AF794943

Mutagenic Impurities

Mutagenic Impurities

Strategies for Identification and Control

Edited by

Andrew Teasdale
AstraZeneca
Cambridge, United Kingdom

WILEY

This edition first published 2022
© 2022 John Wiley & Sons, Inc.

All rights reserved. No part of this publication may be reproduced, stored in a retrieval system, or transmitted, in any form or by any means, electronic, mechanical, photocopying, recording or otherwise, except as permitted by law. Advice on how to obtain permission to reuse material from this title is available at http://www.wiley.com/go/permissions.

The right of Andrew Teasdale to be identified as the author of the editorial material in this work has been asserted in accordance with law.

Registered Office
John Wiley & Sons, Inc., 111 River Street, Hoboken, NJ 07030, USA

Editorial Office
111 River Street, Hoboken, NJ 07030, USA

For details of our global editorial offices, customer services, and more information about Wiley products visit us at www.wiley.com.

Wiley also publishes its books in a variety of electronic formats and by print-on-demand. Some content that appears in standard print versions of this book may not be available in other formats.

Limit of Liability/Disclaimer of Warranty
In view of ongoing research, equipment modifications, changes in governmental regulations, and the constant flow of information relating to the use of experimental reagents, equipment, and devices, the reader is urged to review and evaluate the information provided in the package insert or instructions for each chemical, piece of equipment, reagent, or device for, among other things, any changes in the instructions or indication of usage and for added warnings and precautions. While the publisher and authors have used their best efforts in preparing this work, they make no representations or warranties with respect to the accuracy or completeness of the contents of this work and specifically disclaim all warranties, including without limitation any implied warranties of merchantability or fitness for a particular purpose. No warranty may be created or extended by sales representatives, written sales materials or promotional statements for this work. The fact that an organization, website, or product is referred to in this work as a citation and/or potential source of further information does not mean that the publisher and authors endorse the information or services the organization, website, or product may provide or recommendations it may make. This work is sold with the understanding that the publisher is not engaged in rendering professional services. The advice and strategies contained herein may not be suitable for your situation. You should consult with a specialist where appropriate. Further, readers should be aware that websites listed in this work may have changed or disappeared between when this work was written and when it is read. Neither the publisher nor authors shall be liable for any loss of profit or any other commercial damages, including but not limited to special, incidental, consequential, or other damages.

Library of Congress Cataloging-in-Publication Data Applied for:

[hardback isbn 9781119551218]

Cover Design: Wiley
Cover Image: © theasis/GettyImages

Set in 9.5/12.5pt STIXTwoText by Straive, Pondicherry, India

10 9 8 7 6 5 4 3 2 1

MIX
Paper from responsible sources
FSC www.fsc.org FSC® C013604

Contents

List of Contributors

Ian Ashworth
Chemical Development, Pharmaceutical Technology and Development Operations, AstraZeneca
Macclesfield, UK

Steven W. Baertschi
Baertschi Consulting LLC
Carmel, IN, USA

Joel P. Bercu
Gilead Sciences
Foster City, CA, USA

Tony Bristow
Chemical Development
Pharmaceutical Technology and Development Operations, AstraZeneca
Macclesfield, UK

Michael Burns
Lhasa Limited, Leeds, UK

Adrian Clarke
Novartis Pharma AG
Technical Research and Development
Chemical and Analytical Development
Basel, Switzerland

Stephen Coombes
Chemical Development
Pharmaceutical Technology and Development
Operations, AstraZeneca
Macclesfield, UK

Jason Creasey
Maven E&L Ltd
Stevenage, Herts, UK

Kevin P. Cross
Instem, Columbus, OH, USA

Frank David
RIC-group, Kortrijk, Belgium

Stephen Dertinger
Litron Laboratories, Rochester, NY, USA

Shareen H. Doak
Institute of Life Science
Swansea University Medical School
Singleton Park, UK

David Elder
David P. Elder Consultancy, Hertford, UK

J. Gair Ford
CMC Regulatory Affairs
AstraZeneca, Macclesfield, UK

Robert S. Foster
Lhasa Limited, Leeds, UK

Mark Harrison
Chemical Development
Pharmaceutical Technology and Development
Operations, AstraZeneca
Macclesfield, UK

James Harvey
GlaxoSmithKline R&D
Hertfordshire, UK

Catrin Hasselgren
Genentech Inc., South San Francisco, CA, USA

Gareth J.S. Jenkins
Institute of Life Science
Swansea University Medical School
Singleton Park, UK

George E. Johnson
Institute of Life Science
Swansea University Medical School
Singleton Park, UK

Lucie Lovelle
Novartis Pharma AG
Technical Research and Development
Chemical and Analytical Development
Basel, Switzerland

Melisa J. Masuda-Herrera
Gilead Sciences, Foster City, CA, USA

Glenn Myatt
Instem, Columbus, OH, USA

Christine Mee
Gentronix Ltd, Cheshire, UK

Justin Moser
Global Pharmaceutical Operations Science and Technology
Merck & Co., Inc., West Point, PA, USA

John Nicolette
AbbVie, Inc. North Chicago, IL, USA

Raphael Nudelman
Teva, Netanya, Israel

Michael O'Donovan
Independent Genetic Toxicology Consultant
Nottingham, UK

David J. Ponting
Lhasa Limited, Leeds, UK

Andrew R. Phillips
Pharmaceutical Sciences, R&D
Biopharmaceuticals
AstraZeneca, Macclesfield, UK

Koen Sandra
RIC-Group, Kortrijk, Belgium

Pat Sandra
RIC-Group, Kortrijk, Belgium

David J. Snodin
Xiphora Biopharma Consulting, Bristol, UK

Alan Steven
CatSci Ltd, CBTC2, Cardiff, UK

Andrew Teasdale
Chemical Development
Pharmaceutical Technology and Development
Operations, AstraZeneca
Macclesfield, UK

Adam D. Thomas
Faculty of Health and Applied Science
University of West of England
Bristol, UK

Alejandra Trejo-Martin
Gilead Sciences, Foster City, CA, USA

Michael W. Urquhart
GlaxoSmithKline R&D, Hertfordshire, UK

Gerd Vanhoenacker
RIC-group, Kortrijk, Belgium

Richard V. Williams[†]
Lhasa Limited, Leeds, UK
[†]Deceased

Preface

Since the advent of the first book specifically focused on genotoxic (mutagenic) impurities that I authored and was released in 2010, a considerable amount of progress has been made in terms of management of such impurities. Indeed many of the concepts defined for the first time within the book have ultimately become enshrined in standard practice relating to mutagenic impurities. These include how to conduct a risk assessment, the use of SAR evaluation tools, compound-specific limits, and the use of purge factors to assess carryover risk, to name a few. Indeed most if not all were ultimately captured within ICH M7, which was formalized in 2014.

As a result of this and continued advancement of the science pertaining to many of these and other concepts, it seemed timely to seek to update the book. Little did I or others foresee at the point where this was proposed the dramatic impact of *N*-nitrosamines. By necessity, the book seeks to address not only the changes in general around management of mutagenic impurities, it also seeks to outline a holistic approach to management and control of *N*-nitrosamines, covering all aspects of their management from a safety and quality perspective. As well as this new chapter has been added specifically seeking to examine side reactions that can result in mutagenic impurities, dramatically expanding on the work in the first book that was focused simply on sulfonate esters.

Section 1

The Development of Regulatory Guidelines for Mutagenic/Genotoxic Impurities – Overall Process

1

Historical Perspective on the Development of the EMEA Guideline and Subsequent ICH M7 Guideline

Andrew Teasdale

Chemical Development, Pharmaceutical Technology and Development, Operations, AstraZeneca, Macclesfield, UK

1.1 Introduction

To enable a thorough understanding of the current regulatory position relating to mutagenic impurities (regularly referred to as genotoxic impurities initially) as defined by ICH M7 [1], it is first important to consider the history behind the events that led up to this point and their context. Like many events, the exact point at which concerns relating to the potential presence of mutagenic impurities (MIs) in pharmaceuticals first emerged is difficult to determine. At the time that ICH Q3 guidelines were constructed, specifically ICH Q3A [2], only passing reference was made to compounds of "unusual toxicity" and the potential need for limits tighter than those defined by the guidelines. Although the term "genotoxic" or indeed "mutagenic" is not specifically mentioned, many have taken this to refer to impurities that are mutagenic.

The first public evidence of specific regulatory concern relating to genotoxic impurities was an article published within PharmEuropa in 2000 [3], which drew attention to the potential risk of formation of sulfonate esters resulting from the combination of sulfonic acids in alcoholic solution as part of a salt formation process. At this point this publication was merely a call for "further information," it being part of an attempt to better understand the extent of any risk involved. The publication is now seen as a landmark event, signaling a new era of focus on genotoxic impurity risk assessment and control. It is interesting to reflect on the irony that the "start point" should indeed be sulfonate esters given the long-standing concerns and investigations performed relating to such potential impurities over the subsequent years, much of which are described throughout the chapters within this book.

The first real attempt to generate some form of regulatory framework pertaining to MIs was a position paper relating to genotoxic impurities (GIs) published by the Committee for Proprietary Medicinal Products (CPMP[1]) on behalf of the European Medicines Agency (EMA) Safety Working Party (SWP) for comments in December 2002 [2]. Outlined below is an evaluation of this first draft position paper and an assessment of its later significance in the context of the finalized ICH M7 guideline.

1 Now CHMP.

Mutagenic Impurities: Strategies for Identification and Control, First Edition. Edited by Andrew Teasdale.
© 2022 John Wiley & Sons, Inc. Published 2022 by John Wiley & Sons, Inc.

1.1.1 CPMP – Position Paper on the Limits of Genotoxic Impurities – 2002

1.1.1.1 Scope/Introduction

Within the introduction to the position paper, it was made clear that the need for such guidance was due to the fact that control over levels of genotoxic residues was not adequately addressed through existing ICH guidance.

The position paper consists of a series of sections that addressed the issue of genotoxic impurities from both a toxicological and quality perspective. The key points from those sections are described below.

1.1.1.2 Toxicological Background

Within the position paper, genotoxic compounds were split into two categories:

1) Genotoxic compounds, for which sufficient evidence existed to support a thresholded mechanism. (A thresholded mechanism is one for which a clearly discernible limit exists below which no significant toxicological effect is observed. Several examples were given within the paper of mechanisms of genotoxicity for which a thresholded mechanism may exist, including, for example, topoisomerase inhibition, inhibition of DNA synthesis, and overload of defense mechanisms.)
2) Genotoxic compounds without sufficient evidence for a thresholded mechanism.
   The position paper stated that such thresholds either were unlikely to exist or would be difficult to prove for DNA-reactive chemicals.

This categorization of impurities, on the basis of a mechanistic understanding of toxicological action, has remained in place in the finalized ICH M7 guideline, and the belief that DNA-reactive compounds have no threshold remains widely held. However, as will be explored in a later chapter, there is significant evidence now challenging this for even the most potent of mutagenic carcinogens (Chapter 8).

1.1.1.3 Pharmaceutical (Quality) Assessment

The assumption that some *"in vivo"* genotoxins can damage DNA at any exposure level, and therefore that any level can represent a risk, led to a conservative stance being proposed in terms of quality assessment. It was stipulated that a justification must be provided in relation to the manufacturing process that clearly explained why, for that specific process, the presence of genotoxic impurities was "unavoidable." The position paper also stated that, wherever possible, alternative routes that avoid genotoxic residues should be used and that an applicant was obliged to update the manufacturing process should a safer alternative process be available. If, after these steps had been taken, a risk remained, it was suggested residual levels should be reduced to the level that was "as low as technically feasible." It is interesting to reflect on recent issues relating to contamination of sartans (most notably valsartan [4]) where similar language to that within this preliminary position paper has been used with calls for "nitrosamine free" sartans being requested by some authorities. This is perhaps not entirely unreasonable for the sartans, given that one route to a particular sartan can bring nitrosamine risk and another manufacturing process (or indeed another sartan) can be free from nitrosamine risk.

1.1.1.4 Toxicological Assessment

The guideline made it clear that only after the use of a genotoxic reagent had been justified and every effort had been made to reduce levels should a toxicological assessment be made. Different options were provided by which risk assessments could be carried out, these being through either:

1) **Quantitative risk assessments** – this being essentially based on the linear extrapolation of the dose–response curve from rodent cancer bioassays from a high dose to low dose region.

In this case the low dose recommended being one associated with a 1 in 100 000 risk. (One excess cancer death per 100 000 people exposed to the agent concerned over a lifetime [70 years]).

2) **Uncertainty factor approach** –this approach, one that involves the determination of a no effect level (NOEL) from preclinical studies, along with the subsequent application of uncertainty factors would be appropriate where a threshold-mediated mechanism has been established. Such an approach is consistent with that described within ICH Q3C – Residual Solvents [5].

The position paper in this initial form was a cause of significant concern to the industry. The main concern perhaps related to the safety testing requirements. For many reagents the only safety data available often relates to limited *in vitro* studies, e.g. an Ames test. Such data are unsuitable for establishing a NOEL or for performing a quantitative risk assessment. Thus, to generate data to support the determination of a NOEL or to carry out a quantitative risk assessment as prescribed in the concept paper would require the conduct of further significant *in vivo* studies. This could have resulted in a significant increase in animal studies, something considered potentially unacceptable both at the time and now as efforts are made to refine, reduce, and replace animal experimentation.

Thus, alternatives to this were immediately sought. An alternative approach, previously adopted within other spheres, such as the food arena, was the concept of a "virtually safe dose." This had been developed to deal with low-level contaminants within food. This concept itself was based on the principal of establishing a level at which any new impurity, even if it was subsequently shown to be carcinogenic, would not constitute a significant risk. This paved the way ultimately for the employment within subsequent versions of the guideline of the Threshold of Toxicological Concern (TTC) concept.

1.1.2 Guideline on the Limits of Genotoxic Impurities – Draft June 2004

Significant revisions were made to the original position paper before its rerelease as a draft guideline in June 2004 [6]. The revised guideline struck a carefully considered note. For example, the "as low as technically feasible" terminology used previously was replaced with the ALARP (as low as reasonably practical) principle, a small but in many ways significant shift in emphasis. ALARP does not expect, for example, exploration of unusual or extremely difficult technologies that could be required to be evaluated, irrespective of other impacts (e.g. synthetic efficiency) under "as low as technically feasible (ALATF)." For example, in the context of analysis, ALARP would typically be considered as the application of available standard techniques such as high performance liquid chromatography – mass spectrometry (HPLC-MS), rather than as low as technically feasible that might refer to the need to attempt to apply "state of the art" or even revolutionary experimental approaches. Another important change was the removal of the requirement to introduce an alternative route/process should one "less at risk" be identified. The need to provide justification of the route selected remained.

The most significant change was the acceptance that the concept of elimination of risk in its entirety (zero risk) was often going to be unachievable and therefore an alternative to this principle was required. This led to the adoption of the concept of an acceptable risk level. This acceptable risk was defined as a level sufficiently low that even if the compound in question was ultimately shown to be carcinogenic it would pose a negligible risk to human health. This took the form of the *TTC*. This concept obviates the need to generate extensive *in vivo* data to establish specific limits, by adoption of a conservative generally applicable limit.

The most important aspect of the TTC concept is the derivation of a single numerical limit of 1.5 μg/day based on a lifetime (70 years) exposure resulting in a worst-case excess cancer risk of 1 in 100 000. Within other areas (e.g. food) a 1 in 1 000 000 figure had been applied; this was revised by a factor of 10 in relation to pharmaceuticals to recognize the specific, desired, and otherwise unavailable benefit derived from pharmaceutical treatment. This concept allows an adequate basis of safety and control limits to be established in the absence of specific *in vivo* data on a particular impurity.

The guideline, having established this TTC limit, also stated that, under certain circumstances, higher limits could be established. Such circumstances included short-term exposure, treatment of a life-threatening condition for which no safer alternatives existed, where life expectancy was less than five years, or where the impurity was a known substance for which exposure from other sources (e.g. food) was significantly greater than that associated with exposure from pharmaceuticals. Notably, no fixed alternative limits were provided that could be applied in such instances, perhaps as there are a myriad of potential circumstances where such considerations could apply and thus it was considered that this topic was best left to the assessment of a specific product and a specific risk benefit analysis to agree acceptable limits. It is reasonable that product-specific risk/benefit considerations are applied, and this in many ways supports not establishing fixed acceptable limits in the guideline. Many of these concepts were revisited in the development of ICH M7, this being discussed in depth in Chapter 2.

It should also be noted that impurities considered to be highly potent form part of a cohort of concern and require control to limits lower than the TTC. This concept and its implications are illustrated by the example of *N*-nitrosamines (Chapter 12).

Since the time that the TTC concept was first introduced through this draft guideline, the TTC has come under question principally because of its conservative nature [7]. A detailed evaluation reveals that the TTC derivation process is shaped by the use of the lowest statistically significant TD_{50}s (which can produce a false-carcinogen phenomenon) and by employing linear extrapolation for nonmutagenic carcinogens. Despite such concerns, no effective alternative methodology has been developed, and hence the TTC remains the effective baseline for control of MIs. The 2004 EMA draft guideline [6] and indeed more latter ICH M7 [1] itself explicitly recognize this conservatism. For this reason, the necessity of this specific threshold can be questioned. To understand the importance of the TTC concept, it is imperative to look at it in the context of the initial concept paper. Before the TTC concept was introduced, the primary objective was elimination of risk and only where this was proved impossible could limits be established. However, setting limits would, as already described, require extensive *in vivo* studies. Set in this context the concept of an agreed baseline limit, even if conservative, was valuable in establishing the basis of regulation.

One addition at this point was the widening of the scope of evaluation to include excipients. It is assumed this was to address concerns in relation to some excipients, e.g. modified cyclodextrins (concern over residues of alkylating agents used to modify the cyclodextrin). In many ways, excipients are very similar to existing products in that their safety has been well established through use over an extended period in multiple formulations. In addition, many are used in other areas including the food industry, and thus any exposure related to intake of pharmaceuticals is likely to be small compared to other sources. It is though important to note this presumption of suitable safety based on a history of use is only applicable to well-established excipients. Novel excipients are expected to be assessed in a manner very similar to a new active substance.

At this point in time, there was a lack of any guidance relating to permissible doses during short-term clinical trials. This led, in some instances, to an expectation to meet the 1.5 μg/day lifetime exposure limit, even for very short duration studies. This led to the development of a position paper, outlining a "staged" TTC concept for durationally adjusted control limits. This now well-established concept is described below.

1.1.3 PhRMA (Mueller) White Paper

A Pharmaceutical Research and Manufacturers of America (PhRMA) expert group, led by Lutz Mueller, sought to establish acceptable limits for MIs linked to duration of exposure. This was referred to as a "staged TTC" approach and was based on the established principle that certain types of exposure risk can be defined in terms of cumulative dose [8]. Inherent to this principle is that the risk associated with an overall cumulative dose of a mutagen will be equivalent in terms of risk, irrespective of dose rate and duration. Thus, short-term exposure limits could be based on linear extrapolation from accepted long-term exposure limits.

The group published the outcome of their deliberations in January 2006. The key aspect of this paper, the proposed "staged TTC" limits, is displayed in tabular form in Table 1.1.

It is important to note that included within this proposal is the application of a 1 in 1 000 000 risk factor when calculating limits for durations <12 months, as opposed to the 1 in 100 000 applied in relation to the standard TTC based on lifetime exposure. This precautionary approach was taken in recognition of the fact that during the clinical phase studies are often performed on healthy human volunteers and also that, even for patients at this stage, the therapeutic benefit has often yet to be established.

As well as the staged TTC principle, the paper also proposed a classification system for impurities, defining five classes:

- Class 1: genotoxic carcinogens
- Class 2: genotoxic – carcinogenicity unknown
- Class 3: alerting structure – unrelated to parent
- Class 4: alert related to parent, with associated known toxicological potential
- Class 5: no alerts

This classification and effectively the limits defined within this paper have become the basis of the impurity management system used in ICH M7.

Based on this classification system, the paper defined a strategy for impurity assessment based on the use of structure activity relationships (SARs). SAR evaluation is used as the first stage to give a preliminary evaluation of risk. Thereafter, this can be augmented by the use of safety testing, specifically the Ames test, to determine whether or not the impurity is actually genotoxic. This is particularly true where the impurity is classified as Class 3. Alternatively, one can simply assume the compound in question to be genotoxic on the basis of the prediction and control in line with the appropriate TTC level.

Table 1.1 Proposed allowable daily intakes (μg/day) for potential genotoxic impurity (PGIs) during clinical development, a staged TTC approach depending on duration of exposure.

	Duration of Exposure				
	⩽1 month	**>1–3 month**	**>3–6 month**	**>6–12 month**	**>12 month**
Allowable daily intake (μg/day) for different duration of exposure (as normally used in clinical development)	120[a] or 0.5%[b] whichever is lower	40[a] or 0.5%[b] whichever is lower	20[a] or 0.5%[b] whichever is lower	10[a] or 0.5%[b] whichever is lower	1.5[c] *b*

[a] Probability of not exceeding a 10^{-6} risk is 93%.
[b] Other limits (higher or lower) may be appropriate, and the approaches used to identify, qualify, and control ordinary impurities during development should be applied.
[c] Probability of not exceeding a 10^{-5} risk is 93%, which considers a 70-year exposure.

Such a strategy, often augmented by a science-based impurity purge assessment (incorporating factors such as reactivity of the impurity and downstream process conditions), has become the foundation of most, if not all, control strategies used within the industry (see Chapter 9 for a detailed evaluation of such strategies).

1.1.4 Finalized EMA Guideline on the Limits of Genotoxic Impurities – June 2006

The finalized version of the EMA guideline was issued on 28 June 2006 with an effective date of 1 January 2007 [9]. In terms of the final guideline, a number of key points were addressed, and it would be wrong not to recognize this or to ignore the significant progress made from the original position paper; however, it is equally important to note that concerns remained around several key areas. Outlined below are the key areas that had been addressed and also a reflection on the areas of concern.

The published guidance attempted to clarify how the concepts of the guidance were to be applied to existing substances and products. A concern had been that existing medicines would be required to comply with all aspects of the new guidance. This could have led to there being a perceived shortfall in control strategies or quality for a significant number of medicinal products that had been developed in the years prior to the development of this guidance and furthermore that had proved to be adequately safe across this period. The published guideline included the following specific statement.

"It also relates to new applications for existing products, where assessment of the route of synthesis, process control and impurity profile does not provide reasonable assurance that no new or higher levels of genotoxic impurities are introduced as compared to products currently authorised in the EU concerning the same active substance. The same also applies to variations to existing Marketing Authorisations pertaining to the synthesis. This guideline does, however, not need to be applied retrospectively to authorised products unless there is specific cause for concern." As examined below, there was considerable uncertainty as to what this term meant in practice; there being no clear definition, at least not initially.

In practice this proved extremely difficult to interpret consistently, both from an industry and regulators perspective, particularly in relation to the potential "catch all" phrase "cause for concern." The impact of this uncertainty is explored in detail in the following section.

Another addition within the Recommendations section was advice on the scope of investigations in terms of what impurities should be considered as part of an assessment. The guideline stating:

> As stated in the Q3a guideline, actual and potential impurities most likely to arise during synthesis, purification and storage of the new drug substance should be identified, based on a sound scientific appraisal of the chemical reactions involved in the synthesis, impurities associated with raw materials that could contribute to the impurity profile of the new drug substance and possible degradation products. This discussion can be limited to those impurities that might reasonably be expected based on the knowledge of the chemical reactions and conditions involved.

Although entirely sensible and reasonable on the face of it, in practice this was difficult to interpret consistently.

Another significant change was the exclusion of excipients from the finalized guideline, this having present in the 2004 draft. A separate specific position paper addressing excipients has subsequently been issued jointly by the Quality Working Party (QWP) and SWP within European Medicines Evaluation Agency (EMEA) (and will be discussed later in this chapter).

1.1.4.1 Issues Associated with Implementation

Many of the concepts and principles outlined in the finalized guideline were of real significance in achieving a practical guidance. However, many of the concepts outlined in the guideline also required careful implementation and left certain concerns unaddressed.

1.1.4.1.1 The Relevance of the TTC Concept for Short Durational Exposure

The value of the "threshold of toxicological concern" (TTC) concept is undeniable, but many experts were concerned, and remain concerned, that the maximum daily exposure of 1.5 μg was overly conservative (being based on the combination of several "worst case" assumptions in its derivation) and especially conservative if applied to short duration usage and acute-use therapies. Importantly, the guideline as published did not provide clear guidance on what standards would be expected of investigational medicinal products during the clinical development phases, when controlled and often short duration clinical trials are conducted. It should be unnecessary to apply a control standard applicable to lifetime exposure in such short duration clinical studies, but the guideline gave no specific guidance on what standard would be expected, leaving the implementation of the guideline to be potentially inconsistent. Of primary concern was the lack of any indication as to whether or not the staged TTC concept, as outlined in the Mueller paper [5], was acceptable or not. This led to considerable confusion and uncertainty, which was ultimately resolved with the publication, some 18 months later, of the EMEA staged TTC limits through the SWP Q&A Document [10].

1.1.4.1.2 Application to Existing Products

Similarly, despite the useful focusing of the scope of the applicability of the guideline on "causes for concern" and "significant change" of existing medicinal products, it left unclear what was considered to constitute a "significant cause for concern" or a "significant change."

Did a "cause for concern" exist if an existing impurity in an existing medicine had known genotoxicity (but the medicine concerned had been safely used for many years)? Did a "cause for concern" exist if an existing impurity in an existing medicine had a structural alert for potential genotoxicity but there was no known toxicological findings associated with the impurity?

Did a manufacturing change bring significant new risks if the same route of manufacture was scaled up or conducted at a different site? Did a manufacturing change bring significant cause for concern if process changes were conducted to optimize manufacture that instituted a change in manufacturing chemistry but not a change in the specification of the active substance? What about a change in manufacture of a starting material for active substance manufacture?

Such topics and a lack of clear, specific guidance in the published text left the guideline open to considerable degrees of interpretation and with it the possibility for inconsistent implementation. As a consequence of this uncertainty, a considerable increase in queries linked to existing products was triggered, many asking for a full evaluation of the MI risk, sometimes triggered by even simple variations not linked to the manufacturing process.

So, one can see that even with elements of the guideline that were viewed as "positive," like the TTC concept and the risk-based application to existing products, there were elements of detail that seemed to bring a need for further clarity to support consistent implementation. And of course, there were other aspects of the published guideline that were less well received or were simply not considered in the guideline as it was first published. These too are worthy of consideration.

1.1.4.1.3 Standards Required of Investigational Products

The lack of clear standards that would be expected of investigational products was quickly identified as a gap in the EMA guideline [9]. It could be considered that the original intent of the guideline had been to provide guidance on the management of potentially MIs for marketing applications, not for investigational materials, and thus to make good the "gap" in the ICH impurity guidelines. These ICH guidelines, which provide potential registration requirements for marketing applications, point to a potential need for more rigorous control for some impurity classes (e.g. genotoxic impurities) but do not provide guidance on how to manage such impurities. Given this ICH-driven provenance, one might consider that the Committee for Human Medicinal Products (CHMP) guideline as published was not intended to apply to investigational materials, but like ICH guidelines, to provide potential registration requirements for commercial products. However, the guideline's applicability was ambiguous, and of course with no further specific guidance for investigational materials, it was most likely that the same standards might begin to be applied to investigational materials, even if this was not the initial intent of the expert authors of the original guidance.

1.1.4.1.4 Circumstances that Support Modification of the TTC Limit

As already described the published guideline also contained guidance to the effect that the general TTC limit (1.5 μg/day) could be modified in certain circumstances (e.g. for short duration treatments, particular indications, or patient groups) to provide for modified control of potential MIs in these products. Unfortunately, while this was potentially a very useful aspect of the guidance, the published guideline provided no further specific advice. Similarly, and importantly, there could be some medicines, indications, or patient groups where it might be unnecessary to implement *any* rigorous, "low level" control of potential MIs. For example, if an oncology treatment is itself known to be genotoxic, it would seem unnecessary to control potentially MIs in such an active substance to levels consistent with the TTC. Furthermore, many oncology treatments are used either post or in tandem with cytotoxics during the clinical phase, particularly in advanced stages of the disease. The cytotoxic agent itself poses a significant, but accepted risk, of secondary cancer. This again challenges the value to the patient, of controlling MIs to levels consistent with the TTC.

1.1.4.1.5 Control Requirements When Multiple GIs May Be Present

Given the complex multistep nature of the synthesis of many synthetic Active Pharmaceutical Ingredient (APIs), it is possible for a product to contain more than one potentially MI. Indeed, a study published in 2016 [11] showed that multiple reagents of mutagenic potential are used in a "typical" synthesis. The EMA guidance [9] was not clear on what control expectations would exist when more than one potential MI was likely to be present in the active substance on product. Would each be simply controlled on the basis of individual TTC limits?[2]

This would seem reasonable given the conservative nature of the derivation of the general 1.5 μg/day TTC limit. Or would there be an expectation that the total genotoxic impurity load would be controlled to a total level of 1.5 μg/day or other limit? There might be some scientific basis for implementing such a cumulative control if the impurities were known to be (or likely to be) toxicologically similar, but far less need to do so if the impurities were known to be (or likely to be) toxicologically distinct. These are all interesting and potentially important considerations, but the published guideline provided no detailed guidance on these questions. In terms of the toxicological risk, Bercu et al. reviewed the supporting evidence and demonstrated that with the addition of one to two MIs, a slight but insignificant increase in cancer risk was observed. There was not an increase in cancer risk when comparing structurally related impurities with structurally unrelated impurities. They therefore concluded that there was little evidence to support a view that effect was cumulative at low levels, <5 μg/day [12].

---

2 It is important to recognize that the risk factors in terms of probability of an MI being present at levels of concern are specific to the MI in question, its physicochemical properties, and process conditions.

1.1.4.1.6 *Application to New Marketing Authorisation Approval (MAA) Applications Relating to Existing Products*

One of the specific challenges of the guideline was that it also potentially applied to applicants for generic versions of existing products. On one level, an applicant for a generic medicine might assume that the active substance in their medicine is "out of scope," as clearly such a medicine has a significant preexisting period of use such that its safety is known. However, this assumption relies upon the generic active pharmaceutical ingredient and medicinal product having the same quality and impurity profile as the existing drug substance and drug product. This may on many occasions not be the case, as even if similar chemistry is used, subtleties of manufacture or formulation can lead to potentially significant differences in impurity profile, especially when "significant" is no longer being considered as reflecting levels commensurate with that of the ICH unspecified impurity control limits [2] (e.g. in the order of 0.1% – i.e. parts per thousand) but at the levels of TTC-based controls (which can be in the order of parts per million and indeed for those impurities defined by the cohort of concern, parts per billion).

A particular challenge in terms of development of generic products relates to how can a generic applicant assure themselves they have introduced no new risk factors with respect to previously approved materials? Could this be achieved by simply meeting the preexisting European Pharmacopoeia (EP) monograph for the active substance (if one exists)? In reality even now it is likely that this will *not* be sufficient: monographs rarely include controls on potential MIs at low levels. Thus, many of these potential risks may be "invisible" in terms of the public quality standard; indeed, the recent issues surrounding *N*-nitrosamines clearly highlight this. Maybe the generic applicant could simply test their drug substance against the previously approved drug substance? But what analytical methods should be used? Of course, this lack of transparency relates not only to the generic manufacturer, the regulator charged with assuring the suitability of the new product faces a similar challenge.

Of course, if the generic applicant decided to do a comprehensive and independent risk assessment of their drug substance or drug product and their manufacturing processes and establish TTC-based controls for any potentially MIs (on the basis of structural alerts, etc.), then no doubt the regulatory agencies will be presented with a potentially approvable drug substance, associated specification, and manufacturing process. What even now remains unclear is how will the agency view the previously approved marketing application holders. Issues surrounding valsartan and *N*-nitrosamines show that certainly where the risk is deemed to be a general risk this will very likely lead to a request to test all current approved products for the MIs that the subsequent applicant has determined to be potentially present.

1.1.4.2 Control Expectations for Excipients

When finalized the EMA guideline stated that it did not apply to excipients used in pharmaceutical manufacture, this being addressed by a separate EMA publication [13] (discussed further below). Clearly, some excipients are also manufactured by chemical synthesis and may therefore also be exposed to routes of manufacture that contain reactive and "at risk" reagents and intermediates. Global pharmacopoeias such as the EP and United States Pharmacopeia (USP) contain many synthetic excipients; some like polyethylene glycol (PEG) are polymers of epoxides or use epoxides to derivatize other materials (e.g. cyclodextrins). Epoxides are alkylating materials and hence are potentially mutagenic potential impurities in the excipients. Clearly with excipients often being a more significant percentage in weight terms of a medicine than the active substance, the potential risk associated with excipient impurities might also be of concern. Despite these potential risks, it is clear that many of the excipients have a significant history of safe use, many indeed are listed within the Food and Drug Administration (FDA) generally regarded as safe (GRAS) list [14] and hence the guidance ultimately concluded that there are no significant issues associated with well-precedented excipients given their long-established and demonstrated safety profile.

But what of the potential risk associated with manufacturing process changes related to excipient manufacture? What of the increasing case of a novel excipient being developed? In the case of a novel excipient, it is very likely that the expectations for the assessment of MIs should mirror that of any new chemical entity (NCE).

1.1.4.3 Control Expectations for Natural/Herbal Products

Pharmaceuticals are in the majority of cases well-characterized small molecules manufactured by well-defined chemical synthesis. However, the situation can be quite different in relation to some medicines derived from natural sources. Some of these natural product-derived medicines, including herbal medicines, can be less well-characterized materials and are often complex mixtures that may vary batch to batch in terms of components. Of course, the control of impurities in such medicines is also important and by extension one perhaps should consider whether such medicines too might contain potentially MIs [15]. It is, however, practically impossible to apply the same degree of risk management to the manufacture/isolation of a natural product, nor the same degree of process selection and design. How should one approach the management of potential mutagenic risk in such active substances? The EMA guideline [9] provides no specific guidance and therefore EMA published a separate guideline for such products.

1.1.4.4 Identification of Potential Impurities

As highlighted earlier the guideline noted that risk assessment of manufacturing processes should be undertaken to identify potential MIs and that impurity structures should be risk assessed (using *in silico* systems that link structural motifs to potential toxicological responses). This sounds very reasonable and practicable. However, one could find two “experts” in the field who might draw up two different lists of potential impurities associated with a particular manufacturing process. The challenge is one of defining what is considered reasonable when defining impurities, particularly MIs. Within the framework of the original guideline, it was also unclear as to how many steps within a process should be taken into consideration when performing such an assessment. This was a specific topic returned to when the ICH M7 [1] guideline was elucidated.

1.1.4.5 The Principle of Avoidance

The guidance also contained very specific expectations that the pharmaceutical development efforts should first and foremost “avoid” genotoxic materials or impurities and take every effort to select a manufacturing process that avoids there being potential genotoxic risks associated with the product.

> A justification needs to be provided that no viable alternative exists, including alternative routes of synthesis . . .
>
> If a genotoxic impurity is considered to be unavoidable in a drug substance, technical efforts (e.g. purification steps) should be undertaken to reduce the amount of the genotoxic residues in the final product in compliance with safety needs or to a level as low as reasonably practicable.

These were elements of the guideline that had provoked considerable comment during the drafting process. Assembling drug substances by chemical synthesis is predicated on the combination of simple chemicals into more complex drug substance structures. This synthesis involves chemical reactions often driven by reactive functional groups that as a consequence of their reactivity (e.g. alkylating functionality) can be potentially toxic and indeed potentially genotoxic. Thus, to have "complete avoidance" as the fundamental principle of chemical process development would be extremely problematic. In extremis the effect of such an approach could be that many important, necessary (and well-understood) reactions would suddenly be declared unsuitable or at the very least subject to intense scrutiny. Not only would avoidance be problematic as a fundamental principle but avoidance can also be appreciated to be inherently unnecessary, in risk management terms, when one considers that a manufacturing process can be designed in such a way as to ensure that the residues of these reactive materials are not significantly present in the drug substance.

An important consequence of the intrinsic reactivity of the materials "to be avoided" is that they can easily break down to innocuous materials during isolation of intermediates, for example by hydrolysis. This would mean that one would be being told to avoid a useful synthetic material that would anyway be destroyed and removed during manufacture. This removal would make "avoidance" unnecessary. Furthermore, manufacturing processes can be designed with the removal of potential genotoxic reagents, intermediates, or impurities in mind, either by using such reagents early in a multistep manufacturing process or by designing isolation processes or purification processes specifically to remove materials of concern. Thus, having "avoidance" as a fundamental design criterion for drug substance manufacture could be considered to be an overreaction and extremely precautionary. When all aspects of risk management and scientific understanding are considered, avoidance can be seen to be nonscientific. The risks being avoided can be managed in other scientifically sound ways and furthermore can also be controlled to appropriate levels, if need be, by analytical testing. The primary consideration of the chemical manufacture of drug substances (and medicinal products) should be the safety (and efficacy) of the medicine, and since the adoption of the TTC principle establishes a basis of adequate safety (or acceptable risk), then control strategies and control tests on specifications can be established to "control" the adequate safety of manufactured drug substances without imposing a "ban" on the use of many important reagent and reaction types.

Let's be sure we are absolutely precise and fair to the wording of the guideline. In the guideline, "avoidance" was stated to be a fundamental principle but was not required if the applicant had shown that no other manufacturing process free of attendant genotoxic risk factors could be employed

> A justification needs to be provided that no viable alternative exists . . .
> If a genotoxic impurity is considered to be unavoidable. . .

While on the face of it a seemingly reasonable request, in practice this particular aspect of guidance is in reality a case of "how long is a piece of string?" in terms of the expected extent of such investigations. How many alternative routes of synthesis need to be evaluated and discarded before one can conclude "there is no viable alternative"? How many potential routes should one explore if a drug substance is made by a manufacturing process that uses "risky" reagents like alkylating reagents but contains no trace of the impurity that would have potential genotoxicity? If a route of synthesis not employing "at risk" materials can be shown to be feasible but the drug substance cannot be made economically, or in an environmentally acceptable manner, by that route, should that potential medicine be "avoided"? Having development chemists chasing alternative routes to one medicine is a sure fire way to prevent development chemists developing other medicines. Thus this guidance by placing "avoidance" above "control" could very well have prevented the innovation of new medicines or new manufacturing routes (with improved environmental benefits).

1.1.4.6 The ALARP Principle

The EMA guideline [9] also suggested that if avoidance was not possible, then residues of any mutagenic materials to be used should be removed to a level that was "as low as reasonably practicable" (the so-called ALARP concept). This concept too sounds immediately reasonable, especially in the context of the original request to control to as low as technically feasible, but is flawed when one begins to consider aspects of its implementation. Consider a case when an applicant has developed a process to deliver an active ingredient that contains a measurable, but low, level of a potentially MI. The applicant has established a control strategy in accordance with a TTC-based limit. Should the assessor approve this application or require the applicant to further modify the process to lower the residual level yet further? How much more work would be required to be considered "reasonably practicable"? Can such a judgment be consistently applied, by all assessors, to all applicants? Will some applicants or assessors expect more to be done than others? All such considerations could introduce inconsistencies in what needs to be a level regulatory landscape. Given the conservative nature of the guideline, there should simply be no *need* to further improve quality if a TTC-based control strategy has been established. After all the TTC is considered a virtually safe dose.

1.1.4.7 Overall

Potentially the most troublesome aspect of the EMA guideline [9] was the scope for inconsistent interpretation even in relation to the many apparently "well-developed" concepts. Adding to this challenge of interpretation and consistent implementation are some further gaps that became evident the more one started to consider the kind of scenarios that would be encountered as the guideline was implemented. It is perhaps not surprising that regulators and industry alike struggled to fully understand how to interpret and apply it in its entirety.

To begin to resolve the difficulties described, further work and discussion took place after the final publication of the guidance in both regulatory circles (CHMP SWP and European Directorate for the Quality of Medicines) and in industry and industry trade associations (European Federation of Pharmaceutical Industry Associations – EFPIA) as all parties involved looked to examine these important topics, in depth, principally through a Question and Answer [10] process initiated by the EMA SWP.

1.1.5 SWP Q&A Document

The SWP Q/A document [10] (published 26 June 2008 as EMEA/CHMP/SWP/431994/2007) sought to address several key areas within the original EMA guideline, these included:

1.1.5.1 The Application of the Guideline in the Investigational Phase and Acceptable Limits for GIs Where Applied to Studies of Limited Duration

Through the Q&A document [10], it was clarified that durational adjustments to the TTC limit are acceptable for investigational studies. This approach of extrapolating the lifetime-based TTC limit to shorter duration exposures had originally been proposed by a PhRMA [8] cross-industry workgroup led by Lutz Mueller (Roche) who as described earlier proposed a set of "staged" TTC limits dependent upon study duration. The SWP accepted the principle of such duration-dependent modifications to the TTC but published a set of durational limits that are slightly different from the original PhRMA proposal. These are (see also Table 1.2):

> The acceptable limits for daily intake of genotoxic impurities are 5, 10, 20, and 60 μg/day for a duration of exposure of 6-12 months, 3-6 months, 1-3 months, and less than 1 month, respectively. For a single dose an intake up to 120 μg is acceptable.

Table 1.2 Acceptable limits for MIs based on duration of exposure.

	Duration of exposure				
	Single dose	**≤1 months**	**≤3 months**	**≤6 months**	**≤12 months**
Allowable daily intake (μg)	120	60	20	10	5

> Compared to the proposal of a staged TTC in the Mueller et al (Reg Tox & Pharm, 2006, 44, 198–211) paper these values incorporate a dose rate correction factor of 2 to account for deviations from the linear extrapolation model.

The scientific basis/driver behind the proposal by the SWP to apply a correction factor to the linear model was unclear given the conservative nature of the linear extrapolation model itself; as is the rationale that requires restricting the 120 μg/day to a single dose. This was a topic revisited in the development of ICH M7 [1].

In the published Q/As, the SWP also stated that these modified limits while applicable in the investigational phase only could not be automatically presumed to apply to commercial products that are used for short durations. The applicant for such an acute-use therapy could, however, propose amended control limits in their MAA and the approval of product-specific limits for the commercial product will be established during the review process, considering the full product-specific risk benefit of the product. Like the proposed adjusted durational limits, this became a key topic during development of ICH M7.

1.1.5.2 Application of the Guideline to Existing Products

The EMA guideline [9] limited the application to existing products to "known causes for concern" and to "change management." However, the lack of a definition of what constituted a "cause for concern" was a real shortfall in the guidance. This shortfall had led to difficulty in interpretation and to inconsistent application of the guidance both by regulatory agencies and industry. This led to the SWP looking to provide a clarification, again via the official Q/A publication, that a "cause of concern" is a material with either preexisting or new known genetic toxicology findings (and in their answer the SWP gives one example class of impurity that would be considered as constituting a cause for concern – mesylates and alkyl mesylates).

> If a manufacturing procedure for API remains essentially unchanged a re-evaluation with respect to the presence of potentially genotoxic impurities is generally not needed. However, new knowledge may indicate a previously unknown cause for concern. One example is the mesylate salt drug substances for which a few years ago, a concern regarding the potential for formation of genotoxic alkyl mesylates was raised. This concern resulted in the "Production Statement" requesting a specific evaluation of the potential for formation of these highly toxic products now included as part of the PhEur monographs for all the mesylates salts.

The European Directorate for the quality of Medicines (EDQM) have further extended the clarity on this point by noting, in a PharmEuropa publication [16], that structurally alerting functionality alone does not constitute a cause for concern, without actual toxicology data.

Structural alert does not automatically imply genotoxicity.

Action is needed only where there is study data demonstrating genotoxicity of the impurity. The existence of structural alerts alone is considered insufficient to trigger follow-up measures.

In reality for many this interpretation of cause for concern was not entirely accepted and hence uncertainty remained, again establishing a key area for clarification within ICH M7 [1].

EDQM also addressed this issue through its PharmEuropa publication, discussing the applicability of the guideline to pharmaceutical monographs. The statement made provides a very useful, risk-based approach to managing the application of the guideline to monographed materials. The EDQM article stated that "Substances included in medicinal products authorised in recent years have been thoroughly evaluated for safety and in view of the experience with their use the need for retrospective application of a policy on genotoxic impurities is not considered necessary unless there is specific cause for concern," again emphasizing the role of prior clinical exposure and pharmacovigilance in their management of the existing products. The detailed Appendix table from this EDQM publication provides a variety of tiers of potential change and the action considered necessary to support each "change." This vastly helpful table is reproduced in Table 1.3.

Table 1.3 EDQM decision table for use during elaboration or revision of monographs.

Status	Action
Substance included in a medicinal product authorized after issuance of the CHMP guideline	Monograph should be based on marketing authorization(s)
Substance included in a medicinal product authorized before issuance of the CHMP guideline: no PGI expected from synthetic route	No action needed, monograph based on marketing authorization
Substance included in a medicinal product authorized before issuance of the CHMP guideline: PGI expected from synthetic route of first authorized product – subsequently authorized products (if any) have no expected PGI or same PGI as the first authorized product at same or lower level and no data showing genotoxicity	No action needed during elaboration of monograph (based on marketing authorization), no revision of existing monographs
Substance included in a medicinal product authorized before issuance of the CHMP guideline: PGI expected from synthetic route of an authorized product and data showing genotoxicity of an expected PGI	Monograph should be elaborated or revised based on evaluation by the Competent Authority
Substance included in a medicinal product authorized before issuance of the CHMP guideline: PGI expected from synthetic route of first authorized product and subsequently authorized products have a new expected PGI or same PGI as innovator product at a higher level and data showing genotoxicity of an expected PGI	Monograph should be elaborated or revised based on evaluation of new PGI or higher level of previously known PGI by the Competent Authority
Substance included in a medicinal product authorized before issuance of the CHMP guideline: PGI not expected from synthetic route of first authorized product and subsequently authorized product(s) have a new expected PGI and data showing genotoxicity of an expected PGI	Monograph should be elaborated or revised based on evaluation of new PGI by the Competent Authority

1.1.5.3 Avoidance and ALARP

It was confirmed by the SWP in a Q/A that if MIs associated with a product or substance had been shown to be controlled (i.e. that any MIs associated with the product were controlled to an appropriate safety-based limit), then it was not necessary to conduct further work to further reduce the levels present. Thus, the driver to have ALARP principles drive further process development of an acceptably safe product was unnecessary.

1.1.5.4 ICH Identification Threshold and its Relation to MI Assessment

Another topic addressed within the EMEA Q&A document was the question as to how to relate the ICH identification threshold to unknown impurities and how this related to MI assessments. The answer provided by SWP was to confirm that the identification threshold outlined in ICH Q3a remained appropriate, because the overall quality of drug substance is supported by a well-defined and reasoned risk assessment of the manufacturing process, which serves to identify significant potential major concerns. This focused risk assessment is employed to assure the quality of the drug substance and should mean that the level of risk associated with any unknown impurity present below the identification threshold has a low probability of being potentially mutagenic. As demonstrated by the discovery of *N*-nitrosamines in sartans, this relies very heavily on a detailed knowledge of the chemistry employed; this is explored in detail in Chapter 12.

The EMEA guideline and subsequent Q&A document [10] of course are technically related only to Europe. Up to the end of 2008, the FDA's position remained somewhat unclear although it was clear from podium presentations that the FDA supported the underlying principles, e.g. the TTC of the EMEA guideline. In December 2008 the FDA finally published their draft guideline addressing the topic of genotoxic impurities [17].

1.1.6 FDA Draft Guideline

Drafted in December 2008 but never finalized, unsurprisingly there were significant similarities between this FDA draft guideline and the EMA guideline, certainly in terms of the key principles such as the TTC, the acceptance of a staged approach where study duration is limited, and the use of SAR evaluation.

There were though areas of concern the most significant perhaps was the suggestion of the need to introduce lower limits for different patient populations, specifically pediatric populations. Additional safety factors of 3 and 10 were mentioned and suggested for consideration. The need for this additional level of control for pediatric medicines was unclear when considered in the context of the extremely conservative assumptions that form the basis of the calculated TTC control.

Additionally, other subtler differences also existed, including differences in staged TTC values in relation to very short (less than 14 days) studies; the FDA favoring the extension of the 120 μg/day for the whole of this period.

Taken together the differences between the two guidelines would ultimately present anyone faced with having to comply with both with a challenge. It was the recognition of these challenges and others described in more detail below that led to establishment of an ICH process to deliver what became ICH M7. The development of ICH M7 is discussed in Chapter 2.

1.1.7 Other Relevant Guidance

As well as the main guidelines and supporting documents described, there are a short series of other documents that relate to this area that warrant comment.

It has certainly been interesting to follow publications on two related topics – the control of potential genotoxic (GT) impurities in herbal medicines and a paper considering the degree of risk associated with MIs in pharmaceutical excipients. Each of these is briefly examined below.

1.1.7.1 Excipients

The EMEA's CHMP was requested by the European Commission to provide an opinion on the risks associated with possible presence of carcinogens, mutagens, and substances toxic to health within excipients used as ingredients of medicinal products for human use. The CHMP published this opinion (in a joint paper from the QWP and SWP, published as EMEA/CHMP/SWP/146166/2007 on 18 October 2007) [13]. The conclusion of this evaluation was that the accumulated safety and pharmacovigilance data regarding well-known, established, and standard excipients that meet EP requirements for quality served to provide a generally acceptable proof of acceptable safety and quality.

1.1.8 Herbals

A draft guideline was published by the EMEA's Committee of Herbal Medicine (EMEA/HMPC/107079/2007, dated 31 October 2007) [18] addressing how genotoxic impurity (GTI) management of herbal medicines should potentially be approached. This proposed a very simple approach – a herbal medicine and its source and manufacturing process should simply be evaluated in genotoxicity studies, and if the material is free of genotoxic response, then the impurity profile needs no further risk assessment. This is despite that for genotoxic tests like the Ames test to be effective in determining the true risk of an impurity being genotoxic there needs to be a particular level of exposure to the impurity in the test (250 μg/plate as a stated recommended threshold). An impurity below this threshold in the test cannot be deemed to be non-genotoxic if the test comes back without adverse findings. Thus, the acceptance of such an approach as the basis of defining the safety for herbal medicine is different from EU/ICH M7 expectations for synthetically derived pharmaceuticals, where a deeper (and more stringent) evaluation of risk coming from particular impurities is possible and is mandated.

1.1.9 ICH S9

ICH S9 – Nonclinical evaluation for anticancer pharmaceuticals reached Step 4 in late 2009 [19]. This, as the title suggests, aims to provide guidance relating to the level of safety evaluation required to support the use of anticancer pharmaceuticals particularly in the context of the stage of disease. It suggests a reduced package of testing where treatment is associated with advanced disease, recognizing the inconsequential impact any toxicity-related issues would have in terms of patient risk/benefit.

ICH S9 contains the specific statement:

> It is recognised that impurities are not expected to have any therapeutic benefit, that impurity standards have been based on a negligible risk (e.g. an increase in lifetime risk of cancer of one in 10^5 or 10^6 for genotoxic impurities) and that such standards might not be appropriate for anticancer pharmaceuticals intended to treat advanced stage patients.

In 2018 ICH S9 was augmented by a Q&A document [20]. Within this specifically in relation to ICH M7 the following is tabulated (Table 1.4).

Table 1.4 Extracted text from ICH S9, Q&A document relating to MIs.

4.12	Should impurities exceeding the established qualification limits in ICH Q3A/B be assessed in genotoxicity studies?	**API genotoxic?**	**Impurity exceeds 3A/B qualification threshold?**	**Proposed action**
	When the API is genotoxic?	Yes	No	None
	When the API is non-genotoxic?	Yes	Yes	None
		No	No	None
		No	Yes	Genotoxicity assessment of impurities should be conducted.
4.13	Is ICH M7, giving guidance for the management of MIs, applicable to the patient population covered in the scope of ICH S9?	The scope of ICH M7 specifically states that the guidance does not apply to "drug substances and drug products intended for advanced cancer indications as defined in the scope of ICH S9." Therefore, MIs in products used for treatment of indications under the scope of ICH S9 should be considered for management consistent with the concepts outlined in ICH Q3A/B (see Question 4.12).		

Unlike the earlier EMA guideline, which simply stated that factors such as limited life expectancy and seriousness of disease can be used as the basis for establishing alternative limits, ICH M7 explicitly states that this guideline does not apply to drug substances and drug products intended for advanced cancer indications as defined in the scope of ICH S9.

Another interesting aspect of this is the question as to whether the oncology therapy area is a "special case" or whether the same logic applies to other treatments of life-threatening conditions. To date experience has been that extrapolation to other life-limiting conditions has been difficult in terms of regulatory acceptance.

1.1.10 Conclusions

Despite the substantive progress made in the evolution of the European guidelines [9], it was clear as described above significant challenges remained in terms of interpretation and also alignment with other key guidelines. It is for this reason MIs were adopted as an ICH topic, and it is this process and outcomes that are described in Chapter 2.

References

1 ICH guideline M7(R1) on assessment and control of DNA reactive (mutagenic) impurities in pharmaceuticals to limit potential carcinogenic risk European Directorate for the Quality of Medicines and Healthcare. EMA/CHMP/ICH/83812/2013.

2 ICH Q3A: Impurities in New drug substances (R2), October 2006.

3 European Directorate for the Quality of Medicines and Healthcare (2000). Enquiry: alkyl mesilate (methane sulfonate) impurities in Mesilate salts. *PharmEuropa* 12: 27.

4 Committee for Proprietary Medicinal Products, Position paper on the limits of Genotoxic Impurities, London 18th December 2002, CPMP/SWP/5199/02/draft 2.

5 Committee for Medicinal Products for Human Use (CHMP), Guidelines on the Limits of Genotoxic Impurities, London 23rd June 2004. CPMP/SWP/5199/02.
6 ICH Q3C Residual Solvents (R6), October 2016.
7 Snodin, D.J. and McCrossen, S. (2013). Mutagenic impurities in pharmaceuticals: a critique of the derivation of the cancer TTC (threshold of toxicological concern) and recommendations for structural-class-based limits. *Regul Toxicol Pharmacol* 67: 299–316.
8 Muller, L., Mauthe, R.J., Riley, C.M. et al. (2006). *Regulatory Toxicology and Pharmacology* 44: 198–211.
9 Committee for Medicinal Products for Human Use (CHMP), Guidelines on the Limits of Genotoxic Impurities, London 26th June 2006. CPMP/SWP/5199/02.
10 CHMP SAFETY WORKING PARTY (SWP), Question & Answers on the CHMP Guideline on the Limits of Genotoxic Impurities. London, 17 December 2009. EMEA/CHMP/QWP/251344/2006Q&A.
11 Elder, D.P. and Teasdale, A. (2015). Is avoidance of genotoxic intermediates/impurities tenable for complex, multistep syntheses? *Organic Process Research and Development.* 19: 1437–1144.
12 Bercu, J.P., Hoffman, W.P., Lee, C.y., and Ness, D.K. (2008). *Reg Toxicology and Pharmacology* 51 (3): 270–277.
13 CHMP Scientific article 5(3) Opinion on the potential risks of carcinogens, mutagens and substances toxic to reproduction when these substances are used as excipients of medicinal products for human use. London, 18 October 2007. EMEA/CHMP/SWP/146166/2007.
14 GRAS List: http://www.fda.gov/Food/FoodIngredientsPackaging/GenerallyRecognizedasSafeGRAS/GRASListings/default.htmExcipients
15 Teasdale, A., Elder, D.P., and Snodin, D. (2010). Analytical approaches for the detection of epoxides and hydroperoxides in active pharmaceutical ingredients, drug products and herbals. *Journal of Pharmaceutical and Biomedical Analysis* 51: 1015–1023.
16 EDQM (2008). Potentially genotoxic impurities and European pharmacopoeia monographs on substances for human use. *PharmEuropa* 20 (3): 426–427.
17 Guidance for Industry Genotoxic and Carcinogenic Impurities in Drug Substances and Products: Recommended Approaches Draft, Center for Drug Evaluation and Research (CDER) December 2008.
18 Concept Paper on the Development of a Guideline on the Assessment of Genotoxic Constituents in Herbal Substances/Preparations, Committee for Medicinal Products (CHMP), European Medicines Agency, London, 25 October 2006. (EMEA/HPPC/413271/2006).
19 ICH Topic S9 Non-clinical Evaluation for Anticancer Pharmaceuticals, Step 3, EMEA/CHMP/ICH/646107/2008.
20 ICH S9 guideline on nonclinical evaluation for anticancer pharmaceuticals - questions and answers. EMA/CHMP/ICH/453684/2016

2

ICH M7 – Assessment and Control of DNA Reactive (Mutagenic) Impurities in Pharmaceuticals to Limit Potential Carcinogenic Risk

Andrew Teasdale[1] *and Raphael Nudelman*[2]

[1] *Chemical Development, Pharmaceutical Technology and Development, Operations, AstraZeneca, Macclesfield, UK*
[2] *Teva, Netanya, Israel*

2.1 Introduction

ICH M7 [1] was first introduced in June 2014, undergoing a first revision in July 2017. The guidance itself is a multidisciplinary guideline reflecting the fact that mutagenic impurities are both a quality and safety topic. It focuses on the control of mutagenic impurities and was introduced to address differences that existed between regional guidelines pertaining to genotoxic (mutagenic) guidelines, the principle regional guidelines being the European Medicines Agency (EMA) guideline [2], supplemented by its associated Question and Answer (Q&A) document [3] and the draft Federal Drug Administration (FDA) guideline [4]. No formal guidance was ever published by Pharmaceutical and Medical Devices Agency (PMDA), Japan. A comprehensive overview of the chronological development of the EMA and FDA draft guidelines is described in Chapter 1. These guidelines were introduced to address a perceived gap in existing impurity guidelines ICH Q3A [5] and Q3B [6] relating to management of DNA reactive mutagenic impurities that may arise during the synthesis of the active form and through degradation of either/both the drug substance (DS)/drug product (DP). Although broadly similar within their concept, there were nevertheless sufficient enough differences between the EMA and FDA guidelines to render a global approach challenging.

Clear within the ICH M7 concept paper [7] was a desire to harmonize these guidelines and to specifically address areas of uncertainty. Critical areas identified within the concept paper included:

1) Use of *in silico* structure activity relationship (SAR) tools for the assessment of mutagenic potential.
2) What are acceptable levels of genotoxic impurities during drug development?
3) What are acceptable levels of genotoxic impurities for marketing?
4) Should those impurities be regulated differently that are likely to have threshold effects?
5) Should levels of genotoxic impurities be regulated using a Threshold of Toxicological Concern (TTC) approach?
6) Structurally related genotoxic impurities are likely to have similar mechanisms of action. Should these be summed in calculating a TTC?
7) What process of qualification testing should be followed for impurities that are metabolites?
8) What additional data are needed to support having no special restrictions, or a higher acceptable daily intake (ADI) than the TTC, for a genotoxic impurity?

Mutagenic Impurities: Strategies for Identification and Control, First Edition. Edited by Andrew Teasdale.
© 2022 John Wiley & Sons, Inc. Published 2022 by John Wiley & Sons, Inc.

The following chapter will look in detail at the ICH M7 guideline, examining each aspect of the guideline, seeking in doing so to give a comprehensive overview of its practical implementation. It will also examine whether or not the guideline has succeeded in addressing the critical areas outlined in the concept paper.

2.2 ICH M7

The format and structure of ICH M7 [1] is understandably very similar to that of the preceding regional guidelines, with specific sections focused on quality and safety aspects, although terminology does differ. Crucially, the fundamental concept upon which the guideline is predicated has not changed; the principle of a TTC, i.e. a virtually safe dose (VSD), remains the underpinning principle upon which the guideline is based. In addition to the overall framework, there are additional sections intended to provide definitive guidance in specific areas, these include:

1) Established Products
2) Documentation
3) Case Studies

These and other sections are now described in detail below.

2.2.1 Introduction

The introduction, Section 1 of the guideline, provides an overview of the relationship between the guideline and the closely related quality guidelines, ICH Q3A [5] and Q3B [6]. It also makes clear that the focus of the guideline is on DNA reactive, mutagenic impurities, emphasizing for the first time the specific focus on such impurities as opposed to the more general group of genotoxic impurities. This is a point returned to in more detail in the general principles section.

2.2.2 Scope

The scope (Section 2) seeks to outline when and where the guideline is applicable, in terms of previously established products (predating the introduction of the guideline), anticancer treatments, and therapy class.

2.2.2.1 Established Products

Previous guidelines, especially the EMA [2] guideline, have struggled to address the thorny question of the potential retrospective application of the guideline to existing products. This had led to variable interpretation, often resulting in requests for application to existing products triggered by apparent trivial changes (in the context of effect on levels of mutagenic impurities within the DS/DP) such as change in dose size or formulation volume. A lot of the uncertainty is related to the inclusion within the original EMA guideline of a catchall phrase of "*cause for concern*." Attempts to resolve these uncertainties made via the EMA Q&A process [3] proved only partially successful.

In contrast ICH M7 is clear and concise. Existing products are only in scope where:

1) changes to the DS synthesis result in new impurities or increased acceptance criteria for existing impurities;

2) changes in the formulation, composition, or manufacturing process result in new degradants or increased acceptance criteria for existing degradants;
3) changes in indication or dosing regimen are made which significantly affect the acceptable cancer risk level.

It is hoped that this addresses much, if not all, of the uncertainty that has existed previously. One remaining area is how this applies to products marketed after the introduction of regional guidelines, particularly the EMA guideline, as this was formally issued and became effective in January 2007. Although not stated directly, many have assumed such products, marketed after the introduction of earlier regional guidelines, to be in scope.

2.2.2.2 Anticancer Treatments

A challenging question has long been, what is the relationship between guidelines pertaining to mutagenic impurities and the ICH S9 [8] Guideline – Non-Clinical Evaluation for Anti-Cancer Pharmaceuticals? ICH S9 lays out clear principles relating to the nonclinical (i.e. toxicological) evaluation of impurities present within an anticancer agent. Specific within this is mutagenic impurities, S9 making clear that control to levels defined by EMA, FDA, and now ICH M7 guidelines are not appropriate. ICH M7, within Section 2 (scope), makes it clear that ICH S9 holds primacy.

The only remaining uncertainty relates to what is an acceptable level for a mutagenic impurity? This is not expressly defined. ICH S9 states "Such limits are not appropriate for pharmaceuticals intended to treat patients with advanced cancer, and justifications described above should be considered to set higher limits." Many have interpreted this to mean that you can default to ICH Q3A [5]/Q3B [6]; however, in reality it is likely that limits will be established on a case-by-case basis, with levels typically lower than ICH Q3A/3B but higher than defined in ICH M7 being agreed.

In 2018 ICH S9 was augmented by a Q&A document [9]. Within this specifically in relation to ICH M7 the following is tabulated (Table 2.1):

This reinforces the statement within M7 [1] making the specific point that where ICH M7 is not applicable the concepts outlined in ICH Q3A [5]/Q3B [6] should be applied.

The scope section also addresses the scenario whereby the therapeutic agent in question is itself genotoxic. Importantly and, it is assumed deliberately, the wider term genotoxic is chosen, as opposed to the specific term mutagenic. Hence this encapsulates pharmacologically active agents whose mechanism of action can be described as genotoxic, for example, inhibition of cell proliferation e.g. topoisomerase inhibition. The guideline specifically states that:

> *Exposure to a mutagenic impurity in these cases would not significantly add to the cancer risk of the drug substance. Therefore, impurities could be controlled at acceptable levels for non-mutagenic impurities.*

However, there exists an equally important question. "Does ICH M7 apply to other non-cancer, therapeutic disorders (e.g. rare diseases), where there is a significantly reduced life expectancy, i.e. <2 years (i.e. life threatening disorders, with a poor prognosis and with a similar risk benefit as is seen with cancer indications)?" The logical answer would be yes, but in the absence of an "ICH S9-like" guidance for rare diseases, this will require discussion with authorities on a case-by-case basis.

2.2.2.3 Nature of Therapeutic Agent/Excipients

The guideline makes clear that an assessment of the mutagenic potential is not required for the following types of DSs and DPs:

Table 2.1 Relationship between ICH S9 and ICH M7 [1] [9].

4.12	Should impurities exceeding the established qualification limits in ICH Q3A/B be assessed in genotoxicity studies? When the API is genotoxic? When the API is nongenotoxic?	**API genotoxic?**	**Impurity exceeds 3A/B qualification threshold?**	**Proposed action**
		Yes	No	None
		Yes	Yes	None
		No	No	None
		No	Yes	Genotoxicity assessment of impurities should be conducted.
4.13	Is ICH M7, giving guidance for the management of mutagenic impurities, applicable to the patient population covered in the scope of ICH S9?	The scope of ICH M7 specifically states that the guidance does not apply to "drug substances and drug products intended for advanced cancer indications as defined in the scope of ICH S9." Therefore, mutagenic impurities in products used for treatment of indications under the scope of ICH S9 should be considered for management consistent with the concepts outlined in ICH Q3A/B (see Question 4.12).		

Source: Reproduced from ICH S9 Q&A document.

- Biological/biotechnological, peptide, oligonucleotide, radiopharmaceutical, fermentation products, herbal products, and crude products of animal or plant origin.

Close scrutiny of this highlights some anomalies; certainly many peptides and oligonucleotides are synthetic/semisynthetic in nature, and it is perhaps difficult to rationalize their exclusion (indeed, there was some commentary about these classes in the penultimate draft version of the guideline).

Established excipients are confirmed as also being out of scope. However, the guidance does indicate that it may be applicable for new synthetic excipients. "The safety risk assessment principles of this guideline can be used, if warranted, for impurities in excipients that are used for the first time in a drug product and are chemically synthesized."

The position in relation to extractables and leachables is somewhat ambiguous, while stating that the guidance is not applicable to leachables, it states that "Application of this guideline to leachables associated with drug product packaging is not intended, but the safety risk assessment principles outlined in this guideline for limiting potential carcinogenic risk can be used, if warranted." In practice the use of the TTC in the assessment of mutagenic impurities has become common.

2.2.3 General Principles

Key within this section, (Section 3 of the guideline), is the reaffirmation of the specific focus of the guideline on mutagenic impurities. Indeed, the guideline goes further in making the specific statement that:

> *Other types of genotoxicants that are non-mutagenic typically have thresholded mechanisms and usually do not pose carcinogenic risk in humans at the level ordinarily present as impurities.*

It also makes clear that in assessing mutagenic potential that this can be achieved through a combination of *in silico* SAR evaluation, and where required, with a bacterial reverse mutation assay (Ames test [10]). This is certainly useful in clarifying that *in silico* SAR assessments should focus specifically on mutagenicity. Other SAR models for other toxicological end points such as chromosomal activity and carcinogenicity exist within many of the *in silico* systems utilized; however, these do not need to be specifically applied in the evaluation of impurities; the primary focus is mutagenicity. It can be further concluded that in terms of *in vitro/in vivo* assays that no other test other than a reverse mutation assay is required, precluding in particular a mammalian cell assay defined as part of impurity qualification within ICH Q3A [6].

What is less clear is if there is evidence that an impurity is clastogenic/aneugenic, whether or not this can be ignored, or if limits should be based on an evaluation of the available data – i.e. a limit calculated based on an observed no observable effect level (NOEL) using the permitted daily exposure (PDE) calculation shown in ICH Q3C [11]. In practice it seems logical to use such safety data to set acceptable limits even though such impurities are effectively outside of the scope and would be defined as Class 5 (see Table 2.2).

Also reaffirmed within this section is the fact that the guideline remains, as per earlier regional EMA and FDA guidelines, based on the TTC principle [12–16], and a limit of 1.5 μg/day based on a 1 in 100 000 risk following lifetime exposure (70 years). While clear flaws within the TTC have been identified by Snodin [17] and Delaney [18], what is singularly lacking is any scientifically justified alternative. The TTC is without doubt an overtly conservative interpretation of the risk; one that employs multiple worst-case scenarios, but to date an alternative to it remains elusive. This innate conservatism is even recognized within the guideline itself:

> *The use of a numerical cancer risk value (1 in 100,000) and its translation into risk-based doses (TTC) is a highly hypothetical concept that should not be regarded as a realistic indication of the actual risk. Nevertheless, the TTC concept provides an estimate of safe exposures for any mutagenic compound. However, exceeding the TTC is not necessarily associated with an increased cancer risk given the conservative assumptions employed in the derivation of the TTC value. The most likely increase in cancer incidence is actually much less than 1 in 100,000.*

Finally, within this section a passing reference is made to metabolites. The guideline simply states that where an impurity is also a metabolite its mutagenic potential should be addressed through evaluation of the metabolite. Again this was a specific area highlighted within the concept paper; however, it is difficult to see how this has been meaningfully addressed and sits as perhaps the one remaining area of uncertainty following publication of ICH M7. The issue is thoroughly examined in detail by Dobo et al. [19].

2.2.4 Considerations for Marketed Products

Section 4 of the guideline is intended to be applied to products that were either marketed prior to, or after, the adoption of this guideline. The two types of products are treated differently. For those marketed after implementation, a further set of requirements (Section 8.5 Lifecycle Management) apply.

The need for further evaluation is defined in terms of changes. For those products marketed before the advent of ICH M7, application of the guideline is limited to the criteria defined within Section 4 of the guideline. This itself is divided into four subsections:

- Changes to Drug Substance
- Changes to Drug Product

- Changes to Clinical Use
- Other Considerations

2.2.4.1 Post-approval Changes to Drug Substance, Chemistry, and Manufacturing Controls

Changes to the DS manufacturing route or process should be evaluated to determine whether they result in any new mutagenic impurities or higher acceptance criteria for existing mutagenic impurities. An important point is that the reevaluation of impurities should focus on those stages impacted by the change, and that other stages and their associated impurities unaffected by the change do not require reevaluation.

Another important point is that changes should be assessed from the registered starting material (RSM) onward. However, this should be put into the context of where in the synthesis the RSM is introduced (see ICH Q11 [20]). If the RSM is introduced four or five stages before the active pharmaceutical ingredient (API), then there is likely to be effective purging of any mutagenic impurities in the downstream chemistry, whereas for an RSM introduced in the penultimate stage, such an assessment is very important. This is reflected in Section 5.1. (Relationship between the risk assessment and testing of starting materials):

> *For starting materials that are introduced late in the synthesis of the drug substance (and where the synthetic route of the starting material is known) the final steps of the starting material synthesis should be evaluated for potential mutagenic impurities.*

This may suggest that in the context of an existing product, where a starting material is introduced late in the synthesis and changes are subsequently made to the route/process of the material (i.e. preregistered stages), that in such instances a reevaluation would be necessary.

2.2.4.2 Post-approval Changes to Drug Product Chemistry, Manufacturing, and Controls

In the context of changes to the DP, such as a modified formulation, the guideline makes clear that the focus is specifically on the DP and on any new or increased levels of a mutagenic degradant arising as a result of that change. Critically, it makes clear that changes to the formulated product do not trigger a reevaluation of the route of synthesis of the DS, unless there is a concomitant change.

2.2.4.3 Changes to the Clinical Use of Drug Products

This applies where there is a change in either dose or duration (increase) or a change in indication from a serious or life-threatening condition to an indication for a less serious condition where the existing impurity acceptable intakes (AIs) may no longer be appropriate. An obvious example of the latter would be in instances where an anticancer treatment used in late stage disease for which limits have been based on ICH S9 [8] is then used for either a revised patient population, e.g. adjuvant therapy, or a new non-oncology indication.

Another important point that relates to the innate conservatism of the TTC is the statement that changes to patient population in terms of age do not trigger the need for reassessment and revision of limits. Previously the draft FDA guideline [4] had indicated the need to modify limits for pediatric indications. This has been removed from the ICH M7 guideline.

In addition, the earlier FDA draft guideline had indicated that different limits may be applicable dependent on the planned route of delivery, i.e. oral, pulmonary, etc. Similarly, this has been removed from the ICH M7 guideline.

2.2.5 Other Considerations for Marketed Products

The section relates to where a new impurity is discovered and there is "a cause for concern." Previously, EDQM stated that structural alert alone is insufficient to trigger a concern, there must be new safety data and this point is reiterated in ICH M7: "The existence of impurity structural alerts alone is considered insufficient to trigger follow-up measures, unless it is a structure in the cohort of concern."

New safety data may be generated for a number of reasons and via a variety of routes. One such example is REACH testing, where chemical registration requirements require a battery of safety tests that may include an assessment of mutagenicity.

What is REACH?

REACH is a European Union (EU) regulation concerning the Registration, Evaluation, Authorisation and Restriction of Chemicals. It came into force on 1 June 2007 and replaced a number of European directives and regulations with a single system.

REACH has several aims:

- To provide a high level of protection of human health and the environment from the use of chemicals.
- To make the people who place chemicals on the market (manufacturers and importers) responsible for understanding and managing the risks associated with their use.
- To allow the free movement of substances on the EU market.
- To assess risk, safety testing may be performed.

There are concerns over how new safety data are managed in practice. As a result of REACH, a coupling reagent 1-ethyl-3-(3-dimethylaminopropyl)carbodiimide (EDAC) (see Figure 2.1) was shown to be Ames positive. This chemical was widely used and at the time the data were generated there was no clear guidance as to how should authorities be informed of this new finding. Also uncertain is the impact on marketing authorization, would such a finding lead to immediate suspension of the license for the product(s) impacted or would time be permitted to make the changes needed to ensure control to appropriate levels? In the case of EDAC, the issue was minor due to the instability of the coupling agent, meaning that it was readily purged and thus presented no concern in terms of levels in associated products. Nevertheless, given the wide utility of commodity reagents like EDAC, the impact could be significant, impacting multiple products globally.

What is not entirely clear is how to address the scenario whereby an impurity is detected that based on its' structure belongs to the cohort of concern. It is clear based on the example of *N*-nitrosamines (see Chapter 10) that detection of such an impurity would be a trigger for investigation.

2.2.6 Drug Substance and Drug Product Impurity Assessment

This section (Section 5 of the guideline) is essentially the start point for the overall risk assessment process itself. Critically, in practical terms, this essentially advocates the same approach as previous guidelines. The emphasis remains focused on actual impurities and potential impurities *likely*

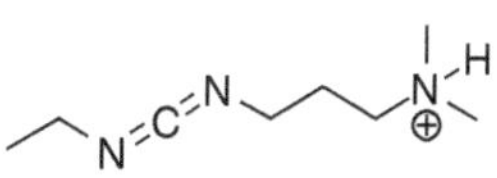

Figure 2.1 EDAC.

to be present in the API/DP. Another important aspect of this section of the guideline is that it also looks to link the ICH M7 [1] to ICH Q3A [5]/Q3B [6] reporting and identification requirements. Actual impurities are defined as those observed in the DS above the ICH Q3A reporting thresholds. Identification of actual impurities is expected when the levels exceed the identification thresholds outlined by ICH Q3A [6]. This confirms the primacy of ICH Q3A in terms of identification thresholds.

In practical terms this means that there is no formal requirement to attempt to identify every possible impurity at a TTC-like level. Such an attempt would be in reality, impractical. Instead, the guideline tacitly accepts the finite but low risk of a mutagenic impurity existing, unidentified, at level below the ICH Q3A identification threshold. This is considered to be an acceptably low risk provided a comprehensive risk assessment is made of the synthetic process, its related impurities, DP manufacturing process, and associated degradants.

2.2.6.1 Synthetic Impurities

Unlike earlier regional guidelines, ICH M7 clearly defines a start point in terms of the risk assessment, indicating that such an assessment may begin at the starting materials. As described above there is a caveat that relates to the proximity of the starting material to the final API. The guideline states that for starting materials that are introduced late in the synthesis of the DS (and where the synthetic route of the starting material is known), the final steps of the starting material synthesis should be evaluated for potential mutagenic impurities. One may consider – what is meant by final steps? Two steps or perhaps three? This is not defined. Despite this uncertainty, the guideline provided by ICH M7 is much clearer in terms of its starting point than previously and should make the process simpler as a consequence.

Another important aspect of this section of the guideline is the clear recognition that proximity to the final API is a key factor in determining risk (i.e. purging capacity of the downstream process), that early stages will, in general, represent a lower risk of carryover than later stages. This point is returned to later in the guideline within the control options section and is again a key, new, aspect in the guideline.

In the context of synthetic impurities, mutagenic impurities can arise via three sources:

1) Mutagenic reagents used deliberately in the synthesis. Many of the common reagents used in the synthesis of the DS are mutagenic. The use of such reagents, i.e. methyl iodide, epichlorohydrin, etc., is effectively unavoidable; it is simply impractical to construct C—C and C—N bonds without the use of such reagents [21].
2) Mutagenic intermediates – often the use of a deliberately formed, highly reactive, intermediate is required – examples include tosylates, hydrazides, and epoxides; such an intermediate being deliberately utilized to effect an efficient synthesis.
3) Side reactions. Perhaps the most difficult to assess, those impurities formed as a result of predictable side reactions. Wherever possible this should be based on existing scientific knowledge. This has been drawn into sharp focus by events surrounding *N*-nitrosamines. The risk of MIs arising from side reactions is the focus of Chapter 11.

2.2.6.2 Degradation Products

Unlike earlier guidelines, ICH M7 specifically addresses the issue of mutagenic degradants. Similar to DS impurities, the guideline makes clear the primacy of ICH Q3B [8] and the identification thresholds for the product. It also makes clear again the need to focus on degradants likely to be present in the final DP. While clearly helpful, this nevertheless does not define how the risk posed

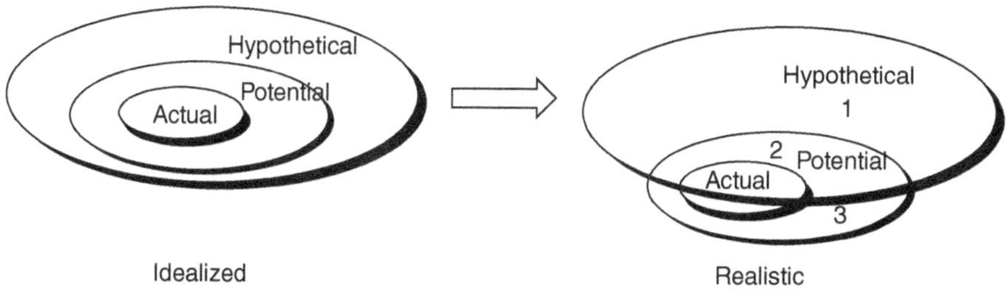

Figure 2.2 Interrelationship between degradant classes.

by degradants should be evaluated. The guideline provides some advice defining degradants in terms of actual degradants and probable degradants. Actual degradants are those observed to form over prolonged storage at ambient temperature, i.e. ICH long-term accelerated storage conditions. Probable degradants are defined in terms of those observed under accelerated conditions, e.g. 40°/75%. Reference is made within the guideline to the use of knowledge derived from stress studies. The evaluation of mutagenic degradants is examined in detail by Baertschi et al. [22]. The use of stress studies to identify major degradative pathways and their associated primary degradants is underpinned by the established relationship between degradants formed under stress conditions to those seen under ambient conditions. Baertschi [23] demonstrated that degradants formed under ambient conditions were a contained "sub-set" of those observed to form under stressed conditions, this is illustrated in Figure 2.2, where the relationship between idealized and realistic degradation knowledge landscapes is considered.

Hypothetical degradants arise from *in silico* and *in cerebro* assessments; potential degradants are observed as major degradation products in stress testing and accelerated stability testing; actual degradation products are those that arise under ICH long-term (real-time) storage conditions.

Another important consideration in the assessment of degradants is what is an appropriate identification threshold within the context of a stress study. The identification threshold defined within ICH Q3B [7] for a DP with a dose of >10 mg – 2 g is set at 0.2% or 2 mg. This is relative to a typical maximal level of degradation of typically 2% total degradation in DP. In the context of a stress study where degradation levels may well exceed 10% total, an adjustment factor of five to the identification threshold would seem pragmatic, raising this to a value of 1.0%. Using such a structured approach will ensure that impurities identified during a stress study are the primary degradants and hence commensurate with the focus of the guideline on degradants likely to be present in the final DP.

Overall in terms of the impurity assessment the impurities for consideration are reflected in Figure 2.3.

This topic is examined in detail in Chapter 14.

2.2.7 Hazard Assessment

The emphasis of the guideline (Section 6) now shifts and focuses on an assessment of the mutagenic potential of impurities identified in the preceding risk assessment. Such an assessment is typically made through the use of *in silico* SAR systems. The guideline defines the need to apply two (Q)SAR methodologies. One methodology should be expert rule based, and the other methodology should be statistical based; however, the guidance does not define which software packages are preferable; this decision is left to the end user. Importantly, it also highlights the need for an expert evaluation of the results.

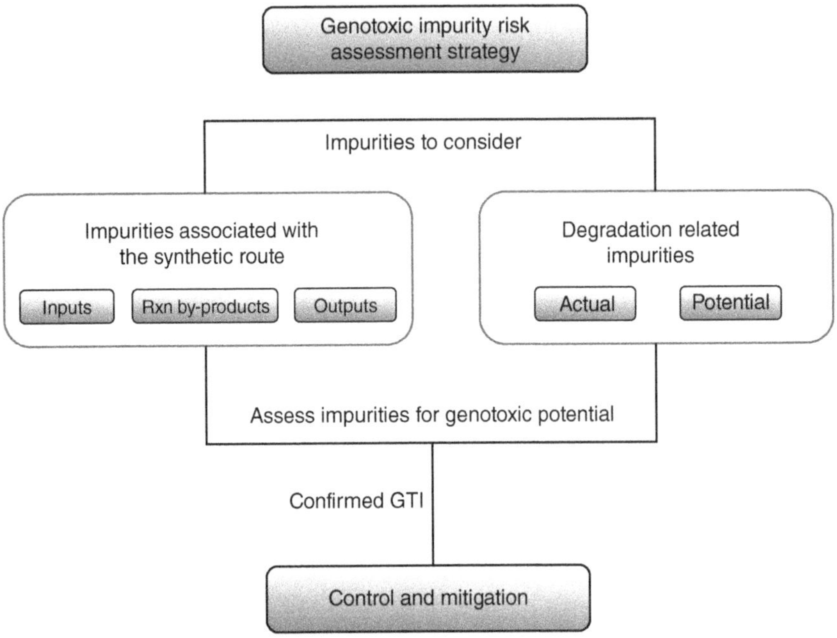

Figure 2.3 Potential sources of mutagenic impurities.

The use of two methodologies throws up a number of different permutations; these include not only situations where predictions are conflictory in nature but also out of domain predictions. These arise where the molecule in question, or at least a significant proportion of the molecule, is not recognized by the training set of the *in silico* tool, and hence it cannot accurately predict its mutagenic potential. In such circumstances, expert evaluation is required to make an overall consensus prediction. Barber et al. [24] examine this in detail. They describe the various scenarios potentially encountered (Figure 2.4), examining how to address conflictory predictions as well as out of domain predictions. Barber et al. provide advice on how to challenge predictions made by both rule based and (Q)SAR systems as well as providing a series of examples that serve to provide effective practical illustration of the key points made within the paper.

In a related study, Green et al. [25] examined the relative predictive performances of popular commercial *in silico* systems. Using a data set of some 801 chemicals and pharmaceutical intermediates, they showed the overall accuracy of each of the systems was generally comparable, ranging from 68 to 73%; however, their studies showed significant differences in sensitivity of each system (i.e. how many Ames positive compounds are correctly identified) results varying between 48 and 68%. The studies did not, however, identify any stand out system or specific combination of rule based/(Q)SAR systems. Perhaps the most significant finding of the studies was the number of contradictory predictions observed when two different methodologies were applied, i.e. those where one system predicted positive and the other did not or the statistical models were not able to make a prediction. Over one-third of all the compounds in this 801 compound data set were seen to give a conflictory prediction. The authors concluded there is clearly a need for expert opinion to be applied to determine the appropriate classification.

Ultimately, the outcome of any such assessment is then classified using the system defined by Mueller et al. [26]. This is shown below in Table 2.2.

A particular challenge with respect to Class 4 compounds is defining structural similarity. Mathematical approaches such as Tanimoto scores may be utilized; however, great care is required

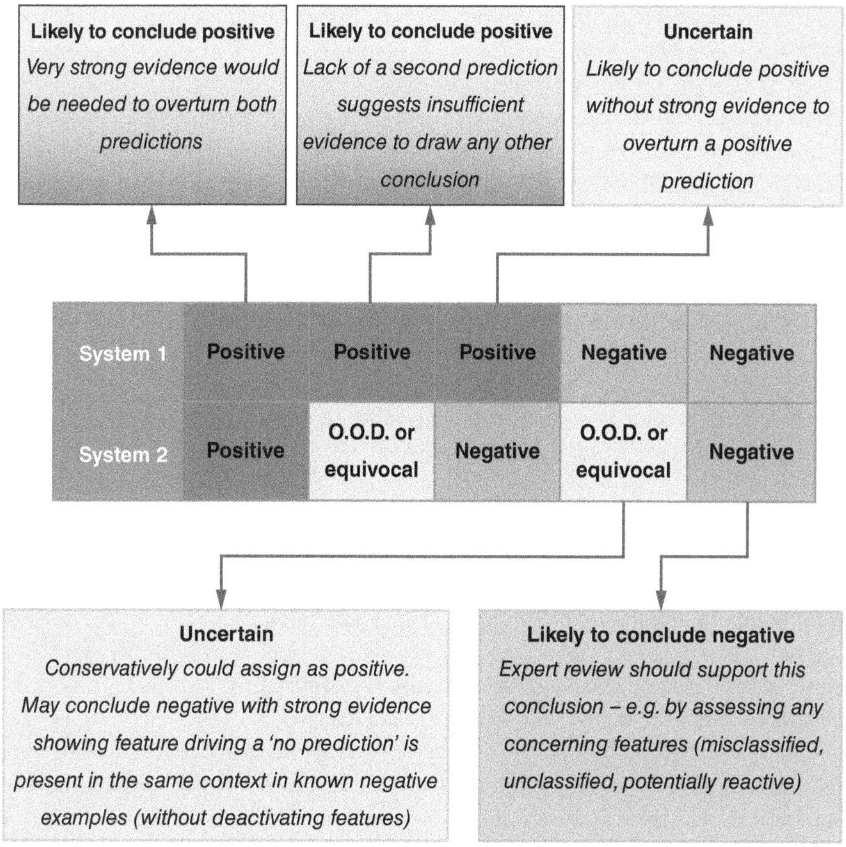

Figure 2.4 Decision matrix when evaluating two *in silico* predictions.

Table 2.2 Impurities classification with respect to mutagenic and carcinogenic potential and resulting control actions.

Class	Definition	Proposed action for control (details in Sections 7 and 8)
1	Known mutagenic carcinogens	Control at or below compound-specific acceptable limit
2	Known mutagens with unknown carcinogenic potential (bacterial mutagenicity positive, no rodent carcinogenicity data)	Control at or below acceptable limits (appropriate TTC)
3	Alerting structure, unrelated to the structure of the DS; no mutagenicity data	Control at or below acceptable limits (appropriate TTC) or conduct bacterial mutagenicity assay; if nonmutagenic = Class 5; if mutagenic = Class 2
4	Alerting structure, same alert in DS or compounds related to the DS (e.g. process intermediates), which have been tested and are nonmutagenic	Treat as nonmutagenic impurity
5	No structural alerts, or alerting structure with sufficient data to demonstrate lack of mutagenicity or carcinogenicity	Treat as nonmutagenic impurity

in their use and similarity cannot simply be defined by a score exceeding a predefined threshold. In all cases it is important to assess the environment, both steric and electronic, in close proximity to the alerting moiety within the impurity in question.

Based on the outcome of the SAR assessment, for those compounds considered a concern, in particular those classified as Class 3, further evaluation in the form of testing may be performed. The earlier scope section of the ICH M7 guideline makes clear that the emphasis is on mutagenic impurities and that the bacterial reverse mutation test (Ames) [10] should be used to follow up any SAR alert.

In addition this section provides an overview of potential *in vivo* follow-up tests that can be utilized in order to investigate further a positive bacterial assay. The tests themselves are described in detail in Table 2.3 (based on Note 3 within the guideline).

The guideline states that such tests can be used to assess the *in vivo* relevance of the positive findings of the *in vitro* bacterial reverse mutation test, suggesting that the results may support the establishment of a compound-specific limit.

2.2.8 Risk Characterization

This section, Section 7 in the guideline, outlines the risk characterization principles used to define acceptable limits for compounds classified in groups 1, 2, or 3, see Table 2.3.

2.2.8.1 Acceptable Intakes Based on Compound-specific Risk Assessments

2.2.8.1.1 Mutagenic Impurities with Positive Carcinogenicity Data (Class 1)

It is important to note that the guideline specifically stipulates that where adequate carcinogenicity data exist it should be used to calculate a compound-specific AI or ADI. It also outlines that the approach should mirror that of the derivation of the TTC itself, i.e. linear extrapolation to a risk

Table 2.3 Tests to investigate the *in vivo* relevance of *in vitro* mutagens (positive bacterial mutagenicity).

in vivo test	Factors to justify choice of test as fit-for-purpose
Transgenic mutation assays	• For any bacterial mutagenicity positive. Justify selection of assay tissue/organ
Pig-a assay (blood)	• For directly acting mutagens (bacterial mutagenicity positive without S9[a])[b]
Micronucleus test (blood or bone marrow)	• For directly acting mutagens (bacterial mutagenicity positive without S9) and compounds known to be clastogenic[b]
Rat liver unscheduled DNA synthesis (UDS) test	In particular for bacterial mutagenicity positive with S9 only; responsible liver metabolite known to be generated in test species used to induce bulky adducts
Comet assay	• Justification needed (chemical class specific mode of action to form alkaline labile sites or single-strand breaks as preceding DNA damage that can potentially lead to mutations) • Justify selection of assay tissue/organ
Others	• With convincing justification

[a] S9 – Supernatant fraction obtained from an organ (usually liver) homogenate and contains cytosol and microsomes. The microsomes component of the S9 fraction contains cytochrome P450 isoforms (Phase I metabolism) and other enzyme activities.
[b] For indirect acting mutagens (requiring metabolic activation), adequate exposure to metabolite(s) should be demonstrated.
Source: Reproduced from ICH M7.

value of 1 in 100 000 analogous in risk terms to the TTC. It also states that other established risk-vassessment practices such as those used by international regulatory bodies may be applied either to calculate AIs or the actual values themselves used. This apparently helpful statement does in fact lead to considerable confusion: should for example the value for a particular compound specified by the US environmental protection agency (EPA) (or other agencies) be simply adopted or in such instances should the available data be evaluated using the linear extrapolation? The guideline provides no clear statement on such a point and nor does it provide any specific example. In practice it would seem appropriate to use the default approach of linear extrapolation where data are available. Dobo et al. [27] recently reported on various ADIs that can be generated for hydrazine, exemplifying the considerable ambiguity that can be found when trying to generate a compound-specific ADI for regulatory use.

Linked to this section is Note 4 where a specific example calculation is provided. The calculation outlines the determination of an AI for ethylene oxide. It is surprising that ethylene oxide was chosen as it is a gas, with good purging potential and of little synthetic utility, making its presence in final product very unlikely. Furthermore there is strong evidence that it is also generated endogenously [28]. In terms of the calculation itself, it is relatively straight forward. Terminal dose (TD_{50}) values are taken from the Carcinogenicity Potency Database (CPDB) for both rat and mouse, with the more conservative value being selected, 21.3 mg/kg/day (rat) and the limit calculated by dividing by 50 000 to adjust to a 1 in 100 000 risk and multiplied by the internationally accepted average human body weight (50 kg), to give an ADI of 21.3 μg/day for lifetime exposure.

On the face of it, this looks relatively straightforward; however, this is a simple example. In reality this is often far more complex. In many cases data are available for multiple carcinogenicity studies, within the CPDB these are combined and reported in terms of the harmonic mean. The studies involved may be of variable quality, e.g. insufficient duration, low animal numbers, and limited number of doses studied. Another important factor for consideration is tumor site and relevancy, for example forestomach tumors in rodents. Such tumors are often associated with local irritation/inflammation and are considered nonrelevant to humans, both from a physiological perspective (humans have no forestomach) and from an exposure perspective; the impurity at the low levels observed within a pharmaceutical product renders it extremely unlikely to result in such irritation.

As a result of the complexity described, a cross-industry initiative was established that looked to develop an addendum table to the guideline. Included within this would be agreed limits for a range of common mutagenic/carcinogenic reagents. In addition to the actual agreed AIs, specific criteria were established to allow for the calculation of limits for other reagents in addition to those captured in the addendum. This ultimately culminated with the publication of the revised guideline in July 2017. The derivation of limits is described in detail in Chapter 7.

2.2.8.2 Acceptable Intakes for Class 2 and Class 3 Compounds

In the absence of specific carcinogenicity data, the ICH M7 guideline outlines AIs based on the TTC. Critically, for both products in clinical development and marketed products, limits are based on duration of exposure, i.e. a staged approach [26]. The recognition that limits should be based on duration of exposure irrespective of phase of development or marketing status is a major step forward as prior to finalization of the guideline the lifetime limit of 1.5 μg/day was routinely applied to marketed products, irrespective of their intended use and likely duration.

The calculation of less-than-lifetime (LTL) AIs is ultimately based on the principle of Haber's law, where concentration (C) × time (T) = a constant (k). The limits are described in Table 2.3.

Within this section of the guideline is a specific section dedicated to clinical development. This proposes that for studies of <14 days in duration that limits only apply to those impurities for which there is actual *in vitro* safety data (Class 2), and that no specific controls are required over levels of impurities where there is only a structural alert (Class 3). While this is an interesting concept that would appear to provide flexibility, it remains to be seen whether or not this is utilized to a significant extent by applicants. In many cases it is unlikely that material will be restricted to use in clinical studies of <14 days duration, in which case in order to utilize material for longer duration studies, compliance with the limits outlined in Table 2.4 will be required for both Class 2 and Class 3 impurities.

In the context of marketed products, a particular challenge is defining duration. Again this is addressed through a specific note in the guideline, (Note 7), taking the form of a table where, based on specific therapeutic areas, the likely treatment durations are discussed. Useful advice is provided within the guideline itself where it is made clear that the duration of use should be defined based on typical use. This recognizes that, in some cases, a subset of the population of patients may extend treatment beyond the marketed products' anticipated duration of use. It states that in such instances the increase in risk is negligible.

It is important to note also the existence of a "quality cap." Because limits for MIs are based on dose, this can create scenarios within low dose products where limits for an MI are very high in concentration terms, well beyond a level acceptable from a quality perspective. The guideline addresses this with the following statement:

> The acceptable intakes derived from compound-specific risk assessments (Section 7.2) can be adjusted for shorter duration of use in the same proportions as defined in the following sections (Section 7.3.1 and 7.3.2) or should be limited to not more than 0.5%, whichever is lower.

2.2.8.3 Multiple Impurities

One of the questions posed within the original ICH M7 concept paper relates to potential synergistic effect of multiple mutagenic impurities, especially where they are closely related. Based on the work of Bercu [29], the guideline concludes that such effects are unlikely at the low levels potentially present and hence permits the presence of up to three mutagenic impurities, to give a total AI as defined in Table 2.5.

Table 2.4 Acceptable intakes for an individual impurity.

Duration of treatment	≤1 month	>1–12 months	>1–10 years	>10 years to lifetime
Daily intake (μg/day)	120	20	10	1.5

Source: Reproduced from ICH M7.

Table 2.5 Acceptable total daily intakes for multiple impurities.

Duration of treatment	≤1 month	>1–12 months	>1–10 years	>10 years to lifetime
Total daily intake (μg/day)	120	60	30	5

Source: Reproduced from ICH M7.

It is important to note that this total only applies to specified Class 2 and Class 3 impurities. In reality most synthetic routes will utilize multiple mutagenic reagents/intermediates; however, due to the innate reactivity of the reagents in question and the downstream process conditions, few actually pose a substantive risk in terms of potential carryover. This is discussed in detail in the control section.

2.2.8.4 Exceptions and Flexibility in Approaches

The final area covered in Section 7 of the guideline addresses various further scenarios where modified limits may apply. Areas include:

1) Where substantive exposure to the impurity may occur either exogenously or endogenously. Perhaps the clearest example of this is formaldehyde.
2) Exceptions to the use of the appropriate AI can be justified in cases of severe disease, reduced life expectancy, late onset but chronic disease, or with limited therapeutic alternatives. In practice this is likely to be applicable where effective control has not been possible, in such instances levels will likely be controlled to as low as reasonably practical (ALARP) and justified following subsequent consultation with regulatory authorities.
3) Where the mutagenic impurity, based on its structure, is considered to likely be a potent carcinogen (cohort of concern), i.e. aflatoxin-like-, *N*-nitroso-, and alkyl-azoxy structures, the guideline states that if these compounds are found as impurities in pharmaceuticals, AIs would likely be significantly lower than the AIs defined in this guideline. This area has of course been brought sharply into focus by issues surrounding *N*-nitrosamines, this being examined from both a quality and safety perspective in Chapters 10 and 7, respectively.

2.2.9 Control Strategy

One of the most significant aspects of ICH M7 [1] is the control section (Section 8). This provides far greater flexibility in terms of strategies to demonstrate absence than are available in the preceding regional guidelines. These are expressed in terms of a series of control options that provide the ability to more widely use chemical/process-based arguments to assess purging and effectively aligns the guideline with the risk-based approaches outlined in ICH Q9 [30]. There are four options:

1) Option 1 – include a test for the impurity in the DS specification with an acceptance criterion at or below the acceptable limit.
2) Option 2 – include a test for the impurity in the specification for an input raw material, starting material or intermediate, or as an in-process control, with an acceptance criterion at or below the acceptable limit.
3) Option 3 – include a test for the impurity in the specification for a raw material, starting material or intermediate, or as an in-process control, with an acceptance criterion above the acceptable limit of the impurity in the DS. Such a limit is justified based on clear understanding of the fate and purge in question within the downstream process negating the need for any additional testing later in the process.
   Such an option can be justified when the level of the impurity in the DS will be less than 30% of the acceptable limit by review of data derived from laboratory-scale spiking experiments.
4) Option 4 – process control. The highly reactive nature of most mutagenic impurities is such that they are unlikely to survive typical downstream processes at levels of concern in the subsequent DS. In order to assess the potential carryover of such impurities requires an understanding of

the effect that process conditions have on residual impurity levels. Where there is sufficient confidence that the level of the impurity in the DS will be below the acceptable limit, then no analytical testing may be required and the impurity does not have to be specified.

Examining these options in more detail many would conclude that the ordering should be reversed. As stated above, most mutagenic reagents are highly reactive by nature or deliberate design. They are used in the construct of the DS to form the molecular skeleton of the compound. This high reactivity is the reason why they are also effectively purged. The challenge has been to be able to assess the level of the risk without resorting to extensive analytical testing. Such an approach was developed by Teasdale et al. [29, 31]; this is examined in detail in Chapter 9. In this approach calculations are based on an evaluation of the specific physicochemical properties of the impurity in question (e.g. reactivity/solubility), relative to the downstream processing conditions they will be exposed to. This is based on the scoring system described in Table 2.6.

This specific approach is referenced directly within the guideline and its application examined in detail in the attached case study.

Examples of the options described are further examined within the guideline through a series of case studies. There are four cases in total: case studies 1 and 2 examine the use of Option 3, case study 3 examines the use of Options 2 and 4 combined, and case study 4 examines the use of Option 4.

Case study 3 (guideline) – reproduced below, is concerning.

The Step 1 intermediate of a five-step synthesis is a nitroaromatic compound that may contain low levels of impurity C, a positional isomer of the Step 1 intermediate and also a nitroaromatic

Table 2.6 Purge values.

Physicochemical parameters	Purge factor
Reactivity	Highly reactive = 100
	Moderately reactive = 10
	Low reactivity/unreactive = 1
Solubility[a]	Freely soluble = 10
	Moderately soluble = 3
	Sparingly soluble = 1
Volatility	Boiling point >20 °C *below* that of the reaction/process solvent = 10
	Boiling point ± 10 °C that of the reaction/process solvent = 3
	Boiling point >20 °C *above* that of the reaction/process solvent = 1
Ionizability	Ionizaion potential of genotoxic impurity (GI) significantly different to that of the desired product[b]
Physical processes – chromatography	Chromatography – GI elutes prior to desired product = 100
	Chromatography – GI elutes after desired product = 10
	Others evaluated on an individual basis

[a] This relates to solubility within the context of a recrystallization/isolation process whereby the impurity in question, if highly soluble, will remain within mother liquors and hence be purged from the desired product.
[b] This relates to a deliberate attempt to partition the desired product/GI between an aqueous and organic layer, typically achieved through the manipulation of pH to change the ionized/unionized state of one of the components.

compound. The amount of impurity C in the Step 1 intermediate has not been detected by ordinary analytical methods, but it may be present at lower levels. The Step 1 intermediate is positive in the bacterial mutagenicity assay. The Step 2 hydrogenation reaction results in a 99% conversion of the Step 1 intermediate to the corresponding aromatic amine. This is confirmed via in-process testing. An assessment of purge of the remaining Step 1 nitroaromatic intermediate was conducted, and a high purge factor was predicted based on purge points in the subsequent Steps 3 and 4 processing steps. Purge across the Step 5 processing step is not expected, and a specification for the Step 1 intermediate at the TTC-based limit was established at the Step 4 intermediate (Option 2 control approach). The positional isomer impurity C would be expected to purge via the same purge points as the Step 1 intermediate and therefore will always be much lower than the Step 1 intermediate itself and therefore no testing is required and an Option 4 control strategy for impurity C can be supported without the need for any additional laboratory or pilot scale data.

In this example during Step 1 the nitro-aromatic is reported to be converted at high yield (99%) conversion to an aromatic amine; further reductions in level are reported to occur at Steps 3 and 4. Certainly, were such a scenario encountered in reality, it would seem logical to conduct a risk assessment based on an evaluation of the purging capacity of the process before simply defaulting to a control strategy based on Option 2.

2.2.9.1 Considerations for Control Approaches

Within the guideline, this subsection of the overall control strategy section provides further clarification as to what is required to support the specific control options in terms of data. Of particular note is the need to take into consideration the effect of scale, i.e. it is important to address the expected scale dependence or independence of any data especially where supporting data are derived from lab or pilot scale manufacture.

If options 3 and 4 cannot be justified, then a test for the impurity on the specification for a raw material, starting material or intermediate, or as an in-process control (Option 2) or DS (Option 1) at the acceptable limit should be included.

Another important point made within this specific section relates to impurities introduced in the last synthetic step; it states that an Option 1 control approach would be expected unless otherwise justified. The caveat that other options may be justified is an important caveat and potentially allows other control options to be justified based on a scientific risk assessment as opposed to simplistic dogma.

Earlier regional guidelines had been predicated on the application of "As Low As Reasonably Practicable" (ALARP) even where levels were below the TTC. They also talked about avoidance of the use of mutagenic reagents. ICH M7 addresses both points directly. It is not necessary, if the level of the mutagenic impurity is below acceptable limits, to implement limits based on the ALARP principle. Similarly, it is not necessary to demonstrate that alternate routes of synthesis have been explored.

Where the ALARP principle is applicable is in cases where efforts to reduce the level of a mutagenic impurity to below the acceptable limit have proven unsuccessful. In this instance, that provided levels are ALARP – a higher limit may be justified based on a risk/benefit analysis.

2.2.9.2 Considerations for Periodic Testing

This section seeks to provide guideline in terms of the use of periodic testing; however, it unfortunately lacks real clarity, failing to provide any definite advice as to when such options are to be applied. A specific concern in the absence of real clarity is that such an approach may be requested even when the control process has been fully justified.

2.2.9.3 Control of Degradation Products

Appearing somewhat out of place the control strategy section also includes specific advice relating to the control of degradation products. The guideline states the need to understand the degradation pathway for a mutagenic degradant and its relevance to the DS and DP manufacturing processes and/or their proposed packaging and storage conditions. It also articulates how either accelerated stability studies (e.g. 40 °C/75% relative humidity, six months) or kinetically equivalent shorter term stability studies at higher temperatures in the proposed commercial package may be used to determine the relevance of the degradation pathway prior to initiating longer term stability studies. This of course makes perfect sense, there being a considerable risk in relying on long-term studies to provide the required assurance of control. Finding out control is inadequate at a late stage, i.e. at the end of long-term stability studies, is likely to have a serious impact. Earlier identification of any potential issue would certainly allow more time to consider formulation/packaging options for control. Again, as with synthetic impurities, if formulation development and packaging design options are unable to control mutagenic degradation product levels to less than the acceptable limit, it may be possible to justify a higher limit based on a risk/benefit analysis.

2.2.10 Lifecycle Management

This section specifically applies to those products approved after the issuance of this guideline. A considerable portion of the section is taken up by what amounts to a description of lifecycle management and the need for an appropriate quality system. It is perhaps not wholly clear as to why this is included as it is not specific to the control of mutagenic impurities.

What is clear and useful is the statement that any proposed change to the manufacturing process should be evaluated for the impact on the quality of DS and DP. This evaluation should be based on understanding of the manufacturing process and should determine if appropriate testing to analyze the impact of the proposed changes is required.

It also stresses the need throughout the lifecycle of the product to reassess if testing is recommended when intended or unintended changes occur in the process, highlighting the need to consider this in particular where there is no routine monitoring (Option 3 or Option 4).

Mention is also made as to the potential use of statistical process control and trending of process measurements, but in the absence of any clear examples, it is not entirely clear as to how this might be utilized.

2.2.11 Documentation

The final section, Section 9 of the guideline, seeks to provide guidance as to the level of information to include within regulatory submissions pertaining to the assessment and control process for mutagenic impurities. This is split into two sections:

2.2.11.1 Clinical Trail Applications

This is further differentiated based on phase:

1) For Phase 1 studies of 14 days or less, the guidance defines the need to provide a description of efforts to mitigate risks of mutagenic impurities focused on Class 1 and Class 2 impurities and those in the cohort of concern.
2) For Phase 1 clinical trials greater than 14 days and for Phase 2a clinical trials, it states that in addition Class 3 impurities that require analytical controls should be included.

3) For Phase 2b and Phase 3 clinical development trials, a list of the impurities assessed by (Q)SAR now should be included, as well as a description of the systems used. Any Class 1, 2, or 3 actual and potential impurities should be described along with details as to how they are controlled. The results of bacterial mutagenicity tests should also be reported.

It is also stated that chemistry arguments may be used in lieu of analytical data for potential impurities that present a low likelihood of being present; this represents in effect Option 4 control. It is assumed therefore that this is applicable irrespective of the phase, despite this not being specifically stated.

2.2.11.2 Common Technical Document (Marketing Application)

The guideline stipulates:

1) The mutagenic impurity classification and rationale for this classification, for actual and potential process-related impurities and degradation products, needs to be provided. Particular emphasis is placed in the guideline on the need to justify the classification of impurities into Class 4 and Class 5 impurities.
2) When bacterial mutagenicity assays were performed on impurities, study reports should be provided for bacterial mutagenicity assays on impurities. What is not clear is whether or not this should be included directly in the submission in the relevant section (would this be within the preclinical section or S3.2 Impurities?) of the dossier or as a supplementary reference.
3) Justification for the overall control strategy and any proposed specification. For Option 3 and Option 4 control approaches, a summary of purge factors and identification of factors providing control (e.g. process steps and solubility in wash solutions) needs to be provided.

Within both later stage clinical applications and the Common Technical Document (CTD), there is an explicit need to properly articulate and justify the classification of impurities based on SAR and any subsequent bacterial mutagenicity data. One potential option would be to include a summary table.

2.2.12 Other Aspects

2.2.12.1 Relationship Between ICH M7 and ICH Q3A

Ahead of development of the guideline the question as to the relationship between guidelines pertaining to mutagenic impurities and ICH Q3A [5] was a considerable concern. These concerns centered on the issue of impurity qualification. For a marketed product, any impurity present at level >0.15% requires qualification. In addition to general toxicological qualification such an impurity requires qualification in terms of its genotoxic potential. It is of course obvious that ICH Q3A predates ICH M7 and indeed earlier regional guidelines; however, the question of genotoxic qualification remains. Of particular concern within this was, not only does ICH Q3A ask that consideration be given to conduct of assessment of mutagenic potential, but it also advises that an assessment of chromosomal activity be also considered.

Note 1 within ICH M7 addresses this specific concern. It states that no further qualification for mutagenic potential is required provided that an SAR assessment of mutagenic potential has been performed on any impurity present at levels either above or below those defined in ICH Q3A/Q3B. This includes the initial use of (Q)SAR tools to predict bacterial mutagenicity. Only in cases where the amount of the impurity exceeds 1 mg daily dose for chronic administration is further evaluation recommended. There seems little logic in the 1 mg limit; however, in reality this is unlikely to be regularly encountered in the context of reactive mutagenic impurities.

The other notes cover the following topics; Notes 2–6 each are examined as part of other detailed chapters.

- Note 2: Ames test – Good Laboratory Practice (GLP)
- Note 3: additional safety tests
- Note 4: compound-specific limits – extrapolation of TD_{50} values
- Note 5: compound-specific limits – class-based limits – monofunctional alkyl halides
- Note 6: thresholded mechanisms

Note 7: derivation of LTL limits

The guideline provides a helpful table that aligns treatment type/disease area to the duration of treatment and hence the applicable durational limit. It should be noted that this is based on median duration not outliers. Many treatments involve multiple and complex patient groups making it impractical to establish the longest potential time small and often specialist subpopulations receive a medicine.

2.3 Conclusions

At the concept stage of the ICH M7 guideline, a series of critical areas were identified, these included:

1) Use of *in silico* SAR tools for the assessment of mutagenic potential.
2) What are acceptable levels of genotoxic impurities during drug development?
3) What are acceptable levels of genotoxic impurities for marketing?
4) Should those impurities be regulated differently that are likely to have threshold effects?
5) Should levels of genotoxic impurities be regulated using a TTC approach?
6) Structurally related genotoxic impurities are likely to have similar mechanisms of action. Should these be summed in calculating a TTC?
7) What process of qualification testing should be followed for impurities that are metabolites?
8) What additional data are needed to support having no special restrictions, or a higher ADI than the TTC, for a genotoxic impurity?

Does ICH M7 adequately address these points? Yes. Certainly most, if not all, of the ambiguity present in earlier guidelines is addressed. Some areas of ambiguity do remain though. Certainly, management of mutagenic metabolites remains an area of uncertainty and some new areas added within the guideline may also prove challenging to interpret; this includes periodic testing and some of the documentation requirements.

2.4 Commentary on ICH M7 Questions and Answers

At the time of writing this commentary, the ICH M7 Q&A (version 29 June 2020) has been signed off as a Step 2 document and released for public consultation. (https://www.ich.org/page/multidisciplinary-guidelines).

The ICH M7 guideline was adopted by ICH in June 2014, and the first addendum (R1) was adopted in May 2017. The Q&A document was developed to provide additional clarification to details having led to differing interpretation by stakeholders, such as justification of control strategy in marketing authorization applications, organization and detail of information on mutagenic impurities in marketing authorization applications, and clarification with regard to (Q)SAR systems. Ultimately, the intention of the Q&A document was to promote further harmonization in using this guidance in regulation of mutagenic impurities in pharmaceuticals.

During the process of preparation of the Q&A document, stakeholders submitted more than 100 questions to the reference as ICH M7 expert reference group. The EWG consolidated related questions, and finally 25 Q&As were included in the Step 2 document. The Q&A document is structured according to the sections in the original guideline.

This commentary will present the relevant section within the original guideline and then discuss the respective Q&As on that section. Some queries will be put forth showing that not all controversial issues have been clarified, and we can only hope that the further steps of this Q&A document will be modified to clarify these contentious topics.

2.4.1 Section 1 – Introduction

The four Q&As in Section 1 relate to the Note 1 in the M7 guideline and clarify the meaning of mutagenic and genotoxic potential, as well as the recommendations for evaluation of impurities present below and above 1 mg.

Note 1 in the M7 guideline states as follows (emphasis added on the dubious phrases): "The ICH M7 Guideline recommendations provide a state-of-the-art approach for assessing the potential of impurities to induce point mutations and ensure that such impurities are controlled to safe levels so that below or above the ICH Q3A/B qualification threshold no further qualification for *mutagenic* potential is required. This includes the initial use of (Q)SAR tools to predict bacterial *mutagenicity*. In cases where the amount of the impurity exceeds 1 mg daily dose for chronic administration, evaluation of *genotoxic* potential as recommended in ICH Q3A/B could be considered. In cases where the amount of the impurity is less than 1 mg, no further *genotoxicity* testing is required regardless of other qualification thresholds."

2.4.1.1 Question 1.1

Question	Answer
Note 1 provides general guidance on the relationship of ICH M7 with ICH Q3A and Q3B. The use of both "mutagenic potential" and "genotoxic potential" in Note 1 is confusing. Are these terms considered interchangeable?	No. The terms "mutagenic potential" and "genotoxic potential" are not interchangeable. Mutagenic potential refers to the ability of a compound to induce point mutations (i.e. bacterial reverse mutation assay), while genotoxic potential refers to both mutagenic and clastogenic potential. ICH M7 focuses specifically on mutagenicity.

Throughout the ICH M7 guideline it is emphasized that the focus is on DNA reactive substances that have a potential to directly cause DNA damage when present at low levels leading to mutations and therefore potentially cause cancer. These types of mutagenic substances are usually detected in a bacterial reverse mutation assay (i.e. the Ames test). The guideline further states that other types of genotoxicants that are nonmutagenic (i.e. clastogenic) typically have threshold mechanisms and usually do not pose carcinogenic risk in humans at the level ordinarily present as impurities. This will be discussed later in Sections 3 and 6.

2.4.1.2 Question 1.2

Question	Answer
What are the expectations for evaluation of the mutagenic potential for an impurity where the amount of impurity is less than or equal to 1 mg daily dose?	In the context of ICH M7, (quantitative) structure activity relationships ((Q) SARs) are considered an appropriate initial evaluation of mutagenic potential of an impurity at a daily dose of ≤1 mg. When a structural alert is identified, a follow-up *in vitro* evaluation (e.g. bacterial reverse mutation assay) could be conducted, or the impurity could be controlled by TTC. Negative results in either evaluation would classify the impurity under Class 5. The result of the bacterial reverse mutation assay overrules the (Q)SAR prediction. Additionally, impurities should not be assigned to Class 5 based solely on the absence of structural alerts by visual evaluation alone. There is an expectation that structural alert assessment will be conducted using (Q)SAR prediction.

The main point made in this Q&A is that (Q)SAR analysis is necessary in order to determine if an impurity has an alert for mutagenicity, and visual evaluation alone (which was conventional prior to the implementation of ICH M7) is currently not acceptable. Here again the emphasis is on the Ames test as the proper assay to perform in case testing is desired, and the results from this test overrule any (Q)SAR prediction.

2.4.1.3 Question 1.3

Question	Answer
If an impurity generates negative predictions in two appropriate (Q)SAR systems and is present at a level less than or equal to 1 mg daily dose, is further genetic toxicity testing recommended?	No. If an impurity generates negative predictions in two appropriate (Q)SAR systems and is present at a level ≤1 mg/day, further genetic toxicity testing is not warranted.

Interestingly, after Q&A 1.1 clearly explaining that the focus is on mutagenicity, the term used in this Q&A is "genetic toxicity testing" and not "mutagenic toxicity testing." Does this mean that an impurity that is above the ICH Q3A/Q3B qualification threshold but has no (Q)SAR alerts and is found at less than 1 mg/day, there is no need for the Ames test or chromosomal aberrations? This is not clear, because the ICH Q3A and Q3B guidelines tell us that in order to qualify an impurity at a level above the qualification threshold you should test for mutagenicity (in the Ames test) and clastogenicity (in the chromosomal aberrations assay) in addition to general toxicity studies (one species, usually 14–90 days). The hope is that this topic will be further clarified in the final version of the Q&A document.

2.4.1.4 Question 1.4

Question	Answer
What are the expectations for evaluation of the genotoxic potential for an impurity where the amount of impurity exceeds 1 mg daily dose?	In cases where the amount of impurity is >1 mg daily dose for chronic administration, regardless of the impurity classification, a minimum screen of genotoxicity studies (point mutation and chromosomal aberration) can be considered.

This Q&A clarifies the situation where even if the impurity is a Class 4 or Class 5, if it is present at above 1 mg/day, you are compelled to perform an Ames test and a chromosomal aberrations assay in order to complete the qualification. The question that arises here is: What is the rationale behind requiring the Ames test in such a situation? If the entire basis of classifying impurities in the ICH M7 guideline is based on (Q)SAR analysis, and an impurity can be classified as being non-mutagenic, then why does this classification not prevail also when the level exceeds 1 mg/day? This is not a logical recommendation and here again the hope is that this topic will be further clarified.

2.4.2 Section 2 – Scope

The single Q&A in Section 2 relates to applicability of the ICH M7 on impurities in semisynthetic DPs.

2.4.2.1 Question 2.1

The relevant section in Section 2 of the guideline states that: "Assessment of the mutagenic potential of impurities as described in this guideline is not intended for the following types of drug substances and drug products: biological/biotechnological, peptide, oligonucleotide, radiopharmaceutical, fermentation products, herbal products, and crude products of animal or plant origin."

Question	Answer
Are semisynthetic DSs and DPs included in the scope of ICH M7?	Yes, for certain cases. If a semisynthetic DS is manufactured using steps that could introduce mutagenic impurities or degradants (e.g. post-modification of a fermentation product or late-stage introduction of a linker), a risk assessment is warranted. The following compounds used in the manufacturing process of semisynthetic DSs and DPs should be considered within the scope of the application of ICH M7: • Chemically synthesized intermediates and actual impurities therein • Reagents

This clarification is very important because there were cases where stakeholders considered, for instance, fermentation products, to be excluded from the scope of the ICH M7 guideline, even if they included synthetic steps following the fermentation process. This Q&A makes it clear that any manufacturing step that may introduce a chemical moiety that can be reactive enough to directly react with DNA brings the product within the scope of the application of ICH M7.

2.4.3 Section 3 – General Principles

The two Q&As in Section 3 relate to impurities which are nonmutagenic carcinogens or mutagenic noncarcinogens.

The relevant section in the guideline states the following (with emphasis on the critical phrase): "The focus of this guideline is on DNA reactive substances that have a potential to directly cause DNA damage when present at low levels leading to mutations and therefore, potentially causing cancer. This type of mutagenic carcinogen is usually detected in a bacterial reverse mutation (mutagenicity) assay. **Other types of genotoxicants that are non-mutagenic typically have**

threshold mechanisms and usually do not pose carcinogenic risk in humans at the level ordinarily present as impurities. Therefore to limit a possible human cancer risk associated with the exposure to potentially mutagenic impurities, the bacterial mutagenicity assay is used to assess the mutagenic potential and the need for controls. Structure-based assessments are useful for predicting bacterial mutagenicity outcomes based upon the established knowledge. There are a variety of approaches to conduct this evaluation including a review of the available literature, and/or computational toxicology assessment."

2.4.3.1 Question 3.1

Question	Answer
Should nonmutagenic, carcinogenic impurities be controlled according to ICH M7?	No. Carcinogens that are negative in the bacterial reverse mutation assay do not have a DNA reactive mechanism of carcinogenicity and therefore are not in scope of the ICH M7 guidance (e.g. acetamide and hydroxylamine).

This Q&A assists stakeholders understand that when impurities are negative in the Ames test but contained carcinogenic effects, they do not need to be controlled within the scope of the ICH M7 and should be controlled according to the ICH Q3A/Q3B guidelines. In such cases the carcinogenicity data may serve as the point of departure for calculation of an appropriate PDE as described in the ICH Q3C guideline.

2.4.3.2 Question 3.2

Question	Answer
Should mutagenic, noncarcinogenic impurities be controlled according to ICH M7?	No. Mutagens that are demonstrated to be noncarcinogenic in appropriate and well-conducted animal bioassays will be treated similarly to Class 5 impurities.

This question clarifies the scenario where the stakeholder may consider a compound as a Class 3 impurity, without taking into account the negative carcinogenicity. The answer to this question tells you that such a compound does not need to be controlled as a mutagenic impurity and that for all matters of control, it can be treated similarly to nonmutagenic impurities. The rationale behind this is because the ultimate concern with mutagenic impurities is their potential to cause cancer, so if it has been shown that the compound is noncarcinogenic, even if it is mutagenic, the primary concern is eliminated and the mutagenic properties are irrelevant.

2.4.4 Section 4 – Considerations for Marketed Products

The one Q&A in this section focuses on the meaning of significant increase in clinical dose of marketed products.

The relevant section in the guideline (4.3 Changes to the Clinical Use of Marketed Products) states the following (with emphasis on the critical phrase): "Changes to the clinical use of marketed products that can warrant a reevaluation of the mutagenic impurity limits include a

significant increase in clinical dose, an increase in duration of use (in particular when a mutagenic impurity was controlled above the lifetime acceptable intake for a previous indication that may no longer be appropriate for the longer treatment duration associated with the new indication), or for a change in indication from a serious or life threatening condition where higher acceptable intakes were justified (Section 7.5) to an indication for a less serious condition where the existing impurity acceptable intakes may no longer be appropriate."

2.4.4.1 Question 4.1

Question	Answer
What does "significant increase in clinical dose" mean in "4.3 Changes to the Clinical Use of the Marketed Products"?	Any increase in dose of the API that would increase any mutagenic impurity to levels above the acceptable limits is considered significant (see tables 2 and 9 and the addendum). In such cases a reevaluation of the mutagenic impurity limits is recommended.

This essentially means that any increase in the dose warrants a re-evaluation of the mutagenic impurity limits.

2.4.5 Section 5 – Drug Substance and Drug Product Impurity Assessment

There are no Q&A drafted on this section.

2.4.6 Section 6 – Hazard Assessment Elements

This section has four Q&As, which relate to the following topics:

- Recommendations for validation and documentation to provide for in-house or not commonly used(Q)SAR models.
- Expectations for qualification of an (Q)SAR "out of domain" or "non-coverage" result to assign an impurity to Class 5.
- Ames negative impurities with positive clastogenicity study results.
- Rationale for follow-up assays in Note 3.

2.4.6.1 Question 6.1

In Section 6 of the guideline, it says that (emphasis on the critical phrase): "A computational toxicology assessment should be performed using (Q)SAR methodologies that predict the outcome of a bacterial mutagenicity assay (Ref. 6). Two (Q)SAR prediction methodologies that complement each other should be applied. One methodology should be expert rule-based and the second methodology should be statistical-based. **(Q)SAR models utilizing these prediction methodologies should follow the general validation principles set forth by the Organisation for Economic Co-operation and Development (OECD).**"

Question	Answer
What information and/or documentation should be provided to regulatory agencies to sufficiently demonstrate validation of (Q)SAR models that are developed in-house or are not commonly used?	Section 6 of ICH M7 states that "(Q)SAR models utilizing these prediction methodologies should follow the general validation principles set forth by the Organization for Economic Co-operation and Development (OECD)" (OECD Validation 2007). In the context of ICH M7, the OECD Principles of (Q)SAR Validation are: 1) A defined end point – the model should be trained using experimental data generated according to the standard OECD protocol for the *in vitro* bacterial reverse mutation assay. 2) An unambiguous algorithm – the algorithm used to construct the model should be disclosed. It should be clear whether the model is considered statistical (constructed via machine learning) or expert rule-based (created from human expert-derived knowledge). 3) A defined domain of applicability – describe whether a test chemical falls within the model's applicability domain and how it is calculated. It should warn the user when the model does not have enough information to make a reliable prediction on a chemical. 4) Appropriate measures of goodness-of–fit, robustness, and predictivity – the model should be evaluated and shown to be sufficiently predictive of bacterial reverse mutagenicity. Standard validation techniques that should be used are recall, cross-validation, and external validation. Evidence that the model has not been over-fit should also be provided. 5) A mechanistic interpretation – is there adequate information to allow an assessment of mechanistic relevance to be made (e.g. specific descriptors)? For any system developed in-house or not commonly used, to demonstrate how each model follows these principles and to understand how a (Q)SAR model was developed and validated, submission of the OECD (Q)SAR Model Reporting Format (QMRF) (OECD QRMF 2017) for each model used should accompany each regulatory submission. A harmonized template for the QMRF was developed by the Joint Research Centre (JRC) and EU Member State authorities. This template summarizes and reports key information on (Q)SAR models, including the results of any validation studies, as well as provides supplementary information on applicability of the model to a given chemical.

Most stakeholders currently use commercial software that have been validated and approved by the regulatory agencies. However, if one were to develop their own in-house software, then the criteria for validation, as described above, need to be rigorously followed.

2.4.6.2 Question 6.2

Section 6 of the guideline says the following with respect to the interpretation of the (Q)SAR predictions: "If warranted, the outcome of any computer system-based analysis can be reviewed with the use of expert knowledge in order to provide additional supportive evidence on relevance of any positive, negative, conflicting or inconclusive prediction and provide a rationale to support the final conclusion."

Question	Answer
When an out of domain or noncoverage result is obtained from one of the two (Q) SAR models as described in ICH M7, can the impurity be classified as a Class 5 impurity?	No. Out of domain or noncoverage is not considered equivalent to Class 5. Additional assessment is warranted. Given that the relationship between chemical structure and DNA reactivity is well understood, it is unlikely that a structure with mutagenic potential would be associated with an out of domain result. However, expert review can provide reassurance in assignment of such impurities to Class 5. Expert review may include one or a combination of the following (Amberg et al. 2019): 1) Comparison to structurally similar analogs for which bacterial reverse mutation assay data are available (read-across approach). 2) Expert review of the chemical structure to determine if there is potential for the chemical to react with DNA. 3) (Q)SAR output from an additional validated model (see Question 6.1) of the same methodology (i.e. expert rule-based or statistical) that generates a prediction that is within its applicability domain.

In most cases it is the statistical (Q)SAR prediction that is out of domain. Comparison to structurally similar compounds will usually occur only when the stakeholder has in-house data that will enable an expert review. In the absence of such data, an expert review by a chemist can determine if the out of domain moiety is sufficiently unreactive (i.e. non-electrophilic) to consider it a Class 5 impurity. Using an additional (Q)SAR software frequently does not solve the problem, because an out of domain moiety in one software is often also not within the database domain of the additional software, and adding on additional prediction tools can sometimes just add more uncertainty rather than clarity.

2.4.6.3 Question 6.3

Section 6 of the guideline further states that (emphasis added): "To follow up on a relevant structural alert (Class 3 in table 1), either adequate control measures could be applied or a bacterial mutagenicity assay with the impurity alone can be conducted. **An appropriately conducted negative bacterial mutagenicity assay (Note 2) would overrule any structure-based concern, and no further genotoxicity assessments would be recommended (Note 1). These impurities should be considered non-mutagenic (Class 5 in table** 1)."

Question	Answer
In a case where an impurity is demonstrated to be negative in an Ames study but positive in a clastogenicity study (e.g. chromosomal aberration test), how would the impurity be classified per the ICH M7 classification system?	If an impurity tests negative in an Ames assay, it is considered a Class 5 impurity. Addressing positive results in a clastogenicity assay is out of scope of ICH M7.

As discussed above in Question 1.1, it is made explicitly clear that if an impurity is negative in the Ames test and even if it is positive in the chromosomal aberrations assay (clastogenic), this impurity is considered a Class 5 impurity.

2.4.6.4 Question 6.4

The continuation of the text quoted above in Question 6.3 says the following (emphasis added): "A positive bacterial mutagenicity result would warrant further hazard assessment and/or control measures (Class 2 in table 1). For instance, when levels of the impurity cannot be controlled at an appropriate acceptable limit, it is recommended that the impurity be tested in an *in vivo* gene mutation assay in order to understand the relevance of the bacterial mutagenicity assay result under *in vivo* conditions. The selection of other *in vivo* genotoxicity assays should be scientifically justified based on knowledge of the mechanism of action of the impurity and expected target tissue exposure (Note 3). *in vivo* studies should be designed taking into consideration existing ICH genotoxicity Guidelines. **Results in the appropriate *in vivo* assay may support setting compound specific impurity limits.**"

Question	Answer
Please clarify the rationale for the tests included under Note 3 as a follow-up to investigate the *in vivo* relevance of Ames mutagen.	If an impurity is positive in the Ames test, an *in vivo* follow-up test with mutagenic end point (mutagenicity) should be used. The other follow-up tests outlined in Note 3 are also acceptable when scientific rationale is provided to support their use. For any of the above tests, adequate exposure should be demonstrated in line with ICH S2.

Regulatory agencies expect a strong argument to overrule a positive Ames test. It is pertinent that the *in vivo* test(s) used to further investigate the relevance of the *in vitro* results be chosen correctly and that adequate exposure be demonstrated. A negative result in a transgenic mutation assay would normally be the strongest evidence to overrule a positive *in vitro* result, but this assay is relatively expensive and lengthy. Other assays are also acceptable, and often times a stakeholder will chose to combine several assays (e.g. *Pig-a* assay, *in vivo* micronucleus assay, and Comet assay) to be absolutely certain that the impurity is nonmutagenic.

2.4.7 Section 7 – Risk Characterization

This section has five Q&As that start with a direct continuation to Q&A 6.4, which relates to Ames positive impurities. Then an interesting question is presented regarding the application of the LTL to AIs derived by linear extrapolation from TD_{50} values but not to PDEs derived for impurities. Then a Q&A relates to the explanations and implications of moving HIV disease from treatment duration of <10 years to lifetime treatment, and finally a Q&A relates to the application of limits for individual impurities when three or more Class 2 and Class 3 impurities are present.

2.4.7.1 Question 7.1

This question relates to the same paragraph from the guideline that was stated above in Question 6.4.

Question	Answer
If an Ames positive impurity is subsequently tested in an appropriate *in vivo* assay and the results are clearly negative, is that sufficient to demonstrate lack of *in vivo* relevance?	Yes. A well-conducted and scientifically justified *in vivo* study (see Question 6.4 in this document) is sufficient to demonstrate lack of *in vivo* mutagenic relevance. If the results of the *in vivo* study are clearly negative, the impurity can be assigned to ICH M7, Class 5.

This Q&A almost states the obvious. The entire intention of Note 3 in the guideline is to provide adequate *in vivo* tests that can overrule a positive Ames test, and if an impurity is determined to be nonmutagenic, then it can be considered a Class 5 impurity.

2.4.7.2 Question 7.2

With regard to mutagenic impurities with positive carcinogenicity data (Class 1 impurities), the guideline states that: "Compound-specific risk assessments to derive acceptable intakes should be applied instead of the TTC-based acceptable intakes where sufficient carcinogenicity data exist. For a known mutagenic carcinogen, a compound-specific acceptable intake can be calculated based on carcinogenic potency and linear extrapolation as a default approach. Alternatively, other established risk assessment practices such as those used by international regulatory bodies may be applied either to calculate acceptable intakes or to use already existing values published by regulatory authorities (Note 4)."

Question	Answer
If an Ames positive impurity is subsequently tested in an appropriate *in vivo* assay and the results are positive, does that support setting compound-specific impurity limits?	No. *in vivo* gene mutation assays are currently not validated to directly assess cancer risk because the end point is mutation and not carcinogenicity (i.e. they are used for hazard identification). Results from these tests could identify mode of action and/or direct further testing strategy to complement the available data for a weight of evidence approach.

The Q&A here clarifies that *in vivo* mutagenicity data cannot be used to set compound-specific impurity limits and that the proper methods to determine AIs for mutagenic carcinogens is the linear extrapolation from the TD_{50} or from the BMDL10 (Benchmark Dose Lower Confidence Limit 10%), as described in Note 4 of the guideline.

2.4.7.3 Question 7.3

The LTL concept, previously known as "staged TTC," is based on the assumption that cancer risk of known carcinogens increases as a function of cumulative dose. The guideline explains that the LTL approach is applied to mutagenic impurities in which the acceptable cumulative lifetime dose is uniformly distributed over the total number of exposure days during LTL exposure. This would allow higher daily intake of mutagenic impurities than would be the case for lifetime exposure and still maintain comparable risk levels for daily and non-daily treatment regimens.

Question	Answer
Can an LTL approach be applied to AIs or PDEs using the same ratio as in Table 2.2?	The LTL approach can be applied to compounds with exposure limits based on the TTC or a compound/class specific AI. However, this approach is not applicable to PDEs. Higher levels of exposure for short-term exposure (30 days or less) may be acceptable on a case-by-case basis.

The Q&A here makes the distinction between limits that derive from TTC or compound/class specific AI and PDEs. The limits that derive from mutagenicity/carcinogenicity data fall under the assumption that cancer risk of known carcinogens increases as a function of cumulative dose; however, the toxicity of other end points that are used in order to calculate PDEs does not necessarily increase as a function of cumulative dose; therefore, LTL is not applicable to PDEs. Conversely, the answer to this question also refers to an option of assigning higher limits than a calculated PDE for short-term dosing (30 days or less), as mentioned in the ICH Q3C and Q3D guidelines. This concept is further elaborated in Harvey et al. (https://doi.org/10.1016/j.yrtph.2016.12.011) where it is proposed to modify the AI of nonmutagenic impurities for clinical studies of less than six months duration, based on the Haber's Law (the same rule upon which the LTL concept is based on).

An interesting example for the use of LTL is the case of monofunctional alkyl chlorides that have a class-specific AI of 10 times the default TTC (Note 5 in the ICH M7 guideline). For a monofunctional alkyl chloride in a DP that is administered for ≤1 month, the theoretical LTL AI can be calculated to be: 15 μg/day × 80 (the ratio between 120 and 1.5 μg/day) = 1200 μg/day. However, this limit now exceeds the 1 mg/day, and thus it may be necessary to perform a screen of genotoxicity studies to qualify such a level (see Question 1.3 above).

2.4.7.4 Question 7.4

This Q&A relates to the AI of HIV treatment that under Note 7 (Table 2.7) of the M7 guideline falls under the LTL category of a treatment duration of >1–10 years where the AI is defined as 10 μg/day. The HIV disease has now been moved from "treatment duration <10 years" to "lifetime" treatment. The Q&A provides further explanation to this change.

Question	Answer
Why was HIV disease moved to the "Treatment duration of >10 years to lifetime" in the clinical use scenarios table? How should this change be implemented?	The treatment duration category was changed because of advances in the clinical treatment of HIV disease. To avoid disruption of supply of HIV drugs already on the market, this change would not be applied to currently marketed products. For example, when a new DS supplier is proposed, the AI would remain at 10 μg/day in cases where the DS produced by this supplier, using the same route of synthesis, is a component of an existing DP marketed in the specific region (see ICH M7 Section 4.1). For regulatory submissions 18 months after the date that the M7 Q&A reached Step 4, the 1.5 μg/day or other appropriate AI would be applied in the following situations: • New DSs and new DPs during their clinical development and subsequent applications for marketing. • Changes to the DS synthesis resulting in new or increased acceptance criteria for existing impurities. • Changes in the formulation, composition, or manufacturing process resulting in new degradation products or increased acceptance criteria for existing degradation products. • Introduction of a new source of the DS through a drug master file (DMF) from a supplier who has not had a previously accepted DMF in the relevant region. • Changes made to a specific synthetic step as described in ICH M7 Section 4.1. • A newly discovered Class 1 or Class 2 impurity, a structure in the cohort of concern, or new relevant impurity hazard data, as described in ICH M7 Section 4.4

Table 2.7 Examples of clinical use scenarios with different treatment durations for applying AIs.

Scenario[a]	Acceptable intake (μg/day)
Treatment duration of ≤1 month: e.g. drugs used in emergency procedures (antidotes, anesthesia, acute ischemic stroke), actinic keratosis, and treatment of lice	120
Treatment duration of >1–12 months: e.g. anti-infective therapy with maximum up to 12 months treatment (HCV), parenteral nutrients, prophylactic flu drugs (~5 months), peptic ulcer, assisted reproductive technology (ART), preterm labor, preeclampsia, presurgical (hysterectomy) treatment, and fracture healing (these are acute use but with long half-lives)	20
Treatment duration of >1–10 years: e.g. stage of disease with short life expectancy (severe Alzheimer's), nongenotoxic anticancer treatment being used in a patient population with longer term survival (breast cancer, CML), drugs specifically labeled for less than 10 years of use, drugs administered intermittently to treat acute recurring symptoms[b] (chronic Herpes, gout attacks, substance dependence such as smoking cessation), macular degeneration, and HIV[c]	10
Treatment duration of >10 years to lifetime: e.g. chronic use indications with high likelihood for lifetime use across broader age range (hypertension, dyslipidemia, asthma), Alzheimer's (except severe AD), hormone therapy (e.g. GH, TH, PTH), lipodystrophy, schizophrenia, depression, psoriasis, atopic dermatitis, COPD, cystic fibrosis, and seasonal and perennial allergic rhinitis	1.5

[a] This table shows general examples; each example should be examined on a case-by-case basis. For example, 10 μg/day may be acceptable in cases where the life expectancy of the patient may be limited, e.g. severe Alzheimer's disease, even though the drug use could exceed 10-year duration.
[b] Intermittent use over a period >10 years but based on calculated cumulative dose it falls under the >1–10 year category.
[c] HIV is considered a chronic indication but resistance develops to the drugs after 5–10 years and the therapy is changed to other HIV drugs.

2.4.7.5 Question 7.5

This Q&A relates to the need of assigning limits to individual impurities when three or more Class 2 and Class 3 impurities are present.

Question	Answer
Does "Table 2.2: Acceptable Intakes for an Individual Impurity" apply when three or more Class 2 or Class 3 impurities are specified in the DS specification?	Yes. In this scenario, a limit for each "Individual Impurity" should be listed in the DS specification as per limits provided in Table 2.2 (for example >10 years to lifetime not more than (NMT) 1.5 μg/day). Additionally, a limit for "Total Mutagenic Impurities" should be listed in the DS specification as per limits provided in Table 2.7 (for example >10 years to lifetime NMT 5 μg/day). As stated in the guidance, compound-specific or class-related acceptable limits (Class 1) and degradation products that form in the DP are excluded from total mutagenic impurity limits.

This Q&A clarifies that in the case where there are three or more Class 2 and Class 3 impurities, the expectation is to assign individual limits for each of the impurities in *addition* to assigning a limit for "Total Mutagenic Impurities."

2.4.7.5.1 Section 8 – Control

The Q&As in this section relate to the four options described in the Control section of the guideline and provide much needed clarifications to scenarios that have been contentious ever since the guideline's publication. There is not much need for commentary to this section as the questions and respective answers are very apprehensible.

The six Q&As focus on:

- Option 4 control strategy.
  - When is it appropriate?
  - Elements recommended when using predictive purge calculations to claim no analytical testing as per Option 4.
- Considerations for control of impurities introduced or formed in the last synthetic step.
- When is periodic verification testing allowed?
- Do level of impurities consistently found <30% TTC in multiple batches justify no testing?
- Batch scales recommended to provide experimental data to support control Options 3 and 4.

#	Question	Answer
1	When is it appropriate to use an Option 4 control strategy?	Use of Option 4 is appropriate when a mutagenic impurity is demonstrated to have a negligible risk of being present in the final DS (e.g. 1% TTC). The risk assessment can be based on scientific principles alone (e.g. impurity reactivity or solubility), calculated purge factors (i.e. predicted), measured purge factors (i.e. spike and purge data), or a combination of these approaches, considering the process-relevant conditions. The acceptability of Option 4 will be assessed by authorities on a case-by-case basis, including additional requests for supporting information. See also Question 8.3 in this document for impurities introduced in the last step.
2	When predictive purge calculations are used for Option 4 control, what elements should be considered?	When using predictive purge calculations for Option 4 control, the following elements should be considered: • Predictive purge calculations should be based on the DS manufacturing process as described in the application and should consider reactivity, solubility, volatility, and other factors of the impurity in each step. The predictive purge calculation should use conservative values and methodology, since predictive purge often does not rely on experimental purge factors. An example predictive purge calculation approach based on scientific principles has been described (Barber et al. 2017). Predictive purge calculations can be paper based or software based. • The amount of information (i.e. impurity reactivity or solubility data, spike and purge data under the process relevant conditions) to justify a predictive purge calculation approach should be guided by knowledge of the manufacturing process, risk to the final DS, and the stage of drug development. • A predictive purge calculation justification submitted in an application could range from a high-level summary to detailed information on the calculation (e.g. scientific justification for individual purge factors) and other supporting data. More detailed information on the calculation is expected when the predicted level of the impurity in the DS approaches the TTC. Even if not submitted, information on how each individual purge factor is derived should be available upon request.

#	Question	Answer
3	What is meant by "for impurities introduced in the last synthetic step, an Option 1 approach would be expected unless otherwise justified" in Section 8.2 Considerations for Control Approaches?	For potential mutagenic impurities introduced or generated in the last synthetic step, given the proximity to the final product, Option 1 is the preferred control strategy. However, Options 2 and 3 control strategies may be possible, for example, when the crude DS is an isolated material that is purified subsequently (e.g. by recrystallization). An Option 4 control strategy for an impurity introduced or generated in the last synthetic step is discouraged and should be reserved for highly reactive species (e.g. thionyl chloride) or materials with low boiling point (e.g. methyl chloride). In case of highly effective purification operations (e.g. chromatography), an Option 4 control approach may also be acceptable for less reactive materials. However, in such cases, the negligible risk of an impurity to be carried to the final product (e.g. 1% TTC) should be justified with experimental data (e.g. spike and purge data under the process-relevant conditions). A justification solely based on calculations (predictions) is not considered sufficient.
4	Is periodic verification testing (i.e. skip testing) allowed for Options 2 and 3 control?	No. Periodic verification testing is not appropriate for Options 2 and 3 control. Periodic verification testing is only discussed as a control strategy for Option 1 control in Section 8.1 of ICH M7. The Option 1 periodic verification testing strategy references ICH Q6A. The Option 1 periodic verification testing concept (per ICH Q6A) should generally be implemented post-approval and applies to testing in the final DS.
5	If test data (i.e. in-process, intermediate, or DS impurity test data) for a potential mutagenic impurity is consistently <30% TTC in multiple batches, is that sufficient to justify no testing of that impurity in the control strategy?	No. Batch data alone demonstrating that a potential mutagenic impurity is consistently <30% TTC is not sufficient to justify no testing of that impurity. Options 1, 2, and 3 should test either at release or upstream in the process. However, if there is negligible risk of the impurity to be present in the DS, an Option 4 control strategy may be considered with appropriate justification. See Questions 8.1 and 8.2 for recommendations on supporting an Option 4 control strategy.
6	What scale considerations are relevant when generating analytical experimental data in support of control Options 3 and 4.	Lab-scale experiments are typically sufficient when generating measured purge factors or when defining in-process control points. These studies should employ the final process as described in the application and should consider the potential impact of scale and equipment-related differences between the laboratory and production environment (e.g. the effects of mixing on impurity levels in heterogeneous systems and the quality of liquid–liquid phase separations). In the case of observed scale dependencies, confirmatory testing on batches manufactured at pilot or commercial scale may be advisable. There is no expectation to perform spiking studies at pilot or commercial scale.

2.4.8 Section 9 – Documentation

The two Q&As in this section clarify if (Q)SAR predictions made earlier in development are still valid for market authorization and provide recommendations for clarity of ICH M7 risk assessment and control strategy with respect to the location in the CTD and details to be provided in Modules 2, 3, and 4. This section also does not require much commentary as it is straightforward.

#	Question	Answer
1	If (Q)SAR predictions are made during drug development, should they be repeated for the marketing application?	(Q)SAR models developed for use under ICH M7 are generally updated regularly with new bacterial reverse mutagenicity assay data and more refined structural alerts. A sponsor is not expected to update their (Q)SAR assessment during drug development unless there is a safety concern such as when newly available bacterial reverse mutagenicity assay data and/or mechanistic knowledge suggest that the prediction is incorrect (see below). It is recommended that the sponsor rerun (Q)SAR predictions prior to the initial marketing application to ensure predictions reflect the most current data available. If the marketing application is later submitted in other regulatory jurisdictions, reassessment may be considered. As an example, in cases where there is reason to question the outcome of a negative prediction (e.g. an aromatic amine is present, but the model gave a negative prediction). Reassessment may also be considered if the predictions made for the initial global marketing application did not use a recent version of the software. In general, predictions generated with models developed prior to ICH M7's publication in 2014 are considered unacceptable.
2	For marketing applications, what content and CTD placement recommendations could improve the clarity of an ICH M7 risk assessment and control strategy?	In Module 2, a brief summary of the ICH M7 risk assessment and control strategy should be included (Sections 2.3 and 2.6). In Module 3, the ICH M7 risk assessment and control strategy should be provided in detail. This type of information is often placed in Section 3.2.S.3.2 Impurities; however, it is sometimes placed in other CTD locations per ICH M4Q guidance. A table summary of the ICH M7 hazard assessment and ICH M7 impurity control strategy is recommended to improve clarity. • Information recommended for an ICH M7 hazard assessment table includes impurity chemical structure, (Q)SAR results (pos/neg predictions, out of domain), bacterial reverse mutagenicity assay results (pos/neg, if available), ICH M7 impurity class (1–5) assignment, and supporting information (e.g. information/links for bacterial reverse mutagenicity assays, literature reports, and (Q)SAR expert analysis). The *in silico* systems used (name, version, and end point) can also be noted. • Information recommended for an ICH M7 impurity control strategy table includes impurity origin (e.g. synthetic step introduced and degradant), ICH M7 class, purge factors (e.g. measured or predicted), ICH M7 control option (1–4), control strategy (i.e. including in-process or compound testing rationale), and supporting information (e.g. information/links for justifications and calculations). The maximum daily dose, TTC, and proposed duration of treatment can also be noted. • Additionally, it is recommended that compound code names be cross-referenced, if Module 3 and Module 4 (including toxicity study reports) use different compound naming conventions. In Module 4, full safety study-related information on impurities (e.g. bacterial reverse mutagenicity assay reports, (Q)SAR reports, genotoxicity test reports, and additional testing) should be included to support the risk assessment and control strategy. This information is often placed in Section 4.2.3.7.6 Impurities (see ICH M4S for additional information) and can be cross-referenced to Module 3 by hyperlinks.

For Sections 10 (Illustrative Examples) and 11 (Glossary) there were no Q&As.

To summarize, critical topics were addressed in the Q&A draft document and these are greatly welcome. The main goal of the EWG was to minimize different interpretation of specific aspects of risk assessment and control strategy of mutagenic impurities. Most of the Q&As achieve this goal; however, several still require further clarification, in particular the requirements around testing for impurities found above the 1 mg/day level.

References

1 ICH M7 – Assessment and control of dna reactive (mutagenic) impurities in pharmaceuticals to limit potential carcinogenic risk. https://www.ema.europa.eu/en/documents/scientific-guideline/ich-guideline-m7r1-assessment-control-dna-reactive-mutagenic-impurities-pharmaceuticals-limit_en.pdf (accessed October 2020).

2 Committee for Medicinal Products for Human Use (CHMP). Guidelines on the Limits of Genotoxic Impurities, London 26th June 2006. CPMP/SWP/5199/02.

3 Chmp Safety Working Party (SWP). Question & Answers on the CHMP Guideline on the Limits of Genotoxic Impurities. London, 17 December 2009. EMEA/CHMP/QWP/251344/2006Q&A.

4 Guidance for Industry Genotoxic and Carcinogenic Impurities in Drug Substances and Products: Recommended Approaches Draft, Center for Drug Evaluation and Research (CDER) December 2008.

5 ICH Q3A (R2). (2006). Impurities in new drug substances ICH. CPMP/ICH/2737/99.

6 ICH Q3B (R2). (2006). Impurities in new drug products ICH. CPMP/ICH/2738/99.

7 Final Concept Paper M7: Assessment and Control of DNA Reactive (Mutagenic) Impurities in Pharmaceuticals to Limit Potential Carcinogenic Risk, dated 27 November 2009. http://www.ich.org/fileadmin/Public_Web_Site/ICH_Products/Guidelines/Multidisciplinary/M7/M7_Final_Concept_Paper_June_2010.pdf (accessed October 2020).

8 ICH S9 – Nonclinical Evaluation for Anticancer Pharmaceuticals. http://www.ich.org/fileadmin/Public_Web_Site/ICH_Products/Guidelines/Safety/S9/Step4/S9_Step4_Guideline.pdf (accessed October 2020).

9 Overview of comments received by EMA on 'Questions and answers – ICH S9 guideline on nonclinical evaluation for anticancer pharmaceuticals – Step 2b' (EMA/CHMP/ICH/453684/2016)

10 Ames, B.N., Durston, W.E., Yamasaki, E.E., and Lee, F.D. (1973). *Proc. Natl. Acad. Sci. U. S. A.* 72: 2423–2427.

11 ICH guideline Q3C (R6) on Impurities: Guideline for Residual Solvents. EMA/CHMP/ICH/82260/2006.

12 Munro, I.C., Ford, R.A., Kennepohl, E., and Sprenger, J.G. (1996). *Food Chem. Toxicol.* 34: 829–867.

13 Munro, I.C., Kennepohl, E., and Kroes, R. (1999). *Food Chem. Toxicol.* 37 (2–3): 207–232.

14 Kroes, K. and Kozianowski, G. (2002, 2002). *Toxicol. Lett.* 127: 43–46.

15 Kroes, R., Renwick, A.G., Cheeseman, M. et al. (2004). *Food Chem. Toxicol.* 42: 65–83.

16 Cheeseman, M.A., Machuga, E.J., and Bailey, A.B. (1999). *Food Chem. Toxicol.* 37: 387–412.

17 Snodin, D.J. and McCrossen, S.D. (2013). *Regul. Toxicol. Pharmacol.* 67 (2): 299–316.

18 Delaney, E.J. (2007). *Regul. Toxicol. Pharmacol.* 49: 107–124.

19 Dobo, K. and Walker, D.. *Genotoxic Impurities, Strategies for Identification and Control.* Wiley. ISBN: ISBN 978-0-470-49919-1.Teasdale Chapter 6

20 ICH Q11. Development and Manufacture of Drug Substances (Chemical Entities and Biotechnological/Biological Entities). http://www.ich.org/fileadmin/Public_Web_Site/ICH_Products/Guidelines/Quality/Q11/Q11_Step_4.pdf (accessed October 2020).

21 Elder, D. and Teasdale, A. (2015). *Org. Process Res. Dev.* 19 (11): 1437–1446.
22 Kleinman, M., Baertschi, S., Teasdale, A. et al. (2015). *Org. Process Res. Dev.* 19 (11): 1447–1457.
23 Steve Baertschi. Assessing and controlling mutagenic impurities from degradation. *Lhasa 2018 Pharmaceutical Industry and Regulators Symposium*, Brazil (24–25 May 2018).
24 Barber, C., Amberg, A., and Custer, L. (2015). *Regul. Toxicol. Pharmacol.* 73 (1): 367–377.
25 Greene, N., Dobo, K., Kenyon, M. et al. (2015). *Regul. Toxicol. Pharmacol.* 72 (2): 335–349.
26 Müller, L., Mauthe, R.J., Riley, C.M. et al. (2006). *Regul. Toxicol. Pharmacol.* 44: 198–211.
27 Ellis, P., Kenyon, M., and Dobo, K. (2013). *Regul. Toxicol. Pharmacol.* 65 (2): 201–213.
28 Bolt, H. (1996). *Biochem. Pharmacol.* 52: 1–5.
29 Teasdale, A., Fenner, S., Ray, A. et al. (2010). *Org. Process Res. Dev.* 14 (4): 943–945.
30 ICH guideline Q9 on quality risk management, September 2015 EMA/CHMP/ICH/24235/2006.
31 Teasdale, A., Elder, D., Chang, S.J. et al. (2013). *Org. Process Res. Dev. 17* (2): 221–230.

3

Control Strategies for Mutagenic Impurities

Andrew Teasdale[1], Michael Burns[2], and Michael Urquhart[3]

[1] *Chemical Development, Pharmaceutical Technology and Development, Operations, AstraZeneca, Macclesfield, UK*
[2] *Lhasa Limited, Leeds, UK*
[3] *GlaxoSmithKline R&D, Hertfordshire, UK*

3.1 Introduction

Since the advent of guidelines relating to mutagenic impurities (MIs), the chronology of which is defined in Chapter 1, it has been necessary for pharmaceutical companies to consider the potential risk posed by MIs within their products. This has therefore driven the need to develop an effective strategy that both identifies and assesses the risk posed by any MI, both those directly related to the synthesis and those resulting from degradation within the formulated product.

In order to synthesize any small synthetic active pharmaceutical ingredient (API) efficiently, it is necessary to build up the molecular structure through the combination of simple structural motifs. This typically involves the formation of carbon-carbon, carbon-nitrogen, and carbon-oxygen bonds. The current status of synthetic methodology [1] is such that this is impractical to achieve without the use of electrophilic species that fall into the broad class of alkylating agents, and hence are a potentially mutagenic impurity (PMI).

Thus, many intrinsically reactive starting materials, intermediates, and reagents used in the synthesis of APIs are potentially mutagenic, and furthermore may present as residual impurities within the API. Although avoidance is generally considered to be the preferable option from a regulator's perspective, there is tacit acceptance of the fact that this is impractical, and hence rather than avoidance, the issue becomes one of control. Indeed, Elder et al. [1] concluded that the average number of registered steps required to synthesize each API was 6 (5.9) and that the average number of reactive intermediates per synthetic route was 4 (4.1), roughly equating to just under one PMI per stage.

Several organizations have published details of their approach to MI risk assessment [2–4] and these are discussed below; all are based on the same general principal.

- First, identify potential impurities that are associated with the synthesis of the active and also potential degradation products. Potential synthetic impurities require expert elicitation, reviewing the synthetic route for what is known and "reasonably" predicted.
- Second, the identified potential impurities are screened for potential mutagenicity, typically through the application of an appropriate (quantitative) structure activity relationship [(Q)SAR] process.
- For those impurities still considered a concern i.e. structurally alerting, an evaluation of the likelihood of the material in question carrying through to the API is undertaken. This should

Mutagenic Impurities: Strategies for Identification and Control, First Edition. Edited by Andrew Teasdale.
© 2022 John Wiley & Sons, Inc. Published 2022 by John Wiley & Sons, Inc.

take into consideration the properties of the compound in question and the downstream process conditions. For those still considered a risk in terms of potential carryover, actual levels may be measured by the development of a suitable analytical method to confirm the impurity is not present at levels that would constitute a concern to the patient (<30% Threshold of Toxicological Concern [TTC]).

- Either prior to, or after, the evaluation of the fate of the impurity (impurities) in question, the actual mutagenicity of the impurity can be confirmed through conduct of an Ames test [5]. This may be followed by further relevant *in vivo* testing (Chapter 6) to further understand the risk.
- Finally, once the evaluation is complete a suitable control strategy may be established; this may range from control based on existing process controls through to control through specification or even modification of the route/process for manufacture of the API. These control options are described in detail below.

The following chapter describes this evaluation process in detail. A structured approach is defined based on the principals of quality by design (International Council for Harmonisation [ICH] Q8 [6]) and risk assessment (ICH Q9 [7]), providing an effective, robust process that identifies and addresses the risk posed by MIs, including recent amendments made to specifically manage the risk posed by *N*-nitrosamines. It examines the scope of such activities and the critical factors to consider when assessing risk. The relationship between analytical and safety testing, as well as the relative timing of such activities is also considered.

The practical application of this process is then demonstrated in several case studies.

3.2 Assessment Process

3.2.1 General

The process begins with the expert elicitation of the synthetic route for postulated and/or known impurities [8]. This is followed by structural assessment of agreed "reasonably predicted" impurities along with other route materials and reagents where appropriate. Once the impurities of potential concern are defined, the potential hazard for the identified impurities needs to be established. An alerting material is considered mutagenic until proven otherwise, and therefore where risks are identified, appropriate quantitation may be required or safety testing to confirm/discharge the risk. Once a risk is confirmed, it leads to the establishment of an appropriate control strategy.

The evaluation process is represented schematically in Figure 3.1.

The process should be flexible; each API synthesis and drug product formulation has its own distinctive features, and where appropriate, the ordering of the steps described may be changed; however, the overall process should generally remain the same.

There is a clear link between the assessment of risk and the permitted level for a MI. Any such evaluation should therefore take into account the phase of development, the intended dose, and likely clinical trial study duration. Permissible limits are based on the "staged TTC" principle. Limits cited within ICH M7 are reproduced in Table 3.1.

ICH M7 [8] also states that values higher than the TTC may be acceptable under certain conditions including short-term exposure, for treatment of a life-threatening condition, when life expectancy is less than five years, or when there is greater exposure from other sources such as food or endogenous metabolism e.g. formaldehyde. This aspect of ICH M7 is examined in greater detail in Chapter 2.

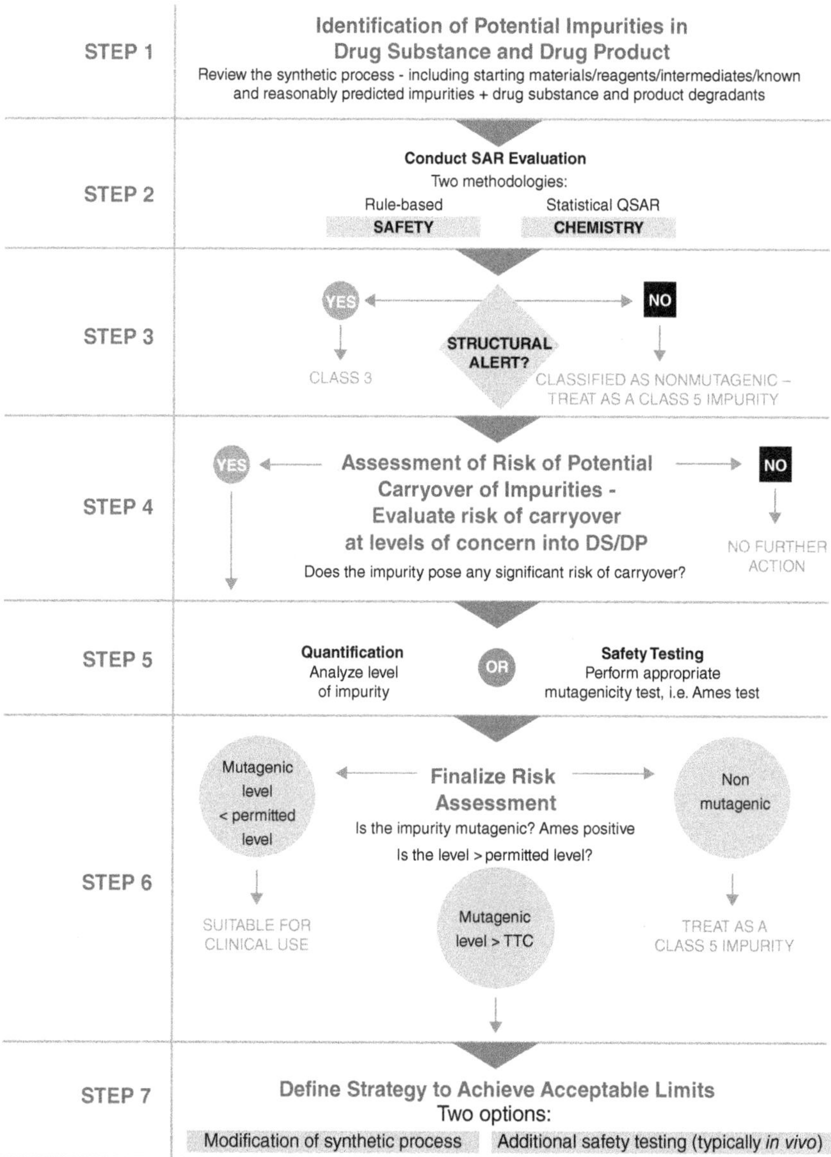

Figure 3.1 Proposed process flow for MI risk assessment for a pharmaceutical product.

Table 3.1 Acceptable intakes for an individual MI.

Duration of treatment	≤1 month	>1–12 months	>1–10 years	>10 years to lifetime
Daily intake (μg/day)	120	20	10	1.5

It is recommended that a permitted limit, e.g. staged TTC, is established in advance of instigating the formal evaluation, with the caveat that this limit will change depending on both time (duration of clinical phase) and dose (absolute level of exposure).

3.2.2 Step 1 – Evaluation of Drug Substance and Drug Product Processes for Sources of Potentially Mutagenic Impurities

The responsibility for this step is likely to fall to the chemists and analysts responsible for the design and development of the API synthetic process, with additional input from formulation development scientist groups who can comment on issues arising from stability and degradation studies, as well as excipient compatibility.

An evaluation of the synthetic route, focused on starting materials, intermediates, reagents, catalysts, and solvents, is carried out to identify materials that could possibly survive the process and present in the API as impurities. It should also include consideration of other potential impurities that may arise from the synthetic route, particularly in the final stages. These could include related substances of the API or intermediates, through to materials derived from interactions between reagents and solvents.

It is recommended to focus on what could be considered "reasonably" predicted. This aspect of the risk assessment has been thrown into sharp focus by the issues surrounding *N*-nitrosamines. This is examined in depth in Chapter 10 where specific issues surrounding *N*-nitrosamines are examined and Chapter 11 that looks in detail at potential side reactions. Throughout the evolution of guidance pertaining to MIs, the scope in terms of what to include in MI risk assessments has been a topic of considerable debate/discussion. For example, the earlier European Medicines Agency (EMEA) guideline contains the following advice:

> As stated in the Q3A guideline, actual and potential impurities **most likely** to arise during the synthesis, purification and storage of the new drug substance should be identified, based on sound scientific appraisal of the chemical reactions involved in synthesis, impurities associated with raw materials that could contribute to the impurity profile of the new drug substance, and possible degradation products. **This discussion can be limited to those impurities that might reasonably be expected based on scientific knowledge of the chemical reactions and conditions involved.**

ICH M7 [8] also addresses this within Section 5.1 of the guideline. The emphasis is focused on actual impurities and potential impurities likely to be present in the API/drug product (DP). Another important aspect of this section of the guideline is that it also looks to link the ICH M7 to ICH Q3A 6/Q3B [9, 10] reporting and identification requirements. For marketed assets, actual impurities are defined as those observed in the drug substance and drug product above the ICH Q3A and ICH Q3B reporting thresholds and identification of actual impurities is expected when the levels exceed the identification thresholds outlined by ICH Q3A and ICH Q3B. This confirms the primacy of ICH Q3A and ICH Q3B in terms of identification thresholds for marketed assets. While ICH Q3A and ICH Q3B do not apply to products within the clinical phases of development, the included identification levels for impurities can be a useful guide for later phase assets. In general for earlier phase products, other impurity identification thresholds have been proposed that recognize the controlled nature of clinical development and the knowledge that most toxicities are dose and dosing duration dependent [11].

PMIs that might be present in API generally fall into the following categories:

- Unreacted contributory materials or intermediates with alerting substructures that have survived processing (for example, an unreacted nitroaromatic functionality within an API due to incomplete hydrogenation or a positional isomer unable to cyclize).
- Substances derived from contributory materials, intermediates, or the API itself that contain an alerting structural motif.

- Unrelated substances formed by combinations of solvents and reagents with each other or with contributory materials or intermediates, *N*-nitrosamines, and sulfonate esters are of course the most well-known and studied examples of this. As highlighted above the issue of MIs generated by side reactions is examined in detail in Chapter 11.

It is important to remember that whatever the root cause of a potential MI, the consideration of factors such as dose, duration, and proximity remain important factors in establishing actual risk.

Such an approach is fully aligned with the key central tenets of ICH Q9 [7] that focuses on the probability of an event occurring, combined with an evaluation of the impact of the event occurring, leading to a consideration of the risk posed. The magnitude of the risk is therefore related to the probability of the PMI being present. The greatest risk is posed by those agents used in the late stages of the API synthesis that possess well-established alerting structural motifs, as there are fewer opportunities for them to be removed during processing, and these should be the main focus of the evaluation.

At an appropriate point in the development of an API, the risk assessment should also include consideration of materials arising from degradation during manufacture or on long-term storage of the API or its formulated product. This review may be based on a combination of factors including expert scientific knowledge and *in silico* predictions, e.g. Zeneth™, of the typical degradation pathways of the API and formulated product based on the chemical structure and literature precedent. The conduct of such assessments is described in detail in Chapter 14.

Having agreed a list of materials, which might comprise degradation products, specified impurities, probable process impurities, including intermediates, reagents, and raw materials as well as "reasonably predicted" impurities based on potential side reactions, these should then be subjected to a formal structural assessment for mutagenicity.

3.2.3 Step 2 – Structural Assessment

Once potential impurities have been identified, the next step is an assessment of their mutagenic potential. The relationship between structure and mutagenicity is discussed in detail in Chapter 4. Such an assessment is typically made through the use of *in silico* SAR systems. ICH M7 [8] defines the need to apply two (Q)SAR methodologies. One methodology should be expert rule-based and the second methodology should be statistical-based; however, the guidance does not define which software packages are preferable; this decision is left to the end user. Importantly, it also highlights the need for an expert evaluation of the results. Again this concept is examined in depth in Chapter 4.

The use of the described methodologies provides numerous permutations, which range from confirmatory, likely to be mutagenic or not, to the *in silico* tools being unable to provide a prediction with varying levels of certainty/uncertainty in between.

It is typical that through the application of *in silico* screening the number of potential impurities that remain of concern will have been drastically reduced, perhaps by close to 90% of those initially part of the assessment.

3.2.4 Step 3 – Classification

Once a structural assessment has been completed, each impurity should be categorized according to its mutagenic hazard. The five-class classification scheme, defined by Müller et al. [12], was ultimately adopted for this purpose, and this is shown in Table 3.2.

It is important to be aware that the SAR evaluation procedures can only be as good as the databases and rule sets that underpin the SAR systems. It is known that there are complexities in the models for some compound classes, for example those relating to anilines and heteroaromatic

Table 3.2 The Mueller five-class classification scheme.

Class	Definition	Proposed action for control (details in section 3.2.6.7)
1	Known mutagenic carcinogens	Control at or below compound-specific acceptable limit
2	Known mutagens with unknown carcinogenic potential (bacterial mutagenicity positive, no rodent carcinogenicity data)	Control at or below acceptable limits (generic or adjusted TTC)
3	Alerting structure, unrelated to the structure of the drug substance; no mutagenicity data	Control at or below acceptable limits (generic or adjusted TTC) or do bacterial mutagenicity assay; If non-mutagenic = Class 5 If mutagenic = Class 2
4	Alerting structure, same alert in drug substance which has been tested and is non-mutagenic	Treat as non-mutagenic impurity
5	No structural alerts, or alerting structure with sufficient data to demonstrate lack of mutagenicity	Treat as non-mutagenic impurity

amines. It may be advisable to treat these cases where expert elicitation becomes essential with the option to consider safety testing (Ames test).

Although *in silico* systems are comprehensive in terms of the compound classes covered, there are nevertheless examples of classes that are not covered and for which there is no closely related data in the underlying database. This point was made by Dobo et al. [3] in respect of heteroaromatic nitro compounds. Hence, it is important for the recipients of the SAR output to scrutinize the findings. If an impurity has no flags for mutagenicity, but is used in the process as an electrophile, then it would be prudent to seek expert judgment with respect to the strength of the underlying data set. In such cases and particularly if the synthetic route is likely to remain the same up to and beyond marketing authorization, further assessment, i.e. an Ames test, may be prudent as this is likely to be ultimately required as part of worker safety expectations.

Evaluation of mutagenic risk can also be augmented by data derived from within the public domain. Indeed, such data forms the basis of the ICH M7 addendum table [8], and the review of common chemicals conducted by Bercu et al. [13]. This topic is explored in detail in Chapter 7.

Data sources include:

- Hazardous Substances Databank (HSDB),
- Chemical Carcinogenesis Research Information System (CCRIS), and
- Integrated Risk Information System (IRS).

These provide an excellent source of safety data for many common chemicals. Another related system is the Berkeley database. Indeed, as described in Chapter 7, it is often possible with common reagents to locate sufficient safety data to allow mutagenic risk to be assessed on a compound-specific basis rather than simply applying the TTC. Until recently these, and a number of other references, were accessible via TOXNET, a searchable database provided by the US Library of Medicine. TOXNET provided access to a series of databases through a common portal. While TOXNET is no longer available, alternative search engines such as TOXPLANET are available. Additionally, Lhasa has now reproduced the Berkeley database, which is accessible via their website. As well as the original Berkeley database, which is no longer maintained, the Lhasa database includes more recent data available since the freezing of the Berkeley database. See Chapter 7 for a detailed overview.

3.2.5 Step 4 – Assessment of Risk of Potential Carryover of Impurities

Once impurities with a potential mutagenic safety concern have been identified by the SAR evaluation process, the next step is to consider the likelihood of them being present in the isolated API, often referred to as impurity fate mapping.

The impurities under consideration are often highly reactive, and hence their removal during downstream processing is facilitated by this intrinsic reactivity. This removal can also occur as a result of a variety of factors including solubility, through extractive processes, or within the isolation solvent, i.e. mother liquors during isolation of the desired product, volatility, etc. For example, acidic and/or basic workup conditions frequently encountered in manufacturing processes may lead to decomposition and/or removal of the material of concern. Similarly, other reagents used in downstream processing may react with the material rendering it nonmutagenic, and thus the resulting impurity can be controlled to levels aligned with ICH Q3A or the appropriate clinical phase. It is important as part of such an assessment to consider the fate and effects with respect to what the downstream product could be "reasonably predicted" to be. While rare, it is possible that a PMI could be converted to another PMI through processing, e.g. oxiranes ring opening with HCl to the chloro-alcohol.

It is important that some consideration should be given to what the impurity might be converted to. Factors that contribute to removal of such impurities are reviewed below in the following section. A more detailed examination is provided in Chapter 9.

Initially, such impurity fate assessments were largely based on the theoretical knowledge and experience of the evaluating chemist. Unfortunately, however compelling the arguments developed, they were viewed as nonquantitative and subjective from a regulatory perspective. Thus, in many cases there is a need to provide further analytical data to substantiate the impurity fate assessment. Hence, a quality by testing (QbT) approach was adopted rather than a quality by design (QbD) approach.

It was against this context that Teasdale et al. [14, 15] looked to define a potentially standardized approach to such assessments. The aim was to assess fate semiquantitatively based on factors linked to the impurity's physicochemical properties (and taking into account those of the API and intermediates) and the process conditions employed in the route of manufacture to the API. Pierson et al. [4] had earlier suggested that an assumption could be made of a 10-fold reduction per synthetic stage. In many cases this would suffice and indeed may even be a cautious estimate of the risk. However, in certain circumstances, for example an unreactive mutagenic reagent or intermediate used in a "telescoped" process (no isolations between stages), this may be too simplistic and may even overestimate the potential purge. For this reason, a more quantitative approach, based on actual process conditions and the physicochemical properties of the MI in question, was sought and is outlined below.

A number of contributory factors have been defined that should be taken into account for such an assessment; these are described in detail in Chapter 9.

3.2.6 Overall Quantification of Risk

As described above the acceptability of chemistry-based arguments to demonstrate purge of MIs was initially only partially successful due to its empirical nature. In order to make a quantitative assessment of the level of carryover of a particular material into an API or downstream intermediate, Teasdale et al. [14, 15] defined a number of mitigating criteria; these are defined in Table 3.4. This scoring system has been widely used [16–18], and the concept is enshrined within ICH M7, aligning with control Options 3 and 4; see Chapter 2 (ICH M7) and Chapter 9 (purge factor concept).

For each mitigating criteria, a purge factor can then be selected according to the characteristics of the material under consideration. The numerical scale has been developed to link individual process steps to the physicochemical properties of the individual impurity in question. Each factor is scored (high-low) in terms of its ability to purge the impurity; thus, the higher the score, the greater the likelihood that the impurity would be purged from the process (Table 3.3).

Table 3.3 Purge factor calculation scoring system.

Physicochemical parameter	Scale of purge factor
Reactivity	Highly reactive = **100**
	Moderately reactive = **10**
	Low reactivity/unreactive = **1**
Solubility	Freely soluble = **10**
	Moderately soluble = **3**
	Sparingly soluble = **1**
Volatility	Boiling point >20 °C below that of the reaction/process solvent = **10**
	Boiling point within ±20 °C of that of the reaction/process solvent = **3**
	Boiling point >20 °C above that of the reaction/ process solvent = **1**
pK_a/pK_b	Ionization potential of PMI significantly different from that desired product (**3–10**)
Physical processes: chromatography	Chromatography: **10–100** based on extent of separation
Physical processes: e.g. other scavenger resins	Evaluated on an individual basis (**3—100**)

Hence, if a material is identified three steps from API, the characteristics of the material concerned should be critiqued with the nature of the three downstream processing stages to understand likelihood of purging. An overall purge factor can be assigned by multiplying the purge factors arising from each separate stage, and based on this value, a decision can be made as to what, if any, further action may be required.

The relationship between predicted purge and required purge was examined by Barber et al. [19]. ICH M7 outlines a series of control options, but it does not provide guidance in terms of how to systematically decide which is an appropriate control option. To address this, Barber et al. developed a regulatory decision tree (Figure 3.2) with detailed description of action limits (Table 3.4) depending on the ratio. The decision tree, and description of action limits, links purge predictions and their relationship with required purge (i.e. purge ratio) with recommendations for control strategy development. It also defines the level of detail and content in terms of supporting data presentation.

The first stage of the process is to establish the required purge for each PMI in question. This is based on factors such as the dose, duration of treatment (this being linked to the permitted dose for an MI), and the starting concentration of the PMI in the process. Where a mole equivalent is used, a start point of 1 000 000 ppm is assumed.

Once this has been done, the next step is to determine the predicted purge factors for each PMI using the identified scoring system. It is important to compare the predicted purge factor for each PMI versus the required purge in order to determine the ratio between the required purge and that which is predicted.

This gives the purge ratio as described in the equation below.

$$\text{Purge ratio}\left(\text{PR}\right) = \frac{\text{Predicted purge factor for PMI}}{\text{Required purge factor}\left(\text{based on the permitted limit}\right)}$$

3.2.6.1 Predicted Purge Factor

Predicated purge factors can be determined either by a paper-based approach or by knowledge-based software (see Chapter 9). Individual purge factors are predicted for each PMI present within

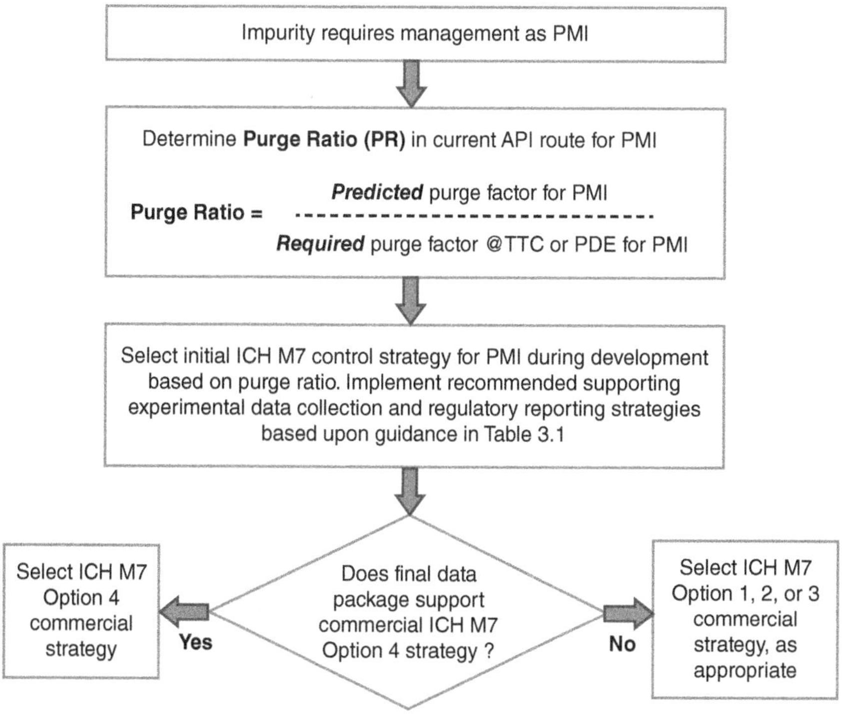

Figure 3.2 (P)MI purge factor decision tree for use under ICH M7.

the process. The purge factors are defined based on the known, or estimated, purging capacity of the downstream chemistry and operations. Based on the outcome of the assessment and comparison to the required purge factor, the resultant purge ratio is then used to determine which ICH M7 control strategy is appropriate and what, if any, further evidence may be required to support the claim that the PMI in question is purged. A systematic process relating the purge ratio to data requirements and regulatory strategy was defined and is illustrated in the decision tree, Figure 3.2.

3.2.6.2 Required Purge Factor

The required purge factor is calculated by dividing the maximum PMI level at a defined point[1] in the process, by the acceptable limit in the final API. The acceptable limit, or acceptable daily intake (ADI), is the level of the PMI associated with negligible risk and is typically based on the TTC, the less than lifetime (LTL) limit, or permitted daily exposure (PDE[2]) as described in ICH M7 R1 [8] (R1 includes the addition of the addendum). The safety-based limit (acceptable daily exposure [ADE] or PDE) is then converted into a concentration limit based on the clinical dose and duration; typically, these are based on the highest-anticipated or highest-approved clinical dose/longest dosing duration.[3]

1 The maximum observed PMI level can be designated by several means. These include: (i) by the amount of PMI introduced to the process, (ii) by the amount of PMI measured at a specific stage in the process, (iii) the amount in the process or by a level allowed by an acceptance criterion such as an assay value in an intermediate, or (iv) a hypothetical amount formed; this final approach is typically used where a PMI is formed by a side reaction.

2 ADIs are based on linear extrapolation of carcinogenicity data; PDEs are used where there is a definable threshold or no-adverse effect level (NOEL).

3 In the case of a marketed product, this may be based on median duration for the class/disease area.

Table 3.4 Relationship between purge factor ratios and regulatory reporting action limits and potential supplementary reporting requirements.

If PR ⩾ 1000×	If 1000 > PR ⩾ 100×	If PR < 100×
Data collection recommendations		
Collection of additional experimental data not recommended for *noncommercial* or *commercial* API routes to support scientific rationale.	Collection of additional non-trace experimental data (solubility, reactivity, and volatility) recommended for both *noncommercial* and *commercial* API routes to support scientific rationale. Collection of additional trace PMI analysis not necessary for *noncommercial* or *commercial* API routes to support scientific rationale.	For *noncommercial* API routes, experimentally measure PMI purging, including trace PMI analyses as appropriate, to support scientific rationale. Note: additional data are expected to support an Option 4 control strategy when PMI purge ratio is <<100×. For *commercial* API routes, detailed experimental fate and purge studies are expected for all PMI to support a commercial Option 4 control strategy.
Regulatory reporting recommendations		
Report "unlikely to persist" or cumulative predicted purge factor and purge ratio for *noncommercial* API routes in regulatory submissions. Replace with summary of key elements of predicted purge factor calculations and purge ratio for *commercial* API routes in regulatory submissions.	Report the cumulative predicted purge factor and purge ratio for *noncommercial* API routes in regulatory submissions. Replace with summary of key elements of predicted purge factor calculations, = purge ratio, and supporting non-trace data on purge properties for *commercial* API routes in regulatory submissions.	Report summary of key elements of predicted purge factor calculations, purge ratio, and supporting non-trace or trace data for *noncommercial* API routes in regulatory submissions. Replace with complete summary of predicted purge factor calculations, purge ratio, and supporting trace and non-trace fate and purge data for *commercial* API routes in regulatory submissions.

3.2.6.3 Purge Ratio

The purge ratio is determined by simply dividing the predicted purge by the required purge. For example, if the predicted purge was 1×10^6 and the required purge was 100, then the purge ratio is 10000, indicating that the PMI in question is anticipated to be removed by the process to levels at least 10000-fold lower than that required to reduce to the acceptable limit established for the individual PMI.

As illustrated in Figure 3.2, the purge ratio can then be utilized to determine the most appropriate ICH M7 control strategy. Based on this ratio, Barber et al. [19] defined a series of action limits incorporating both the purge ratio and the phase of development, see Table 3.4. The action limits recommend the extent of data required to specifically support an Option 4 based control strategy. Clearly, the ratio reflects the extent of risk and hence the data requirements are directly proportionate to this, i.e. that increasing amounts of supporting data are required as the purge ratio decreases. The aim is ultimately to support the voracity of the predicted purge through experimental data, ensuring the robustness of the proposed control strategy.

Barber et al. [19] also set out a series of recommendations on reporting expectations for purge ratio justifications within regulatory submissions. In accord with the phase-dependent data requirements defined within section 9 of the ICH M7 [8] guideline, this covers both clinical development and post-approval, marketing phase.

Experience has shown that in reporting purge factors and the proposed control option, particularly Option 4, transparency is key. It is thus recommended to include predicted purge factors for each key purging step and within each step, each unit operation in the process. That should be augmented by inclusion of any supporting experimental physicochemical data that strengthens and increases confidence in the final prediction.

3.2.6.4 High Predicted Purge

If the predicted purge ratio is $\geq 10^3$ (1000), then Barber et al. proposed that there was no additional specific data collection required. In view of the demonstrated conservative nature of the scoring system derived by Teasdale et al. [14, 15], this conclusion should be valid.

3.2.6.5 Moderate Predicted Purge

If the predicted purge ratio is < 1000 but ≥ 100, then additional data (e.g. reactivity, solubility, and relevant test data) may be required to support the purge argument and the subsequently defined control strategy. This should be assessed on a case-by-case basis and allied to the factors that are most critical to the overall purge of the PMI in question. For example, if the removal of a PMI is predominately due to its solubility, then providing supporting solubility data may be key to underpinning the overall purge factor.

3.2.6.6 Low Predicted Purge

If the predicted purge ratio is < 100, then ICH M7 Option 4 may not be an appropriate strategy unless it can be supported by further substantive experimental data. While it is possible that an Option 4 approach might still be valid in view of the inherent underestimation of the purge estimate approach, a predicted purge of <100 alone was viewed as insufficient to support an Option 4 approach. Therefore, measured purge factors, based on both batch data, and deliberate spiking and purge studies, would be required to support an Option 4 proposal.

3.2.6.7 ICH M7 Control Option 1, 2, or 3

Critically, if the experimentally measured purge factor is insufficient to support an ICH M7 Option 4 control strategy, then the applicant should assess the relative merits of the other ICH M7 control strategies, i.e. Options 1–3 (see Figure 3.2).

3.2.6.8 Step 5 – Further Evaluation

A much abbreviated shortlist of remaining materials of concern is likely to have resulted from having compiled an initial list of potential impurities, identified those that are known or suspected mutagens, and evaluated which of these may potentially be present in API at a level of concern, based on the material characteristics, origin in the process, and ability to survive the process intact.

There are now two ways by which the risk of such remaining potential PMIs may be mitigated.

1) *Safety testing*. Demonstrating that a material is nonmutagenic will allow it to be addressed under ICH Q3A/B.
2) *Analytical testing*. Demonstrating that a material is below the permitted safety limit.

Which approach to take will depend on the specific nature of the project and the impurity concerned and may be influenced by factors such as the availability of pure samples of the material of concern and/or availability of appropriate analytical methodology with which to determine levels.

3.2.6.9 Safety Testing

For any impurity identified as being potentially mutagenic (based on SAR evaluation) and assessed as having a high likelihood of carryover into the API, the next step is often to carry out *in vitro* safety testing.

If *in vitro* testing is selected, it is recommended that the synthesized or isolated impurity is tested for mutagenicity as an individual impurity. However, where this is impractical, then spiked samples or batches of material that contain elevated levels of the impurity of concern may be tested. The latter approach is not generally encouraged by regulatory authorities, and in such cases, an early dialogue with the relevant regulatory authority is recommended. In relation to the material quality, it is recommended to ensure the sample tested is as pure as is practically possible and is fully characterized in terms of the impurities present with the test sample. It is important to ensure that any result from subsequent testing is not confounded by the presence of trace impurities that may themselves elicit a response in the assay but would otherwise not be relevant to the route in which the (P)MI is being assessed. It is therefore important to fully characterize the test sample in terms of impurity profile with a specific emphasis on other potential mutagenic species that could be present, especially any used to synthesize the test sample.

ICH M7, Section 3 general principles, makes clear that the focus of the guideline is on deoxyribonucleic acid (DNA)-reactive substances that have a potential to directly cause DNA damage when present at low levels leading to mutations and therefore, potentially causing cancer. It then states that to detect this type of mutagenic carcinogen a bacterial reverse mutation (mutagenicity), i.e. Ames test, has the necessary sensitivity and specificity.

A positive result in one or more of these tests is generally sufficient evidence to define the impurity as mutagenic, in which case it will be then necessary to adopt the appropriate TTC approach. Occasionally, a thresholded mechanism can be argued based on available safety data; this concept is examined in detail in Chapter 8. If an impurity is found to be negative, it is considered nonmutagenic (qualified for mutagenicity) and can then be treated as a normal impurity under ICH Q3A/B [13, 14].

The mutagenic potential of *in vitro* positive materials may be further evaluated *in vivo*, in order to look to establish the biological relevance of the *in vitro* findings (Table 3.5); this is highlighted in ICH M7, Note 3; and this is examined in detail in Chapter 6.

3.2.7 Quantification of Level Present

For PMIs that have been assessed as having a reasonable likelihood of being present in API at levels of concern, it may be appropriate to attempt to determine the level in parallel with, or *in lieu* of, the safety testing described above. The level of concern will be set by the appropriate limit (ADI, PDE, or TTC), which itself is impacted by factors such as maximum clinical dose and the maximum duration of the proposed trial(s). This in turn will have an effect on the choice of analytical technique.

- *Choice of technique?*
  The nature of the impurity (analyte), the characteristics of the API or intermediate (matrix), and the level to be determined will influence the detection technique employed. Many organizations have developed specific strategies for refining such selections; this is examined in detail in Chapters 12 and 13.
- *Where in the process to test?*
  Testing may be performed on upstream intermediates, API, or drug product as appropriate. It is often desirable to test as close as possible to the point of introduction of a PMI into the process. This approach may permit standard techniques, such as high-performance liquid chromatography (HPLC) with ultraviolet (UV) detection, to be used, if this is allied to spiking experiments demonstrating the removal in the downstream process. Indeed, such an approach aligns with one of the control concepts defined within ICH M7, Section 8.1 of ICH M7 (specifically control Option 3). While development laboratories may be equipped with more sensitive techniques suitable for analysis at the low ppm level, manufacturing quality control laboratories are unlikely to have such facilities.

Table 3.5 Tests to investigate *in vivo* relevance of *in vitro* mutagens.

Note 3 Tests to Investigate the *in vivo* Relevance of *in vitro* Mutagens (Positive Bacterial Mutagenicity)	
in vivo test	Factors to justify choice of test as fit-for-purpose
Transgenic mutation assays	For any bacterial mutagenicity positive. Justify selection of assay tissue/organ
Pig-a assay (blood)	For directly acting mutagens (bacterial mutagenicity positive without S9)[a]
Micronucleus test (blood or bone marrow)	For directly acting mutagens (bacterial mutagenicity positive without S9) and compounds known to be clastogenic[a]
Rat liver Unscheduled DNA Synthesis (UDS) test	In particular for bacterial mutagenicity positive with S9 only Responsible liver metabolite known to be generated in test species used to induce bulky adducts
Comet assay	Justification needed (chemical class specific mode of action to form alkaline labile sites or single-strand breaks as preceding DNA

[a] For indirect acting mutagens (requiring metabolic activation), adequate exposure to metabolite(s) should be demonstrated.
Source: Reproduced from ICH M7.

- *Quantitative assay or limit test.*
  Both types of methods are used in the analysis of MIs. Quantitative tests are useful to furnish data for process development and to support further process modifications to reduce or more consistently control levels of a PMI. Having established a validated process, limit tests are likely to be favored for routine quality control (QC) testing.
  Limit tests are also more likely to be applied to upstream testing at an intermediate stage where they are used in conjunction with demonstrated evidence of further reduction through processing (control Option 3 – ICH M7) [8].
  Quantitative assays are usually applied at the final isolated API, as they provide a measure of true levels of the PMI/MI that would be present in the drug product, and the material would not be administered to patients if measured levels are found to be too high. Since the staged TTC concept for acceptable levels of PMIs/MIs is routinely applied during clinical stages of development, a quantitative test is generally desirable since acceptable levels vary as the clinical program develops. However, limit tests may be appropriate at the API or DP stage if this figure is well below the staged TTC control level.

3.3 Step 6 – Overall Risk Assessment

Once analytical and/or safety test data are available, these are used to finalize the risk assessment.
Possible outcomes include:

- A PMI returns a negative Ames test result and thus no longer requires control as an MI but defaults to ICH Q3 levels of control [13, 14].
- A PMI returns a positive Ames test result, but analytical testing demonstrates adequate process control over levels, i.e. level well below appropriate TTC limit.
- Analytical data demonstrates that a PMI/MI is below a current staged TTC but above future permitted dose duration levels (i.e. where studies are of longer duration or higher doses are needed). In such circumstances this may necessitate a modification of the process to reduce or

eliminate the impurity in question. If the material is potentially mutagenic rather than a known genotoxin, expediting safety testing, with the possibility of a negative Ames test result, would either remove the need for further process development and analytical control at trace levels, or confirm that additional control is required.

- Analytical and safety data reveal an Ames positive material above a staged TTC level for a planned clinical study.[4] In such a scenario, it is likely that the material in question would need to be reprocessed, unless a compelling case could be made for the benefit of the treatment over the risk posed (see ICH S9 [20] for example). In most cases, the process would need to be redeveloped to bring levels of the mutagen in question within the TTC for the envisaged marketed product dose and duration. For this reason it is important that an appropriate procedure to link the assessment to the formal release of material for clinical trial use sits alongside the risk assessment. Similarly, it is also important to understand whether an impurity is above the TTC prior to manufacture of drug product given that once formulated, any reprocessing becomes very difficult and secondary processing equipment could become compromised. For this reason, most companies will have an analytical release process for the drug substance that includes having a completed MI risk assessment and that any described controls for any PMIs have been met. Additionally, conducting the MI risk assessment prior to the drug substance manufacturing campaign will help identify PMIs that may be introduced late within the process, and therefore may not efficiently purge. This will allow additional processing to be engineered into the drug substance manufacturing campaign to reduce the likelihood of failed drug substance batches and reprocessing.

This is not meant to be an exhaustive list but serves to illustrate some of the potential outcomes and likely courses of action in each case.

It should be recognized that the evaluation of mutagenic risk is an iterative process and needs to be updated in line with any process-related changes and/or emerging information relating to impurities, and/or degradants, in drug substance or drug product. Other factors such as a change in the trial duration, trial population, specifically in terms of oncology where the treatment is extended from an initial S9 population to a non-S9 population and/or dose may also require a review of the risk assessment.

3.4 Further Evaluation of Risk – Purge (Spiking) Studies

Alongside the theoretical evaluation of risk described above, there is often the need to examine this experimentally through conducting appropriate purging or spiking experiments. This is most likely required where a moderate to high risk of potential carryover into the API has been defined, i.e. where control Option 4 is not considered appropriate. Spiking refers to the practice of adding in a fixed quantity, or spike, of the material to be tracked, in order to confirm a quantifiable baseline. Purging refers to the extent to which the material in question is removed out of the downstream material or API as a consequence of the processing conditions to which it has been exposed.

3.5 Conclusion

The need to adequately assess the risk posed by MIs, and to limit the level present in API/DP, is clearly established in ICH M7. Within this chapter the authors have looked to define a risk-based

4 Should it be discovered that the level of an MI is above permitted levels in material currently used in clinical trials, then this may lead to suspension of the trial and expedited reporting under 15-day rules to regulatory authorities.

approach to such an assessment, one based on a combination of semiquantitative assessment, allied to analytical results and data from appropriate purging studies. Such an approach should ensure that any actual MI-related risk is clearly identified and managed.

3.6 Case Studies

3.6.1 Case Study 1 – GW641597X

GW641597X was developed as a PPAR-alpha agonist for the treatment of dyslipidemia. Described below is an overview of the MI risk assessment, and while this product predated ICH M7 [8], a useful retrospective assessment in alignment with ICH M7 was performed using current best practice to inform the reader of the specific steps required. The development of the process to GW641597X and a discussion of the MI control strategy has been published [21].

Applying mutagenic, or potentially mutagenic, impurity (PMI) controls in accordance with ICH M7 [8] for chronic dosing allows up to 1.5 μg/day for an individual MI specified in the drug substance or up to 5 μg/day for the total quantity of three or more PMIs that may be specified. A maximum dose of 600 μg/day was predicted for GW641597X, and therefore the "commercial" TTC-based acceptable limit for GW641597X was determined as 2500 μg/g for individual MIs, and for three or more specified MIs, 8333 μg/g would be the maximum total amount.

The first stage for the assessment was to identify potential impurities, this was performed by assessing identified and reasonably predicted drug substance impurities (Figure 3.3), together with assessment of the synthetic process (Figure 3.4) for starting materials, intermediates, and reasonably predicted reaction by-products from the synthetic process.

The identified drug substance impurities and reasonably predicted impurities (derived from route reagents, intermediates, and potential by-products/degradants) were all assessed for potential mutagenicity by (Q)SAR screening, which amounted to > 20 separate structures. In accordance with ICH M7, two methodologies were employed, one expert rule-based software (Derek Nexus v6.0) [22] and one statistics-based software (Leadscope v2.2.1) [23], and all output results were subject to expert assessment [2].

Out of all the materials assessed, only three structures of potential mutagenic or carcinogenic concern were identified, which were the reagents ethyl bromoisobutyrate **2**, hydroxylamine hydrochloride, and the alkyl chloride **8** (Table 3.6).[5]

The remaining compounds were Ames negative, non-SAR alerting, or the equivocal predictions could be refuted following expert review.

The next stage was to assess the probability for these impurities to be present within the drug substance at a level of concern. This was achieved using a paper-based purge calculation using the

---

5 While chloroacetyl chloride and its hydrolysis product chloroacetic acid contain structural features that alert for mutagenicity, both have been found to be nonmutagenic in the Ames test: American Conference of Governmental Industrial Hygienists. (1991). *Documentation of the Threshold Limit Values and Biological Exposure Indices, Cincinnati, Ohio*, 6e. ACGIH, p. 269. European Chemicals Agency, http://echa.europa.eu/ CAS RN 79-11-8, https://echa.europa.eu/registration-dossier/-/registered-dossier/15187/7/7/2.Toluene and denatured ethanol are used within the process to manufacture GW641587X, and these solvents can contain low levels of the known carcinogen benzene. While benzene has an ADI of 20 μg/day, this solvent would need to be controlled to 2 μg/g within GW641597X as per ICH Q3C expectations (ICH guideline Q3C (R6) on impurities: guideline for residual solvents). Typical specifications of benzene in toluene and denatured ethanol are not greater than (NGT) 50–100 and 2 μg/g, respectively (in-house experience). Toluene is used three isolated stages away from GW641597X, and levels of <0.5% solvent (ethanol/methanol) within GW641597X were achieved through drying (drier off-load condition). It is reasonable to anticipate that residual levels of benzene would be purged to significantly below 2 μg/g in drug substance, and this would be confirmed at later phases of development through appropriate testing.

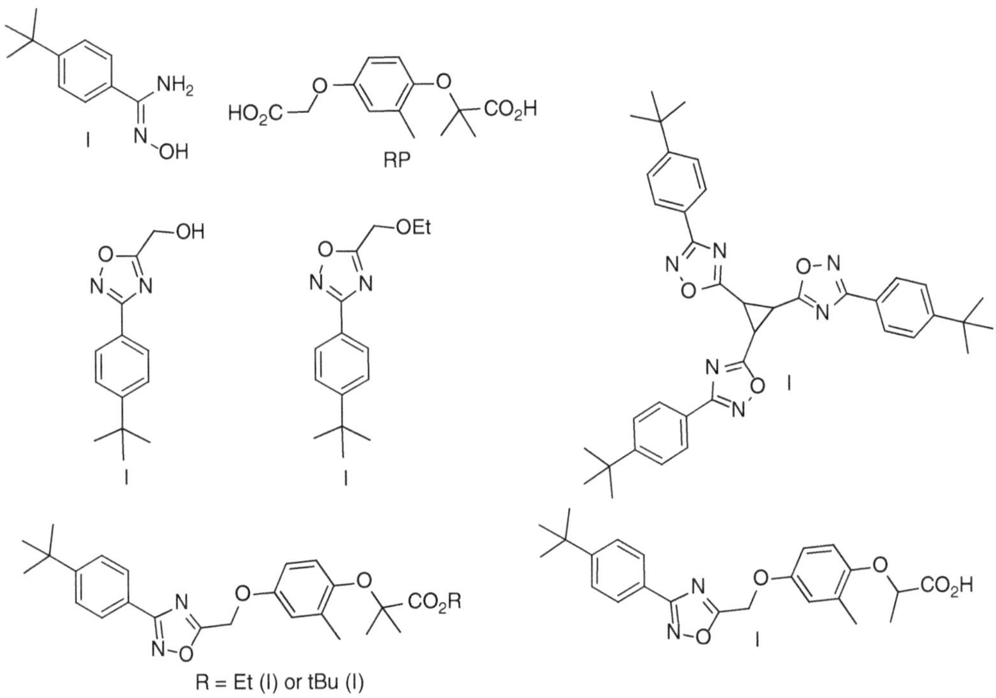

Figure 3.3 Identified (I) and reasonably predicted (RP) impurities within GW641597X drug substance.

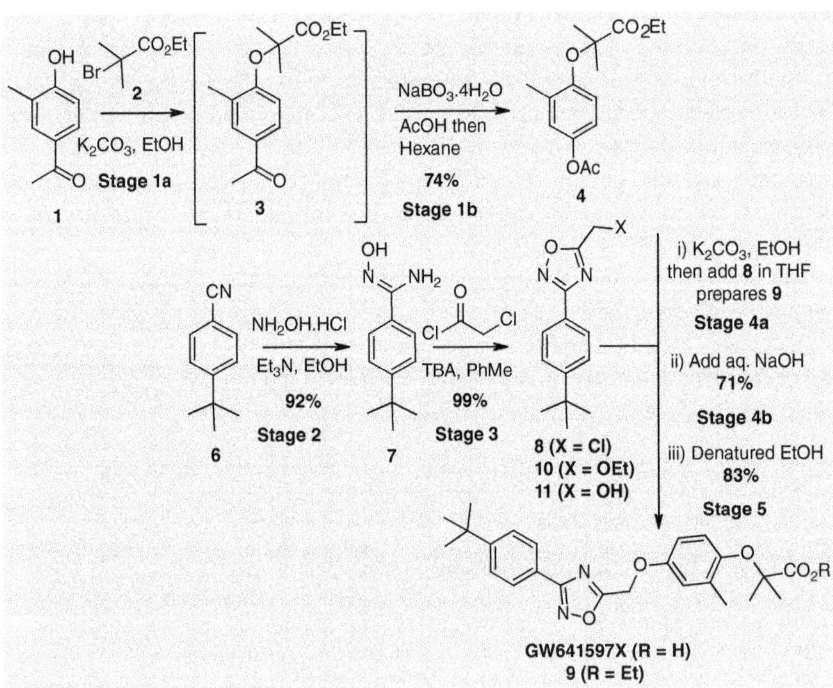

Figure 3.4 Synthetic process to GW641597X.

Table 3.6 Summary of mutagenicity assessment for synthetic process to GW641597X.

Compound	Derek	Leadscope	Ames assay	ICH M7 impurity classification
Ethyl bromoisobutyrate **2**	Positive	Positive	Positive	2
Hydroxylamine hydrochloride	Negative	Negative	Negative	5[a]
Chloromethyloxadiazole **8**	Positive	Positive	Not tested	3[b]

[a] Hydroxylamine is not mutagenic but is carcinogenic in rats and has a permitted daily exposure (PDE) of 23 μg/day. Hydroxylamine is non-SAR alerting using the SAR tools for this case study. A published review of available data considers carcinogenesis to be via a non-thresholded mechanism and as such hydroxylamine can therefore be considered ICH M7 Class 5, i.e. Ref. [19].

[b] Alkyl chloride **8** is a *mono*functional alkyl chloride and should be controlled to a class-specific limit <15 μg/day.

principles established by Teasdale et al. [14, 15], and detailed knowledge of the processing as well as of the material attributes greatly facilitates this process. In each case, once the predicted purge was calculated, this purge was compared to the required purge (based on dose/duration factors) and a purge ratio calculated the magnitude that helps inform likely control strategy for that material [19]. For clarity, each of the potential mutagenic materials is discussed below together with a rationale for its proposed control.

3.6.1.1 Ethyl Bromoisobutyrate 2

Two equivalents of ethyl bromoisobutyrate **2** were charged into the Stage 1a process to achieve complete conversion to product at an acceptable rate of reaction, and as a consequence, there could theoretically be 1 equivalent of **2** remaining on completion of the reaction to prepare **3**.[6]

Where large excesses of reagents are concerned (e.g. >2 eq.), one can generally assume 1 equivalent has been used for reactions that produce > 90% yield of product. For reaction where < 90% conversion is observed, the assessors should consider including the residual reagent that may be remaining through incomplete conversion to product. If stoichiometric or only minor excesses of reagent are used, then it is considered best practice to base the equivalents used, and therefore equivalents left behind, on the yield of product. In this case study, yields were generally significantly over 90% and the decision was taken to assume 1 equivalent of reaction for the reagent of concern.

Given the permitted limit based on a 600 μg/day dose is <2500 μg/g within GW641597X, the required purge was conservatively calculated on the basis of this potential 1 mol excess of **2,** which leads to a purge factor of 400 being required for **2** (1 000 000/2500 = 400).

Ethyl bromoisobutyrate **2** reacts to form the ester **3** in the Stage 1a process. An excess of this reagent is required because **2** polymerizes during the process. A conservative approach was taken such that no reactivity for this side process was assigned and neither was any reactivity of this reagent anticipated within the oxidative Stage 1b process. No solubility can be predicted during the stages to prepare **3** and **4** as there is no formal product isolation stage.[7] Moderate reactivity is predicted within the process to alkylated **4** with **8** as there is opportunity for alkyl bromide **2** to react with the phenolate derived from **4**, as well as potential to form ethers from reaction with the

6 Excess reagents are generally only used to maximize yield for the certain processes because the reagent is otherwise being reduced either through undesired side reactions or through physical processes. If where large excesses are used, it is acknowledged for a high yielding reaction that 1 equivalent of reagent can be assumed to have reacted to form product that leaves the excess reagent requiring assessment.

7 Compound 4 is an oil and was isolated by evaporation, and therefore no solubility score can be assigned for residual alkyl bromide 2.

ethanol solvent during the manufacture of ester **9**. Ethyl bromoisobutyrate **2** is anticipated to react with the aqueous base during the preparation of crude GW641597X, through hydrolysis, polymerization, and potential ether formation. As **2** is an oil, full solubility was anticipated during the isolation and washing processes during the preparation of crude GW641597X and its subsequent purification. Based on this highly conservative approach, the purge ratio (Table 3.7) for **2** is insufficient to justify an ICH M7 Option 4 control strategy by itself as the purge ratio is less than 1000 [19] and, without further information to support an Option 4 control strategy, an alternative ICH M7 control option would be recommended. One option would have been to measure the level of **2** at the completion of Stage 1. Other options can include the use of "prior knowledge," which can be literature reactivity data for similar transformations, with structurally similar compounds, in order to further justify/reinforce an Option 4 proposal. Such data can also be used to assign a higher purge score (reactivity, solubility, etc.) if the original predictions are considered overly conservative when reviewed against a similar reported transformation.

3.6.1.2 Hydroxylamine

Hydroxylamine reacts with benzonitrile **6** during Stage 2 to form the benzimidamide **7** (Figure 3.4). A 2.5 molar excess of the reagent is used to assure reaction completion, and therefore, as much as 1.5 equivalents might remain within the Stage 2 process after formation of **7**.

Hydroxylamine has an ADI of 23 μg/day that equates to a permitted limit of ≤ 38 333 μg/g of GW641597X and correlates to a required purge factor of 39.1 (1 500 000/38 333 = 39.1).

With respect downstream purging, hydroxylamine free base has a low boiling point (58°C), and it is reasonable to expect losses through evaporation during higher temperature processing as well

Table 3.7 Purge predictions for ethyl bromoisobutyrate **2**.

Stage	Reactivity	Solubility	Volatility	Total	Rationale
1a	1	1	1	1	2 M equivalents used where **1** is used to prepare ester **3**. Telescoped process; therefore, no isolation. It was also decided to assume no additional reactivity; therefore, 1 equivalent remains at the end of this stage (conservative).
1b	1	1	1	1	No reactivity or solubility predicted from Stage 1b during the preparation of **4**.
4a	10	1	1	10	Expected to react with the phenolate from **4** during Stage 4a to prepare ester **9**, potential etherification from reaction with ethanol solvent and polymerization.
4b	100	10	1	1000	Anticipated to react with the aqueous base through hydrolysis, polymerization, and potential etherification during the hydrolysis of **9** (Stage 4b). Ethyl bromoisobutyrate **2** is an oil and highly soluble in the isolation solvent.
5	1	10	1	10	Ethyl bromoisobutyrate **2** is an oil and likely to be highly soluble in the isolation solvent.
Predicted purge factor				1×10^5	
(Required purge)				(400)	
				Purge ratio = 250 (would inform ICH M7 Option 1, 2, or 3 control without supporting data, e.g. collection of supporting "non-trace" experimental data)	

as during isolation and drying through volatilization. Despite this, a conservative approach was taken and volatility was not scored for the entire process. Hydroxylamine can be considered to purge through solubility within processes to prepare **7**, **8**, **9**, and GW641597X, and reactivity was anticipated during Stages 3 and 4 for the preparation of **8** and **9** through reaction with chloroacetyl chloride and alkyl chloride **8**, respectively. Even with the application of a highly conservative approach i.e. assuming a starting concentration of 1.5 equivalents and discounting likely volatility, the predicted purge factor was significant, 1×10^8 (Table 3.8). The purge ratio is significantly greater than 1000, which justifies an ICH M7 Option 4 control rationale without recourse to additional experimental data and simply reporting "unlikely to persist" should be sufficient for a regulatory submission as per the published guidance [19].

3.6.1.3 Alkyl Chloride 8

The alkyl chloride **8** is a *mono*functional alkyl chloride[8] and as such can be controlled to a class-specific limit of <15 µg/day; refer Note 5 of ICH M7 [8]. This control limit approximates to 2.5% w/w, but as per ICH M7 section 7.2, the allowable intake of an MI should be capped at 0.5%; therefore, the impurity purge calculations for this material were set against this lower limit. This reagent is charged into the Stage 4 process at 1.15 equivalents, meaning that up to 0.15 equivalents could remain at the end of the reaction assuming a high yielding process, which would need to be reduced to <5000 µg/g of GW641597X, which gives a required purge factor of 30 (150 000/5000 = 30).

Table 3.8 Purge predictions for hydroxylamine.

Stage	Reactivity	Solubility	Volatility	Total	Rationale
2	1	10	1	10	2.5 Equivalents used in Stage 2 – anticipate 1 equivalent would react as reaction is high yielding. Hydroxylamine anticipated to be highly soluble in the isolation solvent (aqueous ethanol). Volatility during isolation and drying expected but not included within the prediction.
3	100	10	1	1000	Expected reaction with chloroacetyl chloride. Anticipated solubility during extraction process (aqueous acid). Volatility during "put and take" drying process expected but not included within the prediction.
4	100	10	1	1000	Expected reaction with alkyl chloride **8** and solubility within isolation solvents (aqueous ethanol). Volatility during isolation and drying expected but not included within the prediction.
5	1	10	1	10	Anticipated solubility within the crystallization/ isolation solvent (denatured ethanol). Volatility during isolation and drying expected but not included within the prediction.
Predicted purge factor				1×10^8	
(Required purge)				(39)	
				Purge ratio = 2.56×10^6 (informs ICH M7 Option 4 control)	

8 Monofunctional alkyl chloride can be controlled to 10 times the default LTL daily intakes according to Note 5 of ICH M7.

The alkyl chloride **8** reacts with the phenolate derived from **4** to provide the ethyl ester **9**. An excess of reagent is used (1.15 eq.) to ensure reaction completion and maximize yield of product. It is assumed that 1 equivalent is required to form the high yield of ethyl ester **9**, and therefore the residual 0.15 molar excess should be assessed.[9] Reactivity for alkyl chloride **8** could be anticipated within Stage 4a through side reactions with the ethanol solvent and base to form the corresponding ethyl ether **10**, but this was not scored. A moderate reactivity was predicted during Stage 4b where hydrolysis to alcohol **11** or ethanolysis to **10** could be reasonably predicted. Solubility was anticipated during product isolations in Stages 4b and 5. The resulting purge ratio (Table 3.9) was insufficient to justify an ICH M7 Option 4 control strategy as the purge ratio is less than 1000 and, without further data to support an Option 4 control strategy, an alternative ICH M7 control option would be recommended [19].

3.6.1.4 Additional Evidence for the Purging of Ethyl Bromoisobutyrate and Alkyl Chloride 8

In both the case of ethyl bromoisobutyrate **2** and the alkyl chloride **8**, purge factors alone were insufficient to allow an Option 4 approach to be defined, and hence it was decided to generate further supporting data to justify the use of Option 4.

3.6.1.4.1 Including "Measured" Purge into the Purge Rationale for Ethyl Bromoisobutyrate 2

Stage 1a and 1b products are generally telescoped through into the Stage 4 process; however, to understand likely levels of the alkyl bromide **2** that might be present, a sample was removed from

Table 3.9 Purge predictions for alkyl chloride **8**.

Stage	Reactivity	Solubility	Volatility	Total	Rationale
4a	1	1	1	1	1 Equivalent of **8** reacts with the phenolate from **4**, and a high yield of the ethyl ester product **9** is formed. This has been reflected in starting equivalents. Anticipated reactivity to form ethyl ether **10** and hydrolysis product **11** have not been included in the prediction.
4b	10	10	1	100	Excess alkyl chloride **8** anticipated to react to form ether **10**, hydrolysis product **11**, and potential *homo* coupled products. Anticipated to be soluble within the aqueous ethanol isolation solvents.
5	1	10	1	10	Solubility expected within the crystallization/isolation solvent (denatured ethanol).
Predicted purge factor				1000	
(Required purge)				(30)	
				Purge ratio = 33 (would inform ICH M7 Option 1, 2, or 3 control without additional data, e.g. experimentally measure PMI purging)	

9 The reaction for deacetylation of acetate **4** proceeds to at least 98% completion, and the subsequent reaction of the intermediate phenolate with **8** to prepare **9** also proceeds to at least 98% conversion. This means that in principle there could be up to 4% of phenolate unavailable for reaction, meaning that an additional 0.04 equivalents of **8** could remain. The authors considered this a negligible difference to their assumption that up to 0.15 molar equivalents would remain, but it is acknowledged that a more accurate representation would be to consider 0.19 molar equivalents for the purge calculations.

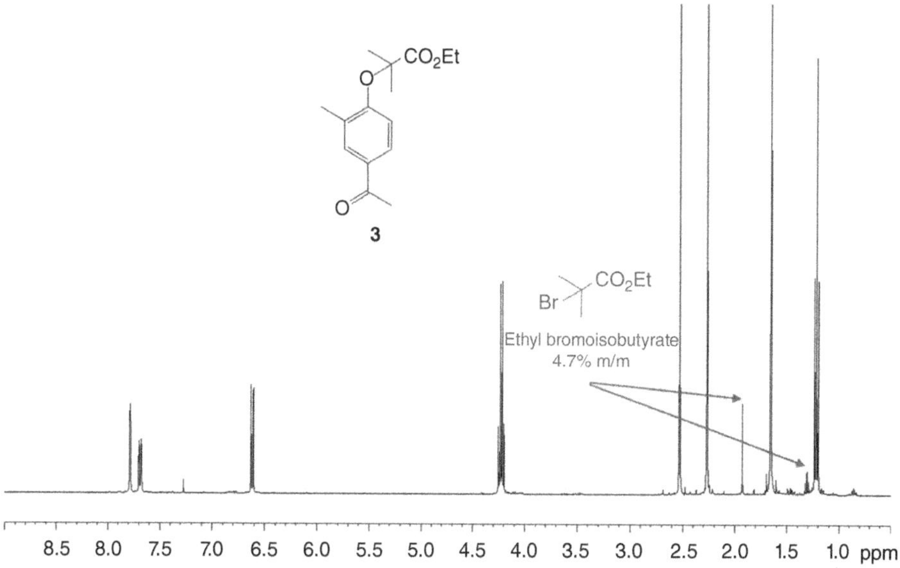

Figure 3.5 ^{1}H NMR of Stage 1a product **3**.

Stage 1a and the reaction solvent removed under vacuum.[10] Subsequent evaluation by ^{1}H NMR (Figure 3.5) confirmed that the alkyl bromide **2** was present at <5% molar ratio (m/m) within this product, rather than the initial conservative estimate of 100% molar ratio based on the 1 molar equivalents excess used for this reagent. This observation aligns with the formation of polymer and allows a more realistic purge prediction to be estimated based on "point of last measurement." Substituting a residual level of ethyl bromoisobutyrate of 5% (50 000 μg/g) into the Stage 4 process, we can now estimate that the predicted purge factor for Stages 4a, 4b, and 5 is 1×10^5 with a required purge factor of 20 (50 000/2500 = 20), which gives a purge ratio of 5000 and fully justifies an ICH M7 [8] Option 4 control. While not necessarily required, checking the ^{1}H NMR of the crude GW641597X (Stage 4b product) confirmed that alkyl bromide **2** is not detected, helping to further validate the predicted purge factor.

3.6.1.4.2 Inclusion of "Measured" Purge into the Purge Rationale for Alkyl Chloride 8

To understand "actual" purge for the alkyl chloride **8**, the Stage 4b product crude GW641597X was checked by ^{1}H NMR (Figure 3.6), which confirmed that the level of **8** was present at approximately 0.2% m/m (0.12% w/w[11]) and is in good alignment with the predicted purge factor of 100 for Stages 4a and b.[12] While the observed level in this typical batch of crude GW641597X is already significantly below the required TTC for a *mono*functional alkyl chloride of 0.5% w/w, checking the ^{1}H NMR for GW641597X derived from the Stage 5 product confirmed that alkyl chloride **8** is no longer observed, giving confidence that it is present at levels <0.10% w/w. These observations fully justify an ICH M7 [8] Option 4 control for alkyl chloride **8**.

10 The boiling point for ethyl bromoisobutyrate is >160 °C.

11 Weight ratio calculated by converting from the molar ratio with the molecular weights for alkyl chloride **8** (250.73) and GW641597X (424.50).

12 The purge prediction for Stage 4a and 4b would indicate that 15 mol% excess of alkyl chloride **8** should reduce to 0.15% m/m.

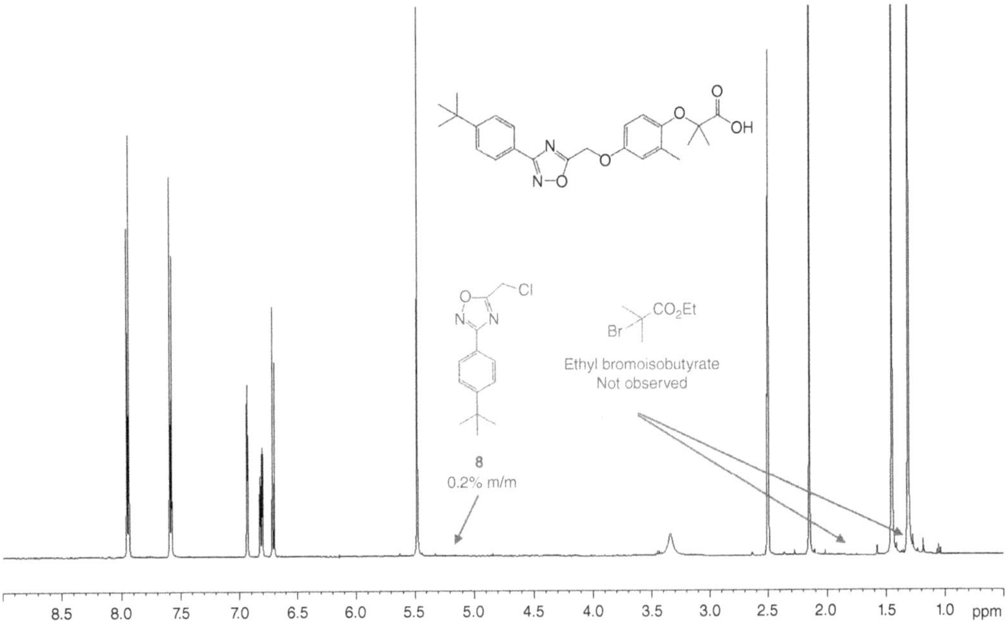

Figure 3.6 ^{1}H NMR of Stage 4b product crude *GW641597X*.

The use of ^{1}H NMR to confirm purging for both ethyl bromoisobutyrate **2** and alkyl chloride **8** helps to verify the purge predictions from the process. They are also useful examples of the expected additional "non-trace" experimental data (solubility, reactivity, and volatility) recommended for both *noncommercial* and *commercial* API routes to support an ICH M7 control Option 4 scientific rationale as advocated by the pharmaceutical consortium within the Barber publication [19].

3.6.2 Proposed ICH M7-aligned Potential Mutagenic Control Regulatory Discussion

Based on the initial purge calculations allied to the additional non-trace analysis, a control summary table (Table 3.10) is presented below.

Further options for control could be considered specifically in the case of chloromethyl oxadiazole **8** where an Ames test could be performed to assess whether or not it is mutagenic.

3.6.3 Case Study 2 – Candesartan

As shown within the previous case study (GW641597X), the normal process for a risk assessment would be to identify the potential impurities within the drug substance and subsequently establish where mutagenicity concerns exist. However, this can also be extended to identify component constituents that may react together to an impurity of concern. By applying ICH M7 [8] control principles to these reactive species, the MI risk and any necessary control strategies can be established.

Candesartan cilexetil was developed as an angiotensin-II receptor antagonist for the treatment of hypertension. Following the discovery of *N*-nitrosodimethylamine (NDMA) in batches of valsartan and subsequently in additional sartans (losartan and irbesartan), it became necessary for all

Table 3.10 Proposed high-level control summary table for potential MIs ethyl bromoisobutyrate **2**, hydroxylamine, and alkyl chloride **8**.

Impurity	Point of potential formation/ introduction and summary of rationale for impurity purging	Required purge and predicted purge	Control
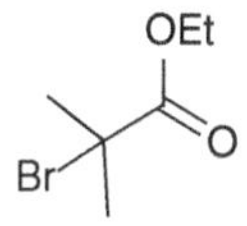 **2**	Starting material in Stage 1a (2 eq.), four steps from drug substance (DS). Consumed to low level (<5%) in Stage 1b; reactive during processing (Stage 4); soluble in isolation solvents (Stages 4 and 5).	Required purge = 20 Predicted purge = 1.0×10^5 Purge ratio = 5000	Option 4 – controlled through chemical reactivity and physical processing.
NH_2OH	Reagent in Stage 2 (2.5 eq.), four steps from DS. Reactive during processing (Stages 2, 3, and 4), highly soluble in isolation solvents (Stages 2, 3, 4, and 5).	Required purge = 39 Predicted purge = 1.0×10^8 Purge ratio = 2.56×10^6	Option 4 – controlled through chemical reactivity and physical processing.
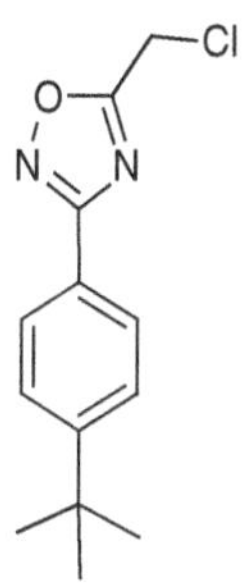 **8**	Starting material in Stage 4 (1.15 eq.), two steps from DS. Confirmed at low level (c. 0.2%) within Stage 4b product following additional reactivity with aqueous base used within the process and solubility within the isolation solvent. Additional solubility anticipated in Stage 5 isolation solvent.	Required purge = 30 Predicted purge =1000 Purge ratio = 33 Measured purge = 75 (Stage 4b) Measured purge ≥ 150 (Stages 4b and 5)	Option 4 – controlled through chemical reactivity and physical processing.

sartan-containing medications to evaluate the risk posed by nitrosamines, which form part of the cohort of concern, in drug products [24]. This case study evaluates the risk of nitrosamine formation and subsequent carryover to the API by examining the fate of the individual components required to generate a nitrosamine impurity.

Candesartan cilexetil is prescribed for chronic use and is therefore subject to lifetime TTCs for any impurities present within the drug product. However, regulatory guidance at the time also indicated that due to the potent mutagenic and carcinogenic potential of some nitrosamines, LTL limits for nitrosamines could not be used. Interim limits for the presence of nitrosamine impurities NDMA and *N*-nitrosodiethylamine (NDEA) were set at 96 and 26.5 ng/day, respectively, based on extrapolation from the respective TD_{50s}. For candesartan, which has a maximum daily dose of 32 mg, this equates to a final API impurity concentration of 3 ppm for NDMA and 0.83 ppm for NDEA.

The most common method by which nitrosamines are formed is through the reaction of secondary or tertiary amines with a nitrosating agent such as sodium nitrite in acidic media. The process used for candesartan incorporates the use of triethylamine and dimethyl formamide (DMF), which are known to contain as contaminants, or decompose into, diethylamine and dimethylamine, respectively. The process additionally utilizes sodium nitrite, thereby introducing a theoretical risk of nitrosamine formation (Figure 3.7). The risk assessment process therefore needs to address

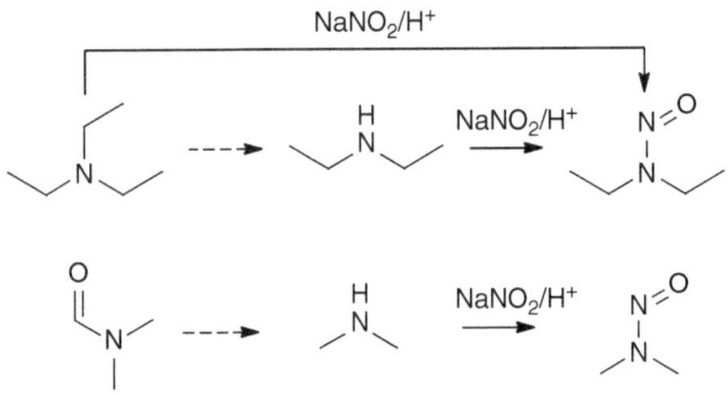

Figure 3.7 Nitrosamine formation pathways from Et_3N and DMF.

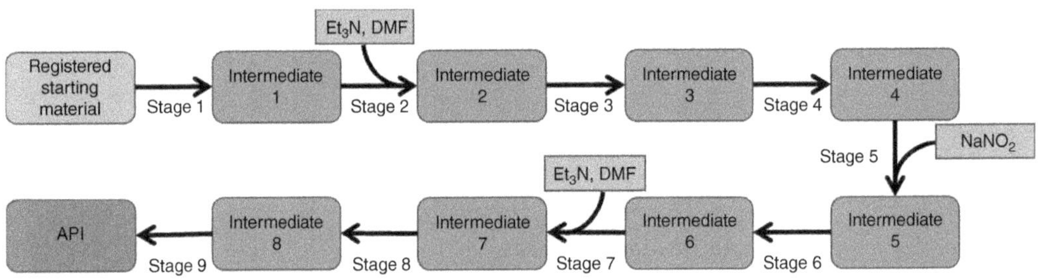

Figure 3.8 Process map of candesartan synthesis.

whether these amines or the parent compounds are likely to be present within the same step (under appropriate conditions) at a level of concern, and thereby identify the degree of risk present for nitrosamine formation and any subsequent removal if formed.

The manufacturing process for candesartan involves a nine-stage synthesis whereby triethylamine and DMF are introduced in Stage 2, whereas nitrite is not present until Stage 5 (Figure 3.8). As a result, the key questions related to assessing the risk of nitrosamines within the synthesis of candesartan are:

1) Do Et_3N, DMF, or their secondary amine degradants persist at an appreciable level into Stage 5?
2) Do traces of $NaNO_2$ persist at an appreciable level through to Stage 7?
3) If yes to either Q1 or Q2, could a nitrosamine formed be expected to be present in the final API?

The use of purge assessments allows the adequate assessment of these questions.

Triethylamine is introduced into the synthesis at Stage 2 through its use as a base at stoichiometric quantities. However, following completion of the reaction, the crude product is obtained through concentration under full vacuum at 70–75 °C, which is able to remove the vast majority of the Et_3N (boiling point 89 °C). While no measured data was available, a conservative estimate based on the volume reduction indicated a level of 5% Et_3N for the crude product after concentration. This value (50 000 ppm) was therefore utilized as the starting concentration for the purpose of purge calculations (Figure 3.9 – step 2). The crude material then undergoes a range of unit processes to afford the clean intermediate 2 presenting a number of mechanisms of purge for Et_3N. An initial extraction with $HCl_{(aq)}$ would result in formation of the corresponding salt ($Et_3N.HCl$), which is known to be highly soluble in water and can therefore reasonably be expected to purge to

Stage Details *Purge of triethylamine during Stage 2*	Reactivity (*H* = 100, *M* = 10, *L* = 1)	Solubility (*F* = 10, *M* = 3, *L* = 1)	Volatility (*H* = 10, *M* = 3, *L* = 1)	Total multiple per stage	Rationale for purge factor values selected
2. Concentration under reduced pressure	Concentration occurs at 70–75 °C at full vacuum; sufficient to distil DMF/TEA. No data on the extent of concentration. Based on this and the reduction in volume a conservative estimate made of residual TEA of 5%				
3. Extraction with H_2O/HCl	1	10	1	10	Purged during the liquid–liquid extraction as the HCl salt.
4. Extraction with $H_2O/NaHCO_3$	1	1	1	1	No purge during the liquid–liquid extraction as converted back to the free base.
5. Concentration under reduced pressure	1	1	3	3	Assigned a PF = 3 for volatility. The boiling point of TEA is 89 °C. The boiling point of EtOAc is 77 °C.
6. Concentration of MeOH	1	1	1	1	Assigned a PF = 1 for volatility. The boiling point of TEA is 89 °C. The boiling point of MeOH is 66 °C.
7. Isolation by filtration	1	10	1	10	Purged in the mother liquors as the free base.
8. MeOH wash	1	10	1	10	Purged in the MeOH wash as the free base.
9. Drying step	1	1	1	1	Assigned a PF = 1 for volatility. The boiling point of TEA is 89 °C. The boiling point of MeOH is 66 °C.
			Total for Stage	3000	

Figure 3.9 Breakdown of purge assignments for Et_3N in the Stage 2 workup processes.

a high degree (Purge factor [PF] = 10) based on solubility as a result of its ionizability (Figure 3.9 – step 3). However, a subsequent basic extraction would not result in a similar purge, as the dominant free base shows excellent solubility in both organic and aqueous solvents making the distribution more even. A purge of 1 is therefore the highest value that can be assigned, despite the likely removal of some Et_3N in this process, to ensure a conservative prediction (Figure 3.9– step 4). Subsequent removal of the EtOAc (boiling point 77 °C) under reduced pressure will result in the co-evaporation/azeotroping of triethylamine given the closeness of their respective boiling points. Utilizing the scoring system, a purge of 3 is scored for this step (Figure 3.9 – step 5), whereas the subsequent uptake and concentration in methanol (boiling point 66 °C) does not warrant application of purge as the boiling point of the triethylamine now exceeds 20 °C above that of the solvent (Figure 3.9 – step 6). Once again, this reiterates the conservative nature of the purge assignment, as some azeotroping is still likely to occur, particularly at reduced pressures where the difference in boiling points will contract to within 20 °C [14, 15]. Following precipitation and filtration of the intermediate, the Et_3N that is both highly soluble in methanol and a liquid itself can safely be considered to remain extensively within the mother liquors. Additionally, the subsequent wash of the filter cake to remove residual mother liquors and surface impurities allows for a further cautious score of 10 based on solubility (Figure 3.9 – steps 6 and 7).

In the workup processes following Stages 3 and 4, the purge of Et_3N is observed through similar mechanisms, reliant on the high degree of solubility in the process solvents and low boiling point. The total predicted purge for triethylamine up to the point of introduction of $NaNO_2$ in Stage 5 is 8.1×10^8 against a required purge of 60 240 to achieve the 0.83 ppm limit for NDEA within the API. Utilizing the approach to reporting for Option 4 strategies developed by Barber et al., this corresponds to a purge ratio of 13 446 for Et_3N (Figure 3.10). At this ratio very little justification would be necessary to demonstrate control of the impurity. In the scenario detailed here, the triethylamine is not the impurity of concern, but the nitrosamine NDEA that may be formed from it. Utilizing the same limit for the parent amine as for the nitrosamine imparts a further degree of conservatism, as quantitative conversion is hugely unlikely to occur, and therefore NDEA formation from Et_3N is demonstrated to be well controlled and suitably de-risked.

A similar assessment was performed for both DMF (potential source of dimethylamine [DMA]) and the amines of concern, assuming their presence within the starting materials or from degradation (Figure 3.11). In each of these cases the degree of purge establishes the risk of carryover into Stage 5 to be low. In the case of DMF, which is approximated at a concentration of 200 000 ppm following the Stage 2 reaction, a target concentration of <1 ppm (below the 3 ppm limit for NDMA) and a predicted purge of 7.3×10^9 equates to a purge ratio of 36 500. This demonstrates the potential for NDMA formation, resulting from DMA formed by the degradation of DMF in Stage 5, to be insignificant and requiring minimal justification.

The purge appraisal of DMA and DEA highlights their greater propensity to be removed, primarily linked to their low boiling points (Figure 3.12). Determining a ratio for these impurities is difficult, as a starting concentration cannot be determined; however, they cannot be present in greater quantities than their parent structures and yet the potential for purge is far greater. As such, any purge ratio derived would be far in excess of those obtained for Et_3N and DMF and therefore posing no appreciable risk to nitrosamine formation.

Purge calculations of amine-related impurities within this synthesis has clearly demonstrated there to be no risk of formation of NDMA or NDEA within Stage 5, as the initial question has been answered – amine impurities and sodium nitrite are not present together within the same stage.

In order to fully de-risk the formation of nitrosamines in the API, the formation of nitrosamines must also be considered within Stage 7, where both Et_3N and DMF are reintroduced into the synthesis. Once again this can be assessed by considering the ability for carryover of one of the reacting components, in this case the $NaNO_2$. The purge assessment of $NaNO_2$ (Figure 3.13) indicated a high degree of purge in the two steps, with a predicted purge of 1×10^6. While the

Impurity	*Stage*	*Reactivity*	*Solubility*	*Volatility*	*Predicted*	*Required*	*Ratio*
Et_3N	Stage 2	1	1000	3	3000		
	Stage 3	1	100	3	300		
	Stage 4	1	100	9	900		
	Overall				8.1×10^8	60 240	13 446

Figure 3.10 Purge calculation summary for Et_3N.

Impurity	*Stage*	*Reactivity*	*Solubility*	*Volatility*	*Predicted*	*Required*	*Ratio*
DMF	Stage 2	1	900	1	900		
	Stage 3	1	100	1	100		
	Stage 4	1	27 000	3	27 000		
	Overall				7.3×10^9	200 000	36 500

Figure 3.11 Purge calculation summary for DMF.

Impurity	*Stage*	*Reactivity*	*Solubility*	*Volatility*	*Predicted*	*Required*	*Ratio*
DMA/DEA	Stage 2		10	300 000	3 000 000		
	Stage 3		100	10 000	1 000 000		
	Stage 4		10	27 000	270 000		
	Overall				8.1×10^9	NA	NA

Figure 3.12 Purge calculation summary for DMA and DEA.

purge ratio for nitrite at this point is only 1, this does not preclude an Option 4-type approach, merely necessitating a greater degree of evidence to support the assessment (see Barber et al. [19]). Purge assessments are conservative due to the limitation in purge value that can be assigned at each step, whereas in practice the true purge may be far greater. If this can be demonstrated, then the application of an Option 4 approach remains valid. To this effect, nitrite testing during Stage 6 found it was not present above 100 ppm (limit of detection [LoD]), thereby confirming the conservatism within the Stage 5 assessment.

Impurity	Stage	Reactivity	Solubility	Volatility	Predicted	Required	Ratio
$NaNO_2$	Stage 5	100	10		1000		
	Stage 6	10	100				
	Overall				1×10^6	1×10^6	1

Figure 3.13 Purge calculation summary for $NaNO_2$.

The use of purge calculations therefore, in conjunction with analytical testing, established the formation of NDEA or NDMA cannot occur to a level of concern within the synthesis of candesartan, as there are sufficient levels of control of the component parts (amines and $NaNO_2$) to ensure that they are never present within the same stage at a concentration of concern, something that can be easily conveyed through a simple schematic (Figure 3.14).

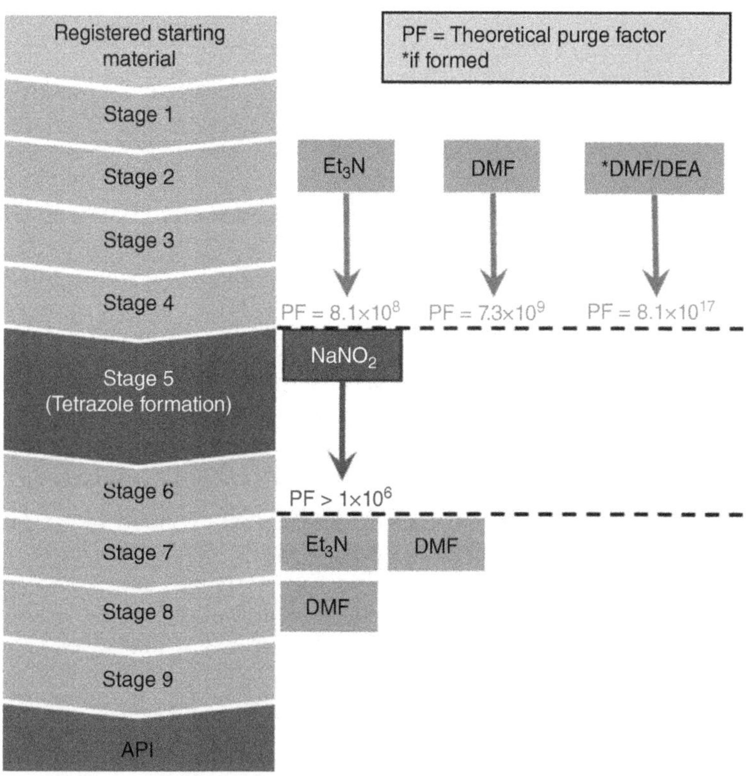

Figure 3.14 Schematic of candesartan process highlighting the purge-based risk assessment for nitrosamine formation and clearance.

The de-risking process described for candesartan was further validated through trace analytical testing for NDMA and NDEA. While no risk of nitrosamine formation was identified within the candesartan synthesis, had the potential for formation been established, the purge principles could have been further exploited to determine the risk of carryover of the nitrosamines themselves into the final API, as any nitrosamine formed would still have the opportunity to be purged and controlled in subsequent stages. In the case of candesartan, a purge assessment of NDMA and NDEA from Stage 5 onward indicates theoretical purge factors of ~10 000 and ~1000, respectively.

In addition, analytical testing of over 100 batches of candesartan have confirmed the absence of NDMA or NDEA above 5 ppb (LoD), thereby validating the expert theoretical assessment that they could not be formed to a level of concern.

References

1 Teasdale, A. and Elder, D.E. (2015). Is avoidance of genotoxic intermediates/impurities tenable for complex, multistep syntheses? *Org. Process Res. Dev.* 19: 1437–1446.

2 Jacobson-Cram, D. and McGovern, T. (2007). Toxicological overview of impurities in pharmaceutical products. *Adv. Drug Deliv. Rev.* 59: 38–42.

3 Dobo, K.L., Greene, N., Cyr, M.O. et al. (2006). The application of structure based assessment to support safety and chemistry diligence to manage mutagenic impurities in actives pharmaceutical ingredients during drug development. *Regul. Toxicol. Pharmacol.* **44**: 282–293.

4 Pierson, D.A., Olsen, B.A., Robbins, D.K. et al. (2009). Approaches to assessment, testing decisions and analytical determination of mutagenic impurities in drug substances. *Org. Process Res. Dev. 13* (2): 285–291.

5 Ames, B.N., Durston, W.E., Yamasaki, E.E., and Lee, F.D. (1973). Carcinogens and mutagens: a simple system combining liver homogenates for activation and bacteria for detection. *Proc. Nat. Acad. Sci. USA* 72: 2423–2427.

6 ICH Q8 (R2). Pharmaceutical Development. http://www.emea.europa.eu/pdfs/human/ich/51881907enfin.pdf (accessed September 2020).

7 ICH Q9. Quality Risk Management. http://www.emea.europa.eu/Inspections/docs/ICHQ9Step4QRM.pdf (accessed September 2020).

8 ICH Guideline M7 (R1). On Assessment and Control of DNA Reactive (mutagenic) Impurities in Pharmaceuticals to Limit Potential Carcinogenic Risk.

9 ICH Q3a (R2). Impurities in New Drug Substances (Revised Guideline). CPMP/ICH/2737/99.

10 ICH Q3B (R2). Impurities in New Drug products (Revised Guideline). CPMP/ICH/2738/99.

11 Harvey J. A, Fleetwood A, Ogilvie R, Teasdale A, Wilcox A.P and Spanhaak S; Management of organic impurities in small molecule medicinal products: deriving safe limits for use in early development *Reg. Tox. Pharmacol.*; 2017, 84, 116 to 123.

12 Müller, L., Mauthe, R.J., Riley, C.M. et al. (2006). A rationale for determining, testing and controlling specific impurities in pharmaceuticals that possess potential for mutagenicity. *Regul. Toxicol. Pharmacol.* 44: 198–211.

13 Bercu, J., Galloway, S.M., Parris, P. et al. (2018). Potential impurities in drug substances: compound-specific toxicology limits for 20 synthetic reagents and by-products, and a class-specific toxicology limit for alkyl bromides. *Regul. Toxicol. Pharmacol.* 94: 172–182.

14 Teasdale, A., Fenner, S., Ray, A. et al. (2010). A tool for the semiquantitative assessment of potentially genotoxic impurity (PGI) carryover into API using physicochemical parameters and process conditions. *Org. Process Res. Dev.* 14: 943–945.

15 Teasdale, A., Elder, D., Chang, S.-J. et al. (2013). Risk assessment of genotoxic impurities in new chemical entities: strategies to demonstrate control. *Org. Process Res. Dev.* 17: 221––230.

16 Elder, D.P., Okafo, G., and McGuire, M. (2013). Assessment of predictivity of semiquantitative risk assessment tool: pazopanib hydrochloride genotoxic impurities. *Org. Process Res. Dev.* 17: 1036–1041.

17 McLaughlin, M., Dermenijan, R.K., Jin, Y. et al. (2015). Evaluation and control of mutagenic impurities in a development compound: purge factor estimates vs measured amounts. *Org. Process Res. Dev.* 19: 1531–1535.

18 Lapanja, N., Zupančič, B., Toplak Časar, R. et al. (2015). A generic industry approach to demonstrate efficient purification of potential mutagenic impurities in the synthesis of drug substances. *Org. Process Res. Dev.* 19: 1524–1530.

19 Barber, C., Antonucci, V., Baumann, J.-C. et al. (2017). A consortium-driven framework to guide the implementation of ICH M7 option 4 control strategies. *Regul. Toxicol. Pharmacol.* 90: 22–28.

20 ICH guideline S9 on nonclinical evaluation for anticancer pharmaceuticals. EMA/CHMP/ICH/646107/2008.

21 Boyer T, Choudary B.M, Edwards A.J, *et al.*; Development of a scalable process for the PPAR-α agonist GW6415 7X incorporating Baeyer–Villiger chemistry and retrospective ICH M7 assessment *Org. Proc. Res. Dev.*; 2020, 16 (12), 1927 to 1939.

22 Derek v6.0 (KB 2018v1.1), Lhasa Ltd.

23 Leadscope v2.2.1 (Salmonella V3, E.Coli-TA102 V1), Leadscope Inc.

24 Sartan medicines: companies to review manufacturing processes to avoid presence of nitrosamine impurities. https://www.ema.europa.eu/en/documents/referral/valsartan-article-31-referral-sartan-medicinescompanies-review-manufacturing-processes-avoid_en.pdf (accessed September 2020).

Section 2

In Silico Assessment of Mutagenicity

4

Use of Structure–Activity Relationship (SAR) Evaluation as a Critical Tool in the Evaluation of the Genotoxic Potential of Impurities

Catrin Hasselgren[1] and Glenn Myatt[2]

[1] *Genentech Inc., South San Francisco, CA, USA*
[2] *Instem, Columbus, OH, USA*

4.1 Introduction

Computational toxicology includes approaches to predict the toxicity of a chemical based solely on its chemical structure [1]. The predictions are generated using computational models that encode complex structure–activity (toxicity) relationships (SAR). These models are referred to as (Quantitative) Structure–Activity Relationship models or (Q)SAR models. They are routinely used to predict the DNA-reactive (mutagenic) potential of impurities since they are both fast and sufficiently accurate for the protection of public health [2]. They are also cost-effective because they often avoid the need to perform a reverse bacterial mutation assay (referred to as an Ames test throughout this chapter) [3] to qualify an impurity as non-mutagenic. Avoiding the time and costs associated with any acquisition of the test material is especially valuable. This can be expensive in cases when the impurity needs to be synthesized in sufficient quantities and purity to enable experimental testing.

The regulatory acceptance of computational toxicology is primarily attributable to the International Council on Harmonization (ICH) M7 guideline ("Assessment and control of DNA reactive (mutagenic) impurities in pharmaceuticals to limit potential carcinogenic risk") [4]. The guideline explicitly defines (Q)SAR approaches as an acceptable regulatory test to assess the mutagenic potential of the impurities. This is the first ICH guideline to define (Q)SAR methods as a regulatory test. Prior to the release of the ICH M7 guideline, a series of position papers [5–7] and guidelines for specific jurisdictions [8–10] were published that supported the development of ICH M7's computational strategy. These papers justified the use of (Q)SAR approaches as sufficiently predictive and discussed the importance of including an expert review as part of the overall assessment [6]. For more background to the ICH M7 guideline, see Chapter 2.

Figure 4.1 describes the overall approach for assessing impurities where each impurity is assigned to one of five classes, numbered 1–5. The first step is to determine whether there are any relevant historical experimental data, including bacterial mutagenicity data as well as carcinogenicity data. After assessing any available experimental study results, it may be possible to assign the impurity to one of three classes: class 1 (where it was determined that the chemical is a mutagenic carcinogen), class 2 (where it is determined that the chemical is a DNA-reactive mutagen with no carcinogenicity data), or class 5 (where it is determined that the chemical is not mutagenic). In the absence of any relevant or adequate data, it is possible to perform a computational assessment to evaluate the mutagenic potential of each impurity. This includes running the impurities through two different

Mutagenic Impurities: Strategies for Identification and Control, First Edition. Edited by Andrew Teasdale.
© 2022 John Wiley & Sons, Inc. Published 2022 by John Wiley & Sons, Inc.

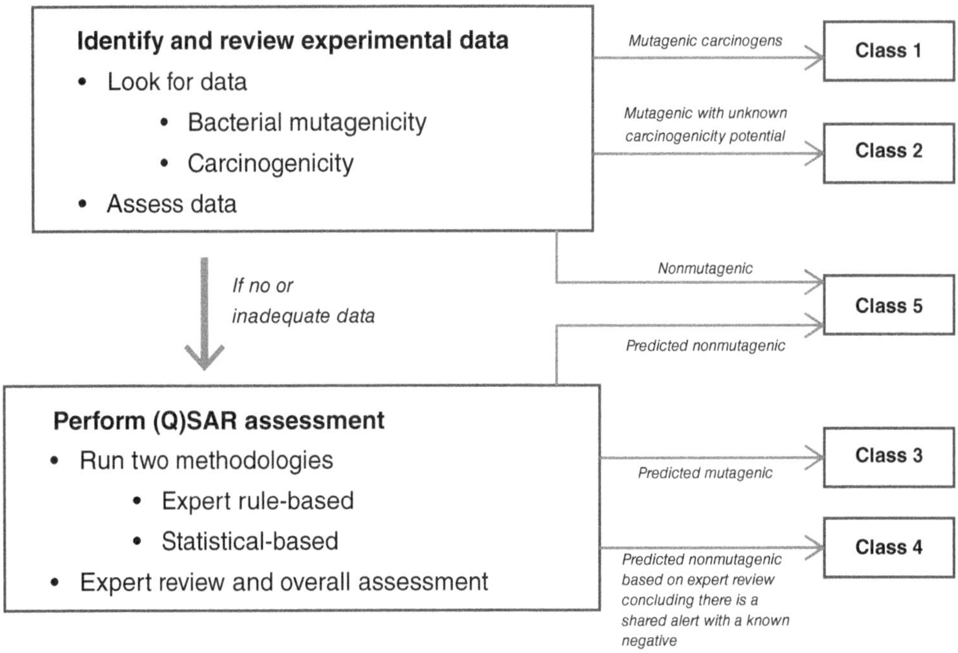

Figure 4.1 Assignment of class 1–5 based on computational models and experimental data.

(Q)SAR methodologies: an expert rule-based and a statistical-based methodology. An overall assessment of the results, including a potential expert review, is then used to determine whether the impurity is predicted to be DNA-reactive mutagenic (assigned to class 3) or predicted to be non-mutagenic (generally assigned to class 5). In situations where there is a mutagenic prediction with a clearly identified structural basis or structural alert, and this alert is also shared with an experimentally determined non-mutagenic chemical, it may be assigned to class 4.

Assignment to class 1, 2, or 3 requires the impurity to be controlled to levels specified in the guideline (including using carcinogenicity data for class 1 chemicals), although a class 3 assignment based on positive computational results alone may be reassessed by conducting an Ames test. In specific situations, the strains to test in may be determined based on the strain response to particular structural features which are present in the impurity. A positive result in an Ames assay may be subsequently followed up by other *in vivo* studies described in Note 3 in the guideline, if desired, although this is not common practice. Assignment to class 4 or 5 designates the impurity as non-mutagenic and no additional work is required under the ICH M7 guideline. The derivation of limits for non-mutagenic impurities is outlined in the corresponding guidelines for non-mutagenic impurities [11, 12].

The following chapter provides an overview of (Q)SAR methodologies related to ICH M7 and outlines the process of assigning and documenting impurities to the five ICH M7 classes, including an expert review.

4.2 (Q)SAR Assessment

4.2.1 Looking-up Experimental Data

The first step in any computational assessment is to search databases and perform literature queries to determine whether any historical data can be used to establish that the impurity is a mutagenic carcinogen (class 1), a mutagenic chemical with unknown carcinogenicity (class 2), or non-mutagenic (class 5). To make this classification, it is important to search for carcinogenicity data as well as bacterial mutagenicity data; however, it is permissible to identify other data that support this designation, such as certain mouse lymphoma study results [2]. A number of databases are often used to support this task with a comprehensive listing provided in Amberg et al. [2]. Some of the more common sources of data include: Chemical Carcinogenesis Research Information System [13], Carcinogenic Potency Database (CPDB) [14], European Chemicals Agency (ECHA) [15], International Agency for Research on Cancer (IARC) [16], Leadscope [17], MultiCASE [18], National Toxicology Program (NTP) [19], and VITIC [20]. A number of public databases, including CCRIS and CPDB, are no longer accessible through ToxNet and hence require alternative search services to access this content.

Once any data has been identified, it is important to review how the experiment was performed to determine whether the study was adequate and hence if the data can be used in the computational assessment. Factors to consider in this evaluation include whether the test was performed according to the OECD protocols (e.g. OECD 471 for Ames) as well as whether any experimental artifact may have contributed to a potentially incorrect result. It will also be important to reference sources such as the ICH M7 addendum, the ICH Q3 guidelines, and Bercu et al. [21], where permissible daily intake calculations have been established. Chapter 7 provides more details on this subject.

4.2.2 (Q)SAR Methodologies

4.2.2.1 Overview

In the absence of adequate experimental data, the ICH M7 guideline recommends the use of two complementary computational toxicology methodologies that predict the results of the bacterial mutagenicity test: (i) expert rule-based and (ii) statistical-based. These two methodologies are often referred to as (Q)SAR approaches with "Q" (standing for quantitative) in parenthesis to include the nonquantitative expert rule-based methodology. The use of two complementary methodologies for such regulatory submissions generally reduces the number of false-negative results [22]. This is desirable since impurities have no benefit to the patient and the regulation is designed to protect patient safety. The ICH M7 guideline also discusses the use of expert review as part of the process, especially when the results are inconclusive. As such, any models used should be transparent and interpretable to support such a review [23].

4.2.2.2 OECD Validation Principles

The ICH M7 guideline states that the computational toxicology methods employed should adhere to the OECD QSAR validation principles [24] which are:

(1) **A defined endpoint:** The defined endpoint is explicitly described in the ICH M7 guideline as the results of the bacterial mutagenicity assay and any selected method should predict only this endpoint.
(2) **An unambiguous algorithm**: The modeling approach should be well documented to support a transparent and independent assessment of the methodology.
(3) **A defined domain of applicability**: The method should indicate when a prediction can be made with sufficient reliability, i.e. when it is within the applicability domain of the model.
(4) **Appropriate measures of goodness-of-fit, robustness, and predictivity**: An assessment of the model's performance based on the widely used Cooper statistics [25] (number of true positives, number of true negatives, number of false positives, number of

false negatives, concordance, sensitivity, specificity, positive predictivity, negative predictivity) using the model's reference/training set, using cross-validation, and ideally using a sufficiently large external dataset will provide an indication of how well a model performs. An external validation analysis for many of the available models has been recently performed using a set of over 12 000 chemicals [26]. The results show good performance for the systems generally used in the assessment of mutagenic or potentially mutagenic impurities (PMIs), previously often referred to as GTIs. This is particularly important since false-negative and false-positive predictions have consequences in terms of costs and time to get drugs to market. A false positive may result in unnecessary Ames testing (and potentially costly synthesis of the impurity) or the implementation of control strategies that are not needed. A false negative may result in patient safety issues further down the drug pipeline.

(5) **A mechanistic interpretation, if possible**: An assessment of potential mechanisms associated with any prediction is useful to support an expert review of the information.

4.2.3 Expert Rule-Based Methodology

The expert rule-based methodology is a predictive approach where chemicals are predicted mutagenic if they contain one or more structural alerts whose presence is likely to result in a mutagenic chemical [22]. An example structural alert is shown in the chemical in Figure 4.2 which contains an aziridine substructure. Expert rule-based systems generate results that are generally easy to interpret; and since they are not automatically derived from any specific database, they can flexibly incorporate information not in the public domain (such as alerts derived from proprietary information).

The process of building an expert system is illustrated in Figure 4.3 and the outcome of this process is a knowledge base containing a collection of structural alerts. One source for alerts is the public literature. One of the first publications presenting such a collection was by Ashby and Tennant [27] who identified 17 chemical classes or alerts believed to be associated with mutagenicity, including alkyl esters of either phosphonic or sulphonic acids, aromatic nitro groups, and others. There are a number of later publications that build on this work and that include both new and overlapping structural alerts. The aziridine alert is, for example, cited in Kazius et al. [28], Bailey et al. [29], and Benigni and Bossa [30, 31]. The public literature often includes information on the mechanistic basis of the published alerts, which is useful to understand the applicability of the alert to a query compound. For example, in Section 4.2.5 of Benigni [31], the aziridine substructure in Figure 4.2 is described as ". . . extremely reactive alkylating agents that may react by ring-opening reactions . . ." ". . . activity of these compounds depends on their ability to act as DNA cross-linking agents, via nucleophilic ring-opening of the aziridine moiety by N7 positions of purines."

Although the literature is a good starting point for the identification of alerting features in chemicals, it has been shown that using literature alerts alone will result is many false-positive and false-negative predictions [32]. For example, the aromatic *N*-oxide alert is an alert cited in the

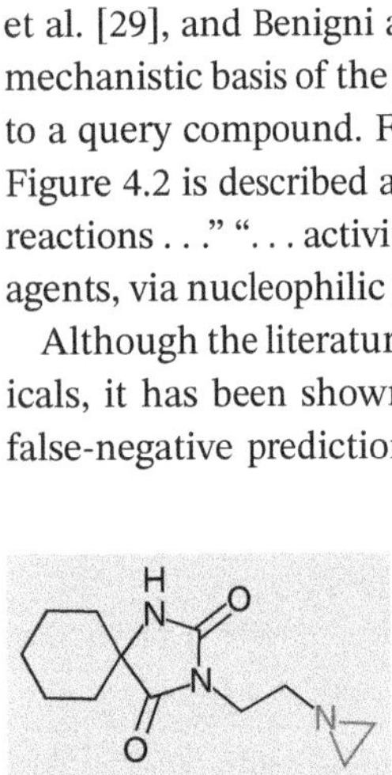

Prediction:	Mutagenic
Alert:	24: aziridine

Figure 4.2 Prediction using an expert alert system.

Figure 4.3 Building an expert alert knowledge base.

original Ashby Tennant publication as well as others; however, a recent analysis of all available data could not establish a sufficiently strong association except for certain subclasses of aromatic *N*-oxides [33]. Hence, any system that is used for ICH M7 purposes (with minimal false positives and false negatives) should consider additional information from public and proprietary databases. This information is used to:

(1) **Qualify alerts**: All literature alerts should be further evaluated to determine if they should be included in the system's knowledge base. Such analysis is usually based on the weight of the supporting evidence, including whether there is a higher number of mutagenic examples matching the alert than you would expect by chance (i.e. statistically significant greater numbers of mutagenic examples), whether there is a plausible mechanism, and whether any matching examples contain other structural moieties that are more likely responsible for the mutagenicity. This analysis should ideally consider both public and proprietary data. An alert is designated as positive when there is clear evidence of an association with bacterial mutagenicity. Alternatively, the alert may be designated an "indeterminate" or "equivocal" category, reflecting any potential uncertainty in determining a clear "positive" or "negative" categorization. This could be due to the activating and deactivating nature of the alerting structural features not being well understood, or there being ambiguous or conflicting data for chemicals containing the particular alert.

(2) **Assigning confidence**: A level of confidence associated with an alert is based on the supporting evidence, such as the precision of the alert derived from the matching examples (i.e. the proportion of mutagenic examples).

(3) **Refining alerts**: Using supporting data, specific structural factors are often identified that both activate and deactivate the alert. Since sharing proprietary information is not always possible, a methodology referred to as SAR fingerprinting provides an opportunity to share knowledge on chemical classes (such as the number of mutagenic or nonmutagenic examples that match a series of substructure searches) without sharing any proprietary information [34]. This information can be pooled across multiple collections and used to refine alerts to improve their predictivity. For example, a proprietary dataset of over 12 000 chemicals was made available by The Division of Genetics and Mutagenesis,

National Institute of Health Sciences (DGM/NIHS) of Japan under confidentiality agreements to (Q)SAR developers to support the improvements of models and alerts [26].

(4) **Identify new alerts**: Alerts not reported in the public literature, such as halo pyridines, can be identified from analysis of both public and proprietary databases.

When making a prediction using an expert rule-based methodology, the alert's rules are compared with the impurity's chemical structure. If one or more alerts match the impurity (i.e. the impurity contains an alert and there is no additional structural or physicochemical reasons that the alert is deactivated) then a potentially mutagenic prediction is made. The prediction reflects how the alert rule has been designated, and, for example, a prediction may be "mutagenic" or "indeterminate." The absence of any structural alerts in the impurity results in a "non-mutagenic" prediction, if there is sufficient supporting information. Figures 4.4 and 4.5 show predictions made for two chemicals using an expert rule-based methodology.

In Figure 4.4, the query chemical was predicted as mutagenic, based on alert #321 (aryl methyl halide). In this example, a defined activating structural fragment matches a substructure of the chemical and although the alert contains a series of mitigating structural factors, none of these factors deactivate the alert.

In Figure 4.5, a prediction of another chemical was made using the expert rule-based methodology. In this example, the chemical is predicted non-mutagenic since no alerts matched it. In compliance

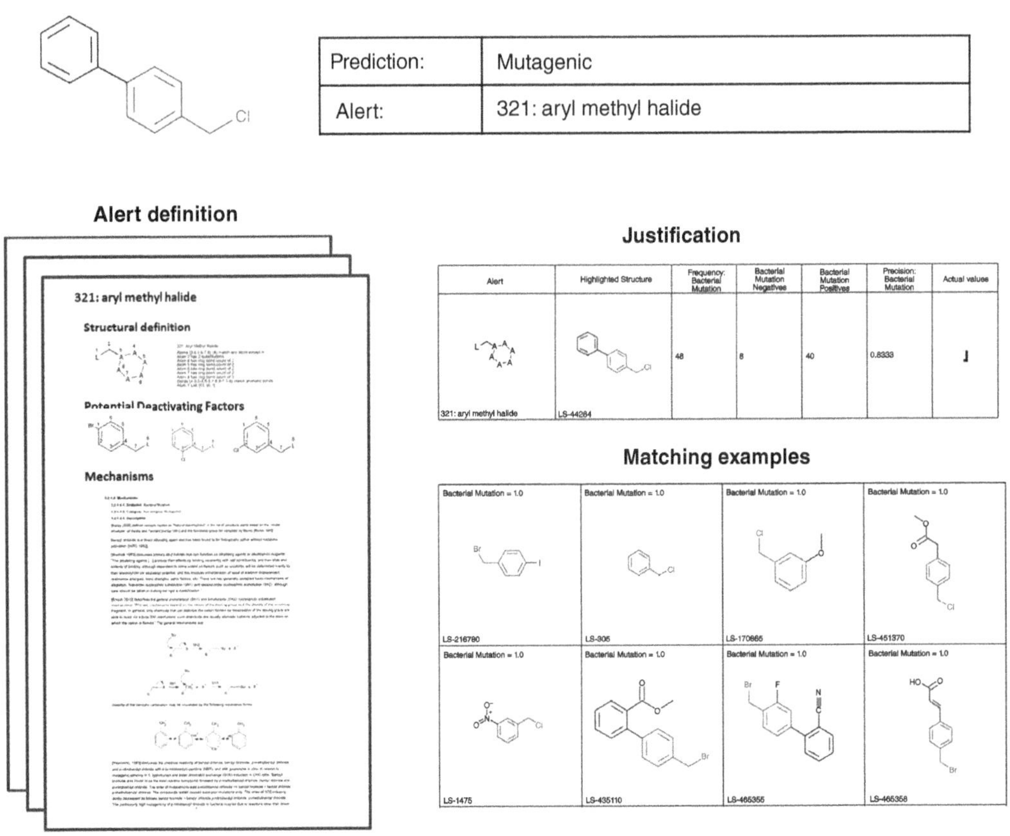

Figure 4.4 Example of a positive prediction using an expert rule-based system.

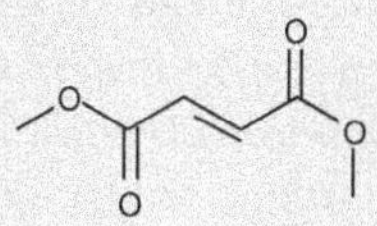

LS-6913

Prediction:	Nonmutagenic
Alert:	No alerts

Chemical analogs

Structure	Similarity	Experimental value-bacterial Mut
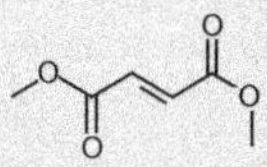 LS-6913	1.0	
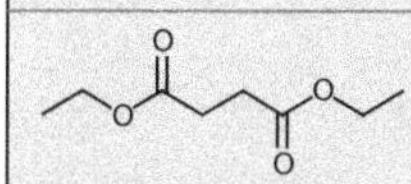 LS-5823	0.8	Negative
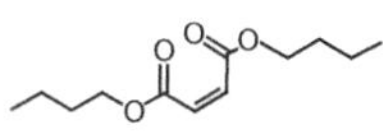 LS-5790	0.8	Negative
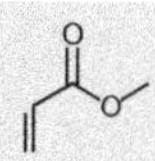 LS-2008	0.71	Negative
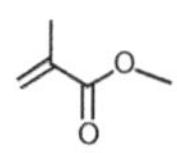 LS-1799	0.71	Negative

Figure 4.5 Example of a negative prediction using an expert rule-based system.

with OECD validation principle number 4, before a chemical may be determined to be non-mutagenic, an assessment of the applicability domain should be performed. In this situation, since there are several close structural analogs in the reference database, there is sufficient evidence to determine the chemical is in the applicability domain and hence it was predicted non-mutagenic.

Such systems should allow for an examination of the underlying information to support any subsequent expert review, including the structural definition of the alerts and mitigating factors, the proposed mechanism, the number of mutagenic and non-mutagenic examples matching the alerts, and the underlying study information.

4.2.4 Statistical-Based Methodology

The second recommended methodology is referred to as statistical-based. This approach uses statistically derived models that represent the complex relationships between the bacterial mutagenicity data and (i) activating (often reactive) features, (ii) deactivating structural features, and (iii) physicochemical properties. The process of building such models is illustrated in Figure 4.6. Since these methods learn from training sets of historical bacterial mutagenicity data, the first step in any process is to put together a set of high-quality data. This includes consolidating any study results where a chemical has been tested more than once and removing any results where the study is determined to be inadequate or conflicting with other studies. The next step is to generate calculated descriptors including the presence or absence of structural features and physiochemical properties to be used in the model. Chemical features that are correlated with the endpoint are included in the model. Interpretable structural features support the subsequent expert review and will be used to explain the model's results. Incorporating features representing external knowledge of DNA-reactive mutagenicity will improve such models and is aligned with best practices in building predictive models. The features are organized into a data matrix, where the calculated chemical descriptors (structural and physicochemical) are the *x*-variables and the bacterial mutagenicity data is the *y*-variable. The descriptors (*x*-variables) are usually weighted based on their associations with the mutagenic or non-mutagenic compounds and those descriptors showing little to no association are often removed. The remaining descriptors are used to build a statistical-based regression/classification model that predicts mutagenicity. This model is conceptually shown in Figure 4.6 relating different weightings of the *x*-variables (chemical descriptors) to the *y*-variable (bacterial mutagenicity).

Figure 4.7 shows an example of how a statistical-based methodology predicts whether a chemical is mutagenic or non-mutagenic. The structural features and molecular properties that were included in the model formula are calculated for the impurity. In Figure 4.7, this includes aromatic nitro, nitro, diazole, polar surface area, etc. The presence or absence of structural features in a test compound, alongside physicochemical properties, is used in the model as the *x*-variables of the statistical formula to predict bacterial mutagenicity (i.e. the *y*-variable). In the example, the result is a probability of a mutagenic outcome and this value (0.73) determines that this chemical is pre-

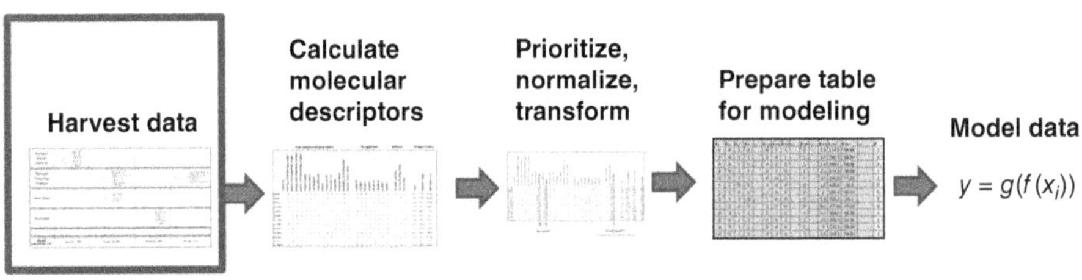

Figure 4.6 Process of building a statistical-based model.

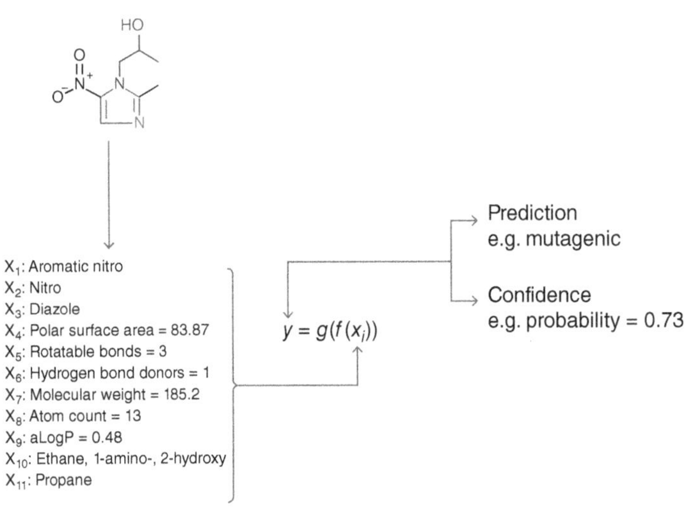

Figure 4.7 Process of making a positive prediction with a statistical-based methodology.

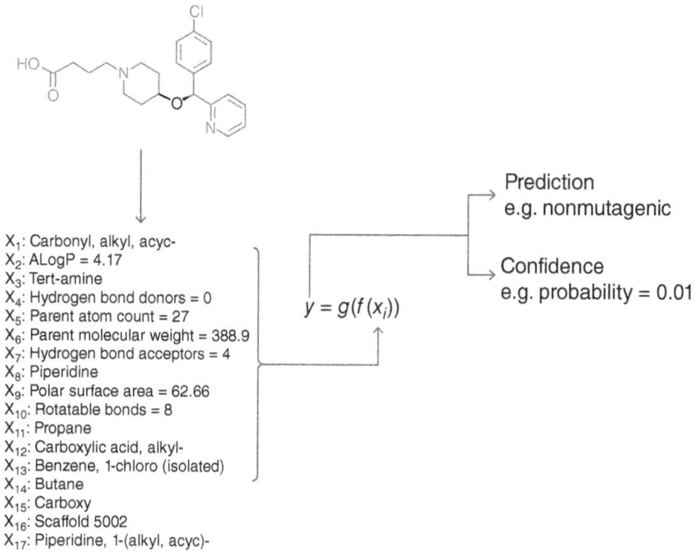

Figure 4.8 Process of making a negative prediction with a statistical-based methodology.

dicted to be mutagenic. An assessment of whether the chemical is within the applicability domain of the model is also required and made as part of the prediction process. Figure 4.7 shows the contribution of the structural features highlighted on the chemical. In addition, atoms and bonds that are colored black indicate that there are no features covering this portion of the molecule used for the prediction. The coloring is an effective means to support expert review of the results. In this example, the coloring allows for rapid identification of the aromatic nitro group which is the primary reason that this chemical is assigned as mutagenic. Moreover, since most of the molecule is color coded (except the methyl group), the model's structural features are taking into consideration practically the whole molecule.

In Figure 4.8, another chemical is predicted using a statistical-based methodology. Again, the features present in the molecule are identified, and are used as x-variables in the model. The computed prediction of bacterial mutagenicity (y-variable) is used to classify the chemical as nonmutagenic with a low probability of 0.01. Again, an assessment of whether the chemical is within the applicability domain of the model is made. A visual inspection indicates that the model's structural features assessed most of the chemical (indicated with the gray coloring).

4.2.5 Applying (Q)SAR Models

In preparation for assessing a set of impurities, an electronic file of the chemical structure(s) for each impurity should be prepared. It is important to consider any necessary structure drawing conventions required by the prediction software [1]. Typically, a single chemical may be drawn using a structure drawing package (such as ChemDraw [35]) and copied into the prediction software. Alternatively, a MOL file [36] or SMILES string [37] may be prepared and copied or loaded into the software. For a batch of chemicals, a list of SMILES strings or an SD file [36] is typically prepared and loaded into the software.

Examples of expert rule-based software routinely used for PMI assessments include the Leadscope Genetox Expert Alerts [17] and Derek Nexus [38]. Common examples of statistical-based methods include the Leadscope Genetox Statistical Suite [17], Sarah Nexus [18], and MultiCASE [18]. These commercial systems have been refined and validated over many years and are familiar to the regulatory agencies. A recent assessment of (Q)SAR systems was performed as part of a collaboration with the NIHS in Japan [26]. Three sets of chemicals totaling 12 140 were provided to (Q)SAR developers over a series of years (2014–2017) with known Ames results. The proportion of positives was 14.4% across the entire set. For the final set of 4409 chemicals, the range of values for certain cooper statistics [25] for the commercial systems discussed was 44.0–59.6% for sensitivity, 82.3–92.8% for specificity, and 77.3–87.3% for accuracy.

Once the chemical(s) are uploaded into the software and the models applied, the programs will typically generate a number of calls (although different systems use different naming convention for their output) along with supporting data to explain how the result was reached. Generally, these calls fall into the following categories:

- **Negative**: A non-mutagenic prediction, where the chemical is within the applicability domain of the model.
- **Positive**: A mutagenic prediction.
- **Indeterminate**: A less certain or equivocal prediction where there is some uncertainty in the supporting data.
- **Out-of-domain**: The (Q)SAR assesses the test chemical to be outside the applicability domain of the model.

Figure 4.9 shows an example of the prediction results for four chemicals (A, B, C, D) using two (Q)SAR methodologies, one statistical-based and one expert rule-based, accessed through the Leadscope Model Applier. Additional model output to support an expert review is presented, including information on the prediction and its confidence, as well as any expert alerts that match. Additionally, a hyperlinked result indicates that the chemical is present in the underlying databases. For example, compound B was predicted to be negative (non-mutagenic) in both methodologies; however, an Ames test had previously been performed and the underlying study data can be accessed via the hyperlink. Figure 4.9 also shows that the tool may provide a consensus prediction as coming from the combination of the statistical-based and expert rule-based results. The consensus is usually based on a conservative approach. For example, compound A of Figure 4.9 is predicted to be indeterminate by the expert alerts and negative by the statistical model (i.e. non-mutagenic), and such results are combined by the tool into an inconclusive consensus outcome. On the other hand, compound C is predicted to be negative (i.e. non-mutagenic) in both methodologies.

4.2.6 Expert Review

4.2.6.1 Overview

As discussed earlier, any prediction software used for impurities assessment should generate additional information to support an expert review. For expert rule-based methods, this includes information reflecting the confidence of a positive or negative prediction (such as the alert's precision), the structural and mechanistic basis of the alert, as well as examples matching the alert or from generally similar chemicals along with the underlying study information and results. For statistical-based methodologies, supporting information includes the structural and physicochemical descriptors used and their contribution to the overall prediction, the underlying data along with full study results, and information on analogs. A review of this background information and other knowledge is used to make an overall classification based on the two methodologies. This

		Statistical Models			Expert Alerts			
Structure	Bacterial Mutation Consensus	Bacterial Mutation v2 Prediction	Bacterial Mutation v2 Positive	Bacterial Mutation v2 Structure	Bacterial Mutation Alerts v6	Bacterial Mutation Alerts v6	Bacterial Mutation Alerts v6 Matched Alerts	Bacterial Mutation Alerts v6 Structure
A 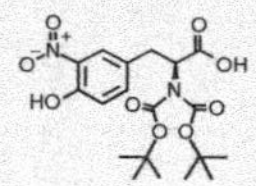 LS Ex4	Inconclusive	Negative	0.131	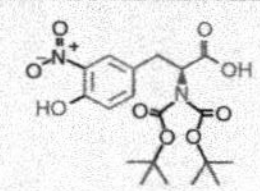 LS Ex4	Indeterminate	0.575	270: aromatic nitro (benzene, 1-hydroxy-,2-nitro- and benzene, 1-hydroxy-,4-nitro-) (Indeterminate)	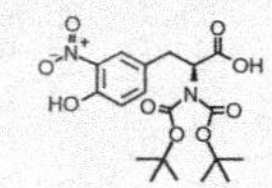 LS Ex4
B 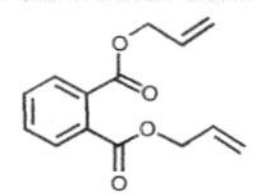 LS-30-1	Negative	Negative	0.19	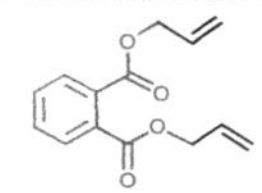 LS-30-1	Negative	0.1462	No Alerts	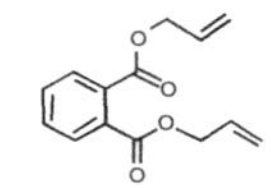 LS-30-1
C 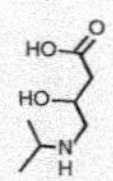 LS_EX1	Negative	Negative	0.0324	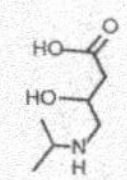 LS_EX1	Negative	0.1462	No Alerts	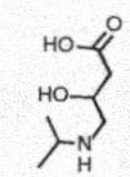 LS_EX1
D 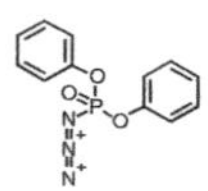 tst3	Positive	Positive	0.958	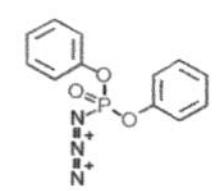 tst3	Positive	0.9878	41: azide (0.99)	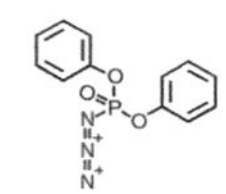 tst3

Figure 4.9 Four examples with predictions from expert rule-based and statistical-based methods.

assessment may also include refuting an individual prediction with sufficient supporting information. In addition, the expert review looks at the totality of the information generated from the methodologies, and other knowledge in order to determine an overall assessment. A series of papers has described this expert review process in detail, including Powley [39], Barber et al. [40], Amberg et al. [2], and Amberg et al. [41]. A number of papers have also covered the SAR around specific classes. These may support an expert review and are summarized in Chapter 5.

The expert review is generally performed by someone with knowledge of how computational models operate, who has a general SAR and mechanistic knowledge of DNA-reactive mutagenicity, as well as sufficient experience in how to assess the supporting bacterial mutagenicity or carcinogenicity data. On occasion, it may be necessary to perform a review of potentially reactive groups, of the potential for steric hindrance as well as metabolic activation, and the influence of physicochemical properties which may require consultation with experts in these areas.

The following section describes different factors and approaches to consider as part of any expert review.

4.2.6.1.1 Looking at Chemical Analog(s)

Chemicals that are sufficiently similar in terms of molecular structure to the impurity are often used as part of an expert review to support a "read-across" of the bacterial mutagenicity data from the analog(s) to the impurity [2]. They support: (i) refuting a prediction, (ii) affirming a prediction, and (iii) providing additional supportive evidence, especially in cases where there is an inconclusive result.

When using chemical analogs to refute a positive prediction, it is possible that searching public database may identify analogs to support such an assessment. However, more commonly, proprietary analogs such as the Active Pharmaceutical Ingredient (API) are often effectively used. In the ICH M7 guideline, a class 4 assignment is possible that uses one or more Ames negative analogs (or API) to refute a positive prediction. To use this type of review, the structural basis for the positive prediction should be known and the alerting substructure should be also present within the analog(s) in the same environment [4].

The example shown in Figure 4.10 demonstrates the reasoning. Compound A was predicted negative by the statistical-based model and indeterminate by the expert rule-based model (based on an alert #270 – aromatic nitro (benzene, 1-hydroxy, 2-nitro and benzene, 1-hydroxy, 4-nitro)). The alert is indeterminate and hence an expert review would be prudent before making a final assessment. A search for chemical analogs (shown in Figure 4.10) identifies a non-mutagenic analog that contains the same alerting structure in the same chemical environment (i.e. ICH M7 class 4). The part of the impurity not covered by the analogs was not considered to be reactive. Therefore, after an expert review, the overall assessment for this chemical is non-mutagenic.

4.2.6.2 Refuting a Statistical-Based Prediction

Statistical-based methods use a training set of known mutagenic and non-mutagenic chemicals to automatically learn from the data. As part of this machine learning process, the statistical-based system determines associations between different chemical descriptors (such as functional groups or physicochemical properties) and both positive and negative bacterial mutagenicity data. These associations are encoded within statistical models. Even though the system will identify statistically significant associations, it may also identify associations that are not biologically meaningful.

One such situation is where a statistical-based model identifies a feature that maps onto a significant number of mutagenic examples. However, it is possible that after a closer examination of the examples, they all contain another structural alert more likely to account for the training set

Structure		Similarity	ChemName	Experimental value - bacterial
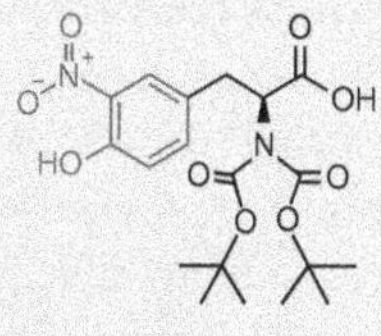 **A**		1.0		
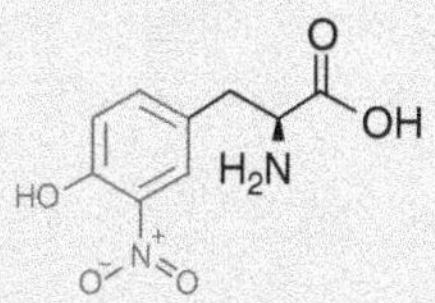 LS-188180		0.58	3-Nitro-L-tyrosine	Negative

Figure 4.10 An assessment by chemical analogs.

chemicals' DNA-reactive mutagenicity. Where a positive prediction is primarily attributable to this coincidentally identified structural feature, the prediction may be refuted based on such an analysis if no other cause exists. Another situation is where again a positive prediction is made and an inspection of the underlying study information might suggest the calls are questionable. This could be due to the studies not having been performed according to the necessary guidelines or there was an experiment artifact that resulted in false-positive experimental results. One such example is the influence of the DMSO solvent on the potential mutagenicity of acyl/sulfonyl halides leading to false positives in the Ames test [42].

Other information related to the selected descriptors and underlying examples may also be taken into account when refuting a positive prediction from a statistical-based model. For example, if the test chemical is more similar to the negative training set examples than the positive ones, or there were not enough examples to make a positive call with sufficient confidence or there were other factors that could deactivate the alerting structure.

4.2.6.3 Mechanistic Assessment

Expert rule-based systems are generally based on information that is in the public domain and alerts are selected based on a series of criteria including a sufficient number of mutagenic examples, a plausible mechanistic basis, and so on. It is possible that an organization has tested similar chemicals or performed mechanistic assessments and this knowledge is not in the public domain. It is also possible that information has been reported that was not used in the development of the expert system. Therefore, any additional information concerning mechanistic factors that both activate or deactivate mutagenicity should be considered in reviewing the results (possibly refuting the prediction). This includes any mechanistic understanding of electronic or steric effects that influence the chemical's mutagenicity.

4.2.6.4 Assessing Lack of Chemical Reactivity

An assessment that the impurity lacks reactive potential supports a number of expert review scenarios. For example, an analysis of the reactivity for out-of-domain predictions can be used as part of an expert review to call the compound negative. Another scenario is to provide additional supportive evidence for a negative prediction. This type of assessment is often performed in consultation with experts or using model output or other computational tools. For example, it may be guided by software that shows what different substructures of the impurities are associated with greater number of mutagenic examples than would be expected by chance, i.e. having a higher prevalence toward occurring in mutagenic compounds.

Compound C is predicted negative in both methodologies, as shown in Figure 4.11. An inspection of the results shows that both the probability and alert precision are low, no alerts are present in the chemical, and the statistical-based model considers all atoms and bonds in its assessment and does not identify any reactive features (i.e. all atoms and bonds are color-coded gray to indicate they have been considered and no potentially reactive groups are identified). Therefore, after an expert review, the non-mutagenic assessment is in agreement with the (Q)SAR model results.

4.2.6.4.1 *Applying a Third Model*

Although it is not specified in the ICH M7 guideline, a practical approach to handling inconclusive or out-of-domain results is to apply another (Q)SAR model. One approach is to apply another commercial model since these models may be generated based on different training sets, using different descriptors, using different algorithms as well as different methodologies for assessing the applicability domain. Another approach is to use models generated from proprietary data (possibly in combination with public datasets). These proprietary-based models may result in fewer out-of-domain chemicals if they contain chemicals within the organization's proprietary chemical space; however, the models will need to be validated and additional information provided to regulatory agencies concerning how they adhere to OECD (Q)SAR validation principles.

4.2.6.4.2 *Assessing the Strength of a Single Prediction*

Both (Q)SAR methodologies have their strengths and weaknesses and will indicate the underlying basis for making any predictions, as discussed earlier. In some situation, one (Q)SAR method may be able to make a prediction with confidence whereas another may not. For example, a statistical-based system may make a prediction with high confidence (such as a high probability score), with

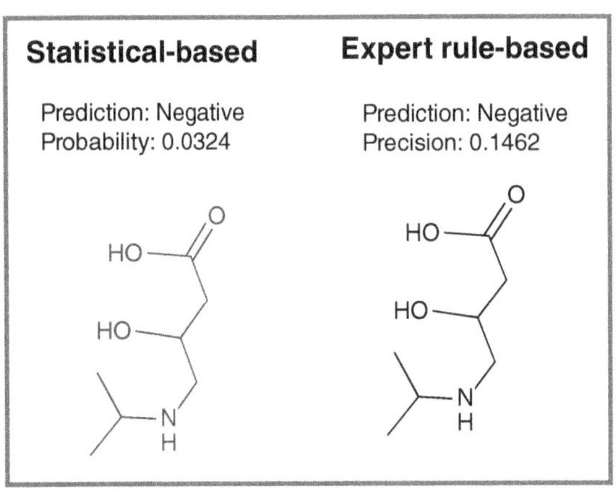

C

Figure 4.11 Assessing the lack of reactive potential through visual inspection.

many analogs that are structurally similar to the impurity, and so on, whereas a rule-based system may not have a sufficient mechanistic basis to make a prediction with confidence. In these situations, an expert review of the totality of the information will support an overall conclusion.

4.2.7 Class Assignment

4.2.7.1 Overview

Figure 4.1 presents a process for assigning the results from the two (Q)SAR models or experimental data to the five ICH M7 classes. Where there is no adequate experimental data for the impurity, the ICH M7 guideline permits the use of two complementary (Q)SAR systems to assign an impurity to either class 3, class 4, or class 5. As illustrated in Figure 4.12, when the results from the two systems are both negative, then the overall assignment is likely to be negative (i.e. class 5). Similarly, when the results from the two systems are positive, then the overall assessment is likely to be positive (i.e. class 3). However, in situations where one or more of the results are either conflicting, out-of-domain, or indeterminate, the class assignment will invariably need to include a robust expert review. If an expert review of all known information can refute a positive/indeterminate result or demonstrate that the impurity lacks reactive potential, then it is possible to assign the impurity as non-mutagenic (i.e. class 4 or 5). However, if this is not possible, then an impurity will likely be assigned to class 3 (mutagenic).

In Amberg et al. [41], an analysis was performed where 15 886 proprietary chemicals with known Ames results (from different pharmaceutical companies) were run against the different (Q)SAR methodologies, including different commercial systems from Leadscope (an Instem company) and Lhasa Limited. These chemicals were similar to those used in impurity assessments. For all combinations of (Q)SAR outcomes, the percentage of mutagens was calculated, representing the likelihood of missing a mutagenic impurity. For example, there were 7978 chemicals (out of the 15 886) that were negative in both the expert rule-based and statistical-based models. 8.1% of those negatively predicted chemicals were in fact Ames positive. A summary of the results is shown in Figure 4.13. This analysis shows the risk of missing a mutagenic impurity from the different results from the two (Q)SAR systems.

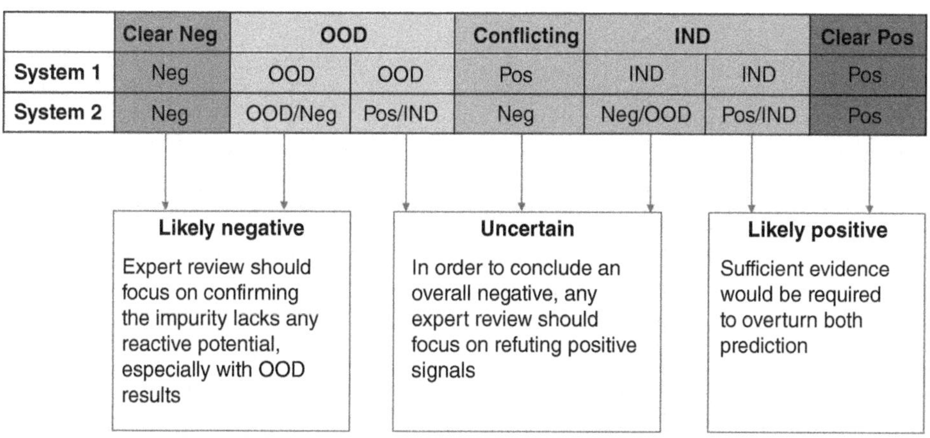

Figure 4.12 Resolving results from different (Q)SAR result. Neg – negative, OOD – out-of-domain, Pos – positive, IND – indeterminate. (*See insert for color representation of this figure*).

Statistical-based	Expert rule-based	Percent mutagens	Expert review
Negative	Negative	<10%	Likely to conclude negative
OOD	Negative or OOD	10%–20%	Expert review proportional to the likelihood of missing a mutagenic impurity
Negative	IND		
IND	Negative or IND	20%–30%	
Positive	IND or Negative		
OOD	IND		
OOD or Negative	Positive	30%–40%	
Positive or IND	Positive	> 50%	Likely to conclude positive

Figure 4.13 Summary of the risk of missing a mutagenic impurity based on an analysis of 15 886 chemicals.

Table 4.1 Results where the two (Q)SAR methodologies were negative.

Statistical-based	Expert rule-based	Percent mutagens
Negative	Negative	8.1% (count = 7978 out of 15 886)

4.2.7.1.1 *Clear Negative Assessment*

When both methods generate a negative (in domain) assessment, a visual inspection of the results may be prudent to determine whether any potentially reactive groups have been overlooked. This minimal and fast assessment is justified since the objective of using a computational assessment is to provide a high-throughput assessment of large numbers of impurities. The chance of obtaining a consensus false-negative prediction is low and expert review generally serves to confirm a negative conclusion. The use of multiple (Q)SAR models under ICH M7 is inherently sensitive and is designed to protect patient safety by minimizing false-negative predictions since a positive prediction from either model results in a class 3 assignment. The risk of missing a mutagenic impurity given two negative predictions was previously calculated to be 6% by Dobo et al. [6] and 8.1% in the currently discussed study [41] based on 7978 examples out of 15 886, as shown in Table 4.1. An expert review has been shown to further reduce the number of missed mutagenic impurities by 5%, as shown by Dobo et al. [6]. Similarly, the US FDA/CDER performed a retrospective analysis of their internal (Q)SAR assessments and determined that expert review overturned only 2% of consensus negative predictions to positive in their consults [41].

Figure 4.14 shows two examples (E and F) that are predicted negative in both (Q)SAR methodologies. The statistical-based methodology predicted E as negative with a probability of 0.264 and the expert rule-based methodology also predicted it to be negative since no alerts matched and the chemical was within the applicability domain of the alert's reference set. Within this particular (Q) SAR system, any alerts or potentially reactive features would be highlighted. The system also highlights those atoms and bonds considered as part of the statistical-based model assessment. In this example, all atoms and bonds (except a single bond shown in black) are considered and no atoms or bonds are highlighted as potentially reactive. Hence, a rapid visual inspection of the results is sufficient to conclude the impurity lacks reactive potential and is therefore assigned to class 5.

The second example (F) is also predicted negative in both (Q)SAR methodologies. Since the chemical contains an aromatic amide functional group, the chemical may be metabolized to a primary aromatic amine. Certain primary aromatic amines are mutagenic and the position of

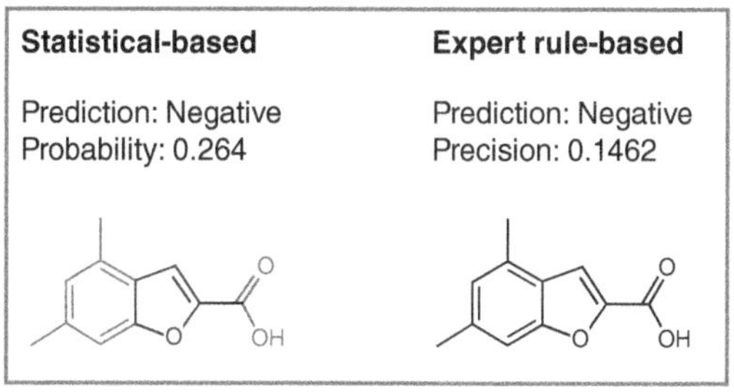

E

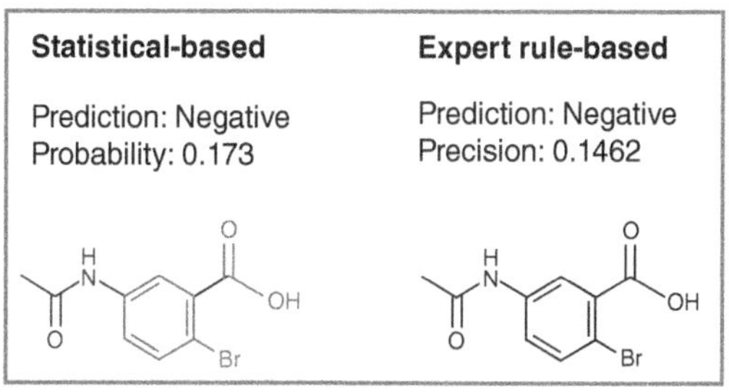

F

Figure 4.14 Two chemicals where the results are negative in both systems.

other functional groups on the aromatic ring relative to the amine influences the chemical's mutagenic potential. Ahlberg et al. 2016 analyzed a series of functional groups in different positions relative to the primary amine to determine whether they are potentially activating. In this example, the carboxylic acid in the meta position and the bromide in the para position are not considered activating. Therefore, even if the chemical undergoes metabolic activation resulting in a primary aromatic amine, the theoretical metabolite is unlikely to be mutagenic given the two substituents. Hence, this impurity is determined to be non-mutagenic and assigned to class 5.

4.2.7.1.2 Clear Positive Assessment

An overall mutagenic assessment is usually made when both methodologies predict the chemical to be mutagenic. This is supported by the data showing that approximately 60% of these chemicals will in fact be positive in the Ames test (based on 1253 examples) as shown in Table 4.2 [41]. Where there is sufficient refuting information, it may be possible to conclude an overall non-mutagenic outcome based on an expert review; however, such a review will need to soundly justify why both prediction results should be overturned.

In Figure 4.15, compound D is predicted positive in both methodologies. The azide functional group is identified as an alert, and is also highlighted in the statistical-based model as responsible for the positive prediction. Azides are a well-known alert for mutagenicity [28] and the expert rule-based model shows that the precision of this alert is around 0.99 (i.e. 99% of chemicals in the reference set with an azide alert are positive). Therefore, the consensus positive conclusion is confirmed and the chemical is assigned to class 3.

4.2.7.1.3 Conflicting Predictions

When the results from the two systems do not agree, i.e. one system predicts the impurity as mutagenic and the other as non-mutagenic, the risk of missing a mutagenic impurity is less than when both systems were positive (Table 4.1) but greater than when both systems were negative (Table 4.2), as shown in Figure 4.13 and Table 4.3. Therefore, an expert review to refute the single positive

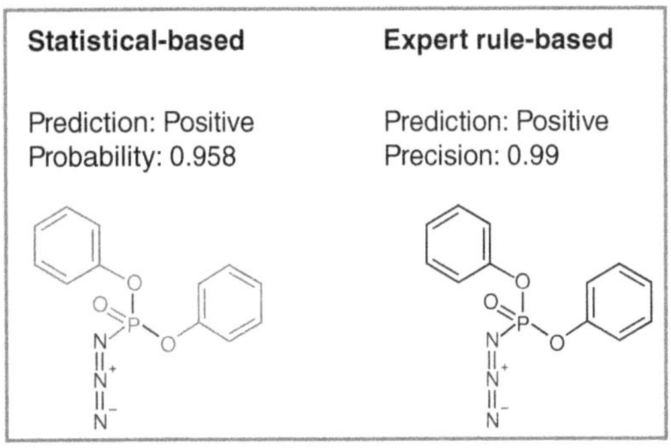

D

Figure 4.15 An example that is positive in the two (Q)SAR methodologies.

Table 4.2 Results when the two methodologies are positive.

Statistical-based	Expert rule-based	Percent mutagens
Positive	Positive	59.7% (count = 1253 out of 15 886)

Table 4.3 Results for conflicting results.

Statistical-based	Expert rule-based	Percent mutagens
Negative	Negative	8.1% (count = 7978 out of 15 886)
Positive	Negative	24.6% (count = 353 out of 15 886)
Negative	Positive	37.5% (count = 499 out of 15 886)
Positive	Positive	59.7% (count = 1253 out of 15 886)

result would be prudent in order to possibly conclude the overall assessment to be non-mutagenic, otherwise the overall assessment would be mutagenic.

Example G's statistical-based results was negative and its expert rule-based result was positive (based on a specific alkyl halide alert), as shown in Figure 4.16. There is additional supporting evidence from the alert system, including a detailed mechanistic basis for the specific alkyl halide alert along with a series of relevant analogs that are mutagenic. Since there is supporting evidence for the positive expert rule-based result, an overall mutagenic assessment is made and the impurity is assigned to class 3.

4.2.7.1.4 Handling Indeterminate predictions

The risk of missing a mutagenic impurity when one of the systems generates an indeterminate result is dependent on the result from the second system. Table 4.4 shows that when the expert rule-based system generates an indeterminate prediction, the risk is dependent on the statistical-based model result: positive (27.7% mutagenic; 155 examples), indeterminate (20.4% mutagenic; 93 examples), and negative (11.8%; 314 examples). Similarly, when the statistical-based result is indeterminate, the risk is dependent on the expert rule-based result: positive (50.6% mutagenic [516 examples]), negative (23.2% mutagenic [668 examples]), and indeterminate (20.4% mutagenic [93 examples]).

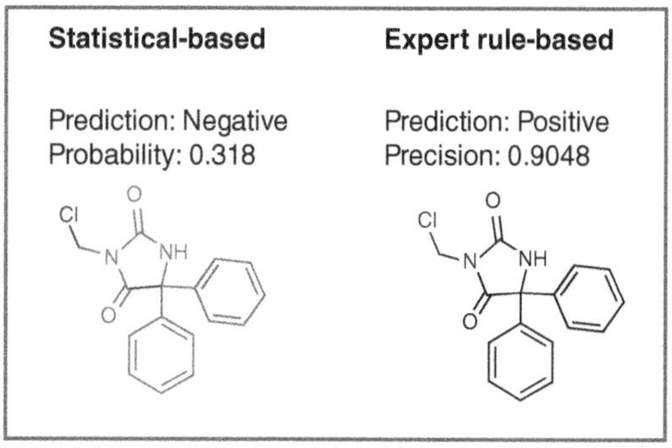

G

Figure 4.16 An example with conflicting (Q)SAR results.

Table 4.4 Results when there is an indeterminate prediction (IND = indeterminate).

Statistical-based	Expert rule-based	Percent mutagens
Negative	IND	11.8% (count = 314, out of 15 886)
IND	IND	20.4% (count = 93, out of 15 886)
IND	Negative	23.2% (count = 668, out of 15 886)
Positive	IND	27.7% (count = 155, out of 15 886)
IND	Positive	50.6% (count = 516, out of 15 886)

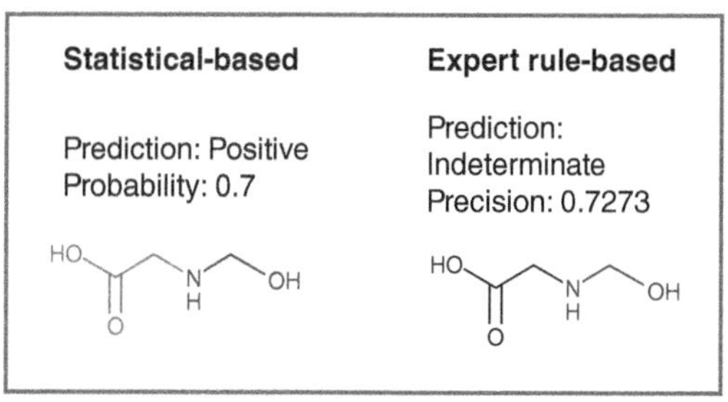

H

Figure 4.17 Example of a prediction where one of the methodologies is an indeterminate prediction.

Example H, shown in Figure 4.17, is predicted positive in the statistical-based methodology and indeterminate in the expert rule-based methodology. An examination of the expert rule-based result indicates that there are some close analogs and a plausible mechanism. Similarly, the statistical-based model is based on similar mutagenic chemicals. Since there is sufficient information supporting a positive prediction, this impurity is assigned to class 3. It should be noted, when there is not sufficient additional information present, the conservative approach is to call the consensus positive.

4.2.7.1.5 Handling Out-of-Domain Results

The risk of missing a mutagenic impurity when one or more of the methodologies is out-of-domain varies depending on the result from the second method. For example, when the statistical-based method is out-of-domain, the risk is dependent on the expert rule-based results: positive (36.1%

Table 4.5 Results where at least one of the methodologies is out-of-domain.

Statistical-based	Expert rule-based	Percent mutagens
OOD	NEG	11.3% (count = 2027, out of 15886)
OOD	IND	28.2% (count = 78, out of 15886)
OOD	POS	36.1% (count = 296, out of 15886)

Table 4.6 Results showing the effect of using the probability score.

Statistical-based	Expert rule-based	Percent mutagens
Negative	Negative	8.1% (7978)
OOD (Pr < 0.2)	Negative	8.8% (1415)
OOD	Negative	11.3% (2027)
OOD (Pr 0.2–0.4)	Negative	17.3% (339)

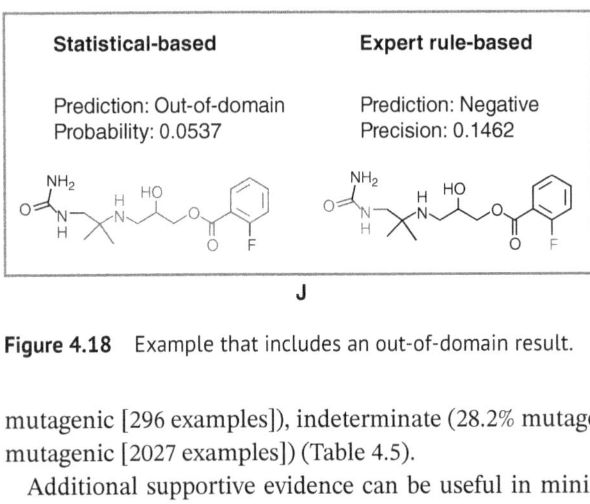

J

Figure 4.18 Example that includes an out-of-domain result.

mutagenic [296 examples]), indeterminate (28.2% mutagenic [78 examples]), and negative (11.3% mutagenic [2027 examples]) (Table 4.5).

Additional supportive evidence can be useful in minimizing the risk of missing a mutagenic impurity. An analysis of a statistical-based model's probability scores for chemicals that were out-of-domain and negative in the expert rule-based model shows that the risk of missing a mutagenic impurity is reduced by taking the probability score into account. A probability score of less than 0.2 results in 8.8% missed mutagens (which is a similar result to two negative predictions) whereas a probability score of 0.2–0.4 would result in 17.3% of missed mutagens (Table 4.6).

For chemical J (shown in Figure 4.18), the expert rule-based result is negative and the statistical-based model is out-of-domain. The statistical-based model also generates a probability score of 0.0537. Since the impurity appears to lack any reactive potential in addition to the low probability score, the risk of missing a mutagenic impurity in this situation is low; therefore, the impurity is assigned to class 5.

4.2.8 Documentation

It will be important to document and share the computational assessments with colleagues across the organization to support any necessary control or testing strategies. This is generally handled using spreadsheets of the (Q)SAR results that includes a preliminary expert review.

Once the list of impurities to assess has been finalized, documentation of the (Q)SAR results, including any expert review along with reports generated from the software for each chemical (preferably), should be generated and archived. Different systems support this step in different ways, for example, (Q)SAR consultation reports are stored at the US FDA/CDER in an internal Leadscope Enterprise database. In addition, a web-based Leadscope client is made available to reviewers allowing them to search the database by chemical structure and retrieve (Q)SAR assessment reports along with other data directly [43].

The US FDA/CDER has also proposed a table of results consisting of the following columns [44]:

- Chemical name
- Chemical structure (with any alert highlighted)
- Expert rule-based prediction
- Statistical-based prediction
- ICH M7 class
- Comments

The table should include all chemicals being assessed as exemplified in Figure 4.19. The comments field can include information on any expert review, particularly if the software calls were overruled. When an Ames assay was run, it will be important to include the study data along with any justification for the use of nonstandard assay formats. Alongside this table should be a short overview of the materials and methods used, including (Q)SAR system, model version number, software version number, and any databases used with version numbers.

Information related to impurities should be included in the submission dossier in the following modules [44]:

- Quality (module 3): Impurity dossier, review of scientific principles/purging claims.
- Nonclinical (module 4): Review of Ames and other compound-specific data.
- Clinical: Dose and duration of exposure; population/s excluded; risk:benefit consideration.

4.3 Discussion

(Q)SAR models are periodically updated with the latest data and knowledge. In fact, during the typical 7–10 year drug development process, a number of different versions of the software may have been developed. Also, during this development process, the synthetic routes may be changed, modifying the list of impurities present. A recent cross-pharma ceutical consortium investigated the issue of how often should such assessment be performed given the changes to the training sets and software versions [45, 46]. They determined that the number of changes from negative/indeterminate/out-of-domain to positive prediction was rare (2%). It is likely that some or all of these could have been identified by expert review, but these changes are important since any newly identified mutagenic impurity may need to be tested or have controlled strategies put in place. To mitigate issues related to a new software version generating new positives, it was proposed to reassess after commercial route of synthesis has been selected and prior to submission. This assessment may be through an expert review of the historical classification or the (Q)SAR software could be rerun over the impurity list.

Executive Summary

Materials and methods

An assessment aligned with ICH M7 was performed on the actual and potential impurities listed in the table below using the following QSAR methodologies and systems:

Expert rule-based methodology and parameters:	Leadscope genetox expert alerts v6 (System: Leadscope Model Applier v2.4.2.1); the domain assessment was turned on
Statistical-based methodology and parameters:	Leadscope Bacterial Mutation statistical-based QSAR model v2; domain assessment was turned on
Genetic toxicity database used for searching:	Leadscope SAR genetox 2019
Rodent carcinogenicity database used for searching:	Leadscope SAR carcinogenicity 2019

These *in silico* methodologies follow the general validation principles set forth by the Organisation for Economic CO-operation and Development (OECD). The Leadscope statistical-based methodologies and the Leadscope SAR databases were developed through a research collaboration agreement with the US FDA.

Results

The following table and notes summarize the results of this QSAR analysis for 4 impurities. Structural alerts and significant model features are highlighted. Alerts and positive model features inred, negative model features in blue-green, and indeterminate in gray.

#	Impurity Structure	Laboratory Data	Expert rule-based system	Statistical-based system	Overall assessment	M7 Class assignment	Additional supportive evidence and comments
1	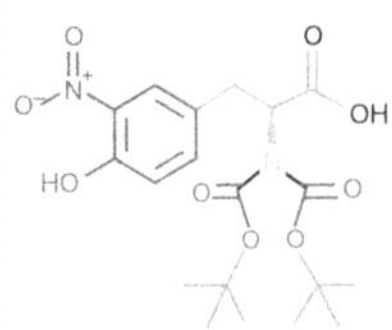 LS Ex4	None	Indeterminate	Negative	Nonmutagenic	4	**Refuted inconclusive *in silico* result.** An inconclusive prediction was refuted based upon a chemical analogs expert review, The compound was reassigned to be negative (class 5). See 'LS Ex4 #1 - Refuting positive *in silico* results. [1-refute]'
2	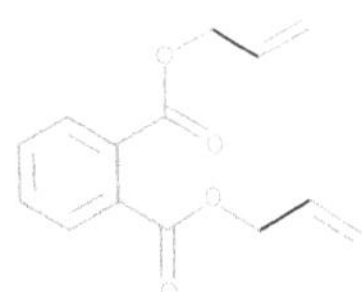 LS-30-1	Class 5- Nonmutagenic Bacterial mutagenesis See 'LS-30-1 #2 - Assess available data. [2-appendix]'	Not performed	Not performed	Nonmutagenic	5	Laboratory data was used determine the classification Class 5 - Nonmutagenic - Bacterial mutagenesis
3	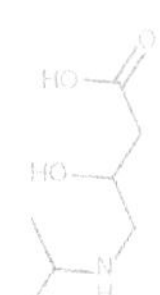 LS_EX1	None	Negative	Negative	Nonmutagenic	5	**Accepted negative *in silico* result.** The impurity lacks obvious reactive potential.
4	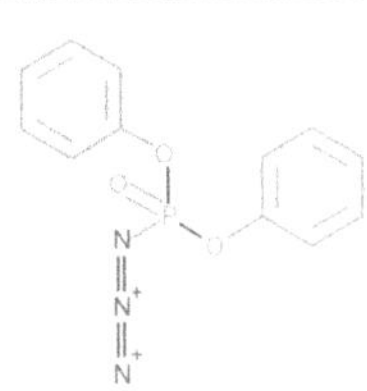 tst3	None	Positive	Positive	Predicted Mutagenic	3	**Accepted negative *in silico* result.** Matching alerts: 41: azide (0.99)

Figure 4.19 Impurities report generated for four chemicals.

4.4 Conclusions

Computational toxicology is a safe, inexpensive, and high-throughput approach to assess the mutagenic potential of impurities. The implementation of (Q)SAR assessments for impurities has shown that computational methods have a significant role to play in the assessment of chemicals. It has also led to improvements in science around the SAR knowledge for chemicals of particular concern in impurities assessment, such as aromatic amines. The predictivity of the (Q)SAR methods has improved as more data become available for building models. Implementation of the ICH M7 guideline has advanced our understanding of what an expert review of (Q)SAR results involves and finally, the successful implementation of (Q)SAR methodologies to support the assessment of PMIs has led to its use in other areas such as the assessment of extractables and leachables.

Acknowledgments

The authors wish to thank Arianna Bassan, Kevin Cross, Candice Johnson, Naomi Kruhlak, Manuela Pavan, and Andy Teasdale for their helpful comments as well as the Japanese NIHS (http://www.nihs.go.jp/dgm/eM7QSAR.html) for providing a number of examples used.

References

1 Myatt, G.J., Ahlberg, E., Akahori, Y. et al. (2018). In silico toxicology protocols. *Regul. Toxicol. Pharmacol.* 98: 1–17. https://doi.org/10.1016/j.yrtph.2018.04.014.

2 Amberg, A., Beilke, L., Bercu, J. et al. (2016). Principles and procedures for implementation of ICH M7 recommended (Q)SAR analyses. *Regul. Toxicol. Pharmacol.* 77: 13–24. https://doi.org/10.1016//j.yrtph.2016.02.004.

3 OECD (1997) Test No. 471: Bacterial Reverse Mutation Test. Section 4, OECD Publishing, Paris. http://dx.doi.org/10.1787/9789264071247-en (accessed 31 August 2020).

4 ICH M7, 2017 (R1). (2017). Assessment and control of DNA reactive (mutagenic) impurities in pharmaceuticals to limit potential carcinogenic risk. https://database.ich.org/sites/default/files/M7_R1_Guideline.pdf (accessed 31 August 2020).

5 Müller, L., Mauthe, R.J., Riley, C.M. et al. (2006). A rationale for determining, testing, and controlling specific impurities in pharmaceuticals that possess potential for genotoxicity. *Regul. Toxicol. Pharmacol.* 44: 198–211. https://doi.org/10.1016/j.yrtph.2005.12.001.

6 Dobo, K.L., Greene, N., and Fred, C. (2012). In silico methods combined with expert knowledge rule out mutagenic potential of pharmaceutical impurities: an industry survey. *Regul. Toxicol. Pharmacol.* 62: 449–455. https://doi.org/10.1016/j.yrtph.2012.01.007.

7 Sutter, A., Amberg, A., Boyer, S. et al. (2013). Use of in silico systems and expert knowledge for structure-based assessment of potentially mutagenic impurities. *Regul. Toxicol. Pharmacol.* 67: 39–52. https://doi.org/10.1016/j.yrtph.2013.05.001.

8 EMA. (2006). Guideline on the limits of genotoxic impurities. https://www.ema.europa.eu/en/documents/scientific-guideline/guideline-limits-genotoxic-impurities_en.pdf (accessed 31 August 2020).

9 EMA. (2010). Questions and answers on the 'guideline on the limits of genotoxic impurities'. https://www.ema.europa.eu/en/documents/scientific-guideline/questions-answers-guideline-limits-genotoxic-impurities_en.pdf (accessed 31 August 2020).

10 FDA. (2008). Draft guidance for industry on genotoxic and carcinogenic impurities in drug substances and products: recommended approaches. https://www.federalregister.gov/documents/2008/12/16/E8-29674/draft-guidance-for-industry-on-genotoxic-and-carcinogenic-impurities-in-drug-substances-and-products (accessed 31 August 2020).

11 ICH Q3A(R2). (2006). International conference on harmonisation of technical requirements for registration of pharmaceuticals for human use ICH harmonised tripartite guideline: impurities in new drug substances Q3A(R2). https://www.ema.europa.eu/en/documents/scientific-guideline/ich-q-3-r2-impurities-new-drug-substances-step-5_en.pdf (accessed 31 August 2020).

12 ICH Q3B(R2). (2006). International conference on harmonisation of technical requirements for registration of pharmaceuticals for human use ICH harmonised tripartite guideline: impurities in new drug products ICH Q3B (R2). https://www.ema.europa.eu/en/documents/scientific-guideline/ich-q-3-b-r2-impurities-new-drug-products-step-5_en.pdf (accessed 31 August 2020).

13 CCRIS. (2011). Accessing chemical carcinogenesis research information system (CCRIS) content from PubChem. https://www.nlm.nih.gov/toxnet/Accessing_CCRIS_Content_from_PubChem.html (accessed 31 August 2020).

14 CPDB. (2011). Carcinogenic Potency Database fact sheet. https://www.toxinfo.io/assets/app_data/pdf/CPDB_FS.pdf (accessed 31 August 2020).

15 ECHA. (2020). European chemicals agency. http://echa.europa.eu/ (accessed 31 August 2020).

16 IARC. (2020). Agents classified by the IARC monographs, Volumes 1–127. http://Monographs.iarc.fr/ENG/Classification/index.php (accessed 31 August 2020).

17 Leadscope. (2020). Leadscope home page. www.leadscope.com (accessed 31 August 2020).

18 MultiCASE. (2020). MultiCASE home page. www.multicase.com (accessed 31 August 2020).

19 NTP. (2020). Chemical effects in biological systems. http://tools.niehs.nih.gov/cebs3/ui/ (accessed 31 August 2020).

20 VITIC. (2020). VITIC product information. http://www.lhasalimited.org/products/vitic-nexus.htm (accessed 31 August 2020).

21 Bercu, J.P., Galloway, S.M., Parris, P. et al. (2019). Potential impurities in drug substances: compound-specific toxicology limits for 20 synthetic reagents and by-products, and a class-specific toxicology limit for alkyl bromides. *Regul. Toxicol. Pharmacol.* 94: 172–182. https://doi.org/10.1016/j.yrtph.2018.02.001.

22 Myatt, G.J., Beilke, L.D., and Cross, K.P. (2016). In silico tools and their application. In: *Reference Module in Chemistry, Molecular Sciences and Chemical Engineering* (ed. J. Reedijk). Elsevier https://doi.org/10.1016/B978-0-12-409547-2.12379-0.

23 Landry, C., Kim, M., Kruhlak, N. et al. (2020). Transitioning to composite bacterial mutagenicity models in ICH M7 (Q)SAR analyses. *Regul. Toxicol. Pharmacol.* 109: 104488. https://doi.org/10.1016/j.yrtph.2019.104488.

24 OECD. (2007). Guidance document on the validation of (quantitative) structure-activity relationships [(Q)SAR] models, OECD Environment Health and Safety Publications Series on Testing and Assessment No. 69 (ENV/JM/MONO(2007)2). http://www.oecd.org/env/guidance-document-on-the-validation-of-quantitative-structure-activity-relationship-q-sar-models-9789264085442-en.htm (accessed 31 August 2020).

25 Cooper, J.A., Saracci, R., and Cole, P. (1979). Describing the validity of carcinogen screening tests. *Br. J. Cancer* 39 (1): 87–89.

26 Honma, M., Kitazawa, A., Cayley, A. et al. (2019). Improvement of quantitative structure–activity relationship (QSAR) tools for predicting Ames mutagenicity: outcomes of the Ames/QSAR International Challenge Project. *Mutagenesis* 34 (1): 3–16. https://doi.org/10.1093/mutage/gey031.

27 Ashby, J. and Tennant, R.W. (1988). Chemical structure, Salmonella mutagenicity and extent of carcinogenicity as indicators of genotoxic carcinogenesis among 222 chemicals tested in rodents by the U.S. NCI/NTP. *Mutat. Res.* 204: 17–115.

28 Kazius, J., Mcguire, R., and Bursi, R. (2005). Derivation and validation of toxicophores for mutagenicity prediction. *J. Med. Chem.* 48: 312–320. https://doi.org/10.1021/jm040835a.

29 Bailey, A.B., Chanderbhan, R., Collazo-Braier, N. et al. (2005). The use of structure activity relationship analysis in the food contact notification program. *Regul. Toxicol. Pharmacol.* 42: 225–235. https://doi.org/10.1016/j.yrtph.2005.04.006.

30 Benigni, R., Bossa, C., Jeliazkova, N. et al. (2008). The Benigni/Bossa Rulebase for mutagenicity and carcinogenicity – a module of toxtree. Technical Report EUR 23241 EN. European Commission – Joint Research Centre.

31 Benigni, R. and Bossa, C. (2011). Mechanisms of chemical carcinogenicity and mutagenicity: a review with implications for predictive toxicology. *Chem. Rev.* 111: 2507–2536. https://doi.org/10.1021/cr100222q.

32 Myden, A., Guesne, S.J., Cayley, A., and Williams, R.V. (2017). Utility of published DNA reactivity alerts. *Regul. Toxicol. Pharmacol.* 88: 77–86. https://doi.org/10.1016/j.yrtph.2017.05.0161.

33 Amberg, A., Anger, L.T., Bercu, J. et al. (2018). Extending (Q)SARs to incorporate proprietary knowledge for regulatory purposes: is aromatic *N*-oxide a structural alert for predicting DNA-reactive mutagenicity? *Mutagenesis* 34: 67–82. https://doi.org/10.1093/mutage/gey020.

34 Ahlberg, E., Amberg, A., Beilke, L.D. et al. (2016). Extending (Q)SARs to incorporate proprietary knowledge for regulatory purposes: a case study using aromatic amine mutagenicity. *Regul. Toxicol. Pharmacol.* 77: 1–12. https://doi.org/10.1016/j.yrtph.2016.02.003.

35 ChemDraw. (2019). ChemDraw products. https://www.perkinelmer.com/category/chemdraw (accessed 31 August 2020).

36 Dalby, A., Nourse, J.G., Hounshell, W.D. et al. (1992). Description of several chemical structure file formats used by computer programs developed at Molecular Design Limited. *J. Chem. Inf. Model.* 32 (3): 244. https://doi.org/10.1021/ci00007a012.

37 Weininger, D. (1988). SMILES, a chemical language and information system. 1. Introduction to methodology and encoding rules. *J. Chem. Inf. Model.* 28 (1): 31–36. https://doi.org/10.1021/ci00057a005.

38 Lhasa. (2020). Lhasa Limited homepage. https://www.lhasalimited.org/products/derek-nexus.htm (accessed 31 August 2020).

39 Powley, M.W. (2015). (Q)SAR assessments of potentially mutagenic impurities: a regulatory perspective on the utility of expert knowledge and data submission. *Regul. Toxicol. Pharmacol.* 71: 295–300. https://doi.org/10.1016/j.yrtph.2014.12.012.

40 Barber, C., Amberg, A., Custer, L. et al. (2015). Establishing best practise in the application of expert review of mutagenicity under ICH M7. *Regul. Toxicol. Pharmacol.* 73: 367–377. https://doi.org/10.1016/j.yrtph.2015.07.018.

41 Amberg, A., Andaya, R.V., Anger, L.T. et al. (2019). Principles and procedures for handling out-of-domain and indeterminate results as part of ICH M7 recommended (Q)SAR analyses. *Regul. Toxicol. Pharmacol.* 102: 53–64. https://doi.org/10.1016/j.yrtph.2018.12.007.

42 Amberg, A., Harvey, J., Czich, A. et al. (2015). Do carboxylic/sulfonic acid halides really present a mutagenic and carcinogenic risk as impurities in final drug products? *Org. Process Res. Dev.* 19 (11): 1495–1506. https://doi.org/10.1021/acs.oprd.5b00106.

43 Cross, K.P., Kim, M.T., and Kruhlak, N.L. (2019). The FDA/CDER chemical dictionary. Poster presented at the 2019 GTA Meeting, Newark, DE.

44 Goodwin, A. (2019). Assessment of potentially genotoxic impurities: best practice for regulatory submissions, 2019 GTA Meeting, Newark, DE.

45 Dobo, K.L. (2019). Does my (Q)SAR assessment need a face lift? 2019 GTA Meeting, Newark, DE.

46 Hasselgren, C., Bercu, J., Cayley, A., et al., Management of pharmaceutical ICH M7 (Q)SAR predictions – the impact of model updates, *Regul. Toxicol. Pharmacol.* 2020; 118:104807.

5

Evolution of Quantitative Structure–Activity Relationships ((Q)SAR) for Mutagenicity

James Harvey[1] *and David Elder*[2]

[1] *GlaxoSmithKline R&D, Hertfordshire, UK*
[2] *David P. Elder Consultancy, Hertford, UK*

5.1 Introduction

Numerous publications have described the performance of quantitative structure–activity relationship ((Q)SAR) approaches for the prediction of the potential mutagenic activity of a new chemical entity in the drug discovery and development setting [1–4]. This approach was subsequently applied to the prediction of the potential mutagenic activity of novel drug substance impurities and drug product degradation products [4]. A comprehensive review of the state of the art of the performance of the various *in silico* (Q)SAR models for the prediction of mutagenicity has also been conducted by Benigni and Bossa [5]. When Teasdale's book on Genotoxic Impurities [6] was first published during 2010, the routine industry practice was to assess drug substance and or drug product impurities in one (Q)SAR system; either an expert system (e.g. DfW – Derek for Windows) or a statistical system (e.g. Leadscope or MCASE) in support of the ICH Q3A and Q3B guidelines to determine whether the said impurity was "unusually toxic." At that time, Glowienke and Hasselgren [7] indicated that both MCASE and DfW "cover the scope of compounds that are typically encountered within synthetic processes." The introduction of the ICH M7 guideline changed industry practice and mandated the use of two complementary (Q)SAR systems; an expert rule-based system and a statistical-based system for the assessment of drug substance and or drug product impurities (for definitions, see [8]). Greene et al. [9] established that a combination of two systems normally increases sensitivity, but typically this is at the expense of specificity as well as a significant increase in conflicting (Q)SAR predictions (c. 25% of all cases). An example of this comparing Derek Nexus and Leadscope is illustrated in Figure 5.1.

However, additional publications demonstrated that the use of two (Q)SAR systems coupled with "expert knowledge," typically increases both sensitivity and specificity [10–12]. This chapter will cover advances made to the overall accuracy and performance of the two main types of *in silico* (Q) SAR platforms (i.e. expert rule or statistical-based) that occurred as a result of the introduction of the ICH M7 guideline that exemplify the huge importance of cross-industry collaborative efforts in ensuring that (Q)SAR systems remain fit for purpose. In addition, strategies for incorporating new knowledge into (Q)SAR systems will be discussed; using pertinent examples, for example, such as **aromatic amines** and **aromatic *N*-oxides**. Furthermore, the principles, procedures, and expectations of an "expert-knowledge" review as defined in ICH M7 (which is frequently used to provide supportive evidence on relevance of any positive, negative, conflicting, or inconclusive (Q)SAR

Mutagenic Impurities: Strategies for Identification and Control, First Edition. Edited by Andrew Teasdale.
© 2022 John Wiley & Sons, Inc. Published 2022 by John Wiley & Sons, Inc.

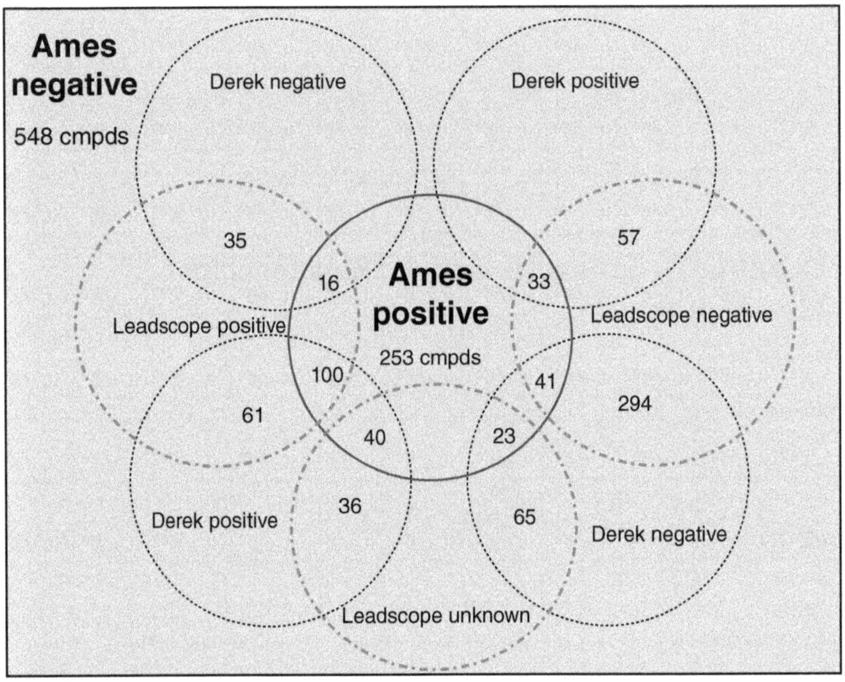

Figure 5.1 Derek Nexus predictions compared with Leadscope modeller predictions for 801 compounds subjected to Ames assay *Source*: Taken from [9].

prediction) will be outlined from the perspective of *in silico* (Q)SAR model builders, model users, and regulators. Finally, a consideration of the future state of the area, such as the adoption of ICH M7 *in silico* (Q)SAR platforms in drug discovery to mitigate against future mutagenic impurity issues that typically occurs late in development, the introduction of rapid refinements to (Q)SAR platforms related to established toxicophores (e.g. *N*-nitrosamines), and an exploration of novel modeling mechanisms to improve access to supporting proprietary mutagenicity data will also be discussed.

5.2 Pre ICH M7 Guideline

Following the introduction of the original Ashby–Tennant SAR framework (see Figure 5.2 – Ashby and Tennant [13]), development of proprietary and non-proprietary *in silico* (Q)SAR models for the prediction of mutagenicity has progressed steadily across various industries with mixed success depending on the particular use case (for review, see Benigni et al. [4]). By way of example, *in silico* (Q)SAR mutagenicity models within the pharmaceutical industry were deployed primarily in the late stages of molecular design or to inform on design aspects of the genotoxicity studies outlined in the ICH S2R genotoxicity testing guidelines. However, within the novel chemical space of the drug discovery setting, mutagenicity (Q)SAR models built with external data sets had limited utility due to the limitations associated with narrow applicability domains [15, 16] (applicability domains commonly being defined as "the response and chemical structure space in which the model makes predictions with a given reliability"). These applicability domain limitations were partially addressed by accessing large data sets of mutagenicity data, careful selection of chemical descriptors and specific modeling algorithms that facilitated further (Q)SAR development [1, 17], and the development of proprietary mutagenicity (Q)SAR models based on a combination of internal and external data

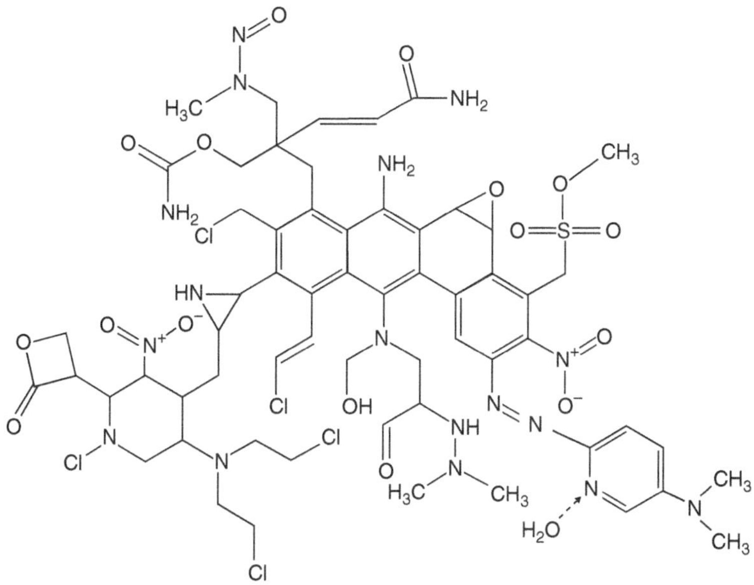

Figure 5.2 Ashby–Tennant super structure. *Source:* Taken from [13, 14].

sets [18]. In contrast, outside of drug discovery, the adoption and deployment of *in silico* (Q)SAR models as a tool for the regulation of chemicals across a range of industries including cosmetics, foods, and industrial was becoming more widely established and was used to inform on testing prioritizations [14, 19]. The adoption has been pragmatic – facilitated by necessity for *in silico* (Q)SAR approaches that offer a valuable low-cost, scientifically robust, and ethical (aligned with efforts to reduce animal testing) screening approach that can be scaled to address the magnitude of current chemical safety testing requirements – it is estimated that 140 million chemical substances have been registered and 100 000 are produced on an industrial scale [3]. Periodically, a significant breakthrough has been made on the basis of a fundamental understanding of the biological mechanisms of mutagenicity (i.e. direct electrophilicity of the chemical or its transformation products – see the boronic acids example) and/or large well-curated public "Ames" bacterial mutagenicity data sets. For example, a data set of *in vitro* genotoxicity assay data with a high degree of experimental reproducibility were made available [20] that facilitated *in silico* (Q)SAR model development and refinement. Additional progress regarding the development of structural alerts was also made via the sharing of proprietary data with (Q)SAR developers, but without the public release of the underlying data (for examples, see Elder et al. [18]).

5.3 Post ICH M7

5.3.1 Evolution of (Q)SAR Platforms

The introduction of the ICH M7 guideline on assessment and control of DNA-reactive (mutagenic) impurities in pharmaceuticals to limit potential carcinogenic risk changed the *in silico* (Q)SAR mutagenicity modeling landscape within the pharmaceutical industry. The regulatory requirement for robust (Q)SAR approaches to support the mutagenic impurity management framework was a major milestone regarding the acceptance of *in silico* (Q)SAR modeling approaches as an alternative to *in vitro* and *in vivo* testing in support of hazard identification and risk assessment to determine the safety of impurities in humans [3]. The ICH M7 guideline also brought an additional dimension to the use of *in silico* (Q)SAR approaches in the drug development setting as it stipulated "the absence of structural alerts from ***two*** complementary (Q)SAR methodologies (expert rule-based and statistical) is sufficient to conclude that the impurity is of no mutagenic concern." The introduction of *in silico* QSAR models in the decision framework outlined in ICH M7 did, however, raise some interesting questions. At that time, *in silico* (Q)SAR models for the prediction of mutagenicity of new chemical entities in drug discovery were typically oriented toward high specificity (to reduce the number of false positives) whereas under ICH M7, the *in silico* (Q)SAR models for the prediction of mutagenicity of impurities needed to be orientated toward high sensitivity (to reduce the number of false negatives). As defined by Landry et al. [21], "in a regulatory environment, high sensitivity and negative predictivity are important characteristics of (Q)SAR models used to support drug safety decisions, thereby minimizing risk to patients" which is important considering that impurities, although an inevitable outcome of drug substance and drug product manufacture, offer no therapeutic benefit to patients. This driver for an increased focus on sensitivity (as opposed to specificity) of the *in silico* approaches provided a sense of urgency which led to modelers, chemists, toxicologists, and regulators across the industry collaborating to ensure the available *in silico* (Q)SAR approaches for predicting mutagenicity were suitably robust to meet the requirements of the ICH M7 guideline.

The commercial *in silico* (Q)SAR vendors, pharmaceutical industry, and regulators addressed these challenges in several ways that resulted in iterative improvements to the available *in silico* (Q)SAR platforms. The major changes included (i) robust negative *in silico* QSAR predictions including gaining a consensus on the meaning of a "negative prediction" in certain (Q)SAR platforms, (ii) development of composite (Q)SAR models that combined the predictions for multiple specific bacterial strains, (iii) large-scale release of proprietary mutagenicity databases to enhance the predictive power of (Q)SAR tools, and (iv) focused data sharing initiatives to improve the accuracy of the (Q)SAR models for specific chemical classes – all of which supported an increase in overall predictive performance and the applicability domains of the major *in silico* (Q)SAR platforms.

5.3.2 Robust Negative *In Silico* (Q)SAR Predictions

Barber et al. [22] and Williams et al. [23] described the integration of a transparent interpretable prediction framework that addressed the robustness of a negative prediction in a (Q)SAR platform. Barber et al. [22] described the development of a novel *in silico* (Q)SAR system that generates positive and negative hypotheses from a self-organizing hypothesis network (SOHN) based on the presence of structural fragments in the query structure which have been associated with activity or inactivity in the training set. Williams et al. [23] discussed the definition of a "negative prediction" from the perspective of a model builder "how can we predict the absence of a signal" and a model user "can I trust this negative prediction for my compound" summarizing the dilemma as "is the absence of a positive prediction sufficient evidence for a negative prediction for my compound?" The timing of this important publication (i.e. after the introduction of ICH M7) was seemingly at odds with the mainstream adoption of these *in silico* (Q)SAR approaches by regulators and industry; however, in reality, this prediction framework was under evaluation by organizations before the ICH M7 guideline was introduced. much earlier. Williams et al. [23] pointed out that whereas positive *in silico* (Q)SAR predictions can be acted upon (via mutagenicity testing of the impurity or control measures in drug product), a negative *in silico* (Q)SAR prediction alone for an impurity inferred a lack of hazard and risk (with no subsequent requirement for additional mutagenicity testing or control measures) and hence a negative prediction carried more weight and conceivably should be held to higher standards than a positive prediction. The authors discussed two approaches that could increase the confidence in negative predictions by examining regions around precedented chemical space associated with mutagenicity, the so-called "predictive space" (e.g. the known substituents that effect mutagenicity of classical mutagenic substructure) and an evaluation of the similarity of a query structure to known mutagens that were incorrectly predicted by *in silico* (Q)SAR approaches. The analysis demonstrated that the "similarity" approach was robust to support negative predictions, but this was not the case for the "predictive space" approach. The resulting decision framework allowed negative predictions to be categorized in an informative way: (1) inactive (i.e. the "query compound is similar to (i) inactive compounds or (ii) active compounds that are well-predicted"), (2) inactive with misclassified features (i.e. "similar to experimentally active compounds that are not predicted to be active"), and (3) inactive with unclassified features, that is "contains unfamiliar features" (i.e. there are no similar compounds in the library). This framework was incorporated into certain commercial (Q)SAR platforms and increased the confidence of model builders and model users in the robustness of a negative prediction and the subsequent consequences.

5.3.3 Development of Composite (Q)SAR Models

Prior to ICH M7, several (Q)SAR platforms have been developed with multiple individual models to determine the potential mutagenicity of a query structure in both *Salmonella typhimuirum* and *Escherichia coli* bacterial strains (and corresponding mutation at the guanine–cytosine [GC] and adenine–thymine [AT] base pairs) that was invaluable in the drug discovery setting to define high-throughput screening cascades. However, within the ICH M7 setting, in practice this actually resulted in the assessment of impurities in three (Q)SAR models by default, i.e. one expert rule model and two statistically based models within a particular QSAR platform for the assessment of both GC and AT mutations in *S. typhimuirum* and *E. coli*, respectively. Landry et al. [21] described the development of composite bacterial mutagenicity (Q)SAR models that combined the data from the *S. typhimuirum* and *E. coli* strains to simplify the ICH M7 impurity assessment process. The authors increased the applicability domains of these composite models following the inclusion of new data sets from government institutions and regulatory authorities (e.g. US FDA (Food and Drug Agency), Japanese NIHS (National Institute of Health Sciences), US EPA (Environmental Protection Agency), US NTP (National Toxicology Program), and the published literature) that effectively doubled the number of compounds in the underlying databases. This process also included the systematic review of 1000+ equivocal and conflicting bacterial mutagenicity results into a binary positive or negative outcome to facilitate (Q)SAR model improvement. The authors described how the composite models improved both the sensitivity and negative predictivity of two of the statistical-based (Q)SAR platforms, i.e. MCASE and Leadscope, including the introduction of modified toxicophores related to aromatic amines, aromatic diazo's, aromatic hydrazines, and vinyl halides. They also demonstrated how the use of three *in silico* (Q)SAR tools (one expert rule model, and two separate composite statistical-based models) could increase the coverage of chemical space encountered in the manufacturing environment, thereby reducing the frequency of out-of-domain impurity assessments. Despite ICH M7 clearly stating that only two complementary (Q)SAR models (one expert rule model and one statistically based model) are required to evaluate impurities, this proposal to use a third validated (Q)SAR model of the appropriate methodology, to address out-of-domain predictions was recently proposed in the ICH M7 Q&A and is a pragmatic way to address the issues frequently experienced during development where impurities related to proprietary chemistry are out-of-domain in one or two of the standard (Q)SAR platforms. In summary, the composite (Q)SAR models were validated against an external data set specific to pharmaceutical chemical space and were shown to have broader applicability domains and improved predictive accuracy for the assessment of mutagenic potential of drug impurities as defined in the ICH M7 guideline.

5.3.4 Expansion of Training Data Sets to Enhance the Predictive Power of (Q)SAR Tools

Honma et al. [3] described a landmark study in which an external validation of the commercial *in silico* (Q)SAR platforms was conducted using a proprietary data set of 12 140 chemical substances from government databases and the subsequent integration of this novel data set to increase the predictive performance and applicability domains. After extensive curation of the proprietary data sets, the bacterial mutagenicity database was assessed in various phases with the data being released and incorporated into the respective (Q)SAR models (in order to improve their performance). The authors concluded that the predictive power of all the (Q)SAR models improved over the course of the study, as compared with the original QSAR models, indicating that this project successfully fulfilled the principal aim of enhancing (Q)SAR performance. (One such example was outlined in additional detail by Chakravarti and Saiakhov [24] who used this specific data from Honma et al. [3] set alongside an equivalent data set from public databases to develop a novel method to generate molecular fingerprints from the structural environment of mutagenicity alerts that enabled the incorporation of activating or deactivating features on a series of chemical classes including primary aromatic amines [PAAs], aromatic nitros, epoxide, and alkyl chlorides all of which are of particular relevance in the ICH M7 setting). Despite these advances, Honma et al. [3] ended on a more sobering note that indicated the biggest challenge in this exercise was not necessarily the modeling but the creation and curation of the underlying database (recognition that certain chemicals were almost certainly incorrectly classified due to nonstandard laboratory practices, reactions between the test compound and vehicle [25], and impurities in the test sample among other factors) and that further benchmark data sets of reliable bacterial mutagenicity data needed to be utilized to develop more accurate *in silico* (Q)SAR models. An observation that has been made by others and that subsequently informed upon more focused data sharing initiatives which is discussed in the following section.

5.3.5 Focused Data Sharing Initiatives on Specific Chemical Classes

Prior to the introduction of ICH M7, there had been several pivotal publications to increase the transparency of the available bacterial mutagenicity data in order to enhance the development of *in silico* (Q)SAR platforms [1, 20]. Pre-competitive collaborations increased the number of focused studies on specific chemical classes, resulting in a rapid increase in publications related to the collaborative improvement and refinement of these *in silico* (Q)SAR approaches. These initiatives were essentially aimed at ensuring the available *in silico* (Q)SAR mutagenicity models had high sensitivity, high specificity, and broad applicability domains that reduce the number of false-negative predictions. This led to a series of improvements and refinements in the various (Q)SAR models that strengthened the *in silico* (Q)SAR led impurity mutagenicity testing framework that is now embedded throughout drug substance and drug product manufacturing processes in the pharmaceutical industry. The numerous cross-industry initiatives [18] can be broadly categorized into three areas: (i) Understanding *in vitro* mechanisms leading to mutagenicity – defining and understanding the mechanisms of mutagenicity that can occur in the standard *in vitro* test systems, (ii) Shared data, Shared progress – the intentional rapid focused sharing of proprietary and nonproprietary mutagenicity testing data for specific chemical classes, and (iii) Novel data mining approaches of proprietary mutagenicity testing data all of which resulted in improvements and refinements of the various (Q)SAR platforms and a robust framework for their application in drug development. At the center of all of these initiatives was a drive to get access to high-quality benchmark data sets of reliable bacterial mutagenicity data for model development.

5.3.5.1 Understanding *In Vitro* Mechanisms Leading to Mutagenicity

Regarding access to chemistry expertise, the inevitable interactions between chemists and toxicologists following the introduction of ICH M7 guideline led to some simple but significant improvements in *in silico* (Q)SAR approaches. A clear example is how the mechanistic understanding of the chemical reactivity of the acyl and sulfonyl chlorides led to a modification to the (Q)SAR predictions for this chemical class [25]. Acyl/sulfonyl halides are very reactive compounds often used as key intermediates in the synthesis of many other important organic compounds, e.g. esters, amides, aromatic ketones (Friedel–Crafts acylation), etc. Routine mutagenicity testing of chemical intermediates for occupational toxicology and worker safety purposes led to the generation of a database of 50+ acyl and sulfonyl chlorides (much of which was in the public domain) which appeared to be representing a compelling case for the development of robust (Q)SAR models in commercial (Q)SAR systems relating to mutagenicity for this chemical class. Amberg et al. [25] conducted a critical review of the data that highlighted that the design of the bacterial mutagenicity assay (Ames Test) and specifically the routine use of DMSO (dimethyl sulfoxide) as a vehicle in the assay (due to its effective solubilizing properties) had compromised the integrity of this data set. This is because chemists have known for a considerable time (>50 years) that acyl and sulfonyl halides react with DMSO via the Pummerer rearrangement reaction to form alkyl halides, i.e. halodimethylsulfides such as chlorodimethyl sulfide (CDMS), which are reported to be strongly mutagenic [26, 27]. This knowledge was unavailable to toxicologists conducting the mutagenicity testing and it would subsequently render the majority of the mutagenicity data for 50+ acyl and sulfonyl chlorides redundant as the compounds in the public domain that had been shown to be mutagenic had been tested using DMSO as the test vehicle – in direct contradiction of the OECD (Organization for Economic Co-operation and Development) guideline for the bacterial mutagenicity assay which states the vehicle should not be suspected of chemical reaction with the test substance. Subsequent retesting of multiple members of this chemical class in alternative vehicles (e.g. acetone, ethanol, water, etc.) provided a very different picture of the mutagenic potential of acyl and sulfonyl chlorides. The consortium collected and re-examined both internally generated mutagenicity data, as well as external published mutagenicity test data. Where appropriate, the consortium retested the appropriate compounds in either the five-strain bacterial mutagenicity test (GSK), or the screening Ames II assay (Sanofi-Aventis), using alternate solvents to DMSO. Interestingly, methane sulfonyl chloride (MsCl) was Ames positive in DMSO, acetonitrile, acetone, and methanol. This could be rationalized due to reaction of MsCl with all of these solvents, forming mutagenic species, i.e. CDMS with DMSO, IPS (isopropyl methane sulfonate) with acetone, MMS (methyl methane sulfonate) with methanol, and MSN (methyl sulfonyl nitrilium) with acetonitrile. A combination of chemistry expertise, a focused literature review of the mutagenicity testing data, and a series of experimental studies provided strong evidence that formation of the mutagen CDMS in DMSO from the test materials which had led to false-positive findings in the mutagenicity tests. The result of this collaborative effort between Sanofi-Aventis and GSK ultimately led to a revision of the available commercial (Q)SAR systems models with respect to the prediction of the mutagenicity of acyl and sulfonyl chlorides [21].

5.3.5.2 Shared Data, Shared Progress

5.3.5.2.1 Boronic Acids

The boronic acid chemical class serves as an example of how (Q)SAR tools evolved to address new challenges primarily as a result of a global network of genetic-toxicologists sharing precompetitive data and knowledge and national conferences to solve an emerging issue. Palladium (or platinum)-catalyzed coupling reactions, e.g. the Suzuki–Miyuara reaction, involving boronic acids or esters (referred to as boronic acids) are key intermediates in the preparation of many active pharmaceutical ingredients (APIs) [28]. Boronic acids can also be used in rhodium-catalyzed C–C coupling reactions [29–31]. The potential mutagenicity of arylboronic acids was first discussed in the public domain during 2010. O'Donovan et al. [32] indicated that, "Boronic acids as a class do not trigger alerts in any publicly available quantitative structure activity relationship ((Q)SAR) program and there was no a priori reason to them." Ellis et al. [33] had implemented data mining of Lhasa's Vitic database [34] containing both public and private, proprietary toxicity information on an extensive number of chemicals (particularly synthetic intermediates). They showed that 3-fluorobenzeneboronic acid was mutagenic in *Salmonella typhimurium* strains TA97a and TA100, and *Escherichia coli* strain WP2uvrA with and without metabolic activation (±S9). They subsequently tested several aryl boronic acids and demonstrated that they were Ames positive. O'Donovan et al. [32] also highlighted a similar finding, showing positive Ames test results on a boronic acid; the test was required for Occupational Safety purposes. The same authors subsequently found that 12/13 of the boronic acids that they tested were positive in a 5-strain Ames test without metabolic activation [32]. Their mutagenic activity (in 12/13 cases) was restricted to a single strain, either TA100 and/or WP2uvrA(pKM101) strains and the mutagenic response in all cases was relatively weak. In the two compounds that were mutagenic in both TA100 and WP2uvrA(pKM101) strains, there was no evidence of DNA-adduct formation using ^{32}P-postlabeling. Based on these findings, the authors speculated that this class of compound represented a novel class of mutagens that did not appear to be active via direct covalent binding to DNA. The authors concluded that the, "mechanism of action remains to be elucidated and it cannot yet be determined whether or not they represent a real genotoxic hazard." Pellizzaro et al. [35] identified that there are electronic and steric factors related to the bacterial mutagenicity of arylboronic compounds, which can be assessed using ^{11}B NMR chemical shifts with a predicted accuracy of 86%. Ciaravino et al. [36] highlighted that four benzoxaboroles and one boronic acid ester produced negative results in the Ames assay, chromosomal aberration, and *in vivo* micronucleus study (rat bone marrow). Furthermore, one of the benzoxaboroles has been studied in mouse and rat two-year bioassays and was not found to have any carcinogenic potential. These results demonstrate that it is possible to design boron-based therapeutic agents with no *in vitro* or *in vivo* mutagenic liabilities. Ames data on "44 commercially available boronic acids, boronic acid derivatives and boron containing reagents" were published by Hansen et al. [37]. The authors discussed trends in the Ames data from a (Q)SAR perspective. The authors postulated that the mechanism of mutagenicity could be oxidative in nature leading to radical-mediated DNA damage. They indicated that "currently available *in silico* computational models were found to provide little value in predicting the Ames assay for boronic acids and derivatives." The authors did make a plea for the publication of additional Ames data that "will lead to improved (Q)SAR models and an understanding of the mechanism of mutagenicity." To address the need to rapidly enhance the existing boronic acid data set in Derek Nexus, a consortium consisting of eight pharmaceutical companies was rapidly established by Lhasa. This resulted in a rapid increase in the data set and enabled testing to be targeted toward specific areas of chemical space where there was poor coverage [38]. This initiative ensured that a larger data set comprising of 90 arylboronic acids and derivatives was rapidly generated, resulting in refinement of the alert in Derek Nexus; this was despite the continuing lack of any clear mechanistic basis for boronic acid mutagenicity [38]. In addition to Ames test data, the consortia agreed to further evaluate the potential mechanism of action through the conduct of other *in vitro* genotoxicity studies "using the data generated in mammalian *in vitro* systems (rodents or human): mutagenicity (mouse lymphoma assay), clastogenicity (chromosome aberration assay, micronucleus test, comet assay) to increase knowledge/understanding." [18] Finally, eight aryl boronic acid/ester compounds that were mutagenic in the Ames test were shown to be non-mutagenic in follow-up *in vivo* studies, i.e. Pig-A, comet assays, and micronucleus test. All orally administered aryl boronic acids show high systemic exposure [38]. The authors suggested that structurally similar aryl boronic acids may be considered non-mutagenic and managed as per ICH Q3A/Q3B guidelines.

In summary, it appears that the majority of aryl boronic acids are *in vitro* mutagens, but that based on a very limited data set, some/all of these are *in vivo* non-mutagenic. Pragmatically, it would seem to be logical to continue to use the (Q)SAR alert for arylboronic acid or derivative (i.e. Derek Nexus Alert – 746) and develop sensitive (but not necessarily specific) ICP-MS methods [39] to monitor and control this class of compounds during early-stage development. After which, the organization can ask itself the key question: Do we proceed with continuing to monitor and control these boronic acid(s) impurities in drug substance of product or do we need to discharge the risk? If it is the latter, then a suite of *in vivo* mutagenicity tests can be applied to demonstrate that these boronic acid(s) are indeed not *in vivo* mutagens.

5.3.5.2.2 Primary Aromatic Amines (PAAs)

In addition to this, there has been a concerted industry-wide effort to share proprietary data to improve the (Q)SAR approaches for predicting the mutagenicity of PAAs. PAAs are often present in starting materials used to synthesize pharmaceuticals and hence may be present as drug impurities [40]. In addition, potential drug metabolites may contain PAAs, especially for drugs containing an aromatic amide/sulphonamide. A number of publications have indicated that PAAs are an alert for DNA-reactive mutagenicity [2, 13, 41–46]. This alert is based on both a statistical association between chemicals containing PAAs and mutagenicity as well as plausible mechanistic basis for the alert. This mechanistic basis is based on a series of steps including cytochrome P450-mediated metabolic activation to a hydroxylamine and eventual formation of a nitrenium ion that forms covalent adducts with DNA, see Figure 5.3.

In addition, other pathways have been proposed [45]. There is a statistical association and a mechanistic rationale that supports the relationship between for the PAAs chemical class and mutagenicity, i.e. approximately 60% of PAAs are reported to be mutagenic [1] and approximately 70% of mutagenic PAAs have been shown to be carcinogenic in rodents [47]. The prediction of the mutagenicity of aromatic amine class has been notoriously challenging given their diversity of physiochemical properties and the factors that can influence the multistep mechanism of mutagenicity (i.e. steric properties that facilitate or impede metabolism to the DNA-reactive nitrenium ion and the stability of the ultimate mutagen, i.e. the nitrenium ion itself). Despite these challenges, given their ubiquitous presence in pharmaceutical molecular design and their subsequent occurrence as impurities in manufacturing processes and as degradants in drug products, there have been several initiatives to improve the *in silico* (Q)SAR models for this chemical class.

Patel et al. [48] described a collaboration initiated by the Royal Society of Chemistry (RSC) that highlighted the challenges related to modeling the molecular properties of aromatic amines that contribute to their propensity for mutagenicity (involving their metabolism and subsequent DNA reactivity) and those associated with compilation of a consistent data set given the inherent variability in the standard protocols for the Ames bacterial mutagenicity assay. Specifically, a cross-pharmaceutical collaborative project managed by Lhasa was started in 2011 with the objective of creating a database of

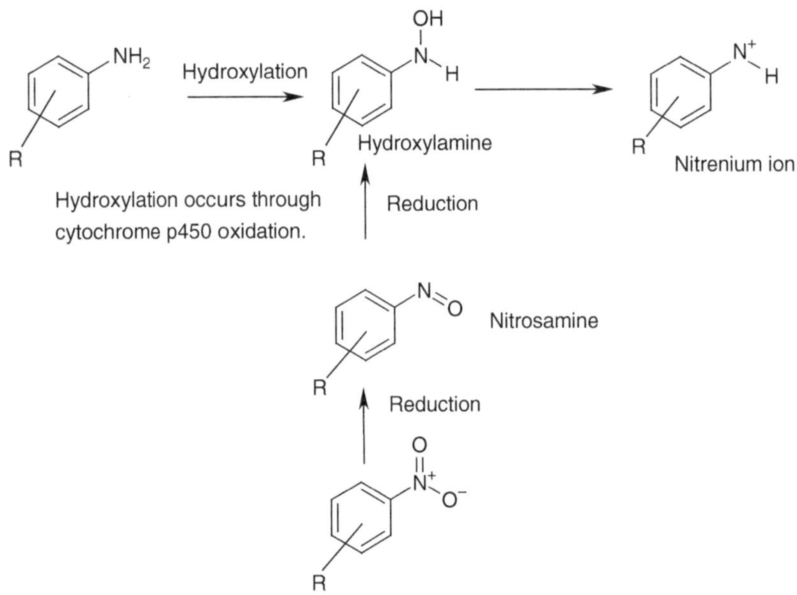

Figure 5.3 Metabolic activation of aromatic amines/aromatic nitro compounds to generate a nitrenium ion.

PAA mutagenicity data that could be used to support regulatory submissions and to augment the performance of these (Q)SAR predictive models [18]. The resulting Consortium for the Investigation of Genotoxicity of Aromatic Amines (CIGAA) looks to enhance both the understanding and predictability of the Ames test for PAAs. This precompetitive data sharing group consisted of 13 pharmaceutical companies who agreed to share data for drug intermediates and precursors, with the aim of constructing a database facilitating the development of improved predictive (Q)SAR tools. The consortium used four (Q)SAR tools in total; three were commercially available and the fourth was a quantum mechanics tool. The current CIGAA database that was published consisted of headline Ames data for 300 PAAs and in total, data for over 650 compounds have now been donated to the consortium [48]. The need for database curation was also highlighted. For example two of the compounds were direct acting mutagens (i.e. not indicative of the metabolism-mediated mutagenic PAA mechanism of action that forms the basis of the (Q)SAR predictions) and furthermore several closely related analogs were reported to be non-mutagenic in the literature. For example, 2′-aminoacetophenone was Ames positive (but predicted to be Ames negative), while its closely related analog, 4′-aminoacetophenone was Ames negative and predicted to be Ames negative. When the Ames-positive data were re-examined, it was found to be positive without metabolic activation in strains TA98/TA100; whereas as described above, PAAs require metabolic activation to induce nitrenium ion formation. Finally, a repeat 5-strain Ames test for 2′-aminoacetophenone was found to be negative across all five strains; supporting the conjecture that the original data was flawed; possibly caused by impurities present in the original sample (for instance, PAAs are often synthesized by catalytic reduction of the corresponding aromatic nitro compounds [49] and aromatic nitro compounds have a higher incidence of mutagenicity than the corresponding PAA). Therefore, (Q)SAR tools with high predictivity can be used to help better curate the test library and highlight those Ames test data that require re-evaluation; which in turn would improve predictivity [3, 49]. Impurities in test materials that result in inaccurate or ambiguous mutagenicity testing data in the training set bedevils the development, improvement, and refinement of silico (Q)SAR models. Honma et al. [3] recently reported on two aromatic amines predicted to be positive by most of the (Q)SAR models, i.e. 17 in total, used in their challenge project, but which were actually recorded as Ames negative. However, on closer examination of the Ames data, both amines showed weakly positive outcomes with metabolic activation in TA98 and TA100 strains. Both PAAs have constituents which are known to enhance mutagenicity (compound 1 is a 4-amino-3-methoxy derivative; whereas, compound 2 is a 2,4-diamino derivative). In summary, the authors concluded that this chemical class remains a challenge in terms of mutagenicity prediction (given that minor structural modifications can substantially influence potential mutagenic activity) and has continued the RSC collaboration to drive data sharing for this chemical class. They are also progressing further model refinement and the use of consensus (Q)SAR approaches to enable consistent predictions of mutagenicity for this important chemical class which is ubiquitous in drug discovery settings.

5.3.6 Novel Data Mining Approaches

5.3.6.1 Case Study: Primary Aromatic Amines (PAAs)

Ahlberg et al. [50] also focused on improving the performance of (Q)SAR approaches for predicting the mutagenicity of PAAs by piloting a novel data mining approach using an "SAR fingerprint" that could be used to extract knowledge from proprietary databases without releasing the chemical structures. The SAR fingerprint approach allowed the authors to predefine a series of substructure search queries (that focused on the relationships between the aromatic amine and other functional groups e.g. ortho, meta, and para substituted anilines and polycyclic and heteroaromatic amines) that were ultimately used to understand factors that increased or decreased the potential mutagenicity of aromatic amine compounds. These searches were then performed across several public and proprietary databases with bacterial mutagenicity data. The number of mutagenic and non-mutagenic examples that match each substructure were identified in each database. These results were then summed across the different collections to generate a total number of mutagenic and non-mutagenic examples corresponding to each substructure search. Although the analysis was performed on a series of proprietary data, no confidential information was shared, only numerical counts of the number of mutagenic and non-mutagenic outcomes matching each substructure. These numerical counts were used to identify subclasses of PAAs which appear to activate or deactivate DNA-reactive mutagenicity based on a statistical analysis. These rules have been encoded into expert alert systems, such as the Leadscope genetic toxicity expert alerts, and may be used to support an expert review of the (Q)SAR results for PAAs, see Figure 5.4. Figures 5.5 and 5.6 summarize some classes that activate PAA mutagenicity and Figure 5.7 illustrates a number of chemical classes that deactivate PAA mutagenicity.

The analysis ultimately demonstrated that anonymized meta data derived from proprietary data sets could be used to improve the prediction of aromatic amine mutagenicity and expand the applicability domain of (Q)SAR models. As a consequence, the use of the SAR fingerprint approach to improve (Q)SAR predictions is now being deployed on a regular basis across industry and is also being used for the rational design of training data sets for (Q)SAR platform development to enhance the prediction of aromatic amines and other chemical classes.

5.3.6.2 Case Study: Aromatic *N*-oxides

A further example of this is highlighted by Amberg et al. [51] who extended the "SAR fingerprint" approach to explore the robustness of the aromatic *N*-oxide structural alert for mutagenicity. Pharmaceutical products may contain aromatic *N*-oxide functional group or may degrade or be metabolized to a chemical containing aromatic *N*-oxides [52, 53]. Any degraded or metabolized product may present challenges in terms of acquiring sufficient quality of the test article, for example, it may not be possible to easily synthesize these chemicals. In addition, reagents may contain aromatic *N*-oxides and may be present as an impurity [54]. As such, it was important to understand

E. Ahlberg et al. / Regulatory Toxicology and Pharmacology 77 (2016) 1–12

At least one activating group				At least one deactiving group				No known activating or deactivating group			
Functional group	ortho	meta	para	Functional group	ortho	meta	para	Functional group	ortho	meta	para
amino(NH2)-	SA	SA	SA	sulfonate-	SD	SD	SD	oxymethyl-	-	-	-
methyl-	SA	SA	SA	sulfonyl-	SD	SD	SD	(2-aminoethyl)-	?	-	-
methoxy-	SA	WA	SA	sulfonamide-	WD	SD	SD	arylamino-	?	-	-
alkylthio-	SA	?	SA	aminocarbonyl-	SD	WD	SD	mercapto-	-	?	?
trifluoromethyl-	SD	SD	SA	aminomethyl-	SD	SD	WD	acetyl-	?	-	?
phenyl-	WA	-	SA	trifluoromethyl-	SD	SD	SA	N-iminomethyl-	?	?	-
heteroaryl-	?	-	SA	carboxylate-	SD	-	SD				
fluoro-	WA	SD	WA	(2-oxyethyl)-	WD	-	WD				
alkenyl-	?	WA	SA	(2-oxyethyl)-	-	SD	WD				
acetamido-	?	SA	SA	carboxylic acid-	-	WD	SD				
3-oxopropyl-	?	SA	-	hydroxymethyl-	?	?	WD				
carbonylamino-	-	WA	-	fluoro-	WA	SD	WA				
hydroxy-	-	SA	WD	chloro-	-	WD	SD				
phenoxy-	-	-	SA	bromo-	-	WD	WD				
aryl thiol-	?	?	SA	iodo-	SD	?	-				
alkoxy-	-	-	WA	nitrile-	SD	-	-				
heteroiminomethyl-	?	?	WA	tert-butyl-	?	WD	?				
((aminophenyl)(phenyl)methanone-	WD	WA	?	trifluoromethoxy-	?	?	WD				
alkylamino-	WD	-	WA	hydroxy-	-	SA	WD				
				alkylamino-	WD	-	WA				
				((aminophenyl)(phenyl)methanone-	WD	WA	?				

- SA Strong activating
- WA Weak activating
- \- No assignment
- WD Weak deactivating
- SD Strong deactivating
- ? Not enough data

Figure 5.4 Substitution pattern analysis illustrating activating and deactivating aniline substitution patterns. (*See insert for color representation of this figure*).

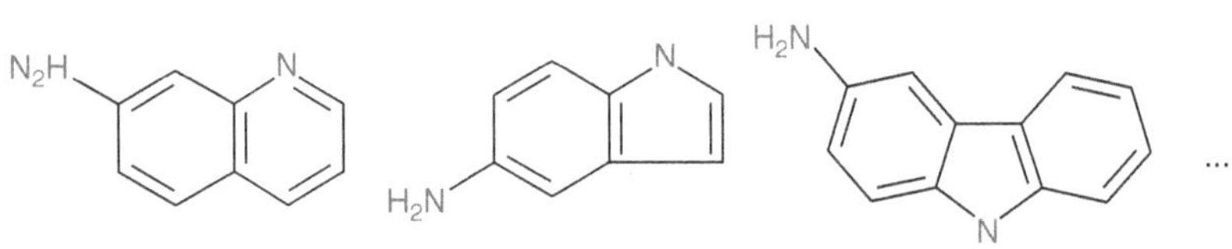

Figure 5.5 Examples of strong activating primary aromatic amines (fused rings).

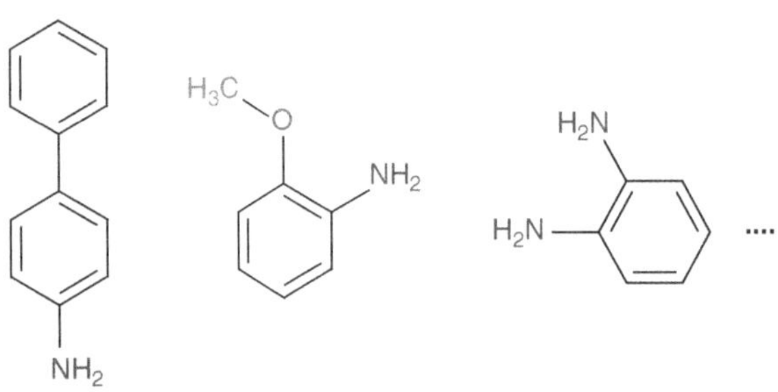

Figure 5.6 Examples of strong activating primary aromatic amines (anilines).

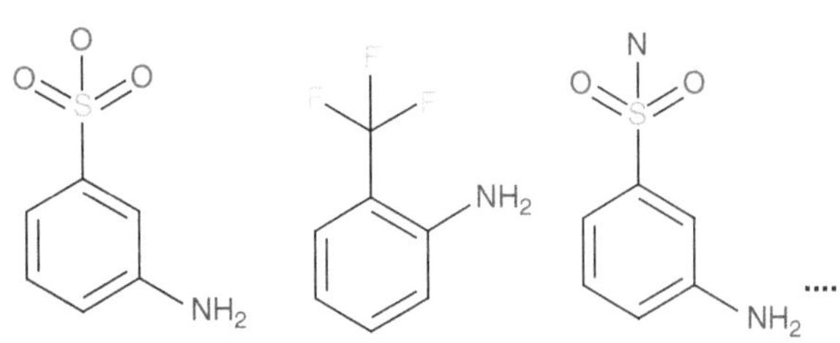

Figure 5.7 Example of strong deactivating primary aromatic amines (anilines).

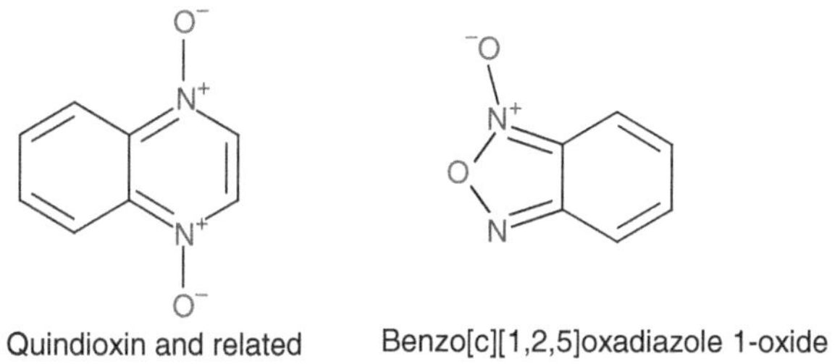

Figure 5.8 Quindioxin and related compounds, and benzo[c][1,2,5]oxadiazole 1-oxide.

DNA-reactive mutagenicity for this class of chemicals to support an assessment of impurities. The seminal Ashby–Tennant paper on "Chemical structure, Salmonella mutagenicity and extent of carcinogenicity as indicators of genotoxic carcinogenesis among 222 chemicals tested in rodents by the U.S. NCI/NTP" included aromatic *N*-oxides as one of the 18 original alerting substructures. However, this general alert has been questioned for lack of empirical and supporting mechanistic explanation [55]. Amberg et al. [51] performed a SAR-fingerprinting exercise to determine whether aromatic *N*-oxides or any subset should be classified as an alert for DNA-reactive mutagenicity. In a similar manner to the PAA exercise, a series of predefined substructures containing aromatic *N*-oxides were searched over a number of public and proprietary databases. The number of mutagenic and non-mutagenic examples matching each substructure search was determined and these results (alongside a thorough analysis of the public data) were used to make the following conclusions:

Two subclasses of aromatic *N*-oxides – (i) quindioxin and related compounds, as well as (ii) benzo[c][1,2,5]oxadiazole 1-oxide, see Figure 5.8, were considered alerts based on the public and proprietary data analysis. However, the general class of aromatic *N*-oxides is not an alert since (i) there was not a sufficient statistical association, (ii) there was a lack of mechanistic rationale for the alert, (iii) all the mutagenic public examples contained another structural alert (including the two subclasses identified earlier) that are more likely responsible for the mutagenicity, and (iv) there was a number of non-mutagenic diverse chemicals containing aromatic *N*-oxides [51].

5.4 Expert Knowledge

In parallel with the series of general improvements and refinements made to *in silico* (Q)SAR models, a series of additional publications set out to define the practical aspects of applying *in silico* tools in industrial practice akin to good laboratory practice [56, 57] as well as more detailed examination of the set of principles and procedures to examine the relevance of any positive, negative, conflicting, or inconclusive (Q)SAR prediction. Myatt et al. [56] described considerations related to the selection of expert rule-based and statistical-based (Q)SAR models based on their performance, applicability domain, and model complementarity in relation to the specific use case, e.g. high sensitivity vs. high specificity. This was referred to as "good *in silico* practice". Myatt et al. [56]described how in addition to OECD-compliant (Q)SAR models there was a requirement for reproducible, transparent, and standardized procedures that enabled (Q)SAR results to be reviewed rapidly and thoroughly by, for example, regulatory agencies. Myatt et al. [56] described a process by which a user could select, assess, and integrate *in silico* predictions alongside experimental data for defined toxicological effects or mechanisms that would improve the reliability and confidence of the overall *in silico* and *in vitro*/*in vivo* toxicological assessments.

Hasselgren et al. [57] further defined the practical aspects of model application and described how *in silico* (Q)SAR approaches could also inform the user as to the robustness of the reported *in vitro* mutagenicity and/or *in vivo* genotoxicity data. Specifically, Hasselgren et al. [57] described how a combination of robust complementary (expert rule-based and statistical-based) *in silico* (Q)SAR predictions and expert review could be used to augment partial or incomplete *in vitro* mutagenicity data sets (with a weak Klimisch[1] scores) or to challenge a positive *in vitro* mutagenicity finding (with a strong Klimisch score) without a (Q)SAR-based rationale for the reported mutagenicity. Using a series of case studies the combined consideration of *in silico* (Q)SAR approaches and *in vitro* mutagenicity data produced a "reliability score" that provided a consistent decision framework that contributed to a "weight of evidence" that facilitated robust interpretation of nonproprietary and proprietary mutagenicity test data.

Amberg et al. [12] referred to a presentation by Powley [58] that demonstrated the scale of the problem by analyzing the frequency of out-of-domain and indeterminate predictions in routine regulatory submissions (c. 18% over a two-year period). A review of these submissions indicated a variety of approaches had been taken to address the uncertainty around the "absence of structural alerts" categorisation including assuming the impurity was potentially mutagenic and controlling it in drug product accordingly, conducting mutagenicity testing of the said impurity, application of expert knowledge to understand the *in silico* (Q)SAR predictions and the use of a third (Q)SAR system to clarify the (Q)SAR prediction. Clearly, simply assuming that impurities with either out of domain and/or indeterminate predictions are presumptive "positive" predictions, and hence potential mutagens, would have significant impact on the operational aspects of ICH M7 compliance if applied across the drug development process. Therefore, the application of an "expert review" (the value of which had been demonstrated previously [59–62] to aid the interpretation of positive, negative, and indeterminate results of the (Q)SAR predictions (expert rule-based and statistical-based), based on the analysis of underlying training set and/or database analogs (including the associated experimental data), has become a routine part of ICH M7 *in silico* (Q)SAR model interpretation.

Amberg et al. [63] provided some guiding principles and procedures and a series of case studies to manage out-of-domain and indeterminate (Q)SAR predictions that occur as part of the ICH M7-recommended (Q)SAR analyses. (i) The use of structurally and toxicologically meaningful analogs from additional databases. For example, comparisons of cyclohexyldiphenylphosphine oxide and cyclohexyldiphenylphosphine that were out-of-domain in conventional (Q)SAR systems, with the closely related structures triphenylphosphine and triphenylphosphine oxide that were Ames negative, (ii) Assessment of nonreactive groups, for example, the addition of protective groups to facilitate synthetic chemistry processes, such as tertbutyloxycarbonyl (BOC) which show a lack of potential mutagenic, but which may place the impurity out-of-domain of the *in silico* (Q)SAR tools), where their presence within an impurity may change the prediction from negative to out-of-domain, (iii) A critical review of the relevance of the data sets contributing to indeterminate (Q)SAR predictions. (iv) The use of a third complementary (Q)SAR models based on proprietary data to increase the applicability domains of the overall (Q)SAR evaluation [21].

The assessment by Amberg et al. [63] of the overarching data set that summarized the likelihood of misclassifying a mutagenic impurity as non-mutagenic based on the different combinations of *in silico* (Q)SAR predictions (e.g. where one prediction was either out-of-domain or indeterminate

1 The Klimisch score is a method of assessing the reliability of toxicological studies, mainly for regulatory purposes, that was proposed by H.J. Klimisch, M. Andreae, and U. Tillmann of the chemical company BASF in 1997. Generally, only Klimisch scores of 1 or 2 can be used by themselves to cover an endpoint. However, Klimisch score 3 and 4 data can still be used as supporting studies or as part of a weight of evidence approach.

as compared with two negative predictions) clearly demonstrated the value of consistent expert review. However, they concluded that ultimately the performance of the (Q)SAR models would continue to improve only as a result of the incorporation of data and knowledge from proprietary databases and collaborations, which would in turn decrease the frequency of out-of-domain and indeterminate results.

5.5 Future Direction

Clearly the introduction of the ICH M7 guideline and the adoption of an *in silico* (Q)SAR framework that could be used *in lieu* of actual mutagenicity studies for hazard identification and risk assessment resulted in major advances in the predictive performance of the two main types of *in silico* (Q)SAR platforms used within the pharmaceutical industry. These advances were the result of a combination of innovative modeling approaches and increased access to high-quality curated proprietary mutagenicity databases. From an operational perspective, the orientation of the *in silico* (Q)SAR models toward a high sensitivity (to reduce the number of false negatives) inevitably led to an increase in the number of false-positive predictions which was addressed via the introduction of an additional review process using "context-dependent expert knowledge," the value of which has been clearly demonstrated in multiple publications. However, ideally iterative improvements to the *in silico* (Q)SAR platforms would continue to codify this "expert knowledge" that would eventually decrease still further the requirement for human intervention in the interpretation of the in-silico predictions.

Interestingly the "expert-knowledge" within the pharmaceutical industry has indirectly addressed some of these issues around out-of-domain and indeterminate predictions as well as the nuances around human interpretation of positive predictions, by moving some of the more challenging *in silico* (Q)SAR scenarios earlier in development or even into discovery. By deploying the *in silico* (Q)SAR platforms much earlier in the molecular design stage of discovery (to assess metabolic and degradation products), false-positive (Q)SAR predictions can be readily circumvented by deprioritizing molecules in the discovery stage, *in vitro* bacterial mutagenicity testing at a stage that does not have significant development costs (e.g. large-scale chemical synthesis campaigns) and in addition of out-of-domain molecules can be tested for bacterial mutagenicity and the results incorporated into the (Q)SAR databases. Many pharmaceutical companies have adopted the *in silico* (Q)SAR approaches defined in ICH M7 to assess the potential mutagenicity of potential impurities, degradants, and importantly associated metabolites, of early candidates within the medicinal chemistry stages of drug discovery. Several publications discuss how these approaches can be valuable to address the potential mutagenicity of PAAs that could potentially form from the parent drug molecule via either metabolism (i.e. oxidation or hydrolysis) or via physiological degradation (i.e. acid-catalyzed hydrolysis) of amide, urea, or sulphonamide bonds [18, 64].

The medicinal chemistry approach is typically to either remove the "embedded" PAA for other structural subunits known to be safer or "by exhaustively profiling of the putative aromatic amine metabolites" [64, 65]. A recent report assessed the challenge of circumventing the formation of an "embedded" 4-aminobiphenyl moiety, which was optimal for desired pharmacological activity, but 4-aminobiphenyls are usually strongly mutagenic. Hamann et al. [66] found trace levels of a PAA metabolite generated via hydrolytic metabolism. This aniline fragment was shown to be mutagenic in an Ames assay. Subsequent concomitant optimization for target activity and avoidance of mutagenicity from the PAA fragment led to the discovery of a pharmacologically improved molecule without any mutagenic potential. As described earlier, PAAs are believed to generate mutagens after *N*-hydroxylation, to yield an electrophilic species, i.e. nitrenium ion, which can react with

DNA. In the case of 4-aminobiphenyl, it was believed that the more planar the PAA structure the more resonance stabilized would be the resulting nitrenium ion. Using *in silico* approaches, Leach et al. [67] linked the potential mutagenic activity with the dissociation energy for the formation of the nitrenium ion from the corresponding hydroxylamine. A series of 142 4-aminobiphenyl molecules were assessed using quantum mechanical calculations to link their dissociation energies with structure. The additional drug-like properties of molecular weight and calculated Log P were also included in the in-silico calculations. 26/142 of the molecules were identified as being of interest. The authors found that adding a halogen, i.e. fluorine and particularly chlorine in the 3-position (i.e. meta), of the 4-amino ring, reduced mutagenicity in the Ames test (resulting in some Ames-negative molecules), presumably by forcing the two ring systems into a nonplanar conformation due to electronic and steric considerations. This allows the mutagenic liability to be designed out of the molecule using (Q)SAR approaches [64].

Zeller et al. [47] described the implementation of a discovery strategy known as "Aware, Avoid, Assess" that is deployed in many pharmaceutical companies at the early molecular design stage to address complications that can occur when there is an overlap in enzyme-mediated drug metabolism pathways with chemical degradation pathways. The deployment of the ICH M7-aligned *in silico* (Q)SAR and subsequent testing paradigm to address the potential formation of potential mutagenic PAA's from parent drug molecules allowed early resolution of complex protracted risk assessments that had previously been encountered late in development when human metabolism data are generated alongside long-term drug product stability data [47]. In the case studies presented, Zeller described the termination of new candidate drug molecule based on the "Aware, Avoid, Assess" approach which demonstrated the renewed impact of the *in silico* (Q)SAR approaches adopted for the ICH M7 framework in the drug discovery lead optimization setting.

While these pragmatic small changes to the drug discovery process can be very effective, removing the need for human intervention in the ICH M7 process during the later more expensive stages of development, they should be considered a parallel operational activity. Fortunately, there is still a continued drive to improve the sensitivity and specificity of the available *in silico* (Q)SAR mutagenicity models as commercial vendors will continue to innovate and drive predictive performance of the *in silico* (Q)SAR platforms in support of the ICH M7 impurities framework. One relatively recent development that may prove to be particularly impactful is the development of a federated machine learning approach that can enable continued *in silico* (Q)SAR model development from proprietary databases without the requirement to physically share proprietary *in vitro* mutagenicity data. Hanser et al. [68] described a machine learning approach that could be used to combine different sources of knowledge to build improved structure–activity relationship models and ultimately build a platform along with algorithms to facilitate knowledge discovery and produce interpretable and accurate predictions, i.e. SOHN. This approach has been demonstrated to improve the prediction of mutagenicity [22] and hERG (Human Ether-a-go-go-Related Gene) activity [69]. However, when combined with a novel knowledge transfer model (known as a student–teacher model), the approach is able to leverage the knowledge distributed across corporate data without the disclosure of confidential information. In short, the approach involves the development of multiple "teacher" models based on proprietary data from different institutions which are then used to "label" a thoroughly curated and well-defined public data set. These resulting labeled public data set is then used to develop a "student" model that leverages the combined knowledge of multiple institutions [70]. In short, the approach would allow unparalleled "learning" from proprietary *in vitro* mutagenicity data if done on a routine basis would ensure that *in silico* (Q)SAR platforms leverage all the available high-quality mutagenicity data that has been generated without incurring any issues around confidentiality.

Until very recently, one may have considered the continued development of the *in silico* (Q)SAR platforms for the prediction of mutagenicity that supports the ICH M7 process to be without merit. However, in 2019, following the detection of certain *N*-nitrosamine impurities in a series of established drug products, a series of regulatory recommendations were introduced related to the control of *N*-nitrosamine impurities in drug products (see Chapter 10). While the recommendations initially focused on the control of a series of well-established mutagenic and carcinogenic *N*-nitrosamines, the scope was extended to the entire *N*-nitrosamine chemical class (ranging from simple alkyl nitrosamines to complex nitrosated drug substances). In the latter case, i.e. *N*-nitroso valsartan, it is worth highlighting that these compounds are typically Ames negative – perhaps reflecting the influence of steric factors on the outcome of mutagenicity. This represented a particular challenge as somewhat surprisingly the *in silico* (Q)SAR models related to this chemical class were not well developed in the commercial platforms (a combination of their rare occurrence in the pharmaceutical industry and the disparate legacy *in vitro* mutagenicity data sets that include a variety of study designs). As a consequence, commercial (Q)SAR model developers have combined with industrial partners to study available aspects of *N*-nitrosamine mutagenicity to enable *in silico* (Q)SAR platform. One based on the established mechanism of action (mutagenicity being dependent on hydrolysis of the alpha-carbon to the amine to form the reactive diazonium metabolite) and the examination of a series of structural features that are reported to reduce or inhibit mutagenicity and carcinogenicity [71–74]. These activities have involved the conduct of additional mutagenicity studies with a series of *N*-nitrosamines at a rapid pace that are being used to progress the development of expert rule-based and statistical-based (Q)SAR prediction methodologies for this chemical class (see Chapters 6 and 10). These activities should establish whether novel untested *N*-nitrosamine impurities can be accurately classified as per the ICH M7 framework, i.e. class 3 (Structural alert – predicted to be mutagenic) or class 5 (no structural alerts – predicted to be non-mutagenic). In addition, the (Q)SAR platforms are likely to evolve again as regulators, industry and (Q)SAR vendors attempt to introduce "read across" approaches to define "class-specific acceptable intakes" for class 1 - mutagenic N-nitrosamines that reflect their predicted carcinogenic potential. This rapid response to understand the hazard and risk associated with the *N*-nitrosamine chemical class would not have been possible without the aforementioned precompetitive collaborations that enabled the development of the ICH M7 *in silico* (Q)SAR framework and will be used to support any future ICH M7 revisions regarding the control of *N*-nitrosamine impurities in drug products.

References

1 Kazius, J., McGuire, R., and Bursi, R. (2005). Derivation and validation of toxicophores for mutagenicity prediction. *J. Med. Chem. 48* (1): 312–320.

2 Benigni, R. and Bossa, C. (2011). Mechanisms of chemical carcinogenicity and mutagenicity: a review with implications for predictive toxicology. *Chem. Rev.* 111 (4): 2507–2536.

3 Honma, M., Kitazawa, A., Cayley, A. et al. (2019). Improvement of quantitative structure-activity relationship ((Q)SAR) tools for predicting Ames mutagenicity: outcomes of the Ames/(Q)SAR international challenge project. *Mutagenesis* 34: 3–16.

4 Benigni, R., Bassan, A., and Pavan, M. (2020). In silico models for genotoxicity and drug regulation. *Expert Opin. Drug Metab. Toxicol.* 16 (8): 651–662.

5 Benigni, R. and Cecilia Bossa, C. (2019). Data-based review of QSARs for predicting genotoxicity: the state of the art. *Mutagenesis* 34: 17–24.

6 Teasdale, A. (2010). *Genotoxic Impurities, Strategies for Identification and Control* (ed. A. Teasdale). Hoboken, NJ: Wiley.

7 Glowienke, S. and Hasselgren, C. (2010). Structure activity relationship (SAR) evaluation as a critical tool in the evaluation of the genotoxic potential of impurities. In: *Genotoxic Impurities, Strategies for Identification and Control* (ed. A. Teasdale), 97–120. Hoboken, NJ: Wiley.
8 Barber, C., Hanser, T., Judson, P. et al. (2017). Distinguishing between expert and statistical systems for application under ICH M7. *Regul. Toxicol. Pharmacol.* 84: 124–130.
9 Greene, N., Dobo, K.L., Kenyon, M.O. et al. (2015). A practical application of two in silico systems for identification of potentially mutagenic impurities. *Regul. Toxicol. Pharmacol.* 72 (2): 335–349.
10 Dobo, K.L., Greene, N., Fred, C. et al. (2012). In silico methods combined with expert knowledge rule out mutagenic potential of pharmaceutical impurities: an industry survey. *Regul. Toxicol. Pharmacol.* 62: 449–455.
11 Araya, S., Lovsin-Barle, E., and Glowienke, S. (2015). Mutagenicity assessment strategy for pharmaceutical intermediates to aid limit setting for occupational exposure. *Regul. Toxicol. Pharmacol.* 73 (2): 515–520.
12 Amberg, A., Beilke, L., Bercu, J. et al. (2016). Principles and procedures for implementation of ICH M7 recommended (Q)SAR analyses. *Regul. Toxicol. Pharmacol.* 77: 13–24.
13 Ashby, J. and Tennant, R.W. (1988). Chemical structure, Salmonella mutagenicity and extent of carcinogenicity as indicators of genotoxic carcinogenesis among 222 chemicals tested in rodents by the U.S. NCI/NTP. *Mutat. Res.* 204 (1): 17–115.
14 Worth, A., Fuart-Gatnik, M., Lapenna, S., et al. (2011). The use of computational methods in the toxicological assessment of chemicals in food: current status and future prospects. https://op.europa.eu/en/publication-detail/-/publication/49b38592-9a5c-4b98-aacf-4b66d71d8377/language-en (Accessed 06 October 2020).
15 Cariello, N.F., Wilson, J.D., Britt, B.H. et al. (2002). Comparison of the computer programs DEREK and TOPKAT to predict bacterial mutagenicity. Deductive Estimate of Risk from Existing Knowledge. Toxicity Prediction by Computer Assisted Technology. *Mutagenesis* 17: 321–329.
16 Snyder, R.D. and Smith, M.D. (2005). Computational prediction of genotoxicity: room for improvement. *Drug Discov. Today* 10: 1119–1124.
17 Votano, J.R., Parham, M., Hall, L.H. et al. (2004). Three new consensus QSAR models for the prediction of Ames genotoxicity. *Mutagenesis* 19 (5): 365–377.
18 Elder, D., White, A., Harvey, J. et al. (2015). Mutagenic impurities: precompetitive/competitive collaborative and data sharing initiatives. *Org. Process Res. Dev.* 19 (11): 1486–1494.
19 Marchant, C.A. (2012). Computational toxicology: a tool for all industries. *Wiley Interdiscip. Rev.: Comput. Mol. Sci.* 2: 424–434. https://doi.org/10.1002/wcms.100.
20 Hansen, K., Mika, S., Schroeter, T. et al. (2009). Benchmark data set for in silico prediction of Ames mutagenicity. *J. Chem. Inf. Model.* 49 (9): 2077–2081.
21 Landry, C., Kim, M.T., Kruhlak, N.L. et al. (2019). Transitioning to composite bacterial mutagenicity models in ICH M7 QSAR analyses. *Regul. Toxicol. Pharmacol.* 109: 104488.
22 Barber, C., Cayley, A., Hanser, T. et al. (2016). Evaluation of a statistics-based Ames mutagenicity QSAR model and interpretation of the results obtained. *Regul. Toxicol. Pharmacol.* 76: 7–20.
23 Williams, R.V., Amberg, A., Brigo, A. et al. (2016). It's difficult, but important, to make negative predictions. *Regul. Toxicol. Pharmacol.* 76: 79–86.
24 Chakravarti, S.K. and Saiakhov, R.D. (2019). Computing similarity between structural environments of mutagenicity alerts. *Mutagenesis* 34: 55–65.
25 Amberg, A., Harvey, J.S., Czich, A. et al. (2015). Do carboxylic/sulfonic acid halides really present a mutagenic and carcinogenic risk as impurities in final drug products? *Org. Process Res. Dev.* 19 (11): 1495–1506.
26 Boyle, R.E. (1966). The reaction of dimethyl sulfoxide and 5-dimethylaminonaphthalene-1-sulfonyl chloride. *J. Org. Chem.* 31: 3880–3882.

27 Thea, S. and Cevasco, G. (1988). A novel reaction of benzoyl chlorides in dimethyl sulfoxide. *J. Org. Chem.* 53: 4121–4122.

28 Hooshmand, S.E., Heidari, B., Sedghi, R., and Varma, R.S. (2019). Recent advances in the Suzuki–Miyaura cross-coupling reaction using efficient catalysts in eco-friendly media. *Green Chem.* 21: 381–405.

29 Nadeu, C., Aly, S., and Belyk, K. (2011). Rhodium-catalyzed enantioselective addition of boronic acids to *N*-benzylnicotinate salts. *J. Am. Chem. Soc.* 133 (9): 2878–2880.

30 Allsop, G.L., Cole, A.J., Giles, M.E. et al. (2009). A convergent synthesis of AR-C123196 utilising reaction between a cyclic carbonate and a phenol for aryl alkyl ether formation. *Org. Process Res. Dev.* 13: 751–759.

31 Brock, S., Hose, D.R.J., Mosely, J.D. et al. (2008). Development of an enantioselective, kilogram scale, rhodium catalyzed 1,4-addition. *Org. Process Res. Dev.* 12: 496–502.

32 O'Donovan, M.R., Mee, C.D., Fenner, S. et al. (2011). Boronic acids – a novel class of bacterial mutagen. *Mut. Res./Gentox. Toxicol. Environ. Mutagen.* 724: 1–6.

33 Ellis, K.M., Cheung, J., and Ackerman, J. (2010). Abstract No. 77. Aryl boronic acids: potentially mutagenic synthetic intermediates. *Mutagenesis*, 25(6), 657. In: Abstracts of the 33rd Annual Meeting of the United Kingdom Environmental Mutagen Society, 12–14 July 2010 at the University of Derby, UK. *Mutagenesis* 2010, 25(6), 631–658.

34 Lhasa. Vitic. A chemical database and information management system, offering rapid access to high-quality data from both published and proprietary sources. https://www.lhasalimited.org/products/vitic.htm# (Accessed 06 October 2020).

35 Pellizzaro, M.L., Covey-Crump, E.M., Fisher, J. et al. (2015). Investigating a relationship between the mutagenicity of aryl boronic acids and 11B NMR chemical shifts. *Chem. Res. Toxicol.* 28 (7): 1422–1426.

36 Ciaravino, V., Plattner, J., and Chanda, S. (2013). an assessment of the genetic toxicology of Novel Boron-containing therapeutic agents. *Environ. Mol. Mutagen.* 54 (5): 338–346.

37 Hansen, M.M., Jolly, R.A., and Linder, R.J. (2015). Boronic acids and derivatives – probing the structure-activity relationships for mutagenicity. *Org. Process Res. Dev.* 19 (11): 1507–1516.

38 Masuda-Herrera, M.J., Dobo, K.L., Kenyon, M.O. et al. (2019). *in vivo* mutagenicity testing of arylboronic acids and esters. *Environ. Mol. Mutagen.* 60 (9): 766–777.

39 Patel, I., Venkatramani, C.J., Stumpf, A. et al. (2017). Trace analysis of potentially mutagenic boronic acids and esters in drug substance by ICP-MS. *Org. Process Res. Dev.* 21 (2): 182–186.

40 McCarren, P., Springer, C., and Whitehead, L. (2011). An investigation into pharmaceutically relevant mutagenicity data and the influence on Ames predictive potential. *J. Chem.* 3: 51.

41 Bailey, A.B., Chanderbhan, R., Collazo-Braier, N. et al. (2005). The use of structure activity relationship analysis in the food contact notification program. *Regul. Toxicol. Pharmacol.* 42: 225–235.

42 Benigni, R., Bossa, C., Jeliazkova, N.G., Netzeva, T.I., and Worth, A.P. (2008). The Benigni/Bossa Rulebase for mutagenicity and carcinogenicity - a module of Toxtree. Technical Report EUR 23241 EN. European Commission – Joint Research Centre. http://toxtree.sourceforge.net/carc.html (Accessed 06 October 2020).

43 Enoch, P. and Cronin, M.T.D. (2010). A review of the electrophilic reaction chemistry involved in covalent DNA binding. *Crit. Rev. Toxicol.* 40: 728–748.

44 Enoch, S. and Cronin, M. (2012). Development of new structural alerts suitable for chemical category formation for assigning covalent and non-covalent mechanisms relevant to DNA binding. *Mutat. Res.* 743: 10–19.

45 Shamovsky, I., Ripa, L., Borjesson, L. et al. (2011). Explanation for main features of structure genotoxicity relationships of aromatic amines by theoretical studies of their activation pathways in CYP1A2. *J. Am. Chem. Soc.* 133: 16168–16185.

46 Galloway, S.M., Reddy, M.V., Mcgettigan, K. et al. (2013). Potentially mutagenic impurities: analysis of structural classes and carcinogenic potencies of chemical intermediates in pharmaceutical syntheses supports alternative methods to the default TTC for calculating safe levels of impurities. *Regul. Toxicol. Pharmacol.* 66: 326–335.

47 Zeller, A., Brigo, A., Brink, A. et al. (2020). Genotoxicity assessment of drug metabolites in the context of MIST and beyond. *Chem. Res. Toxicol.* 33 (1): 10–11.

48 Patel, M., Kranz, M., and Munoz-Muriedas, J. (2018). A pharma-wide approach to address the genotoxicity prediction of primary aromatic amines. *Comput.Toxicol.* 7: 27–35.

49 Orlandi, M., Brenna, D., Harms, R. et al. (2018). Recent developments in the reduction of aromatic and aliphatic nitro compounds to amines. *Org. Process Res. Dev.* 22 (4): 430–445.

50 Ahlberg, E., Amberg, A., Beilke, L.D. et al. (2016). Extending (Q)SARs to incorporate proprietary knowledge for regulatory purposes: a case study using aromatic amine mutagenicity. *Regul. Toxicol. Pharmacol.* 77: 1–12.

51 Amberg, A., Anger, L.T., Bercu, J. et al. (2019). Extending (Q)SARs to incorporate proprietary knowledge for regulatory purposes: is aromatic *N*-oxide a structural alert for predicting DNA-reactive mutagenicity? *Mutagenesis* 34 (1): 67–82.

52 Strege, M.A., Osborne, L.M., Hetrick, E.M. et al. (2015). Assessing the risk of formation of potential genotoxic degradation products in a small molecule kinase inhibitor drug substance and drug product. *Org. Process Res. Dev.* 19: 1458–1464.

53 Mfuh, A.M. and Larionov, O.V. (2015). Heterocyclic *N*-oxides – an emerging class of therapeutic agents. *Curr. Med. Chem.* 22: 2819–2857.

54 Chandrasekhar, S., Reddy, C.R., Rao, R.J., and Rao, J.M. (2002). Efficient and chemoselective deoxygenation of amine *N*-oxides using polymethylhydrosiloxane. *Synlett*: 0349–0351. https://doi.org/10.1055/s-2002-19751.

55 Snodin, D.J. and McCrossen, S.D. (2013). Mutagenic impurities in pharmaceuticals: a critique of the derivation of the cancer TTC (Threshold of Toxicological Concern) and recommendations for structural-class-based limits. *Regul. Toxicol. Pharmacol.* 67: 299–316.

56 Myatt, G.J., Ahlberg, E., Akahor, Y. et al. (2018). In silico toxicology protocols. *Regul. Toxicol. Pharmacol.* 96: 1–17.

57 Hasselgren, C., Ahlberg, E., Akahori, Y. et al. (2019). Genetic toxicology in silico protocol. *Regul. Toxicol. Pharmacol.* 107.

58 Powley, M. (2017). Presentation made at the 2017 Genetic Toxicity Association annual meeting "(Q)SAR Evaluation of Potentially Mutagenic Impurities: Regulatory Experience with Out of Domain Results."

59 Sutter, A., Amberg, A., Boyer, S. et al. (2013). Use of in silico systems and expert knowledge for structure-based assessment of potentially mutagenic impurities. *Regul. Toxicol. Pharmacol.* 67: 39–52.

60 Barber, C., Amberg, A., Custer, L. et al. (2015). Establishing best practise in the application of expert review of mutagenicity under ICH M7. *Regul. Toxicol. Pharmacol.* 73: 367–377.

61 Powley, M.W. (2015). QSAR assessments of potentially mutagenic impurities: a regulatory perspective on the utility of expert knowledge and data submission. *Regul. Toxicol. Pharmacol.* 71: 295–300.

62 Kruhlak, N. (2017). A regulatory perspective on the use of (Q)SAR to evaluate potential genotoxic impurities. In: *Presentation at American College of Toxicology (ACT) Annual Meeting*, November 2017, Palm Springs, CA.

63 Amberg, A., Andaya, R.V., Angera, L.T. et al. (2019). Principles and procedures for handling out-of-domain and indeterminate results as part of ICH M7 recommended (Q)SAR analyses *Regul. Toxicol. Pharmacol.* 102: 53–64.

64 Birch, A.M., Groombridge, S., Kaw, R. et al. (2012). Rationally designing safer anilines: the challenging case of 4-aminobiphenyls. *J. Med. Chem.* 55: 3923–3933.

65 Block, M.H., Boyer, S., Brailsford, W. et al. (2002). Discovery and optimization of a series pf carbazole ureas as NPY5 antagonists for the treatment of obesity. *J. Med. Chem.* 45: 3509–3523.

66 Hamann, L.G., Manfredi, M.C., Sun, C. et al. (2007). Tandem optimization of target activity and elimination of mutagenic potential in a potent series of *N*-aryl bicyclic hydantoin-based selective androgen receptor modulators. *Bioorg. Med. Chem. Lett.* 17 (7): 1860–1864.

67 Leach, A.G., Cann, R., and Tomasi, S. (2009). Reaction energies computed with density functional theory correspond with a whole organism effect: modelling the Ames test for mutagenicity. *Chem. Commun.*: 1094–1096.

68 Hanser, T., Barber, C., Rosser, E. et al. (2014). Self organising hypothesis networks: a new approach for representing and structuring SAR knowledge. *J. Chem.* 8 (6): 21.

69 Hanser, T., Steinmetz, F.P., Plante, J. et al. (2019). Avoiding hERG-liability in drug design via synergetic combinations of different (Q)SAR methodologies and data sources: a case study in an industrial setting. *J. Chem.* 2 (11): 9.

70 Johnston, L.A. and Hanser, T. (2020). Building secondary pharmacology models: a novel approach to proprietary data transfer – Recording Lhasa Limited. https://www.lhasalimited.org/publications/building-secondary-pharmacology-models-a-novel-approach-to-proprietary-data-transfer-recording/6967 (Accessed 06 October 2020).

71 Guttenplan, J.B. (1987). *N*-Nitrosamines bacterial mutagenesis and *in vitro* metabolism. *Mutat. Res.* 186: 81–134.

72 Helguera, A.M., González, M.P., Cordeiro, M.N.D.S. et al. (2007). Quantitative structure carcinogenicity relationship for detecting structural alerts in nitroso-compounds. *Toxicol. Appl. Pharmacol.* 221: 189–202.

73 Helguera, A.M., Cordeiro, M.N.D.S., Pérez, M.A.C. et al. (2008). Quantitative structure carcinogenicity relationship for detecting structural alerts in nitroso-compounds: species: rat; sex: male; route of administration: water. *Toxicol. Appl. Pharmacol.* 231: 197–207.

74 Helguera, A.M., Pérez-Machado, C., and Cordeiro, M.N.D.S. (2010). Quantitative structure-activity relationship modelling of the carcinogenic risk of nitroso compounds using regression analysis and the TOPS-MODE approach. *SAR QSAR Environ. Res.* 3-4: 277–304.

Section 3

Toxicological Perspective on Mutagenic Impurities

6

Toxicity Testing to Understand the Mutagenicity of Pharmaceutical Impurities

Andrew Teasdale[1], John Nicolette[2], Joel P. Bercu[3], James Harvey[4], Stephen Dertinger[5], Michael O'Donovan[6], and Christine Mee[7]

[1] *Chemical Development, Pharmaceutical Technology and Development, Operations, AstraZeneca, Macclesfield, UK*
[2] *AbbVie, Inc. North Chicago, IL, USA*
[3] *Gilead Sciences, Foster City, CA, USA*
[4] *GlaxoSmithKline R&D, Hertfordshire, UK*
[5] *Litron Laboratories, Rochester, NY, USA*
[6] *Independent Genetic Toxicology Consultant, Nottingham, UK*
[7] *Gentronix Ltd., Cheshire, UK*

6.1 Introduction

The purpose of this chapter is to describe the toxicological test methods that may be used to understand the mutagenicity of either actual or potential impurities. The testing itself is predicated by ICH M7 [1]. First, it is important to understand the different types of genetic toxicity, which is further defined in Hasselgren et al. [2] and OECD [3]. The three main mechanisms are (quoted from Ref. [2]):

1) Mutagenicity: Permanent, transmissible changes in the DNA that result from the induction of DNA adducts, insertions, inversions, and small deletion.
2) Clastogenicity: Structural chromosomal damage leading to sections of a chromosome being duplicated, deleted, or rearranged.
3) Aneugenicity: Numerical chromosomal abnormalities (aneuploidy) where an abnormal number of chromosomes is generated, often by disruption of the microtubule apparatus necessary for the orderly segregation of chromosomes during nuclear division.

ICH M7 makes clear that the focus of the guideline is on DNA-reactive, mutagenic impurities, emphasizing specific focus on such impurities as opposed to the more general group of genotoxic impurities. This is a point returned to in more detail in the general principles section. Key within the general principle section of ICH M7 (Section 3 of the guideline) is the reaffirmation of the specific focus of the guideline on mutagenic impurities. Indeed, the guideline goes further in making the specific statement that:

Other types of genotoxicants that are non-mutagenic typically have thresholded mechanisms and usually do not pose carcinogenic risk in humans at the level ordinarily present as impurities.

It also makes clear that in assessing mutagenic potential that this can be achieved through a combination of *in silico* structure activity relationship (SAR) evaluation, and where required, with

Mutagenic Impurities: Strategies for Identification and Control, First Edition. Edited by Andrew Teasdale.
© 2022 John Wiley & Sons, Inc. Published 2022 by John Wiley & Sons, Inc.

a bacterial reverse mutation assay (Ames test [4]). This is used to classify impurities; Table 6.1 from ICH M7 is reproduced below.

This is returned to in more detail in Section 6 of the guideline – Hazard assessment elements in ICH M7. This specifically states that:

> To follow up on a relevant structural alert (Class 3 in Table 6.1 – ICH M7 [1]), either adequate control measures could be applied or a bacterial mutagenicity assay with the impurity alone can be conducted. An appropriately conducted negative bacterial mutagenicity assay (Note 2, ICH M7) would overrule any structure-based concern, and no further genotoxicity assessments would be recommended (Note 1). These impurities should be considered non-mutagenic (Class 5 in Table 6.1). A positive bacterial mutagenicity result would warrant further hazard assessment and/or control measures (Class 2 in Table 6.1). For instance, when levels of the impurity cannot be controlled at an appropriate acceptable limit, it is recommended that the impurity be tested in an *in vivo* gene mutation assay in order to understand the relevance of the bacterial mutagenicity assay result under *in vivo* conditions. The selection of other *in vivo* genotoxicity assays should be scientifically justified based on knowledge of the mechanism of action of the impurity and expected target tissue exposure (ICH M7 [1] Note 3). *in vivo* studies should be designed taking into consideration existing ICH genotoxicity guidelines. Results in the appropriate *in vivo* assay may support setting compound specific impurity limits.

The tests themselves are described in detail in Table 6.2 (based on Note 3 within the guideline). The guideline states that such tests can be used to assess the *in vivo* relevance of the positive findings of the *in vitro* bacterial reverse mutation test.

Table 6.1 Impurities classification with respect to mutagenic and carcinogenic potential and resulting control actions.

Class	Definition	Proposed action for control (details in Sections 7 and 8)
1	Known mutagenic carcinogens	Control at or below compound-specific acceptable limit
2	Known mutagens with unknown carcinogenic potential (bacterial mutagenicity positive, no rodent carcinogenicity data)	Control at or below acceptable limits (appropriate TTC)
3	Alerting structure, unrelated to the structure of the drug substance; no mutagenicity data	Control at or below acceptable limits (appropriate TTC[a]) or conduct bacterial mutagenicity assay; If non-mutagenic = Class 5 If mutagenic = Class 2
4	Alerting structure, same alert in drug substance or compounds related to the drug substance (e.g. process intermediates), which have been tested and are non-mutagenic	Treat as non-mutagenic impurity
5	No structural alerts or alerting structure with sufficient data to demonstrate lack of mutagenicity or carcinogenicity	Treat as non-mutagenic impurity

[a] Threshold of toxicological concern.
Source: Reproduced from ICH M7.

Table 6.2 Tests to investigate the *in vivo* relevance of *in vitro* mutagens (positive bacterial mutagenicity).

In vivo test	Factors to justify choice of test as fit-for-purpose
Transgenic mutation assays	• For any bacterial mutagenicity positive. Justify selection of assay tissue/organ
Pig-a assay (blood)	• For directly acting mutagens (bacterial mutagenicity positive without S9[a])[b]
Micronucleus test (blood or bone marrow)	• For directly acting mutagens (bacterial mutagenicity positive without S9) and compounds known to be clastogenic[b]
Rat liver unscheduled DNA synthesis (UDS) test	• In particular for bacterial mutagenicity positive with S9 only • Responsible liver metabolite known • to be generated in test species used • to induce bulky adducts
Comet assay	• Justification needed (chemical class specific mode of action to form alkaline labile sites or single-strand breaks as preceding DNA damage that can potentially lead to mutations • Justify selection of assay tissue/organ
Others	• With convincing justification

[a] S9 – supernatant fraction obtained from an organ (usually liver) homogenate contains cytosol and microsomes. The microsomes component of the S9 fraction contain cytochrome P450 isoforms (phase I metabolism) and other enzyme activities.
[b] For indirect acting mutagens (requiring metabolic activation), adequate exposure to metabolite(s) should be demonstrated. *Source:* Reproduced from ICH M7.

While ICH M7(R1) [1] guidelines emphasize DNA-reactive, mutagenic compounds, it then states in Note 1: "In cases where the amount of the impurity exceeds 1 mg daily dose for chronic administration, evaluation of genotoxic potential as recommended by ICH Q3A/B could be considered." Then, ICH Q3A/B [5, 6] states "consider patient population and duration of use and consider conducting studies for point mutation and chromosomal aberration." Guidance for mutagenicity is covered in the ICH M7 guidance, but chromosomal aberrations (a clastogenic effect) are not.

Clastogenicity may be a consideration as part of the ICH guidelines, but ultimately the results of the studies has minimal impact on the control of impurities. As stated above, the concern for DNA-reactive substances is to cause cancer at low levels, which other genotoxicity can have threshold-related effects. Thus, while impurities may be tested in certain circumstances for clastogenicity, these assays will be de-emphasized in the chapter due to the higher weight given to mutagenicity for the control of impurities.

Because of its importance as the primary screen, the bacterial reverse mutation assay and its variants are considered in greatest detail in this review. In addition, *in vivo* mutation assays, potentially used when the impurity cannot be controlled to low levels, (e.g. Threshold of Toxicological Concern [TTC]) are also be described.

6.2 *In Vitro* Genotoxicity Tests

6.2.1 Background

The ability of chemicals to cause mutations was first realized during the Second World War when Auerbach and Robson showed mustard gas to be mutagenic in *Drosophila* [7, 8]. Mutagenicity has also been studied in bacteria for over 50 years with the initial work aiming to understand the function of

DNA after it had been found to be the hereditary material in all organisms. However, it was not until the early 1970s that the possible use of bacterial tests to predict carcinogenicity was seriously considered and, in 1973, this led Bruce Ames to publish his paper titled "Carcinogens are mutagens" [9]. A few years afterward, Lijinsky [10] described this as "An overenthusiastic interpretation of the available evidence," but the field of what became to be known as genetic toxicology had become established.

The use of bacteria for screening chemicals for potential mutagenicity and, hence, carcinogenicity is based on the observation that the primary structure of DNA is the same throughout the living world. There is a considerable amount of evidence that DNA damage in germ cells can cause heritable genetic defects and in somatic cells can be critical in both the initiation of cancer and subsequent steps in the progression of the disease. A large range of different sorts of genetic damage may have adverse effects on living organisms, but induction of mutation in bacteria is taken as a sensitive indicator that an agent may be capable of causing damage to DNA. The nature of the damage may not be the same as those causing cancer or birth defects in humans, but mutagenicity in bacteria does indicate that a chemical has the intrinsic ability to interact with DNA and modify its function, not simply to destroy it.

Early on, genetic toxicologists came to appreciate that many carcinogens require metabolism to produce the DNA-reactive, electrophilic species, and it was shown that binding of polycyclic aromatic hydrocarbons to DNA was dependent on metabolism by microsomes [11, 12]. It was also clear that the bacteria used for mutation tests had limited capacity for metabolism, and liver homogenates were used to overcome this. To increase the metabolic capability, liver homogenates (S9) were prepared from rats that had been treated with enzyme inducers, initially phenobarbital, and the activity of several carcinogens including aflatoxin B_1, benzo(a)pyrene, and benzidine were shown to require S9 activation [9]. Subsequently, inducers with a broader spectrum of induction have been used including polychlorinated biphenyls (PCBs) such as Aroclor 1254 [13] or a combination of phenobarbital and β-naphthoflavone. The S9 mix is supplemented with cofactors for nicotinamide adenine dinucleotide phosphate (NADPH) generation so it is very efficient at cytochrome p450-mediated phase 1 metabolism; however, phase 2 metabolism is generally very poor unless cofactors for conjugation are also added. Similar S9 systems are used in all *in vitro* genotoxicity test systems.

6.2.2 Bacterial Reverse Mutation or "Ames" Test

The bacterial mutation assay examines mutation in specific strains of *Salmonella typhimurium* and *Escherichia coli* constructed to detect a range of mutagens and is commonly referred to as the "Ames" test after Professor Bruce Ames (University of California, Berkeley) who developed the *Salmonella* strains [13]. Technically, however, this term is incorrect if any of the *E. coli* strains are included since these were developed independently by Green and Muriel [14]. However, the term Ames is used in this book chapter as it is frequently used to refer to the modern bacterial reverse mutation assay.

A standard test uses five different strains, and the ICH S2(R1) guideline [15] gives the following options.

- *S. typhimurium* TA1535
- *S. typhimurium* TA98
- *S. typhimurium* TA100
- *S. typhimurium* TA1537 *or* TA97 *or* TA97a
- *S. typhimurium* TA102 *or E. coli* WP2 *uvrA or* WP2 *uvrA* (pKM101)

The *S. typhimurium* and *E. coli* strains have mutations in the histidine and tryptophan operons respectively, and it is the reversion of these mutations that is measured by the ability of colonies to grow in medium lacking these amino acids. The target sequences in the mutations mean the strains detect agents acting through different mechanisms, *viz.*

Table 6.3 Description of bacterial strains commonly used for the bacterial reverse mutation assay.

	Amino acid marker			Other relevant mutations		
Strain	**Mutation (his or trp)**	**Type of mutation**	**Main DNA target**	**Cell wall**	**DNA repair**	**Plasmid**
***Salmonella typhimurium* LT2**						
TA1535	*hisG46*	Base pair substitution	GC	rfa	*uvrB*	None
TA1537	*hisC3076*	Frameshift	GC	rfa	*uvrB*	None
TA98	*hisD3052*	Frameshift	GC	rfa	*uvrB*	pKM101
TA100	*hisG46*	Base pair substitution	GC	rfa	*uvrB*	pKM101
TA97a	*hisD6610*	Frameshift	GC	rfa	*uvrB*	pKM101
TA102	*hisG428*	Base pair substitution	AT	rfa	+	pKM101 and pAQ1
***Escherichia coli* WP2**						
***uvrA*/ pKM101**	*trpE*	Base pair substitution	AT	+	*uvrA*	pKM101
pKM101	*trpE*	Base pair substitution	AT	+	+	pKM101

Source: S. typhimurium sequences taken from Mortelmans and Zeiger [16].

The various mechanisms are described in Table 6.3.

In addition, the strains have other characteristics influencing their response to mutagens: all the *S. typhimurium* strains except TA102 are DNA repair deficient with the *uvrB* gene deleted, and the *E. coli* strain has a similar deletion, *uvrA*; TA98, TA100, and the *E. coli* strain all contain the pKM101 plasmid, conferring error-prone repair; all the strains also have deficient lipopolysaccharide walls allowing greater permeability to test agents.

Compounds are tested in both the presence and absence of metabolic activation comprising the S9 fraction of livers from rats pretreated with enzyme inducers, supplemented with cofactors for NADPH generation. The enzyme inducers are used to increase the activity of a wide range of cytochrome p450s. Originally, PCBs such as Aroclor 1254 were used but, subsequently, a combination of phenobarbital and β-naphthoflavone was found to be an effective alternative to PCBs [17, 18]; either inducing regime is now accepted. In the standard plate-incorporation test, bacteria and test compound, with or without S9, are mixed with agar before plating onto agar plates, incubated for two to three days, and then numbers of revertant colonies are scored. In the preincubation method, bacteria and test compound ±S9 are incubated together for 20–60 minutes before mixing with agar and plating as before. Either the preincubation or plate-incorporation is considered acceptable under Organization for Economic Co-operation and Development (OECD) 471 guidelines [19], with some notable considerations. Some compounds can be more efficiently detected using the preincubation method such as short chain aliphatic nitrosamines, divalent metals, aldehydes, azo-dyes and diazo compounds, pyrrolizidine alkaloids, allyl compounds, and nitro compounds [20]. Alternative procedures to the standard preincubation and plate-incorporation tests are highlighted in OECD 471 [19] such as azo-dyes, diazo compounds, gases/volatile compounds, and glycosides. These deviations from standard procedures while acceptable and preferred in certain cases should be scientifically justified and documented. Experimental procedures for the

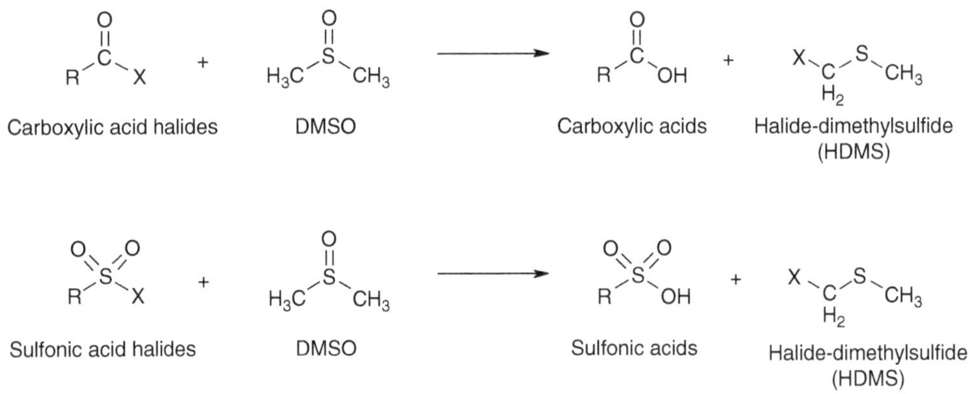

Figure 6.1 Reaction of carboxylic acid/sulfonic acid halides with DMSO.

assay are described in greater detail by Maron and Ames [21] and also specifically by Mortelmans and Zeiger [16] for the *Salmonella* strains and Mortelmans and Riccio [22] for *E. coli*.

A test for regulatory submission should comply with ICH S1 (R1) [23], ICH S2(R1) [15], and OECD Guideline 471, and the highest level used is 5000 μg/plate unless limited by toxicity to the indicator strains, or by solubility in either the initial solvent, routinely dimethyl sulfoxide (DMSO), or the aqueous agar medium. Although DMSO is the default organic solvent used in most laboratories, several others have been shown to be compatible including dimethyl formamide, acetonitrile, acetone, and 95% ethanol [21]. Solvent compatibility is a critical requirement. The solvent/vehicle could chemically react with the test substance, impact the survival of bacteria, or reduce metabolic (i.e. S9) activity influencing the results of the study [22]. One example in Amberg et al. [24] reported that both sulfonyl and acyl chlorides react with DMSO to generate *in situ* mutagenic alkylating agent, see Figure 6.1. Note this is discussed in more detail in Chapter 5. Clearly, the presence of alkylating agent formed *in situ* can lead to a false positive test result. As highlighted in Chapter 10, there are concerns over the use of DMSO when testing short chain aliphatic nitrosamines due to a reduction in metabolic activation [25].

A flowchart illustrating the Ames test is shown in Figure 6.2.

The data outcome is reported as the number of revertant colonies per plate. Individual plate counts, mean number of revertant colonies per plate, the standard deviation of tested concentrations, and positive and negative controls are reported. Positive results are based on fold increase over background, where the cutoffs are strain dependent and based on historical data for the lab. Other methods that can help with determining a positive response are a concentration-related increase, reproducibility, and statistical methods. A positive result in a single strain, with or without activation, is sufficient to conclude a compound is positive. A clear positive negative response does not need to be repeated. Equivocal or weak positive results can be clarified using follow-up testing based on modifying experimental conditions, such as testing concentrations closer to the equivocal range.

ICH M7(R1) [1] highlights some important considerations for pharmaceutical impurities. The bacterial reverse mutation assay should be performed in compliance with good laboratory practice (GLP) regulations; however, deviation from GLPs is allowed and can be documented in the study report. One very common example is that the test article is not usually prepared or analyzed in compliance with GLP regulations. Typically, a certificate of analysis (COA) is provided by the manufacturer, which is appended to the report. It is important that the COA reports purity and appearance, as these are critical factors in interpreting the results of a study. It is recommended that overall purity should

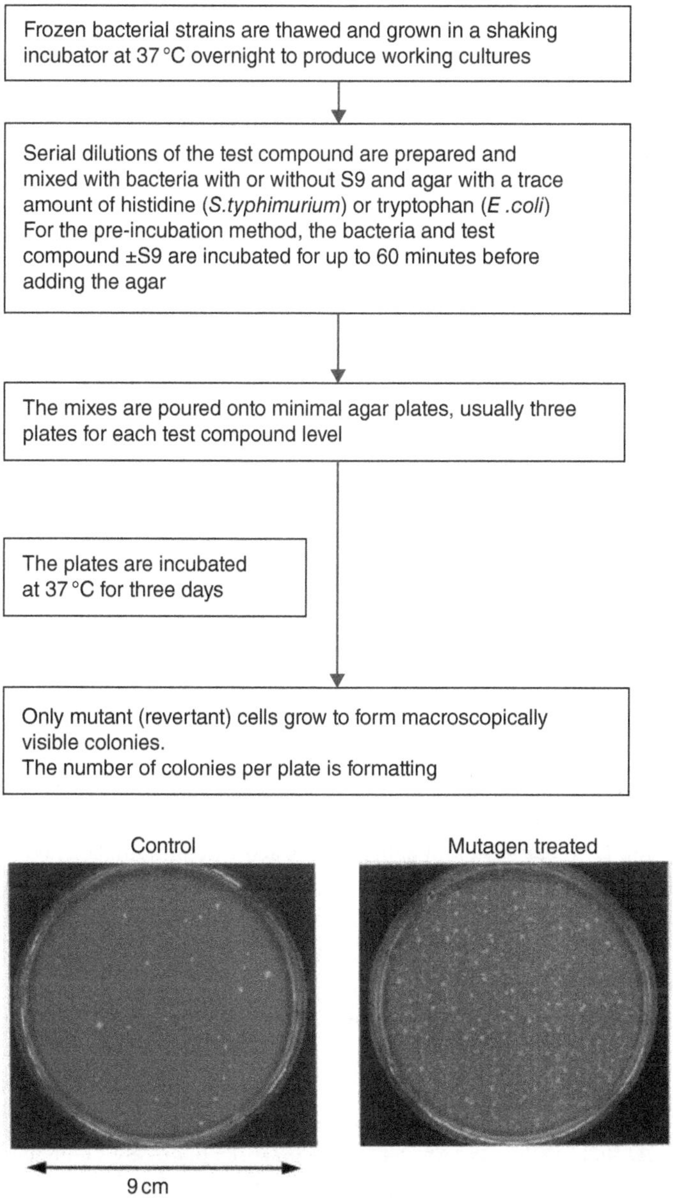

Figure 6.2 Flowchart representing the standard Ames test.

be ~98%, but knowledge of the impurities in test article is important, especially when those impurities are themselves mutagenic. This may influence whether to further repurify the test article. Dose-formulations analysis is not typically performed with impurities.

6.2.3 Modifications to the Standard Ames Test

It is presumed that, for impurities, a five-strain Ames test performed according to OECD 471 [19] and ICH S2(R1) [15] guidelines will be accepted by regulatory authorities worldwide. However, to assess mutagenicity where compound supply is limited, ICH M7 provides flexibility in the Ames test design to accommodate this situation and still be considered acceptable for regulatory submission. Several modifications in the scale of the Ames test have been explored, and various high-throughput screens that often were established for early pharmaceutical candidate screening have been developed, which may be applicable for impurity safety testing. These are considered below, but experience has shown that the number of impurities is not enough to require a screen with genuinely high-throughput capability and supply of test material is not usually limiting.

Since compound availability can be critical in early discovery, methods to reduce the amount of material for screening Ames tests have been investigated in the 30 years since the original methods were published. Approximately 300 mg is required for a single five-strain test performed to regulatory submission standards (without allowing material for formulation analysis), and this amount can only be reduced in the following ways: restricting the number of strains, limiting the highest test level, reducing the number of plates per dose, reducing the size of the plates, and performing the test only in the presence of S9. If modifying the standard OECD 471 [19] method, the strains selected and top concentration should be scientifically justified and documented.

This book chapter will not highlight all the different variations of the Ames test as they are not commonly performed on impurities. For more information on mutagenicity screening tests, especially modifications to the Ames assay, please refer to Escobar et al. [26]. The six-well Ames assay will be highlighted as it is gaining more regulatory acceptance for use with pharmaceutical impurities [27].

6.2.3.1 Six-well Ames Assay

The six-well test reduces the test article need by reducing the surface area for growth (to 9.2 cm [2]) and the volume of reagents to one-fifth typically used in the full-sized plate test. The wells are large enough that the assay can be counted with most automated counters. Based on work by Brooks [28], the assay was first described by Diehl and colleagues [29], showing results of select strains (TA98, TA100, TA102, TA1535, WP2 *uvrA*$^-$) in 91 compounds test in the six-well format and standard plate testing (Diehl et al. [29]). Later, Nicolette [27] and colleagues tested 24 compounds in both formats using five OECD tester strains (adding TA1537 instead of TA102) in two laboratories with 100% concordance. Since this assay follows the same steps and uses the same reagents, it could be adapted in most Ames testing labs. Briefly, 4–5 ml of bottom agar is added to the wells as the base agar. Test concentrations typically up to 50 mg/ml for solvents such as water or DMSO are prepared. A 20–25 μl of overnight bacterial culture, 100 μl of S9-mix or phosphate buffered saline (PBS) (for nonactivated conditions), and 20 μl of test article formulation are added to a tube containing 0.5 ml molten top agar supplemented with histidine, biotin, and tryptophan at 0.5 mM each. The contents of the tube are poured onto the prepared well and gently swirled. After pouring of all wells of a plate, the plate is set aside to allow the top agar to solidify. The plates are incubated for 48–72 hours as is done in the standard plate test, and then colonies can be counted. Data can be analyzed for dose-related increases that either are statistically significant from controls or exceed two- or threefold control values. Mean control colony counts have been published showing that miniaturization leads to average vehicle control colony counts between four- and sevenfold fewer than standard plates. This assay has been shown to have a high qualitative correlation to the standard plate test (Diehl et al. [29]; Nicolette et al. [27]) requiring only between 75 and 85 mg of test item. The six-well test can be evaluated with most colony counters where the aperture can be reduced or with an external camera setup, facilitating the scoring of wells.

6.2.4 Test Strategy

The objective of a strategy is to ascertain whether an impurity is mutagenic and, in some instances, in accord with ICH M7 conduct further investigation to ascertain *in vivo* relevance. Although ICH

M7 defines a series of tests that may be conducted following a positive Ames test, it doesn't define a specific path in terms of testing i.e. what tests will ultimately mitigate a positive Ames test. The expectation appears to be that this would need to be addressed on a case-by-case basis and will therefore very likely require specific engagement with regulatory agencies.

The first stage of the assessment of a potential mutagenic impurity (PMI) is to determine whether it contains a structural alert for mutagenicity. This can be done by literature review but, more typically, a computational toxicology assessment is made. There are multiple systems, both rule-based and statistical systems, that can be utilized, and this is explored in depth in Chapter 4.

If it is decided that a structural alert for mutagenicity requires testing, a bacterial mutation assay is the usual first screen indeed as described above; ICH M7 [1] makes this very clear. Negative results will be sufficient to dismiss the alert.

In terms of the Ames assay, requirements are defined in ICH M7, Note 2; this is reproduced below; importantly, this opens up the possibility to still utilize data where full compliance to the required standard has not been achieved; this is very likely to refer to historical data and potentially avoids unnecessary repeat testing if scientifically justified:

To assess the mutagenic potential of impurities, a single bacterial mutagenicity assay can be carried out with a fully adequate protocol according to ICH S2(R1) and OECD 471 guidelines [15, 19]. The assays are expected to be performed in compliance with Good Laboratory Practices (GLP) regulations; however, lack of full GLP compliance does not necessarily mean that the data cannot be used to support clinical trials and marketing authorizations. Such deviations should be described in the study report. For example, the test article may not be prepared or analysed in compliance with GLP regulations. In some cases, the selection of bacterial tester strains may be limited to those proven to be sensitive to the identified alert. For impurities that are not feasible to isolate or synthesize or when compound quantity is limited, it may not be possible to achieve the highest test concentrations recommended for an ICH-compliant bacterial mutagenicity assay according to the current testing guidelines. In this case, bacterial mutagenicity testing could be carried out using a miniaturized assay format with proven high concordance to the ICH-compliant assay to enable testing at higher concentrations with justification.

As highlighted in Note 2, there are options when the availability of the isolated impurity is limited and a full OECD 471-compliant test may not be possible. As described, miniaturized tests are being evaluated by OECD for routine use for mutation testing, but can be used when compound is limited and justification provided. Testing specific strains or conditions if described in the output of *in silico* evaluation as being sensitive for the structural concern is another option. Lastly, if the only way to assess an impurity is as part of the active pharmaceutical ingredient (API), testing the impurity "spiked" into the API could be considered. It is recognized that the level must be such that most genotoxic agents would actually give positive results if present as an impurity. Based on literature results for approximately 450 mutagens, it appears that about 85% are detected at concentrations of 250 μg/plate or less [30]. Therefore, it was proposed that an impurity may be evaluated as part of the API as long as it is present to give a minimum concentration of 250 μg/plate i.e. 5% if the API can be tested to the limit level of 5000 μg/plate. This has not been accepted for general use for Ames testing, but where testing with the API is the only viable option, this methodology could be justified.

If a PMI gives a positive result in the Ames test, it is generally likely that the default option would be to control it to the TTC limits. This is likely to be the case where the result is consistent with the *in silico* alert and the known mechanism. There are a very few examples such as sodium azide that are clear bacterial mutagens but are not genotoxic in mammalian cells *in vitro* or *in vivo* because of differences in metabolism [31], but it is unlikely that a similar explanation could be found for a novel impurity. Although the default is that mutagenic carcinogens have no threshold, this is being increasingly challenged; Chapter 8 looks at this in depth. Indeed, threshold and safety margins were established even for the DNA-reactive mutagenic carcinogen, ethyl methanesulfonate (EMS), after it was found to be an impurity in Viracept®, and these allowed a human risk assessment to be made [32]. However, this required a significant number of *in vitro* and *in vivo* studies with a compound for which there was already a large amount of data. This work was essential to perform the risk assessment after patients had been exposed, and it would generally be impracticable to generate similar data prospectively for an impurity in a drug in clinical development.

Although a negative Ames test will frequently be the only test used to qualify an impurity, there are occasions where as highlighted in ICH M7 (i.e. >1 mg/day for chronic administration) further testing may be performed, despite there being no specific guidance in terms of how testing interrelates at least in terms of a formal decision tree. However, these tests typically have minimal impact on the control of impurity levels compared to the Ames assay.

In conclusion, negative results in an adequately conducted Ames test should be sufficient to qualify a structurally alerting impurity for most regulatory authorities worldwide. Similarly, a clearly mutagenic impurity will almost certainly need to be controlled to the relevant TTCs at each stage of development, except for those potent agents for which specific lower limits may be required i.e. cohorts of concern (COC). However, whether or not limits between these two extremes can be justified for chemicals that show some activity in *in vitro* tests but for which *in vivo* data are available is not clear. It is essential that feedback is obtained from regulatory authorities for such examples in order for a generally accepted strategy to be developed.

6.3 *In Vivo* Mutation Assays

In ICH M7(R1) Note 3 discusses the use of *in vivo* data to understand the *in vitro* relevance of mutagens (Table 6.2). Such *in vivo* assays can be important when an impurity cannot be controlled to the TTC. The most common assays to test for impurities are the transgenic mutation and *Pig-a* assays. The *Pig-a* assay is performed with stock, nonproprietary rodent models; is less expensive; and data can generally be generated more quickly than the transgenic mutation assay. Since the *Pig-a* assay strictly reports on mutagenic activity that has occurred in the bone marrow compartment, the validity of a negative (non-mutagenic) test result requires evidence that systemic exposure to the presumed mutagenic species was attained. In some cases, for an Ames positive compound solely with metabolic activation, the reactive metabolite is unknown so determining systemic exposure is not possible.

The advantage of the transgenic rodent (TGR) mutation assay is that it can be used for any impurity positive for bacterial mutations, and many different kinds of tissues can be analyzed for mutagenicity. This is very useful especially for determining the effect of mutagenicity at the site of contact (e.g. stomach or small intestines), site of metabolic activation (e.g. liver), germ cells, etc. Multiple tests can be combined together, such as adding comet and *Pig-a* together to understand DNA damage at the site of first contact and site of metabolic activation in addition to determining mutagenicity following systemic exposure. The following sections will discuss most of the *in vivo* mutagenicity assays with the exception of the UDS assay since it is not commonly performed for impurities.

6.3.1 *In Vivo Pig-a* Gene Mutation Assay

The phosphatidylinositol glycan class A (*Pig-a*) gene is 17 kb in length with six exons [33]. As reported by Kawagoe and colleagues, the gene's function and X-chromosome location are highly conserved across mammalian species [34]. *Pig-a* is one of more than two dozen genes involved in glycosylphosphatidylinositol (GPI) anchor biosynthesis. GPI anchors attach a host of specific proteins to the cell surface of cells, for example CD55, CD59, and CD24. Unlike other genes that contribute to GPI anchor synthesis, only the *Pig-a* gene normally exists as one functional copy. Therefore, a single inactivating mutation is sufficient to ablate expression of all GPI-anchored proteins on the cell surface. As first described by Araten and colleagues [35], this explains why the absence of cell surface antigens such as CD55 and/or CD59 and/or CD24 on hematopoietic cells represents a reliable phenotypic reporter of *Pig-a* mutation.

The initial reports of rodent blood cell-based mutation assays based on GPI anchor deficiency first appeared in 2008 [36, 37]. These early studies used flow cytometric analyses to study chemical-induced mutation in rodents, and they took advantage of the same basic principle – fluorescent antibody(s) against GPI-anchored protein marker(s) can distinguish between GPI anchor-proficient and GPI anchor-deficient cells. Whereas wild-type cells are fluorescent, cells lacking GPI-anchored proteins are nonfluorescent and therefore presumed to be *Pig-a* mutants. These early proof-of-principle studies showed increased frequencies of *Pig-a* mutant phenotype erythrocytes in rodents exposed to potent mutagens.

An overview of the assay is presented in Figure 6.3.

Since this time, collaborative interlaboratory efforts have systematically expanded the number of chemicals studied, including weak mutagens and chemicals thought to have little or no genotoxic activity [38]. Furthermore, methodological improvements, especially the use of immunomagnetic separation prior to flow cytometric analysis, are now routinely applied in order to interrogate much greater numbers of reticulocytes [39] or total erythrocytes and reticulocytes [40, 41] per blood sample (see Figure 6.4). These advances have provided investigators with the means to reliably enumerate *Pig-a* mutant cells with high precision and accuracy, despite their ordinarily low frequency.

Left panel: the instrument calibration standard is prepared each day of analysis and contains approximately 50% mutant-mimicking cells (i.e. erythrocytes that were not incubated with anti-CD59-PE) and approximately 50% wild-type erythrocytes (i.e. blood that was fully processed). The instrument calibration standard contains enough events with a full range of phycoerythrin (PE)-fluorescence intensities to optimize photo multiplier tubes (PMT) voltages and compensation settings, and it also provides a means to rationally and consistently set the position of the vertical demarcation line that discriminates mutant phenotype erythrocytes (left) from wild-type erythrocytes (right). Center panel: blood from a mutagenized rat, pre-column analysis. Pre-column analyses are used to determine reticulocyte frequency, reticulocyte to Counting Bead ratio, and total erythrocyte to Counting Bead ratio. Center panel: blood from the same mutagenized rat, post-column analysis. This sample was depleted of wild-type erythrocytes via immunomagnetic separation, and the mutants were further enriched with a subsequent centrifugation step. The numbers of mutant phenotype reticulocytes and mutant phenotype erythrocytes are directly determined from this sample. The denominators, that is, the total number of reticulocytes and erythrocytes, are derived from the pre-column cell to bead ratios and the number of Counting Beads observed in the post-column sample. With immunomagnetic separation in conjunction with flow cytometry, it is possible to analyze > 3 × 10 [6] reticulocytes and > 150 × 10 [6] erythrocytes per sample for the *Pig-a* mutant phenotype.

In parallel to the work described above, other efforts have utilized DNA sequencing to ensure that the GPI anchor-deficient phenotype is indeed due to *Pig-a* mutation. This component of assay validation has taken several forms. Kimoto et al. [42] used a flow cytometer to sort ethyl nitrosourea (ENU)-induced bone marrow erythroids that exhibited a GPI anchor-deficient phenotype. Complementary DNA (cDNA) sequencing demonstrated that the CD24-negative cells were indeed

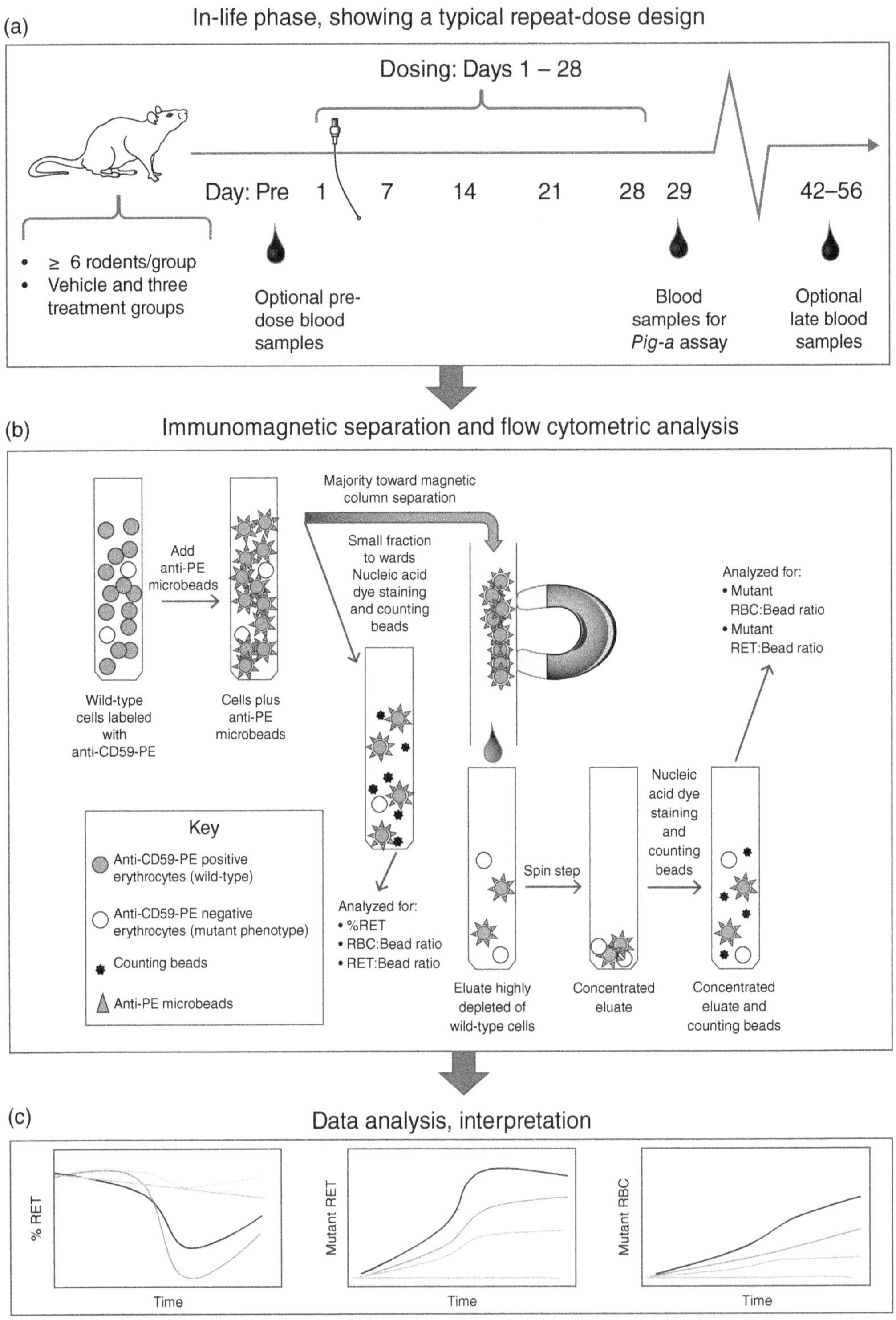

Figure 6.3 Overview of a rodent *Pig-a* study.

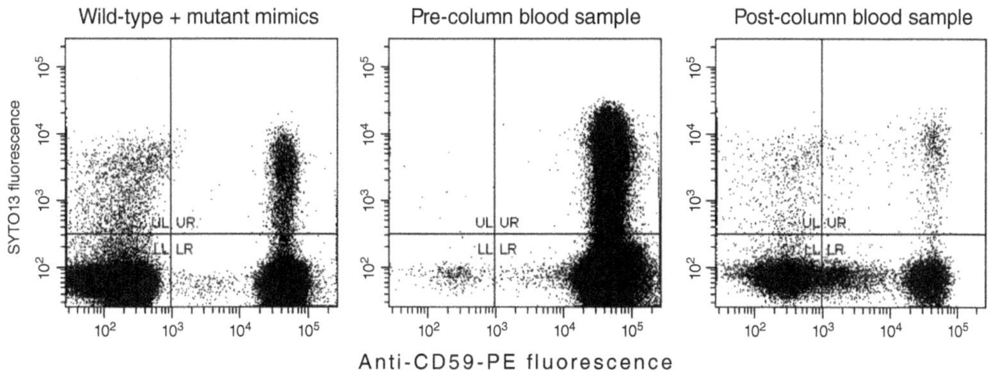

Figure 6.4 Flow cytometric scoring of *Pig-a* mutant phenotype erythrocytes.

Pig-a mutants, with base-pair substitutions typical of ENU spectra. Other sequencing work has considered *N*-ethyl-*N*-nitrosourea- and 7,12-dimethylbenz[a]anthracene-treated rats and in some cases other sources of hematopoietic cells, for example T-lymphocytes from the spleen [43, 44]. Additional support linking GPI anchor deficiency to gene mutation has come from experiments with lymphoblastoid cell lines. This work has been performed with human TK6 [45, 46] and mouse L5178Y cells [47, 48]. For TK6 cells, besides *PIG-a*, the *PIG-l* locus contributes to the frequency of GPI anchor-deficient cells. This is an expected result that is related to the heterozygous status of the *PIG-l* gene in TK6 cells examined to date [45].

An important characteristic of *in vivo Pig-a* assays is that mutant phenotype erythrocytes generally accumulate with repeat exposures. This explains why repeat dosing study designs, for example 28 consecutive days, are often employed – they help maximize assay sensitivity. While shorter-term treatment schedules, for example three consecutive days, are common in the literature, and retrospective analyses suggest that these tests detect nearly all the mutagens that have been found to be active in 28-day studies, it is generally accepted that a 28-day treatment schedule is the preferred design when one is considering the *in vivo* relevance of an *in vitro* mutagenic substance.

Whatever treatment schedule is employed, it is important to consider that different erythrocyte populations manifest the mutant phenotype over different time periods, and this information needs to be considered when selecting the most appropriate blood harvest time(s). As explained in more detail below, current best practices incorporate an evaluation of both immature erythrocytes (IE) (reticulocytes, or RETs) as well as total erythrocytes (red blood cells [RBCs]) for the GPI anchor-deficient phenotype. For this reason, it is necessary to include time point(s) that provide enough manifestation time for effects to be seen in both populations.

Mutant RETs frequencies rise relatively rapidly upon mutagen exposure, over the course of one or two weeks, and can be thought of as a leading indicator of mutation. Genotoxicant-induced increases to the mutant RBC frequency requires several weeks and is therefore a lagging indicator of mutation. Taking into consideration the disparate kinetics of mutant RETs and RBCs, it is possible to limit 28-day repeat dose studies to one postexposure blood collection time, for example day 29 (i.e. one day after cessation of treatment). Note that in addition to the postexposure time points indicated above, some investigators routinely evaluate pretreatment blood samples. This approach can be valuable for ensuring high statistical power, since animals with unusually high/outlier mutation frequencies can be identified and deselected from a study.

A 2013 International Workshop on Genotoxicity Testing (IWGT) Workgroup reviewed existing *Pig-a* literature, identified data gaps, and prioritized assay validation efforts [49]. Largely based on this report, the Health and Environmental Sciences Institute Genetic Toxicology Technical Committee (HESI-GTTC) consortium recommended and organized the effort to move forward with the development of an OECD Test Guideline. Since this time, a detailed review paper and a retrospective validation report have been approved by OECD member nations. Therefore, at the time of this writing, a Mammalian Erythrocyte *Pig-a* Gene Mutation Assay Test Guideline is officially under development.

While the development of an OECD Test Guideline continues at the time of this writing, it should be noted that the *Pig-a* assay has already gained significant regulatory attention and acceptance. For example, both the ICH M7 guideline on the Assessment and Control of DNA Reactive (Mutagenic) Impurities in Pharmaceuticals [1] as well as the Environmental Protection Agency's Office of Pesticide Programs [50] have suggested the use of the *Pig-a* assay for safety evaluation.

One instructive ICH M7-centric use case that centered on the *in vivo Pig-a* assay has been published by a consortium of scientists and involved a series of arylboronic compounds [51]. These chemicals are ubiquitous intermediates in the synthesis of many pharmaceuticals. It was therefore concerning when many of these compounds were found to exhibit mutagenic activity in a bacterial test system [52]. Given the extensive use of these compounds across the industry, these pharmaceutical scientists collaborated to evaluate a set of eight arylboronic compounds in a series of rat *Pig-a* studies.

The structures included aryl- and hetero arylboronic acids and esters and substituents with varying degrees of steric hindrance and electron-withdrawing characteristics. Exposures occurred for 28 consecutive days. Each of the eight studies included assessments of *Pig-a* mutation as the primary end point, while some of the studies included comet and micronucleus assays as well.

The results were clear – arylboronic compounds that were mutagenic *in vitro* were not found to be mutagenic (or otherwise genotoxic) *in vivo*, despite the high systemic exposures that were achieved in these studies [51]. These results have important implications for these and other arylboronic acids and esters with similar scaffolds. Rather than necessitating the control of these agents below the TTC as described in the ICH M7 guidance document, these *in vivo* mutation data support managing them in accordance with the ICH Q3A/Q3B guideline [5, 6].

6.3.2 Rodent Micronucleus Test

Micronucleus tests can be applied to any population of dividing cells either *in vivo* or *in vitro*. Micronuclei are formed from chromosome fragments, or whole chromosomes, left behind during the anaphase stage of mitosis and, therefore, may result from both clastogenic (chromosome breakage) and aneugenic (loss of whole chromosomes) events. They are visualized microscopically in cells that have gone through division as discrete small bodies of chromatin in the cytoplasm, and specialized staining techniques can distinguish whether they contain chromosome fragments or whole chromosomes.

The rodent micronucleus test provides an *in vivo* method for detecting agents that interfere with mitotic cell division [53] and examines rapidly dividing erythropoietic cells that are exposed to plasma levels of drug and metabolites. Erythroblasts expel their nuclei a few hours after the last mitotic division, but micronuclei remain in the cytoplasm (see Figure 6.5).

The most recent version of the guideline describes three treatment schedules and associated sampling times.

Quoting the guideline:

a) Animals are treated with the test chemical once. Samples of bone marrow are taken at least twice (from independent groups of animals), starting not earlier than 24 hours after treatment,

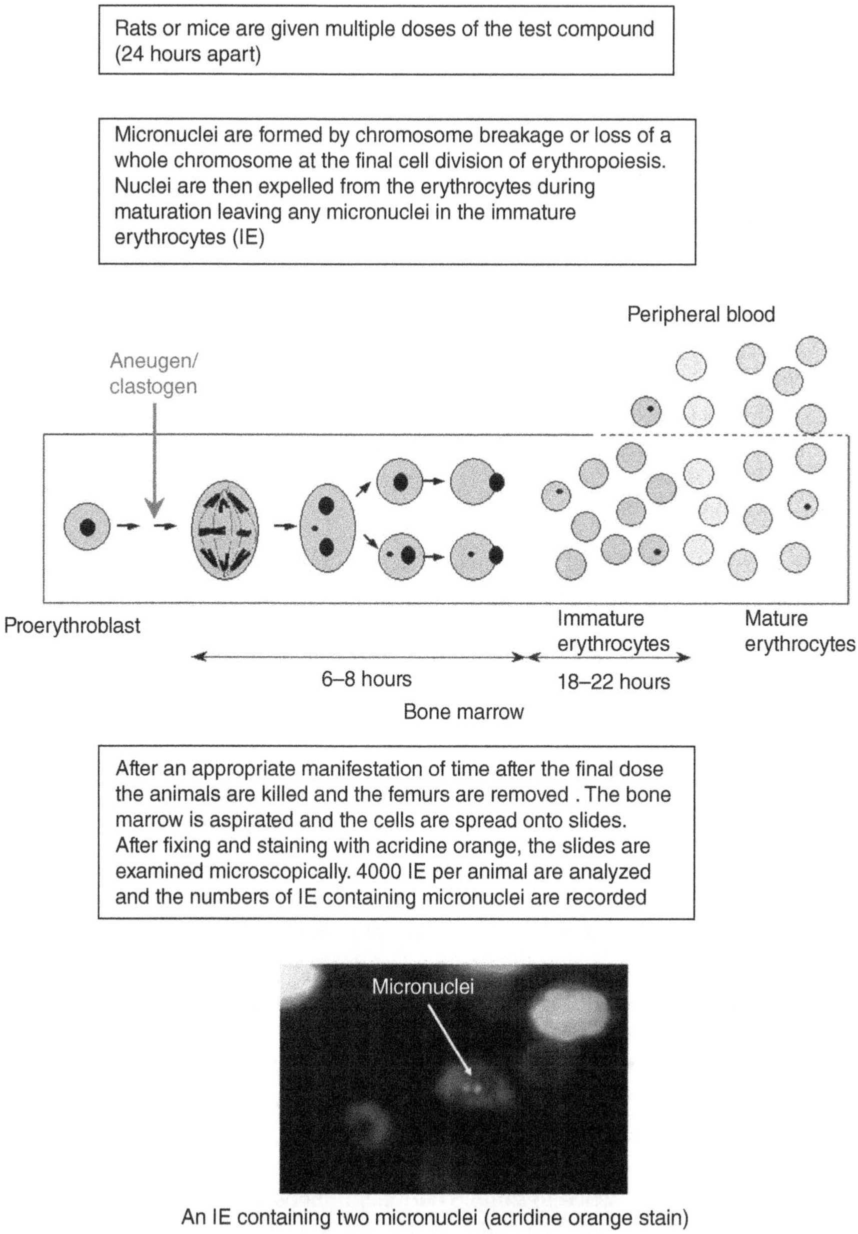

Figure 6.5 The rodent bone marrow micronucleus test. (*See insert for color representation of this figure*).

but not extending beyond 48 hours after treatment with appropriate interval(s) between samples, unless a test substance is known to have an exceptionally long half-life. The use of sampling times earlier than 24 hours after treatment should be justified. Samples of peripheral blood are taken at least twice (from the same group of animals), starting not earlier than 36 hours after treatment, with appropriate intervals following the first sample, but not extending beyond 72 hours. At the first sampling time, all dose groups should be treated and samples collected for analysis; however, at the later sampling time(s), only the highest dose needs to be administered. When a positive response is detected at one sampling time, additional sampling is not required unless quantitative dose-response information is needed. The described harvest times are a consequence of the kinetics of appearance and disappearance of the micronuclei in these two tissue compartments.

b) If two daily treatments are used (e.g. two treatments at 24 hour intervals), samples should be collected once between 18 and 24 hours following the final treatment for the bone marrow or once between 36 and 48 hours following the final treatment for peripheral blood [54]. The described harvest times are a consequence of the kinetics of appearance and disappearance of the micronuclei in these two tissue compartments.
c) If three or more daily treatments are used (e.g. three or more treatments at approximately 24 hour intervals), bone marrow samples should be collected no later than 24 hours after the last treatment and peripheral blood should be collected no later than 40 hours after the last treatment [55]. This treatment option accommodates combination of the comet assay (e.g. sampling two to six hours after the last treatment) with the micronucleus test and integration of the micronucleus test with repeated-dose toxicity studies. Accumulated data suggested that micronucleus induction can be observed over these wider timeframes when three or more administrations have occurred [56].

After sampling the bone marrow, slides must then be stained to allow mature and immature erythrocytes (IE) to be differentiated in order to score micronuclei in the latter. All the stains used rely on detecting the residual RNA in the IE. For the mouse, the species originally used for the micronucleus test, the most common stain was May–Grunwald/Giemsa [53] that identifies IE by their polychromatic appearance, hence the alternative term for IE, "polychromatic erythrocytes" (PCE). However, rat bone marrow preparations contain mast cell granules that stain similarly to micronuclei with May–Grunwald/Giemsa, thus precluding its use. Consequently, modified staining procedures with hematoxylin and eosin [57] or fluorescent dyes were introduced to allow the rat to be used routinely. Acridine orange [58] is now the most commonly used stain for rats and is equally applicable to mice; the RNA in IE fluoresces orange in contrast to the green of DNA in the micronuclei. Whatever stain is used, the frequency of micronuclei in IE (PCE) is scored and the ratio of IE to mature erythrocytes is used as an indicator of bone marrow toxicity.

Either the rat or the mouse can be used, and for drug substances the default species is often the rat in order to relate to other toxicology and toxicokinetic data. However, for impurities, this is unlikely to be a factor and either species is equally acceptable.

A test for regulatory submission must comply with OECD Guideline 474 [59], and the highest dose must be the maximum tolerated up to a limit of 2000 mg/kg. A flowchart illustrating the rodent bone marrow micronucleus test is shown in Figure 6.5.

Although the rodent micronucleus assay was originally performed by examining the bone marrow, it has now become common practice to sample peripheral blood and perform analyses on circulating IE. For instance, the ICH S2(R1) guideline [15] states that "the measurement of micronucleated immature (e.g. polychromatic) erythrocytes in peripheral blood is an acceptable alternative in the mouse, or in any other species in which the inability of the spleen to remove micronucleated erythrocytes has been demonstrated, or which has shown adequate sensitivity to detect clastogens/aneuploidy inducers in blood." While rats are known to remove micronucleated erythrocytes from peripheral blood, it has been established that micronucleus induction by a range of clastogens and aneugens can be detected in blood reticulocytes [60, 61]. Thus, rat blood may be used for micronucleus analysis so long as methods are used to ensure that the most immature fraction of newly formed reticulocytes are analyzed [62, 63], and the sample size is sufficiently large to provide appropriate statistical power, given the lower micronucleus levels in rat blood than bone marrow [64].

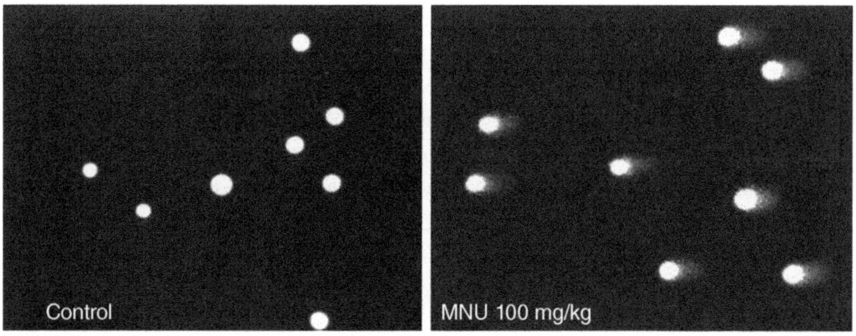

Figure 6.6 The comet assay. Hepatocyte nuclei from rats given an oral dose of methyl nitrosourea (MNU) showing DNA damage.

There are several compelling reasons why peripheral blood has increasingly been considered a preferred tissue compartment for conducting *in vivo* micronucleus tests. First, the time and effort needed to collect peripheral blood is considerably lower compared to bone marrow specimens. Second, animals do not have to be euthanized at sample time(s). This provides greater flexibility and opportunities to combine the end point with other (geno)toxicity studies. Finally, peripheral blood is highly compatible with automated scoring techniques, including flow cytometry and image analysis. Indeed, OECD Test Guideline 474 [59] considers validated automated scoring techniques [62, 65] the preferred means of acquiring micronucleated cell frequencies.

6.3.3 Rodent "Comet" Assay

The single-cell gel electrophoresis or "comet" assay is a method used to evaluate the ability of a test compound to cause DNA strand breaks and alkali labile sites and can be applied *in vitro* or *in vivo* to virtually any eukaryotic cell population following exposure to any genotoxic material(s) that can be obtained as a single-cell suspension [66–68]. The comet assay is now accepted by regulatory authorities (OECD 489) as the second *in vivo* study [69] in place of the rat liver UDS assay, which is now generally considered to lack sensitivity. As the second *in vivo* test, only the liver would be examined routinely but other tissues including peripheral lymphocytes, stomach, and bone marrow are routinely examined when needed.

After single-cell suspensions have been prepared from the relevant tissue, they are embedded in agarose gel on glass microscope slides and lysed to rupture the cell membranes, extract the nuclear proteins, and leave supercoiled DNA. The DNA is unwound in strong alkaline buffer and then electrophoresed [66]. DNA damage is detected as an increase in the migration of DNA resulting from changes in the conformation and molecular weight of DNA and is measured as the amount of DNA present in the comet tail (Figure 6.6). The method has been used for the detection of DNA damage in cells exposed to chemical and physical agents under *in vitro* and *in vivo* conditions [66, 70–72].

The comet assay requires samples to be analyzed at both three to six hours (or at C_{max}, if known). To minimize the number of animals used, doses are given on three consecutive days and the tissues collected and processed typically three to six hours after the last dose (on the third day) day. If it is necessary to perform both the comet and bone marrow micronucleus tests, then it is possible to combine them.

6.3.4 Transgenic Rodent (TGR) Mutation Assay

Within regulatory genetic toxicology testing, the development of the *in vivo* TGR mutation models allowed the detection and quantification of somatic mutations in multiple tissues, which provided direct insight into a compound's potential carcinogenic mechanism [73, 74]. The various TGR mutation assays were developed by introducing multiple copies of surrogate neutral bacterial reporter transgenes into the genome of rodents (Figure 6.7) that could subsequently be recovered and assessed for mutation *in vitro* in a bacterial host using conventional *in vitro* biochemical techniques [75]. A series of comprehensive reviews have been written that describe the origins and the general characteristics of the various TGR mutation assays in use today (e.g. Muta™Mouse λLacZ, Big Blue® λLacI, CII, gpt-delta, and λLacZ plasmid) and robust descriptions of the biochemical methods used to measure and verify the induction of *in vivo* mutation by a test chemical [75, 76]. These publications have provided consensus on study design and statistical analysis resulting in standardized protocols that have maximized the capacity of the various TGR mutation assays to detect the *in vivo* mutagenicity of known DNA-reactive carcinogens in array of target organs. The protocols have recommendations related to the route of administration, study duration, and treatment times (i.e. ≤28 days); recommendations on the selection of tissues including specific sampling times for certain tissues based on factors such as cell proliferation rate (e.g. +3 days postdosing for somatic tissues and +28 days for germ cells); and procedures for confirmatory DNA sequencing [76]. In short, this collaborative academic and industry approach to gain consensus on protocols and study design provided confidence in the robustness of the TGR mutation assay and ultimately allowed the development of the OECD 488 "Test Guideline on Transgenic Rodent Gene Mutation Assays" [77–81].

Considerable discussion regarding the biological responses of the TGR mutation assay has also occurred within these core publications; these include but are not limited to: differences in the mutation and repair rates of the transgene versus endogenous genes; spontaneous mutation frequencies across tissues and their effect on the sensitivity of the assay; concordance between target tissues in TGR mutation assay and associated rodent carcinogenicity assays for exemplar DNA-reactive carcinogens; variation between specific tissues to reach peak mutation frequency; and the effects of biological processes (e.g. clonal expansion) on the mutation frequency following subchronic (12+ weeks) treatment times [76, 77]. These exhaustive publications also provide a valuable database of the 200+ compounds that have been evaluated in the TGR mutation assay

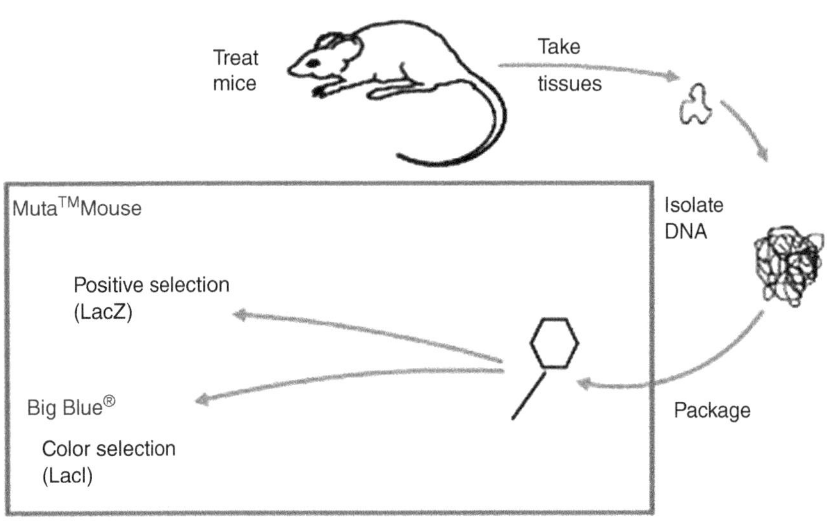

Figure 6.7 The experimental procedure for Big Blue® and Muta™Mouse assays. *Source:* Taken from [75].

that cover a broad range of chemical classes including aromatic amines, alkylating agents, aziridines, hydrazines, propiolactones, aromatic nitros, polycyclic aromatic hydrocarbons, epoxides, *N*-nitrosamines, azos, and aflatoxin-like compounds that are representative of industrial chemicals, exemplar DNA-reactive research compounds, environmental and food contaminants, pharmaceuticals, pesticides, and natural products [77, 78]. Although the compound set could be considered unbalanced from a rodent carcinogenicity perspective (the ratio of rodent carcinogens to noncarcinogens is approximately 3:1), a detailed review of the data indicated that the TGR mutation assay is highly sensitive (i.e. there is a high probability that a chemical with a positive result in the TGR mutation assay is a rodent carcinogen). Although the data indicate there is a low probability that a chemical with a negative result in the TGR mutation assay is a rodent noncarcinogen, this is no different to the bacterial mutation assay itself [77]. Of note, the data also indicated that the *true* positive predictivity (i.e. the number of mutant phenotypes identified by the biochemical selection techniques that were subsequently confirmed as true mutations by DNA sequencing) was extremely high (i.e. the TGR mutation assay correctly identifies *in vivo* mutagens). This concordance analysis has demonstrated that the TGR mutation assay is robust (being qualitatively and quantitively reproducible under standardized conditions), and hence the TGR mutation assay is widely considered to provide valuable insight into the relationships between DNA damage, mutation, and carcinogenicity [77]. This consensus position [77–80] and the finalization of the OECD 488 guideline [81] facilitated deployment of the TGR mutation assay within the field of regulatory genetic toxicology testing paradigms to ascertain the relevance of positive findings in both *in vitro* bacterial and mammalian assays, thereby allowing an *in vivo* assessment of the potential mechanism of action (i.e. mutagenic or non-mutagenic) of suspected carcinogens and ultimately improving the cancer risk assessment process [15]. The subsequent introduction of the ICH M7 impurities guideline for the control of DNA-reactive (mutagenic) impurities in pharmaceuticals provided a unique setting for the use of the *in vivo* TGR mutation assay given the recognition that impurities in drug substance and drug products will often be DNA reactive due to the nature of synthetic chemistry processes [1]. Specifically, the ICH M7 outlined a series of tests that could be used to investigate the *in vivo* relevance of *in vitro* bacterial mutagens (i.e. the TGR mutation, *Pig-a*, micronucleus, UDS, and comet assays) the "results of which could support setting alternative compound specific impurity limits in final drug product". The ICH M7 guideline indicated that the *in vivo* TGR mutation assay could be used to investigate the *in vivo* relevance of any bacterial mutagenicity positive that provided a strong rationale for the use of *in vivo* TGR mutation assay, as opposed to the use of a surrogate end point (e.g. chromosome damage), and the gene mutation end point could be assessed in any target tissue. In short the assay was considered to be appropriate to investigate the *in vivo* relevance of *any* positive finding bacterial reverse mutation assay [1, 76].

Under the ICH M7 framework, the results of the TGR mutation assay could be used in a categorical manner, to determine whether an impurity that was considered to be a "Class 2" mutagenic impurity based on an *in vitro* bacterial mutagenicity data alone could be confirmed as a "Class 2" mutagenic impurity or recategorized as "Class 5" non-mutagenic impurity based on a positive or negative response in the assay, respectively. Furthermore, the results of the TGR mutation assay can support a weight of evidence approach as to whether *in vivo* mammalian mutagen has a biological threshold (Figure 6.8) with an associated no-observed-adverse effect level (NOAEL). Specifically, if the mechanism of action of an *in vitro* and *in vivo* mutagenic impurity is shown to have practical biological threshold, data from the *in vivo* TGR mutation assay could then be used to derive compound-specific impurity limits in final drug product based on the permitted daily exposure (PDE) principles outlined in ICH Q3C [83]. This latter approach was taken following the discovery of the Class 2 mutagenic impurity EMS in the Viracept drug product following a production

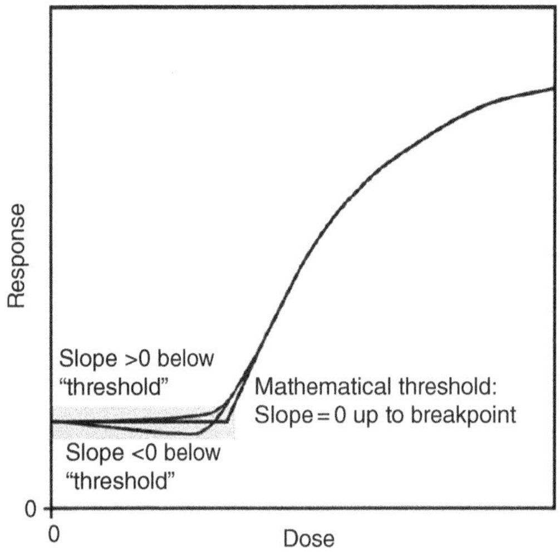

Figure 6.8 Schematic representation of three "threshold" dose–response curves. *Source:* Taken from [82].

accident [82]. EMS is an alkylating agent, a model mutagen, and a presumptive rodent carcinogen, and the default assumption would be that the entity induced mutation via a direct non-thresholded mechanism of action that would be associated with a linear dose–response curve. However, subsequent *in vitro* mechanistic studies and *in vivo* studies in the Muta™Mouse TGR mutation assay indicated that EMS-induced mutation via a direct but thresholded mechanism of action (a biological threshold related to saturation of DNA repair systems) that was confirmed mathematically by the demonstration of a nonlinear dose–response curve [84, 85]. As a consequence, a threshold-type risk assessment was conducted in which the TGR mutation assay data that included the derivation of (a) a PDE for EMS of 10 mg/day based on a cross-species exposure extrapolation from the TGR mutation assay to humans using pharmacokinetic/pharmacodynamic (PK/PD) modeling approaches and (b) a more conservative PDE for EMS of 104 μg/day using the principles and safety factors outlined in the ICH Q3C guideline [83]. In the particular case study, the TGR mutation assay data allowed the levels of EMS in Viracept drug product to be placed into context and provided valuable insight into the concerns raised by the incident related to patient safety.

In summary the TGR mutation assay is widely considered to provide valuable insight into the *in vivo* relevance of *in vitro* bacterial mutagenicity findings and has made a significant contribution to the understanding of the relationships between DNA damage, mutation, and carcinogenicity. In relation to ICH M7, given its value in terms of "setting alternative compound specific impurity limits in final drug product," the TGR mutation assay is considered by many to be the most robust approach to take when managing late stage impurity issue.

6.4 Conclusions

Assessing the mutagenicity of impurities following an *in silico* SAR evaluation is primarily resolved by the bacterial reverse mutation (i.e. Ames) assay. If an Ames positive compound cannot be controlled to the TTC, then the sponsor should consider conducting an *in vivo* mutagenicity study. The TGR assay can be conducted for nearly all Ames positive compounds but is much more resource intensive in terms of personnel, scientific expertise and compound needs, and extra time (~6 months) is needed to generate transgenic animals. Other *in vivo* mutagenicity assays can be conducted more rapidly, but the appropriateness of the assay for an impurity is dependent on the Ames response and potential mechanism for mutagenicity. It is less often to perform DNA binding assays for impurities, but they can be useful biomarkers of exposure and to further understand the mechanism of mutagenicity.

Glossary

Aneuploidy: Small increases or decreases in the modal number of chromosomes in a cell or organism. It may arise spontaneously or be induced by an aneugen.

Buffy coat: Is the fraction of an anticoagulated blood sample after density gradient centrifugation that contains most of the white blood cells and platelets.

Chromatid: The two halves into which a chromosome is longitudinally divided at mitosis. These are held together at the centromere and part from each to become daughter chromosomes at mitosis.

Chromatin: The component of the nucleus that contains the genetic material; it describes the chromosomes visible at mitosis and the more diffuse arrangement of the genetic material in the interphase cell.

Chromosome: The individual thread-like structures in the cell nucleus comprising double helices of DNA complexed with proteins and RNA. They carry the genetic information in a linear array of functional units (genes).

Clastogen: An agent that produces structural breakage of chromosomes, usually detectable by light microscopy.

Deletions: Remove one or more nucleotides from the DNA and may cause frameshifts; large deletions can cause loss of most or all of a gene.

DNA strand breaks: Single- or double-strand scissions in DNA.

Excision repair: DNA excision repair is used when only one of the strands of the DNA helix has a defect and the other strand is used as a template to repair the damage. There are a number of excision repair mechanisms i.e. base excision repair, nucleotide excision repair, and mismatch repair each responding to different types of DNA damage.

Frameshift mutations: A mutation (change in the genetic code) in which one base or two adjacent bases are added to (inserted in) or deleted from the nucleotide sequence of a gene. This may lead to an altered or truncated protein.

Gene conversion: Results from recombination; DNA sequence information is transferred from one DNA helix, which is not altered, to another helix, the sequence of which is altered.

Gene mutation: A detectable permanent change within a single gene or its regulating sequences. The changes may be point mutations, frameshift mutations, insertions, or deletions.

Insertions: Add one or more nucleotides into the DNA and may alter the gene product by affecting messenger ribonucleic acid (mRNA) splicing or causing frameshifts.

Mitosis: The process by which a cell nucleus divides into two daughter nuclei with chromosome numbers and genetic makeup identical to the parent cell.

Nondisjunction: An error at mitosis that results in the two daughter cells not receiving the correct number of chromosomes so that both become aneuploid.

Operon: A unit of genetic transcription comprising adjacent structural genes and a promoter region at one end, where the transcription of the structural genes into messenger RNA begins.

Point mutation: Change in the genetic code, usually confined to a single DNA base pair.

Polyploidy: A multiple of the total chromosome complement.

Recombination: Breakage and balanced or unbalanced rejoining of DNA.

Translocation: A chromosome translocation is an abnormality caused by a rearrangement of parts between nonhomologous chromosomes; it can result in loss of gene function or altered gene expression.

References

1 ICH M7. Assessment and control of DNA reactive (mutagenic) impurities in pharmaceuticals to limit potential carcinogenic risk. https://www.ema.europa.eu/documents/scientific-guideline/ich-guideline-m7r1-assessment-control-dna-reactive-mutagenic-impurities-pharmaceuticals-limit_en.pdf (accessed September 2020).
2 Overview of the set of OECD genetic toxicology test guidelines and updates performed in 2014–2015, Series on Testing and Assessment Nr. 238. Paris, 2016.
3 Hasselgren, C., Ahlberg, E., Akahori, Y. et al. (2019 Oct). Genetic toxicology in silico protocol. *Regul. Toxicol. Pharmacol.* 107: 104403.
4 Ames, B.N., Durston, W.E., Yamasaki, E.E., and Lee, F.D. (1973). *Proc. Natl. Acad. Sci. USA* 72: 2423–2427.
5 ICH Q3A Impurities in New Drug Substances. https://www.ich.org/page/quality-guidelines (accessed September 2020).
6 ICH Q3B Impurities in New Drug Products. https://www.ich.org/page/quality-guidelines (accessed September 2020).
7 Auerbach, C. and Robson, J.M. (1947). The production of mutations by chemical substances. *Proc. R. Soc. Edinb. Biol.* 62: 271–283.
8 Auerbach, C. and Robson, J.M. (1947). Tests of chemical substances for mutagenic action. *Proc. R. Soc. Edinb. Bio.* 62: 284–291.
9 Ames, B.N., Durston, W.E., Yamasaki, E., and Lee, F.D. (1973). Carcinogens are mutagens: a simple system combining liver homogenates for activation and bacteria for detection. *Proc. Natl. Acad. Sci. U. S. A.* 72: 2423–2427.
10 Lijinsky, W. (1979). A view of the relationship between mutagenesis and carcinogenesis. *Environ. Mol. Mutag.* 14 (Suppl. 16): 78–84.
11 Grover, P.L. and Sims, P. (1968). Enzyme catalysed reactions of polycyclic aromatic hydrocarbons with deoxyribonucleic acid and protein in vitro. *Biochem. J.* 110: 159–160.
12 Gelboin, H.V. (1969). A microsome-dependent binding of benzo[a]pyrene to DNA. *Cancer Res.* 29: 1272–1276.
13 Ames, B.N., McCann, J., and Yamasaki, E. (1975). Methods for detecting carcinogens and mutagens with the *Salmonella*/mammalian microsome mutagenicity test. *Mutat. Res.* 31: 347–364.
14 Green, M.H.L. and Muriel, W.J. (1976). Mutagen testing using TRP^+ reversion in *Escherichia coli*. *Mutat. Res.* 38: 3–32.
15 ICH Topic S2A. (1995). Genotoxicity: guidance on specific aspects of regulatory genotoxicity tests for pharmaceuticals. *International Conference on Harmonisation of Technical Requirements for Registration of Pharmaceuticals for Human Use*. Geneva.
16 Mortelmans, K. and Zeiger, E. (2000). The Ames *Salmonella*/microsome mutagenicity assay. *Mutat. Res.* 455: 29–60.
17 Matsushim, T., Sawamura, M., Hara, K., and Sugimura, T. (1976). A safe substitute for polychlorinated biphenyls as an inducer of metabolic activation systems. In: *in vitro Metabolic Activation in Mutagenesis Testing* (eds. F.J. De Serres, J.R. Fouts, J.R. Bens and R.M. Philpot), 85–88. Amsterdam: Elsevier/North Holland.
18 Ong, T.-M., Mukhtar, M., Wolf, C.R., and Zeiger, E. (1980). Differential effects of cytochrome p450 inducers on promutagen activation capabilities and enzymatic activities of S9 from rat liver. *J. Environ. Pathol. Toxicol.* 4: 55–65.
19 Guideline for the Testing of Chemicals No. 471. (1997). *Bacterial Reverse Mutation Test. Organisation for Economic Cooperation and Development*. Paris.

20 Gatehouse, D., Haworth, S., Cebula, T. et al. (1994). Recommendations for the performance of bacterial mutation assays. *Mutat. Res.* 312: 217–233.

21 Maron, D.M. and Ames, B.N. (1983). Revised methods for the Salmonella mutagenicity test. *Mutat. Res.* 113: 173–215.

22 Mortelmans, K. and Riccio, E.S. (2000). The bacterial tryptophan reverse mutation assay with *Eschericia coli* WP2. *Mutat. Res.* 455: 61–69.

23 ICH S1 (R1) guideline on the need for carcinogenicity studies of pharmaceuticals. https://www.ich.org/ (accessed September 2020).

24 Amberg, A., Harvey, J.S., Czich, A. et al. (2015). Do carboxylic/sulfonic acid halides really present a mutagenic and carcinogenic risk as impurities in final drug products? *Org. Process Res. Dev.* 19: 1495–1506.

25 Yahagi, T., Nagao, M., Seino, Y. et al. (1977). *Mutat. Res.* 48: 121–130.

26 Escobar, P.A., Kemper, R.A., Nicolette, J. et al. (2013). Bacterial mutagenicity screening in the pharmaceutical industry. *Mutat. Res.* 752 (2): 99–118.

27 Nicolette, J., Dakoulas, E., Pant, K. et al. (2018 Dec). A comparison of 24 chemicals in the six-well bacterial reverse mutation assay to the standard 100-mm Petri plate bacterial reverse mutation assay in two laboratories. *Regul. Toxicol. Pharmacol.* **100**: 134–160.

28 Brooks, T.M. (1995). The use of a streamlined bacterial mutagenicity assay, the MINISCREEN. *Mutagenesis* 10: 447–448.

29 Diehl, M., Willaby, S., and Snyder, R. (2000). Comparison of the results of a modified miniscreen and the standard bacterial reverse mutation assays. *Environ. Mol. Mutagen.* 35: 72–77.

30 Kenyon, M.O., Cheung, J.R., Dobo, K.L., and Ku, W.W. (2007 Jun). An evaluation of the sensitivity of the Ames assay to discern low-level mutagenic impurities. *Regul. Toxicol. Pharmacol.* 48 (1): 75–86. https://doi.org/10.1016/j.yrtph.2007.01.006. Epub 2007 Feb 4.

31 Arenaz, P., Hallberg, L., Mancillas, F. et al. (1989). Sodium azide mutagenesis: inability of mammalian cells to convert azide to a mutagenic intermediate. *Mutat. Res.* 227: 63–67.

32 Müller, L., Gocke, E., Lavé, T. et al. (2009). Ethyl methanesulfonate toxicity in Viracept – a comprehensive human risk assessment based on threshold data for genotoxicity. *Toxicol. Lett.* 190: 317–329.

33 Iida, Y., Takeda, J., Miyata, T. et al. (1994). Characterization of genomic *PIG-A* gene: a gene for glycosylphosphatidylinositol-anchor biosynthesis and paroxysmal nocturnal haemoglobinuria. *Blood* **83**: 3126–3131.

34 Kawagoe, K., Takeda, J., Endo, Y., and Kinoshita, T. (1994). Molecular cloning of murine *Pig-a*, a gene for GPI-anchor biosynthesis, and demonstration of interspecies conservation of its structure, function, and genetic locus. *Genomics* **23**: 566–574.

35 Araten, D.J., Nafa, K., Pakdeesuwan, K., and Luzzatto, L. (1999). Clonal populations of hematopoietic cells with paroxysmal nocturnal hemoglobinuria genotype and phenotype are present in normal individuals. *Proc. Natl. Acad. Sci. U. S. A.* **96**: 5209–5214.

36 Bryce, S.M., Bemis, J.C., and Dertinger, S.D. (2008). *in vivo* mutation assay based on the endogenous *Pig-a* locus. *Environ. Mol. Mutagen.* **49**: 256–264.

37 Miura, D., Dobrovolsky, V.N., Kasahara, Y. et al. (2008). Development of an *in vivo* gene mutation assay using the endogenous *Pig-A* gene: I. Flow cytometric detection of CD59-negative peripheral red blood cells and CD48-negative spleen T-cells from the rat. *Environ. Mol. Mutagen.* **49**: 614–621.

38 Shemansky, J.M., McDaniel, L.P., Klimas, C. et al. (2019). *Pig-a* gene mutation database. *Environ. Mol. Mutagen.* 60: 759–762.

39 Kimoto, T., Chikura, S., Suzuki, K. et al. (2011). Further development of the rat *Pig-a* mutation assay: measuring rat *Pig-a* mutant bone marrow erythroids and a high throughput assay for mutant peripheral blood reticulocytes. *Environ. Mol. Mutagen.* **52**: 774–783.

40 Dertinger, S.D., Bryce, S.M., Phonethepswath, S., and Avlasevich, S.L. (2011). When pigs fly: Immunomagnetic separation facilitates rapid determination of *Pig-a* mutant frequency by flow cytometric analysis. *Mutat. Res.* **721**: 163–170.

41 Dertinger, S.D., Phonethepswath, S., Weller, P. et al. (2011). Interlaboratory *Pig-a* gene mutation assay trial: Studies of 1,3-propane sultone with immunomagnetic enrichment of mutant erythrocytes. *Environ. Mol. Mutagen.* **52**: 748–755.

42 Kimoto, T., Suzuki, K., Kobayash, X.M. et al. (2011). Manifestation of *Pig-a* mutant bone marrow erythroids and peripheral blood erythrocytes in mice treated with *N*-ethyl-*N*-nitrosourea: direct sequencing of *Pig-a* cDNA from bone marrow cells negative for GPI-anchored protein expression. *Mutat. Res.* **723**: 36–42.

43 Revollo, J., Pearce, M.G., Petibone, D.M. et al. (2015). Confirmation of *Pig-a* mutation in flow cytometry-identified CD48-deficient T-lymphocytes from F344 rats. *Mutagenesis* **30**: 315–324.

44 Revollo, J.R., Crabtree, N.M., Pearce, M.G. et al. (2016). Mutation analysis with random DNA identifies (MARDI) catalogs *Pig-a* mutations in heterogeneous pools of CD48-deficient T cells derived from DMBA-treated rats. *Environ. Mol. Mutagen.* **57**: 114–124.

45 Nicklas, J.A., Carter, E.W., and Albertini, R.J. (2015). Both *PIGA* and *PIGL* mutations cause GPI-a deficient isolates in the TK6 cell line. *Environ. Mol. Mutagen.* **56**: 663–673.

46 Krüger, C.T., Fischer, B.M., Armant, O. et al. (2016). The *in vitro PIG-A* gene mutation assay: glycosylphosphatidylinositol (GPI)-related genotype-to-phenotype relationship in TK6 cells. *Arch. Toxicol.* **90**: 1729–1736.

47 Wang, Y., Revollo, J., McKinzie, P. et al. (2018). Establishing a novel *Pig-a* gene mutation assay in L5178YTk+/- mouse lymphoma cells. *Environ. Mol. Mutagen.* **59**: 4–17.

48 Bemis, J.C., Avlasevich, S.L., Labash, C. et al. (2018). Glycosylphosphatidylinositol (GPI) anchored protein deficiency serves as a reliable reporter of *Pig-a* gene mutation: Support from an *in vitro* assay based on L5178YTK+/- cells and the CD90.2 antigen. *Environ. Mol. Mutagen.* **59**: 18–29.

49 Gollapudi, B.B., Lynch, A.M., Heflich, R.H. et al. (2015). The *in vivo Pig-a* assay: a report of the International Workshop on Genotoxicity Testing (IWGT) Workgroup. *Mutat. Res.* **783**: 23–35.

50 US EPA OPP (US Environmental Protection Agency, Office of Pesticide Programs). (2012). Advances in genetic toxicology and integration of *in vivo* testing into standard repeat dose studies. http://www.epa.gov/pesticides/science/integrating-gentox-studies.html (accessed September 2020).

51 Masuda-Herrera, M.J., Dobo, K.L., Kenyon, M.O. et al. (2019). *in vivo* mutagenicity testing of arylboronic acids and esters. *Environ. Mol. Mutagen.* **60**: 766–777.

52 O'Donovan, M.R., Mee, C.D., Fenner, S. et al. (2011). Boronic acids-a novel class of bacterial mutagen. *Mutat. Res.* **724**: 1–6.

53 Schmid, W. (1975). The micronucleus test. *Mutat. Res.* 31: 9–15.

54 Higashikuni, N. and Sutou, S. (1995). An optimal, generalized sampling time of 30 +/- 6 h after double dosing in the mouse peripheral blood micronucleus test. *Mutagenesis* 10 (4): 313–319.

55 Hayashi, M., JT, M.G., Gatehouse, D.G. et al. (2000). in vivo rodent erythrocyte micronucleus assay. II. Some aspects of protocol design including repeated treatments, integration with toxicity testing, and automated scoring. *Environ. Mol. Mutagen.* 35 (3): 234–252.

56 MacGregor, J.T., Wehr, C.M., Henika, P.R., and Shelby, M.D. (1990). The in vivo erythrocyte micronucleus test: measurement at steady state increases assay efficiency and permits integration with toxicity studies. *Fundam. Appl. Toxicol.* 14 (3): 513–522.

57 Pascoe, S. and Gatehouse, D. (1986). The use of a simple haemotoxylin and eosin staining procedure to demonstrate micronuclei within rodent bone-marrow. *Mutat. Res.* 164: 237–243.

58 Hayashi, M., Sofuni, T., and Ishidate, M. (1983). An application of acridine orange fluorescent staining to the micronucleus test. *Mutat. Res.* 120: 241–247.

59 Guideline for the Testing of Chemicals No. 474. (1997). Mammalian Erythrocyte Micronucleus Test. *Organisation for Economic Cooperation and Development, Paris.*

60 Wakata, A., Miyamae, Y., Sato, S. et al. (1998). Evaluation of the rat micronucleus test with bone marrow and peripheral blood: summary of the 9th Collaborative Study by CSGMT/JEMS.MMS. *Env. Mol. Mutag.* 32: 84–100.

61 Hamada, A., Sutou, S., Morita, T. et al. (2002). Evaluation of the rodent micronucleus assay by a 28-day treatment protocol: summary of the 13th collaborative study by the collaborative study group for the micronucleus test (CSGMT)/Environmental Mutagen Society of Japan (JEMS)-Mammalian Mutagenicity Study Group (MMS). *Environ. Mol. Mutag.* 37: 93–110.

62 Hayashi, M., MacGregor, J.T., Gatehouse, D.G. et al. (2007). *in vivo* rodent erythrocyte micronucleus assay. III. Validation and regulatory acceptance of automated scoring and the use of rat peripheral blood reticulocytes with discussion of non-hematopoietic cells and a single dose-level limit test. *Mutat. Res.* 627: 10–30.

63 MacGregor, J.T., Bishop, M.E., McNamee, J.P. et al. (2006). Flow cytometric analysis of micronuclei in peripheral blood reticulocytes. II. An efficient method of monitoring chromosomal damage in the rat. *Toxicol. Sci.* 94: 92–107.

64 Kissling, G.E., Dertinger, S.D., Hayashi, M., and MacGregor, J.T. (2007). Sensitivity of the erythrocyte micronucleus assay: dependence on number of cells scored and inter-animal variability. *Mutat. Res.* 634: 235–240.

65 Hayashi, M., MacGregor, J.T., Gatehouse, D.G. et al. (2000). *in vivo* rodent erythrocyte micronucleus assay. II. Some aspects of protocol design included repeated treatments, integration with toxicity testing and automated scoring. *Environ. Mol. Mutag.* 35: 234–252.

66 Singh, N.P., McCoy, M.T., Tice, R.R., and Schneider, E.L. (1988). A simple technique for quantitation of low levels of DNA damage in individual cells. *Exp. Cell Res.* 175: 184–191.

67 Ostling, O. and Johnson, K.L. (1984). Micro-electrophoretic study of radiation-induced DNA damage in individual mammalian cells. *Biochem. Biophys. Res. Commun.* 175: 184–191.

68 McKelvey-Martin, V.J., Green, M.H.L., Schmezer, P. et al. (1993). The single cell gel electrophoresis assay (Comet assay): a European review. *Mutat. Res.* 288: 47–63.

69 Brendler-Schwaab, S., Hartmann, A., Pfuhler, S., and Speit, G. (2005). The in vivo comet assay: use and status in genotoxicity testing. *Mutagenesis* 20: 245–254.

70 Tice, R.R., Andrews, P.W., Hirai, O., and Singh, N.P. (1991). The single cell gel (SCG) assay: an electrophoretic technique for the detection of DNA damage in individual cells. In: *Biological Reactive Intermediates IV, Molecular and Cellular Effects and Their Impact on Human Health*, vol. 157 (eds. C.R. Witmer, R.R. Snyder, D.J. Jollow, et al.), 164. New York: Plenum Press.

71 Singh, N.P., Tice, R.R., and Schneider, E.L. (1991). A microgel electrophoresis technique for the direct quantitation of DNA damage and repair in individual fibroblasts cultured on microscope slides. *Mutat. Res.* 252: 289–296.

72 Tice, R.R., Strauss, G.H.S., and Peters, W.P. (1992). High-dose combination alkylating agents with autologous bone marrow support in patients with breast cancer: preliminary assessment of DNA damage in individual peripheral blood lymphocytes. *Mutat. Res.* 271: 101–113.

73 Heddle, J.A., Dean, S., Nohmi, T. et al. (2000). in vivo transgenic mutation assays. *Environ. Mol. Mutagen.* 35: 253–259.

74 Thybaud, V., Dean, S., Nohmi, T. et al. (2003). in vivo transgenic mutation assays. *Mutat. Res.* 540: 141–151.

75 Dean, S. (2012). *Genetic Toxicology. Methods in Molecular Biology (Methods and Protocols)*, vol. 817 (eds. J. Parry and E. Parry). New York, NY: Springer.

76 Lambert, I.B., Singer, T.M., Boucher, S.E., and Douglas, G.R. (2015). Detailed review of transgenic rodent mutation assays. *Mutat. Res.* 590: 1–280.

77 Detailed review paper on transgenic rodent mutation assays. (2009). Series on testing and assessment. Number 103. http://www.oecd.org/ ENV/JM/MONO (2009) 7 Part 2: Annexes to the detailed review paper on transgenic rodent mutation Assays. Number 103. http://www.oecd.org/ENV/JM/MONO(2009)29 (accessed September 2020).

78 Transgenic rodent somatic and germ cell gene mutation assays: retrospective performance assessment. Number 145. http://www.oecd.org/ENV/JM/MONO(2011)20 (accessed September 2020).

79 Test No. 488: Transgenic Rodent Somatic and Germ Cell Gene Mutation Assays. www.oecd-ilibrary.org (accessed September 2020).

80 Draft TG 488: Transgenic Rodent Somatic and Germ Cell Gene Mutation Assays. www.oecd-ilibrary.org (accessed September 2020).

81 OECD (2020). *Test No. 488: Transgenic Rodent Somatic and Germ Cell Gene Mutation Assays*, OECD Guidelines for the Testing of Chemicals, Section 4. Paris: OECD Publishing https://doi.org/10.1787/9789264203907-en.

82 Lutz, W.K. (2009). **The** Viracept (nelfinavir)--ethyl methanesulfonate case: a threshold risk assessment for human exposure to a genotoxic drug contamination? *Toxicol. Lett.* 190: 239–242.

83 ICH guideline Q3C (R6) on impurities: guideline for residual solvents. www.ich.org (accessed September 2020).

84 Gocke, E., Ballantyne, M., Whitwell, J., and Müller, L. (2009). MNT and MutaMouse studies to define the in vivo dose response relations of the genotoxicity of EMS and ENU. *Toxicol. Lett.* 190: 286–297.

85 Gocke, E. and Müller, L. (2009). Considerations regarding a permitted daily exposure calculation for ethyl methanesulfonate. *Toxicol. Lett.* 190: 330–332.

7

Compound- and Class-Specific Limits for Common Impurities in Pharmaceuticals

Joel P. Bercu[1], Melisa J. Masuda-Herrera[1], Alejandra Trejo-Martin[1], David J. Snodin[2], Kevin P. Cross[3], George E. Johnson[4], James Harvey[5], Robert S. Foster[6], David J. Ponting[6], and Richard V. Williams[6,†]

[1] *Gilead Sciences, Foster City, CA, USA*
[2] *Xiphora Biopharma Consulting, Bristol, UK*
[3] *Instem, Columbus, OH, USA*
[4] *Institute of Life Science, Swansea University Medical School, Singleton Park, UK*
[5] *GlaxoSmithKline R&D, Hertfordshire, UK*
[6] *Lhasa Limited, Leeds, UK*

7.1 Introduction

The synthesis of pharmaceuticals involves the use of many reactive reagents, solvents, and starting materials, and the resulting drug substances can contain a cariety of by-products. In pharmaceutical development, active measures are put in place to minimize patient exposure to these compounds as impurities. The process chemistry is designed and refined not only to synthesize the drug substance but also to develop processes that can purge impurities. In addition, the drug substance is formulated and packaged in such a way to minimize degradation products as well as drug substance–excipient interaction impurities.

However, exposure to impurities and degradation products is unavoidable. The main process for removal of impurities is crystallization [1]. Crystallization can be inefficient with respect to reducing impurities to diminishingly low levels while using a large supply of raw materials and generating an excess of waste. In addition, there is a balance between removing impurities and affecting other quality attributes. In the case of degradation, this process can be innate to a drug substance and drug product. The rate of degradation can be slowed through formulation and packaging, but these measures cannot eliminate degradation. Therefore, impurities and degradation products should be controlled to acceptable levels, but 100% purity for a drug substance or drug product is impractical.

In analytical chemistry there is no such thing as "zero exposure." Analytical techniques are simply constrained by their levels of detection and quantitation. Modern analytical methodology can be used to detect impurities and degradation products at extremely low concentrations of exposure, but this will not make the product inherently safer. A more relevant question is determining the level of safe exposure to impurities and degradation products. This can be used to guide analytical chemistry on the appropriate limit of detection (LOD) and limit of quantitation (LOQ).

[†] Deceased

Mutagenic Impurities: Strategies for Identification and Control, First Edition. Edited by Andrew Teasdale.
© 2022 John Wiley & Sons, Inc. Published 2022 by John Wiley & Sons, Inc.

Table 7.1 Classes of mutagenic and/or carcinogenic impurities.

Class	Description	Toxicology limit
1	Mutagenic carcinogen	Compound-specific AI[a] or PDE[b]
2	Mutagen without carcinogenicity data	PDE or class-specific AI or TTC[e]
3	Predicted positive in (Q)SAR[c] for mutagenicity, alert not contained in the drug substance (or compounds related to the drug substance; e.g. process intermediates); no mutagenicity data[d]	Class-specific AI or TTC[e]
4	Predicting positive in (Q)SAR but the alerting region in a similar chemical – space is contained in the drug substance or compounds related to the drug substance and was negative for mutagenicity	Treat as nonmutagenic impurity
5	Predicted negative in (Q)SAR for mutagenicity; or tested for mutagenicity and not positive; or mutagenic but negative for carcinogenicity	Treat as nonmutagenic impurity

[a] AI – acceptable intake.
[b] PDE – permissible (permitted) daily exposure.
[c] (Q)SAR – (quantitative) structural activity relationship.
[d] Mutagenicity generally refers to bacterial reverse mutation assay. However, other mutagenicity assays such as *in vivo* mutagenicity can be used as well.
[e] TTC – Threshold of Toxicological Concern (for carcinogenic activity).

Toxicologists play a vital role working with analytical and process chemistry providing targets for exposures to impurities and degradation products. Unfortunately, for most drug substances and products, there is little toxicology information on their impurities and degradation products. This is because the chemical space for new pharmaceuticals is novel. There are processes outlined in International Council for Harmonisation (ICH) Q3A(R2) and Q3B(R2) guidelines to handle novel impurities in drug substances and products [2, 3]. Within the guidelines, it states that lower limits are needed if an impurity is unusually toxic or potent. The ICH M7(R1) guideline describes methods for identifying and controlling impurities and degradation products that are mutagenic and so may be potential or known carcinogens [4].

There are five classes of impurities described in ICH M7(R1) and highlighted in Table 7.1. The type of toxicology limit is dependent on the class of impurity. For a Class 1 impurity, i.e. mutagenic carcinogen, the carcinogenicity data is available to generate a "compound-specific" limit. It is typically assumed in the absence of contrary data that mutagenic carcinogens do not have a threshold. In this case, an acceptable intake (AI) is developed based on linear low-dose extrapolation from carcinogenicity data. In some cases, a threshold-related mode of action (MOA) is related to the compound's toxicity and thus a permissible (or permitted) daily exposure (PDE) can be developed based on applying adjustment factors (AF) (also referred to as modifying or uncertainty factors) to a no-observed effect level (NOEL)/no-observed adverse effect level (NOAEL) or lowest-observed effect level (LOEL)/lowest-observed adverse effect level (LOAEL).

Class 2 impurities are mutagenic impurities that do not have carcinogenicity information. Therefore, developing an AI based on linear low-dose extrapolation is not possible. Thus, the Threshold of Toxicological Concern (TTC) concept is applied, which is a dose that has a high probability of not exceeding a 1 in 100 000 excess risk of cancer even when the carcinogenic potency is unknown [5, 6]. In some cases a class-specific AI can be developed for the mutagenic compound when no carcinogenicity information is available [4]. A class-specific AI is a limit based on an analysis of carcinogenic potency of compounds with a specific structural alerting feature. Finally, a PDE can be developed based on mutagenicity data. Dose-response information and a threshold for mutagenicity can be developed from an *in vivo* mutagenicity assay [7, 8]. Examples of *in vivo* mutagenicity assays can include the transgenic rodent or *Pig-a* assays [9–11].

Class 3 impurities are predicted to be positive in the bacterial reverse mutation assay. Similar to Class 2 impurities, the carcinogenic potency of Class 3 impurities is unknown. Thus, the TTC is used to generate a toxicology limit or a class-specific limit could be applied depending on the type of structural alert.

Class 4 and 5 impurities are not considered mutagenic or carcinogenic. In these cases, they are treated like a nonmutagenic impurity and they can be qualified in nonclinical toxicology studies conducted with the parent molecule [2, 3] or alternatively, if the data exists on the entity itself, a PDE can be developed. If this is not possible, then the ICH Q3A/Q3B qualification thresholds can be applied, noting that these limits can be modified based on scientific rationale and level of concern, including drug class effects and clinical experience [2, 3]. While Class 4 and 5 impurities are out of scope for ICH M7(R1), they can often be confused as Class 1, 2, or 3 impurities given that some compounds have structurally alerting features for mutagenicity but are not mutagenic and/or carcinogenic.

Efforts have been made to document class and compound-specific AIs or PDEs of potential impurities in drug substances. The goal of these efforts is to harmonize approaches for toxicology-based risk assessment for common impurities in pharmaceuticals. This chapter will describe these activities in further detail, discuss case studies, and provide some perspectives for further development.

For each compound-specific AI or PDE, a monograph was developed to document the rationale for the compound-specific limit. The monographs were developed to express key information on each chemical. The following are subsections and types of information developed in each monograph. While the focus of this book is on mutagenic impurities, compound-specific AIs and PDEs are also relevant for nonmutagenic or noncarcinogenic impurities especially those that are commonly confused as carcinogenic or mutagenic.

7.2 Monograph Development

To develop an AI or PDE for impurities, the health hazards should be listed for the impurity. Prior to developing the monograph, a literature search should be performed to identify available data and information for the monograph. Given the large number of databases that are available, it is useful to search using a meta-database that searches multiple databases at once and can generate data from multiple sources or databases of summarized toxicity information of chemicals. Examples of commonly used, but not all-inclusive, databases used to search for toxicology data are shown in Table 7.2. Some databases are proprietary, while others are based on publicly available information. Databases included are meta-databases or databases that have summary toxicology information for a compound. In most cases the toxicology data on test chemicals in these databases has been generated to a standard that meets international guidelines (e.g. Organisation for Economic Co-operation and Development [OECD]), and in certain cases the data has even been summarized and subjected to rigorous peer review that rapidly facilitates the generation of compound-specific limits. Where this has not happened, the quality of the toxicology data will always require further assessment specifically to decide upon its reliability, relevance, and adequacy (basic review principles also known as the "Klimisch score") before it can be used to generate compound-specific limits [12].

7.2.1 Exposure to the General Population

In ICH M7(R1), higher toxicology limits could be justified when human exposure is greater from other sources such as food or endogenous metabolism [4]. The general population exposure to a specific compound may occur through the environment (via air and water), food and food additives, cosmetics, occur naturally, or be generated in human, animal, or plant metabolism. The monograph takes into consideration these sources of exposure and compares with the potential exposure of the

Table 7.2 Examples of commonly searched toxicity databases.

Database	Description	Link or source
CDC ATSDR	Toxicology summary profiles for chemicals	https://www.atsdr.cdc.gov/toxprofiledocs/index.html
CEBS database	Toxicity information generated by NTP	https://manticore.niehs.nih.gov/cebssearch
COSMOS database	Curated toxicity information for chemicals	http://www.cosmostox.eu/what/COSMOSdb/
CPDB	Database for carcinogenicity risk potency estimates (no longer being updated)	https://files.toxplanet.com/cpdb/index.html
Echemportal	Meta-database toxicology search	https://www.echemportal.org/echemportal/index.action
ELSIE	Proprietary toxicity database for extractables and leachables	http://www.elsie.org/
EU Food Additive Database	Information on food additives approved for use in food in the EU and their conditions of use	https://ec.europa.eu/food/safety/food_improvement_agents/additives/database_en
Food Database	Information about compounds found in food	https://foodb.ca/
HERA database	Toxicology assessments of cleaning substances	https://www.heraproject.com/RiskAssessment.cfm
Human Metabolome Database	Information about compounds (i.e. metabolites) in the human body such as blood levels and excretion	https://hmdb.ca/
ICH Q3C, Q3D, and M7	Contain AIs/PDEs for common solvents, metals, and mutagenic/carcinogenic compounds	www.ich.org
JECDB	Japanese chemical database for toxicity information	http://dra4.nihs.go.jp/mhlw_data/jsp/SearchPageENG.jsp
Inchem database	Searches summary toxicity assessments provided by IPCS	http://www.inchem.org/#/search
Leadscope Toxicity Database	Curated proprietary toxicity data	http://www.leadscope.com/toxicity_database/
Lhasa Carcinogenicity Database	Database for carcinogenicity risk potency estimates	https://www.lhasalimited.org/Initiatives/lhasa-carcinogenicity-database.htm
PPRTV database	Toxicity values for chemicals assessed under USEPA Superfund	https://hhpprtv.ornl.gov/
Registered Substances Database	Summary of toxicology information provided to ECHA under REACH	https://echa.europa.eu/information-on-chemicals/registered-substances

Toxnet	Meta-database for toxicology information	https://toxnet.nlm.nih.gov/
Toxplanet	Proprietary meta-database for toxicology information	www.toxplanet.com
USEPA Pesticide Reregistration Status	Contains pesticide chemical-related information including toxicology testing	https://archive.epa.gov/pesticides/reregistration/web/html/status.html
USEPA IRIS	USEPA summary toxicity assessments for chemicals	https://www.epa.gov/iris
Vitic	Proprietary curated database of toxicology information. Also contains AI/PDEs for impurities	https://vitic.lhasalimited.org/vitic/query

AI – acceptable intake, ATSDR – Agency for Toxic Substances and Disease Registry, CDC – Centers for Disease Control and Prevention, CEBS – Chemical Effects in Biological Systems, CPDB – Carcinogenic Potency Database, ECHA – European Chemicals Agency, ELSIE – Extractables and Leachables Safety Information Exchange, EU – European Union, HERA – Human and Environmental Risk Assessment, ICH – International Conference on Harmonization, IPCS – International Programme on Chemical Safety, IRIS – Integrated Risk Information System, JECDB – Japan Existing Chemical Database, NTP – National Toxicology Program, PDE – permitted (permissible) daily exposure, PPRTV – Provisional Peer-Reviewed Toxicity Values, REACH – Registration Evaluation and Authorization of Chemicals, USEPA – United States Environmental Protection Agency.

compound as an impurity in drug substances or products to develop toxicology limits. The exposure in the general population is generally quantitated if feasible, but it is not possible in all cases.

7.2.2 Mutagenicity/Genotoxicity

Available mutagenicity data is reviewed giving more weight to *in vitro* and *in vivo* assays considered sufficiently validated according to ICH S2(R1) that use experimental protocols recommended by the OECD guidelines [13]. The bacterial reverse mutation (Ames) assay in accordance with OECD 471 and ICH S2(R1) is specifically evaluated to determine if the impurity should be considered Class 2 under ICH M7(R1) [4, 13, 14]. In cases where there is no available data to evaluate the potential mutagenicity of an impurity, a (quantitative) structure activity relationship ((Q)SAR) from two complementary methodologies (one statistically and the second rules-based) could be used to predict the outcome in the bacterial reverse mutation assay [4]. Other mutagenicity assays may be considered relevant for ICH M7 classification on a case-by-case basis. Other genotoxicity end points such as clastogenicity, aneugenicity, deoxyribonucleic acid (DNA) damage, etc., while less relevant to ICH M7(R1), should be collected based on *in vitro* and *in vivo* studies to fully understand the genotoxic hazards of the impurity [15].

7.2.3 Noncarcinogenic Effects

In this section, accessible toxicity studies via different routes of administration are reviewed. Oral is generally the most relevant route of administration, but given that pharmaceuticals may be administered intravenously, subcutaneously, intramuscularly, by inhalation, etc., other routes should also be considered.

Acute toxicity is evaluating doses that will cause serious adverse effects following a single short-term administration [16]. In this case, single or multiple doses may result in adverse effects that occur rapidly, including mortality. The major parameters measured are lethal oral/dermal dose or airborne concentration (LD_{50}, LC_{50}), but other subtler effects can be measured as well.

Repeat-dose toxicity studies measure adverse effects following subacute, subchronic, and chronic exposure. Detailed end points are measured in these studies, which typically include clinical observations, body weight, food/water consumption, hematology/clinical biochemistry, pathology, gross necropsy, and histopathology [17]. More weight is given to studies following regulatory guidelines (e.g. ICH/OECD) and conducted under good laboratory practices (GLPs) [18].

Reproductive toxicity studies focus on adverse effects on sexual function or fertility [16]. Effects can include alterations of male and female reproduction function or performance or adverse effects in reproductive organs.

Developmental toxicity studies determine the adverse effects on offspring of pregnant animals [16]. Interference with fetal development could occur prenatally, during gestation, or postnatally. Understanding whether there is fetal toxicity or teratogenicity is important when evaluating the severity of toxicity. Also important is whether the developmental toxicity occurred in the presence of maternal toxicity, which suggests that developmental effects were influenced by the health of the mother.

7.2.4 Carcinogenic Effects

Carcinogenicity evaluates tumor formation following chemical exposure, whether through a mutagenic or nonmutagenic MOA, and its relevance to humans [16, 19–22]. Evaluations and classifications from organizations such as the International Agency for Research on Cancer (IARC) or United States Environmental Protection Agency (USEPA) are provided to understand weight of evidence on carcinogenicity information. Available carcinogenicity studies are evaluated and summarized [23] with a discussion of each study considering:

- Adequacy of the experimental design and conduct (e.g. animal species, strain, sex, number per group, age at start of treatment, route of exposure, dose levels, duration of exposure, survival, and information on tumors)
- Statistical significance of the observed tumor response
- Tumor progression (e.g. benign to malignant or preneoplastic to neoplastic lesions)
- Dose–response relationships
- Lesions in the different sexes or species

7.2.5 Mode of Action (MOA) and Assessment of Human Relevance

To determine the appropriate data used to calculate the toxicology limit, an understanding of the toxicological dose–response relationship and possible MOA for the type of tumors seen in animals is evaluated. For a compound that is mutagenic in short-term studies, and no other MOA can be determined, it is reasonably assumed that the impurity would be a Class 1, mutagenic carcinogen.

There are compounds where the induction of tumors in rodents is through a nonmutagenic mechanism. In these cases there is sufficient evidence of a nonlinear dose-response curve for which a threshold or point of departure (PoD) below which carcinogenicity is not expected [24]. In some cases, the carcinogenesis may be a rodent-specific phenomenon, which is not relevant to humans as rodents are more sensitive than humans [20–22].

Compounds that are mutagenic can also exhibit a threshold depending on the MOA [8, 25]. In these cases, the *in vivo* mutagenic outcome may be used as the PoD to set the PDE.

7.2.6 Toxicokinetics

Toxicokinetic data providing quantitative information on the compound's absorption, distribution, metabolism, and elimination are used to extrapolate based on route of administration used in the toxicity study to the relevant pharmaceutical route of administration [26]. Toxicokinetics is also important when determining interspecies differences from animals to human and human variability [27]. Finally, toxicokinetics can be used to extrapolate data from daily to intermittent exposure or vice versa [26, 28].

7.2.7 Regulatory/Published Limits

Regulatory organizations have derived toxicology-based limits to established exposure levels that can be tolerated without adverse health effects. These limits have different names: reference dose (RfD), acceptable daily intake (ADI), tolerable daily intake (TDI), or acceptable daily exposure (ADE), depending on the reporting agency. Regulatory limits are calculated based on toxicity data from studies in animals and epidemiological data using different methodologies. They are valuable sources and are considered when developing compound-specific limits. Regulatory/published limits are located on the search sites listed in Table 7.2. Literature data should still be reviewed to determine if new toxicity information has been generated following the regulatory/published limit.

7.3 Derivation of the Compound-specific Limit

Compound-specific limits are protective for a lifelong exposure, for relevant routes of exposure, and to be protective of all patient populations including sensitive subpopulations. Therefore, the limits are calculated using acute, subacute, subchronic, chronic, reproductive, developmental, genotoxicity, and/or carcinogenicity exposure data.

7.3.1 PoD Selection

The PoD is the dose that is selected to derive a compound-specific limit [29]. The selection of the PoD takes into consideration the critical effect, MOA, and dose-response assessment. The critical effect is the most sensitive end point, or end point that would result in the lowest limit relevant to humans.

7.3.2 Limited Data Sets

For compounds that are deemed nonmutagenic and that do not have sufficient carcinogenicity or toxicity data to derive a PDE, the qualification thresholds from the ICH Q3A(R2) guidelines (0.15% or 1 mg/day, whichever is lower for an impurity in a drug substance of up to 2 g/day) are considered health protective based on a review of databases with subchronic and chronic toxicity studies [30]. In cases of compounds with positive results in the bacterial reverse mutation assay with insufficient or no carcinogenicity information to derive a compound-specific limit, the TTC-based AI of 1.5 μg/day is considered health protective for lifetime exposure (except for impurities defined as being part of the cohort of concern) [4].

7.3.3 PDE Development

In cases where the toxicity has a threshold (e.g. noncarcinogens, nonmutagens), the PoD is selected that includes the NOAEL/NOEL, LOAEL/LOEL, or modeled estimate benchmark dose (BMD) [29]. It is ideal to target no-effect, but it is also acceptable to target a no-adverse effect or lowest-adverse effect. The PDE can be calculated by applying appropriate AFs.

$$\frac{\mathrm{NO(A)EL} \times 50\,\mathrm{kg}}{\mathrm{F1} \times \mathrm{F2} \times \mathrm{F3} \times \mathrm{F4} \times \mathrm{F5}} = \mathrm{PDE}$$

Where the following AFs are used to account for:
F1 = Interspecies variability
F2 = Variability between individuals
F3 = Short-term to chronic studies extrapolation
F4 = Cases of severe toxicity
F5 = No NO(A)EL was established

F1 and F2 can be replaced for chemical-specific adjustment factors (CSAFs) if there is sufficient toxicokinetic and toxicodynamic data available [26, 27, 31]. Appropriate applications of AFs are listed in guidance documents and the published literature [4, 28, 32–34].

7.3.4 AI Development

In cases where a threshold cannot be identified (e.g. mutagenic carcinogens, where a threshold cannot be determined), linear extrapolation from the carcinogenic end point is used to determine the negligible excess risk over background (background is about 1 in 3), which is 1 in 100 000 for pharmaceutical impurities [4]. The common way to calculate a 1 in 100 000 excess risk of cancer is using the TD_{50} (dose resulting in a 50% increase in tumors over background) from the most sensitive tumor site and sex in animals (if relevant to humans) from available carcinogenicity studies [35, 36]. Precalculated TD_{50} values can be found in the Carcinogenic Potency Database (CPDB) and Lhasa Carcinogenicity Database (Table 7.2). The AI is calculated by the following equation:

AI (10^{-5} excess risk of cancer) = [TD_{50} (mg/kg/day)/50 000] × 50 kg

Table 7.3 LTL application to ICH M7 Class 1, 2, and 3 impurities.

Duration of treatment	≤1 month	>1–12 months	>1–10 years	>10 years lifetime
Daily intake for mutagenic impurities (μg/day)	120	20	10	1.5
Daily intake class- or compound-specific limit (μg/day)	80 × AI	13.3 × AI	6.7 × AI	AI

Other cancer potency estimates can be used to calculate the AI such as $BMDL_{10}$ (estimated 10% response over background at the lower 95th percent confidence interval). Provided that adequate data exists, the $BMDL_{10}$ modeling approach could be a more relevant PoD than the TD_{50} for subsequent linear extrapolation as it takes into account the shape of the dose-response of the low-dose region with rodent carcinogenicity data [37, 38].

7.3.5 Class-specific Limit

For mutagenic compounds with unknown carcinogenic potential, which are structurally similar to a well-defined class of known carcinogens, a class-specific limit AI could be considered (see Section 7.5).

7.3.6 Less than Lifetime (LTL) AIs

Compound- and class-specific limits are developed based on lifetime exposure. Oftentimes, exposure to pharmaceuticals is not chronic and less-than-lifetime (LTL) limits can be developed in those cases. Table 7.3 represents the LTL limits as recommended by ICH M7(R1) [4].

7.4 Examples of Published Compound-specific Limits

Compound-specific toxicology limits were derived in a collaborative publication by pharmaceutical sponsors and guidelines for industry for compounds that are commonly used in pharmaceutical manufacturing (Table 7.4). In the Bercu et al. [39] publication, compound-specific toxicology limits were derived for 20 widely used synthetic reagents and common by-products used in the synthesis of drug substances, which can also appear as potential impurities in the final drug substances [39]. Based on the available data, PDEs, AIs, or limits based on ICH Q3A qualification thresholds were established for each compound. In the ICH M7 Addendum, AIs or PDEs were derived for 14 chemicals that are considered to be mutagens and/or carcinogens using linear extrapolation from TD_{50}s (from the CPDB), threshold MOA, estimates of endogenous production, or TTC [4].

7.4.1 Mutagenic Carcinogens

There are 18 potential impurities listed in Table 7.4 that are ICH M7 Class 1 compounds or mutagenic carcinogens. Out of the 18 impurities, limits for the majority of compounds (13) were derived using a TD_{50} to determine an AI. Limits for five compounds, acetaldehyde, EMS, formaldehyde, hydrogen peroxide, and vinyl acetate, had a PDE derived instead. All five of these compounds exhibit a nonlinear dose–response curve for mutagenicity and/or carcinogenicity, and, in addition, acetaldehyde, formaldehyde, and hydrogen peroxide have high levels of endogenous/exposure in the general population. Vinyl acetate's PDE was based on its interconversion to acetaldehyde. Therefore, PDEs were developed for these ICH M7 Class 1 impurities with MOAs, which exhibit a threshold dose-response and high levels of endogenous/background exposure.

Table 7.4 Published compound-specific limits.

Compound	CAS number	*In vitro* mutagen	Rodent carcinogen	ICH M7 class	Limit (AI/PDE)	Source
Acetaldehyde	75-07-0	Yes (mammalian)	Yes	1	PDE = 2 mg/day	[39]
Acetamide	60-35-5	No	Yes	5	PDE = 7.1 mg/day	[39]
Acrolein	107-02-8	Yes	No	5	PDE = 50 μg/day $PDE_{Inhalation}$ = 7 μg/day	[39]
Acrylonitrile	107-13-1	Yes	Yes	1	AI = 6 μg/day	[4]
p-Aminophenol	123-30-8	Yes	No	5	PDE = 2 mg/day	[39]
Aniline	62-53-3	No (bacterial), but positive in some genotoxicity assays	Yes	5[a]	PDE = 720 μg/day	[4]
Benzyl chloride	100-44-7	Yes	Yes	1	AI = 41 μg/day	[4]
Bis(chloromethyl)ether	542-88-1	Yes	Yes	1	AI = 0.004 μg/day	[4]
t-Butyl chloride	507-20-0	No	Inadequate information	5	Maintain below ICH Q3A qualification thresholds	[39]
p-Chloroaniline	106-47-8	Yes	Yes	1	AI = 34 μg/day	[4]
1-Chloro-4-nitrobenzene	100-00-5	Yes	Yes	1	AI = 117 μg/day	[4]
p-Cresidine	120-71-8	Yes	Yes	1	AI = 45 μg/day	[4]
dimethyl sulfate	77-78-1	Yes	Inadequate information	2	AI = 1.5 μg/day	[4]
Dimethylcarbamoyl chloride	79-44-7	Yes	Yes	1	AI = 5 μg/day $AI_{Inhalation}$ = 0.6 μg/day	[4]
Epichlorohydrin	106-89-8	Yes	Yes	1	AI = 3 μg/day	[39]
Ethyl chloride	75-00-3	Yes	Yes	1	AI = 1810 μg/day	[4]
1-Ethyl-3-(3-dimethylaminopropyl) carbodiimide (EDAC)	1892-57-5	Yes (not *in vivo*)	Not tested	5	Maintain below ICH Q3A qualification thresholds. Lower limits may be appropriate following topical exposure.	[39]
Ethyl methanesulfonate (EMS)	62-50-0	Yes	Yes	1[b]	PDE = 1 mg/day	[39] and updated, this publication

Formaldehyde	50-00-0	Yes	Oral: no Inhaled: yes	1	PDE = 10 mg/day $PDE_{Inhalation}$ = 7.2 mg/day	[39]
Glycidol	556-52-5	Yes	Yes	1	AI = 4 μg/day	[4]
Hydrazine	302-01-2	Yes	Yes	1	AI = 39 μg/day $AI_{Inhalation}$ = 0.2 μg/day	[4]
Hydrogen peroxide	7722-84-1	Yes	Yes	1	PDE = 68 mg/day (based on endogenous estimate)	[4]
1-Hydroxy-7-azabenzotriazole (HOAt)	39968-33-7	No	Not tested	5	Maintain below ICH Q3A qualification thresholds.	[39]
Hydroxylamine	7803-49-8	No	Yes	5	PDE = 23 μg/day	[39]
Mesityl oxide	141-79-7	No	Not tested	5	PDE = 2.1 mg/day	[39]
Methyl bromide	74-83-9	Yes	No	5	PDE = 2.2 mg/day $PDE_{Inhalation}$ = 467 μg/day	[39]
Methyl chloride	74-87-3	Yes	Yes	1	AI = 1361 μg/day	[4]
Methyl iodide	74-88-4	No	Yes	5	PDE = 375 μg/day	[39]
p-Nitrophenol	100-02-7	No	No	5	PDE = 5 mg/day	[39]
O-(7-azabenzotriazol-1-yl)-*N*,*N*,*N*′,*N*′-tetramethyluronium hexafluorophosphate (HATU)	148893-10-1	No	Not tested	5	Maintain below ICH Q3A qualification thresholds.	[39]
Styrene	100-42-5	Yes	Yes	1	AI = 154 μg/day	[39]
Triphenylphosphine	603-35-0	No	Not tested	5	PDE = 250 μg/day	[39]
Triphenylphosphine oxide	791-28-6	No	Not tested	5	PDE = 200 μg/day	[39]
Vinyl acetate	108-05-4	Yes (mammalian)	Yes	1	PDE = 2 mg/day	[39]

[a] Not mutagenic in bacterial cells and was carcinogenic via a nonmutagenic MOA.
[b] Updated to ICH M7 Class 1 based on more recent analysis of data in this publication.

7.4.2 Nonmutagenic Carcinogens

There are four examples of nonmutagenic carcinogens in Table 7.4. These potential impurities were labeled as ICH M7 Class 5 compounds since a compound must be a known mutagen for designation as Class 1. In each of these cases, a PDE was developed for the impurity. For acetamide, aniline, and hydroxylamine, PDEs were determined on the basis of their carcinogenicity data. In the case of methyl iodide, carcinogenicity was considered threshold-based (thyroid tumors), where rats were more susceptible than humans to the compound's carcinogenic effects [40, 41]. The PDE for methyl iodide was based on noncarcinogenic lesions of the esophagus, salivary gland, and stomach following one-year oral exposure in the dog [42].

7.4.3 Mutagenic Noncarcinogens

There are three potential impurities that are mutagenic but not carcinogenic. These impurities are ICH M7 Class 5 impurities as carcinogenicity is the more biologically relevant outcome. A PDE can be calculated for these impurities based on noncarcinogenic effects. However, it is often incorrectly believed that if these compounds are mutagenic, then the AI should be based on the TTC. In the case of acrolein, *p*-aminophenol, and methyl bromide, the PDEs were developed based on noncarcinogenic effects for these mutagenic, but not carcinogenic, compounds.

7.4.4 Nonmutagenic Compounds

It is important to identify when potential impurities are nonmutagenic; as emphasized in ICH M7(R1), a structural alert alone is generally insufficient to label an impurity as mutagenic, and in other cases there may be conflicting or equivocal data that require an evaluation based on weight of evidence. There are six potential impurities that are structurally alerting but are nonmutagenic. These six compounds are aniline, *t*-butyl chloride, hydroxylamine, methyl iodide, mesityl oxide, and *p*-nitrophenol. Therefore, they are classified as ICH M7, Class 5 compounds. PDEs were provided for 5 of the 6 compounds, while *t*-butyl chloride had inadequate data to derive a PDE and thus the ICH Q3A/B thresholds were considered appropriate (Table 7.1). HOAt had some conflicting mutagenicity data but was determined to be nonmutagenic based on the weight of evidence. This was also controlled to ICH Q3A/B thresholds.

7.4.5 Mutagenic *In vitro* but not *In vivo*

1-Ethyl-3-(3-dimethylaminopropyl) carbodiimide (EDAC) was an example of a mutagenic compound that tested positive in the bacterial reverse mutation assay but was nonmutagenic *in vivo*. According to the ICH M7(R1) guidance document, for positive bacterial mutagens, additional hazard assessments and/or control measures may be applied. In cases where an impurity cannot be controlled to an appropriate acceptable limit, the impurity may be tested in an *in vivo* gene mutation assay to "understand the relevance of the bacterial mutagenicity assay result under *in vivo* conditions." [4] The ICH M7(R1) guidance document lists several *in vivo* assays and the corresponding scientific justifications to determine the most appropriate assay for use, which includes mechanism of action of the impurity and expected target tissue exposure. The results of the *in vivo* assay would overrule the *in vitro* result and can be used to set a compound-specific limit (Table 7.5).

EDAC yielded positive results *in vitro* in the bacterial mutagenicity assay and induced micronuclei in Chinese hamster ovary (CHO) cells. However, EDAC was negative when tested orally up to the maximum tolerated dose (300 mg/kg/day) in rats daily for 28 days for erythrocyte *Pig-a*

Table 7.5 Tests to investigate the *in vivo* relevance of *in vitro* mutagens (positive bacterial mutagenicity) [4].

In vivo test	Factors to justify choice of test as fit-for-purpose
Transgenic rodent gene mutation assays	For any bacterial mutagenicity positive, justify selection of assay tissue/organ
Pig-a gene mutation assay	For directly acting mutagens (bacterial mutagenicity positive without S9)[a]
Micronucleus test	For directly acting mutagens (positive without S9) and compounds known to be clastogenic[a]
Rat liver unscheduled DNA synthesis (UDS) test	• In particular for bacterial mutagenicity positive with S9 only • Responsible liver metabolite known to be generated in test species used to induce bulky adducts
Comet assay	• Justification needed (chemical class-specific MOA to form alkaline labile sites or single-strand breaks as preceding DNA damage that can potentially lead to mutations) • Justify selection of assay tissue/organ
Others	With convincing justification

[a] For indirect acting mutagens (requiring metabolic activation), adequate exposure to metabolite(s) should be demonstrated.

mutations, micronuclei in peripheral blood, or the liver comet assay [43]. Therefore, EDAC is treated as a nonmutagenic compound with insufficient carcinogenicity data, and limits in accordance with ICH Q3A/Q3B thresholds are recommended.

7.4.6 Route of Administration-specific Limits

There are five impurities with specific limits based on route of administration: acrolein, dimethylcarbamoyl chloride, formaldehyde, hydrazine, and methyl bromide. ICH M7(R1) states: "The above risk approaches are applicable to all routes of administration and no corrections to acceptable intakes are generally warranted. Exceptions to consider may include situations where data justify route-specific concerns that should be evaluated case-by-case." A separate route-specific limit was warranted if the AI or PDE was significantly lower by the inhalation route compared to the systemic limit. In all the five cases, an inhalation route-specific limit has been determined because exposure via the lung provided direct localized or enhanced toxicity.

7.5 Class-specific Limits

7.5.1 Alkyl Chlorides

Alkyl chlorides are commonly used in the synthesis of chemicals and pharmaceutical ingredients because they react with a range of nucleophiles. However, owing to this marked electrophilic character, alkyl chlorides also have the potential to react with DNA, which can lead to carcinogenesis. Brigo and Müller [44] previously analyzed the carcinogenicity and mutagenicity data of 27 alkyl chloride compounds [44]. Upon evaluation of the TD_{50} values for alkyl chlorides, it was determined that the monofunctional (i.e. only a single alerting feature for mutagenicity) alkyl chlorides were less potent carcinogens (i.e. TD_{50} above 15 mg/kg/day) than those containing additional alerts. The TD_{50} values ranged from 36 to 1810 mg/kg/day and considering a conservative class-specific potency reference point of 36 mg/kg/day would result in an AI 10 times greater than the ICH M7(R1) default TTC of 1.5 μg/day. Therefore, a class-specific toxicology limit of 15 μg/day was derived for monofunctional alkyl chlorides. This class-specific limit was adopted within the ICH M7(R1) guideline [4].

7.5.2 Alkyl Bromides

Alkyl bromides, which have similar chemical reactivity as alkyl chlorides, are also used in the synthesis of chemicals and pharmaceutical. In the Bercu et al. [39] publication, a comparable methodology was employed, where the carcinogenicity data of 26 alkyl bromides was analyzed to derive a class-specific toxicology limit [39]. The derived acceptable limit was 15 μg/day for monofunctional alkyl bromides. Unlike monofunctional alkyl chlorides, the class-specific limit is not included in ICH M7 guidelines to date.

7.5.3 *N*-Nitrosamines

N-Nitrosamines are a heightened concern among regulators and pharmaceutical companies because of their potential to form as low-level contaminants [45–47]. *N*-Nitrosamines were first detected in sartans, thought to form from the use of sodium nitrite as a quenching agent (used to deactivate sodium azide) in the manufacture of tetrazole ring systems (Chapter 10). *N*-Nitrosamines were also detected in pioglitazone, ranitidine, nizatidine, and metformin, which led to the European Medicines Agency (EMA) requiring that *N*-nitrosamine risk assessments be performed on every marketed product in an aggressive timeline [48, 49]. Other regulatory agencies have required similar risk assessments [50–52]. The United States Food and Drug Administration (USFDA) has emphasized that manufacturers have a responsibility to understand the mechanism involved in the formation and carryover of *N*-nitrosamine impurities, as well as developing appropriate analytical methods for their detection (see Chapter 12). The USFDA has also published methods to guide industry for analytical detection of *N*-nitrosamine impurities. Both industry and health authorities have expended significant chemistry effort to understand the potential sources and formation of *N*-nitrosamines in drug substances and drug products, with industry guidance available to help address risk in accordance with ICH M7(R1) principles (see Chapter 10) [45].

This heightened concern related to the presence of *N*-nitrosamines in sartan drug products also resulted in the conduct of a specific pharmacovigilance study where the cancer risk associated with exposure to *N*-nitrosodimethylamine (NDMA)-contaminated valsartan products was evaluated. A study with 5510 Danish patients using valsartan was followed for a median of 4.6 years with groups that were considered exposed to NDMA and those not exposed to NDMA [53]. The exposure to valsartan contaminated with NDMA was not associated with a markedly increased risk of cancer (adjusted hazard ratio of 1.09, 95% confidence interval 0.85–1.41). This study provides further evidence that the ICH M7 impurity control framework is conservative to protect for a theoretical risk in patients (i.e. as stated with ICH M7, the numerical cancer risk value of 1 in 100 000 and its translation into risk-based doses is a hypothetical concept that should not be regarded as a realistic indication of the actual risk). However, the study is limited by the duration of follow-up and number of patients to truly estimate excess cancer risk in a large number (>100 000) of patients.

Therefore, to prevent further incidents and to guide future risk assessments for the presence of *N*-nitrosamines, a key element of any strategy is the generation of compound- and/or class-specific limits for *N*-nitrosamines.

7.5.3.1 Regulatory Limits for *N*-Nitrosamines

Given the acute need to address the risk of *N*-nitrosamines in pharmaceuticals, regulatory agencies have provided temporary interim (provisional) limits for *N*-nitrosamine impurities (Table 7.6) [49,

Table 7.6 Regulatory interim AIs for *N*-nitrosamine compounds.

Compound (CAS number)	Structure	Temporary AI (ng/day)	AI toxicology derivation
N-Nitrosodimethylamine (NDMA) [62-75-9]	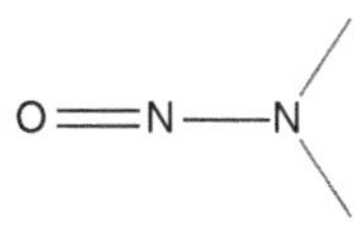	96	Extrapolation from TD_{50}
N-Nitrosodiethylamine (NDEA) [55-18-5]	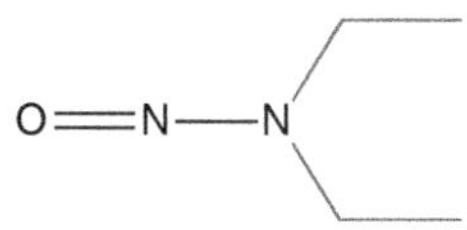	26.5	Extrapolation from TD_{50}
4-[(Methyl)(nitroso)amino] butanoic acid (NMBA) [61555-55-4]	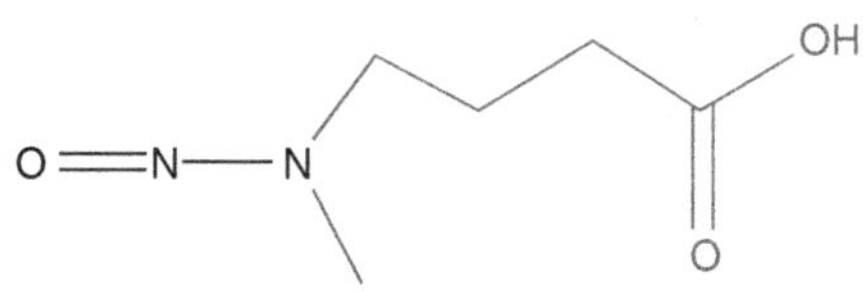	96	SAR comparison with NDMA
N-Nitrosodiisopropylamine (NDIPA) [601-77-4]	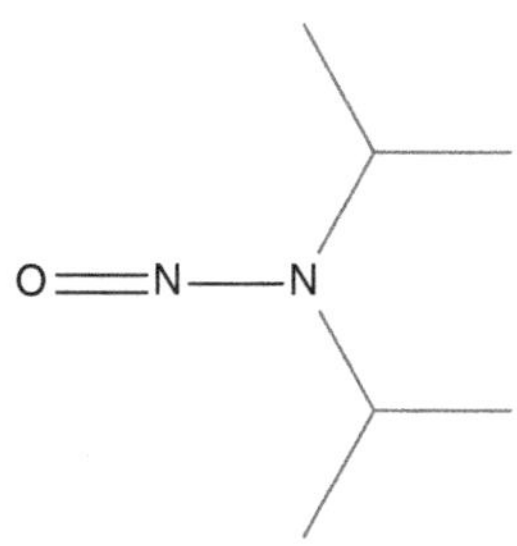	26.5	SAR comparison with NDEA
N-Nitrosoethylisopropylamine (NEIPA) [16339-04-1]	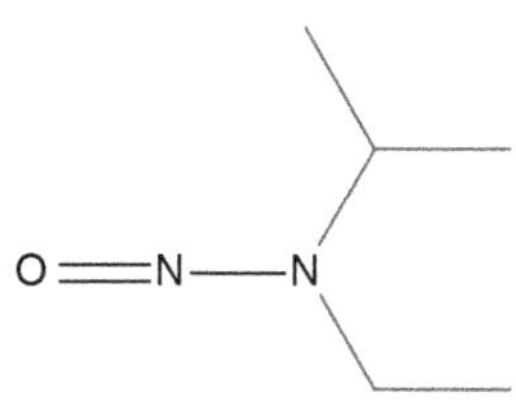	26.5	SAR comparison with NDEA
1-Methyl-4-nitrosopiperazine (MeNP) [16339-07-4]	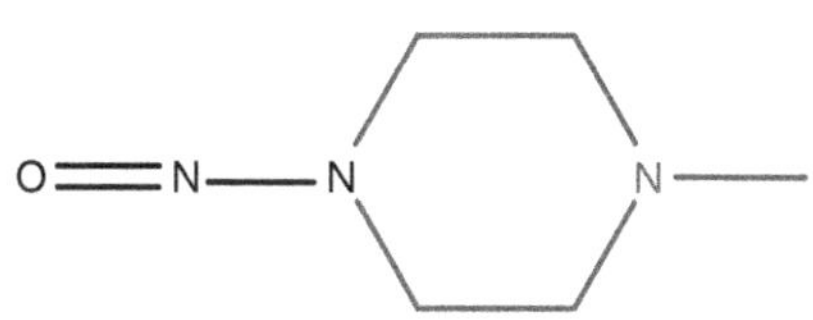	26.5	SAR comparison with NDEA
N-Nitrosodibutylamine (NDBA) [924-16-3]	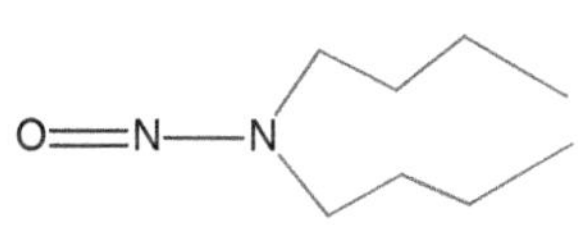	26.5	SAR comparison with NDEA
N-Nitrosomethylphenylamine (NMPA) [614-00-6]	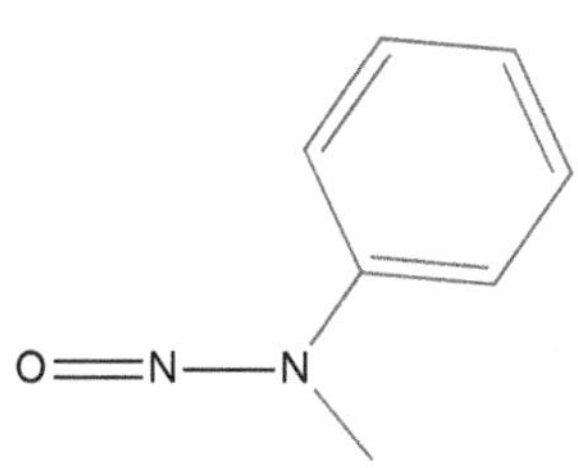	34.3	Extrapolation from TD_{50}

Source: Interim limits taken from [49, 54].

51, 54, 55]. NDMA, *N*-nitrosodiethylamine (NDEA), and *N*-nitrosomethylphenylamine (NMPA) AIs were developed based on linear extrapolation from their respective TD_{50}s to determine a 1 in 100 000 excess risk of cancer. The 4-[(methyl)(nitroso)amino]butanoic acid (NMBA) AI was based on its SAR with NDMA. *N*-Nitrosodiisopropylamine (NDIPA), *N*-nitrosoethylisopropylamine (NEIPA), and *N*-nitrosodibutylamine (NDBA) AIs were based on their SAR comparison with NDEA. EMA had also calculated NDMA AIs using benchmark dose lower confidence limit 10% ($BMDL_{10}$) methods in accordance with ICH M7(R1). The $BMDL_{10}$ values ranged from 0.029 to 0.043 mg/kg/day based on total rat liver tumors to develop corresponding AIs of 145–215 ng/day. While this approach was endorsed by the expert consultation, EMA favored the use of the more conservative TD_{50} calculation [55–57].

The International Programme on Chemical Safety (IPCS) of the World Health Organization (WHO) has also evaluated the carcinogenicity of NDMA [58]. A TD_{05} (dose level that results in a 5% increase in tumor incidence over background) of 34 μg/kg/day was derived based on the development of biliary cystadenomas in female rats. USEPA has developed a cancer oral slope factor of 51 $(mg/kg/day)^{-1}$ for NDMA based on liver tumors observed in female rats [59]. The USEPA also developed an oral slope factor of 150 $(mg/kg/day)^{-1}$ for NDEA based on rat female liver tumors [60]. Converting the oral slope factor to an AI (see Table 7.10 for the NDMA and NDEA AIs) is done by the following formula:

$$\text{AI (ng/day)} = 0.00001/\text{Oral Slope Factor (mg/kg/day)}^{-1} \times 50\,\text{kg} \times 10^6\,\text{ng/1 mg}$$

The Scientific Committee on Consumer Safety (SCCS) of the European Commission (EC) has developed risk values (T_{25} [25% tumor response over background], $BDML_{10}$, and TD_{50}) for a variety of *N*-nitrosamines (Table 7.7) [61]. For the T_{25} analysis, NDMA was the most potent carcinogen, but NDEA was the most potent carcinogen in the TD_{50} and $BMDL_{10}$ analyses. It may be assumed that the T_{25} and $BMDL_{10}$ may be simply half or five times less than the TD_{50}, respectively, but this is usually not the case based on the way each estimate is calculated.

7.5.3.2 Additional Proposed Limits for *N*-Nitrosamines

The temporary limits provided by the regulatory agencies have been developed based on conservative, yet simple, linear extrapolation from TD_{50}s and an SAR analysis to extrapolate based on read-across to NDEA and NDMA values.

The BMD described in a 2019 EMA report resulted in AIs of 145–215 ng/day for NDMA [55]. BMD methodology was employed previously to develop a TDI for NDMA [62]. Their techniques

Table 7.7 EC SCCS tumor potency estimates for *N*-nitrosamines [61].

N-Nitroso	T_{25} (mg/kg/day)	$BMDL_{10}$ (mg/kg/day)	TD_{50} (mg/kg/day)
Dimethylamine	0.058	0.027	0.0959
Diethylamine	0.085	0.018	0.0265
Morpholine	0.094	0.7	0.109
Methyl-*n*-dodecylamine	0.46	NA	0.537
Bis(2-hydroxypropyl)amine	0.54	NA	0.846
Pyrrolidine	0.57	0.16	0.799
Diethanolamine	2.09	0.73	3.17

T_{25} based on mean estimates; NA – not available.

involved mathematical modeling on the incidence data of rat hepatocellular carcinomas and hemangiosarcomas. The $mBMD_{0.05}$ (modified BMD at a 5% extra risk) was generated to understand dose-response modeling at the sub-experimental region and a large safety factor (3000–5000) to generate a TDI of 4–9.3 ng/kg/day or 200–465 ng/day for a 50 kg person. Finally, another $BMDL_{10}$ of 62 μg/kg/day was developed for NDMA based on liver cell tumors in rats by Dybing et al., [63], which resulted in an AI of 310 ng/day for a 50 kg person [63].

The reason cited by EMA for non-adoption of the BMD approach is the lack of international harmonized calculation methodology [48]. However, organizations such as European Food Safety Authority (EFSA) and the UK's Committee on Carcinogenicity have provided such international guidance in order to standardize the use of BMD in cancer risk assessment [37, 38]. These expert-committee examples could provide the framework to use BMD in the context of ICH M7(R1) guidance.

There has also been another approach to developing PDEs for NDMA and NDEA, which correlates data on mutagenic thresholds of these carcinogens with corresponding carcinogenicity information [64]. This work is important as it shows how understanding low-dose mutagenicity/carcinogenicity data can provide more realistic limits than the default linear extrapolation from the TD_{50}. In these cases, the PDE based on carcinogenic threshold (rat liver tumors) would be 6.2 μg/day for NDMA and 2.2 μg/day for NDEA, which is significantly higher than their corresponding temporary AIs developed by EMA (see Chapter 7). Also calculated, based on the incidence of liver tumors in rat cancer bioassays, were a $BMDL_{10}$ of 0.06 mg/kg/day for NDMA and 0.022 mg/kg/day for NDEA.

7.5.3.3 *N*-Nitrosamine Exposure in the General Population

N-Nitrosamines, particularly NDMA, are known contaminants at low levels in foodstuffs, including cured meats, dairy products, certain vegetables, and drinking water [46]. There are several restrictions on nitrite used in meat curing processes to reduce nitrite, but this cannot be completely removed [65]. Extensive literature reviews have been performed on *N*-nitrosamine exposures in food, beverages, tobacco, and personal care products [66, 67]. More than 24 *N*-nitrosamine compounds have been monitored routinely for exposure from the different sources. Gushgari and Halden [66], based on their literature review of analytical exposure data of *N*-nitrosamines, estimated average worldwide concentrations of total nitrosamines to be 40 ng/l in water, 6.7 ng/g in food and beverages, 16 100 ng/g in tobacco, and 1500 ng/g in personal care products [66]. Behavioral choices impact total *N*-nitrosamine exposure, from a lower bound of 1900 ng/day, to an upper bound of 25 000 ng/day, with tobacco being the largest contributor to total *N*-nitrosamines. Lee, [67], provided a literature review of *N*-nitrosamine exposures in processed meat and poultry products [67]. The estimated mean levels of volatile *N*-nitrosamines can vary, ranging from not detected to 35.6 μg/kg total *N*-nitrosamines. The most frequent of volatile amines were NDMA, *N*-nitrosopyrrolidine, and *N*-nitrosopiperidine. When grilling meat, air-sample levels of NDMA ranged from 69.4 to 906 ng/m^3, which means that a person could potentially breathe 69–900 ng NDMA in air while grilling meat assuming a breathing volume of 1 m^3 (this would be ~1 hour of grilling using ICH Q3C assumptions) [68]. A survey was performed on 387 diet study samples in 7 of the largest cities of South Korea for 7 different types of *N*-nitrosamines [69]. The result of study was that NDMA and NDEA were the most frequently detected in agricultural food products. For example, a maximum level of 6.01 μg/kg NDMA and 1.53 μg/kg NDEA were observed in fresh vegetables, and 6.21 μg/kg NDMA and 0.6 μg/kg NDEA were observed in fruit. In addition, NDMA and NDEA were found in meats, oils, and seasonings, with up to 1.54 μg/kg NDMA and 7.9 μg/kg NDEA in sausage, up to 2.83 μg/kg NDMA in sunflower oil, and up to 13.48 μg/kg NDMA and 1.01 μg/kg NDEA in seasonings (artificial flavored black pepper and salt). In Taiwan, up to 16.3 ng/l of NDMA was observed in drinking water samples [70]. While NDMA and NDEA are highlighted here, many of the foods/drinking water contained low-level concentrations of multiple different *N*-nitrosamines.

In addition, it has been estimated that exposure to *N*-nitrosamines from endogenous formation could be higher than from exogenous exposure [71–74]. There are different mechanisms for the formation of endogenous *N*-nitrosamines from the ingestion of nitrosating compounds, which can interact with (secondary) dietary amines via enzymatic or acid-catalyzed nitrosation reactions (in the stomach) [72, 74]. Vegetables are the predominant source of nitrate in the diet, and oral bacteria can reduce dietary nitrate to nitrite. The mean endogenous production of NDMA has been estimated to be from 100 μg/day to nearly 2500 μg/day, which was based on a biological-based modeling analysis and dietary sources of amines and nitrate [72]. NDMA has been measured in the blood of adult human volunteers at mean concentrations of 20–600 ng/kg body weight [72]. NDMA has been found in the urine of volunteers ranging from 11 to 385 ng NDMA eliminated/24 hours. In one study, NDMA was observed in blood samples of all but 1 of 51 samples in 23 human volunteers with a mean concentration of 0.5 μg/l [75]. After a test meal of bacon, spinach, bread, and beer, the concentration of NDMA increased with the estimated total body burden by about 40–50 μg; this is based on increased blood concentration and pharmacokinetic parameters measured in rabbits. Therefore, exogenous and endogenous exposures for *N*-nitrosamines can exceed exposure via pharmaceutical impurities.

7.5.3.4 Developing a Class-specific Limit for *N*-Nitrosamines

N-Nitroso compounds are part of the so-called cohort of concern (COC) or compounds that are considered highly potent or toxic and the AI could possibly be below the TTC. *N*-Nitroso compounds comprise a variety of additional structural types such as *N*-nitrosamides or *N*-nitrosoureas; all are generally positive for carcinogenicity in rodent bioassays; however, the focus of this discussion will be *N*-nitrosamines. A class-specific limit is needed for *N*-nitrosamines for those potential impurities where a compound-specific limit cannot be calculated. For any new *N*-nitrosamine, the principles of ICH M7 should be applied. Specifically, (i) the focus of the assessments should be on DNA-reactive *N*-nitrosamines that are positive in the bacterial reverse mutation assay (see Note 2 of the ICH M7 guideline). The bacterial reverse mutation assay is recognized to be an appropriate test for the detection of *N*-nitrosamines [13, 14, 76]. The sponsor should ensure that the bacterial reverse mutation assay is appropriately designed for the detection of the potential mutagenicity of an *N*-nitrosamine (i.e. standard bacterial strains with a consideration of appropriate test vehicle, pre-incubation protocols, and appropriate metabolic activation as per OECD 471). (ii) In the case of an identified DNA-reactive *N*-nitrosamine with robust rodent carcinogenicity data, safety limits based on the compound-specific AI (defined by the most appropriate modeling approach on the most robust carcinogenicity data) should be used to define an acceptable limit in final drug product (see Note 4 of the ICH M7 guideline) [56]. (iii) In the case of an identified DNA-reactive *N*-nitrosamine without reliable rodent carcinogenicity data, safety limits based on compound-specific AIs derived from carcinogenicity data on closely related structures should be used to define an acceptable limit in the final drug product (see Section 7.5 of the ICH M7 guideline).

The EMA Safety Working Party (SWP) developed a class-specific of 18 ng/day for nitrosamines where a compound-specific limit cannot be developed and an SAR based on analog structures cannot be made [57]. This was derived by analyzing the TD_{50} distribution of the Lhasa Carcinogenicity Database [77] and selecting the lower 95th percentile at a 1 in 100 000 excess cancer risk. If using TD_{50}s from the CPDB, the derived lower 95th percentile would be 44 ng/day [5].

EMA provided guidance that SAR read-across can be used for *N*-nitrosamines to develop AIs [57]. This is especially important for *N*-nitrosamines that may be formed from the drug substance where no carcinogenicity data exists. Developing a read-across framework for *N*-nitrosamines can be

Figure 7.1 General structure of an *N*-nitrosamine.

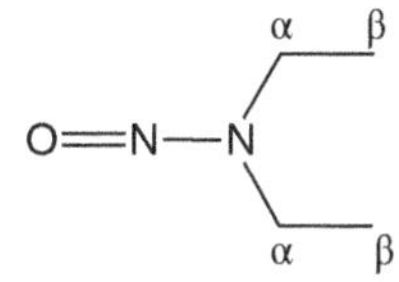

determined by researching structurally similar compounds and understanding structural aspects that may contribute to carcinogenic potency, including potential toxicokinetics. A general description of *N*-nitrosamines is provided in Figure 7.1. *N*-Nitrosamines are indirect acting mutagens requiring cytochrome P450 (CYP2E1) activation by hydroxylation of the carbon atom in the α-position relative to the nitroso moiety [78]. Further reactions produce a diazonium ion that is thought to be the proximate mutagen. α-Oxidative metabolism is the general pathway for mutagenicity, but there are also several alternative routes not involving α-oxidation, which are reviewed in Guttenplan [79]. Nonetheless, this principal mechanism is essential to understanding the SAR for mutagenicity and carcinogenicity potential of *N*-nitrosamines (Figure 7.2).

The diazonium ion formed by elimination of an aldehyde moiety from the hydroxylated nitrosamine undergoes an S_N1-type alkylation with DNA [80]. For NDMA, N7-methylguanine makes up about 70% of the DNA adducts and O6-methylguanine is also formed but significantly less (10-fold) than N7-methylguanine. The N7-methylguanine is repaired slowly via DNA glycosylases, and O^6-methylguanine is repaired by methylguanine DNA methyltransferase (MGMT), a suicide enzyme. Overwhelming DNA repair probably explains why a threshold and a sublinear dose-response have been observed for NDMA liver carcinogenicity in the rat [81, 82].

Mutagenesis is dependent on several factors highlighted by Guttenplan, [79], which includes (i) metabolic activation, (ii) cellular permeation, and (iii) formation of DNA adducts and their potential for DNA repair [79]. Metabolic activation to the diazonium ion can be reduced if certain structural features are present [79, 83–86]. Simple alkyl nitrosamine molecules are easily hydroxylated to form the pre-mutagenic species. Mutagenic potency tends to be reduced dramatically when the total number of carbons in the alkyl substituents shown in Figure 7.1 exceeds 12–14 [79]. However, carcinogenicity has been observed with the long-chain alkyl groups such as *N*-nitroso-methyl-*n*-dodecylamine. The carcinogenic response may involve a different mechanism compared to that shown in Figure 7.2 [87, 88]. It has been suggested that these long-chain alkyl substituents are initially activated via ω-oxidation and then later metabolized by successive β-oxidation, mimicking fatty acid metabolism [89]. If the α-position is blocked as in the case for a *t*-butyl substituent, this type of *N*-nitrosamine is expected not to be an alkylating agent *in vivo* and will likely be noncarcinogenic (or in some cases weakly carcinogenic) [79, 87, 90]. Even in the presence of α-hydrogen substituents, substitution at the α- and β-positions can still reduce mutagenic and carcinogenic potency of *N*-nitrosamines [91, 92]. In the case of NDIPA, there is a partial steric blockage that reduces its mutagenic and carcinogenic potency (which conflicts with the EMA limit derived by read-across to NDEA). Further testing of sterically blocked or inhibited *N*-nitrosamines using current guidelines for bacterial mutation assays (i.e. OECD 471 guidelines) would be useful to further validate these SAR features.

Hydroxyl-, carboxy-, and cyano-groups increase the polarity of the molecule that facilitates excretion and reduces metabolism [84–86]. This has shown an associated correlation with reduced carcinogenic potency. However, increased polarity for *N*-nitrosamines does not result in a universal reduction in carcinogenic potency. When comparing carcinogenic potency with log P (measured as JPogP, a consensus predictor learning from disparate methods for high accuracy), no immediate correlation was observed (Figure 7.3). However, more exploration on outliers or certain subclasses is needed to see if correlations could improve. When a meta-analysis was performed, the most

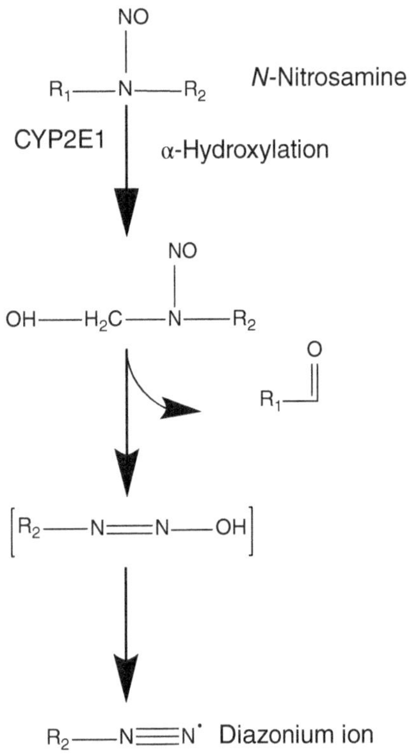

Figure 7.2 General *N*-nitrosamine metabolism to reactive metabolite.

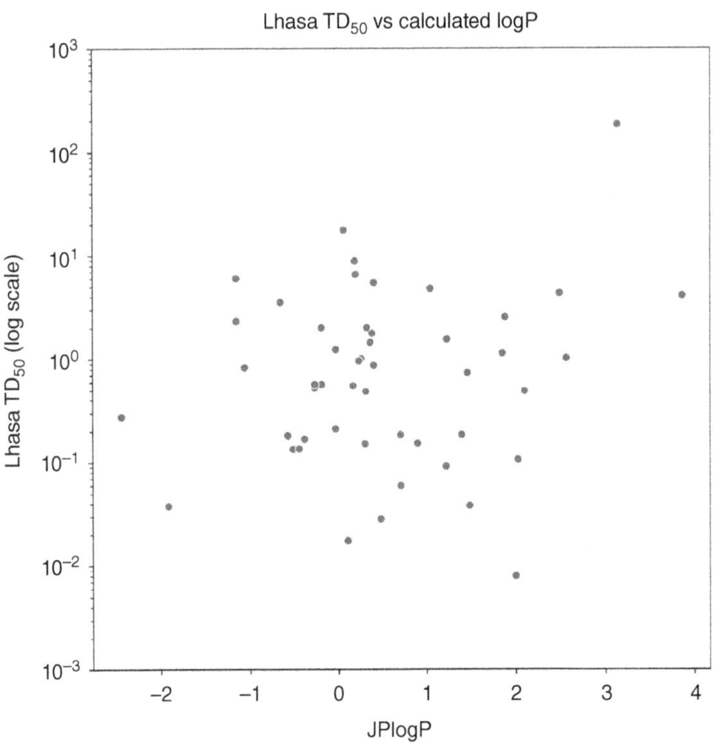

Figure 7.3 Comparison of Lhasa TD_{50}s and Log P (as JPogP) [93].

sensitive organ site for *N*-nitroso-alkylamines following oral exposure was the liver, indicating mutagenicity/carcinogenicity at the site of first-pass metabolism [94]. In contrast, most of the alcohol, keto, or carboxy *N*-nitrosamines were lower-potency bladder carcinogens indicating reduced first-pass metabolism by the liver to the reactive metabolite. Out of the 15 alcohol, keto, or carboxy *N*-nitrosamines analyzed by Buist et al. [94], 6 were considered negative for carcinogenicity and 8 were about approximately10-fold less potent than NDEA. One exception was *N*-nitrosobutyl-4-hydroxylbutylnitrosamine, which is a potent bladder carcinogen in rats via the formation and excretion of *N*-nitrosobutyl-3-carboxypropylamine [89, 95]. *N*-Nitrosodiphenylamine also may work through an alternative carcinogenic mechanism (such as trans nitrosation) as it is nonmutagenic and produces urinary bladder tumors in the rat with a low carcinogenic potency (Table 7.8).

A sufficiently large body of carcinogenicity data is available on *N*-nitrosamines to generate structural comparisons to gain an understanding of the likely carcinogenic potency of any untested *N*-nitrosamine. The data is from a variety of sources, and study design quality for carcinogenic potency analysis differs per compound. A series of *N*-nitrosamines and their carcinogenic potency were collected to develop class-specific limits for subclasses of *N*-nitrosamines (Table 7.8). Only mono *N*-nitroso compounds were selected for which there was at least one carcinogenicity study with at least two dose groups. Both the Gold TD_{50} (from the CPDB – Table 7.2) and Lhasa TD_{50} (from the Lhasa Carcinogenicity Database – Table 7.2) harmonic mean values were reported for comparison of carcinogenic potency. As a reference point, TD_{50}s 0.15 and 1.5 mg/kg/day correspond to AIs of 0.15 and 1.5 μg/day, respectively. Derivation of the Lhasa TD_{50} is very similar to the Gold TD_{50} except that the former TD_{50} has different criteria for inclusion of data, for example excluding compounds that have only a single dose group or the use of lifetable data (tumor incidence tracked over time instead of terminal sacrifice) [77]. Also, the Lhasa Carcinogenicity Database is being maintained and updated, whereas the CPDB data was last supplemented in 2007. In nearly all cases, the two TD_{50} values were within 2× of each other, showing good precision given that the data sets spanned four orders of magnitude. The only exception was *N*-nitrosopyrrolidine, where the Lhasa TD_{50} was not calculated for mice because only a single dose group was used.

Table 7.9 lists *N*-nitrosamines that were tested for carcinogenicity in animals with negative results. These compounds are listed in the CPDB/Lhasa Carcinogenicity Database. All but *N*-nitroso cimetidine had testing limited to a single species. Included were top doses for all studies for the compounds as reported in the Lhasa Carcinogenicity Database. For many of the *N*-nitrosamines, the top dose in the study was low so it is difficult to determine if these compounds are weak carcinogens or if testing at higher doses would have resulted in a tumorigenic response. Nonetheless, these negative results are indicative of a noncarcinogenic or weak response.

EMA limit of 18 ng/day is conservative and health protective but extremely challenging from an analytical perspective. For example, if targeting 10% of the limit in a 1 g/day dose (recommended by EMA for omission of a specification [49]) would require a 1 ppb analytical limit. When combined with the stated requirement to focus testing on the drug product as opposed to the drug substance, this becomes an even greater challenge as the complexity of the drug product matrix further challenges the ability to detect and quantify at the 1 ppb level.

Data on NDMA and NDEA may, however, be used to develop a revised class-specific limit for *N*-nitrosamines. Defining the AI or PDE for NDMA and NDEA is critical as they are some of the best studied compounds of the *N*-nitrosamine class and are highly potent carcinogens. AIs or PDEs developed for NDMA or NDEA are listed in Table 7.10. The important assumption in defining AIs/PDEs for NDMA and NDEA is the existence of a threshold. If assuming a threshold dose-response for mutagenicity and corresponding carcinogenicity, then the current class-specific AI recommended

Table 7.8 Carcinogenic potencies of selected *N*-nitrosamines.

N-Nitroso	CAS number	Structure	Bacterial mutation result[a]	TD_{50} Gold[b] (mg/kg/day)	TD_{50} Lhasa[c] (mg/kg/day)	Species (tumor sites)[d]
Methyl-2-phenylethylamine	13256-11-6	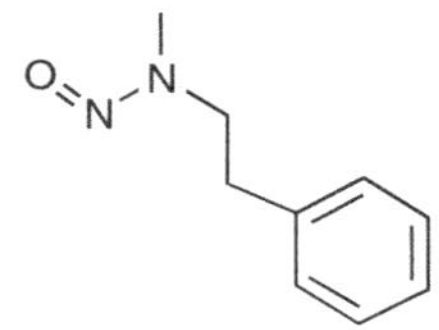	Pos	0.01	0.008	Rat
Methyl-2-oxopropylamine	55984-51-5	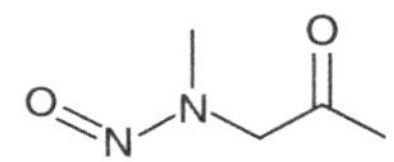	Pos	0.017	0.018	Rat
Diethylamine	55-18-5	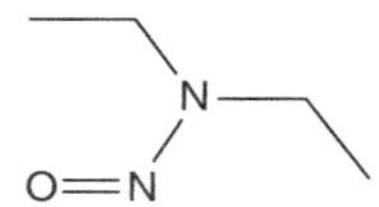	Pos	0.0265	0.018	Rat
Heptamethyleneimine	20917-49-1	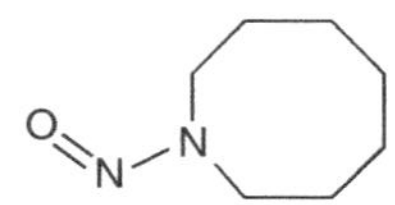	Pos	0.038	0.038	Rat
1,2,5,6-Tetrahydropyridine	55556-92-8	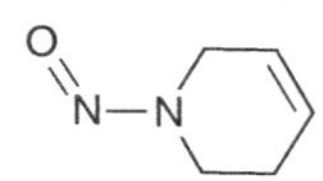	Pos	0.06	0.06	Rat
Ethylurethane	614-95-9	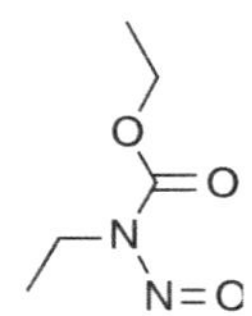	Pos	0.09	0.066	Rat
Dimethylamine	62-75-9	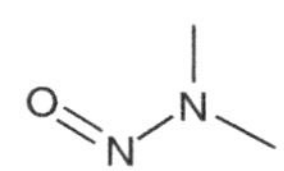	Pos	0.096	0.177	Rat

4-(Methylnitrosoamino)-1-(3-pyridinyl)-1-butanone; 4-(*N*-Methyl-*N*-nitrosamino)-1-(3-pyridyl)-1-butanone (NNK)	64091-91-4	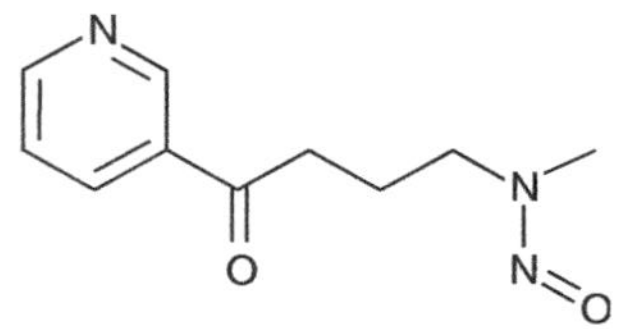	Pos	0.1	0.142	Rat
Morpholine	59-89-2	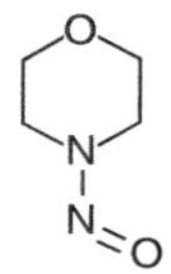	Pos	0.109	0.135	Rat
N-Methylaniline	614-00-6	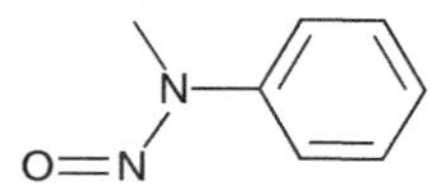	Pos	0.142	0.106	Rat
3,4,5-Trimethylpiperazine	75881-18-4	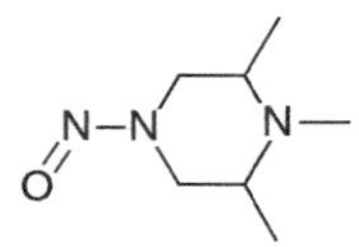	NA	0.151	0.153	Rat
Bis-(2-oxopropyl)amine	60599-38-4	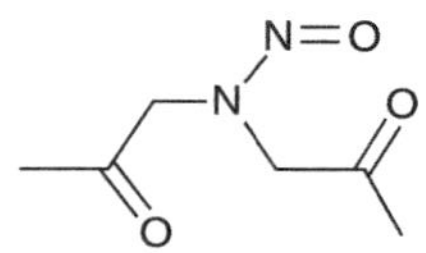	Pos	0.491	0.184	Rat
Pyrrolidine	930-55-2	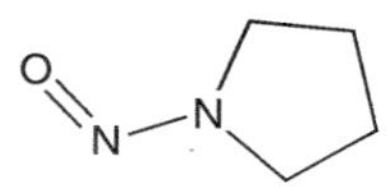	Pos	0.679	2.02	Mouse[e] Rat[f]
2,2′-Dihydroxydipropyl	53609-64-6	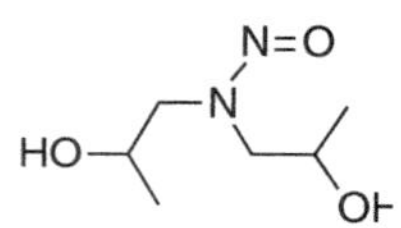	Pos	0.846	1.78	Rat

(Continued)

Table 7.8 (Continued)

N-Nitroso	CAS number	Structure	Bacterial mutation result[a]	TD_{50} Gold[b] (mg/kg/day)	TD_{50} Lhasa[c] (mg/kg/day)	Species (tumor sites)[d]
Caffeidine	145438-96-6	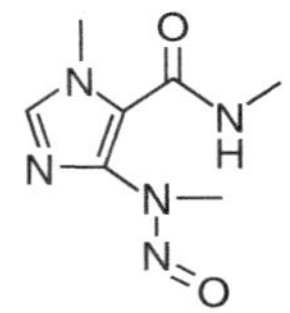	Neg	1	1.01	Rat
Amylurethane	64005-62-5	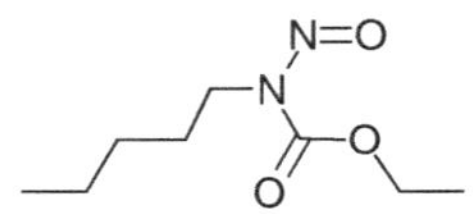	NA	1.01	0.339	Rat
Piperidine	100-75-4	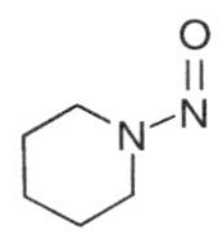	Pos	1.3	1.12	Mouse[e] Rat[f]
(2,2,2-Trifluoroethyl)ethylamine	82018-90-4	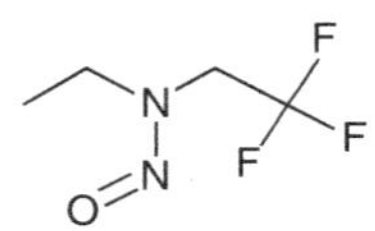	Pos	2.52	2.55	Rat
Diethanolamine	1116-54-7	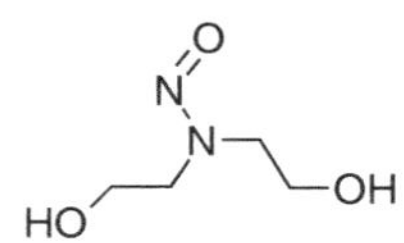	Pos	3.17	3.38	Rat
Di-*n*-pentylamine	13256-06-9	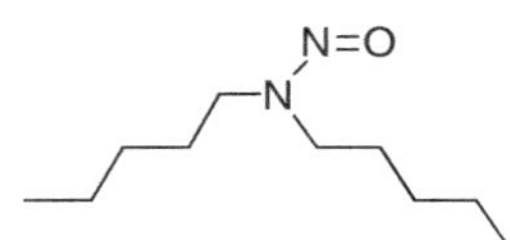	Pos	4.03	4.09	Rat
Thiomorpholine	26541-51-5	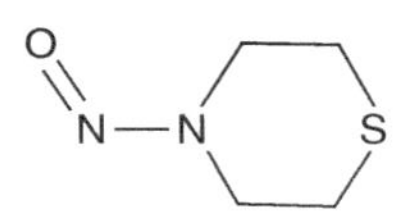	Pos	5.39	3.54	Rat

2,3-Dihydroxypropyl-2-hydroxyethylamine	89911-78-4	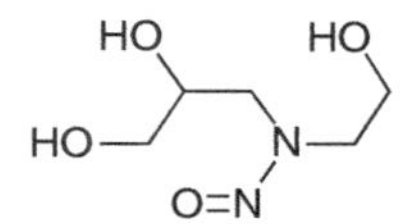	Pos	5.98	6.04	Rat
Piperazine	5632-47-3	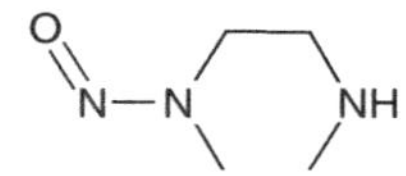	Pos	8.78	6.04	Rat
Diphenylamine	86-30-6	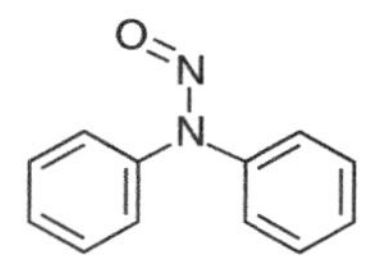	Neg[g]	167	NA[h]	Rat

NA – not available; Pos – positive; Neg – negative.

Compounds with only single-dose studies were excluded from the table.

[a] Ames call determined by consensus calls from Leadscope and Lhasa databases.

[b] Carcinogenicity Potency Database and TD_{50} represent the harmonic mean of positive results in a single species. https://files.toxplanet.com/cpdb/index.html.

[c] Lhasa Carcinogenicity Database and TD_{50} represent the harmonic mean of positive results in a single species. https://carcdb.lhasalimited.org/carcdb-frontend/#.

[d] Most sensitive species of rats and mice.

[e] Most sensitive species in CPDB database.

[f] Most sensitive species in Lhasa database.

[g] Negative in Leadscope database. Two positive results in non-OECD guideline strains (TA104, TA2638) were listed in Lhasa database, set against 70 negative results in OECD strains.

[h] Lhasa TD_{50} not calculated because lifetable was used to calculate the Gold TD_{50}, which is excluded from Lhasa analyses.

Table 7.9 Selected *N*-nitrosamines that were not carcinogenic when tested in animals.

N-Nitroso or *N*,*N*′-dinitroso[a]	CAS number	Structure	Bacterial mutation result[b]	Species tested[c]	Top dose tested (mg/kg/day)[d]
Bis(2,2,2-trifluoroethyl) amine	625-89-8	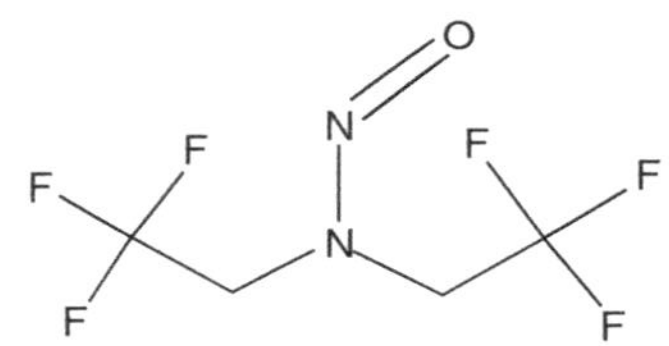	Neg[e]	Rat	0.61
Cimetidine	73785-40-7	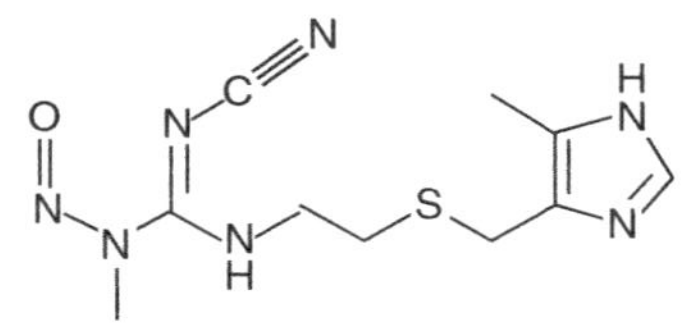	Pos	Mouse Rat	Mouse (226) Rat (71.7)
5,6-Dihydrothymine	62641-67-2	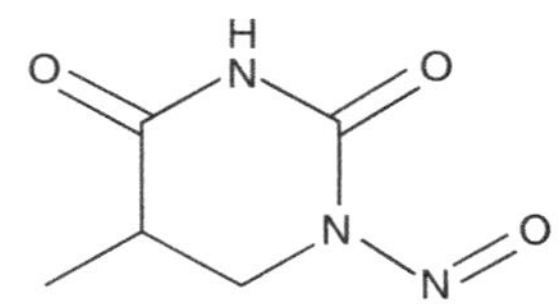	Pos	Rat	3.2
Guvacoline	55557-02-3	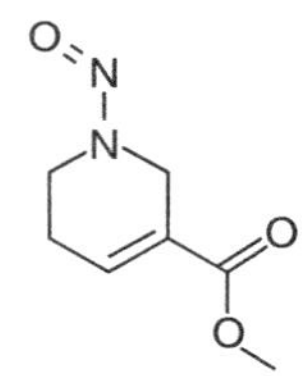	Pos	Rat	1.0
Hydroxyproline	30310-80-6	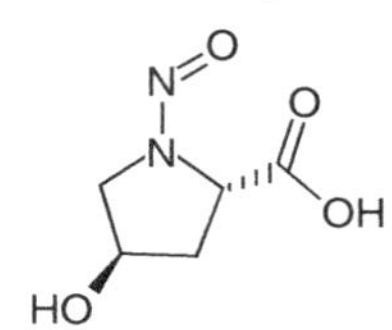	Neg[f]	Rat	4.42

Iminodiacetic acid	25081-31-6	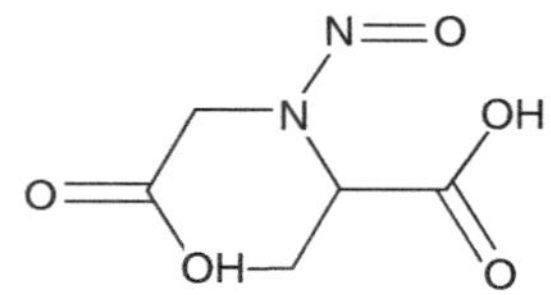	Neg	Rat	56.5
Pentamethylenetetramine	101-25-7	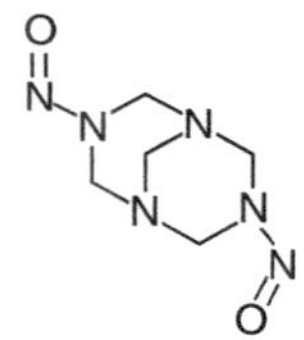	Pos	Rat	1.86
Pipecolinic acid	4515-18-8	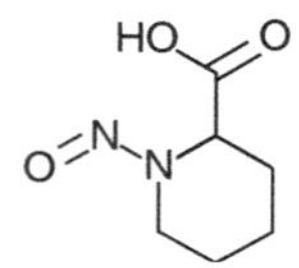	Pos	Rat	4.42
L-Proline	7519-36-0	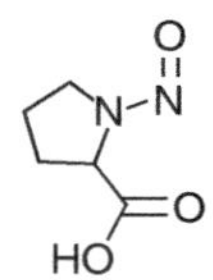	Neg[f]	Rat	4.42

Pos – positive; Neg – negative.
[a] Dinitroso included for *N*, *N′*-dinitrosopentamethylenetetramine (CASRN# 101-25-7).
[b] Ames call determined by consensus calls from Leadscope and Lhasa databases.
[c] As listed in https://carcdb.lhasalimited.org/carcdb-frontend/.
[d] Top dose from all studies listed for that species in https://carcdb.lhasalimited.org/carcdb-frontend/.
[e] Not a five-strain assay but negative in TA1535 with and without rat S9 metabolic activation.
[f] Only tested in a TA1535 with and without rat S9 metabolic activation (plate incorporation) data from Stoltz and Sen, 1977 [96].

Table 7.10 AI and PDE comparisons from different approaches for NDMA and NDEA.

Source	Cancer potency estimate	NDMA AI or PDE (ng/day)	NDEA AI or PDE (ng/day)
CPDB	TD_{50}	96	26.5
Lhasa Carcinogenicity Database	TD_{50}	177	18
[58]	TD_{05}	340	NA
[64]	PDE Threshold	6200	2200
[64]	$BMDL_{10}$	300	110
[55]	$BMDL_{10}$	145–215	NA
[61]	T_{25}	116	170
[61]	$BMDL_{10}$	135	90
[59, 60]	Cancer Potency Slope Factor	10	3
[62]	TDI from $mBMD_{0.05}$	200–465	NA
[63]	$BMDL_{10}$	310	NA

AIs assume a 1 in 100 000 excess risk of cancer and a 50 kg person.
NA – not available.

by EMA would likely be overly conservative (Chapter 7). A threshold for mutagenicity and carcinogenicity could be a valid assumption since *in vivo* mutagenic potency seemed to correlate well with carcinogenic potency for seven *N*-nitroso compounds [94]. This concept was also supported by an analysis of 15 general chemicals (NDMA was included) where *in vivo* mutagenicity measures correlated well with carcinogenicity [97]. In the case of NDMA and NDEA, $BMDL_{50}$s (BMD_{50} is based on the critical effect size for the assay used) calculated for *in vivo* mutagenicity studies were at or below their corresponding $BMDL_{10}$s for carcinogenicity [64]. This indicates that for NDEA and NDMA no tumorigenic response is expected unless the mutagenic threshold is exceeded. The USEPA cancer risk estimates are significantly below other limits, but this may be due to additional allometric scaling applied by USEPA and cancer potency model used (Weibull model), which is not included per ICH M7(R1) guidelines [59, 60]. The $BMDL_{10}$ estimates provided the highest AIs when assuming low-dose linear extrapolation. The UK's Committee on Carcinogenicity analyzed the different cancer potency estimates, i.e. TD_{50}, T_{25}, and $BMDL_{10}$ [37]. They concluded that while the TD_{50} is a pragmatic tool to understand carcinogenic potency, $BMDL_{10}$ is a preferred measure if it can be calculated, due to its superiority in modeling robust dose-response data. Given that one chronic study tested 15 doses and 4080 rats for NDMA and NDEA carcinogenicity (2040 rats for each compound), it would be prudent to perform a more robust dose-response assessment [98].

There could be cases where it may be more appropriate to apply the TTC of 1.5 μg/day for certain *N*-nitrosamines as they are not considered as part of the higher-potency class. This can be achieved via read-across or (Q)SAR modeling as appropriate. Adding inactivating features to both sides of the amine (i.e. R_1 and R_2 see Figure 7.4) such as phenyl, hydroxy, carboxy, keto, cyano, fluoro, and large chain alkyl groups (>5 carbons) can diminish the carcinogenic potency of the *N*-nitrosamine by either reducing metabolism or facilitating excretion. This feature was also demonstrated by the

Figure 7.4 *N*-Nitrosamine representation for potency subclassifications.

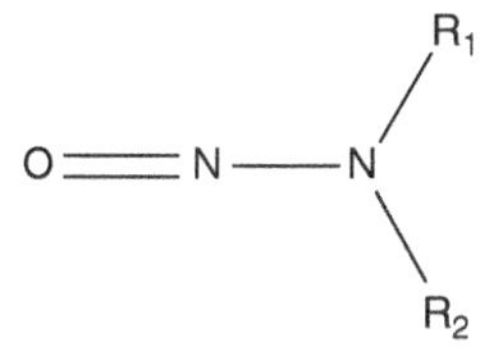

lower potency of *N*-nitroso compounds of diethanolamine; 2,2′-dihydroxydipropyl; dipentylamine; 2,3-dihydroxypropyl-2-hydroxyethylamine; or diphenylamine. Fluorination as in (*N*-nitroso-2,2,2-trifluoroethyl) diethylamine can significantly reduce carcinogenic potency likely due to increased metabolic stability of the compound [99]. Certain cyclic *N*-nitrosamines are of lower carcinogenicity potency as in piperazine, thiomorpholine, and piperidine. Simple alkyl amines can still be highly potent carcinogens even with the introduction of inactivating groups such as in *N*-nitroso derivatives of methyl-2-phenylethylamine, methylaniline, and polar groups such as *N*-nitrosomethyl-2-oxopropylamine. These findings are consistent with carcinogenicity estimates by Buist et al. [94], mentioned above. Further SAR analysis would help determine when it would be appropriate to use the TTC versus the class-specific limit for *N*-nitrosamines.

Consistent with ICH M7(R1), nonmutagenic and/or noncarcinogenic *N*-nitrosamine compounds should be controlled to the ICH Q3A/Q3B thresholds. Examples of this are *N*-nitroso cimetidine or *N*-nitrosamines with a *t*-butyl substituent at the α-position of the amine. There may be other examples of where structural features can used to predict the mutagenicity or carcinogenicity of *N*-nitrosamines, and additional analysis should be performed to help explore this further.

In conclusion, there has been a significant increase in regulatory guidance on the control of *N*-nitrosamines based on the recent contamination events in pharmaceuticals. Limits have been provided in regulatory guidance documents, which are conservative and health protective but may also be a challenge, indeed impractical from an analytical point of view. There are therefore at the time of writing efforts to better understand the class-specific limit for *N*-nitrosamines and further understand the SAR for the carcinogenic potency of *N*-nitrosamines.

7.5.4 Arylboronic Acids and Esters

Arylboronic acids and esters are commonly found as impurities in pharmaceuticals due to their use in the Suzuki cross-coupling reaction used extensively in drug substance synthesis [100, 101]. This impurity class is alerting for bacterial mutagenicity based on positive results in the absence of metabolic activation for a number of compounds from a series of arylboronic acids and derivatives [102, 103]. However, important to PDE/AI setting and the ICH M7 classification of these impurities is that arylboronic acids and their related esters that have been tested *in vivo* are not considered mutagenic.

Eight arylboronic acids and esters representing a variety of chemical scaffolds were tested in repeat-dose *in vivo* mutagenicity studies to investigate the *in vivo* relevance of the *in vitro* positive findings [104]. The results showed that the arylboronic compounds were not mutagenic in *in vivo* assays despite being mutagenic *in vitro*, suggesting that strict control of arylboronic compounds using the TTC is not necessary (Table 7.11). Although a class-specific toxicology limit could not be derived in the absence of repeat-dose toxicity data, the data suggests that ICH Q3A/B thresholds would be sufficiently conservative for arylboronic compounds with similar chemical scaffolds.

Table 7.11 In vivo results for arylboronic acids and esters.

Compound	CAS number	Structure	Rat dosing (route; duration; doses)	*In vivo* measurements	Result
3,5-Difluorophenylboronic acid (DFPBA)	156545-07-2	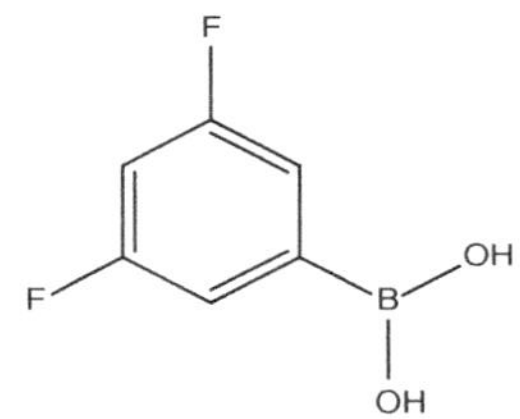	Oral 28 days 0, 30, 100, 300[a] mg/kg/day	*Pig-a* (RBCs, RETs) MN-RET Comet (Liver)	Negative
Phenylboronic acid (PBA)	98-80-6	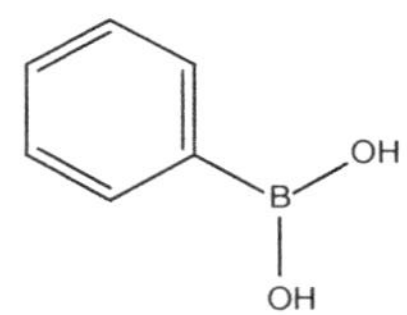	Oral 28 days 0, 30, 100, 250[a] mg/kg/day	*Pig-a* (RBCs, RETs) MN-RET Comet (duodenum, liver)	Negative
3-Quinoline boronic acid (QBA)	191162-39-7	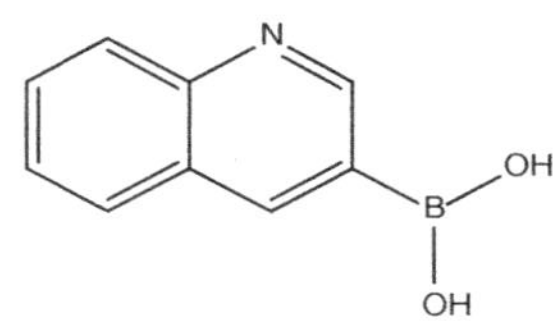	Oral 28 days 0, 250, 500, 1000 mg/kg/day	*Pig-a* (RBCs, RETs) MN-RET Comet (duodenum, liver)	Negative
Imidodicarbonic acid, 2-[6-(4,4,5,5-tetramethyl-1,3,2-dioxaborolan-2-yl)-2-pyrazinyl]-, 1,3-bis(1,1-dimethylethyl) ester (IDCPBE)	1400668-06-5	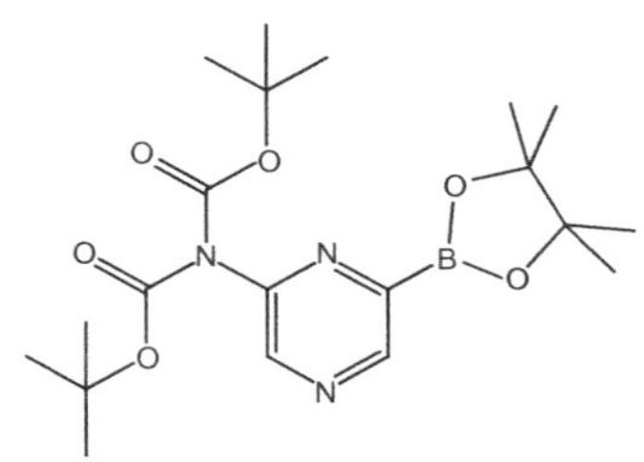	Intravenous[b] 28 days 0, 3, 10, 30[c] mg/kg/day	*Pig-a* (RBCs, RETs)	Negative

Pyrimidinyl boronic acid (PyBA)	109299-78-7	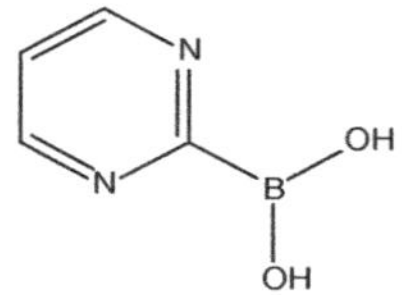	Oral 28 days 0, 100, 3000, 1000/750[d] mg/kg/day	*Pig-a* (RBCs, RETs) MN-RET Comet (duodenum, liver)	Negative
4-Methoxyphenyl boronic acid (MPBA)	5720-07-0	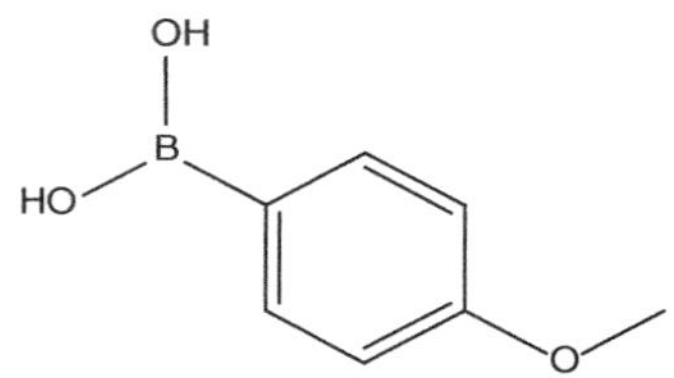	Oral 28 days 0, 250, 500, 700[a] mg/kg/day	*Pig-a* (RBCs, RETs) MN-RET Comet (duodenum, liver)	Negative
Thiomorpholine, 4-[[4-(4,4,5,5-tetramethyl-1,3,2-dioxaborolan-2-yl)phenyl]methyl]-, 1,1-dioxide (TTDPMD)	1092563-25-1	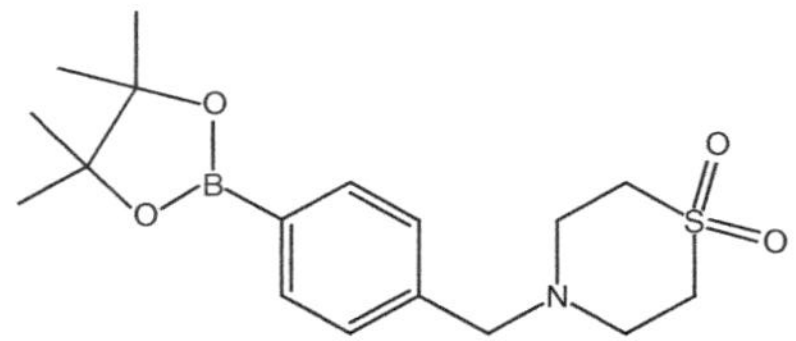	Oral 28 days 0, 100, 300, 1000 mg/kg/day	*Pig-a* (RBCs, RETs) Comet (duodenum, liver)	Negative
Boronic acid, [4-[(1,1-dioxido-4-thiomorpholinyl)methyl]phenyl] (BADTMP)	747413-23-6	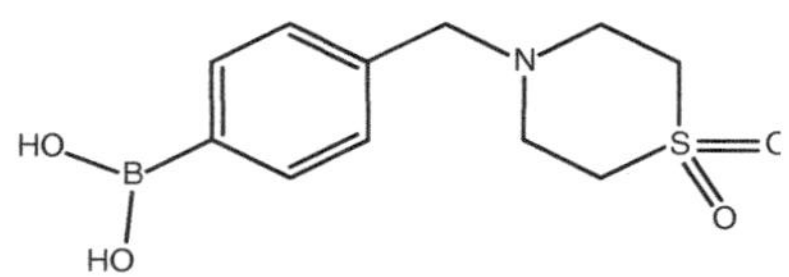	Oral 28 days 0, 100, 300, 1000 mg/kg/day	*Pig-a* (RBCs, RETs) Comet (duodenum, liver)	Negative

MN-RET – micronuclei in reticulocytes; RBCs – red blood cells; RETs – reticulocytes.
All compounds were tested positive in a bacterial mutation assay except for TTDPMD and BADTMP, which have not been tested in accordance with OECD 471 methodology (i.e. 5 strains up to 5 mg/plate).
[a] Maximum tolerated dose (MTD).
[b] Intravenously administered due to instability of compound.
[c] Maximum feasible dose.
[d] Dose lowered to 750 mg/kg/day on day 19 due to unexpected mortality.

7.6 EMS Case Study and Updated Toxicity Analysis

EMS is a case study of a mutagenic impurity that resulted in high patient exposure and required a subsequent thorough investigation of its toxicity. In this section we will discuss some of the issues that caused EMS exposure in patients and the subsequent testing. We also propose a PDE based on an analysis of the toxicity data set.

In summer of 2007, several patients had reported a bad odor and adverse reactions such as nausea to 250 mg nelfinavir mesylate tablets [105]. As a result, the pharmaceutical producer performed a root-cause analysis to determine what was causing the odor and nausea. It was determined that the source of the odor was the presence of EMS, which was measured up to 2300 ppm.

The manufacturing process of nelfinavir mesylate involved careful addition of an equimolar amount of methanesulfonic acid (MSA) to nelfinavir free base suspended in ethanol. Spray drying of the ethanolic solution was then performed to isolate nelfinavir mesylate. The key source of the EMS was as a contaminant in MSA stored in a holding tank. The tank had previously been cleaned with an ethanol-containing product, and EMS was slowly formed over several months by the reaction between MSA and residual ethanol in the tank. The MSA employed was of an ultrapure grade and so, in normal circumstances, it was possible to react nelfinavir base with this MSA and then isolate pure nelfinavir mesylate by spray drying (to avoid solvate formation). (Impurity data on numerous previous batches of nelfinavir mesylate confirmed its high purity.) The use of spray drying for isolation purposes precluded the possibility of impurity purging as would have been the case if a conventional process of filtration/washing of precipitated mesylate salt were employed. As a result, holding tanks were removed and disposable containers of MSA were used instead.

Mesylate is a very popular choice of salt in the pharmaceutical industry due to its chemical properties and experience with the salt. Mesylate exists as a salt in many currently marketed compounds [106]. The formation of alkyl-sulfonate esters in pharmaceutical syntheses has been further explored in detail demonstrating that the high exposure to EMS was unique to the nelfinavir mesylate scenario, which was subsequently corrected. Under normal pharmaceutical processing conditions, ester formation from the alkyl-sulfonic acid is unlikely since it is thermodynamically unfavored [107, 108]. When adding an equimolar of the active ingredient (base) with the sulfonic acid, proton transfer to form an acid salt occurs instantaneously precluding any side reactions leading to ester formation.

Even though it has been well documented that potential formation of alkyl-sulfonate esters like EMS is highly unlikely during pharmaceutical syntheses, there is still a perception of a safety concern due to their innate hazards not necessarily the risk from exposure (see also Chapter 11) [108]. The following section describes toxicity data of EMS and derives a PDE to be compared with exposure estimates.

7.6.1 Potential for Human Exposure

No data available.

7.6.2 Mutagenicity/Genotoxicity

The mutagenicity/genotoxicity data for EMS has been reviewed [109, 110]. EMS is considered mutagenic *in vitro* and *in vivo* by a threshold-based mechanism.

EMS is an alkylating agent to DNA nucleotides. It targets highly nucleophilic centers of the *N*-atoms such as the N^7-guanine atom, with about 58–65% of the profiled DNA adducts at that site [25, 109, 110]. EMS has been reported to be positive in the bacterial reverse mutation assay with and without rat liver S9 metabolic activation [111]. In addition, genotoxic effects have been observed in viruses/phages, fungal, plant, insect, and mammalian cells [109].

In mammalian cells, a threshold was observed for mutation effects and chromosomal damage in human lymphoblastoid cells [25]. High sampling sizes and narrowly spaced doses were used to clearly delineate a NOEL. The LOEL for gene mutations at the hypoxanthine-guanine phosphoribosyl transferase (hprt) locus was 1.4 μg/ml, and 1.2 μg/ml was the NOEL. The LOEL for induction of micronuclei was also 1.4 μg/ml and the NOEL was 1.35 μg/ml.

The data in support of a threshold-related mechanism is related to the types of ethylation lesions EMS produces and their repair [25, 110]. N^7-Ethylation (the predominant alkylation site) is repaired by base excision repair (BER), while O^6-ethylguanine (a minor alkylation site ~2%) is repaired by MGMT and to some extent by nucleotide excision repair. BER is an efficient repair process, yet with mistakes, while MGMT is an error-free repair mechanism, which can become saturated at higher levels as MGMT undergoes suicide inhibition upon action and its resynthesis becomes rate limiting for error-free repair. At lower doses, the EMS DNA adducts are efficiently repaired by BER and to a lesser extent MGMT. In contrast, at doses above a threshold, MGMT is overwhelmed resulting in cellular mutagenesis.

In order to understand the NOEL for mutations of EMS *in vivo*, a 28-day oral study was performed with the transgenic MutaMouse® model [112]. This study employed many dose levels – several of which were selected at or below the suspected threshold to enrich the data at and below the NOEL. Gene mutation frequencies in the bone marrow, liver, and gastrointestinal tract were measured. In the study, the NOEL for gene mutations was 25 mg/kg/day for the bone marrow and gastrointestinal tract and 50 mg/kg/day for the liver. At the same time, it was shown that the NOEL for EMS did not decrease with cumulative dosing. After four weeks of constant dosing of 12.5 mg/kg/day, or a 350 mg/kg cumulative dose, the mutant frequency did not increase over control levels for the time points sampled from beginning to end of study. In addition, a high acute single dose of EMS of 350 mg/kg showed a significant increase in mutation frequency over controls. Therefore, the total daily dose and not the cumulative dose are important for the frequency of EMS mutations. This was clearly different for the other tested agent, ethylnitrosourea (ENU), for which the mutation frequency cumulatively increased over the four-week period and was higher than the mutation frequency following a high single acute dose. This supports the assumption for EMS that saturation of an error-free repair mechanism is rate limiting for the NOEL. In a micronucleus (MN) study, mice were administered EMS by oral gavage for seven days. Doses up to 80 mg/kg/day did not raise the MN-polychromatic erythrocyte (PCE) frequency compared to control levels, but at higher doses an increase was evident, with saturation of the effect at 260 mg/kg/day. Several other *in vivo* studies for clastogenicity and mutagenicity have been performed with EMS, with a similar dose–response curve and the minimal effect dose was considered 50–100 mg/kg/day in these studies (summarized in [110]).

EMS was also tested in the male gpt-delta transgenic mouse model. EMS oral doses were 5–100 mg/kg/day administered by oral gavage for 28 days. Mutation frequencies were measured in the liver, lung, bone marrow, kidney, small intestine, and spleen and *Pig-a* mutation frequencies in peripheral blood reticulocytes (RETs) and total red blood cells (RBCs) [113]. Micronucleated RETs were also measured in peripheral blood. The no-observed genetic effect levels (NOGELs) for different end points measured are listed in Table 7.12. In the context of the transgenic models, the NOGEL will be considered synonymous to a NOEL.

A 28-day oral gavage study in Sprague Dawley rats was performed with EMS, which assessed the frequency of *Pig-a* mutations with doses ranging from 6.25 to 100 mg/kg/day [114]. At Day 55, an increase in mutations was observed in both RBCs and RETs with a LOEL of 25 mg/kg/day and a NOEL of 12.5 mg/kg/day. The threshold using a hockey-stick model (model of the dose–response curve that determines the inflection point or threshold of toxicity) was estimated to be 22 mg/kg/day.

Table 7.12 Genotoxic effects observed in 28-day oral male gpt-delta transgenic mouse model.

End point	NOGEL (mg/kg/day)[a]
Lung mutations	5
Kidney mutations	5
Pig-a mutations in RETs	5
Spleen mutations	13
Bone marrow mutations	20
Pig-a mutations in total RBCs	20
Small intestine mutations	20
Micronuclei (measured on Day 13)	20
Liver	55
Micronuclei (measured on Day 29)	55

[a] NOGEL that is equivalent to a NOEL.

Rats and mice are documented to be more susceptible to saturation of MGMT repair and thus related mutagenesis than humans [115, 116]. Levels of MGMT DNA and protein have been measured in human, rat, and mouse cells of a variety of organs. In all species, the liver contains the most MGMT and human cells contain more MGMT than rats and mice. This suggests that rats and mice would be more susceptible to overwhelming the MGMT-related repair of adducts at lower EMS doses than humans.

7.6.3 Noncarcinogenic Effects

The oral LD_{50} of EMS in mice is 470 mg/kg [117]. Pharmacokinetics of EMS have been characterized in mice, rats, and cynomolgus monkeys, following intravenous administration, and orally in mice [118]. Following intravenous administration, the half-life was short in mice (10–24 minutes) and longer in rats and monkeys (2.5–5 hours). Following oral dosing in mice, peak plasma concentrations were obtained within 10 minutes of administration with about 63–71% oral bioavailability when compared to the intravenous route of administration.

Male and female Wistar rats were administered EMS with repeated oral doses of 20, 60, and 180/120 mg/kg for 28 days [119]. Decreases in food consumption and body weight were dose-limiting primary effects. Adverse target organ effects at doses ≥60 mg/kg/day included effects characterized by depression of cell proliferation (e.g. hematopoiesis and spermatogenesis) and changes suggestive of reduced metabolism and/or some physiological imbalances (e.g. thymolymphatic system and thyroid gland) without signs of inflammatory or necrotic lesions other than single cell necrosis of germ cells. The 20 mg/kg dose was considered well tolerated and was determined to be the NOAEL.

EMS has been evaluated for effects on the germ cells of experimental animals and embryo-fetal development [110]. Germ cell genotoxicity in male rats and mice has resulted in an increase in pre-and postimplantation and conceptus loss. Treatment of female mice with EMS during embryogenesis can cause significant developmental toxicity [120]. In mice, 150 mg/kg dosed intraperitoneally on gestation day 11 resulted in teratogenicity, while 100 mg/kg was the NOEL.

7.6.4 Carcinogenicity

EMS is classified as an agent possibly carcinogenic to humans (Group 2B) by the IARC [121]. EMS has induced lung, renal, brain, abdominal, genital, mammary, and thymic tumors in mice and/or rats (Table 7.13). However, each study has various deficiencies (e.g. limited reporting, doses, duration of exposure, number of animals studied, and end points evaluated) and was not performed to current standards, which limit their interpretation and utility for cancer potency estimation.

During the period of 1 July 2006–30 June 2007, nelfinavir (oral HIV medication) was accidently contaminated with EMS resulting in an international recall [105]. Estimated worst-case patient exposure to EMS in some manufactured lots of nelfinavir during that period of time was 920 ppm. In a maximum dose of 2500 mg/day, this would result in an exposure of about 2.3 mg EMS. A follow-up study in patients exposed to nelfinavir during the risk period (1063 patients) found that the adjusted risk ratio for cancer was not different than those exposed outside the risk period [132].

7.6.5 Regulatory and/or Published Limits

Based on the EMS 28-day MutaMouse study, a PDE of 104 μg/day was developed [133]. This was based on dividing a NOEL of 25 mg/kg/day by a total safety factor of 12 000. The authors commented that the safety factors are likely overestimated due to conservative assumptions and made suggestions for further refinement.

7.6.6 Permitted Daily Exposure

The PDE for EMS was calculated from the mutagenic effects observed *in vivo*. Mutagenicity has been measured in the lung, kidney, spleen, bone marrow, peripheral blood cells, and gastrointestinal tract. The most sensitive site for mutagenicity was in the gpt-delta mouse lung, kidney, and RETs (*Pig-a*) with a NOEL/NOGEL of 5 mg/kg/day.

Mutation data was analyzed using the PROAST BMD model in accordance with the EFSA guidance [38]. A benchmark response (BMR) of 50% was selected as the critical effect size for the mutagenicity assay. $BMDL_{50}$s (BMD 50% response at the lower 95th confidence interval) were derived for EMS based on the most sensitive organ sites using model averaging (Table 7.14) [134, 135]. The lowest $BMDL_{50}$ calculated was 3 mg/kg/day for *Pig-a* mutant RETs, which is below the NOGEL of 5 mg/kg/day. However, this estimate is somewhat limited given that saturation was observed at 13 mg/kg/day and thus higher dose groups were removed from the analysis. Nonetheless, 3 mg/kg/day was used in the calculation of the PDE.

The PDE calculation is: (NOEL × body weight adjustment (kg))/F1 × F2 × F3 × F4 × F5

The following safety factors as outlined in ICH Q3C have been applied to determine the PDE for EMS:

F1 = 4 (Mouse to man. According to WHO guidance [31], if mice are more susceptible to the toxicodynamic effect, which is MGMT DNA repair, then a reduced factor of 4 can be applied in cases where the pharmacokinetic differences are unknown.)

F2 = 10 (Interindividual variability)

F3 = 1 (Study duration. While the PDE was based on 28 days of dosing, dose fractionation experiments indicate that NOEL would unlikely decrease with longer-term dosing. Also, the half-life of EMS is relatively short, 10 minutes to 5 hours in nonclinical species, indicating that systemic accumulation over time is unlikely. Therefore, a reduced factor was applied.)

Table 7.13 EMS – details of carcinogenicity studies.

Study	Animals/dose group	Duration/ exposure	Controls	Doses	Most sensitive tumor site/type/sex	TD_{50} (mg/kg/day)
[122]	36/sex not reported Carworth farm webster (CFW/D) mice (mortality in 14/36 mice)	Single ip	29 Mice (sex and vehicles not reported)	**1**:372 mg/kg[a]	Lung adenomas	NA[b]
[123]	31/male RF mice	Single ip	52 Male mice injected ip with 0.3 ml saline	**1**:175 mg/kg	Lung tumors (unspecified)/Male	NA[b]
[124]	24/female Wistar rats	ip on days 0, 2, 9	Not reported	**1**:275 mg/kg[c]	Renal carcinomas/ Female	NA[b]
[124]	22/sex not reported rats (strain not reported)	Single ip	Not reported	**1**:350 mg/kg	Brain ependymoma in 1/22 animals	NA[b]
[125]	31/male, 33/female Sprague Dawley rats with or without nephrectomy	1/week ip for three weeks	20/sex ip 1 ml saline	**1**:333 mg/kg[d]	Lung carcinoma, abdominal wall adenocarcinoma[e]	NA[b]
[126]	20/sex/group Wistar rats	Single ip	20/sex ip 1 ml saline	**3**:100, 200, and 300 mg/kg	Kidney epithelial and mesenchymal/Male – low incidence at all doses with no apparent dose-response[f]	NA[b]
[127]	4–9/female Albino rats/group with or without partial hepatectomy	Single ip	Not reported	Doses ranged from 100 to 350 mg/kg[g]	Malignant mesenchymal tumor of genital tract/ Female	NA[b]
[128]	48/sex not reported BALB/c newborn mice	Single sc	70 – no treatment 28–0.02 ml 1% aqueous gelatin 63 – Arachis oil Sex not reported	**1**:100 mg/kg[h]	Lung adenomas or adenocarcinomas	NA[b]

[129]	40/Female Wistar King A	12 weeks Drinking water	10 female (tap water)	**1**:372 mg/kg/day[i]	Mammary carcinomas	NA[b]
[130]	50–60/Sex not reported BALB/c newborn mice	1–5 days sc	50–60 – 0.02 ml Arachis oil	**1**:200 mg/kg[j]	Lung adenomas	NA[b]
[131]	40/Female C57BL mice	Single ip	Not reported	**1**:400 mg/kg	Thyomas (not statistically significant)	NA[b]

NA – not applicable; ip – intraperitoneal; sc – subcutaneous.
[a] 3 mmol/kg reported in study. 0.003 mol/kg × 124 g/mol = 372 mg/kg.
[b] NA because the study was not conducted according to current guidelines (e.g. limited number of animals, dose groups, or dosing days or insufficient reporting).
[c] 27.5 mg dissolved in 1 ml saline and injected ip to 100 g rats (reported weight in study).
[d] 33 mg dissolved in 1 ml saline and injected ip to 100 g rats (reported weight in study).
[e] Sexes not differentiated per organ site.
[f] Incidence is 0/40, 3/38, 0/40, and 2/40 rats at 0, 100, 200, and 300 mg/kg, respectively.
[g] Four partially hepatectomized groups (4–8 rats/group) and two intact animals (9 rats/group) and doses were reported as a range for each group.
[h] 100 μg ip for a 1 g mouse (average weight newborn mouse).
[i] 0.001 mol/l reported in the study. 0.001 mol/l × 124 g/mol × 0.31 water (from study)/0.1 kg (from study) = 372 mg/kg/day.
[j] 200 μg ip for a 1 g mouse (average weight newborn mouse).

Table 7.14 BMD estimates for EMS for the most sensitive sites in male gpt-transgenic mouse model (NOGEL – 5 mg/kg/day).

Tissue/end point	$BMDL_{50}$ (mg/kg/day)
Lung	5.6
Kidney	5.3
Pig-a RETs	3.0

$BMDL_{50}$ generated using PROAST version 67.0. proastweb.rivm.nl.

F4 = 3 (Severe toxicity. The end point measured was mutagenicity, which is a precursor and more sensitive effect than carcinogenicity or embryo-fetal toxicity where these severe effects would occur at higher doses, and a nonlinear mechanism of mutagenicity has been demonstrated.)

F5 = 1 (Because a $BMDL_{50}$ is considered the NOEL for genetic effects for EMS.)

Lifetime PDE = 3 mg/kg/day × 50 kg/(4 × 10 × 1 × 3 × 1)

Lifetime PDE = 1 mg/day

7.7 Extractables and Leachables

Leachables and extractables are impurities in pharmaceuticals but occur from product-contact with equipment or container closer systems and do not develop in the synthesis of a pharmaceutical drug substance. Leachables are the more relevant impurities than extractables from a patient safety perspective since they are generated under appropriate storage conditions with the drug product. However, extractable studies are performed under harsher solvent conditions, under a shorter time frame. ISO-10993-17 provides guidance for the development of allowable limits for leachables from medical devices [136]. The process involves developing a tolerable intake (TI) for cancer or noncancer end points, which is similar to the derivation of a PDE or AI, but with some differences in AFs (called uncertainty factors and overall modifying factor in ISO-10993-17) or defining a risk-specific number. The differences are beyond the scope of this discussion; however, the TI derived using ISO-10993-17 standards should be considered similar to the AI/PDE approach. It should also be noted that ISO-10993-17 is under development, likely updating the guidance using more modern toxicology risk assessment techniques. In parallel, extractables and leachables have now been adopted as an ICH topic, ICH Q3E [137].

Many pharmaceutical sponsors currently use the AI/PDE methodology to derive acceptable limits for leachables. A publication was developed from several pharmaceutical industry sponsors, which reviewed the safety evaluation practices for extractables and leachables, included processes for developing PDEs, and adopted AI methodology from ICH M7 for mutagenic carcinogens [138]. The latter publication provided two case study examples for mono(2-ethylhexyl) phthalate (MEHP) and Irganox 1076, both common leachables from polymeric materials. MEHP was negative in *in vitro* and *in vivo* genotoxicity assays, and although carcinogenicity data was not available for MEHP, several carcinogenicity studies were available for its parent compound, DEHP (di-(2-ethylhexyl)phthalate), which indicated that DEHP is not a concern for human-relevant carcinogenesis. Therefore, a PDE for MEHP was derived based on the most sensitive non-tumor end point, which was the LOAEL from a mouse developmental toxicity, yielding a PDE_{oral} of 146 μg/day and a $PDE_{parenteral}$ of 117 μg/day (parenteral PDE based on estimated 80% oral absorption). Irganox 1076 was negative in *in vitro* and *in vivo* genotoxicity assays, and there was no evidence for an increased tumor incidence in oral toxicity studies with mice or rats. A PDE_{oral} of 3 mg/day and a $PDE_{parenteral}$ of 1 mg/day (parenteral PDE based on estimated 30% oral absorption) were derived based on the NOAEL from a repeat-dose 28-day oral rat study. Pharmaceutical sponsors are developing more guidance and methodology to determine the safety of leachables/extractables to further harmonize the efforts across the industry (Parris et al. – publication in process).

7.8 Lhasa AI/PDE Database for Impurities

There is a data-sharing effort led by Lhasa Limited in order to collect high-quality AI/PDE monographs for pharmaceutical impurities (https://www.lhasalimited.org/Initiatives/aipde.htm). The monographs are stored in a proprietary database called Vitic and are only available to companies that participate in the data-sharing initiative (Table 7.2). The calculation of the AIs and PDEs are consistent with current ICH guidelines listed below. The monographs are generated by toxicologists in the industry and peer reviewed by two other toxicologists who participate in the data-sharing initiative. This effort results in time and cost savings. It takes time to research and thoroughly investigate the toxicity data, and it is not cost-effective to have 20 or more different pharmaceutical sponsors each develop their own monograph. Also, it helps with harmonization of efforts, because pharmaceutical sponsors peer review each other's reports, and best practices are shared that will improve the overall development of AIs/PDEs.

7.9 Conclusions and Future Directions

To further the harmonization effort and share best practices for determination of compound-specific limits, a workshop was conducted at the 2018 Genetic Toxicology Association meeting (Drewe et al. – publication in progress). It is clear from this workshop that each impurity's toxicity data set can provide unique challenges and differences can occur in terms of data interpretation. The data used to develop the AI or PDE is oftentimes incomplete resulting in uncertainty. It is important to have experts in the field to develop compound-specific limits that have followed a rigorous peer-review process.

The use of in *vivo* mutation data to derive PDEs is evolving. There have been advances where this data can be employed for risk assessment purposes such as the derivation of compound-specific limits [8]. EMS is an example highlighted here about how a PDE can be derived for an *in vivo* mutagen. Compound-specific data was used to determine the appropriate use of AFs and BMD modeling was applied. However, more work is needed to understand the appropriate selection of the PoD and AFs for other *in vivo* mutagenic compounds.

Compound-specific limits should not necessarily be static as illustrated in revised limits for some residual solvents provided in ICH Q3C. Impurity science will no doubt continue to advance and so it is appropriate to incorporate new developments into risk assessment techniques. It is impractical and unnecessary to update an AI or PDE on a daily or monthly basis, but one may consider revisiting the evaluation when new data is available.

Monofunctional alkyl chlorides and bromides are currently the only two structurally alerting moieties with scientific evidence for class-specific limits. It will be important to explore more class-specific limits to avoid always defaulting to the TTC, which assumes the impurity is a potent carcinogen. It may be possible to use computational techniques, or additional toxicity data may become available to determine more accurate associations between structure and cancer potency [139–142]. An initial analysis in the SAR for *N*-nitrosamines demonstrates that class-specific limits can be developed, which is promising in light of the recent contamination events with certain pharmaceuticals.

Technology for toxicology will also continue to advance. Such technology including computational toxicology, genomics, adverse outcome pathways, high-throughput screening, organ on a chip technology, etc. These new technologies are expected to enable the generation of more sophisticated data sets on chemicals and potentially reduce animal testing. Tomorrow's advances in the field of toxicology may result in new methodology used to derive compound-specific limits. Ideally, these new advances and methodology will help toxicologists reduce the level of uncertainty in the development of compound-specific limits.

7.10 Acknowledgments

Thank you to Ernest Lee, Jared Evans, and Eric Dowdy who looked over sections of the chapter and provided helpful feedback. Finally, this chapter is dedicated to Richard Williams, a brilliant and inspirational scientist, mentor, and friend, whose passion for and impact on predictive toxicology will be an eternal legacy.

Research reported in this publication was supported by the National Institute of Environmental Health Sciences of the National Institutes of Health under Award Number R43ES026909. The content is solely the responsibility of the authors and does not necessarily represent the official views of the National Institutes of Health.

References

1 Teasdale, A., Elder, D., and Harvey, J. (2018). Impurities in new drug substances and new drug products. In: *ICH Quality Guidelines: An Implementation Guide* (eds. A. Teasdale, D. Elder and R.W. Nims), 167–198. Hoboken, NJ: Wiley.

2 International Conference on Harmonisation (ICH) (2006). Impurities in new drug susbtances. Q3A(R2).

3 International Conference on Harmonisation (ICH) (2006). Impurities in new drug products. Q3B(R2).

4 International Conference on Harmonisation (ICH) (2017). Assessment and control of DNA reactive (mutagenic) impurities in pharmaceuticals to limit potential carcinogenic risk. M7(R1).

5 Kroes, R., Renwick, A.G., Cheeseman, M. et al. (2004). Structure-based thresholds of toxicological concern (TTC): guidance for application to substances present at low levels in the diet. *Food Chem. Toxicol.* 42: 65–83.

6 Muller, L., Mauthe, R.J., Riley, C.M. et al. (2006). A rationale for determining, testing, and controlling specific impurities in pharmaceuticals that possess potential for genotoxicity. *Regul. Toxicol. Pharmacol.* 44: 198–211.

7 Johnson, G.E., Soeteman-Hernandez, L.G., Gollapudi, B.B. et al. (2014). Derivation of point of departure (PoD) estimates in genetic toxicology studies and their potential applications in risk assessment. *Environ. Mol. Mutagen.* 55: 609–623.

8 MacGregor, J.T., Frotschl, R., White, P.A. et al. (2015). IWGT report on quantitative approaches to genotoxicity risk assessment II. Use of point-of-departure (PoD) metrics in defining acceptable exposure limits and assessing human risk. *Mutat. Res. Genet. Toxicol. Environ. Mutagen.* 783: 66–78.

9 Gollapudi, B.B., Lynch, A.M., Heflich, R.H. et al. (2015). The in vivo Pig-a assay: a report of the International Workshop On Genotoxicity Testing (IWGT) Workgroup. *Mutat. Res. Genet. Toxicol. Environ. Mutagen.* 783: 23–35.

10 Organization for Economic Co-operation and Development (OECD). (2013). Test No. 488: transgenic rodent somatic and germ cell gene mutation assays.

11 Heflich, R.H., Dertinger, S.D., Dobrovolsky, V.N., et al. (2019). The in vivo erythrocyte Pig-a gene mutation assay: Part 1: detailed review paper and performance assessment. *For the Organisation for Economic Cooperation and Development Working Group of the National Coordinators of the Test Guidelines Programme.*

12 Klimisch, H.J., Andreae, M., and Tillmann, U. (1997). A systematic approach for evaluating the quality of experimental toxicological and ecotoxicological data. *Regul. Toxicol. Pharmacol.* 25: 1–5.

13 International Conference on Harmonisation (ICH). (2011). Guidance on genotoxicity testing and data interpretation for pharmaceuticals intended for human use. S2(R1).

14 Organization for Economic Co-operation and Development (OECD). (1997). Test No. 471: bacterial reverse mutation test.

15 Hasselgren, C., Ahlberg, E., Akahori, Y. et al. (2019). Genetic toxicology in silico protocol. *Regul. Toxicol. Pharmacol.* 107: 104403.

16 United Nations (UN). (2017). *Globally Harmonized System of Classification and Labelling of Chemicals (GHS).* 7 revised edition.

17 Organization for Economic Co-operation and Development (OECD). (2018). Test No. 408. Repeat dose 90-day oral toxicity study in rodents.

18 Organization for Economic Co-operation and Development (OECD). (1998). OECD series on principles of good laboratory practice and compliance monitoring.

19 Smith, M.T., Guyton, K.Z., Gibbons, C.F. et al. (2016). Key characteristics of carcinogens as a basis for organizing data on mechanisms of carcinogenesis. *Environ. Health Perspect.* 124: 713–721.

20 Cohen, S.M., Klaunig, J., Meek, M.E. et al. (2004). Evaluating the human relevance of chemically induced animal tumors. *Toxicol. Sci.* 78: 181–186.

21 Meek, M.E., Bucher, J.R., Cohen, S.M. et al. (2003). A framework for human relevance analysis of information on carcinogenic modes of action. *Crit. Rev. Toxicol.* 33: 591–653.

22 Cohen, S.M., Meek, M.E., Klaunig, J.E. et al. (2003). The human relevance of information on carcinogenic modes of action: overview. *Crit. Rev. Toxicol.* 33: 581–589.

23 Cogliano, V.J., Baan, R.A., Straif, K. et al. (2004). The science and practice of carcinogen identification and evaluation. *Environ. Health Perspect.* 112: 1269–1274.

24 United States Environmental Protection Agency (USEPA). (2005). Guidelines for Carcinogen Risk Assessment.

25 Doak, S.H., Jenkins, G.J., Johnson, G.E. et al. (2007). Mechanistic influences for mutation induction curves after exposure to DNA-reactive carcinogens. *Cancer Res.* 67: 3904–3911.

26 Reichard, J.F., Maier, M.A., Naumann, B.D. et al. (2016). Toxicokinetic and toxicodynamic considerations when deriving health-based exposure limits for pharmaceuticals. *Regul. Toxicol. Pharmacol.* 79 (Suppl. 1): S67–S78.

27 Meek, M.E., Renwick, A., Ohanian, E. et al. (2002). Guidelines for application of chemical-specific adjustment factors in dose/concentration-response assessment. *Toxicology* 181–182: 115–120.

28 International Conference on Harmonisation (ICH). (2019). Guideline for elemental impurities. Q3D(R1).

29 Bercu, J.P., Morinello, E.J., Sehner, C. et al. (2016). Point of departure (PoD) selection for the derivation of acceptable daily exposures (ADEs) for active pharmaceutical ingredients (APIs). *Regul. Toxicol. Pharmacol.* 79 (Suppl. 1): S48–S56.

30 Harvey, J., Fleetwood, A., Ogilvie, R. et al. (2017). Management of organic impurities in small molecule medicinal products: deriving safe limits for use in early development. *Regul. Toxicol. Pharmacol.* 84: 116–123.

31 World Health Organization (WHO). (2005). Chemical-specific adjustment factors for interspecies differences and human variability: guidance document for use of data in dose/concentration-response assessment.

32 Sussman, R.G., Naumann, B.D., Pfister, T. et al. (2016). A harmonization effort for acceptable daily exposure derivation - considerations for application of adjustment factors. *Regul. Toxicol. Pharmacol.* 79 (Suppl. 1): S57–S66.

33 International Conference on Harmonisation (ICH). (2016). Impurities: guideline for residual solvents. Q3C(R7).

34 Escher, S.E., Mangelsdorf, I., Hoffmann-Doerr, S. et al. (2020). Time extrapolation in regulatory risk assessment: the impact of study differences on the extrapolation factors. *Regul. Toxicol. Pharmacol.* 112: 104584.

35 Peto, R., Pike, M.C., Bernstein, L. et al. (1984). The TD_{50}: a proposed general convention for the numerical description of the carcinogenic potency of chemicals in chronic-exposure animal experiments. *Environ. Health Perspect.* 58: 1–8.

36 Sawyer, C., Peto, R., Bernstein, L., and Pike, M.C. (1984). Calculation of carcinogenic potency from long-term animal carcinogenesis experiments. *Biometrics* 40: 27–40.

37 Committee on Carcinogenicity (COC). (2019). Committee on carcinogenicity of chemicals in food, consumer products and the environment. COC Guidance Statement G05 –version 1.1.

38 European Food Safety Authority (EFSA) (2017). Update: use of the benchmark dose approach in risk assessment. *EFSA J.* 15: 1–41.

39 Bercu, J.P., Galloway, S.M., Parris, P. et al. (2018). Potential impurities in drug substances: compound-specific toxicology limits for 20 synthetic reagents and by-products, and a class-specific toxicology limit for alkyl bromides. *Regul. Toxicol. Pharmacol.* 94: 172–182.

40 Hill, R.N., Crisp, T.M., Hurley, P.M. et al. (1998). Risk assessment of thyroid follicular cell tumors. *Environ. Health Perspect.* 106: 447–457.

41 International Agency for Research on Cancer (IARC). (2001). *IARC Monographs on the Evaluation of Carcinogenic Risks to Humans*, vol. 79. Some Thyrotropic Agents.

42 UK CLP Competent Authority. (2013). CLH Report for Iodomethane.

43 Custer, L. (2002). *Impact of New Mutagenicity Data on Marketed Drugs: A Case Study*. Presentation at Genetic Toxicology Association.

44 Brigo, A. and Müller, L. (2011). Development of the threshold of toxicological concern concept and its relationship to duration of exposure. In: *Genotoxic Impurities: Strategies for Identification and Control* (ed. A. Teasdale). Hoboken, NJ: Wiley.

45 Teasdale, A. (2020). Regulatory highlights. *Org. Process Res. Dev.* 24: 12–16.

46 Snodin, D.J. and Elder, D.P. (2019). Short commentary on NDMA (*N*-nitrosodimethylamine) contamination of valsartan products. *Regul. Toxicol. Pharmacol.* 103: 325–329.

47 United States Food and Drug Administration (USFDA). Information about nitrosamine impurities in mediations. https://www.fda.gov/drugs/drug-safety-and-availability/information-about-nitrosamine-impurities-medications (accessed 2 March 2020).

48 European Medicines Agency (EMA). (2019). Information on nitrosamines for marketing auhtorization holders. EMA/189634/2019.

49 European Medicines Agency (EMA). (2020). Questions and answers for marketing authorisation holders/applicants on the CHMP Opinion for the Article 5(3) of Regulation (EC) No 726/2004 referral on nitrosamine impurities in human medicinal products. EMA/409815/2020.

50 Health Canada. (2020). Nitrosamines in pharmaceuticals. Health Canada stakeholder informational webinar.

51 Swissmedic. (2019). Potential nitrosamine contamination: request to perform a risk evaluation. https://www.swissmedic.ch/swissmedic/en/home/news/mitteilungen/aufforderung-zlinhaberinnen-ham.html (accessed 22 Febraury 2020).

52 United States Food and Drug Administration (USFDA). (2020). Guidance for Industry. Control of nitrosamine impurities in human drugs.

53 Pottegard, A., Kristensen, K.B., Ernst, M.T. et al. (2018). Use of *N*-nitrosodimethylamine (NDMA) contaminated valsartan products and risk of cancer: Danish nationwide cohort study. *BMJ* 362: k3851.

54 European Medicines Agency (EMA). (2019). Temporary interim limits for NMBA, DIPNA, EIPNA, impurities in sartan blood pressure medicines. EMA/351053/2019 rev 1.

55 European Medicines Agency (EMA). (2019). Assessment Report: Referral under Article 31 of Directive 2001/83/EC: Angiotensin-II-receptor antagonists (sartans) containing a tetrazole group. EMA/217823/2019.

56 European Medicines Agency (EMA). (2020). Questions and answers on "Information on nitrosamines for marketing authorisation holders." EMA/CHMP/428592/2019 Rev. 3.

57 European Medicines Agency (EMA). (2020). Assessment Report. Nitrosamine impurities in human medicinal products. Procedure under Article 5(3) of Regulation EC (No) 726/2004. EMA/369136/2020.

58 World Health Organization (WHO). (2002). *N*-Nitrosdimethylamine. Concise International Chemical Assessment Document 38.

59 United States Environmental Protection Agency (USEPA). (1987). *N*-Nitrosodimethylamine; CASRN 62-75-9: Integrated Risk Information System (IRIS).

60 United States Environmental Protection Agency (USEPA). (1987). *N*-Nitrosodiethylamine; CASRN 55-18-5: Integrated Risk Information System (IRIS).

61 Eurpean Commission (EC) Scientific Committee on Consumer Safety (SCCS). (2012). Opinion on nitrosamines and secondary amines in cosmetic products.

62 Fitzgerald, D.J. and Robinson, N.I. (2007). Development of a tolerable daily intake for *N*-nitrosodimethylamine using a modified benchmark dose methodology. *J. Toxicol. Environ. Health A* 70: 1670–1678.

63 Dybing, E., O'Brien, J., Renwick, A.G., and Sanner, T. (2008). Risk assessment of dietary exposures to compounds that are genotoxic and carcinogenic--an overview. *Toxicol. Lett.* 180: 110–117.

64 Johnson, G.E. (2020). Webinar: quantitative analysis of in vivo mutagenicity dose-response data for risk assessment and regulatory decision-making: A case study of alkylnitrosamines.

65 European Food Safety Authority (EFSA) (2017). Re-evaluation of potassium nitrite (E 249) and sodiumnitrite (E 250) as food additives. *EFSA J* 15: 1–157.

66 Gushgari, A.J. and Halden, R.U. (2018). Critical review of major sources of human exposure to *N*-nitrosamines. *Chemosphere* 210: 1124–1136.

67 Lee, H.S. (2019). Literature compilation of volatile *N*-nitrosamines in processed meat and poultry products - an update. *Food Addit. Contam. Part A Chem. Anal. Control Expo. Risk Assess.* 36: 1491–1500.

68 Kim, H., Tcha, J., Shim, M.Y., and Jung, S. (2019). Dry-heat cooking of meats as a source of airborne *N*-nitrosodimethylamine (NDMA). *Atmos.* 10: 1–14.

69 Park, J.E., Seo, J.E., Lee, J.Y., and Kwon, H. (2015). Distribution of seven *N*-nitrosamines in food. *Toxicol. Res.* 31: 279–288.

70 Fan, C.C. and Lin, T.F. (2018). *N*-nitrosamines in drinking water and beer: detection and risk assessment. *Chemosphere* 200: 48–56.

71 Zeilmaker, M.J., Bakker, M.I., Schothorst, R., and Slob, W. (2010). Risk assessment of *N*-nitrosodimethylamine formed endogenously after fish-with-vegetable meals. *Toxicol. Sci.* 116: 323–335.

72 Hrudey, S.E., Bull, R.J., Cotruvo, J.A. et al. (2013). Drinking water as a proportion of total human exposure to volatile *N*-nitrosamines. *Risk Anal.* 33: 2179–2208.

73 Krul, C.A., Zeilmaker, M.J., Schothorst, R.C., and Havenaar, R. (2004). Intragastric formation and modulation of *N*-nitrosodimethylamine in a dynamic in vitro gastrointestinal model under human physiological conditions. *Food Chem. Toxicol.* 42: 51–63.

74 Fristachi, A. and Rice, G. (2007). Estimation of the total daily oral intake of NDMA attributable to drinking water. *J. Water Health* 5: 341–355.

75 Gough, T.A., Webb, K.S., and Swann, P.F. (1983). An examination of human blood for the presence of volatile nitrosamines. *Food Chem. Toxicol.* 21: 151–156.

76 Thresher, A., Foster, R., Ponting, D.J. et al. (2020). Are all nitrosamines concerning? A review of mutagenicity and carcinogenicity data. *Regul. Toxicol. Pharmacol.* 116: 104749.

77 Thresher, A., Gosling, J.P., and Williams, R. (2019). Generation of TD_{50} values for carcinogenicity study data. *Toxicol. Res. (Camb).* 8: 696–703.

78 Yang, C.S., Yoo, J.S., Ishizaki, H., and Hong, J.Y. (1990). Cytochrome P450IIE1: roles in nitrosamine metabolism and mechanisms of regulation. *Drug Metab. Rev.* 22: 147–159.

79 Guttenplan, J.B. (1987). N-nitrosamines: bacterial mutagenesis and in vitro metabolism. *Mutat. Res.* 186: 81–134.

80 Souliotis, V.L., Henneman, J.R., Reed, C.D. et al. (2002). DNA adducts and liver DNA replication in rats during chronic exposure to *N*-nitrosodimethylamine (NDMA) and their relationships to the dose-dependence of NDMA hepatocarcinogenesis. *Mutat. Res.* 500: 75–87.

81 Fukushima, S., Gi, M., Kakehashi, A. et al. (2016). Qualitative and quantitative approaches in the dose-response assessment of genotoxic carcinogens. *Mutagenesis* 31: 341–346.

82 Fukushima, S., Wei, M., Kakehashi, A., and Wanibuchi, H. (2010). Thresholds for genotoxic carcinogenicity: evidence from mechanism-based carcinogenicity studies. In: *Cancer Risk Asessment: Chemical Carcinogenesis, Hazard Evaluation, and Risk Quantification* (eds. H. Ching-Hung and T. Stedeford), 207–221. Wiley.

83 Guttenplan, J.B. (1987). Structure-activity relationships in metabolism and mutagenicities of *N*-nitrosamines. *IARC Sci. Publ.* 84: 129–131.

84 Helguera, A.M., Cordeiro, M.N., Perez, M.A. et al. (2008). Quantitative structure carcinogenicity relationship for detecting structural alerts in nitroso-compounds: species: rat; sex: male; route of administration: water. *Toxicol. Appl. Pharmacol.* 231: 197–207.

85 Helguera, A.M., Gonzalez, M.P., Cordeiro, M.N.D., and Perez, M.A. (2007). Quantitative structure carcinogenicity relationship for detecting structural alerts in nitroso-compounds. *Toxicol. Appl. Pharmacol.* 221: 189–202.

86 Helguera, A.M., Perez-Machado, G., Cordeiro, M.N., and Combes, R.D. (2010). Quantitative structure-activity relationship modelling of the carcinogenic risk of nitroso compounds using regression analysis and the TOPS-MODE approach. *SAR QSAR Environ. Res.* 21: 277–304.

87 Lijinsky, W. (1987). Structure-activity relations in carcinogenesis by *N*-nitroso compounds. *Cancer Metastasis Rev.* 6: 301–356.

88 Lijinsky, W., Saavedra, J.E., and Reuber, M.D. (1981). Induction of carcinogenesis in Fischer rats by methylalkylnitrosamines. *Cancer Res.* 41: 1288–1292.

89 Okada, M., Suzuki, E.M., and Mochizuki, M. (1976). Possible important role of urinary *N*-methyl-*N*(3-carboxypropyl)nitrosamine in the induction of bladder tumors in rats by *N*-methyl-*N*-dodecylnitrosamine. *Gann* 67: 771–772.

90 Magee, P.N. and Lee, K.Y. (1964). Cellular injury and carcinogenesis. Alkylation of ribonucleic acid of rat liver by diethylnitrosamine and *n*-butylmethylnitrosamine in vivo. *Biochem. J.* 91: 35–42.
91 Lee, S.Y. and Guttenplan, J.B. (1981). A correlation between mutagenic and carcinogenic potencies in a diverse group of *N*-nitrosamines: determination of mutagenic activities of weakly mutagenic *N*-nitrosamines. *Carcinogenesis* 2: 1339–1344.
92 Lijinsky, W. and Taylor, H.W. (1979). Carcinogenicity of methylated derivatives of *N*-nitrosodiethylamine and related compounds in Sprague-Dawley rats. *J. Natl. Cancer Inst.* 62: 407–410.
93 Plante, J. and Werner, S. (2018). JPlogP: an improved logP predictor trained using predicted data. *J. Chem.* 10: 61.
94 Buist, H.E., Devito, S., Goldbohm, R.A. et al. (2015). Hazard assessment of nitrosamine and nitramine by-products of amine-based CCS: alternative approaches. *Regul. Toxicol. Pharmacol.* 71: 601–623.
95 Okada, M., Suzuki, E., and Hashimoto, Y. (1976). Carcinogenicity of *N*-nitrosamines related to *N*-butyl-*N*-(4-hydroxybutyl)nitrosamine and *N,N*,-dibutylnitrosamine in ACI/N rats. *Gann* 67: 825–834.
96 Stoltz, D.R. and Sen, N.P. (1977). Mutagenicity of five cyclic *N*-nitrosamines: assay with Salmonella typhimurium. *J. Natl. Cancer Inst.* 58: 393–394.
97 Hernandez, L.G., Slob, W., van Steeg, H., and van Benthem, J. (2011). Can carcinogenic potency be predicted from in vivo genotoxicity data?: a meta-analysis of historical data. *Environ. Mol. Mutagen.* 52: 518–528.
98 Peto, R., Gray, R., Brantom, P., and Grasso, P. (1991). Effects on 4080 rats of chronic ingestion of *N*-nitrosodiethylamine or *N*-nitrosodimethylamine: a detailed dose-response study. *Cancer Res.* 51: 6415–6451.
99 Gillis, E.P., Eastman, K.J., Hill, M.D. et al. (2015). Applications of Fluorine in medicinal chemistry. *J. Med. Chem.* 58: 8315–8359.
100 Galloway, S.M., Vijayaraj Reddy, M., McGettigan, K. et al. (2013). Potentially mutagenic impurities: analysis of structural classes and carcinogenic potencies of chemical intermediates in pharmaceutical syntheses supports alternative methods to the default TTC for calculating safe levels of impurities. *Regul. Toxicol. Pharmacol.* 66: 326–335.
101 Magano, J. and Dunetz, J.R. (2011). Large-scale applications of transition metal-catalyzed couplings for the synthesis of pharmaceuticals. *Chem. Rev.* 111: 2177–2250.
102 Hansen, M.M., Jolly, R.A., and Linder, R.J. (2015). Boronic acids and derivatives-probing the structure-activity relationships for mutagenicity. *Org. Process Res. Dev.* 19: 1507–1516.
103 O'Donovan, M.R., Mee, C.D., Fenner, S. et al. (2011). Boronic acids-a novel class of bacterial mutagen. *Mutat. Res.* 724: 1–6.
104 Masuda-Herrera, M.J., Dobo, K.L., Kenyon, M.O. et al. (2019). In vivo mutagenicity testing of arylboronic acids and esters. *Environ. Mol. Mutagen.* 60 (9): 766–777.
105 Gerber, C. and Toelle, H.G. (2009). What happened: the chemistry side of the incident with EMS contamination in Viracept tablets. *Toxicol. Lett.* 190: 248–253.
106 Saal, C. and Becker, A. (2013). Pharmaceutical salts: a summary on doses of salt formers from the Orange Book. *Eur. J. Pharm. Sci.* 49: 614–623.
107 Snodin, D. and Teasdale, A. (2015). Mutagenic alkyl-sulfonate impurities in sulfonic acid salts: reviewing the evidence and challenging regulatory perceptions. *Org. Process Res. Dev.* 19: 1465–1485.
108 Snodin, D.J. (2019). Elusive impurities-evidence versus hypothesis. Technical and regulatory update on alkyl sulfonates in sulfonic acid salts. *Org. Process Res. Dev.* 23: 695–710.

109 Sega, G.A. (1984). A review of the genetic effects of ethyl methanesulfonate. *Mutat. Res.* 134: 113–142.

110 Gocke, E., Burgin, H., Muller, L., and Pfister, T. (2009). Literature review on the genotoxicity, reproductive toxicity, and carcinogenicity of ethyl methanesulfonate. *Toxicol. Lett.* 190: 254–265.

111 Chemical Carcinogenesis Research Information System (CCRIS). (2009). Ethyl methansulfonate. https://toxnet.nlm.nih.gov/cgi-bin/sis/search2/r?dbs+ccris:@term+@rn+62-50-0 (accessed September 2020).

112 Gocke, E. and Muller, L. (2009). in vivo studies in the mouse to define a threshold for the genotoxicity of EMS and ENU. *Mutat. Res.* 678: 101–107.

113 Cao, X., Mittelstaedt, R.A., Pearce, M.G. et al. (2014). Quantitative dose-response analysis of ethyl methanesulfonate genotoxicity in adult gpt-delta transgenic mice. *Environ. Mol. Mutagen.* 55: 385–399.

114 Dobo, K.L., Fiedler, R.D., Gunther, W.C. et al. (2011). Defining EMS and ENU dose-response relationships using the Pig-a mutation assay in rats. *Mutat. Res.* 725: 13–21.

115 Gerson, S.L., Trey, J.E., Miller, K., and Berger, N.A. (1986). Comparison of O^6-alkylguanine-DNA alkyltransferase activity based on cellular DNA content in human, rat and mouse tissues. *Carcinogenesis* 7: 745–749.

116 Pegg, A.E. (1990). Mammalian O^6-alkylguanine-DNA alkyltransferase: regulation and importance in response to alkylating carcinogenic and therapeutic agents. *Cancer Res.* 50: 6119–6129.

117 Registry of Toxic Effects of Chemical Substances (RTECS). (2018). Methanesulfonic acid, ethyl ester.

118 Lave, T., Birnbock, H., Gotschi, A. et al. (2009). in vivo and in vitro characterization of ethyl methanesulfonate pharmacokinetics in animals and in human. *Toxicol. Lett.* 190: 303–309.

119 Pfister, T. and Eichinger-Chapelon, A. (2009). General 4-week toxicity study with EMS in the rat. *Toxicol. Lett.* 190: 271–285.

120 Platzek, T., Bochert, G., and Meister, R. (1995). Embryotoxicity induced by alkylating agents: 9. Low dose prenatal-toxic risk estimation of ethylmethanesulfonate based on no-observed-adverse-effect-level risk factor approach, dose-response relationships, and molecular dosimetry. *Teratog. Carcinog. Mutagen.* 15: 81–92.

121 International Agency for Research on Cancer (IARC). (1974). Ethyl methanesulfonate: summary of data reported and evaluation.

122 Frei, J.V. (1971). Tumour induction by low molecular weight alkylating agents. *Chem. Biol. Interact.* 3: 117–121.

123 Clapp, N.K. (1973). Carcinogenicity of nitrosamines and methanesulphonate esters given intraperitoneally, in RF mice. *Int. J. Cancer* 12: 728–733.

124 Swann, P.F. and Magee, P.N. (1969). Induction of rat kidney tumours by ethyl methanesulphonate and nervous tissue tumours by methyl methanesulphonate and ethyl methanesulphonate. *Nature* 223: 947–949.

125 Hrushesky, W., Sampson, D., and Murphy, G.P. (1972). Carcinogenicity of ethylmethanesulfonate. *J. Natl. Cancer Inst.* 49: 1077–1083.

126 Montesano, R., Mohr, U., Magee, P.N. et al. (1974). Additive effect in the induction of kidney tumours in rats treated with dimethylnitrosamine and ethylmethanesulphonate. *Br. J. Cancer* 29: 50–58.

127 Craddock, V.M. and Frei, J.V. (1976). Induction of tumours in intact and partially hepatectomized rats with ethyl methanesulphonate. *Br. J. Cancer* 34: 207–209.

128 Roe, F.J., Mitchley, B.C., and Walters, M. (1963). Tests for carcinogenesis using newborn mice: 1,2-benzanthracene, 2-naphthylamine, 2-naphthylhydroxylamine and ethyl methane sulphonate. *Br. J. Cancer* 17: 255–260.

129 Ueo, H., Takaki, R., Yamagami, H. et al. (1979). High incidence of rat mammary carcinoma by oral administration of ethyl methanesulphonate. *Cancer Lett.* 7: 79–84.

130 Walters, M.A., Roe, F.J., Mitchley, B.C., and Walsh, A. (1967). Further tests for carcinogenesis using newborn mice: 2-naphthylamine, 2-naphthylhydroxylamine, 2-acetylaminofluorene and ethyl methane sulphonate. *Br. J. Cancer* 21: 367–372.

131 Frei, J.V., Swenson, D.H., Warren, W., and Lawley, P.D. (1978). Alkylation of deoxyribonucleic acid in vivo in various organs of C57BL mice by the carcinogens *N*-methyl-*N*-nitrosourea, *N*-ethyl-*N*-nitrosourea and ethyl methanesulphonate in relation to induction of thymic lymphoma. Some applications of high-pressure liquid chromatography. *Biochem. J.* 174: 1031–1044.

132 Boettiger, D.C., Sabin, C.A., Grulich, A. et al. (2016). Is nelfinavir exposure associated with cancer incidence in HIV-positive individuals? *AIDS* 30: 1629–1637.

133 Muller, L. and Gocke, E. (2009). Considerations regarding a permitted daily exposure calculation for ethyl methanesulfonate. *Toxicol. Lett.* 190: 330–332.

134 Wills, J.W., Johnson, G.E., Battaion, H.L. et al. (2017). Comparing BMD-derived genotoxic potency estimations across variants of the transgenic rodent gene mutation assay. *Environ. Mol. Mutagen.* 58: 632–643.

135 Slob, W. (2017). A general theory of effect size, and its consequences for defining the benchmark response (BMR) for continuous endpoints. *Crit. Rev. Toxicol.* 47: 342–351.

136 International Organization for Standardization (ISO). (2002). 109993-17. Biological evaluation of medical devices. Part 17: Establishment of Allowable Limits for Leachable Substances.

137 International Conference on Harmonisation (ICH). (2020). Concept Paper: Guideline for Extractables and Leachables (E&L). ICH Q3E.

138 Broschard, T.H., Glowienke, S., Bruen, U.S. et al. (2016). Assessing safety of extractables from materials and leachables in pharmaceuticals and biologics - current challenges and approaches. *Regul. Toxicol. Pharmacol.* 81: 201–211.

139 Bercu, J.P., Morton, S.M., Deahl, J.T. et al. (2010). In silico approaches to predicting cancer potency for risk assessment of genotoxic impurities in drug substances. *Regul. Toxicol. Pharmacol.* 57: 300–306.

140 Contrera, J.F. (2011). Improved in silico prediction of carcinogenic potency (TD_{50}) and the risk specific dose (RSD) adjusted Threshold of Toxicological Concern (TTC) for genotoxic chemicals and pharmaceutical impurities. *Regul. Toxicol. Pharmacol.* 59: 133–141.

141 Venkatapathy, R., Wang, C.Y., Bruce, R.M., and Moudgal, C. (2009). Development of quantitative structure-activity relationship (QSAR) models to predict the carcinogenic potency of chemicals I. Alternative toxicity measures as an estimator of carcinogenic potency. *Toxicol. Appl. Pharmacol.* 234: 209–221.

142 Wang, N.C., Venkatapathy, R., Bruce, R.M., and Moudgal, C. (2011). Development of quantitative structure-activity relationship (QSAR) models to predict the carcinogenic potency of chemicals. II. Using oral slope factor as a measure of carcinogenic potency. *Regul. Toxicol. Pharmacol.* 59: 215–226.

8

Genotoxic Threshold Mechanisms and Points of Departure

George E. Johnson[1], Shareen H. Doak[1], Gareth J.S. Jenkins[1], and Adam D. Thomas[2]

[1] Institute of Life Science, Swansea University Medical School, Singleton Park, UK
[2] Faculty of Health and Applied Science, University of West of England, Bristol, UK

8.1 Introduction to Genotoxic Dose Responses

In genetic toxicology, induced deoxyribonucleic acid (DNA) damage is used as a surrogate for cancer risk, as it is well known that cancer is induced by genetic aberrations. Therefore, genetic damage (mutation) is often used as a short-term biomarker of cancer risk. In terms of dose responses, while linearity has been assigned to genotoxic dose responses (see below), the dose relationship with cancer induction is much less clear. This is partly due to the expense and number of animals required to investigate this aspect thoroughly. However, it has been suggested that cancer is induced in a nonlinear fashion for non-genotoxic carcinogens and in a linear manner for genotoxic carcinogens. This is based on mutation response being regarded as linear, but this has recently been challenged in various studies and reviews, where nonlinear models provided more suitable fits to *in vivo* gene mutation and chromosome damage end points [1–4]. This is further supported due to nonlinearity reflecting human carcinogenesis where a number of genetic hits are known to be required to induce cancer. Indeed, linearity in the cancer response would lead to tumor formation occurring much more frequently in the human population. More recently, Bailey et al. [5] demonstrated, in a "mega-animal" study, that cancer was induced by dibenzopyrene (DBP) in a decidedly nonlinear manner using trout as a model organism, which meant that very large animal numbers (>40 000 individual fish) could be readily studied. It is now accepted by leading experts that "at non-toxic dosages, thresholds exist for the induction of experimental cancer by all types of carcinogens" [6].

8.1.1 The Linear Default Position for Genotoxic Carcinogens

Genotoxic agents that are DNA reactive and direct acting have long been assumed to display linear dose responses [7]. Genotoxic agents that require metabolic activation, while having the confounding effect of the kinetics of enzymatic activation, are still also assumed to induce linear responses. However, due to this need for metabolic activation, and the considerable variation in metabolic enzyme levels between different organisms and organs, most research into dose–response relationships for genotoxicants has focused on DNA-reactive genotoxicants that do not require metabolic activation.

In the linear model, DNA damage induction is directly proportional to dose; hence, the implication that there are no "safe" genotoxic doses in terms of DNA damage (and hence by extension, cancer). This linear model has been implemented partly due to the precautionary principle, designed

Mutagenic Impurities: Strategies for Identification and Control, First Edition. Edited by Andrew Teasdale.
© 2022 John Wiley & Sons, Inc. Published 2022 by John Wiley & Sons, Inc.

to protect human populations from exposure to potential genotoxicants. This linear concept has been controversial and has recently been challenged [2, 3, 8], as it assumes a binary situation where chemicals are either genotoxic or not, but does not account for the effect of dose. As pointed out by Paracelsus in the sixteenth century, "only the dose permits something not to be poisonous."

In the case of indirect acting genotoxicants, which have non-DNA targets (e.g. aneugens targeting the mitotic spindle and agents interacting with DNA-modifying enzymes), nonlinear relationships would logically be assumed to apply, and indeed, "thresholds" have now been generally accepted for such genotoxicants [9–12]. Hence, the presentation of solid, experimental evidence has already successfully altered the default linear position on the dose–response relationship for some chemicals. However, for direct acting DNA damaging genotoxicants, linear models are still assumed to apply. The linear model was supported by some early dose-response data, which was often carried out using high doses of genotoxicants. High doses of test agents have traditionally been used in genotoxicity testing (approaching or exceeding the LD_{50} values *in vivo* or *in vitro*) to ensure that DNA damaging effects are identified in the available tests (due to test sensitivity constraints). As it has widely been assumed that genotoxic effects are induced in a linear manner, these high dose values are then extrapolated back to the low dose region and the conclusion linearity accepted, without solid experimental evidence at low doses. The implications emanating from the linear model for genotoxicants can be wide reaching and can have a significant impact on the development of a new pharmaceutical. This can often lead to the need to control a genotoxic impurity down to very low (ppm) levels and can, in some instances, even lead to clinical delays that significantly affect the development of the product concerned. Even without this, controlling such impurities, down to the levels required by the application of the "Threshold of Toxicological Concern" (TTC), can be a significant technical challenge.

8.1.2 Theoretical Evidence for Rejecting the Linear Approach

The main argument against a linear dose response for genotoxicants is the presence of natural cellular defenses, which have evolved to cope with our daily exposure to unavoidable genotoxicants. Humans are constantly exposed to genotoxic substances like cytosolic oxidative agents, sunlight, dietary amines, inhaled hydrocarbons, and many others. Low-level exposures to these genotoxicants have occurred throughout evolutionary time and have led to the development of efficient homeostatic defenses to protect organisms against the deleterious mutagenic consequences. DNA repair is one such homeostatic defense mechanism that may impact heavily on the outcome of genotoxicant exposure. Indeed, even simple bacteria have intricate defenses (like DNA repair) against genotoxicants. As multicellular organisms, humans have several tiers of protection against DNA damage induction.

Hence, it is almost inconceivable that genotoxicants cause DNA damage in a manner proportional only to dose. This is due, in part, to the failure of the genotoxicant to readily access and react and damage the DNA of a target tissue. Even in a simple cell culture system, it is unlikely that true linearity will be seen due to extracellular and intracellular interactions between the genotoxicant and non-DNA biomolecules, as well as membrane-based exclusion. Furthermore, once in the nucleus, the DNA damage induced by the genotoxicants must overcome the homeostatic protection of DNA repair, to produce permanent DNA sequence alterations resulting in phenotypic changes.

Over recent decades, DNA repair has been shown to function in a complex and, in some cases, in an inducible manner to control the genetic stability of the host cell's genome [13]. Several overlapping DNA repair pathways exist (e.g. base excision repair, nucleotide excision repair, homologous recombination, and mismatch repair); each is responsible for repairing specific DNA damage types (small and bulky adducts, DNA strand breaks, and base mismatches). Hence, it is likely that DNA repair will impact directly on the linearity of genotoxic dose responses by removing DNA damage toward a nonlinear response. This is particularly true at low doses of genotoxicant exposure, as DNA repair has evolved to deal with constant low-level DNA damage induction. At higher doses DNA repair may be saturated and hence may not be able to remove newly damaged DNA bases. There is some evidence that DNA adduct formation accrues in a linear fashion [14–16], although recent evidence challenges this [17]. Nevertheless, it is likely that fixed mutations (point mutations and chromosome mutations) will not follow linearity. One complication with comparing the DNA adduct data to the mutation data (point mutation or chromosome mutation) is that DNA adducts can be detected to a level that is three to four orders of magnitude more sensitive than that to which DNA mutations can currently be measured. Furthermore, another complication with this comparison is that DNA mutations (point mutations in particular) are detected in specific gene sequences (hprt, tk, lacZ, Lazl, etc.), whereas current DNA adduct measurements do not define the location of the adducts in the genome. Given that cells have evolved efficient measures to keep gene coding sequences damage free at the expense of noncoding regions [18], it is not possible to currently say if DNA adducts accrue in a linear fashion in the coding sequences, which form the basis for most genotoxicity tests. Therefore, it is not straightforward to currently compare the dose relationships between adducts and mutations.

8.1.3 *In Vitro* Experimental Evidence for Threshold Mechanism

Up until recently, most of the *in vitro* experimental evidence demonstrating nonlinear (thresholded) dose responses were available for indirect genotoxicants or non-DNA reactive agents. As their dose response and underlying mechanisms of action are well understood, threshold mechanisms are now largely accepted for these compounds [19]. Indirect acting compounds generally target non-DNA biomolecules, possessing functional redundancy. These genotoxicants must therefore damage multiple targets, before a significant adverse effect arises. Hence, low concentrations of these compounds can be tolerated by cells.

Classical examples of indirect acting (non-DNA reactive) genotoxic compounds that have clear threshold mechanisms are the spindle poisons. These compounds damage tubulin monomers, which are in excess within cells. Although damage to these biomolecules will arise at low concentrations, it is not sufficient to disrupt microtubule formation, and hence chromosome segregation at mitosis is unaffected. Only when a substantial degree of damage is induced will malformation of the spindle apparatus occur, which in turn leads to aneuploidy through mis-segregation of the replicated chromosomes. Using the micronucleus (MN) assay coupled with fluorescence *in situ* hybridization (FISH), points of departure (PoD) have been established *in vitro* for several chemicals with this underlying biological mechanism, including the aneugens colchicine, mebendazole, carbendazim, and nocodazole [19, 20]. More recently another class of indirect acting genotoxicants, the topoisomerase II (topo II) poisons, has also been shown to display nonlinear dose responses *in vitro* [21, 22]. Topo II enzymes play an essential role in modulating DNA tension and topology during replication, transcription, and repair. Through transiently cutting both DNA strands, these enzymes relieve torsion at key places along the genome before the DNA is resealed. Agents that poison topo II enzyme action often lead to DNA strand breakage during DNA replication by stabilizing the DNA–topo II complex, which possibly inhibits strand religation. Again, these enzymes represent a redundant target within the cells, hence low concentrations of topo II inhibitors can be tolerated by cells as the excess type II topoisomerase molecules ensure that the required level of activity is maintained. However, at higher doses (above the threshold) a genotoxic compound will significantly disrupt more topo II enzymes to have a greater effect on net activity, leading to strand breaks (clastogenicity).

In contrast to the above examples, DNA-reactive chemicals have largely been assumed to have a non-thresholded dose response as they directly induce DNA lesions that can potentially be fixed as point mutations or chromosomal aberrations. However, as mentioned earlier, homeostasis in mammalian cells allows them to adapt to environmental insults, meaning that a range of low doses should have a biologically insignificant effect. Nonetheless, experimental evidence for the existence of thresholds for agents that directly damage DNA is considerably more limited than for indirect acting agents, although some such studies are now starting to emerge. It is important to point out that only large data sets with high-quality data can truly be used to confirm that threshold mechanisms exist. This is discussed below.

One of the first reports to comprehensively demonstrate that direct acting genotoxicants could exhibit PoD with threshold mechanisms for mutation induction and chromosome breakage *in vitro* focused on a set of four well-known genotoxic and carcinogenic alkylating agents: methyl methanesulfonate (MMS), *N*-methyl-*N*-nitrosourea (MNU), ethyl methanesulfonate (EMS), and *N*-ethyl-*N*-nitrosourea (ENU) [23–28]. Given that MMS and EMS were known carcinogens and were often used as positive controls in genotoxicity testing, this evidence clearly refuted the linear model, which was assumed to apply for all direct acting genotoxicants. This is further supported through the MNU and ENU PoD and the elucidation of the DNA repair mechanism supporting this dose response. These potent alkylnitrosoureas were previously common positive experimental controls and are also considered super-mutagens. Therefore, clear PoDs, accompanied with an explanatory threshold mechanism (Figure 8.1) for these substances, are highly relevant when considering how to risk assess potent genotoxic carcinogens, and that the default linear approach may not always be suitable.

At present, only a limited amount of good quality *in vitro* evidence is available for the existence of genotoxic thresholds and this relates to a small number of chemicals. Although thresholds are well accepted for indirect acting (non-DNA reactive) genotoxicants, more evidence is required for

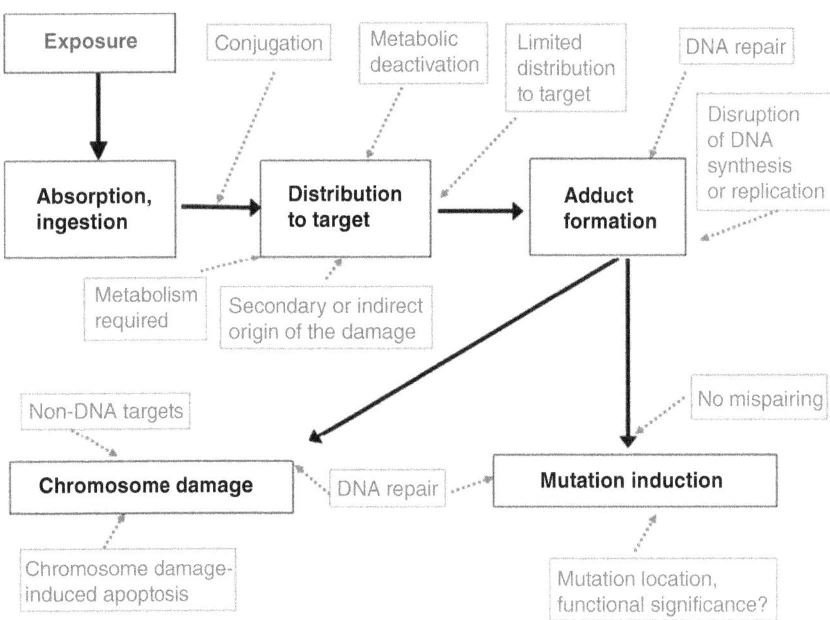

Figure 8.1 Flowchart illustrating potential mechanisms underlying genotoxic thresholds.

direct acting agents, and this data needs to be supported with a clear biological explanation for the shape of the respective dose–response curve. Further investigation is therefore required on a compound-by-compound basis, at least in the short term until sufficient data across different chemical classes has been compiled.

The International Workshops on Genotoxicity Testing (IWGT) stated that *in vitro* genetic toxicity data should be used for potency and mechanistic data and not for derivation of PoD [3]. Following work in this area, benchmark dose (BMD) analysis was used to assess the potency of a range of similar acting genotoxicants, and the BMD confidence interval (CI) provided an excellent measure of potency, in a ranked format. This is the recommended analysis for potency assessment of *in vitro* data [20] (Figure 8.2).

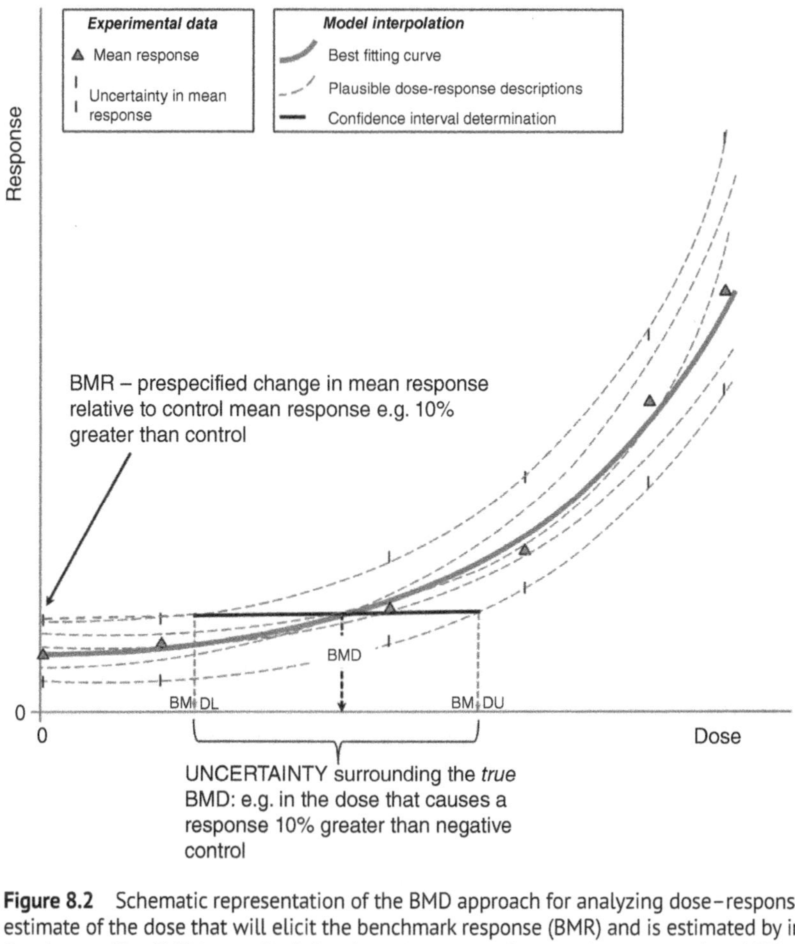

Figure 8.2 Schematic representation of the BMD approach for analyzing dose–response data. The BMD is an estimate of the dose that will elicit the benchmark response (BMR) and is estimated by interpolation from the fitted curve. The BMR is usually defined as a percentage increase in response (e.g. 10%) relative to control, with this BMR adjustable to any desired response level. The uncertainties in the data can be taken into account by calculating a CI for the BMD. Conceptually, one may imagine that, by varying the parameters in the model, different curves can be generated, and for those that are considered compatible with the data (dashed curves) BMDs could be established. Together, they comprise values that make up the CI for the BMD. As an approximate conceptual illustration, the horizontal black line segment intersecting the range of plausible curves results in the BMDL and BMDU, the lower and upper 90% confidence bounds of the BMD estimate, respectively. The width of this interval (expressed as the ratio BMDU to BMDL) therefore represents the BMD precision [20, 29].

8.1.4 In Vivo Evidence for Genotoxic Thresholds

In 2009, a comprehensive analysis of the genotoxic dose responses for two alkylating agents (EMS and ENU) was reported [30]. This research was instigated as a result of the contamination of the drug Viracept® (nelfinavir mesylate) with EMS. This occurred as a consequence of a good manufacturing practice (GMP) failure that resulted in a reaction between methane sulfonic acid and ethanol, which was inadvertently mixed in a tank used in the manufacturing process. This *in vivo* genotoxicity analysis formed the basis of an assessment of the risks posed to patients as a result of their exposure to the contaminated product. The *in vivo* data compared dose responses for both point mutation (LacZ locus in three tissues) and chromosome aberration (bone marrow micronuclei) between ENU and EMS across a range of low doses, which were spaced closely together. This seminal piece of work confirmed the previous *in vitro* threshold for EMS (but not ENU) for both MN induction and LacZ mutation induction *in vivo* [30]. Furthermore, through comprehensive pharmacokinetic studies, a cross species analysis allowed extrapolation from the genotoxic PoD observed in the mice to a corresponding level in humans. Hence, a "threshold" for EMS has been identified, providing a potential precedent for future safety assessment of other genotoxic carcinogens that may also possess threshold mechanisms and PoD. In the seminal assessment from [30–33] chromosome breakage in bone marrow and mutation induction in bone marrow, gut and liver are plotted against dose and no-observed effect levels (NOELs) calculated for each end point at 25 mg/kg/day [30] (Figure 8.3). Bercu et al. [34] reanalyzed EMS in addition to calculating compound-specific toxicology limits for 19 other widely used synthetic reagents and common by-products that are potential drug impurities. This was an extensive and highly regarded analysis that provides guidance and useable permitted daily exposure (PDE) and acceptable intake (AI) metrics for these substances.

A comparable but larger incident has occurred more recently. In 2018, a global recall of valsartan-containing preparations was initiated by the regulatory bodies of 22 countries. This was because the synthesis impurity *N*-nitrosodimethylamine (NDMA) was found in certain valsartan products. NDMA

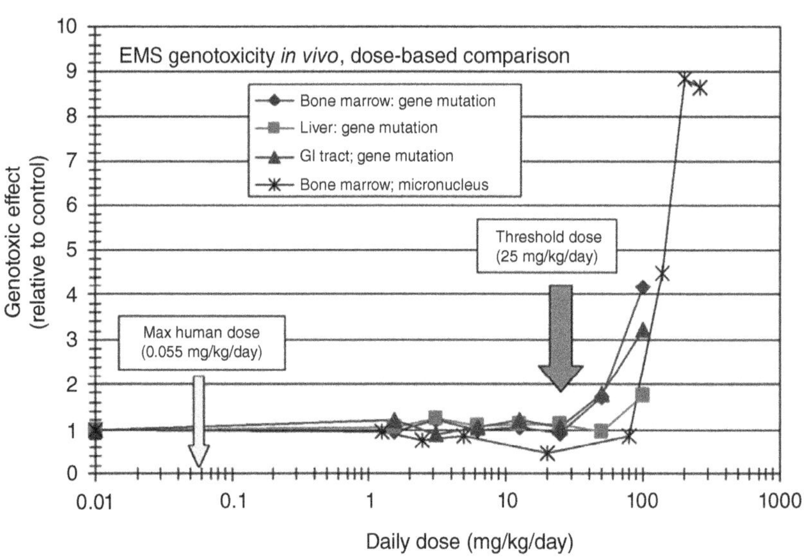

Figure 8.3 EMS-induced thresholded dose response *in vivo*. *Source:* Data taken from Gocke et al. [33] and includes both chromosome breakage data and LacZ mutation data. Here the data are plotted on a log-linear graph to allow better visualization of the data. In the publication by Gocke et al. [33], linear–linear graphs are also plotted. (*See insert for color representation of this figure*).

is a well-studied compound that along with being found in other sartan pharmaceuticals also occurs in other products, for example, 2.5 μg/kg is found in cured meats, approximately 4 μg is present per cigarette smoked, and within beer that has a technical guideline value of 0.5 μg/kg [35]. NDMA is an alkylnitrosamine and a known genotoxic carcinogen. The most comprehensive and relevant dose response for NDMA was the cancer bioassay in the liver of rats [36]. The similar acting alkylnitrosamine *N*-nitrosodiethylamine (NDEA) was also found as a pharmaceutical impurity, and the most relevant data for this structurally related compound are from the same rodent cancer bioassay [36] and the transgenic rodent (TGR) assay in gpt-delta rats [37]. The liver was also the most sensitive tissue for gene mutations in mice [38] supporting the known involvement of metabolic activation [39]. NDMA and NDEA are structurally related, DNA-reactive genotoxic carcinogens that alkylate DNA at numerous nucleophilic sites, producing a spectrum of DNA alkyl adducts, just like the four aforementioned alkylating agents [40, 41]. These agents are both clastogenic (able to produce chromosome breaks) and mutagenic (able to induce gene point mutations); their propensity for either one broadly depends on the ratio of 7-alkylguanine (7-AlkG) and O^6alkylguanine (O^6AlkG) adducts induced, although, in biology, there is some crossover [42–44]. NDMA and NDEA show similar adduct spectra to ENU and MNU, the other well-characterized nitrosoureas, and are, therefore, also considered potent point mutagens, due to the high reactivity with exocyclic oxygen (O^6) of guanine and the relatively high levels of O^6AlkG produced as a result [45]. When left unrepaired, O^6AlkG has enormous miscoding potential during replication and transcription. It is able to form only two hydrogen bonds like its purine analogue adenine, instead of three as with unmodified guanine. As a result, the polymerase eventually fixes the O^6AlkG adduct as a GC>AT transition mutation [46–48].

This process of mutagenesis has been experimentally linked to the mutation and oncogenic transformation of rat sarcoma virus (RAS) oncogenes, leading to mammary and pancreatic cancers in animal models [49–53]. On the other hand, O^6AlkG adducts are directly repaired via a protein called methylguanine DNA methyltransferase (MGMT) [54–57]. In a suicide reaction, MGMT transfers the alkyl group from O^6AlkG to a cysteine residue within its active site. This renders the MGMT protein nonfunctional and becomes marked for proteolytic degradation [58]. However, the guanine residue is restored, and the wild-type DNA sequence remains unchanged. With a lack of evidence of MGMT inducibility, following genotoxic attack, the level of protection against O^6AlkG-mediated mutagenesis, therefore, is limited (saturable) and depends upon the constitutive level of MGMT within each cell. In fact, MGMT loss, through aberrant methylation, has been found to be a significant step in sporadic colorectal cancer [59, 60], most likely as a result of increased susceptibility to mutation. MGMT plays a pivotal role in cellular sensitivity to mutagenesis, via O^6AlkG, and can, therefore, influence the dose response and PoD for both gene mutation and cancer [2, 61–67]. This parallels what has been shown for EMS. Therefore, the same justification for using the PDE could be made.

The linear risk assessment approaches are a suitable way to ensure the safety of the exposed population when little else is known about the repair mechanisms and dose response. However, in this instance there is a better option within the ICH M7 [12]. This relies on the following statement, "the existence of *mechanisms* leading to a dose response that is *non-linear* or has a *practical threshold* is increasingly recognized, not only for compounds that interact with non-DNA targets but also for DNA-reactive compounds, whose effects may be modulated by, for example, rapid detoxification before coming into contact with DNA, or by *effective repair* of induced damage. The regulatory approach to such compounds can be based on the identification of a No-Observed Effect Level (**NOEL**) and use of uncertainty factors (see ICH Q3C(R5)), to calculate a permissible daily exposure (**PDE**) when data are available."

For NDMA and NDEA, the gene mutation dose responses for both compounds are modulated by MGMT within the Ames mutation test [68]. NDMA and NDEA do not react with DNA directly, because metabolic activation is needed to form the DNA-reactive molecules; however, for the purposes of risk assessment, they are regarded as DNA-reactive carcinogens [69]. For the PDE approach to be implemented for DNA-reactive carcinogens, there is a requirement to show that low levels of NDMA- and NDEA-induced damage are modulated by a biological mechanism. A NOEL from a relevant *in vivo* test is then required; however, a NOEL based on pairwise testing suffers from numerous disadvantages compared to the BMDL, which is a comparable but more precise PoD metric [2, 3]. A poorer data set with larger within-group variation should lead to a lower PoD, as this would protect the population from the uncertainty in the data set. This is the case with the BMD approach. However, for the NOEL, the PoD becomes higher with these poorer data, with the assumption that the compound is safer, but this is an artifact and entirely based on the within-group variation. It is essential to use the most relevant and extensive data set to derive the PoD. For NDMA, the best dose-response data for *in vivo* mutation were from Jiao et al. [38] and Gollapudi et al. [70]. For NDEA, transgenic gene mutation data from gpt-delta rats were the most extensive and suitable for analysis [37]. The Peto et al. [36, 71] cancer bioassay study was also hugely extended to increased precision in the dose response for NDMA and NDEA, with 240 animals per dose, at 15 doses, and with 4 × 240 animals as the vehicle control ($n = 4080$). These data were considered to be excellent for further assessment [72] (Table 8.1).

The PDE derived by Johnson et al. [72] is not considered for regulatory purposes and is presented as case studies. There could be justification to use genetox derived PDEs in this case if a new TGR mutation study was carried out at 28 days with repeat dosing, with more animals used to expand the number of doses tested and potentially more replicates as well. Regulatory bodies recommend deriving PoD from multiple studies and to select the most relevant, conservative, and precise metric. In this case, preference would still be to use the cancer-derived PDE, due to cancer being the apical end point for alkylnitrosamines and due to the extensive cancer bioassay data sets for NDMA and NDEA [36, 71].

There are now numerous examples of DNA-reactive substances that have nonlinear dose responses. Many of these are from standard study designs using the BMD approach, but there are also extensive dose responses for use in defining BMD CI with high precision. One of the best-case studies to date

Table 8.1 BMD CI and PDE calculated using *in vivo* mutation and cancer bioassay dose-response data by Johnson et al. [72].

Parameter	NDMA $PDE_{mutation}$	NDMA PDE_{cancer}	NDEA $PDE_{mutation}$	NDEA PDE_{cancer}
End point modeled	Increase in *lacI* Mutant frequency (MF) for Big Blue® rats, liver	Increased incidence of liver cell tumors in rats	Increase in *got* MF for liver in F344/gpt-delta rats	Increased incidence of liver cell tumors in rats
Data source	Gollapudi et al. [70]	Peto et al. [36, 71]	Akagi et al. [37]	Peto et al. [36, 71]
CES (% increase over mean controls)	50	10	50	10
BMD (mg/kg/day)	0.37	0.088	0.011	0.035
BMDU (mg/kg/day)	2.34	0.107	0.028	0.046
BMDL (mg/kg/day)	0.06	0.062	0.004	0.022
Composite uncertainty factor	5000	500	5000	500
PDE (μg/person/day)	0.6	6.2	0.04	2.2

is for benzo(a)pyrene (BaP), which has a huge dose response using TGR and MN. There are also extensive dose responses in *in vivo Pig-a* and MN assays [73–77]. BaP is a genotoxic carcinogen that requires metabolic activation into BaP diol epoxide (BPDE). It is, therefore, most potent in the liver for both mutation and for cancer. The dose responses for both end points have been statistically modeled and provide comparable BMD CI. The dose responses for both end points are not in line with the linear model. BPDE is mutagenic, through bulky DNA adducts that are repaired by nucleotide excision repair (NER). Translesion synthesis (TLS) has also been linked with the dose response for BPDE, and DNA repair is, therefore, intrinsically linked to the PoD defined using these data.

8.2 Threshold Mechanisms

It is becoming increasingly apparent that there are a number of biological explanations that argue against the assertion that all genotoxicants have linear dose responses, as mammalian cells have numerous homeostatic mechanisms that provide protection. Humans are complex multicellular organisms, and therefore, have several tiers of protection against DNA damage. Many of these mechanisms were summarized in the quantitative workgroup of the IWGT 2014 [3] as: critical involvement of non-DNA targets, contribution of DNA repair mechanisms, detoxification of enzymes involved in DNA synthesis or replication, chemical reactivity or properties unlikely to occur *in vivo*, inadequate uptake of toxicokinetics (limiting distribution to target), mutational spectrum in tumor genes being similar to those in untreated animals, structural similarities to similar threshold-acting chemicals, secondary or indirect origin of the observed damage, and species- and tumor-specific nongenotoxic mode of action.

With regard to indirect- and direct-DNA damaging agents, assuming they are able to gain access to the cellular environment, there is a significant disparity in the potential mechanism underlying their dose responses due to the differences in their biomolecular targets. Indirect acting genotoxicants do not damage DNA directly, but instead, interfere with DNA replication, transcription, or repair processes. Nuclear division is a multifactorial process involving a large number of components required for DNA synthesis; subsequent, equal partitioning of the replicated chromosomes; and several signal transduction-based checkpoints, to monitor the mitosis process. Indirect acting agents may therefore target mechanical components vital for segregation of replicated chromosomes (e.g. microtubules and centrosomes), DNA replication (e.g. topoisomerase enzymes, DNA polymerases, and imbalanced nucleotide pools), or DNA repair proteins (e.g. glycosylases, polymerases, endonucleases, and ligases), subsequently resulting in structural and numerical chromosomal abnormalities [78, 79]. Consequently, these agents have to damage multiple targets before a significant adverse biological effect occurs; a prime example being spindle poisons (e.g. nocodazole) that will only induce nondisjunction and chromosome loss at a dose that damages a sufficient number of tubulin monomers to disrupt microtubule assembly and disrupt the appropriate formation of the spindle apparatus [80]. Low doses of a spindle poison lead to slight damage to the microtubules, which is of no consequence in terms of mutagenesis, and at higher doses, modest damage and nondisjunction occur with more serious damage leading to chromosome loss and the formation of micronuclei at even higher doses [80, 81]. Indirect acting agents are, therefore, well accepted to have a threshold mechanism due to the redundancy of their targets.

In contrast, DNA-reactive genotoxicants have long been assumed to have a linear dose response because of their capacity to directly damage the macromolecule by inducing either strand breakages or DNA lesions such as adducts. This linear model is still largely assumed as a precaution for these agents and direct acting genotoxins will only be accepted as having nonlinear dose responses if accompanying data on their mechanism of action supports the shape of the dose–response curve. Although data is starting to emerge indicating that not all direct acting genotoxicants have linear dose responses [2, 3, 8], evidence explaining the biological basis for these responses is limited. However, theoretically, DNA repair processes may be the key mechanism for low exposure doses resulting in no significant increase in mutagenic effects.

The ability of a cell to repair its DNA following damage is a vital means for maintaining the genetic stability of the cell's genome. However, at higher genotoxicant doses, these DNA repair pathways may become saturated and hence be unable to efficiently remove all the damage induced. Currently, the evidence to indicate that DNA repair is indeed responsible for nonlinear dose responses following exposure to direct acting genotoxicants is limited, but data supporting this theory is starting to emerge in relation to alkylating agents. Theoretically, it is likely that DNA repair will be strongly involved in influencing the shape of the respective dose–response curves. In the case of alkylating agents, N7AlkG, N3AlkA, and O^6AlkG all have specific repair mechanisms associated with their removal (e.g. base excision repair [BER] involving methyl purine glycosylase [MPG] and MGMT, respectively), while O^2AlkT and O^4AlkT are very poorly corrected. Consequently, at low MMS exposure levels, DNA repair is likely to be primarily responsible for the efficient removal of the N7G, N3A, and O^6G DNA alkyl adducts resulting in a NOEL, while failing to fully remove all the damage at higher concentrations, due to saturation of enzymatic activity, resulting in a LOEL and subsequent dose-dependent increases thereafter. With respect to the nitrosoureas, it was previously considered possible that at low exposure levels, alkyl adducts at N7G, N3A, and O^6G may be repaired, but those at the O^2T and O^4T positions would persist, giving rise to the more linear dose response observed [23, 24]. However, even these more potent mutagens have now been shown to have clear PoD and threshold mechanisms [2]. The importance of the thymine adducts has been shown by Guttenplan and colleagues [82] who reported that at low ENU doses mutations at AT sites predominated (as a result of O^2– and O^4– ethylT), but as concentrations increased, more mutations were induced at GC sites due to saturation of O^6-ethylG repair. The situation is also very similar for ethylene oxide (EO), with the adduct spectrum, mutation spectrum, and genotoxic potency very comparable to EMS and MMS [83].

These repair pathways are, therefore, very important in governing DNA protection against alkylating agents, and it is only when they fail to remove lesions that the opportunity arises for the damage to become fixed as permanent point mutations or chromosomal aberrations. For example, it has been shown that in MGMT-deficient human cells eightfold more GC mutations are observed following exposure to the alkylating agent ENU due to increased persistence of O^6-ethylG [84].

With regard to low doses of alkyl methanesulfonates, BER may be responsible for removing the 7AlkG lesions induced by MMS and EMS, while MGMT may cope with the low levels of O^6-AlkG, thus resulting in their NOELs for the induction of chromosomal damage and point mutations, respectively. In support of this theory, previous genotoxic studies *in vitro* have highlighted a potential role for MGMT in altering the shape of dose responses. In both bacterial and mammalian cells, MGMT knockouts result in more linear-shaped mutational dose responses in contrast to the sublinear shapes in the wild-type cells [65, 85]. Thus, the basal level of MGMT expression may be responsible for repairing O^6MeG lesions at low doses, before they are fixed as permanent base substitutions [23], but becomes overwhelmed and saturated at higher doses due to the increased demand for repair of high O^6methylG levels. Furthermore, when MGMT was inhibited using O^6-benzylguanine, the PoD for mutation moved to a lower concentration, to further support this mechanism of action [66]. However, this was not the case with regard to chromosomal damage as this LOEL for MMS is lower than for point mutations at 0.85 μg/ml. The evidence therefore indicates that O^6-AlkG is unlikely to be primarily responsible for the clastogenicity observed at these lower doses at this sampling timepoint, because despite the MGMT upregulation observed at 1 μg/ml, at this concentration significant chromosomal damage can already be detected. Hence, removal of O^6AlkG by MGMT appears to have limited influence on the chromosomal damage threshold for MMS, suggesting that O^6AlkG is not a key clastogenic adduct at early timepoints posttreatment. Consequently, it seems that at low doses O^6MeG lesions are responsible for MMS mutagenicity, while 7MeG may be the predominant cause of MMS clastogenicity. It may also be possible that MGMT is the more dominant means of adduct removal at low doses (Figure 8.4), while at higher exposure levels alternative DNA repair pathways are triggered in the cell by the genotoxic responses. Indeed, this has been observed in resistance to the chemotherapeutic agent 1,3-bis(2-chloroethyl)-1-nitrosourea (BCNU) [86]. Furthermore, 7AlkG and O^6AlkG are also substrates for mismatch repair (MMR) and NER, respectively, particularly when MGMT is overloaded [87–91], but further analysis is required to substantiate this theory with regard to monofunctional alkylating agents.

It, therefore, appears that the threshold mechanisms shown experimentally, with clear PoDs, for the direct acting genotoxicants MMS and EMS are probably due to efficient repair of the lesions induced at low doses. Although data to support this theory is now emerging, further evidence is required to clearly demonstrate the interplay between DNA damage profiles and repair pathways for

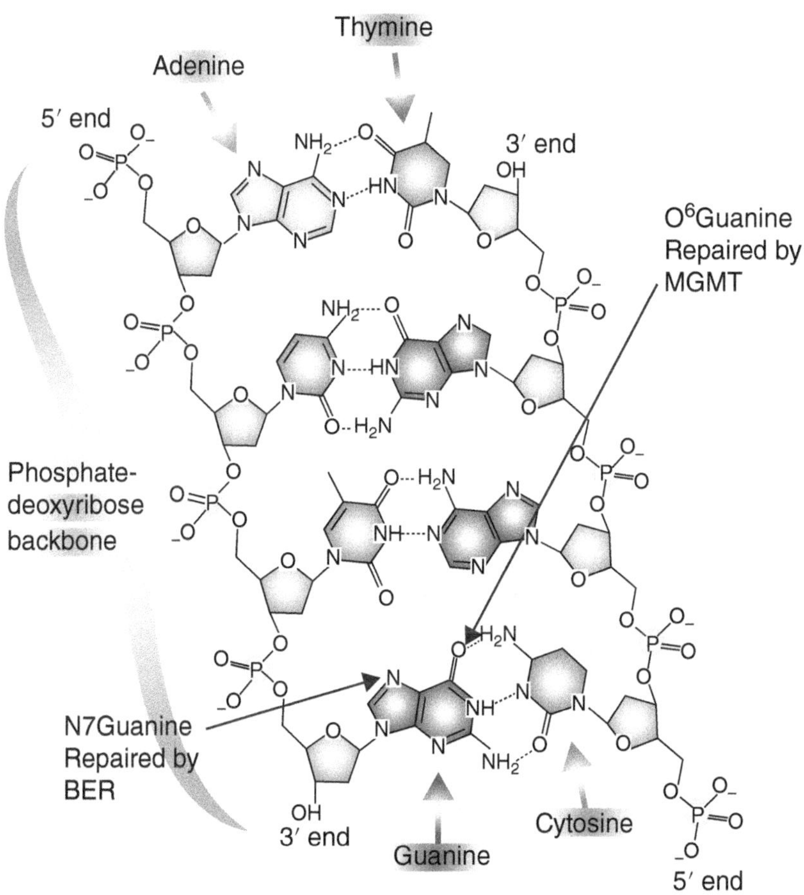

Figure 8.4 DNA adduct locations in DNA coupled to DNA repair processes involved in their correction.

a wider range of chemicals. A full understanding of the biological mechanisms of action imposing genotoxic dose responses is fundamental to accepting the plausibility of these threshold mechanisms in order to support derivation of PoD (Figure 8.1).

8.2.1 Statistical Assessment of Dose Response Data Sets

In order to assess the dose response of genotoxicants, it is imperative that the most sensitive tests available are employed to achieve sufficient power to calculate the PoD. For the purposes of this chapter, a threshold is defined as the point below which there was no dose response. It is only used when discussing the underlying mechanism to result in the PoD. When discussing statistically defined metrics usable for PoD, we will refrain from using the term threshold. This is because an infinite sample size is required to define a threshold using statistical approaches, as without this a nonlinear model will always provide a suitable or better fit to the data. If there is a recommendation to use broken stick (breakpoint dose) modeling in order to define a threshold dose, then there are some important things to consider. A fundamental requirement for such a data analysis is to start with high-quality dose-response data. Poor-quality data is not amenable to such statistical modeling. In order to achieve high-quality data, sensitive, relevant, and robust tests must be used, and adequate replicates are needed to ensure that spurious effects are avoided. Closely spaced doses are also necessary to better define any changes in the shape of the dose response. The more closely spaced the doses, the better the resolution using the breakpoint dose modeling tools.

Once the dose response has been defined, then further statistical analyses can be used to determine the NOEL and LOEL. One approach is the Dunnett's test. This is a widely used and robust method for dose-response data. Dunnett's test is a multiple comparison post hoc method, performed after a one-way analysis of variance (ANOVA) by comparing all treatment data versus the control. Trend tests for homogeneity of variance and normal distribution must be used prior to analysis with the Dunnett's test [2]. In order to pass the trend tests, the data can be transformed using methods such as log or square root transformation, or if the data fails the trend tests, a Dunnett's T3 or Dunn's (nonparametric) test can be used in order to define the NOEL.

The BMD approach is a more advanced statistical approach that models the data using statistical models. A small but measurable increase above the background is then defined with CIs. The lower bound (BMDL) is used as the PoD. The BMD approach is the preferred method for defining PoD for genetic toxicity data [2–4]. BMDS is the statistical software that can be downloaded for free from Environmental Protection Agency (EPA) [92], and PROAST can be downloaded from National Institute for Public Health and the Environment (Dutch agency) (RIVM) or used online at https://proastweb.rivm.nl. These tools can be used to define the BMD CI. Precision in the PoD is calculated by the ratio in the BMD CI, with lower ratios showing increased precision, and higher ratios showing low precision. A rule of thumb is that a ratio of under 2 is excellent, under 10 is good, under 100 is suitable for use, and over 100 is not suitable for use as a PoD. This rule of thumb can help in determining when to consider expanding a study for increased precision or when a BMD CI is not appropriate for use as a PoD.

8.2.2 Extrapolation from One Chemical to Another

Currently, genotoxic PoDs with threshold mechanisms need to be identified on a compound by compound basis. As the number of chemicals with accepted threshold mechanisms is still very low, it is currently not possible to extrapolate from known threshold effects for one chemical to other chemicals within a similar chemical class. As yet, the best proof of a genotoxic threshold mechanism for a DNA-reactive compound is for EMS, as both *in vitro* and *in vivo* evidence supports this [30, 93]. The sister compound, MMS, also exhibits a comparable dose response (Figure 8.5) with clear mechanisms and PoD *in vitro* and *in vivo*. This is also true for the alkyl nitrosoureas ENU and MNU [1, 2] (Figure 8.6). Additional alkylating agents could also be considered to act via the same mechanism with PoD for gene mutation and/or chromosome damage.

8.2.3 Extrapolation of Threshold Mechanisms and PoDs to Populations

Given that DNA repair appears to be centrally involved in the existence of genotoxic threshold mechanisms for DNA-reactive substances, and also given that there is population level variation in DNA repair genes, it can be assumed that this genetic variation may alter susceptibility to genotoxicants in the low-dose region. Indeed, it has previously been suggested that defining a genotoxic threshold for a population might be impossible, due to genetic variation among individuals [94]. There is certainly the possibility that individuals will exist in a population who are more sensitive to a genotoxicant due to possession of a DNA repair variant protein with lower than average efficiency.

Unpicking the haplotypes (combinations) of DNA repair gene polymorphisms, which influence susceptibility to mutation and cancer, is likely to be a complex process. This is discussed in more detail elsewhere [41]. Theoretically, if mismatch repair (MMR) was found to be the main mediator of a threshold response to a genotoxicant and safe exposures levels in humans were calculated based on a threshold dose in proficient models, then patients with hereditary nonpolyposis colorectal cancer

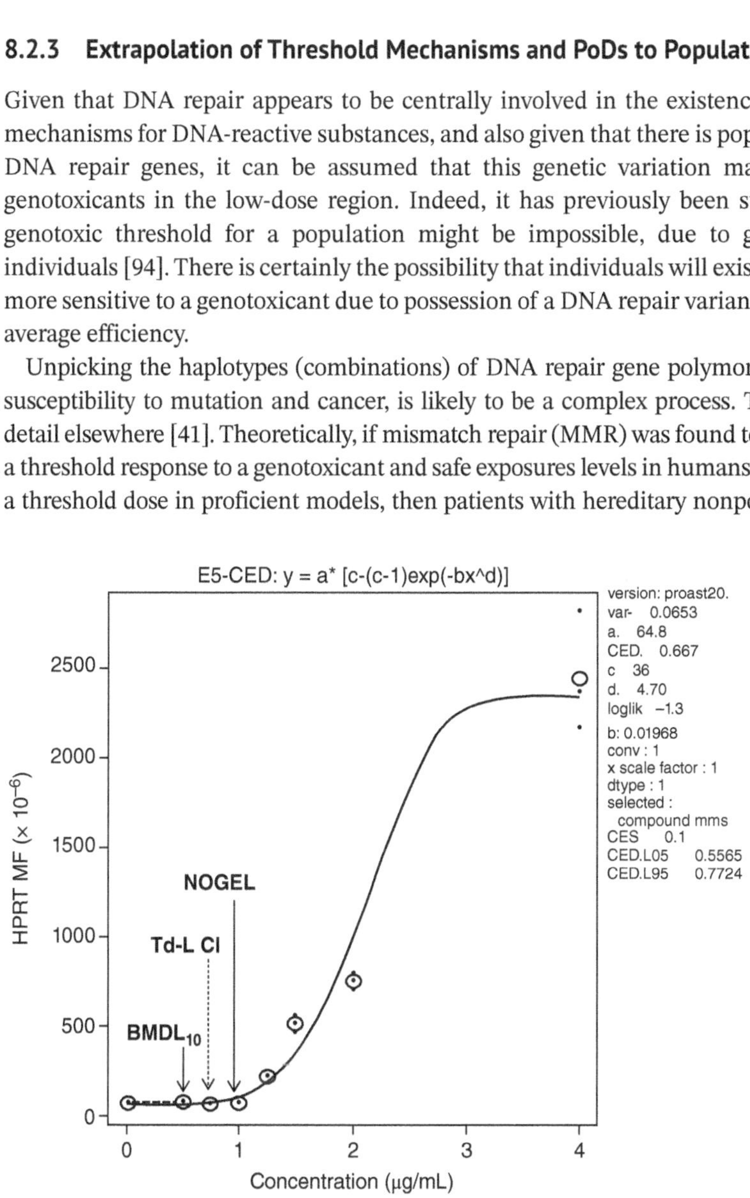

Figure 8.5 Benchmark dose–response modeling results for HPRT gene mutations induced by MMS *in vitro* in AHH-1 cells. Benchmark dose–response modeling. The three parameters NOGEL, Threshold dose lower confidence interval (TD-LCI), and BMDL10 are shown in the graph. CED is the critical effect dose or BMD; CES is the critical effect size, in this case 10% or 0.1; CED-L05 is the lower CI of the BMD or BMDL; CED-L95 is the upper CI of the BMD or the BMDU. For this example, model E5-CED had the highest log-likelihood and provided the best representation of the data [1].

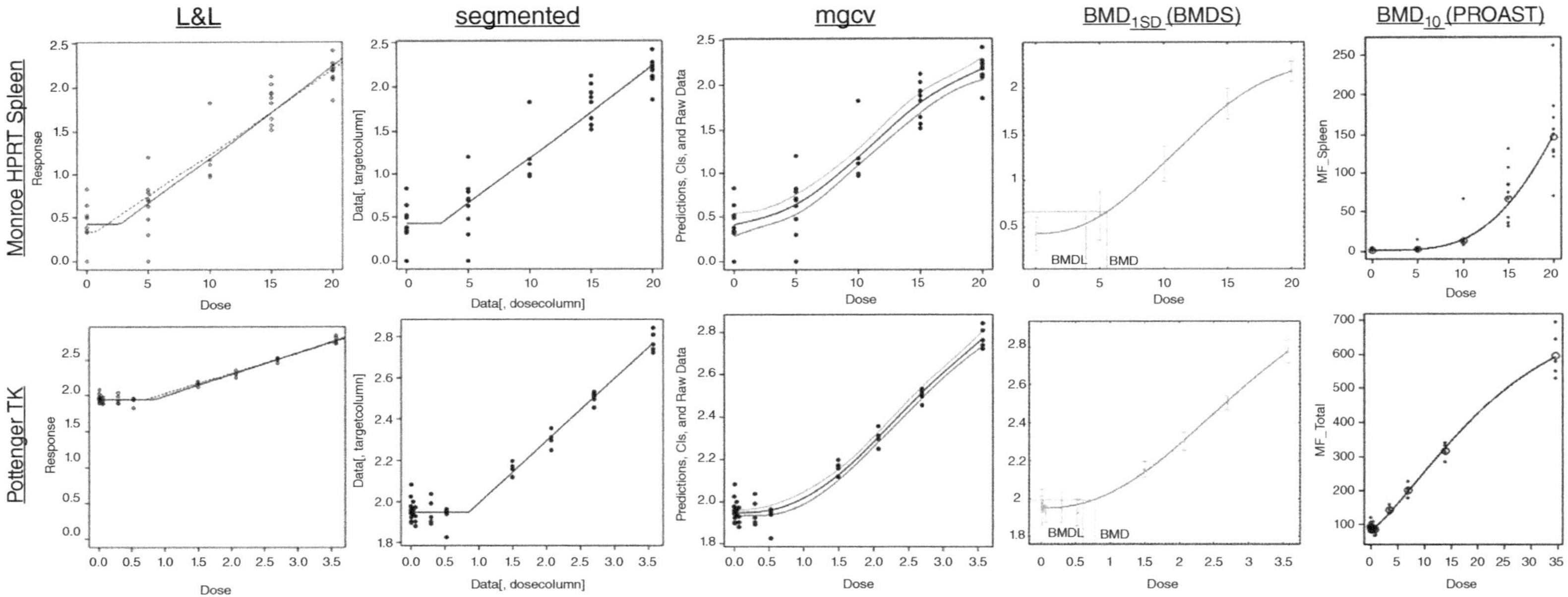

Figure 8.6 Example of breakpoint dose, slope transition dose, and BMD modeling results for two MNU gene mutation data sets. L&L, Lutz, and Lutz hockey stick/bilinear approach for defining a breakpoint dose (BPD); segmented, bilinear model for defining a BPD; mgcv, smoothing regression spline for defining a slope transition dose (STD) using "drsmooth" in R version 3.0.2; BMD_{1SD} (BMDS), BMD 1 standard deviation using BMDS; BMD_{10} (PROAST), BMD 10 using PROAST [2].

(HNPCC; a deficiency in MMR) would be more susceptible to mutation and perhaps cancer, while the general population would be tolerant of exposure. In short, the threshold dose calculated in this example would not necessarily apply for these HNPCC individuals. Of course, the fact that MMR (and indeed other repair processes) can unwittingly facilitate chromosome damage induction complicates this view as efficient MMR can drive chromosome damage [57] and may represent a double-edged sword in mutagenesis. Moreover, due to the overlapping specificity of the different DNA repair processes, other functional counterpart pathways may well compensate the reduced efficiency of one pathway. Therefore, failure of more than one pathway may be necessary to radically alter any population level threshold dose. Repair haplotypes may better define susceptible individuals, allowing a better understanding of risk assessment in terms of exposure to genotoxic carcinogens. Furthermore, while it is accepted that MGMT and BER are centrally involved in the repair of alkyl DNA damage, NER can also contribute to this repair effort as shown by increased alkylation sensitivity in NER-deficient cells [88]. This is particularly true for larger alkyl groups (ethyl, isopropyl) compared to methyl groups [57]. For more complex adducts, TLS components, such as polymerase kappa, can also contribute to the dose response and can also be considered as a threshold mechanism [95].

Certainly, safety factors used in risk assessment need to take account of genetic variation in DNA repair genes (currently included for interindividual variation in general), particularly when risk assessing a genotoxicant with a possible threshold mechanism. Mechanistic studies to better understand the biological basis are essential, as they highlight the key protective factors (like DNA repair) that contribute to the dose response and PoD. In fact, before suggesting safe exposure limits to known genotoxicants, adequate characterization of any protective mechanisms should be undertaken in order to inform the risk assessments necessary. The identification of these protective processes and the genes involved can then lead to the search for susceptible groups before setting safe exposure levels. This may require confirmatory genotoxicity testing in appropriate model systems (e.g. DNA repair deficient) to examine the likely risks to populations with deficiencies in key DNA repair pathways.

8.3 Conclusions

Genotoxic threshold mechanisms exist and have now been well described for a small number of chemicals. The biological basis involves multiple cellular and extracellular mechanisms (e.g. functional redundancy and DNA repair). Understanding the biological mechanism behind the dose response and PoD is fundamental to accepting that these responses are genuine and also can be useful in highlighting susceptible subpopulations.

However, at this point in time, there are insufficient data to apply this on a wider basis to genotoxic carcinogens in general. Thus, individual investigations are required. It is, however, hoped that by the conduct of further studies, across a range of different chemical classes, that a more thorough understanding of the existence of threshold mechanisms will be gained and that this will ultimately be used to refine current methodologies resulting in an approach that better reflects the actual level of risk.

References

1 Gollapudi, B.B., Johnson, G.E., Hernandez, L.G. et al. (2013). Quantitative approaches for assessing dose-response relationships in genetic toxicology studies. *Environ. Mol. Mutagen. 54*: 8–18.

2 Johnson, G.E., Soeteman-Hernandez, L.G., Gollapudi, B.B. et al. (2014). Derivation of point of departure (PoD) estimates in genetic toxicology studies and their potential applications in risk assessment. *Environ. Mol. Mutagen. 55*: 609–623.

3 MacGregor, J.T., Frötschl, R., White, P.A. et al. (2015). IWGT report on quantitative approaches to genotoxicity risk assessment II. Use of point-of-departure (PoD) metrics in defining acceptable exposure limits and assessing human risk. *Mut. Res. Genet. Toxicol. 783*: 66–78.

4 MacGregor, J.T., Frötschl, R., White, P.A. et al. (2015). IWGT report on quantitative approaches to genotoxicity risk assessment I. Methods and metrics for defining exposure-response relationships and points of departure (PoDs). *Mut. Res. Genet. Toxicol. 783*: 55–65.

5 Bailey, G.S., Reddy, A.P., Pereira, C.B. et al. (2009). Nonlinear cancer response at ultralow dose: a 40800-animal ED(001) tumor and biomarker study. *Chem. Res. Toxicol. 22*: 1264–1276.

6 Kobets, T. and Williams, G.M. (2019). Review of the evidence for thresholds for DNA-Reactive and epigenetic experimental chemical carcinogens. *Chem. Biol. Interact. 301*: 88–111.

7 Henderson, L., Albertini, S., and Aardema, M. (2000). Thresholds in genotoxicity responses. *Mutat. Res. 464*: 123–128.

8 White, P.A. and Johnson, G.E. (2016). Genetic toxicology at the crossroads-from qualitative hazard evaluation to quantitative risk assessment. *Mutagenesis 31*: 233–237.

9 COM. (2000). Thresholds for aneugens: extrapolation of data from somatic cells to germ cells. Statement COM/00/S2. In: Health, D. o. (ed.) *Committee on Mutagenicity of Chemicals in Food, Consumer Products and the Environment.*

10 COM. (2007). Statement: Benzimidazoles: An approach to defining a common aneugenic grouping. COM/07/S3. Committee on Mutagenicity of Chemicals in Food, C. P. a. t. E., Ed.

11 COM. (2018). Statement on the quantitative approaches to the assessment of genotoxicity data. Committee on Mutagenicity of Chemicals in Food, C. P. a. t. E., Ed.

12 ICHM7. (2017). M7 (R1): Assessment and Control of DNA Reactive (Mutagenic) Impurities in Pharmaceuticals to Limit Potential Carcinogenic Risk. Guideline, I. H. T., Ed. ICH, pp. 1–116. http://www.ich.org/fileadmin/Public_Web_Site/ICH_Products/Guidelines/Multidisciplinary/M7/M7_R1_Addendum_Step_4_31Mar2017.pdf.

13 Ames, B.N., Shigenaga, M.K., and Gold, L.S. (1993). DNA lesions, inducible DNA repair, and cell division: three key factors in mutagenesis and carcinogenesis. *Environ. Health Perspect. 101* (Suppl 5): 35–44.

14 Perera, F.P. (1988). The significance of DNA and protein adducts in human biomonitoring studies. *Mutat. Res. 205*: 255–269.

15 Zito, R. (2001). Low doses and thresholds in genotoxicity: from theories to experiments. *J. Exp. Clin. Cancer Res. 20*: 315–325.

16 Swenberg, J.A., Fryar-Tita, E., Jeong, Y.-C. et al. (2008). Biomarkers in toxicology and risk assessment: informing critical dose-response relationships. *Chem. Res. Toxicol. 21*: 253–265.

17 Kraus, A., McKeague, M., Seiwert, N. et al. (2019). Immunological and mass spectrometry-based approaches to determine thresholds of the mutagenic DNA adduct O(6)-methylguanine in vivo. *Arch. Toxicol. 93*: 559–572.

18 Hanawalt, P.C. (1994). Transcription-coupled repair and human disease. *Science 23*: 1957–1958.

19 Elhajouji, A., Lukamowicz, M., Cammerer, Z., and Kirsch-Volders, M. (2011). Potential thresholds for genotoxic effects by micronucleus scoring. *Mutagenesis 26*: 199–204.

20 Wills, J.W., Johnson, G.E., Doak, S.H. et al. (2016). Empirical analysis of BMD metrics in genetic toxicology part I: in vitro analyses to provide robust potency rankings and support MOA determinations. *Mutagenesis 31*: 255–263.

21 Lynch, A., Harvey, J., Aylott, M. et al. (2003). Investigations into the concept of a threshold for topoisomerase inhibitor-induced clastogenicity. *Mutagenesis 18*: 345–353.

22 Dertinger, S.D., Kraynak, A.R., Wheeldon, R.P. et al. (2019). Predictions of genotoxic potential, mode of action, molecular targets, and potency via a tiered multiflow(R) assay data analysis strategy. *Environ. Mol. Mutagen.* https://doi.org/10.1002/em.22274.

23 Doak, S.H., Brüsehafer, K., Dudley, E. et al. (2008). No-observed effect levels are associated with up-regulation of MGMT following MMS exposure. *Mutat. Res. 648*: 9–14.

24 Doak, S.H., Jenkins, G.J., Johnson, G.E. et al. (2007). Mechanistic influences for mutation induction curves after exposure to DNA-reactive carcinogens. *Cancer Res. 67*: 3904–3911.

25 Johnson, G.E., Doak, S.S., and Jenkins, G.J. (2010). Quantitative aspects of DNA reactive genotoxins assessed in vitro. *Mutagenesis 25*: 56.

26 Thomas, A., Doak, S., Jenkins, G. et al. *Can O6-Methylguanine methyltransferase (MGMT) protect DNA from low doses of Methyl Nitrosourea?* P116. Barcelona: European Environmental Mutagen Society.

27 Thomas, A. and Johnson, G. (2015). DNA repair and its influence on points of departure for alkylating agent genotoxicity. In: *Thresholds of Genotoxic Carcinogens: from Mechanism to Regulation* (eds. T. Nohmi and S. Fukushima). Springer.

28 Zair, Z.M., Jenkins, G.J., Doak, S.H. et al. (2011). *N*-methylpurine DNA glycosylase plays a pivotal role in the threshold response of ethyl methanesulfonate-induced chromosome damage. *Toxicol. Sci. 119*: 346–358.

29 Wills, J.W., Long, A.S., Johnson, G.E. et al. (2016). Empirical analysis of BMD metrics in genetic toxicology part II: in vivo potency comparisons to promote reductions in the use of experimental animals for genetic toxicity assessment. *Mutagenesis 31*: 265–275.

30 Gocke, E., Ballantyne, M., Whitwell, J., and Muller, L. (2009). MNT and MutaMouse studies to define the in vivo dose response relations of the genotoxicity of EMS and ENU. *Toxicol. Lett. 190*: 286–297.

31 Gocke, E. and Muller, L. (2009). in vivo studies in the mouse to define a threshold for the genotoxicity of EMS and ENU. *Mutat. Res. 678*: 101–107.

32 Gocke, E. and Wall, M. (2009). in vivo genotoxicity of EMS: statistical assessment of the dose response curves. *Toxicol. Lett. 190*: 298–302.

33 Müller, L., Gocke, E., Lavé, T., and Pfister, T. (2009). Ethyl methanesulfonate toxicity in Viracept--a comprehensive human risk assessment based on threshold data for genotoxicity. *Toxicol. Lett. 190*: 317–329.

34 Bercu, J.P., Galloway, S.M., Parris, P. et al. (2018). Potential impurities in drug substances: compound-specific toxicology limits for 20 synthetic reagents and by-products, and a class-specific toxicology limit for alkyl bromides. *Regul. Toxicol. Pharmacol. 94*: 172–182.

35 Abdel-Tawab, M., Gröner, R., Kopp, T., Meins, J., and Wübert, J. (2018). ZL findet NDMA in Tabletten. www.pharmazeutische-zeitung.de (accessed 25 July 2018).

36 Peto, R., Gray, R., Brantom, P., and Grasso, P. (1991). Dose and time relationships for tumor induction in the liver and esophagus of 4080 inbred rats by chronic ingestion of *N*-nitrosodiethylamine or *N*-nitrosodimethylamine. *Cancer Res. 51*: 6452–6469.

37 Akagi, J., Toyoda, T., Cho, Y.M. et al. (2015). Validation study of the combined repeated-dose toxicity and genotoxicity assay using gpt delta rats. *Cancer Sci. 106*: 529–541.

38 Jiao, J., Douglas, G.R., Gingerich, J.D., and Soper, L.M. (1997). Analysis of tissue-specific lacZ mutations induced by *N*-nitrosodibenzylamine in transgenic mice. *Carcinogenesis 18*: 2239–2245.

39 Yamazaki, H., Inui, Y., Yun, C.H. et al. (1992). Cytochrome P450 2E1 and 2A6 enzymes as major catalysts for metabolic activation of *N*-nitrosodialkylamines and tobacco-related nitrosamines in human liver microsomes. *Carcinogenesis 13*: 1789–1794.

40 Beranek, D.T. (1990). Distribution of methyl and ethyl adducts following alkylation with monofunctional alkylating agents. *Mutat. Res. 231*: 11–30.

41 Jenkins, G.J., Doak, S.H., Johnson, G.E. et al. (2005). Do dose response thresholds exist for genotoxic alkylating agents? *Mutagenesis 20*: 389–398.

42 Quiros, S., Roos, W.P., and Kaina, B. (2010). Processing of O^6-methylguanine into DNA double-strand breaks requires two rounds of replication whereas apoptosis is also induced in subsequent cell cycles. *Cell Cycle 9*: 168–178.

43 Ensminger, M., Iloff, L., Ebel, C. et al. (2014). DNA breaks and chromosomal aberrations arise when replication meets base excision repair. *J. Cell Biol. 206*: 29–43.

44 Thomas, A.D. (2020). Biological basis for threshold responses to methylating agents. *Chem. Res. Toxicol.* 33 (9): 2219–2224.

45 Arimoto-Kobayashi, S., Kaji, K., Sweetman, G.M., and Hayatsu, H. (1997). Mutation and formation of methyl- and hydroxylguanine adducts in DNA caused by *N*-nitrosodimethylamine and *N*-nitrosodiethylamine with UVA irradiation. *Carcinogenesis 18*: 2429–2433.

46 Green, C.L., Loechler, E.L., Fowler, K.W., and Essigmann, J.M. (1984). Construction and characterization of extrachromosomal probes for mutagenesis by carcinogens: site-specific incorporation of O^6-methylguanine into viral and plasmid genomes. *Proc. Natl. Acad. Sci. 81*: 13–17.

47 Loechler, E.L., Green, C.L., and Essigmann, J.M. (1984). in vivo mutagenesis by O^6-methylguanine built into a unique site in a viral genome. *Proc. Natl. Acad. Sci. U. S. A. 81*: 6271–6275.

48 Warren, J.J., Forsberg, L.J., and Beese, L.S. (2006). The structural basis for the mutagenicity of O^6-methyl-guanine lesions. *Proc. Natl. Acad. Sci. 103*: 19701–19706.

49 Parsa, I., Marsh, W.H., and Sutton, A.L. (1980). An in vitro model of pancreas carcinogenesis: two-stage MNU--TPA effect. *Am. J. Pathol. 98*: 649–662.

50 Zarbl, H., Sukumar, S., Arthur, A.V. et al. (1985). Direct mutagenesis of Ha-ras-1 oncogenes by *N*-nitroso-*N*-methylurea during initiation of mammary carcinogenesis in rats. *Nature 315*: 382–385.

51 Pazos, P., Lanari, C., Meiss, R. et al. (1991). Mammary carcinogenesis induced by *N*-methyl-*N*-nitrosourea (MNU) and medroxyprogesterone acetate (MPA) inBALB/c mice. *Breast Cancer Res. Treat. 20*: 133–138.

52 Prior, I.A., Lewis, P.D., and Mattos, C. (2012). A comprehensive survey of RAS mutations in cancer. *Cancer Res. 72*: 2457–2467.

53 Westcott, P.M.K., Halliwill, K.D., To, M.D. et al. (2015). The mutational landscapes of genetic and chemical models of Kras-driven lung cancer. *Nature 517*: 489–492.

54 Pegg, A.E., Dolan, M.E., Scicchitano, D., and Morimoto, K. (1985). Studies of the repair of O^6-alkylguanine and O^4-alkylthymine in DNA by alkyltransferases from mammalian cells and bacteria. *Environ. Health Perspect. 62*: 109–114.

55 Pegg, A.E., Scicchitano, D., and Dolan, M.E. (1984). Comparison of the rates of repair of O^6-alkylguanines in DNA by rat liver and bacterial O^6-alkylguanine-DNA alkyltransferase. *Cancer Res. 44*: 3806–3811.

56 Pegg, A.E. and Byers, T.L. (1992). Repair of DNA containing O^6-alkylguanine. *FASEB J. 6*: 2302–2310.

57 Kaina, B., Christmann, M., Naumann, S., and Roos, W.P. (2007). MGMT: key node in the battle against genotoxicity, carcinogenicity and apoptosis induced by alkylating agents. *DNA Repair 6*: 1079–1099.

58 Xu-Welliver, M. and Pegg, A.E. (2002). Degradation of the alkylated form of the DNA repair protein, O^6-alkylguanine-DNA alkyltransferase. *Carcinogenesis 23*: 823–830.

59 Shen, L., Kondo, Y., Rosner, G.L. et al. (2005). MGMT promoter methylation and field defect in sporadic colorectal cancer. *J. Natl. Cancer Inst. 97*: 1330–1338.

60 Svrcek, M., Buhard, O., Colas, C. et al. (2010). Methylation tolerance due to an O^6-methylguanine DNA methyltransferase (MGMT) field defect in the colonic mucosa: an initiating step in the development of mismatch repair-deficient colorectal cancers. *Gut 59*: 1516–1526.

61 Becker, K., Dosch, J., Gregel, C.M. et al. (1996). Targeted expression of human O(6)-methylguanine-DNA methyltransferase (MGMT) in transgenic mice protects against tumor initiation in two-stage skin carcinogenesis. *Cancer Res. 56*: 3244–3249.

62 Becker, K., Gregel, C., Fricke, C. et al. (2003). DNA repair protein MGMT protects against *N*-methyl-*N*-nitrosourea-induced conversion of benign into malignant tumors. *Carcinogenesis 24*: 541–546.

63 Becker, K., Gregel, C.M., and Kaina, B. (1997). The DNA repair protein O^6-methylguanine-DNA methyltransferase protects against skin tumor formation induced by antineoplastic chloroethylnitrosourea. *Cancer Res. 57*: 3335–3338.

64 Becker, K., Thomas, A.D., and Kaina, B. (2013). Does increase in DNA repair allow "Tolerance-to-Insult" in chemical carcinogenesis? Skin tumor experiments with MGMT-overexpressing mice. *Environ. Mol. Mutagen.* n/a-n/a.

65 Kaina, B., Fritz, G., Ochs, K. et al. (1998). Transgenic systems in studies on genotoxicity of alkylating agents: critical lesions, thresholds and defense mechanisms. *Mutat. Res. 405*: 179–191.

66 Thomas, A.D., Jenkins, G.J., Kaina, B. et al. (2013). Influence of DNA repair on nonlinear dose-responses for mutation. *Toxicol. Sci. 132*: 87–95.

67 Fahrer, J., Frisch, J., Nagel, G. et al. (2015). DNA repair by MGMT, but not AAG, causes a threshold in alkylation-induced colorectal carcinogenesis. *Carcinogenesis 36*: 1235–1244.

68 Arimoto-Kobayashi, S., Ohta, K., Yuhara, Y. et al. (2015). Mutagenicity and clastogenicity of extracts of Helicobacter pylori detected by the Ames test and in the micronucleus test using human lymphoblastoid cells. *Mutagenesis 30*: 537–544.

69 EMEA. (2018). CHMP overall rapporteur's preliminary assessment report: Angiotensin-II-receptor antagonists (sartans) containing a tetrazole group medicinal products. EMEA/H/A-31/1471 Agency, E. M., Ed.

70 Gollapudi, B.B., Jackson, K.M., and Stott, W.T. (1998). Hepatic lacI and cII mutation in transgenic (lambdaLIZ) rats treated with dimethylnitrosamine. *Mutat. Res. 419*: 131–135.

71 Peto, R., Gray, R., Brantom, P., and Grasso, P. (1991). Effects on 4080 rats of chronic ingestion of *N*-nitrosodiethylamine or *N*-nitrosodimethylamine: a detailed dose-response study. *Cancer Res. 51*: 6415–6451.

72 Johnson, G.E., Dobo, K., Gollapudi, B. et al. (2021). Permitted daily exposure limits for noteworthy N-nitrosamines. *Environ. Mol. Mutagen.* https://doi.org/10.1002/em.22446.

73 Guerard, M., Johnson, G., Dertinger, S. et al. (2017). Dose-response relationship of temozolomide, determined by the Pig-a, comet, and micronucleus assay. *Arch. Toxicol. 91*: 2443–2453.

74 Zeller, A., Tang, L., Dertinger, S.D. et al. (2016). A proposal for a novel rationale for critical effect size in dose–response analysis based on a multi-endpoint in vivo study with methyl methanesulfonate. *Mutagenesis* 31: 39–253.

75 Asano, N., Torous, D.K., Tometsko, C.R. et al. (2006). Practical threshold for micronucleated reticulocyte induction observed for low doses of mitomycin C, Ara-C and colchicine. *Mutagenesis 21*: 15–20.

76 Bemis, J.C., Labash, C., Avlasevich, S.L. et al. (2015). Rat Pig-a mutation assay responds to the genotoxic carcinogen ethyl carbamate but not the non-genotoxic carcinogen methyl carbamate. *Mutagenesis 30*: 343–347.

77 Bemis, J.C., Wills, J.W., Bryce, S.M. et al. (2016). Comparison of in vitro and in vivo clastogenic potency based on benchmark dose analysis of flow cytometric micronucleus data. *Mutagenesis 31*: 277–285.

78 Crebelli, R. (2000). Threshold-mediated mechanisms in mutagenesis: implications in the classification and regulation of chemical mutagens. *Mutat. Res. 464*: 129–135.

79 Kirsch-Volders, M., Vanhauwaert, A., Eichenlaub-Ritter, U., and Decordier, I. (2003). Indirect mechanisms of genotoxicity. *Toxicol. Lett. 140–141*: 63–74.

80 Elhajouji, A., Van Hummelen, P., and Kirsch-Volders, M. (1995). Indications for a threshold of chemically-induced aneuploidy in vitro in human lymphocytes. *Environ. Mol. Mutagen. 26*: 292–304.

81 Elhajouji, A., Tibaldi, F., and Kirsch-Volders, M. (1997). Indication for thresholds of chromosome non-disjunction versus chromosome lagging induced by spindle inhibitors in vitro in human lymphocytes. *Mutagenesis 12*: 133–140.

82 Guttenplan, J.B. (1990). Mutagenesis by *N*-nitroso compounds: relationships to DNA adducts, DNA repair, and mutational efficiencies. *Mutat. Res. 233*: 177–187.

83 Pottenger, L.H., Boysen, G., Brown, K. et al. (2018). Understanding the importance of low-molecularweight (ethylene oxide- and propylene oxide-induced) DNA adducts and mutations in risk assessment: insights from 15 years of research and collaborative discussions. *Environ. Mol. Mutagen.* https://doi.org/10.1002/em.22248.

84 Bronstein, S.M., Cochrane, J.E., Craft, T.R. et al. (1991). Toxicity, mutagenicity and mutational spectra of *N*-ethyl-*N*-nitrosourea in human cell lines with different repair phenotypes. *Cancer Res. 51*: 5188–5197.

85 Sofuni, T., Hayashi, M., Nohmi, T. et al. (2000). Semi-quantitative evaluation of genotoxic activity of chemical substances and evidence for a biological threshold of genotoxic activity. *Mut. Res. Gen. Toxicol. Environ. Mutagen. 464*: 97–104.

86 Bobola, M.S., Berger, M.S., and Silber, J.R. (1995). Contribution of O^6-methylguanine-DNA methyltransferase to resistance to 1,3-(2-chloroethyl)-1-nitrosourea in human brain tumor-derived cell lines. *Mol. Carcinog. 13*: 81–88.

87 Glaab, W.E., Tindall, K.R., and Skopek, T.R. (1999). Specificity of mutations induced by methyl methanesulfonate in mismatch repair-deficient human cancer cell lines. *Mutat. Res. 427*: 67–78.

88 Ophet Vel, C.W., van Hees Stuivenberg, S., Van Zeeland, A.A., and Jansen, J.G. (1997). Effect of nucleotide excision repair on HPRT mutations in rodent cells exposed to DNA ethylating agents. *Mutagenesis 12*: 417–424.

89 Plosky, B., Samson, L., Engelward, B.P. et al. (2002). Base excision repair and nucleotide excision repair contribute to the removal of N-methylpurines from active genes. *DNA Repair 1*: 683–696.

90 Drablos, F., Feyzi, E., and Aas, P.A. (2004). Alkylation damage in DNA and RNA – repair mechanisms and medical significance. *DNA Repair 3*: 1389–1407.

91 Wyatt, M.D. and Pittman, D.L. (2006). Methylating agents and DNA repair responses: methylated bases and sources of strand breaks. *Chem. Res. Toxicol. 19*: 1580–1594.

92 EPA (2020). Benchmark Dose Software (BMDS) version 3.2 - download. https://www.epa.gov/bmds/benchmark-dose-software-bmds-version-32-download (accessed 23rd Aug 2021).

93 Gocke, E., Muller, L., and Pfister, T. (2009). EMS in Viracept--initial ('traditional') assessment of risk to patients based on linear dose response relations. *Toxicol. Lett. 190*: 266–270.

94 Lutz, W.K. (2000). A true threshold dose in chemical carcinogenesis cannot be defined for a population, irrespective of the mode of action. *Hum. Exp. Toxicol. 19*: 566–568. discussion 571-2.

95 Nohmi, T. and Fukushima, S. (2016). *Thresholds of Genotoxic Carcinogens, 1*e. *From Mechanisms to Regulation*. Elsevier.

Section 4

Quality Perspective on Genotoxic Impurities

9

Mutagenic Impurities – Assessment of Fate and Control Options

Michael W. Urquhart[1], *Andrew Teasdale*[2], *and Michael Burns*[3]

[1] *GlaxoSmithKline R&D, Hertfordshire, UK*
[2] *Chemical Development, Pharmaceutical Technology and Development, Operations, AstraZeneca, Macclesfield, UK*
[3] *Lhasa Limited, Leeds, UK*

9.1 Introduction/Background

Since the publication of the EMEA position paper in 2002 [1] on limits for genotoxic impurities, it has been necessary for pharmaceutical companies to consider the potential risk posed by genotoxic (now more correctly referred to as mutagenic) impurities within their products. This has, therefore, driven the need to develop an effective strategy that both identifies and assesses the risk posed by any mutagenic impurity (MI) within the process used to manufacture the API, and any subsequent mutagenic degradants within the final formulated product. ICH M7 [2] outlines the anticipated steps involved in such a risk assessment, including the implementation of an effective control strategy.

In order to synthesize APIs efficiently, it is necessary to build up the molecular structure through the combination of simple structural motifs. This typically involves the formation of carbon–carbon, carbon–nitrogen, and carbon–oxygen bonds. The current status of synthetic methodology is such that this is impractical to achieve without the use of electrophilic species that fall into the broad class of alkylating agents and are hence potentially mutagenic.

Thus, many intrinsically reactive starting materials, intermediates, and reagents used in the synthesis of APIs are potentially mutagenic, and furthermore may be present as residual impurities within the API. Although avoidance is generally considered to be the preferable option from a regulator's perspective, there is tacit acceptance of the fact that this is generally impractical, and hence rather than avoidance, the issue becomes one of control [3].

Given that most syntheses will necessarily involve the use of materials that are mutagenic, then it is critical to demonstrate effective control to appropriate levels. In the following chapter, we specifically examine the control options and their alignment to ICH M7, focusing in depth on the development and use of purge factor calculations [4, 5].

Once impurities with a potential mutagenic safety concern have been identified by the SAR evaluation process (Chapter 4), the next step is to consider the likelihood of the impurity being present in the isolated API, often referred to as impurity fate mapping.

The impurities under consideration are by nature often highly reactive, hence their removal during downstream processing is typically facilitated by this intrinsic activity. Acidic and/or basic work-up conditions frequently encountered in manufacturing processes may lead to decomposition

Mutagenic Impurities: Strategies for Identification and Control, First Edition. Edited by Andrew Teasdale.
© 2022 John Wiley & Sons, Inc. Published 2022 by John Wiley & Sons, Inc.

and/or removal of the material of concern. Similarly, other reagents used in downstream processing may react with the material converting it to non-mutagenic by-products which ultimately need to be considered using ICH Q3A for general impurities. Another factor is solubility; many of the mutagenic impurities are present within a synthetic scheme by design. In consciously selecting such a reagent, it is done on the basis that it provides an efficient and effective means by which the molecular scaffold of the molecule can be established. As part of this route design, the associated process is further developed to maximize factors such as reactivity, yield, etc. Solubility of the reagents is thus an important factor and hence many of the selected mutagenic reagents will be, by necessity, soluble within the solvents selected for the stages within the manufacturing process. As a result, the removal of a MI can often be achieved as a result of its intrinsic solubility within a manufacturing stage(s), through retention within the mother liquors when the desired intermediate(s), and ultimately API, are isolated in a solid form by crystallization.

This of course is a concept well understood by chemists and analysts. In the case of MIs, and indeed non-mutagenic impurities, chemists look at the impurity's structure, consider its reactivity, and gauge whether it can survive in the downstream process. Unfortunately, however compelling the arguments that are developed, they have often been viewed as nonquantitative from a regulatory perspective. Thus, in many cases, there is a need to provide further analytical data to substantiate the impurity fate assessment. Hence, a quality by testing (QbT) approach is adopted rather than a quality by design (QbD) approach.

The challenge was whether a systematic approach could be developed that would allow risk to be estimated in a quantifiable way which would lead to MI control through process understanding (QbD) rather than confirmation of absence through QbT.

It should be possible to assess fate semiquantitatively based on factors linked to the impurity's physicochemical properties (and taking into account those of the API and intermediates), and the process conditions employed in the route of manufacture to the API. Pierson et al. [6] suggested that an assumption could be made of a 10-fold reduction per synthetic stage. In many cases this would suffice and indeed may even be a cautious estimate of the risk. However, in certain circumstances, for example, an unreactive mutagenic reagent or intermediate used in a "telescoped" process (where there are no isolations between stages), this may be too simplistic and not provide a sufficiently accurate assessment of risk. For this reason, a more quantitative approach based on actual process conditions and the physicochemical properties of the MI in question was sought and is outlined below.

In 2010, the concept of Purge Factor calculations was introduced by Teasdale [4]. This defined a number of contributory factors that should be taken into account for such an assessment, and these are described below.

9.2 Reactivity

As already described, many of the potentially mutagenic impurities (PMIs) or MIs that are likely to be of concern are intrinsically reactive. In extreme cases, they are so reactive that there would be little practical value in monitoring their presence in the outcome of the reaction, for example, acyl halides. Any residual analyte would also be effectively eliminated through procedures such as an aqueous quench or even a simple water wash of the resulting product. As discussed within Chapter 12 (analysis of MIs), the very reactivity of such reagents makes analysis challenging. For example, thionyl chloride, which is explosively reactive with water, would rapidly hydrolyze during any aqueous-based analytical quantification ruling out analysis using aqueous-based High Performance Liquid Chromatography (HPLC). The absence of any carbon atoms effectively rules out gas chromatography (GC), certainly with flame ionization as the detection technique.

Even in processes where there is a likelihood that some residue of the PMI may remain in an intermediate, for many such compounds there is a high probability of purging in a subsequent process stage due to anticipated reactivity or solubility. Consider, for example, an alkyl halide used early in a synthesis; should a further alkylating agent be used downstream in the process, any residual quantities of the initial alkyl halide are highly likely to be further consumed in the latter nucleophilic stage (see Figure 9.1). In the example case presented here, allyl bromide is intrinsically more reactive than the second alkylating agent chloropropylamine. In this process, any residual allyl bromide remaining after stage 1 may carry through into stage 3 where, if still present, would be expected to react to produce an allylated analog of the desired product.

Therefore, whenever assessing the risk posed by a PMI/MI, it may be inappropriate to simply group together similar compounds across a process and treat as an equivalent risk for no other reason than their functional group, particularly where differences in their reactivities would be expected. Factors such as the reactivity of the PMI/MI and process conditions need to be taken into consideration. Purge calculations clearly should demonstrate this.

On the basis of chemical reactivity, the original paper by Teasdale proposed that reactivity could be placed into one of three categories (see Table 9.1).

Figure 9.1 Example reaction scheme highlighting the fate of two mutagenic impurities and their differing purge capacities.

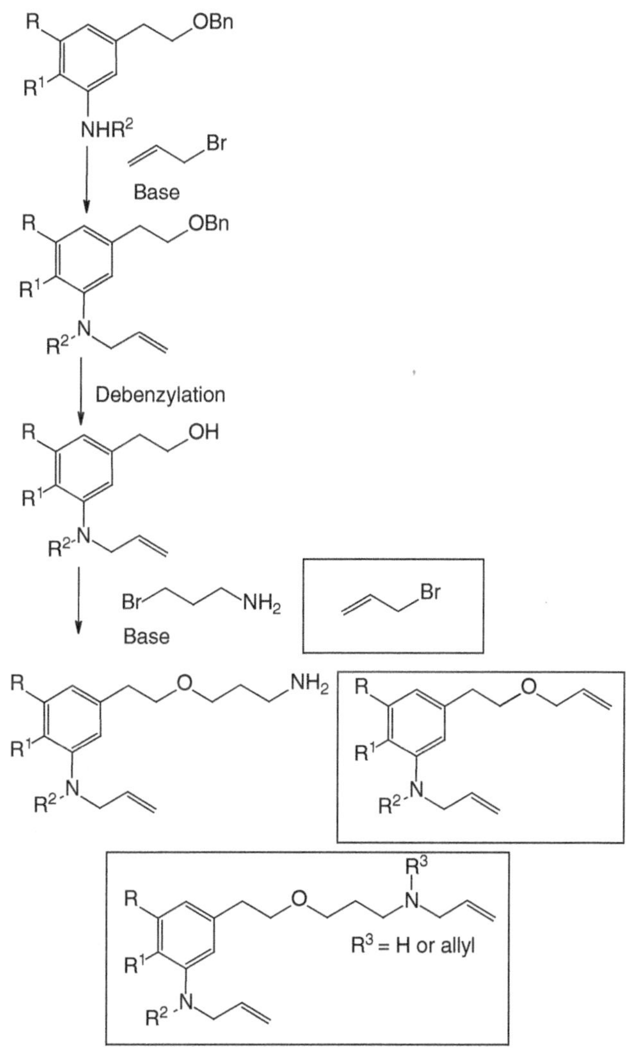

Table 9.1 Potential general reactivity classes.

Reactivity class	Mutagenic groups
Highly reactive[a]	Epoxides/Aldehydes/Sulfonate esters/Acyl halides/Hydrazines
Moderate reactivity	N or S mustards/Michael Reactive Acceptors/allylic halides, Primary alkyl halides/Aziridines
Low reactivity	Amino aryls, Nitro compounds, Purines or Pyrimidines, Carbamates, hindered alkyl halides (secondary/tertiary)

[a] Susceptible to attack by a wide range of potential nucleophiles.

9.2.1 Reactivity Classification

Table 9.1 was intended to be used simply as a guideline; however, as will be discussed below, the knowledge of the reactivity of the classes defined has increased exponentially and in the case of the *in silico* system Mirabilis®, this is reflected in a knowledge base that underpins the system and is used to make predictions.

9.3 Solubility – Isolated Stages

Many of the reagents/intermediates that are highly electrophilic (and hence often mutagenic) are introduced into the synthetic process at those stages specifically designed to optimize the yield and product quality. A critical factor in most processes is that reactants are physically able to react with one another; in practical terms, this is best achieved by the reactants being in solution. By inference, this means that the mutagenic reactant in question is likely to be highly soluble in the reaction solution selected for the process. Thus, should the product from the process be isolated as a solid, then the mutagenic reactant should remain in the reaction mother liquors and thus be removed during the filtration. Its removal would be maximized through having designed an efficient de-liquoring process and augmented by washing the cake with a solvent in which the reagent is freely soluble and the product is not. The de-liquoring process is simply to remove any impurity trapped as a result of inefficient removal of solvent. A wash cannot disrupt the solid form and therefore any impurity trapped within the product's crystal lattice would not be purged. Often, isolation may involve some form of solvent replacement; where this is the case, the solubility of the mutagenic reagent should be considered in the replacement solvent in order to define the appropriate purge factor. In such scenarios, it is possible to transpose solubility data in one solvent and use this to predict solubility in another solvent. Solubility can also be readily generated within a lab environment. The solubility for common materials, like the MI 2-chloropropane, are also readily captured within reagent encyclopedias like (i) the Merck Index, (ii) Scifinder, and (iii) the CRC Handbook of Chemistry and Physics. For example, Scifinder and Reaxys capture solubility, with associated supporting references, within the physical properties section. There is also considerable data that provide at least an indication in terms of implicit solubility of common reagents, i.e. a reagent is seen in literature to have been reacted at X molar concentration in solvent Y. This is an inferred minimum solubility, which can be confirmed if evaluation of available experimental data confirms solution phase chemistry, which can subsequently be used in an evaluation, particularly when at impurity levels.

Where no data exist, then of course solubility may be measured; however, where the MI in question is formed *in situ* and no isolated material is available, one further option is to use surrogate data. An example of this is presented in Figure 9.2. In this example, a predicted MI, phenoxazine **1**,

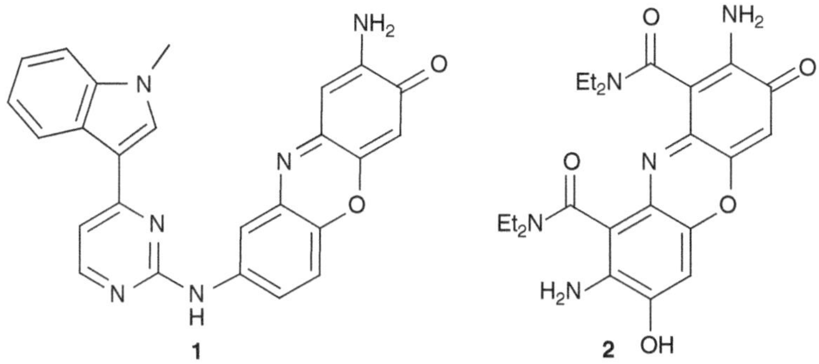

Figure 9.2 Comparative structures of phenoxazines.

was generated *in situ*, and was sufficiently structurally different to the parent compound such that it was not deemed possible to base a prediction of solubility on the parent. In this instance, the approach taken was to conduct a literature search focused on the fused ring; this yielded the phenoxazine **2** for which there was quite extensive solubility data. While it is acknowledged that phenoxazine **1** may not have identical solubility to that of phenoxazine **2**, this approach will give an indication of the type of solubility profile one might expect for phenoxazine 1 where no other information is available.

9.4 Recrystallization

Perhaps one of the most effective ways in which to remove impurities from API or intermediates, including those that are potentially mutagenic, is recrystallization. This involves selecting a solvent, or solvent system, in which the API or intermediate is highly soluble when hot and virtually insoluble when cold. The impure API or intermediate is dissolved in the smallest practical volume of the solvent at an elevated temperature. The hot solution is then typically filtered, to remove any impurities insoluble in the hot solvent, particularly for clinically destined API, as a GMP against extraneous solid contamination. The filtered solution is then allowed to cool under carefully controlled conditions until the product crystallizes out of the cooling solvent. Impurities that are more soluble in the cold solvent remain in solution. The product is then isolated by filtration, washed, and dried, leaving impurities in the filtrate (mother liquors). Other Solid–Liquid separation processes such as precipitation/slurry/re-slurry/filtration and wash sequences should also be taken into consideration and their effect in terms of reduction of remaining levels of the impurity of concern.

The process can be further refined through the introduction of seed crystals (previously isolated product material); although typically used to control or modify a specific property, for example, morphic form, it can help to improve the selectivity of the recrystallization.

9.4.1 Solubility – Liquid/Liquid Partitioning

Another means of purification that is related to solubilities is partitioning. Aromatic amines are perhaps the most obvious example of potentially mutagenic materials that contain an ionizable group. The majority of APIs, and some intermediates, will be potentially ionizable. Where this is the case, and where there is a difference in the ionizability of the mutagen of concern with the desired product(s) in which it is potentially present, it should be possible to reduce the level of the former by manipulation of the pH of the aqueous phase and extraction into an organic solvent.

Reduction of a nitro compound to an amine is an example where such a process would be very effective. While both the nitro and aniline possess a common potentially mutagenic metabolic intermediate (precursor to a nitrenium ion), it may be possible to remove one from the other through an appropriate extraction process, thus reducing the mutagenic liability caused by both materials.

Where there is a need to remove excess nitro compound, this may be achieved very effectively through a liquid/liquid extraction, by employing a 2-phase system with an acidic aqueous layer. Any non-ionizable nitro compound would be retained in the organic layer whereas the ionized amine may have sufficient aqueous solubility to reside in the aqueous layer. The organic layer could then be separated and the product amine then isolated through basification and back extraction of the aqueous phase into a new organic solvent.

Ashworth and Meadows [7] examined a Buchwald–Hartwig coupling reaction between an aryl bromide **3** and 4-methyl piperazine **4** to make aniline **5** and the extraction curve for the three materials as pH was varied (Figure 9.3).

All three species have similar pK_as, but it was observed that the desired product was extracted from the organic phase easily, relative to the starting aryl bromide as the pH was lowered. This was surprising given the similarity of their pK_as. The product's extraction curve was also significantly steeper than expected. A consideration of the relevant equilibria, including all three of the product's pK_as, led to the development of a general model for the liquid–liquid extraction behavior of ionizable molecules, which depends upon the fraction of the molecule in the neutral form, f_N, at the extraction pH (Figure 9.4).

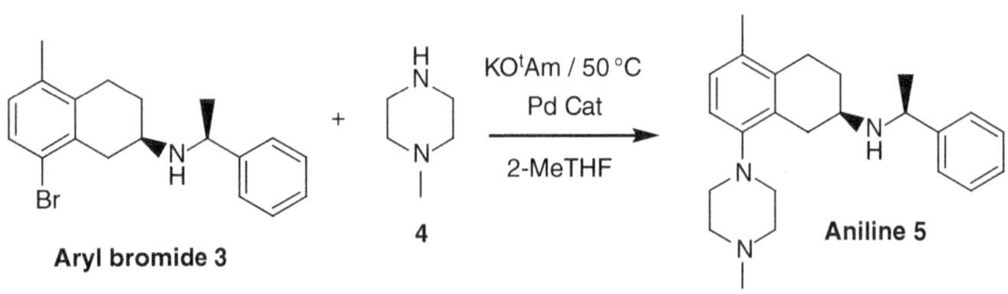

Figure 9.3 Buchwald–Hartwig coupling of an aryl bromide (3) to 4 methyl piperazine (4).

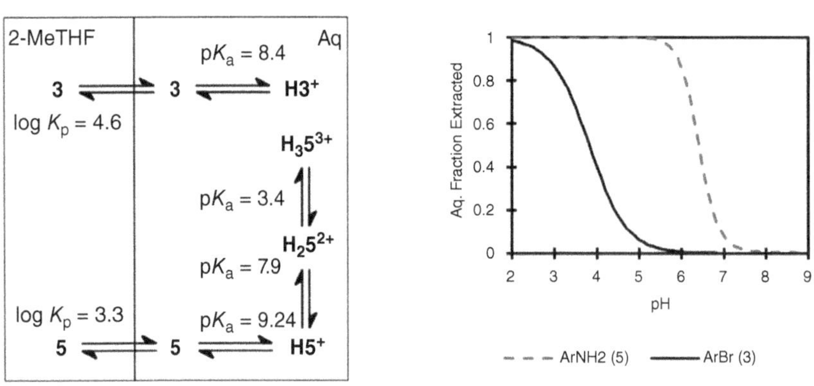

$$\text{Aq. Fraction Extracted} = \frac{1}{1 + K_p V_R f_N} \quad \text{where } K_p = \frac{[X]_{org}}{[X]_{aq}} \text{ and } V_R = \frac{V_{org}}{V_{aq}}$$

Figure 9.4 Extraction curves and equilibria for Aryl Bromide 3 and Aniline 5 and a general model for the liquid–liquid partition of ionizable molecules.

This model can be used to model extraction behavior to determine whether an impurity, particularly a PMI/MI, could be removed through liquid/liquid extraction.

In addition to classical liquid/liquid extractions, solid-phase extraction (SPE) can also be employed. By exploiting both the physical properties (ionizability) and chemical properties (polarity), SPE may confer an advantage over liquid/liquid extraction. The variety of stationary phases available, which can separate analytes according to different chemical properties, is another favorable characteristic of SPE. Most stationary phases are based on silica, and increasingly this has been modified by attachment of a specific functional group. Modifying functional groups include

- hydrocarbon chains of varying length (for reversed phase SPE),
- quaternary ammonium or amino groups (for anion exchange), and
- sulfonic acid or carboxyl groups (for cation exchange).

The main drawback of SPE has been its practicality at a manufacturing scale.

9.5 Volatility

A number of the mutagenic materials likely to be encountered within a typical synthetic process are volatile, including low molecular weight alkyl halides, aldehydes, and nitrogen or sulphur halo-ethyl "mustards." Distillation is frequently used to lower, or completely remove, the volume of reaction solvent present, and this can be effective in reducing, or eliminating, any residual mutagenic material, dependent on the volatility of the genotoxin relative to the boiling point of the solvent. Using this approach, a justification for removal of benzene to <2 ppm in drug substance, as per ICH Q3C expectations, can be put forward entirely based on purge, assuming there was sufficient processing from where benzene was potentially present as an impurity.

9.6 Chromatography

The technique of chromatography offers a range of options to remove or reduce a potential MI from API or an intermediate. Techniques range from simple "filtration" through a silica bed, to preparative liquid chromatography.

Preparative chromatography is typically performed in normal phase mode, that is the use of a polar (usually silica) stationary phase and a nonpolar mobile phase (organic solvent system). The reason for the use of normal phase mode is due to the practical need to isolate the compound in question; normal phase uses volatile solvents that are easily removed.

Preparative HPLC is a now a standard technique within the Pharmaceutical industry for the reduction or removal of impurities, with multi-kilo capability available in the larger companies, as well as a range of contract manufacturers. Improvements in the quality and range of stationary phases, as well as the supporting hardware, have greatly increased the scope of this technique, and there are few separations that cannot be achieved in this fashion.

The removal of a MI can be considered a subset of the standard chromatographic challenge of impurity removal and typically will be approached in the same way. Additional considerations for PMIs may exist, for example, "what is the stability of the PMI?" or "is there any risk of producing additional PMIs?" from the systems being considered. Indeed, with respect to the first consideration, Welch et al. [8] (Merck USA) have published work on the removal of an unwanted oxime via its high reactivity to a resin/packing material which can simply be stirred with the reaction solution, and removed by filtration, or recycled though a preparative, or at line, column.

Most analytical chromatography is now performed in reverse phase mode, which can be used for purification, but the difficulty in removing aqueous-based solvent systems limits the applicability on a manufacturing scale, although it is possible to employ freeze drying as a means of solvent removal. Another potential alternative is super-critical fluid chromatography (SFC); this has the advantage of a readily removable eluent (CO_2).

Bandichhor et al. [9] reported the purification of rizatriptan, a serotonin 5-HT receptor agonist. A mutagenic dimer impurity generated in the synthesis could not be removed to an acceptable limit by conventional processes such as fractional crystallization and recrystallization. A reverse phase method was developed using careful pH and ionic strength modification to increase the selectivity between the rizatriptan and the mutagenic dimer. The mutagenic dimer was strongly retained on the column and the loading optimized to give maximum productivity without any appreciable breakthrough of the mutagenic dimer into the product. Here, the retention of the rizatriptan was kept to a minimum and the mutagenic dimer washed off between injections. The authors reported a decrease in the level of the mutagenic dimer from c. 40 000 ppm to 40–80 ppm (yield >95%). Full details of the method are available in the paper.

9.7 Other Techniques

In addition to the techniques described above, there are a variety of other "niche" techniques that can be applied. Two examples are activated charcoal and resins.

9.7.1 Activated Charcoal

Activated charcoal is used in a variety of industries including the water industry and alcoholic beverage (e.g. Vodka, Rum) industry to remove a range of impurities. Activated charcoal is, however, a complex material in terms of its physicochemical properties and the effect of charcoal is very difficult to predict. In some circumstances, and often in combination with a recrystallization process, it can prove to be very effective in removing certain species, particularly colored impurities. While historically, purification using charcoal has required addition of the solid matrix to the recrystallization medium, it is noteworthy that when activated carbon treatments are used at larger scale, considerable care is needed because its use in bulk increases the risks of dust exposure and dust explosions. Vessel abrasion can also become an issue leading to reactor staining or contamination requiring extensive cleaning post the purification operation. For this reason, filtrations through activated charcoal filter discs or cartridges are now considered safer practice. Examples of the available technologies are 3M's *ZetaCarbon*®, *Zeta Plus*™ SP, and *CUNO*™ CTG as well as Graver Technologies *E-Pak*® for use with small laboratory chemistry through pilot plant to manufacturing applications [10, 11].

9.7.2 Scavenger Resins

Polymer scavengers are functionalized polymers that are designed to react with and bind reagents and by-products. The concept is analogous to that of other extraction and partitioning techniques, the MI of interest binds to the resin and can thus be removed by filtration as the desired product remains in solution. Such resins have found widespread applicability in the combinatorial chemistry arena [12].

Some of the types of resin available and their potential applicability are described in Table 9.2.

The potential use of such resins for the successful removal of methyl sulfonate esters has been reported; [13] however, related ethyl and isopropyl esters were only partially removed. Nevertheless, the authors concluded that the use of such resins showed some potential and suggested that this could be extended to other classes of genotoxins, e.g. alkyl halides.

Table 9.2 Examples of scavenging resins and their application.

Scavenger (functional group)	Structure	Application
Benzaldehyde	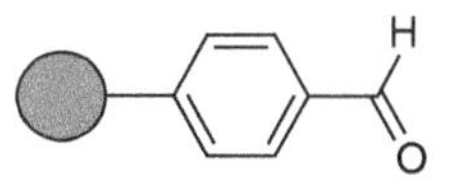	Scavenges nucleophiles including primary amines, hydrazines
Isocyanate	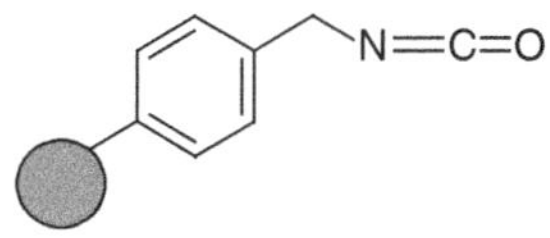	Scavenges nucleophiles including amines
Amine	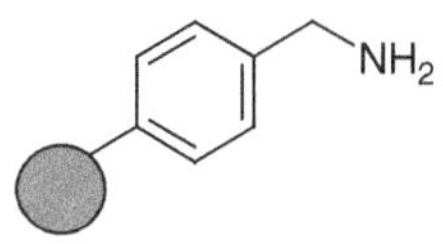	Scavenges acid chlorides, sulfonyl chlorides, and miscellaneous electrophiles
Thiophenol	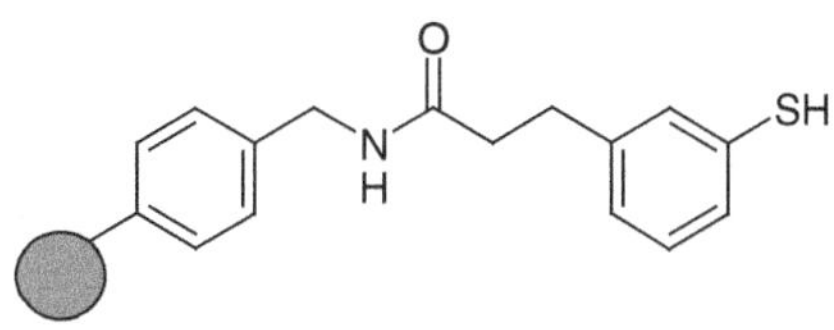	Scavenges alkylating agents, e.g. alkyl halides
Trisamine	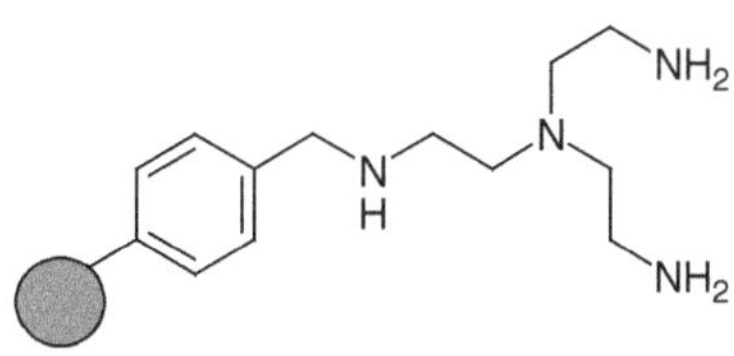	Scavenges acid chlorides, sulfonyl chlorides, and miscellaneous electrophiles
Hydrazide	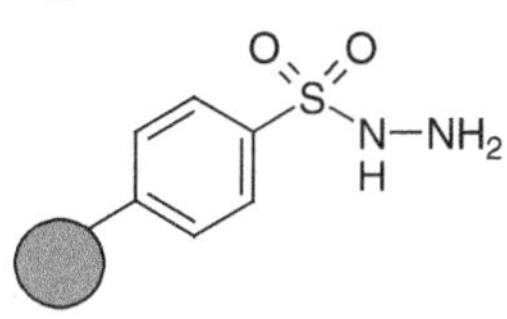	Scavenges aldehydes

A drawback to date of such polymer-based resins has been their stability in aggressive organic solvents such as tetrahydrofuran (THF). Leaching of the monomer has been observed; hence, a procedure to remove PMIs may lead to the potential contamination of the product with another material. Depending on the nature of the monomer in question, this could introduce a bigger problem, and this factor has largely precluded the use of such resins in large-scale synthetic processes. Another issue is the potential environmental impact of using a polymeric scavenger to remove a small mw impurity unless there is potential for recycling of these polymers.

9.8 Overall Quantification of Risk

In order to make a quantitative assessment of the level of carryover of a particular material into an API, or downstream intermediate, a number of mitigating criteria were selected by Teasdale et al. [4] and are defined in Table 9.3. For each mitigating criteria, a purge factor can then be selected according to the characteristics of the material under consideration. The numerical scale has been developed to link individual process steps to the physicochemical properties of the specific impurity in question. Each factor is scored (high–low) in terms of its ability to purge the impurity, thus the higher the score, the greater the likelihood that the impurity would be purged from the process.

Table 9.3 Purge factor scoring system.

Physicochemical parameter	Scale of purge factor
Reactivity	Highly reactive = **100**
	Moderately reactive = **10**
	Low reactivity/unreactive = **1**
Solubility	Freely soluble = **10**
	Moderately soluble = **3**
	Sparingly soluble = **1**
Volatility	Boiling point >20 °C below that of the reaction/ process solvent = **10**
	Boiling point within ±20 °C of that of the reaction/ process solvent = **3**
	Boiling point >20 °C above that of the reaction/ process solvent = **1**
pK_a/pK_b	Ionization potential of GTI significantly different from that desired product – treated as for solubility (**3–10**)
Physical processes: chromatography	Chromatography: **10–100** based on the extent of separation
Physical processes: e.g. other scavenger resins	Evaluated on an individual basis (**3–100**)

The scoring factor defined in the original 2010 paper continues to be the basis of the purge factor concept [4]. The use of the concept has subsequently become widespread and multiple organizations have assessed the utility of the concept to predict the purge factor for PMIs in a given drug substance synthesis and on the basis of such predictions, define an appropriate control strategy. Critical in the use of any such model is the accuracy of predictions in terms of comparison between the predicted purge and the experimentally measured purge. Such comparisons have now been made and published for various classes of PMIs under diverse reaction conditions [5, 15–18], see Table 9.4. These studies have demonstrated a strong correlation between predicted and experimentally measured purge factors, with an overall systematic conservative bias in predicted purge of at least an order of magnitude with respect experimental results. In each case, the authors applied the approach described by Teasdale et al. [4]. to predict the likely purge factors and then compared these predictions to experimentally measured purge factors. Importantly, the scoring system was found to consistently underpredict the actual purge. In some examples illustrated in table 9.4, specifically those involving very high predicted values, the measured purge appears to less than that predicted for the process step. In reality this is a consequence of the limitation of the analytical method used with respect limit of detection for the MI rather than over prediction of the tool. This was designed into the original system by Teasdale to ensure a conservative outcome, one that deliberately underestimates the true purge factor, thus ensuring an inbuilt "safety net."

9.9 Alignment to ICH M7 – Control Options

A critical aspect of ICH M7 is the flexibility to demonstrate control of MIs in a flexible manner utilizing a range of control options presented within the guideline. There are four options, each of which is described below using the specific text within ICH M7.

Table 9.4 Examples of published purge values – predicted compared with measure illustrating the consistent conservatism within the tool [19–22].

API	Impurity	Predicted Purge	Measured Purge
MK-8876 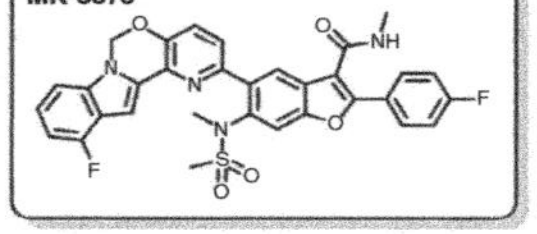	EDCI	1×10^{10}	>50,000
	MeI	1×10^{6}	$>1 \times 10^{5}$
		1×10^{5}	$>2 \times 10^{5}$
	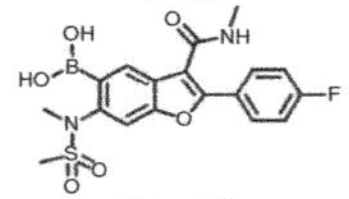	30,000	$>1 \times 10^{6}$
	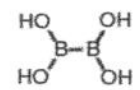	1000	$>2.5 \times 10^{5}$
	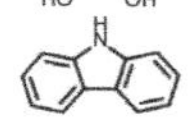	100	>375
Naloxegol oxalate 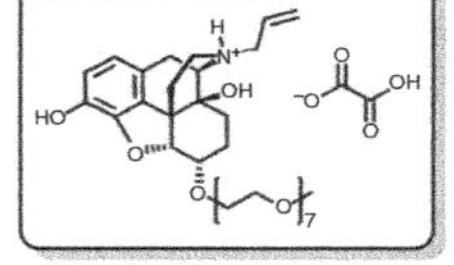	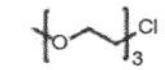	100	>5000
	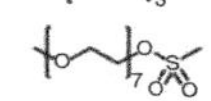	1×10^{5}	$>2 \times 10^{5}$
	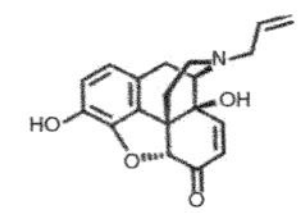	1000	>200
	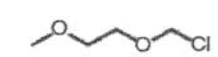	10,000	>5000
Verubecstat 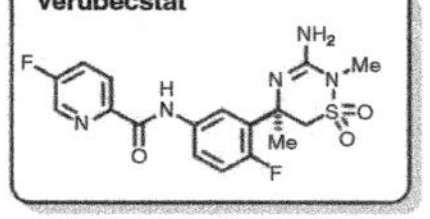	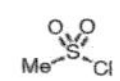	100,000	>60,000
	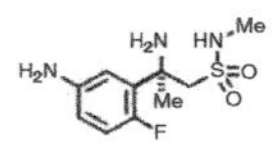	3	16
Camicinal 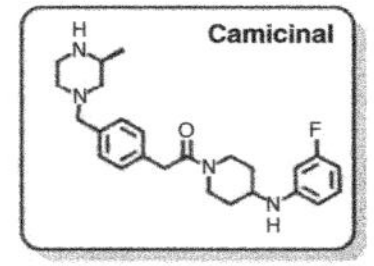	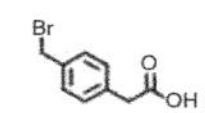	8.1×10^{5}	$>1 \times 10^{5}$
GW641597X 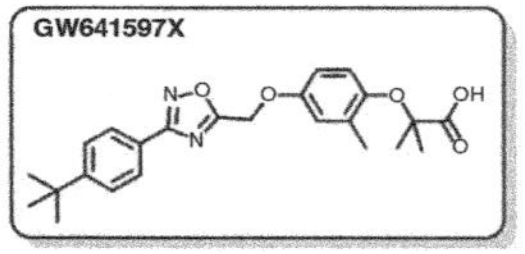	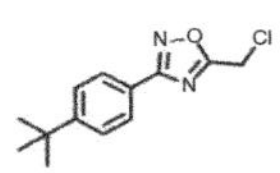	1000	>150
Vortioxetine 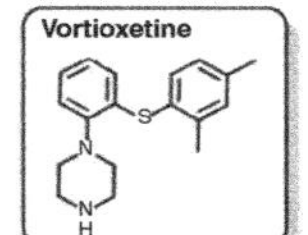	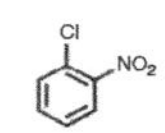	1×10^{9}	8×10^{11}
	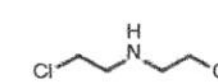	2.7×10^{4}	3×10^{5}

Option 1 – "Include a test for the impurity in the drug substance specification with an acceptance limit." As highlighted, many MIs are highly reactive and this reactivity is in many cases likely to result in the effective purging of an MI within the process downstream of its introduction. Moreover, the reactivity may also result in significant challenges in the analysis of the MI in question. The development and operation of complex analytical methods with commensurate low ppm limits, or even lower limits, within a quality control environment is not necessarily desirable and in many cases is typically only considered where critical to the overall control strategy. As a control option, it is in general rare to utilize Option 1 because mutagenic impurities are not commonly introduced within the final stages of a manufacturing route thus offering opportunities for purge and control through Options 3 or 4. Instances where an Option 1 control may be needed include (i) very late introduction of a PMI with little opportunity to purge or (ii) production of a PMI during the final-stage DS process, possibly through degradation, therefore potentially relying on strict process controls for the control strategy of the critical quality attribute of concern.

Option 2 – "Include a test for the impurity in the specification for a raw material, starting material, or intermediate, or as an in-process control, with an acceptance criterion at or below the acceptable limit."

Again for the reasons described above, in relation to Option 1, this is rarely a required option.

Option 3 – "Include a test for the impurity in the specification for a raw material, starting material, or intermediate, or as an in-process control, with an acceptance criterion above the acceptable limit coupled with demonstrated understanding of fate and purge and associated process controls that assure the level in the drug substance is below the acceptable limit without the need for any additional testing."

The use of an Option 3 approach results in the establishment of acceptance criteria based on higher than acceptable levels at an intermediate stage for PMIs. It is based on an understanding of "fate and purge" and related process controls that ultimately assure the level of the impurity in the API is always below the acceptable limit, without additional downstream testing or an API specification. This approach is particularly useful in establishing criteria for MIs in starting materials, defining an effective standard without having to employ highly sensitive and complex analytical techniques in the analysis of what may often be commodity chemicals.

Option 4 – "Understand process parameters and impact on residual impurity levels (including fate and purge knowledge) with sufficient confidence that the level of the impurity in the drug substance will be below the acceptable limit such that no analytical testing is recommended for this impurity (i.e. the impurity does not need to be listed on any specification). A control strategy that relies on process controls *in lieu* of analytical testing can be appropriate if the process chemistry and process parameters that impact levels of mutagenic impurities are understood and the risk of an impurity residing in the final drug substance above the acceptable limit is determined to be negligible. **In many cases, justification of this control approach based on scientific principles alone is sufficient.**"

An Option 4 approach provides applicants with the potential to establish appropriate chemical purging arguments for a PMI, without the need to provide definitive analytical data or to include a specification either at an intermediate or API stage. The justification of an Option 4 control strategy can thus be based on scientific principles and process knowledge alone.

Of these ICH M7 options, Options 3 and 4 rely upon an intrinsic understanding of the physicochemical properties of the impurity in question and the process conditions in order to assess their likely purge within the process. These options very clearly align to the purge concepts and the language used within ICH M7 correlates with the language used within the initial Teasdale et al. publications. Indeed, ICH M7 takes the unusual step for such guidance by including a direct reference to the Teasdale et al. 2013 paper [5].

The semiquantitative purge factor prediction method developed by Teasdale et al. [4] aligns with ICH M7 Option 4, is in widespread use across the pharmaceutical industry, and has gained regulatory acceptance.

9.10 Control Option Selection

While ICH M7 outlines a series of control options as described, it does not provide guidance in terms of how to systematically decide on which is an appropriate option and in particular what level of supporting information might be required. In order to address this, a cross industry consortium, aligned to the development of Mirabilis™, published a proposed process defining how control option selection could be directly aligned to predicted purge factors, or more specifically to the ratio between predicted purge and required purge [14]. The consortium developed a regulatory decision tree (Figure 9.5) with detailed description of action limits (Table 9.5), depending on the ratio. The decision tree and description of action limits link purge predictions, and their relationship with required purge (i.e. purge ratio), with recommendations for control strategy development. It also defines the level of detail and content in terms of supporting data presentation.

The first stage of the process is to establish the required purge for each PMI in question. This is based on factors such as the dose and duration of treatment which is linked to the permitted daily dose for an MI, and the starting concentration of the MI in the process. Where a mole equivalent is used, a start point of 1 000 000 ppm is assumed.

Once this has been done, the next step is to determine the predicted purge factors for each MI using the defined scoring system and then compare the predicted purges for each MI with the required purge. This determines the ratio between the required purge and that which is predicted.

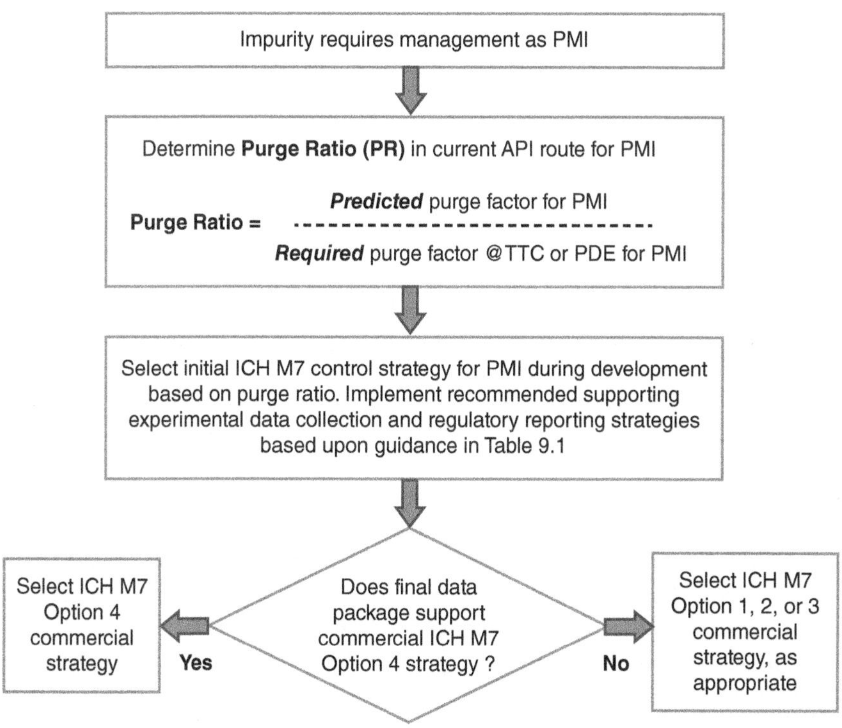

Figure 9.5 Consortium PMI Purge Factor Decision Tree for use under ICH M7.

Table 9.5 Relationship between Purge Factor ratios and Regulatory Reporting Action Limits and potential supplementary reporting requirements.

If PR ⩾ 1000×	If 1000 > PR ⩾ 100×	If PR < 100×
Data collection recommendations		
Collection of additional experimental data not recommended for *noncommercial* or *commercial* API routes to support scientific rationale	Collection of additional non-trace experimental data (solubility, reactivity, and volatility) recommended for both *noncommercial* and *commercial* API routes to support scientific rationale. Collection of additional trace PMI analysis not necessary for *noncommercial* or *commercial* API routes to support scientific rationale	For *noncommercial* API routes, experimentally measure PMI purging, including trace PMI analyses as appropriate, to support scientific rationale. Note: Additional data are expected to support an Option 4 control strategy when PMI Purge Ratio <<100×. For *commercial* API routes, detailed experimental fate and purge studies are expected for all PMI to support a commercial Option 4 control strategy
Regulatory reporting recommendations		
Report "unlikely to persist" or cumulative predicted purge factor and Purge Ratio for *noncommercial* API routes in regulatory submissions. Replace with summary of key elements of predicted purge factor calculations and Purge Ratio for *commercial* API routes in regulatory submissions	Report the cumulative predicted purge factor and Purge Ratio for *noncommercial* API routes in regulatory submissions. Replace with summary of key elements of predicted purge factor calculations, Purge Ratio, and supporting non-trace data on purge properties for *commercial* API routes in regulatory submissions	Report summary of key elements of predicted purge factor calculations, Purge Ratio, and supporting non-trace or trace data for *noncommercial* API routes in regulatory submissions. Replace with complete summary of predicted purge factor calculations, Purge Ratio, supporting trace and non-trace fate, and purge data for *commercial* API routes in regulatory submissions

This gives the Purge Ratio as described in the equation below.

$$\text{Purge Ratio}(\text{PR}) = \frac{\text{Predicted purge factor for PMI}}{\text{Required purge factor}(\text{based on the permitted limit})}$$

9.10.1 Predicted Purge Factor

Predicated purge factors can be determined either by a paper-based approach or by knowledge-based software (see Section 9.10.11). Individual purge factors are predicted for each PMI present within the process. Such PMIs may be present deliberately through their introduction as a reagent and/or formation as a reactive intermediate at a specific stage. PMIs may also be unintended as a result of an undesired side reaction, e.g. *N*-Nitrosamine formation resulting from the interaction between trace levels of secondary amines and nitrite (see Chapters 10 and 11). The defined purge factors are based on the known or estimated purging capacity of the downstream chemistry and operations. Based on the outcome of the assessment, and comparison to the required purge factor, the resultant Purge Ratio is then used to determine which ICH M7 control strategy is appropriate and what, if any, further evidence may be required to support the claim that the PMI in question is purged. A systematic process relating the purge ratio to data requirements and regulatory strategy was defined and is illustrated in the decision tree, Figure 9.4.

9.10.2 Required Purge Factor

The required purge factor is calculated by dividing the maximum PMI level at a defined point[1] in the process, by the acceptable limit in the final API (see Section 2.2). The acceptable limit, or acceptable daily intake (ADI), is the level of the PMI associated with negligible risk, and is typically based on the Threshold of Toxicological Concern (TTC), the Less Than Lifetime limit (LTL), or Permitted Daily Exposure (PDE[2]) as described in ICH M7 R1 (R1 includes the addition of the addendum). The safety-based limit (ADE or PDE) is then converted into a concentration limit based on the clinical dose and duration, typically these are based on the highest anticipated or highest approved clinical dose/longest dosing duration.[3]

9.10.3 Purge Ratio

The Purge Ratio is determined by dividing the predicted purge by the required purge. For example, if the predicted purge was 1×10^6 and the required purge was 100, then the Purge Ratio is 10 000, indicating that the PMI in question is anticipated to be removed by the process to levels at least 10 000 fold lower than the acceptable limit established for the individual PMI.

As illustrated in Figure 9.4, the Purge Ratio can then be utilized to determine the most appropriate ICH M7 control strategy. Based on this ratio, Barber et al. [14] defined a series of action limits, these also being related to both the purge ratio and the phase of development, see Table 9.5. The action limits recommend the extent of data required to specifically support an Option 4-based control strategy. Clearly, the ratio reflects the extent of risk and hence the data requirements are directly proportionate to this such that increasing amounts of supporting data are required as the Purge Ratio decreases. The aim is ultimately to support the voracity of the predicted purge through experimental data, ensuring the robustness of the proposed control strategy.

Barber et al. also set out a series of recommendations on reporting expectations for Purge Ratio justifications within regulatory submissions. In accordance with the phase-dependent data requirements defined with ICH M7, Section 9 of the guideline, this covers both clinical development and post approval, marketing phase.

Experience has shown that in reporting purge factors and the proposed control option, particularly Option 4, transparency is key. It is thus recommended to include predicted purge factors for each key purging step and within each step, each unit operation in the process. This can be augmented by

---

1 The maximum observed PMI level can be designated by several means. These include (i) By the amount of PMI introduced to process, (ii) by the amount of PMI measured at a specific stage in the process, (iii) the amount in the process or by a level allowed by an acceptance criterion such as an assay value in an intermediate, or (iv) a hypothetical amount formed, this final approach is typically used where a PMI is considered to have been formed by a side reaction.

2 ADIs are based on linear extrapolation of carcinogenicity data, PDEs are used where there is a definable threshold or No Adverse Effect Level (NOEL).

3 In the case of a marketed product, this may be based on median duration for the class / disease area or can involve discussions with an appropriate clinician to understand highest anticipated dose and dosing duration.

inclusion of any supporting experimental physicochemical data that strengthens and increases confidence in the final purge particularly when the overall purge ratio is considered low.

9.10.4 High Predicted Purge

If the predicted Purge Ratio is ≥ 10^3 (1000), then Barber et al. [14] proposed that there was no additional specific data collection required. In view of the demonstrated conservative nature of the scoring system derived by Teasdale et al., this conclusion should be valid (Figure 9.4).

9.10.5 Moderate Predicted Purge

If the predicted Purge Ratio is < 1000 but ≥ 100, then additional data (e.g. reactivity, solubility, relevant test data, etc.) may be required to support the purge argument and the subsequently defined control strategy. This should be assessed on a case-by-case basis and allied to the factors that are most critical to the overall purge of the PMI in question. For example, if the removal of a PMI is predominately due to its solubility, then providing supporting solubility data may be key to underpinning the overall purge factor. Particularly, where scoring for solubility can be a gross underestimate of the actual situation when compared to measured values [23]. In terms of reactivity literature evidence of reactivity/mechanistic understanding (such data underpin predictions within Mirabilis and are directly accessible within the system Knowledge base) may be utilized potentially with specific non-trace data, i.e. reaction monitoring.

9.10.6 Low Predicted Purge

If the predicted purge ratio is <100, then ICH M7 Option 4 may not be an appropriate strategy unless it can be supported by further substantive experimental data. While it is possible that an Option 4 approach might still be valid in view of the inherent underestimation of the purge estimate approach, a predicted purge of <100 alone was viewed as insufficient to support an Option 4 approach. Therefore, measured purge factors, based on both batch data and also deliberate spiking and purge studies would be required to support an Option 4 proposal.

A recent manuscript discussing a second-generation synthesis to the antimalarial agent atovaquone **7** demonstrated that process purging of potential mutagenic impurities can be realized even when they are introduced in the later stages of a process (Figure 9.6) [23]. The authors highlighted that the predicted purge for the PMI unsaturated ketone **6** was insufficient by itself to justify an ICH M7 Option 4 control. Following an investigation into the actual observed reactivity and solubility for the conju-

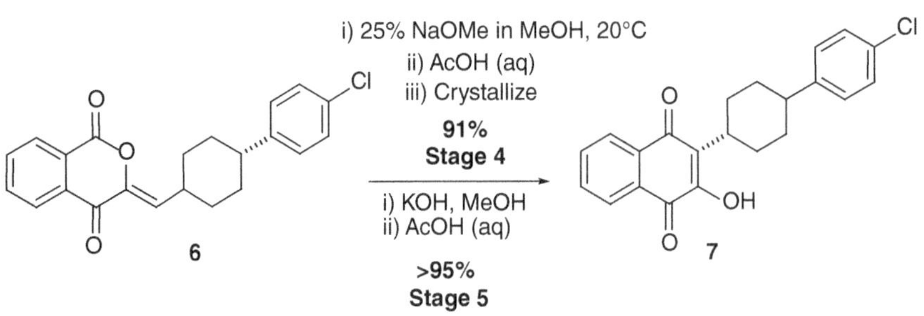

Figure 9.6 Stages 4 and 5 within the second generation manufacture of atovaquone.

gated ketone **6**, coupled with appropriate spiking and purge experiments demonstrated that an option 4 control was fully justifiable when supported with the additional experimental observations.

9.10.7 ICH M7 Control Option 1, 2, or 3

Critically, if the experimentally measured purge factor is insufficient to support an ICH M7 Option 4 control strategy, then the applicant should assess the relative merits of the other ICH M7 control strategies, i.e. Options 1–3 (see Figure 9.4).

9.10.8 Representative Data to be Supplied in Regulatory Submission Under an ICH M7 Control Strategy

Through substantive use of purge factor calculation, industry has been able to tacitly understand what is required in terms of detail from a regulatory perspective. It is very clear that simply providing a total purge factor or purge ratio, however large, is insufficient to demonstrate that a thorough risk evaluation has been performed. This is entirely understandable, in order to understand the risk the applicant must provide the reviewer with sufficient data to allow them to understand how the risk was assessed and the key elements of the risk assessment itself. Within the Barber et al paper [14] the consortium provided commentary on the level of detail and a potential format for presentation. The consortium also proposed that data to support the ICH M7 control strategy should cover the following: Commentary was also provided on the level of detail and a potential format for presentation.

For each PMI, the point at which it is introduced, or formed, should be described as well as its initial starting concentration.

This initial concentration may be defined, measured, or predicted based on appropriate justification (e.g. theoretical impurities).

For each stage, and for each PMI, a summary of the cumulative purge should be provided; this should include a total purge for the stage and record of the specific purge for each unit operation.

In most cases the reactivity of a PMI results in it reacting with another component in the process effectively rendering it safe, for example, an alkylating agent will react with a nucleophile to form C–N, C–O, C–S bonds removing the alkylating agent from the process. However, in some cases, PMIs may be converted into other related PMIs, e.g. Hydrazine to a alkyl hydrazine, as a consequence of the downstream process, where this is the case then both the original PMI and the newly introduced PMI should both be tracked.

Throughout, the level of detail provided is dependent upon the overall magnitude of the Purge Ratio (Table 9.5); however, the focus should be to summarize the key downstream chemistry steps primarily responsible for PMI purging. This should also include relevant supporting evidence; this can be either deliberately generated supporting data or reference to data within peer-reviewed journals.

9.10.9 Summary of PMI Purging Across the Synthetic Route

In order for reviewers to be able to judge the appropriateness of both the purge assessment and subsequent control strategy, it is important to provide a summary of the PMIs, their origins, initial levels, and expected fate and purges throughout the manufacturing process. Within the Barber paper, the consortium provided a visible example of a potential format for such a data summary (Table 9.6) and a similar format is also provided in the recent report detailing the route development to GW641597X (Table 9.7) [21].

Table 9.6 Data to support a regulatory submission that applies the ICH M7 Control Strategy – High-level summary.

PMI	Summary	
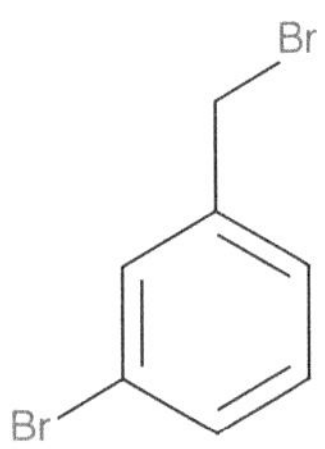	Point of introduction	Stage 1 of 5
	Initial concentrationExplanation	5000 ppm at end of stage 1 *Detected in crude isolate at 0.5%*
	Required Purge Factor	67 [5000/75][a]
	Predicted Purge Factor	10 000
	Purge Ratio	149
Impurity 1 Impurity 1	Recommended ICH M7 Control Strategy	**ICH M7 Option 4; Additional non-trace data recommended to support.**
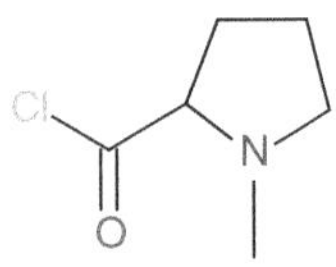	Point of introduction	Stage 2 of 5
	Initial concentrationExplanation	1 000 000 ppm at start of stage 2 *1 eq. added to reaction*
	Required Purge Factor	13 333 [10^6/75][a]
Impurity 2 Impurity 2	Predicted Purge Factor	3×10^{18}
	Purge Ratio	2×10^{14}
	Recommended ICH M7 Control Strategy	**ICH M7 Option 4; no additional data required**
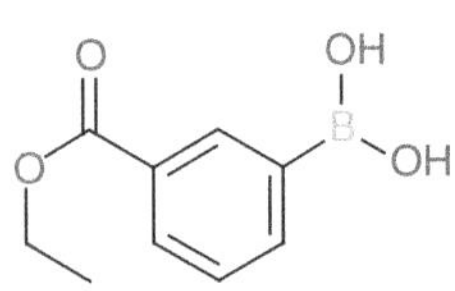	Point of introduction	Stage 3 of 5
	Initial concentrationExplanation	20 000 ppm at end of stage 3 *Max theoretical level is 20 000 ppm at crude reaction (stage 3) based upon 1 eq. used and 98% yield (⇒ max residual impurity is 2%)*
Impurity 3 Impurity 3	Required Purge Factor	267 [20 000/75][a]
	Predicted Purge Factor	100
	Purge Ratio	2
	Recommended ICH M7 Control Strategy	**ICH M7 Option 1, 2 or 3**

[a] Based upon a daily dose of 20 mg of API and a TTC limit of 1.5 μg/day, the PMIs must be limited to (1.5 μg/20 mg =) 75 ppm (ICH M7).

Table 9.6 provides an example of a high-level summary of the supporting information underpinning the control strategy for the designated PMIs which may be used in regulatory submissions. This table shows the introductory point/origin of each PMI in the synthetic process together with a summary of the required and predicted purges. Initial PMI levels are included together with any specific measurements made to support the required purge factor. These measurements could include specific physicochemical information supporting volatility purging (e.g. b.p. of a volatile PMI), solubility purging (e.g. the PMI is liquid and is miscible with all common solvents), or direct experimental measurement of the PMI. The table includes the control strategy for each PMI including any proposed specification limits and the stage where they will be applied (applicable to ICH M7, Options 1–3).

Table 9.7 High-level control summary table for potential mutagenic impurities within the manufacture of GW641597X.

Impurity	Point of potential formation/ introduction and summary of rationale for impurity purging	Required purge and predicted purge	Control
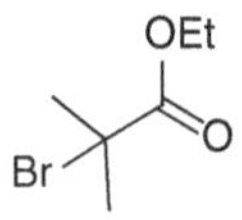	Starting material in stage 1a (2 eq.), 4 steps from drug substance (DS) Consumed to low level (<5%, 50 000 ppm) in stage 1b; reactive during processing (Stage 4); soluble in isolation solvents (Stages 4 and 5)	1 000 000 ppm at start. 1 eq. added to reaction Required Purge = 20 Predicted purge = 1.0×10^5 Purge ratio = 5000	Option 4 – Controlled through chemical reactivity and physical processing.
NH_2OH	Reagent in stage 2 (2.5 eq.), 4 steps from DS Reactive during processing (Stages 2, 3, and 4), highly soluble in isolation solvents (Stages 2, 3, 4, and 5)	Reagent in stage 2 (2.5 eq.), Required purge = 39 Predicted purge = 1.0×10^8 Purge ratio = 2.56×10^6	Option 4 – Controlled through chemical reactivity and physical processing.
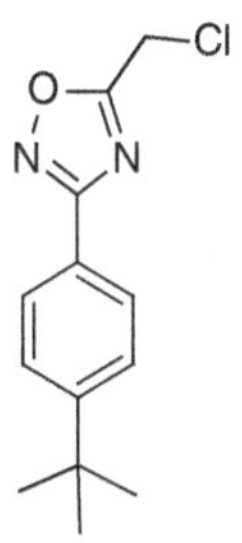	Starting material in stage 4, 2 steps from DS Confirmed at low level (c. 0.2%) within stage 4b product following additional reactivity with aqueous base used within the process and solubility within the isolation solvent. Additional solubility anticipated in stage 5 isolation solvent	Confirmed at low level (c. 0.2%) within stage 4b Required purge = 6 Predicted purge = 1000 Purge ratio = 167 Measured purge = 75 (Stage 4b) Measured purge ≥ 150 (Stage 4b and 5)	Option 4 – Controlled through chemical reactivity and physical processing.

A similar approach to reporting ICH M7 MI option 4 control rationales to regulators has been reported and is provided in Figure 9.7 [21].

9.10.10 Details of Individual Impurity Purging Through the Subsequent Downstream Chemistry

While Tables 9.6 and 9.7 provide high-level summaries of the PMIs and the proposed control strategies, it may also be appropriate to show a more detailed derivation of the overall predicted purge factors when supporting an Option 4 strategy for a PMI present in a commercial API route, especially when more moderate Purge Ratios are predicted. Table 9.8 is proposed to provide a breakdown of the predicted (or measured) purge factors through each stage of the synthesis including handling steps such as isolation and purification, from the point of introduction through all remaining stages which may also be included along with Table 9.6 in a regulatory submission. Any specific studies undertaken to support a predicted purge factor can be included – such as spike-purge experiments or experimentally determined solubilities. References to supporting expert commentary, analysis, or measured data from analogs can be usefully provided here whether from primary literature or from a shared available knowledge base such as that provided with Mirabilis. Again, it is important to retain some conservatism with predictions; therefore, while the amine derived from the reductive amination within stage 5 might be expected to have some reactivity

				Residual impurity of concern						
Reactions	Anticipated reagent	Anticipated conditions	Comments	Alkyl iodides	Alkyl bromides	Alkyl chlorides	N, S-mustards	Halo alkenes	Boronic acids	Epoxides
N-Alkylation	Primary alkyl bromide	Excess R-X with respect to amine, solvent		10	10	1	100	1	1	10
	Primary Iodo	Excess R-X with respect to amine, solvent		10	1	1	100	1	1	10
	Primary Chloro	Excess R-X with respect to amine, solvent		100	10	10	100	1	1	100
Reductive *N*-alkylation	Aldehyde + borohydride		Assumes that product of the reaction is unreactive with respect to R-X	1	1	1	1	1	1	10
N-Arylation (Buchwald)	Aryl bromide/catalyst			100	100	100	100	10	10	100
Amide *N*-alkylation	Deprotonation and alkyl bromide			100	10	1	100	1	1	100
Aniline *N*-alkylation	Alkyl bromide			100	10	1	100	1	1	100
Heteroaryl *N*-alkylation	Alkyl bromide			100	10	1	100	1	1	100
O-alkylation	Base and alkyl bromide			100	10	1	100	1	1	100
S-alkylation	Base (NaOH) alkyl bromide			100	10	1	100	1	1	1

Figure 9.7 A screenshot of part of the reaction grid for common reactive intermediate classes for some alkylation reaction types.

toward impurity 1, this has not been scored because of a lack of process knowledge within the software to reinforce a prediction (Table 9.8).

9.10.11 Development of a Knowledge Base Expert In Silico System

The principle of the purge concept is predicated on the use of prior knowledge, which, in terms of the PMI, can include its physicochemical properties and the processing conditions deployed in the synthetic process to manufacture the API to predict the fate of the PMI in question. When assessed manually, the approach is underpinned by the expertise of the individual, or group, that performs the risk assessment. Prior knowledge is effectively their knowledge of their process allied to their fundamental understanding of chemistry.

Expert systems are used widely in pharmaceutical discovery and development across a range of applications, including the prediction of synthetic reactions [24–26], kinetic modeling of reactions [27], route design via retrosynthetic analysis [28, 29], prediction of toxicity [30–33], drug metabolism [31], and chemical degradation [34, 35]. There was realization that purge factor calculations could be similarly made *in silico* based on a knowledge base and that the utility and consistency of this predictive approach would be further augmented by development of a standardized *in silico* system. The purpose of this section is to describe the principles that underpin the development of a semiautomated computer-based system and the outcomes of the development.

A cross-industry consortium of seven pharmaceutical companies was established in December 2013, to facilitate the development of an *in silico* system, subsequently named Mirabilis™. The consortium has since grown to include 21 companies. The initial objective was to capture and harmonize industry best practices related to purge predictions based on the concept published by Teasdale et al. [4, 5] As highlighted, the suitability, robustness, and acceptability of the concept have been established. Thus, the system was developed based on the exact same principles and concepts of the published paper-based tool, including the scoring system. The ultimate aim was to provide systematic models and processes, based on a comprehensive data set, referred to as the knowledge base and to utilize this to facilitate and standardize the prediction of purges, augmenting the existing expertise of the chemist. A key prerequisite was to retain the simplicity of the original paper-based approach, while enhancing the efficiency, transparency, and consistency of purge predictions.

In order to make effective use of prior knowledge, a team at Astra Zeneca created a "reaction grid" covering all major PMI/MI classes and matrixed them with identified major chemical reactions/transformations [36]. The reaction grid is illustrated in Figure 9.7.

Table 9.8 Detailed purge calculations for an impurity.

Impurity 1 **(impurity class = Primary alkyl bromide)**	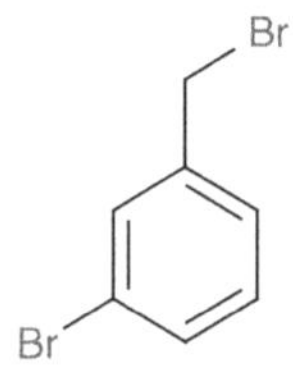		**Total predicted purge = 10 000**
Stage	**Process**	**Predicted purge**	**Comments**
1	Work-up – Liquid–Liquid Extraction (Neutral)	1	No purge expected
	Purification – Recrystallization	**10**	Impurity is a solvent-miscible oil and stage 1 product isolated as an HCl salt a solid
	Total predicted purge for stage 1	10	
2	Reaction – N-Acylation of amine	10	*Default value – ref knowledge base*
	Work-up – Liquid–Liquid Extraction (Neutral)	1	No purge expected
	Total predicted purge for stage 2	10	
3	Reaction – Suzuki coupling	10	*Default value – ref knowledge base*
	Work-up – Liquid–Liquid Extraction (Neutral)	1	No purge expected
	Total predicted purge for stage 3	10	
4	Reaction – Reduction of ester to aldehyde	1	*Default value – ref knowledge base*
	Work-up – Liquid–Liquid Extraction (Neutral)	1	No purge expected
	Total predicted purge for stage 4	1	
5	Reaction – Reductive amination	1	*Default value – ref knowledge base*
	Work-up – Liquid–Liquid Extraction (Neutral)	1	No purge expected
	Purification – Recrystrallization	**10**	Impurity is a solvent-miscible oil and stage 5 product isolated as an HCl salt a solid
	Total predicted purge for stage 5	10	

The grid was shaded based on existing knowledge, the lighter the shading the greater the confidence of the prediction. **It is important to stress that in circumstances where there is no knowledge, or no specific evidence of purge, a conservative approach is taken and no reaction is assumed. This principle remains a key principle of Mirabilis™.** Also included was a further category marked in blue – this referring to PMIs which as a result of a chemical transformation are converted to another related PMI. Such PMIs must also be tracked. An example would include alkylation of hydrazine to yield an alkyl hydrazine.

This reaction grid was further developed within the Mirabilis consortium through a process referred to as expert elicitation. Expert elicitation is a defined form of knowledge sharing [36] based on a consortium-wide consensus view of the purging potential of prioritized classes of PMIs, in various reaction scenarios. Each of the original seven consortium members reviewed the original reaction grid and their expertise was incorporated. Sample results of expert elicitation for five impurity/reaction pairs are shown in Table 9.9. If five or more members agreed on a particular reactivity purge factor, this was considered to be a consensus for that value. However, where no consensus could be reached, or where there was a gap in knowledge, the most conservative value was adopted to ensure the conservative nature of the approach.

A number of PMI classes were revealed to be only partially covered by existing knowledge, this included arylboronic acids, hydrazines, and aromatic amines. A detailed review of the use of common classes of PMIs within multistep synthetic pathways [3] shows that alkyl bromides, hydrazines, and aromatic amines are commonly used reagents in the later stages of API synthesis and arylboronic acids are also seeing increased usage. A consequence of this overlap between commonly used reagents and a lack of appropriate knowledge instigated a significant program of work from within the consortium to address the identified gaps in knowledge.

The first two rows illustrate a consensus call. The third row shows a minor variance in opinion (these were only observed for adjacent factors) and the fourth and fifth rows show uncertainty, and thus conservative calls were made.

In Mirabilis, the concept of the reaction grid has been further developed and incorporated into the system as a reactivity matrix (Figure 9.8). Each cell within the matrix describes the purge factor for a single impurity/reaction class combination. Another major advantage of an *in silico* tool is the ability to retain and visually present supporting information. In the case of Mirabilis, this takes the form of information displayed by the system pertaining to the reactivity purge factor, an executive summary of key points, a table of dependencies which may affect reactivity (and thus purge), and additional scientific comments including relevant references and examples, and experimental data if available (Figure 9.9).

Table 9.9 Sample results from expert elicitations [18].

		Number of experts assigning call				
Impurity	**Reaction type**	**Reactivity = 100**	**Reactivity = 10**	**Reactivity = 1**	**Reactivity = unknown**	**Assigned reactivity**
Epoxide	*S*-Alkylation	6	1	0	0	100
Aromatic nitro compound	Reductive *N*-alkylation	0	0	7	0	1
Epoxide	*N*-Arylation	4	3	0	0	10
Michael-reactive acceptor	Amide *N*-alkylation	0	2	1	4	1
Arylboronic acid	*N*-Arylation	2	0	0	5	1

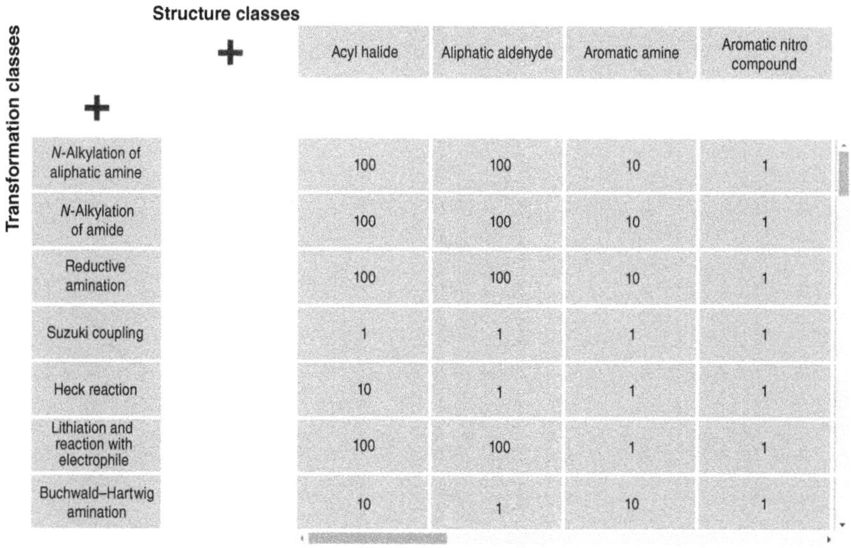

	Acyl halide	Aliphatic aldehyde	Aromatic amine	Aromatic nitro compound
N-Alkylation of aliphatic amine	100	100	10	1
N-Alkylation of amide	100	100	10	1
Reductive amination	100	100	10	1
Suzuki coupling	1	1	1	1
Heck reaction	10	1	1	1
Lithiation and reaction with electrophile	100	100	1	1
Buchwald–Hartwig amination	10	1	10	1

Figure 9.8 Illustration of the reactivity matrix within Mirabilis.

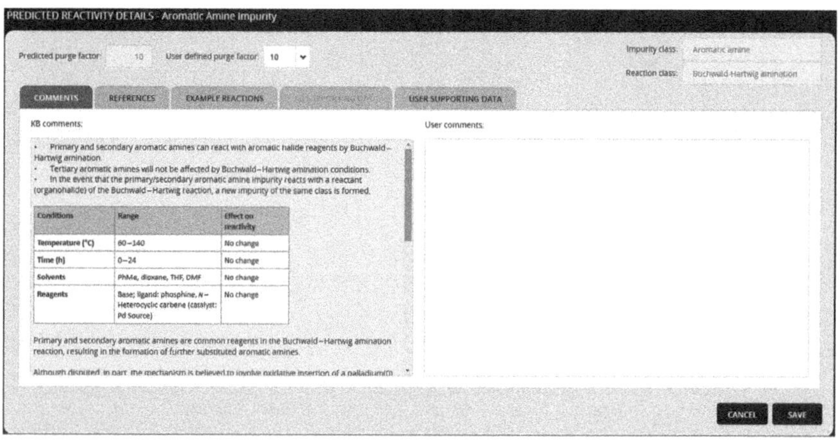

Figure 9.9 Example of additional supporting information displayed within Mirabilis.

Such supporting data are key in underpinning the individual prediction, while their transparency is also vital as it allows any user, both applicant and reviewer direct and visible access to these key data.

9.10.12 Experimental Work to Assess Reactivity

Following the identification of gaps in the knowledge surrounding key PMI structural classes, an experimental workgroup was created and a protocol developed to determine reactivity, to address the aforementioned knowledge gaps. In the first instance, for each PMI class studied, a screening assay was conducted to quickly assess the reactivity of simple PMIs under various reaction conditions. Reactions which were very fast provide evidence to support a reactivity purge value of 100. Reactions where the PMI displayed little or no reactivity provide evidence to support a reactivity purge value of 1. Reactions that did not fall into either of these categories were studied further to collect data in order to develop a kinetic model which describes the reaction behavior of the PMI under specific reaction conditions and exemplified with a range of boronic acids (Table 9.11) [38]. This kinetic information was then used to understand the degree of consumption of the impurity in a particular transformation of interest and to assign a purge value.

Table 9.10 Experimental reactivity protocol.

	Reaction type	Reagent	Solvent	Reactive?
1	Reduction	H_2 Pd/C	Dioxane	No
2		$NaBH_4$	MeOH, THF, DCM	No
3		$LiAIH_4$	THF	No
4		DIBAL-H	THF, DCM	No
5	Oxidation	H_2O_2	DCE, DCM, CH_3CN	Yes
6		Peracetic acid	DCM	Yes[a]
7		Oxone	CH_3CN, H_2O, H_2O:CH_3CN	Yes[b]
8		TEMPO	DCM	Yes[c]
9	Acids	Aq HCI	CH_3CN, THF	No
10		Conc. H_2SO_4	H_2O	No
11		Aq H_2SO_4	H_2O, Dioxane, CH_3CN	No
12		HBr/HOAc	DCM	No
13	Bases	Aq $NaHCO_3$	CH_3CN	No
14		10% NaOH	CH_3CN, Dioxane. H_2O	No
15		50% NaOH	H_2O	Yes
16		DBU	CH_3CN, DCE	No
17	Amide bond formation	CDI (with benzoic acid)	DCM	No
18		EDAc/HOPO (with benzoic acid)	DMF	No
19		Benzoyl chloride	THF	No
20	Nucleophiles	MeOH	THF	No
21		Benzyl amine	THF	No
22	Other reagents	$SOCI_2$	DCE	No
23		NCS	DCE	No
24		NCS/7EA	DCE	No
25		NBS	DCE	Yes[d]
26		Boc_2O/TEA	THF	No
27		TMSCI/TEA	THF	No
28	Cross-coupling	RuPhos-Pd complex (25 mol %), K_2CO_3,THF/H_2O		?
29		Pd_2dba_3 (12.5 mol%), $PtBu_3HBF_4$ (25 mol%), TEA, THF		?

[a] Reaction was complete within five minutes at −78 °C.
[b] Reaction was complete within five minutes at 2.5 °C.
[c] Reaction was complete within five minutes at 2.8 °C.
[d] Reaction was complete within five minutes at 3.2 °C.

Table 9.10 shows the outcome of the specific modeling of a series of boronic acids. What is clear from this work is that under the conditions when reaction does occur then the purge would be significantly higher than 100, again illustrating the conservatism built into the scoring system devised by Teasdale et al. [4].

9.11 Utilizing Mirabilis for a Purge Calculation

The user can generate purge predictions within Mirabilis using the following process:

- Enters the full synthetic scheme capturing each chemical transformation which allows any PMIs formed in the process to be captured, and tracked, as well as highlighting any PMIs formed through transformation of one PMI to another, e.g. hydrazine to hydrazide.
- Selects the impurities they wish to track and the software seeks to determine which class(es) the impurity belongs to. The user confirms selection of the appropriate class(es). Although these are reactive motifs that generally behave as a class, reactivity does not necessarily correspond to mutagenic activity; this is assessed as a separate activity.
- Selects each chemical transformation in turn. The software then seeks to assign the reaction class automatically which is then confirmed by the user.
- Once both the impurity and reaction classes are defined, the extent to which an impurity may be potentially consumed during a reaction is determined. The software uses the knowledge base to assign the purge factor and provides all relevant information contained within to the user. Thus, when patterns are matched, both for the structural motif of the impurity and for the reaction conditions of a synthetic step, the respective purge factor prediction is presented (Figure 9.10).

The user defines the unit operations performed during each step within the synthesis (e.g. reaction, work-up, purification). These are selected from a predefined list to ensure consistency in terminology.

Within Mirabilis, the operations performed during the synthesis are organized into stages and steps. A step is defined as any operation: reaction, work-up, or purification. A stage typically consists of multiple operations, for example, a reaction step, followed by one or more work-up and/or purification steps. Each step can be assigned purge factors from one or more purge parameters, dependent on the nature of the unit operations employed within the step.

The software directs the user to which parameters are appropriate for a given step (for examples, see Figure 9.11).

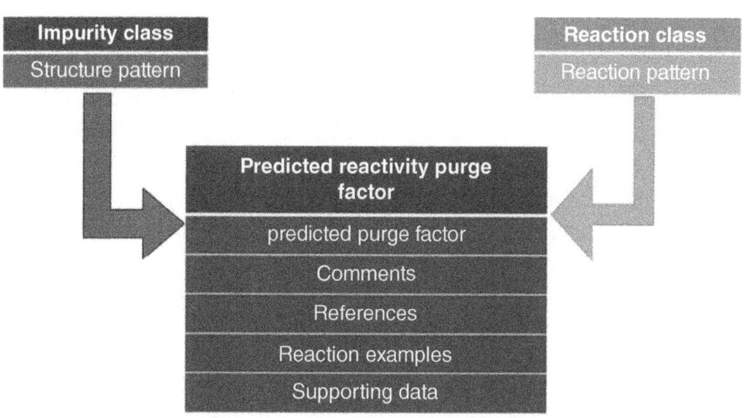

Figure 9.10 Identifying the appropriate purge prediction in Mirabilis.

Steps	Reactivity	Solubility	Volatility	Other
Buchwald-Hartwig amination - Aromatic amine	10		1	
Workup Liquid-Liquid Extraction (Neutral)	1	10		
Workup - Wash		10		
Workup - Drying			3	
Purification - Chromatography				100
Other - Other	1	1	1	1

Figure 9.11 Examples of restrictions on purge parameters for various steps within a stage from Mirabilis.

The other physicochemical properties that can impact the purge of a PMI, such as solubility, volatility, etc., are entered by the user based on their expert knowledge or literature precedent. These are combined with the reactivity purge to give an overall predicted purge value for that PMI. It is noteworthy that while "Ionizability" is not included, this factor is still scored but is captured as "solubility" for that unit operation, etc.

In terms of reactivity where the predictions are made based on the knowledge base, i.e. system defined, the user is able to change the predicted value; however, in doing so, they must record the reason for the change and this change is highlighted both within the system and within subsequent reports.

9.11.1 Utility of In Silico Predictions

Burns et al. [19] recently examined the use of Mirabilis through a series of case studies (Figure 9.12). These case studies compared predictions made within Mirabilis to experimental data and showed over the course of a multistage synthesis that Mirabilis consistently underpredicts cumulative purge versus experimentally measured purge, demonstrating the intentional conservative baseline which Mirabilis provides in purge calculations. The case studies also demonstrated the potential for impurities to be purged far beyond the levels of detection used in trace analyses. In these circumstances, the value of Mirabilis purge predictions is to provide a realistic, but still conservative, approximation of the true magnitude of MI purge that is not bounded by analytical sensitivity limitations, providing additional critical insight into actual process performance.

9.11.1.1 Case Study – Camicinal [38]

Gastroparesis, or chronic delayed gastric emptying without mechanical obstruction, affects about 40% of patients with type 1 diabetes and up to 30% of patients with type 2 diabetes. Diabetic gastroparesis (DGP) typically causes nausea, vomiting, early satiety, bloating, and postprandial fullness. These symptoms can be extremely troubling and result in poor quality of life.

The motilin (GPR38) receptor agonist Camicinal (GSK962040) with gastroprokinetic activity was in development for treating conditions which have reduced gastric motility, i.e. Parkinson's disease, intensive care unit enteral nutrition, and diabetes.

There was a plan to use Camicinal to assist enteric feeding and DGP using oral liquid and tablet products respectively. The dose was estimated as 50 to 125 mg with a dosing duration of less than 1 year which led to a TTC of 0.016% or 160 ppm.

An initial MI risk assessment was conducted to determine potential impurities based on the observed and tentatively identified impurities (Tables 9.12 and 9.15) as well as what could be reasonably predicted from evaluation of the registered synthetic route as well as synthetic route to the piperazine starting material **8** (Figures 9.13 and 9.14).

Mirabilis Purge Report

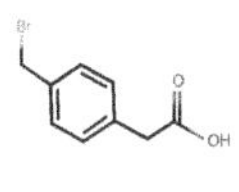

Impurity ID: 4-(Bromomethyl)phenylacetic acid
Impurity Class: Primary alkyl bromide
Total Predicted Purge: 810000

Process	Assigned Purge*	Comments
Stage: Stage 7		
Reaction - *N*-Alkylation of aliphatic amine	100 (10)	Reactivity: Primary reactant consumed through the reaction.
Total Predicted Purge For Stage:	**100**	
Stage: Stage 8		
Reaction - Amide coupling	10 (1)	Reactivity: Reactivity primarily occurs through the carboxylic acid motif, Supported through HPLC monitoring.
Workup - Liquid–liquid extraction (base)	3	Solubility: Impurity soluble in precipitation solvent
Total Predicted Purge For Stage:	**30**	
Stage: Stage 9		
Reaction - *N*-Boc deprotection and *t*-butyl ester hydrolysis	1	No purge expected
Workup - Quench	10	Reactivity: Bromide undergoes substitution with hydroxide under quench conditions.
Workup - Precipitation	3	Solubility: Impurity soluble in precipitation solvent
Total Predicted Purge For Stage:	**30**	
Stage: Stage 10		
Reaction - Unassigned reaction		
Purification - Recrystallization	3	Solubility: Impurity soluble in crystallization solvent
Total Predicted Purge For Stage:	**3**	
Stage: Stage 11		
Reaction - Unassigned reaction		
Purification - Recrystallization	3	Solubility: Impurity soluble in crystallization solvent
Total Predicted Purge For Stage:	**3**	

where predicted values have been changed by the reviewer, the original values are shown in parentheses.

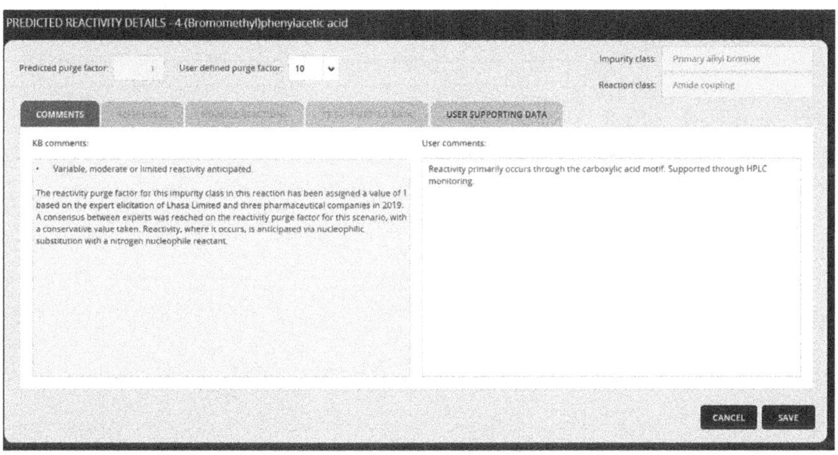

Figure 9.12 Top: Purge table in Mirabilis report. Bottom: Predicted reactivity details in Mirabilis.

Once this overall picture of known and reasonably predicted impurities was established, they were subject to *in silico* analysis, which showed there to be four structures of concern:

Aniline **2** alerts for mutagenicity but was confirmed Ames negative.

Alkyl bromide **7** and alkyl chloride **13** alert as potential alkylating agents.

Chloroacetyl chloride **12** alerted as a potential alkylating agent but was confirmed non-mutagenic.

In addition, degradation was assessed, which highlighted three materials for assessment from hydrolysis and oxidation (Figure 9.16). No materials from forced degradation studies alerted for

Table 9.11 Experimentally derived purge factors for oxidation of boronic acids by peroxide [38].

Arylboronic acid	Predicted purge factor (in the absence of PhSMe)	Experimentally derived purge factor (in the presence of PhSMe)	Standardized reactivity purge factor assignment
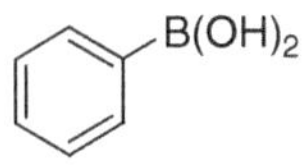	2.1×10^5	1.4×10^4	100
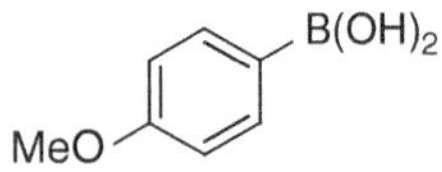	3.3×10^8	6.5×10^6	100
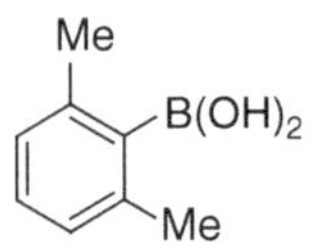	1.2	1.1	1
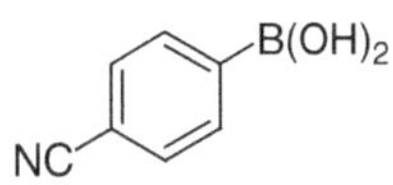	153	5058	100
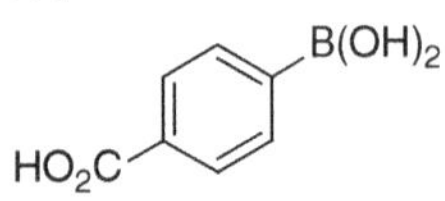	1253	9170	100
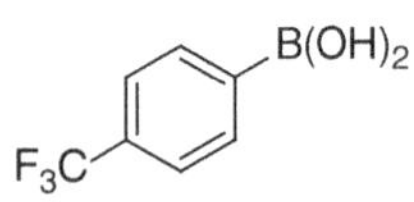	1640	506	100

Table 9.12 Specified impurities for Camicinal (GSK962040).

Identifier	Structural formula	Description	Typical level observed
RRT 0.68	No structure determined	Process impurity	Up to 0.10% area
RRT 0.93	No structure determined	Process impurity	Up to 0.10% area
RRT 0.94	No structure determined	Process impurity	Up to 0.25% area
RRT 1.07	No structure determined	Process impurity	Up to 0.10% area
RRT 1.25	No structure determined	Process impurity	Up to 0.10% area
RRT 1.37	No structure determined	Process impurity	Up to 0.20% area
4	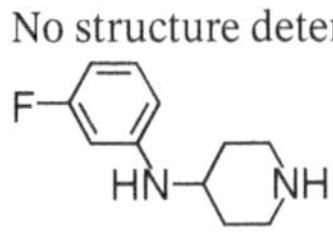	Starting material	Up to 0.10% area
18	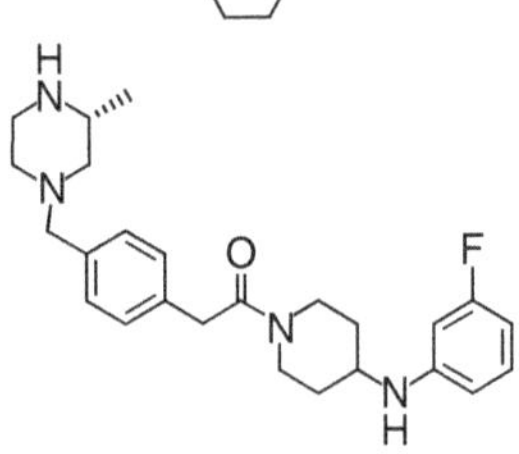	Enantiomer of DS	<0.05% area
10	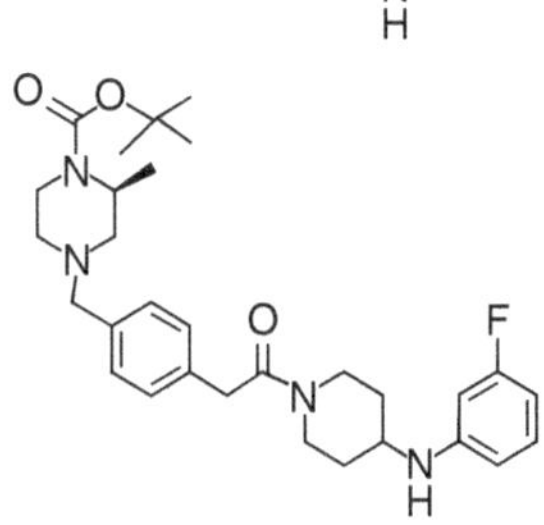	Intermediate	0.05% area

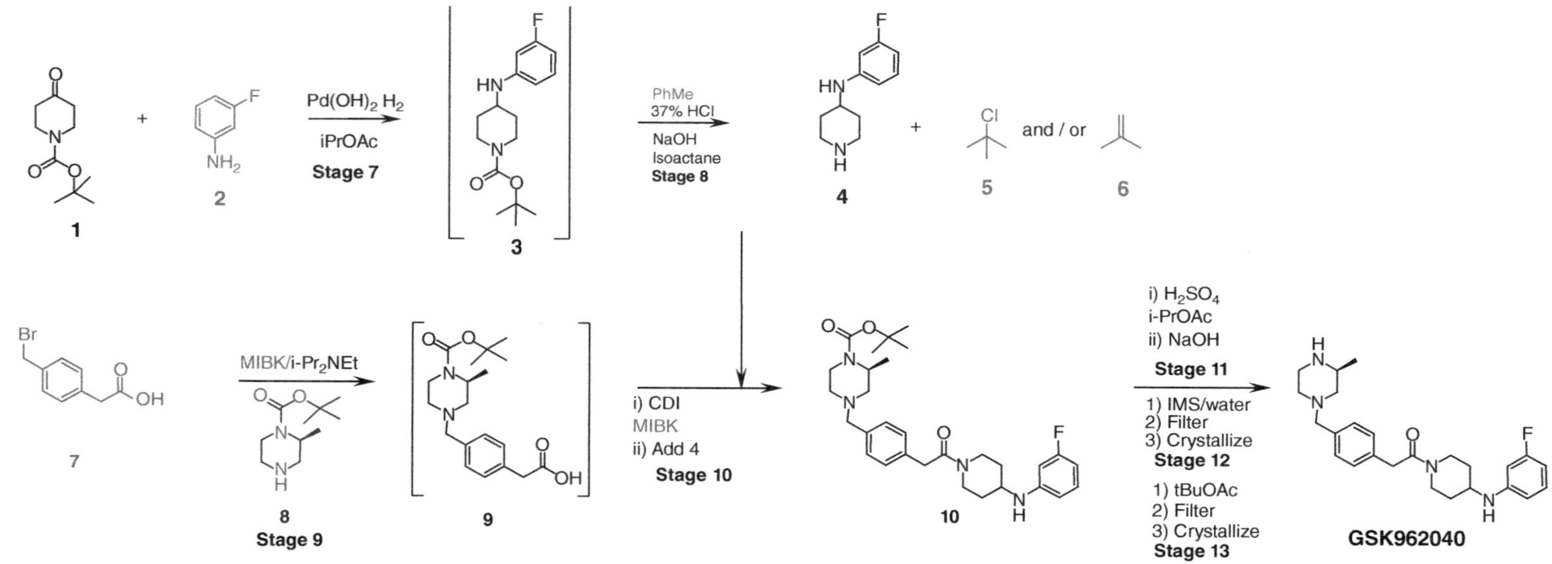

Figure 9.13 Synthetic process to Camicinal (GSK962040).

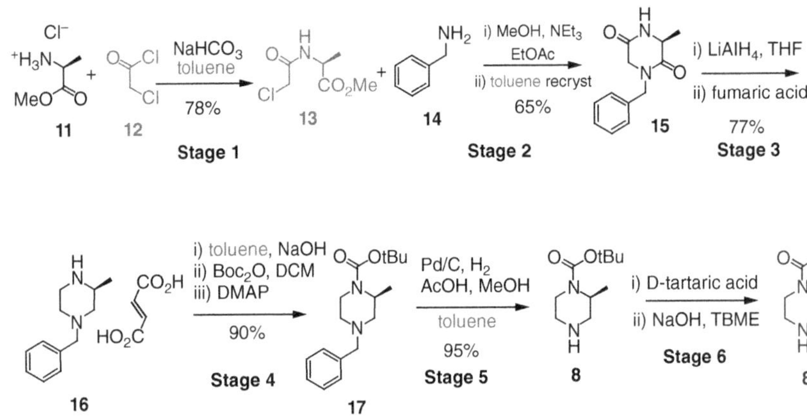

Figure 9.14 Synthetic process to the proposed registered starting material tert-butyl (*S*)-2-methylpiperazine-1-carboxylate (8).

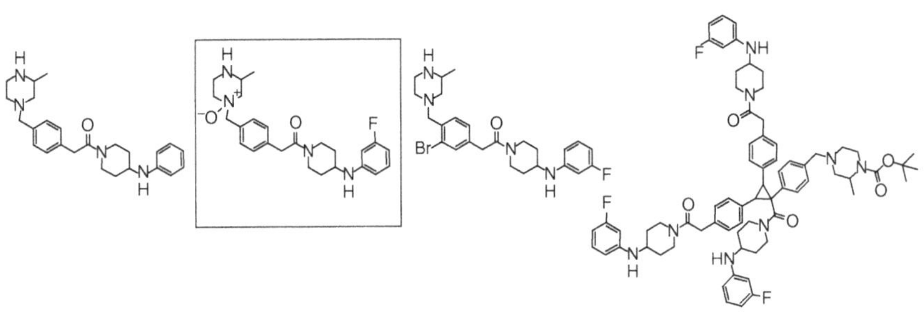

Figure 9.15 Further tentatively assigned and "non-alerting" impurities within Camicinal (GSK962040).

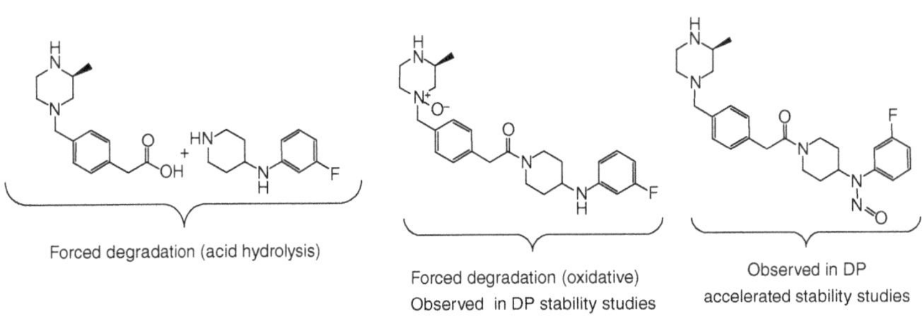

Figure 9.16 Observed degradants from Camicinal forced DS degradation and DP stability studies.

mutagenicity. A new degradant was observed during accelerated stability of the drug product which was the nitrosamine of the drug substance. This nitrosamine was confirmed as non-mutagenic through returning a negative Ames test.

A summary of the potential materials of concern is shown in Table 9.13 and, based on this, the next stage was to assess the purging of the alkyl halides **7** and **13** as only these materials would require control.

Detailed purge factor calculations were determined using Mirabilis (e.g. Figure 9.12) and a summary of these are provided in Table 9.14.

Table 9.13 Summary of mutagenic impurity assessment from Camicinal (GSK962040) review.

Potential mutagenic impurities	Origin	(Q)SAR assessment		Bacterial mutagenicity (Ames Assay)	Ames report reference	ICH M7 class
		Expert knowledge system	QSAR system			
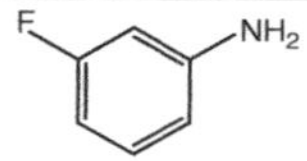	Synthetic intermediate	Negative	Indeterminate	Negative	Internal study	5
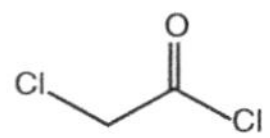	Synthetic intermediate	Positive	Positive	Negative	HSDB	5
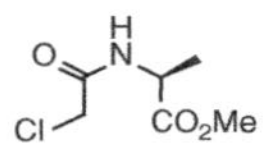 **13**	Synthetic intermediate	Negative	Positive	Not tested	Not applicable	3
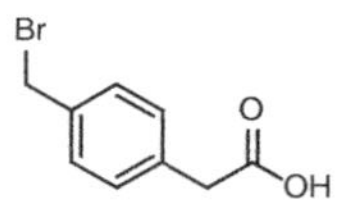 **7**	Synthetic intermediate	Positive	Indeterminate	Not tested	Not applicable	3
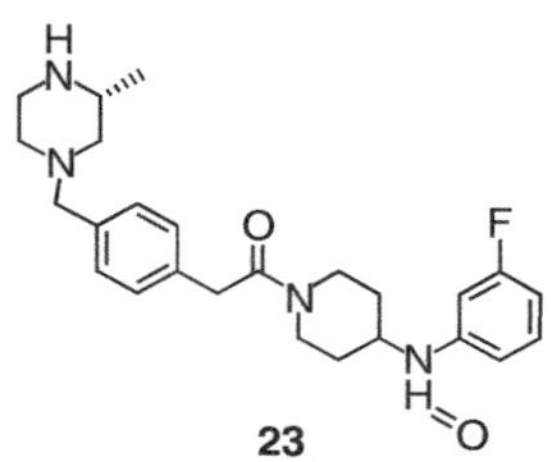 **23**	DP degradant	Positive	Positive	Negative	Internal study	5

Table 9.14 Summary of purge rationales for PMIs **7** and **13** from the Camicinal route of manufacture.

Impurity	Point of potential formation/ Introduction and summary of rationale for impurity purging	Required purge and predicted purge	Control
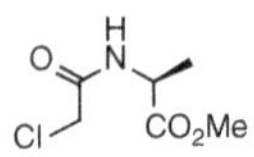 **13**	Starting material in Stage 2 (1 eq), 10 steps from DS consumed to low level (<2%) in Stage 2; expected to be reactive during processing (Stages 3–6, 9, 11); expected to be soluble in isolation solvents (Stages 2–6, 10–13). Assigned as a mono-functional alkyl chloride.	Required purge = 625 Predicted purge = 2.0 $\times 10^{12}$ Purge ratio = 3.2×10^9	Option 4 – Controlled through chemical reactivity and physical processing.
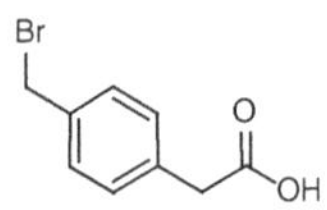 **7**	Starting material in Stage 9 (1 eq), 5 steps from DS consumed to low level (<1%) in Stage 9; expected to be reactive during processing (Stages 10, 11); expected to be soluble in isolation solvents (Stages 10–13).	Required purge = 6250 Predicted purge = 8.1×10^5 Purge ratio = 130 Measured purge $\geq 1 \times 10^5$ (Confirmed <10 ppm in DS) Purge ratio $\geq$ 16 (Measured/Required)	Option 4 – Controlled through chemical reactivity and physical processing.

Comparison of required and predicted purge factors gave a ratio of 130 for impurity **7** (based on $6250/(8.1 \times 10^5)$). Given the ratio was <1000 in accordance with the Barber paper [13], further supporting data were deemed necessary which was achieved by testing the API and confirming levels in DS had been <10 ppm. Given that the TTC for this product was estimated as being 160 ppm for an individual MI, the confirmed level within the drug substance justifies an option 4 control strategy for this impurity. In the case of impurity **13,** where the ratio was >>1000, Option 4 approach was taken without a requirement for any supporting data.

References

1 Carcinogenic potential | European Medicines Agency https://www.ema.europa.eu/en/carcinogenic-potential (accessed 29 September 2020).

2 ICH (2014). ICH Guideline M7 – assessment and control of DNA reactive (mutagenic) impurities in pharmaceuticals to limit potential carcinogenic risk. *Guideline*. https://doi.org/http://www.ich.org/fileadmin/Public_Web_Site/ICH_Products/Guidelines/Multidisciplinary/M7/M7_R1_Addendum_Step_4_31Mar2017.pdf (accessed September 2020).

3 Elder, D.P. and Teasdale, A. (2015). Is avoidance of genotoxic intermediates/impurities tenable for complex, multistep syntheses? *Org. Process Res. Dev.* *19* (11): 1437–1446. https://doi.org/10.1021/op500346q.

4 Teasdale, A., Fenner, S., Ray, A. et al. (2010). A tool for the semiquantitative assessment of potentially genotoxic impurity (PGI) carryover into API using physicochemical parameters and process conditions. *Org. Process Res. Dev.* *14* (4): 943–945. https://doi.org/10.1021/op100071n.

5 Teasdale, A., Elder, D., Chang, S.-J. et al. (2013). Risk assessment of genotoxic impurities in new chemical entities: strategies to demonstrate control. *Org. Process Res. Dev.* *17* (2): 221–230. https://doi.org/10.1021/op300268u.

6 Pierson, D.A., Olsen, B.A., Robbins, D.K. et al. (2009). Approaches to assessment, testing decisions, and analytical determination of genotoxic impurities in drug substances. *Org. Process Res. Dev. 13* (2): 285–291. https://doi.org/10.1021/op8002129.

7 Ashworth, I.W. and Meadows, R.E. (2018). A general liquid–liquid partitioning equation and its consequences: learning from the PH dependent extraction of a pharmaceutical intermediate. *J. Org. Chem. 83* (7): 4270–4274. https://doi.org/10.1021/acs.joc.8b00309.

8 Welch, C.J., Biba, M., Drahus, A. et al. (2003). Selective removal of a pharmaceutical process impurity using a reactive resin. *J. Liq. Chromatogr. Relat. Technol. 26* (12): 1959–1968. https://doi.org/10.1081/JLC-120021763.

9 Bandichhor, R., Reddy L.A., Chackraborty, S. et al. (2009). Preparative chromatography technique in the removal of isostructural genotoxic impurity in Rizatriptan: use of physicochemical descriptors of solute and adsorbent. *Org. Process Res. Dev. 13* (4): 683–689. https://doi.org/10.1021/op9000093.

10 Bennett, A. (2013). Pharmaceuticals and fine chemicals: filtration and separation in the diverse fine chemical sectors. *Filtr. Sep. 50* (6): 30–33. https://doi.org/10.1016/S0015-1882(13)70240-4.

11 Burke, A.J., Marques, C.S., Turner, N.J., and Hermann, G.J. (2018). Catalytic process design: the industrial perspective. *Act. Pharm. Ingred. Synth.* 17: 31–73. https://doi.org/doi:10.1002/9783527807253.ch2.

12 Hermkens, P.H.H., Ottenheijm, H.C.J., and Rees, D. (1996). Solid-phase organic reactions: a review of the recent literature. *Tetrahedron 52* (13): 4527–4554. https://doi.org/10.1016/0040-4020(96)00216-5.

13 Lee, C., Helmy, R., Strulson, C., Removal of Electrophilic Potential Genotoxic Impurities Using Nucleophilic Reactive Resins. *Organic Process Research and Development* 14(4): 1021–1026.

14 Barber, C., Antonucci, V., Baumann, J.-C. et al. (2017). A consortium-driven framework to guide the implementation of ICH M7 option 4 control strategies. *Regul. Toxicol. Pharmacol. 90*: 22–28. https://doi.org/10.1016/J.YRTPH.2017.08.008.

15 Elder, D.P., Okafo, G., and McGuire, M. (2013). Assessment of predictivity of semiquantitative risk assessment tool: Pazopanib hydrochloride genotoxic impurities. *Org. Process Res. Dev. 17* (8): 1036–1041. https://doi.org/10.1021/op400139z.

16 Lapanja, N., Zupančič, B., Toplak Časar, R. et al. (2015). A generic industry approach to demonstrate efficient purification of potential mutagenic impurities in the synthesis of drug substances. *Org. Process Res. Dev. 19* (11): 1524–1530. https://doi.org/10.1021/acs.oprd.5b00061.

17 Teasdale, A., Elder, D., and Fenner, S. (2011). Strategies for the evaluation of genotoxic impurity risk. In: *Genotoxic Impurities: Strategies for Identification and Control* (ed. A. Teasdale), 221–248. New Jersey: Wiley https://doi.org/10.1002/9780470929377.ch9.

18 Szekely, G., Amores de Sousa, C.M., Gil, M. et al. (2015). Genotoxic impurities in pharmaceutical manufacturing: sources, regulations, and mitigation. *Chem. Rev. 115* (16): 8182–8229. https://doi.org/10.1021/cr300095f.

19 Burns, M.J., Ott, M.A., Teasdale, A. et al. (2019). New semi-automated computer-based system for assessing the purge of mutagenic impurities. *Org. Process Res. Dev. 23* (11) https://doi.org/10.1021/acs.oprd.9b00358.

20 McLaughlin, M., Dermenjian, R.K., Jin, Y. et al. (2015). Evaluation and control of mutagenic impurities in a development compound: purge factor estimates vs measured amounts. *Org. Process Res. Dev. 19* (11): 1531–1535. https://doi.org/10.1021/acs.oprd.5b00263.

21 Boyer, T., Choudary, M.B., Edwards, J.A. et al. (2020). Development of a scalable process for the PPAR-α agonist GW641597X incorporating Baeyer–Villiger chemistry and retrospective ICH M7 assessment. *Org. Process Res. Dev. 24* (3): 371–386. https://doi.org/10.1021/acs.oprd.9b00385.

22 Kragelj Lapanja, N., Zupančič, B., Toplak Časar, R. et al. (2018). Evaluation of starting materials for PMIs (potentially mutagenic impurities): a vortioxetine case study. *Org. Process Res. Dev. 22* (2): 125–135. https://doi.org/10.1021/acs.oprd.7b00239.

23 Urquhart, M.W.J., Bardsley, B., Edwards, A.J. et al. (2018). Managing emerging mutagenicity risks: late stage mutagenic impurity control within the Atovaquone second generation synthesis. *Regul. Toxicol. Pharmacol. 99*: 22–32. https://doi.org/https://doi.org/10.1016/j.yrtph.2018.08.004.

24 Höllering, R., Gasteiger, J., Steinhauer, L. et al. (2000). Simulation of organic reactions: from the degradation of chemicals to combinatorial synthesis. *J. Chem. Inf. Comput. Sci. 40* (2): 482–494. https://doi.org/10.1021/ci990433p.

25 Socorro, I.M. and Goodman, J.M. (2006). The ROBIA program for predicting organic reactivity. *J. Chem. Inf. Model. 46* (2): 606–614. https://doi.org/10.1021/ci050379e.

26 Socorro, I.M., Taylor, K., and Goodman, J.M. (2005). ROBIA: a reaction prediction program. *Org. Lett. 7* (16): 3541–3544. https://doi.org/10.1021/ol0512738.

27 Gasteiger, J., Bauerschmidt, S., Burkard, U. et al. (2002). Decision support systems for chemical structure representation, reaction modeling, and spectra simulation. *SAR QSAR Environ. Res. 13* (1): 89–110. https://doi.org/10.1080/10629360290002253.

28 Law, J., Zsoldos, Z., Simon, A. et al. (2009). Route designer: a retrosynthetic analysis tool utilizing automated retrosynthetic rule generation. *J. Chem. Inf. Model. 49* (3): 593–602. https://doi.org/10.1021/ci800228y.

29 Huang, Q., Li, L.-L., and Yang, S.-Y. (2011). RASA: a rapid retrosynthesis-based scoring method for the assessment of synthetic accessibility of drug-like molecules. *J. Chem. Inf. Model. 51* (10): 2768–2777. https://doi.org/10.1021/ci100216g.

30 Matthews, E.J., Kruhlak, N.L., Benz, R.D. et al. (2008). Combined use of MC4PC, MDL-QSAR, BioEpisteme, leadscope PDM, and Derek for windows software to achieve high-performance, high-confidence, mode of action-based predictions of chemical carcinogenesis in rodents. *Toxicol. Mech. Methods 18* (2–3): 189–206. https://doi.org/10.1080/15376510701857379.

31 Marchant, C.A., Briggs, K.A., and Long, A. (2008). In silico tools for sharing data and knowledge on toxicity and metabolism: Derek for Windows, Meteor, and Vitic AU – Marchant, Carol A. *Toxicol. Mech. Methods 18* (2–3): 177–187. https://doi.org/10.1080/15376510701857320.

32 Faulkner, D., Rubin Shen, L.K., De La Rosa, V.Y. et al. (2017). Chapter 3 tools for green molecular design to reduce toxicological risk. In: *Computational Systems Pharmacology and Toxicology*, 36–59. The Royal Society of Chemistry https://doi.org/10.1039/9781782623731-00036.

33 Patlewicz, G., Jeliazkova, N., Safford, R.J. et al. (2008). An evaluation of the implementation of the cramer classification scheme in the Toxtree Software. *SAR QSAR Environ. Res. 19* (5–6): 495–524. https://doi.org/10.1080/10629360802083871.

34 Baertschi, S.W. (2011). *Pharmaceutical Stress Testing: Predicting Drug Degradation II Edition*. https://doi.org/10.1080/03639040500529234.

35 Parenty, A.D.C., Button, W.G., and Ott, M.A. (2013). An expert system to predict the forced degradation of organic molecules. *Mol. Pharm. 10* (8): 2962–2974. https://doi.org/10.1021/mp400083h.

36 Carey, J.S., Laffan, D., Thomson, C., and Williams, M.T. (2006). Analysis of the reactions used for the preparation of drug candidate molecules. *Org. Biomol. Chem. 4* (12): 2337–2347. https://doi.org/10.1039/B602413K.

37 Elder, D.P., White, A., Harvey, J. et al. (2015). Mutagenic impurities: precompetitive/competitive collaborative and data sharing initiatives. *Org. Process Res. Dev. 19* (11): 1486–1494. https://doi.org/10.1021/acs.oprd.5b00128.

38 Betori, R.C., Kallemeyn, J.M., and Welch, D.S. (2015). A kinetics-based approach for the assignment of reactivity purge factors. *Org. Process Res. Dev. 19* (11): 1517–1523. https://doi.org/10.1021/acs.oprd.5b00257.

39 Urquhart, M. W. (2020). The mutagenic impurity risk assessment process. Motilin (GPR38 Agonist) Camicinal GSK962040B. https://www.lhasalimited.org/publications/gsk-mi-risk-assessment-process-camicinal-at-al-2020-ich-m7-india-roadshow/6566 (accessed September 2020).

10

N-Nitrosamines

Andrew Teasdale[1], Justin Moser[2], J. Gair Ford[3], and Jason Creasey[4]

[1] *Chemical Development, Pharmaceutical Technology and Development, Operations, AstraZeneca, Macclesfield, UK*
[2] *Global Pharmaceutical Operations Science and Technology, Merck & Co., Inc., West Point, PA, USA*
[3] *CMC Regulatory Affairs, AstraZeneca, Macclesfield, UK*
[4] *Maven E&L Ltd, Stevenage, Herts, UK*

10.1 Background

On 6 June 2018, Zhejiang Huahai Pharmaceuticals was informed by a customer of an unexpected impurity in the manufacturer's VALSARTAN API. After an initial investigation, on 20 June 2018, Zhejiang Huahai sent a letter to its customers informing them of the presence of "a previously unknown impurity that may have genotoxic potential." Zhejiang requested that they immediately put on hold the use of its valsartan API. Shortly after Zhejiang Huahai contacted its customers again, stating that the impurity in question was *N*-Nitrosodimethylamine (NDMA) and that this was likely to be "process related." Little did anyone at the time envisage the events that would unfold as a consequence of this incident. Given the seriousness of this incident, the fact that NDMA belongs to the "cohort of concern" referred to in ICH M7 [1], it led in July 2018 to the EMA reporting [2] that investigations into this contamination had been instigated with the result that national authorities in Europe had initiated the recall of products containing the active Valsartan. Almost immediately the FDA [3, 4] and other authorities followed suit.

Initially, neither the root cause nor the specific levels were reported; however, this action set off a chain of events that led to similar notifications of recall, across multiple authorities/regions and several Sartan products. Moreover, the announcement of the discovery of NDMA was quickly followed by reports of the detection of another *N*-Nitrosamine – *N*-Nitrosodiethylamine (NDEA). See Table 10.1 for overview of events (Figure 10.1).

These events understandably triggered many authorities to launch formal investigations, including the European Commission (EC). The EC thus initiated a procedure in July 2018 [5] pursuant to Article 31 of Directive 2001/83/EC. Since this point there have continued to be reports of further concerns relating to other *N*-Nitrosamines. These events, potentially questioning the quality of effectively an entire class of drugs, i.e. Sartans, were a major concern; in particular, the continued identification of new contamination issues. The perception was that current controls and assessment processes have failed to address these potential quality concerns.

A chronology of these events and others relating to *N*-Nitrosamines are illustrated in Figure 10.2.

The chemistry of both formation and potential removal of *N*-Nitrosamines is summarized below, a more detailed description can be found in Chapter 11.

Mutagenic Impurities: Strategies for Identification and Control, First Edition. Edited by Andrew Teasdale.
© 2022 John Wiley & Sons, Inc. Published 2022 by John Wiley & Sons, Inc.

Table 10.1 Initial events relating to Sartan contamination in mid-2018.

Date	Authority
3 August 2018	The Taiwan Food and Drug Administration alerted regulators worldwide of the discovery of NDMA in valsartan APIs manufactured by two other companies, Zhejiang Tianyu and Zhuhai Rundu Pharma.
30 August 2018	Zhejiang Huahai confirmed the presence of a second *N*-Nitrosamine, *N*-Nitrosodiethylamine (NDEA), in some batches of its valsartan API.
14 September 2018	BfARM reported trace amounts of NDEA, in another Sartan, Losartan, from Hetero Labs.
17 September 2018	EDQM reported traces of NDEA in Irbesartan from another API manufacturer, Aurobindo Pharma Limited.

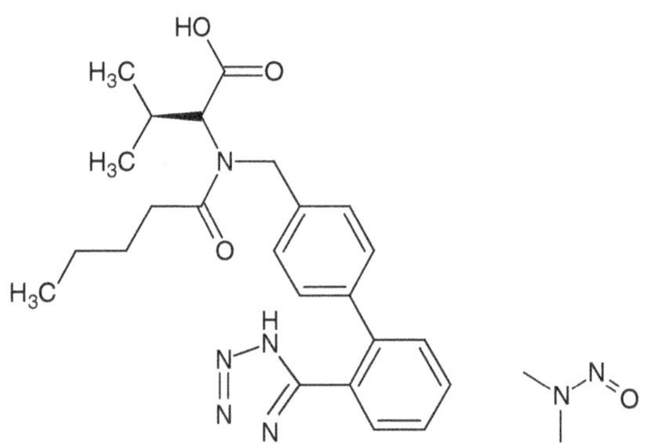

Figure 10.1 Valsartan and *N*-Nitrosodimethylamine (NDMA).

10.2 Generation of *N*-Nitrosamines

Formation of *N*-nitroso impurities require two components to react together: a nitrosating agent (e.g. NO^+ derived from nitrite) and a secondary amine (Figure 10.3). Primary amines can also nitrosate but the nitrosamine formed is not stable and tautomerizes to form a diazo hydroxide, which decomposes rapidly to form a reactive alkyl diazonium ion.

The critical aspect in understanding the manufacturing risk inherent in the manufacture of certain Sartans is the realization that sodium nitrite (as nitrous acid) is a reagent routinely used to quench azides and azides are used in the generation of the tetrazole ring common within the Sartan family of active substances.

In the case of the reported Sartan contaminations, it was not obvious at first assessment of the process how these specific components came together. Even though NDMA was found as the original issue in some valsartan supplies, the root cause was not immediately clear as dimethylamine (DMA) was not used in the manufacture of Valsartan. However, it was later reported that a change in the valsartan process (see Figure 10.4) was the root cause. Introduced to improve process efficiency, the change, an approved variation, involved the replacement of the azide reagent used to form the tetrazole ring, replacing tributyl tin azide with the more reactive and less toxic sodium azide. The use of sodium azide required the use of a quenching agent, this being sodium nitrite, Figure 10.5. The change also saw a change in solvent from Xylene to dimethylformamide (DMF).

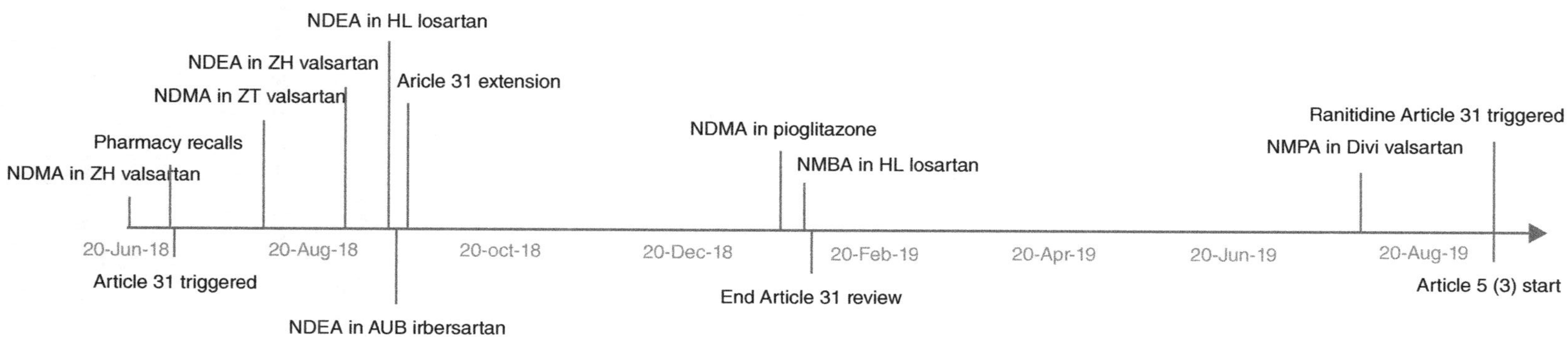

Figure 10.2 Timeline of events relating to *N*-Nitrosamine contamination of pharmaceuticals – June 2018–August 2019.

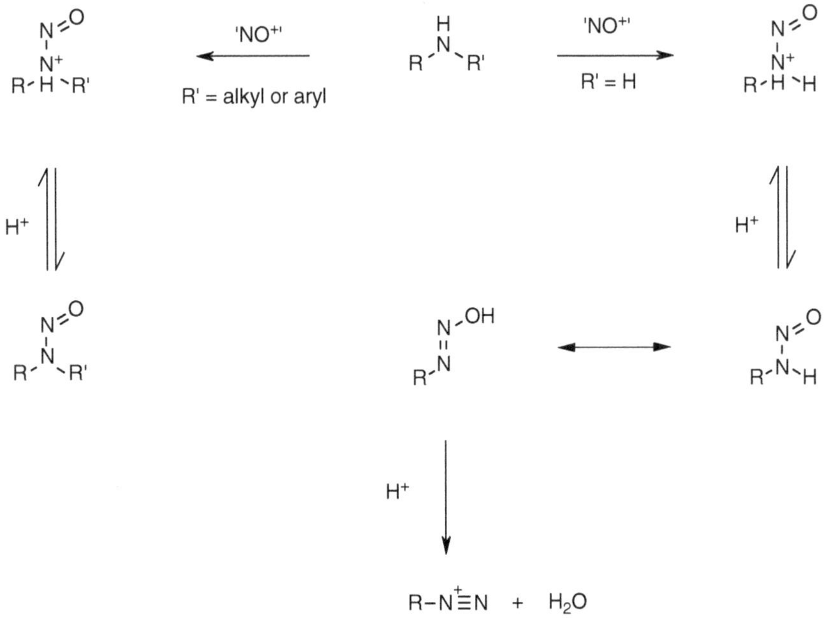

Figure 10.3 Comparison of the nitrosation of secondary and primary amines.

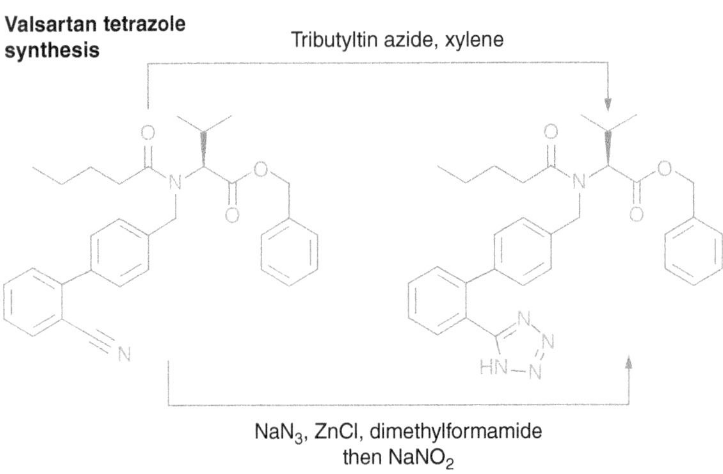

Figure 10.4 Comparison of new revised process, based on use of sodium azide to original process based on use of tributyl tin azide.

$NaN_3 + NaNO_2 + 2\ H+ \rightarrow N_2 + N_2O + H_2O$

Figure 10.5 Quenching of sodium azide using sodium nitrite.

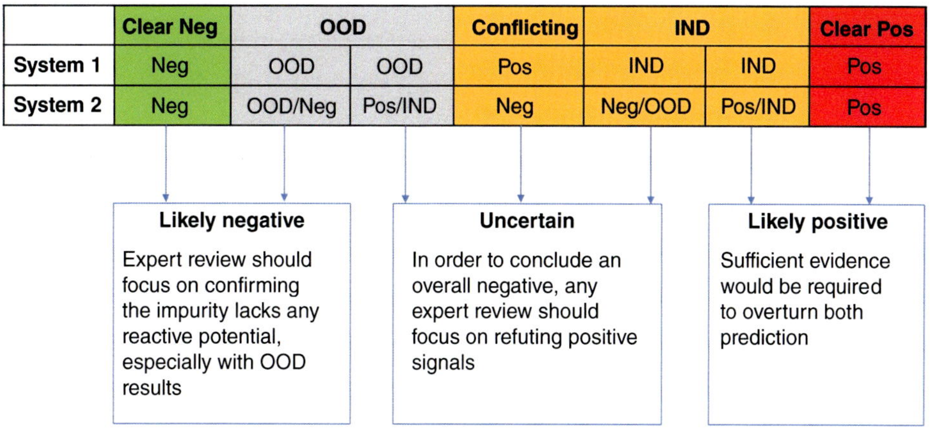

Figure 4.12 Resolving results from different (Q)SAR result. Neg – negative, OOD – out-of-domain, Pos – positive, IND – indeterminate.

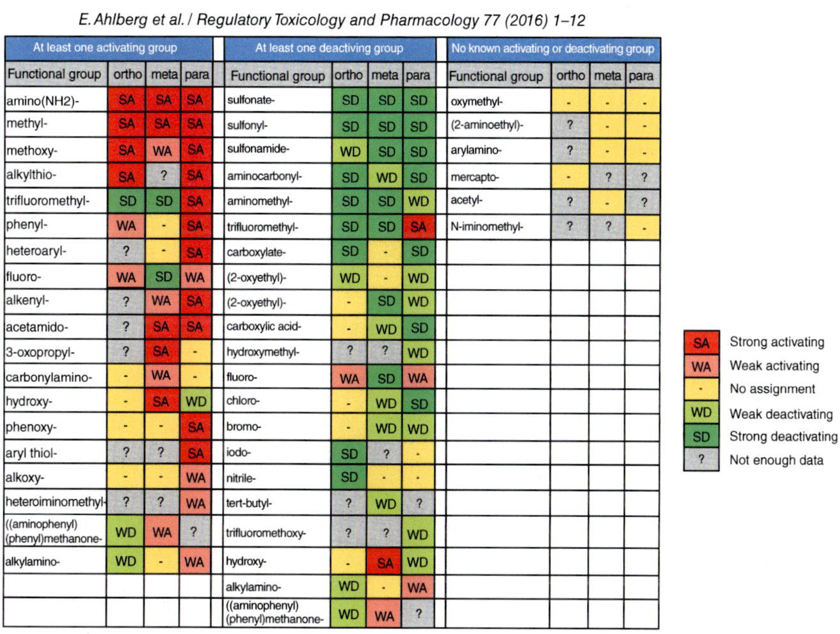

E. Ahlberg et al. / Regulatory Toxicology and Pharmacology 77 (2016) 1–12

At least one activating group				At least one deactiving group				No known activating or deactivating group			
Functional group	ortho	meta	para	Functional group	ortho	meta	para	Functional group	ortho	meta	para
amino(NH2)-	SA	SA	SA	sulfonate-	SD	SD	SD	oxymethyl-	-	-	-
methyl-	SA	SA	SA	sulfonyl-	SD	SD	SD	(2-aminoethyl)-	?	-	-
methoxy-	SA	WA	SA	sulfonamide-	WD	SD	SD	arylamino-	?	-	-
alkylthio-	SA	?	SA	aminocarbonyl-	SD	WD	SD	mercapto-	-	?	?
trifluoromethyl-	SD	SD	SA	aminomethyl-	SD	SD	WD	acetyl-	?	-	?
phenyl-	WA	-	SA	trifluoromethyl-	SD	SD	SA	N-iminomethyl-	?	?	-
heteroaryl-	?	-	SA	carboxylate-	SD	-	SD				
fluoro-	WA	SD	WA	(2-oxyethyl)-	WD	-	WD				
alkenyl-	?	WA	SA	(2-oxyethyl)-	-	SD	WD				
acetamido-	?	SA	SA	carboxylic acid-	-	WD	SD				
3-oxopropyl-	?	SA	-	hydroxymethyl-	?	?	WD				
carbonylamino-	-	WA	-	fluoro-	WA	SD	WA				
hydroxy-	-	SA	WD	chloro-	-	WD	SD				
phenoxy-	-	-	SA	bromo-	-	WD	WD				
aryl thiol-	?	?	SA	iodo-	SD	?	-				
alkoxy-	-	-	WA	nitrile-	SD	-	-				
heteroiminomethyl-	?	?	WA	tert-butyl-	?	WD	?				
((aminophenyl)(phenyl)methanone-	WD	WA	?	trifluoromethoxy-	?	?	WD				
alkylamino-	WD	-	WA	hydroxy-	-	SA	WD				
				alkylamino-	WD	-	WA				
				((aminophenyl)(phenyl)methanone-	WD	WA	?				

SA Strong activating
WA Weak activating
\- No assignment
WD Weak deactivating
SD Strong deactivating
? Not enough data

Figure 5.4 Substitution pattern analysis illustrating activating and deactivating aniline substitution patterns.

Mutagenic Impurities: Strategies for Identification and Control, First Edition. Edited by Andrew Teasdale.
© 2022 John Wiley & Sons, Inc. Published 2022 by John Wiley & Sons, Inc.

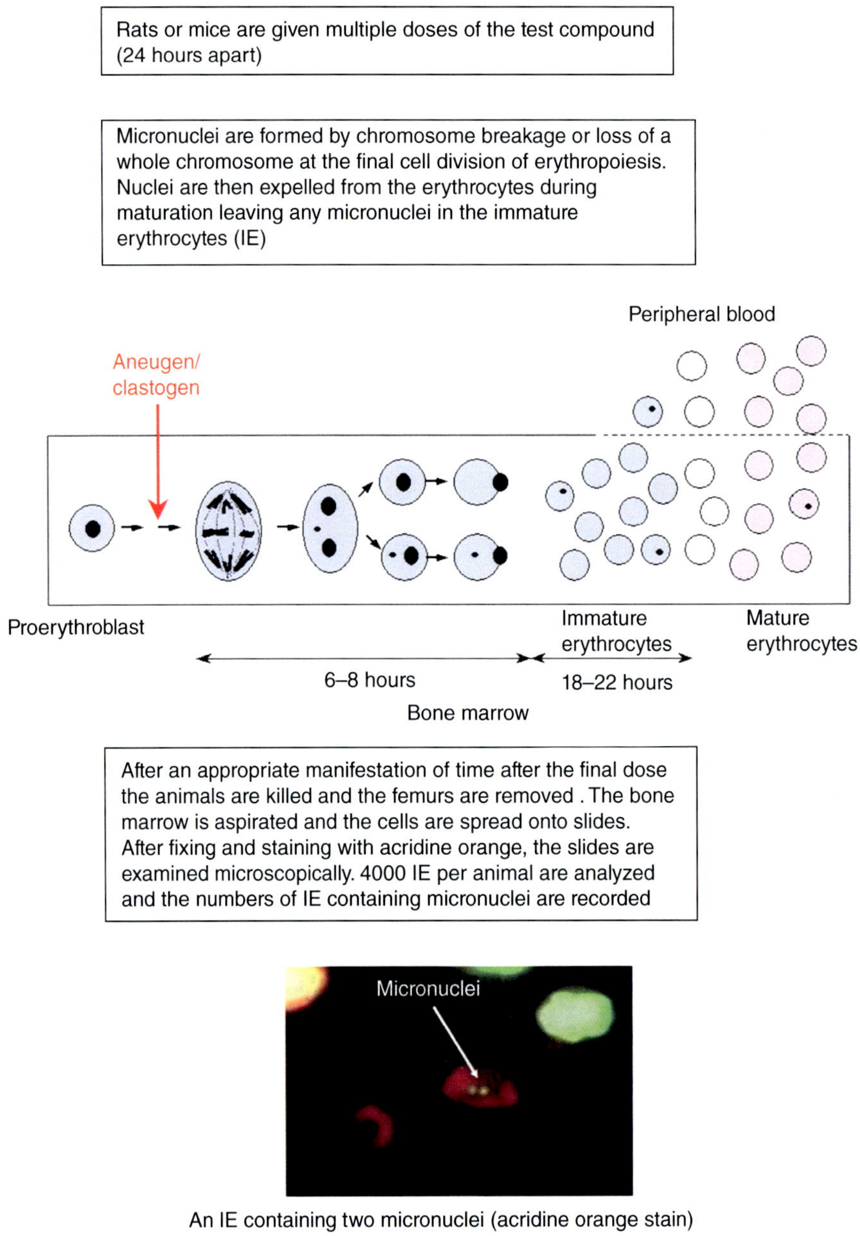

Figure 6.5 The rodent bone marrow micronucleus test.

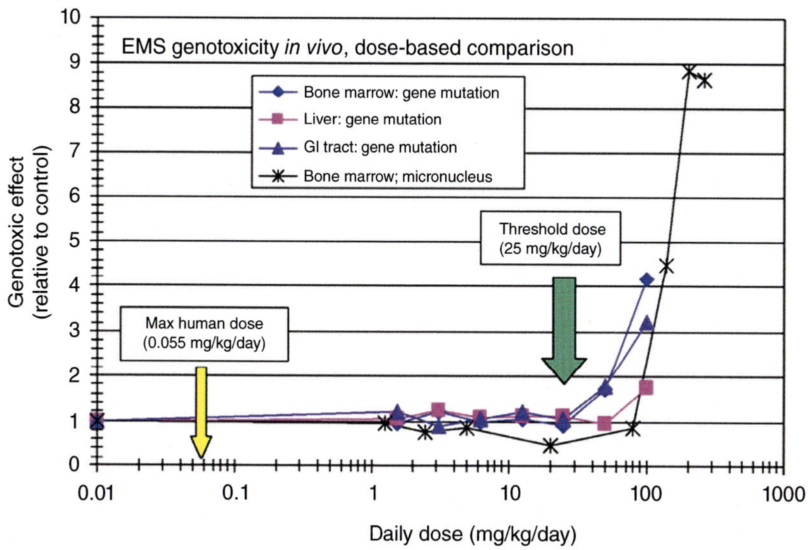

Figure 8.3 EMS-induced thresholded dose response *in vivo*. *Source*: Data taken from Gocke et al. [33] and includes both chromosome breakage data and LacZ mutation data. Here the data are plotted on a log-linear graph to allow better visualization of the data. In the publication by Gocke et al. [33], linear–linear graphs are also plotted.

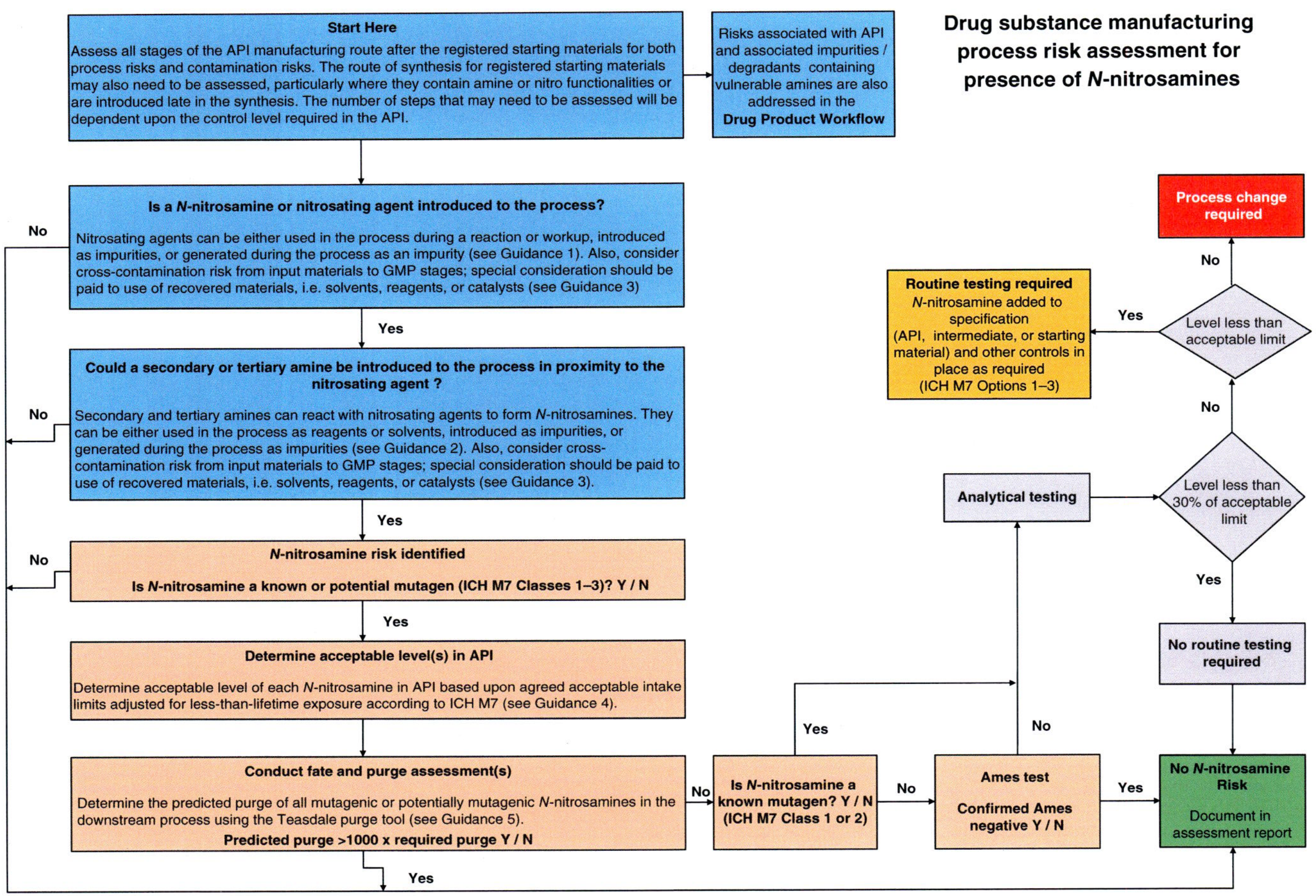

Figure 10.10 EFPIA drug substance risk assessment decision tree.

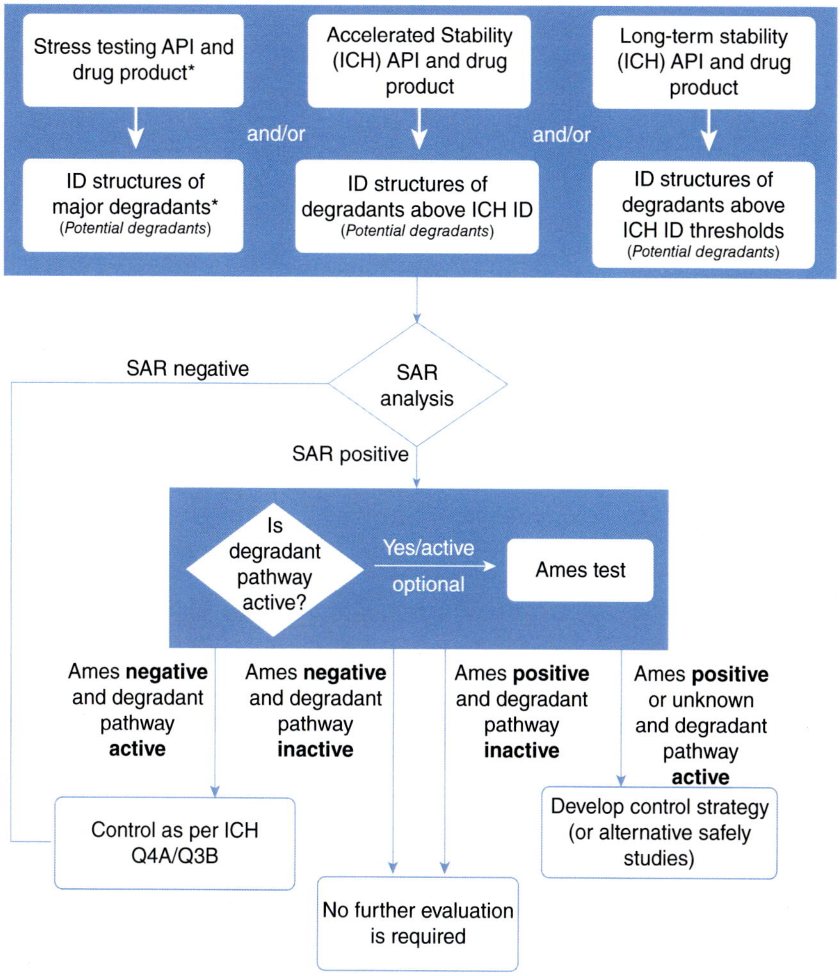

Figure 14.2 Proposed process flow for assessing degradation products in the DS and DP.

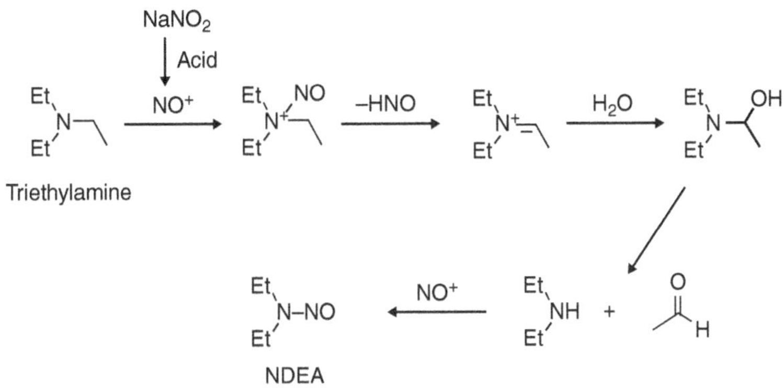

Figure 10.6 Generation of *N*-Nitrosodiethylamine (NDEA) from Triethylamine. NO^+ has been used to represent the nitrosating agent; however, the actual species depends upon the conditions and is likely to be N_2O_3 under quench conditions where the nitrite concentration is quite high.

Under acidic conditions, DMF will hydrolyze to yield DMA (and formic acid), and DMA is also a known impurity in DMF. Hence, the source of the two required agents, the nitrite and the secondary amine, is now clear. It should though be noted in terms of obviousness, the root cause is an impurity/degradant (DMA) of an impurity (DMF).

The mechanistic understanding of the reaction between a secondary amine and a nitrosating agent also serves to potentially explain the observation of NDEA, in some Sartan products. In this case, the secondary amine present is diethylamine, and the most likely source of this is as an impurity in the triethylamine used in the earlier synthetic steps. Triethylamine can also undergo nitrosative dealkylation to give diethylamine via the mechanism shown below (Figure 10.6).

With this understanding, it becomes obvious that such risks emanate from the presence of a secondary or tertiary amine within the reaction system, allied to the presence of a nitrosating agent. For example, in the case of the reported presence of *N*-Nitroso-methyl-4-aminobutyric acid (NMBA) in Losartan, it was generated due to the formation of *N*-methyl-4-aminobutyric acid during solvent recycling due to hydrolysis of the *N*-methyl-2-pyrrolidinone (NMP) solvent and subsequent reaction of the freed secondary amine with nitrite.

The result of the combination of these concerns triggered regulatory action from authorities, this took the form of direct, specific product-related requests to marketing authorization holders (MAHs) and in Europe centralized action via the Article 31 process.

10.3 Article 31

Pursuant to Article 31 of Directive 2001/83/EC, the European Commission initiated a procedure on 5 July 2018 [5]. The investigation has now been finalized and the report was submitted to the European Commission; this report contained as series of recommendations. It is important to note that these were specifically intended to apply to Sartans. The full outcome and implications of this are examined later in Section 10.6.2.

At the time of initiation, there were significant concerns over what was being requested, as this marked a significant shift away from the principles outlined within ICH M7 [1]. The guideline defines the need to control mutagenic impurities below the threshold of toxicological concern, or in the case where specific carcinogenicity data are available, limits should be based on the principles outlined in note 4. It should be noted that the guideline also defines a cohort of concern, this relating to classes of impurities where their apparent potency is such that the TTC itself cannot be applied. The carcinogenicity of *N*-Nitrosamines is examined in detail in Chapter 7. Nowhere though in ICH M7 [1] is it stated that other principles enshrined in ICH M7 such as linear extrapolation of carcinogenicity data is discounted when addressing compounds such as *N*-Nitrosamines belonging to the cohort of concern, indeed the interim limits described in the Article 31 [5] are based on that exact principle. The divergence comes after the two-year period where limits were proposed to shift to a blanket limit of 30 ppb. The exact reason for this is not entirely clear. It may be supposed that the failure of some Sartan suppliers to manage this risk appropriately by suitable risk assessment, process design, and analytical control approaches has led to this imposition of "avoidance" rather than "control" based approaches to their manufacture. It is interesting to reflect on the origins of guidance pertaining to mutagenic impurities, this being examined in detail in Chapter 1. Avoidance principles were an integral part of early drafts of regulatory guidance before the development of ICH M7 but were ultimately replaced by risk assessment and TTC-based controls in ICH M7. A recurring, but in the view of the authors an incorrect view, is that ICH M7 failed to address issues with Sartans contamination and that this apparent oversight resulted in the failure to set suitable controls on quality. As a consequence of this misplaced perspective, it has often led to a view that total avoidance of risk is required to regulate quality. What remains unclear is whether the blanket limit associated with Article 31 remains valid in relation to Sartans or is superseded by later guidance, particularly the Revised Q&A process issued in August 2020 [6]. Informal communication from EDQM strongly indicates that revision of limits and alignment with Article 5(3) is very likely.

It is important though to note that the development of Sartans predates even the earliest of guidance relating to mutagenic impurities and thus it seems incongruous to expect that drugs developed before the establishment of the guidance should comply with it.

So what would the Sartans situation look like through the lens of ICH M7? It is important to understand that the risk in terms of nitrosamine levels in Sartan drug substance depend on a series of factors. These include:

- Proximity of the tetrazole ring formation to the final API in the synthetic route. *N*-Nitrosamines formed earlier in a synthesis are more likely to be purged through factors such as reactivity, solubility, etc., than those formed at a late stage in the synthesis. The more processing steps between the point of introduction and the final API the lower the risk of impurity carryover in general terms. This correlates with no detectable levels of NDMA or NDEA being reported in Candesartan, Losartan, or Olmesartan by the CHMP (Overall rapporteur's preliminary assessment report [7]).
- Proximity and concentration of reactants. *N*-Nitrosamines can only be formed if secondary amines and nitrite are present simultaneously and under particular conditions, thus the risk can be eliminated by ensuring that this is avoided.
- Reagent quality. As described, DMF and triethylamine may contain secondary amines, DMA, and diethylamine, respectively. Secondary amines can also form in situ, e.g. decomposition of DMF. The use of purified materials will minimize the risk of nitrosamine formation.
- Solvent recycling; *N*-nitroso-*N*-methyl-4-aminobutyric acid (NMBA) has been observed in Losartan, this is likely to have been generated from the formation of *N*-methyl-4-aminobutyric acid during solvent recycling due to hydrolysis of NMP solvent. Again, the control of material quality will minimize the risk of nitrosamine formation.

While the failure to identify a risk as significant as the nitrosamine presence in some Sartans was clearly undesirable, as outlined above, the formation of *N*-Nitrosamines is ultimately understandable. Given this understanding, this risk can be mitigated through an appropriate control strategy without the need for complete redesign of the manufacturing process or revision of ICH M7 [1].

Many of the issues associated with the Sartans could in reality be defined as GMP failures, poor oversight of change control, poor batch record keeping, and poor oversight over the quality of input materials, e.g. recycled solvents and by extension the inspection procedures that oversee these. It should be noted that the changes made to the Valsartan and other Sartan manufacturing processes were formally approved and thus this also points to issues with the approval process for variations, perhaps highlighting deficiencies in terms of technical oversight.

10.4 Further Issues – Cross Contamination and Ranitidine

Unfortunately, the issue of *N*-Nitrosamines has continued to bedevil the industry. Initially, it was thought to be limited to the synthesis of Sartans, through the use of sodium nitrite as a quenching agent in the manufacture of tetrazole ring systems. The issue has become further complicated by the discovery of certain *N*-Nitrosamines in some Sartans that do not correlate with the specific route and process employed in the manufacture of the Sartan concerned. Matters have been complicated still further by the discovery of detectable levels of *N*-Nitrosamines in non Sartans, specifically Pioglitazone where the European Medicines agency reported detection of NDMA [8]. The situation became even more complex when it was reported in September 2019 that detectable levels of NDMA were seen in Ranitidine [9, 10]. This led to widespread withdrawal of Ranitidine products globally and submission of a citizen's petition [11] calling for the removal of all Ranitidine products from the US market. Like many issues relating to *N*-Nitrosamines, the exact reason for the presence of *N*-Nitrosamines, specifically NDMA within Ranitidine at the time of writing still remain somewhat unclear; however, it is believed that there is a correlation between crystal form of the active in the solid phase and levels of NDMA observed [12].

The public citizen's petition proposed suspension of all medicines containing Ranitidine, claiming that hundreds of thousands of ppb had been seen using an FDA method; however, this was later challenged by the FDA who recommended that drug makers follow its lead and use a low-heat method of testing the antacids and not the high-temperature method used by a "third-party laboratory," which it stated *actually generates NDMA* and as a result misleadingly exhibits much greater levels of the impurity. FDA later reported the results of its own investigations [13], commenting that the levels of NDMA in ranitidine and nizatidine are similar to the levels you would expect to be exposed to if you ate common foods like grilled or smoked meats. Despite this, as noted above, the FDA eventually in April 2020 issued notice of the withdrawal of all Ranitidine-based products [14].

This example, particularly the results reported by a third-party laboratory, highlights some of the challenges associated with the trace analysis of *N*-Nitrosamines. Further issues relating to analysis of N Nitrosamines are examined in depth in Chapter 13.

As a consequence of the discovery of NDMA in Ranitidine, and also of reports of trace level *N*-Nitrosamine contamination of Pioglitazone, the European Medicines Agency issued through EMA's human medicines committee (CHMP) a request that as a matter of precaution that MAHs for human medicines containing chemically synthesized active substances review ALL their synthetic medicines for the possible presence of nitrosamines and test all products at risk; this was instigated under the banner of the Article 5(3) process [15]. This stated that if nitrosamines were detected in any of their medicines, MAHs must inform authorities promptly so that appropriate regulatory actions can be taken.

A notice to this effect was posted by the EMA in September 2019 providing MAHs with information on the actions they should take. To support this, the EMA also issued a questions-and-answers document, available on EMA's website; as described below, this guidance has continued to evolve. The whole process including advice on reporting templates and reporting mechanisms was also published [16]. This process and reported outcomes is examined in depth later in the chapter, see Section 10.6.1.

10.4.1 Article 5(3) and Associated Q&A Document

The EMA press release stated that MAHs are responsible for ensuring that *every batch* of their finished product is of satisfactory quality, including the active substances and other ingredients used to make them, e.g. excipients, raw materials including solvents, water, etc.

Companies were specifically advised to take the following steps:

Evaluate possibility of nitrosamines being present in every concerned medicine within six months.

1) Prioritize evaluations, starting with medicines more likely to be at risk of containing nitrosamines.
2) Take into account findings from CHMP's review of Sartans.
3) Notify authorities of outcome of risk evaluations.
4) Test products at risk of containing any nitrosamines.
5) Immediately report detection of nitrosamines to authorities.
6) Apply for necessary changes to marketing authorizations to address nitrosamine risk.
7) Complete all steps within three years, prioritizing high-risk products.

This is split into 3 steps:

Step 1 Risk evaluation: MAHs should perform risk evaluation of their medicinal products containing chemically synthesized API. MAHs were asked to prioritize products in order to establish the sequence in which their products are to be evaluated, based on the principles defined within ICH M7 [1] and ICH Q9 [17]. To assist in this process, further guidance was provided within a Q&A document [6]. It was made clear that for products identified as high priority, the risk evaluation should be done immediately.

This initially stipulated that the risk evaluation of all products should be concluded at the latest within six months of the publication of the notification, the deadline initially being 26 March 2020. MAHs were required to inform the concerned Competent Authorities when the risk evaluation was concluded. It also stated that if at the end of Step 1 a risk of presence of nitrosamines was identified as a result of the evaluation, the MAH should proceed to Step 2.

One of the challenges in addressing the requirements outlined is *what is a risk* and *when would Step 2 be instigated*? Risk factors are examined below in detail; however, the boundary between risk assessment and any testing done in the context of this and formal Step 2 testing has remained somewhat unclear, as has the definition of "no risk" and how this fits into a risk-based approach. A binary approach of risk/no risk is too simplistic and can lead to the compounding of a series of highly improbable scenarios where in reality this is simply not feasible.

Step 2 If the risk evaluation conducted in step 1 identified a risk of the presence of nitrosamines, the EMA document stated that confirmatory testing should be carried out and any changes this indicated were needed should be concluded at the latest within three years of the publication of the notification.

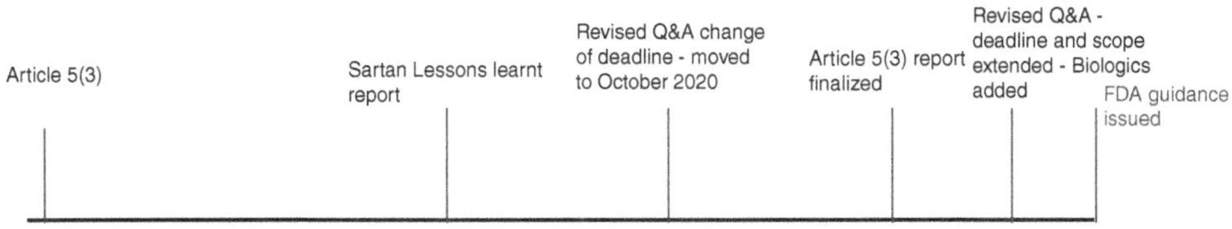

Figure 10.7 Chronology of events from September 2019 to September 2020.

It also stated that MAHs should inform the competent authorities immediately if tests confirm the presence of a nitrosamine impurity irrespective of the amount detected.

Step 3 Updating of registered control strategy. This relates to the implementation of changes within the manufacturing procedure, should a risk be identified in Step 1 and confirmed in Step 2.

Similar requests were made at a similar time by Health Canada and other agencies; however, the scope and timelines were somewhat different. Within the Health Canada request, there was more emphasis on the drug product-related risk. Even more significant though was that the timeline for the health Canada request [18] is to complete Step 2 within two years, not three years as defined by the EMA.

Another authority to publish specific guidance was Swiss Medic [19]. Again, there were differences compared with the EMA request. The two most significant are Swissmedic defined a limit of 30 ppb LOQ, unlike EMA where no limits were originally defined, and again like Health Canada the timeline for Step 2 was set at two years.

Following the issuance of the Article 5(3) [16] request and the initial Question and Answer document, there has been continued evolution of the guidance, scope, and timing of requirements, see Figure 10.7. This is illustrated below and examined in depth in Section 10.6.

The impact of the changes over the period concerned has been significant, leading to a feeling of trying to hit a moving target in terms of what is required. Significant changes include:

- Publication of revised limits, culminating in the most recent version of the EMA Q&A document where a default limit in the absence of specific safety data for an *N*-Nitrosamine of 18 μg/day has been specified.
- Timelines, initially six months, this has been ultimately extended in Europe and a number of other authorities to 18 months for completion of Step 1. It should be noted though the FDA guideline, published in September 2020, gives a deadline of six months, there being no pre-existing requirements in the United States. Even then the deadlines for Europe and the United States are misaligned, the United States requiring completion by the beginning of March 2021, Europe 31 March.
- Change of Scope, Europe, in the Article 5(3) report and latest Q&A document, the scope was extended to include Biologics, despite recognizing the very low risk. This is examined later this chapter (Section 10.5.6).
- Another change in scope, or at the very least, emphasis is the highlighting of additional risk factors other than those described above pertaining specifically to the chemistry. These include:
  - Formulation risks – this relating to the presence of nitrites within certain excipients and the potential for reaction within the formulated product during manufacture and subsequent packaging and storage where there are traces of secondary amines present.
  - Related to the above is the risk of contamination during packaging, specifically the use of Nitrocellulose-based lidding foils.
  - And the alleged ubiquitous risk posed by water and the presence of nitrites within it.

Of course, despite these challenges, risk assessments must be performed and this process is now examined.

10.5 How to Assess the Risk Posed in Pharmaceuticals

Clearly, there is need to examine the general risk across all current medicines but how should this be done in practice? Part of the issue is the lack of facts and the separation of these from hypotheses. Outlined below is an overview of the current risk factors and the extent of knowledge at the present time.

10.5.1 Drug Substance

Certainly, one irrefutable fact is that Nitrosating agents will readily react with secondary amines. Multiple *N*-Nitrosamines have been reported; this is not a surprise as any secondary amine, present in a reaction system with a Nitrosating agent will form an *N*-Nitrosamine. This reaction is well known and well understood even if much of this understanding was gathered several decades ago. Importantly armed with this knowledge it is possible to predict the levels of *N*-Nitrosamines formed under some conditions. The chemistry of *N*-Nitrosamine formation is studied in detail in Chapter 11.

10.5.1.1 Where do Nitrites Come Within Drug Substance Come From?

The most obvious source is the use of Sodium nitrite to quench sodium azide as in the formation of tetrazole ring systems. Sodium nitrite is also used in the generation of a diazonium intermediate in the formation of other ring systems, such as triazoles. In such scenarios, control can be potentially exercised several ways:

- Avoid the presence of any secondary or tertiary amine in the process in stages where nitrite is used or may be present.
- Remove any nitrite in stages following its deliberate use. The high aqueous solubility of nitrite makes this relatively straightforward of nitrite itself; however, the removal of other nitrosating species should be carefully considered. The use of nitrous acid scavengers may ameliorate this.
- Destroy any *N*-Nitrosamine formed. At present, the understanding of such processes is limited and this is a key area for further examination, but it offers the real potential to deliberately eliminate traces of *N*-Nitrosamines [20].

10.5.1.2 What Other Sources Are There?

10.5.1.2.1 Nitrite in Water, NOx in Nitrogen

Pertaining to water, a crucial question is, *at the levels present in water is this an issue*? As outlined, the kinetics of such systems are well understood and can be modeled. One relevant scenario is the use of drinking water that just meets the WHO guideline acceptable limit (NMT 3 mg/l ~6.5×10^{-5} M nitrite) in a process containing traces of a secondary amine such as dimethylamine.

Such risks are examined in a detailed review by Ashworth et al. [21] Both the mechanistic and kinetic aspects of amine reactivity were used as the basis of an assessment of the risk that traces of nitrite in the water used during active pharmaceutical ingredient (API) manufacturing could give rise to significant levels of *N*-Nitrosamines. It was concluded that the levels of nitrite typically found in potable water used for API manufacture are very low (<0.01 mg/l) and will not give rise to significant levels of *N*-Nitrosamines through reaction with basic secondary amines ($pK_a > 9.5$) in the majority of cases. However, the presence of less basic amines ($pK_a < 9.5$), elevated processing temperatures, or low pH conditions in combination with significantly higher levels of nitrite have the potential to generate levels of *N*-Nitrosamines that could lead to significant quantities being present in the isolated API if the downstream processing does not provide an adequate purge. A key point is that the kinetic models described may be used to risk assess specific situations or processes. It also provides an evaluation of the risk posed by tertiary alkylamines. These can nitrosate via a dealkylative process, which is significantly slower than secondary amine nitrosation. This concludes they do not represent a risk of *N*-Nitrosamine formation under conditions where there is no significant risk of secondary amine nitrosation. This and other risk factors in terms of formation are examined in detail in Chapter 11.

However, the situation regarding waste water is more complex. It is evident that oxidation of unsymmetrical dimethylhydrazine (UDMH) (high volatility at 63 °C, also very polar and miscible in water link) to NDMA is a known pathway and happens even in the presence of air. The most likely channel to UDMH is again through degradation of DMF to DMA, and then amination (ammonium hydroxide, NaOCl), see Figure 10.8, with an electrophilic source of ammonia, e.g. monochloramine NH_2Cl.

The specific risk is associated with amidic aprotic solvents such as DMF and their combination with chlorinating or oxychlorinating media.

The factors leading to generation of *N*-Nitrosamines is discussed in detail in Chapter 11.

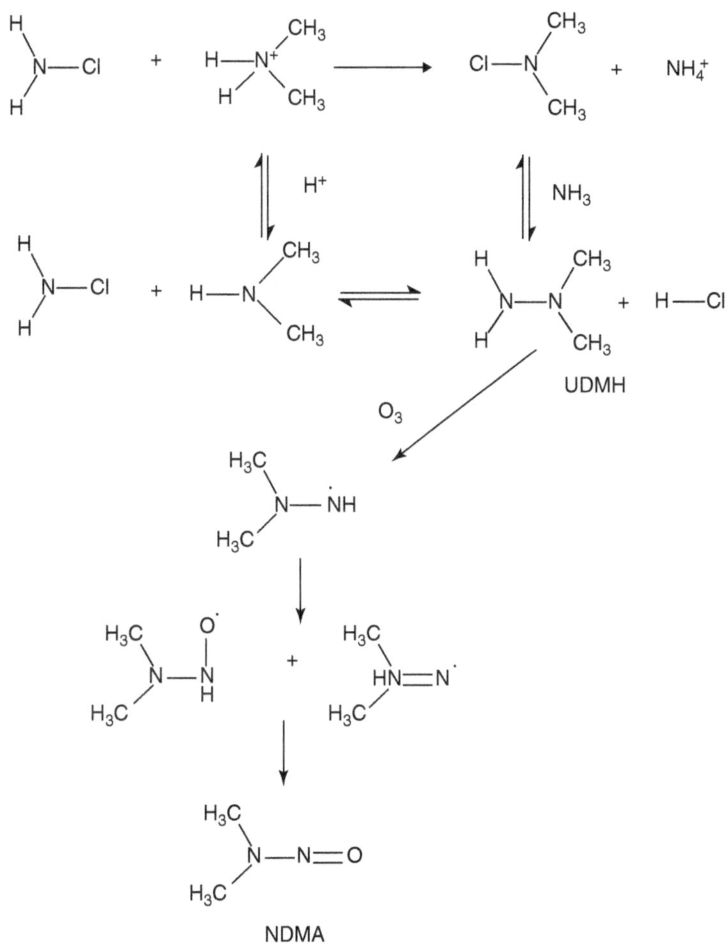

Figure 10.8 Postulated reaction scheme for NDMA formation via UDMH. [22]

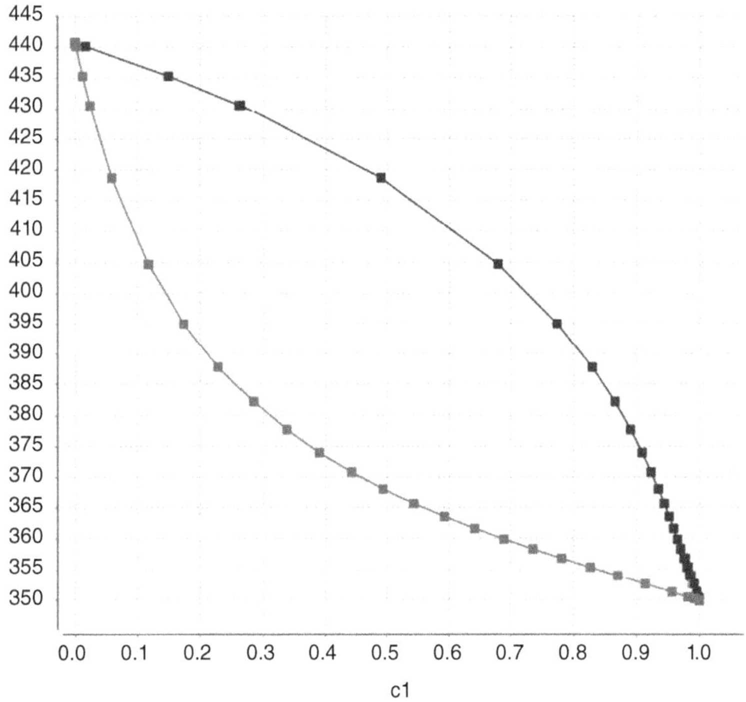

Figure 10.9 Binary VLE diagrams (constant pressure at 1 atm). Mass fraction solvent (EtOAc) on *X* axis; temperature (Kelvin) on *Y* axis – EtOAc and dimethyl nitrosamine.

10.5.1.3 Other Factors Associated with Drug Substance Synthesis

10.5.1.3.1 Extrinsic Contamination

As described in the Sartans lessons learnt report and above in this chapter, several Sartans were reported to contain detectable levels of *N*-Nitrosamines that, based on the specific chemistry, would not have been expected to contain them. *So why were they there?* The finger clearly points to cross contamination but *arising how and where?* are critical questions. There are obvious reasons why this is a concern, both in terms of the associated safety risk and the apparent lack of GMP. The extent to which this is a genuine risk across the industry is not clear.

Several materials have been implicated, which include solvents and reagents. In the case of solvents, specific concerns relate to recycled solvents. Again, there is a vital need to understand the true nature of such issues. Detailed modeling of the distillation of common solvents, e.g. ethyl acetate, shows no evidence of an azeotrope and that when contaminated with NDMA that a correctly operated distillation would in fact remove the NDMA (this remaining in the residue after distilling off and recycling the Ethyl acetate), see Figure 10.9.

How does this then correlate with the circumstantial evidence available at present around solvent recycling? Although not proven, there appears to be a strong causal link between contaminated Sartans and serious GMP failures at both the manufacturers and their solvent recyclers.

10.5.2 Process to Assess Drug Substance-Related Risk

In order to address drug substance-related risk, it is important to have a systematic process. An attempt was made through a cross-industry collaboration to define such a process in the form of a decision tree, Figure 10.10.

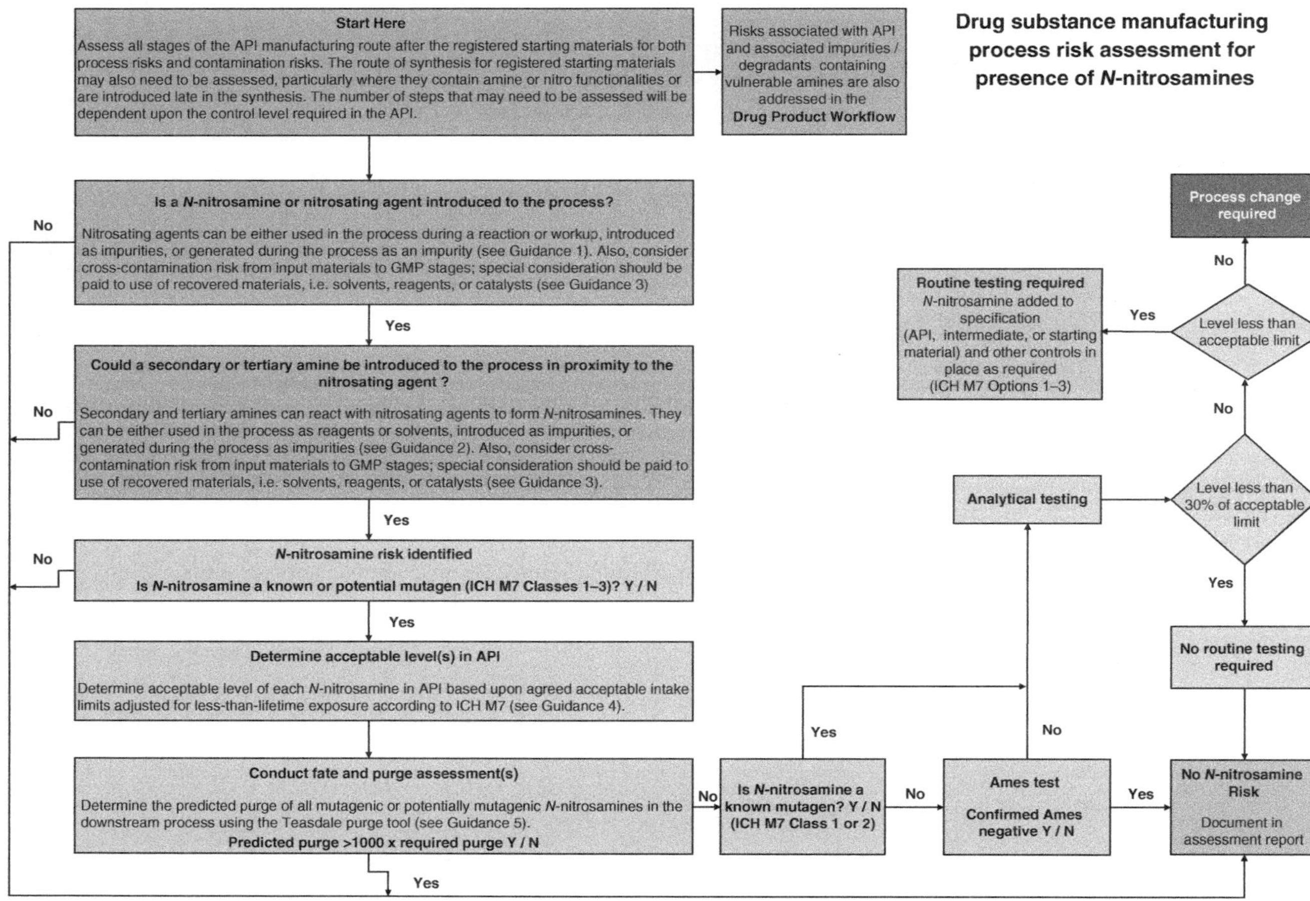

Figure 10.10 EFPIA drug substance risk assessment decision tree. (*See insert for color representation of this figure*).

Also critical to such a process are overall quality aspects; adherence to GMP, presence of an appropriate quality system aligned to principles of Q10 [23] audit performance, management of change, etc. Factors clearly lack in several of the reported cases of *N*-Nitrosamine contamination. Often, API manufacture is carried out by an external contractor, requirements for this are examined in the Q and A document [6], specifically question 16, reproduced:

Q16: What are the responsibilities of MAHs for APIs with CEPs or ASMFs?

MAHs/Applicants, manufacturing authorization holders and API manufacturers should work together and take precautionary measures to mitigate the risk of presence of nitrosamines during the manufacture and storage of all medicinal products containing chemically synthesized APIs.

MAHs/Applicants must ensure that appropriate and robust risk evaluations are carried out by the relevant manufacturing authorization holders and API manufacturers (including ASMF or CEP holders) in accordance with Article 46 of Directive 2001/83/EC.

This is often achieved in practice through the use of detailed questionnaires sent by the MAH to the API manufacturer.

Arguably the most significant risk factor is that of recycled solvents, an approach to this is highlighted below in Figure 10.11.

10.5.3 Drug Product-Related Risk

10.5.3.1 Related Risks of Contamination and Formation in Drug Products

Another potential source of *N*-Nitrosamines is their formation during the drug product formulation manufacturing process and/or during product shelf life. This potential risk arises from the presence of a nitrosating agent in the drug product formulation along with presence of secondary amines or sources that can react to yield secondary amines. Nitrosating agents typically originate from nitrites present in the drug product formulation. The manufacturing processes involved in production of drug products aim at creating well-mixed blends and final dosage forms which could enable nitrosating agents and secondary amines to react and form *N*-Nitrosamines. In addition, these manufacturing processes induce stresses (e.g. temperature, moisture, mechanical forces) that can promote reaction and/or influence the kinetics of *N*-Nitrosamine formation. *N*-Nitrosamine chemistry in solution state is well developed for which the mechanisms and reaction chemistry kinetics can be leveraged to help in drug product risk assessments [21]. Published kinetic models can be utilized to estimate formation risk in drug product solution manufacturing processes and solution drug products. However, much less is known and documented about *N*-Nitrosamine formation in the manufacturing and shelf life of solid drug products. This section therefore aims to discuss information available to evaluate *N*-Nitrosamine formation risk in drug products and the associated manufacturing processes as well as discuss what risk factors should be considered. Finally, given the recent emergence of this field of research, a summary of considerations for future research will be provided.

The first step one must take is to determine the sources of *N*-Nitrosamine impurities and predecessor reactants that could enter the drug product manufacturing process. Certainly, a thorough analysis of the drug substance risks is a logical first step. *N*-Nitrosamine impurity risk in the drug substance could have direct impact on the drug product. However, an additional consideration should be made as to whether the drug substance may carry a reactive amine source. There exist various types of amines that can nitrosate to form *N*-Nitrosamines. Secondary and tertiary amines can react to form *N*-Nitrosamines. Additionally, even quaternary ammonium ions have been reported to nitrosate under forcing conditions [21]. However, it is known that secondary amines pose the highest risk of reaction with nitrosating agents to form *N*-Nitrosamines and therefore

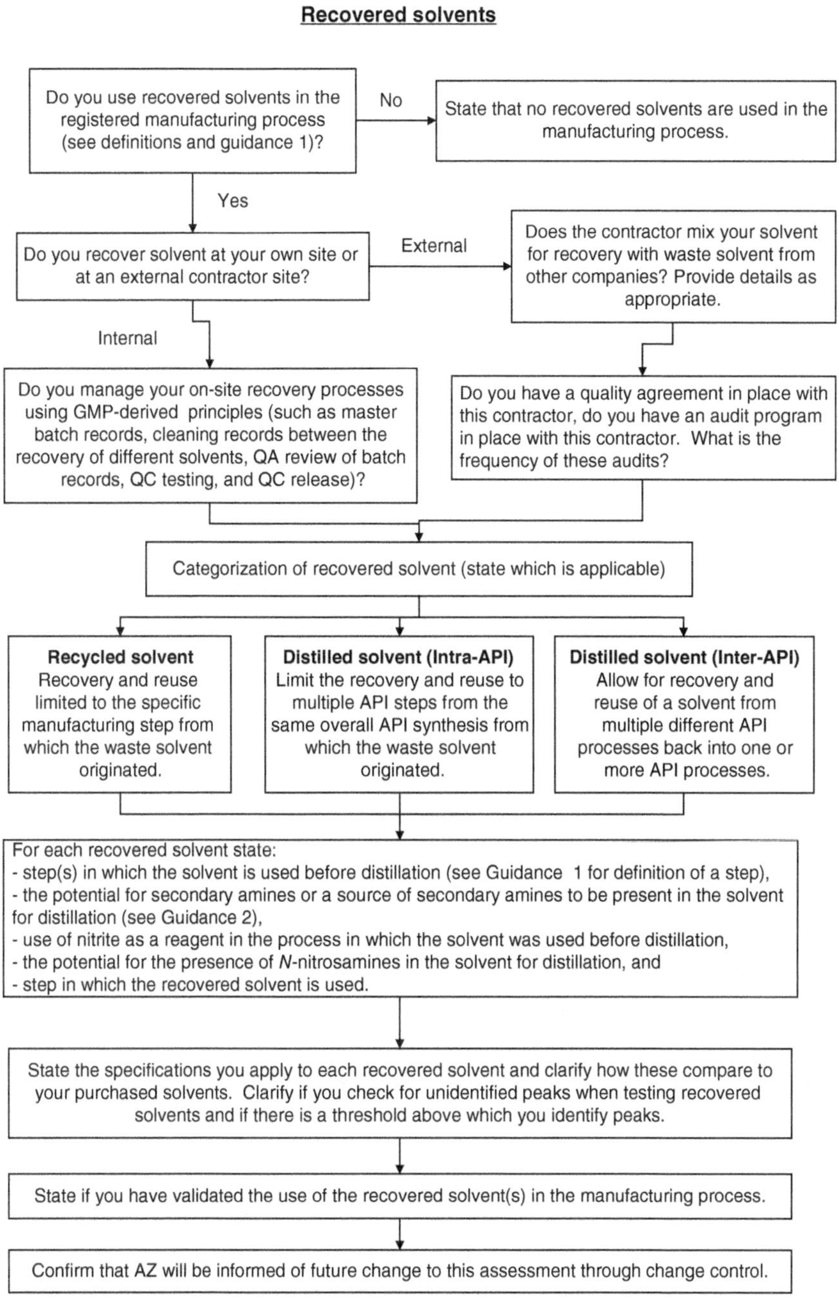

Figure 10.11 Solvent recycling decision tree.

should be a focal point. Specific secondary amine reagents/solvents used in drug substance synthesis, like DMA and diethylamine, may end up in trace quantities in the final drug substance and are typically limited by ICH Q3A [24] or specified in the drug master file. An evaluation of the drug substance process and specifications is a useful first step to determine the likelihood of the presence and quantities of secondary amines. In addition to process solvent impurities, the chemistry of the drug substance should be analyzed to determine the likelihood of degradants that could be reactive. In addition, some smaller molecular weight tertiary amide, common process solvents, have been known to readily hydrolyze to form secondary amines that can then go onto form *N*-Nitrosamines. A well-known example is *N,N*-Dimethylformamide (DMF) [25] used in the tetrazole chemistry for Valsartan. Finally, an evolving area of research involves the study of secondary amine functionality in the drug substance structure itself and whether it can nitrosate to form a *N*-nitroso form of the drug substance. The source and presence of nitrosating agents in the drug product formulation must also be assessed. Nitrosating agents can come in numerous forms and some common ones are: nitrites (e.g. sodium nitrite, $NaNO_2$) and nitrous acid (HNO_2), nitric oxide (NO), nitrosyl halides (e.g. ClNO, BrNO), dinitrogen trioxide (N_2O_3), dinitrogen tetroxide (N_2O_4), and organic nitrites (e.g. t-BuONO). It has become apparent that the nitrite presence is the most abundant concern for nitrosating source in drug products. This section will focus on nitrite as the key nitrosating agent given they have the highest prevalence in drug product formulations. Prior research has shown that many common excipients used in drug products contain nitrite levels that can often be >1 ppm (4) [26]. Considering the acceptable daily intake (ADI) values of the documented *N*-Nitrosamines range from 26.5 to 96 ng/day; the level of nitrite that could be present in the drug product is in far excess of what is needed to react and form *N*-Nitrosamine levels beyond those ADI values. As a result, there is a large amount of research ongoing to study the nitrite levels in excipients used in drug product formulations. One example of such research is happening within an *N*-Nitrosamine team chartered under the International Consortium for Innovation and Quality (IQ). An additional known source of nitrite in drug product manufacturing is from the water used.

However, most drug product manufacturing processes use purified water or in the cases of sterile products, water for injection, which are expected to have very low and negligible levels of nitrite [27]. Once the sources and worst-case concentrations of secondary amines and nitrite are known, a good next step is to calculate what the theoretical maximum *N*-Nitrosamine content could be in the drug product. Equation 10.1 lays out this calculation. MW refers to the molecular weight. One must account for maximum unit doses that could be prescribed to a patient in a given day and multiply the value from Equation 10.1 by this number to get a daily value. If the calculated value is less than the ADI of the *N*-Nitrosamine of interest, there is no risk present to the patient assuming known or conservative estimates were used for the total limiting reagent in the calculation.

Equation 10.1 Maximum *N*-Nitrosamine per unit dose based on limiting reagent

$$\text{max nitrosamine}\left(\frac{\text{ng}}{\text{unit dose}}\right)=\frac{\text{limiting reagent per dose}\left(\mu g\right)}{\text{MW}\left(\text{limiting reagent}\right)}\times \text{MW}\left(\text{nitrosamine}\right)\times 1000$$

If the theoretical maximum amount of an *N*-Nitrosamine of interest exceeds the ADI, the next step is to dive deeper into the risks carried by the drug substance, excipients, and manufacturing process. As part of the response to Article 5(3), pharmaceutical manufacturers have queried suppliers of excipients to understand the risks associated with direct presence of *N*-Nitrosamines as well as the common reactants, such as amines and nitrites. A comprehensive, cross-industry summary is not yet available. However, examination of a subset of the industry-wide data being gathered suggests that no excipients have been identified that carry an *N*-Nitrosamine impurity risk. The work on excipient impurity risks is being consolidated and should be shared in future publications. The likely reason is that manufacturing processes of excipients typically do not use

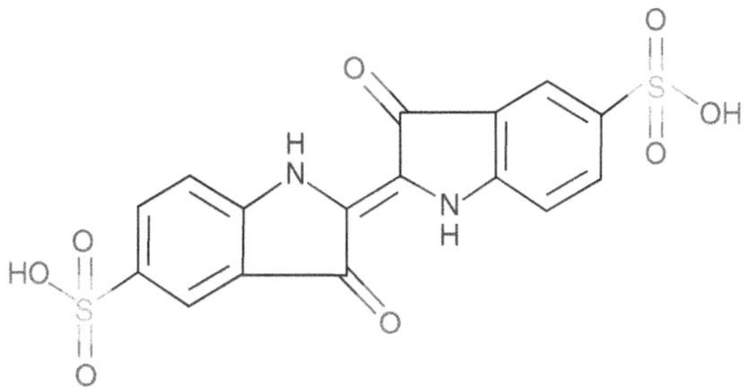

Figure 10.12 Structure of FD&C Blue #2/Indigo carmine aluminum lake.

secondary amines or sources thereof, and for this matter, excipients typically do not carry a potential amine impurity of concern. However, there are a few exceptions. As the industry is learning more about risks that excipients carry, a few excipients have been identified which warrant additional consideration. Povidone can carry *N*-Vinylpyrrolidone as an impurity [28]. Another example is FD&C blue #2 indigo carmine aluminum lake dye which has secondary amine functionality in the structure (Figure 10.12). However, the carbon atoms next to the amine nitrogen lack an alpha-hydrogen and therefore, even if it nitrosates to an associated *N*-Nitrosamine, the *N*-Nitrosamine is not expected to be mutagenic.

Excipients do commonly carry a risk of nitrite presence. Prior research studying reactive impurities in excipients showed that nitrite impurity is present in common excipients at trace or ppm levels [26]. Recent ongoing studies have shown similar findings that nitrite is present at various levels in common excipients (Table 10.2). Also, this work gives early indication that nitrite content can vary to some extent between lots and suppliers for a given excipient as evidenced by the results for corn starch. However, additional work is needed in drug product formulations to understand if this variation has a meaningful effect on *N*-Nitrosamine formation risk. Given the commonality of these excipients in many different types of drug products, it is likely that a nitrite source is present in most drug products. There is a collaborative effort ongoing among numerous pharmaceutical companies to build a database of nitrite content in commonly used excipients.

In addition to understanding the sources and risk levels in the drug product components themselves, the type of manufacturing process and induced stressors can play a significant role in the risk of *N*-Nitrosamine formation. The following considerations of risk rankings are based on the

Table 10.2 Nitrite results for common excipients in drug product formulations.

Excipients	Nitrite ion amount (μg/g)	LOD (μg/g)	LOQ (μg/g)
Corn starch (supplier 1, lot 1)	0.252	N/A	0.05
Corn starch (supplier 1, lot 2)	0.498	N/A	0.05
Corn starch (supplier 2)	0.088	N/A	0.05
Pre-gelatinized starch	0.562	N/A	0.05
Sodium citrate dihydrate	Non-Detected	0.07	N/A
Hypromellose	Non-Detected	0.08	N/A
Silicon dioxide	Non-Detected	0.03	N/A
Magnesium stearate	0.71	N/A	0.05

limited experimental data available to date and understanding of general reactivity in pharmaceutical drug products. Liquid-based processes and drug products where the drug substance, assuming it carries the reactive amine source, is dissolved or suspended are considered highest risk given the high mobility of the reactants and probability for reaction. Creams and lotion products have some reduced risk due to elevated viscosity and resistance to reactant mobility. Amorphous products may have elevated concern as the drug, or amine source, is present at an elevated energy state and can support or promote higher reactivity in solid state [29]. Wet granulation is also seen as higher risk based on the levels of water present. Nitrite is not an effective nitrosating agent itself. It must protonate to form nitrous acid which can readily happen in low pH, aqueous media [30]. It is a solution-phase process. Nitrite on its own is not a nitrosating agent, it must either:

1) React to form a nitrosating agent "XNO^{+}" ($X = NO^{2-}$, Cl^{-}, Br^{-}, H_2O, etc.) or
2) react with an amine that has been activated to form an iminium ion.

Nitrite is very water soluble and could readily dissolve during the process. The water distributed among powders in wet granulation may be an effective means to distribute nitrite among the reactive amine source in the granulating mixture. In addition, with the assumption that mobility in the water phase is a key factor to the nitrosation risk, we should consider how this homogenized blend may concentrate the reactants together during granule drying. Figure 10.13 outlines the stages of this wet granulation process to visualize the stages of powder agglomeration, distribution of water, and how the nitrite may concentrate in the final granule drying process.

Dry processed, compacted drug products are of lower concern given the moisture level remains lower throughout the manufacturing process, and nitrite contained within the excipients is not given the potential to dissolve in the granulating fluid and distribute as it may during wet granulation processes. Finally, dry powder blends produced as the final dosage form represent the lowest risk given they are not compacted and maintain more separation of the reactants to minimize probability of nitrosation. Additional process and formulation factors that may influence the degree of nitrosation include product temperature during manufacture, content uniformity of nitrite and reactive amine in the drug product, effective contact area of reactants in the formulation, p*Ka* of the reactive amine, and pH in the formulation. The content uniformity and effective contact area of the reactants may be more relevant in solid drug products where mobility is highly limited. The p*Ka* of the amine is important as it describes the fraction of amine available in the reactive freebase form to nitrosate [21]. Amines with lower p*Ka*, like morpholine, will nitrosate

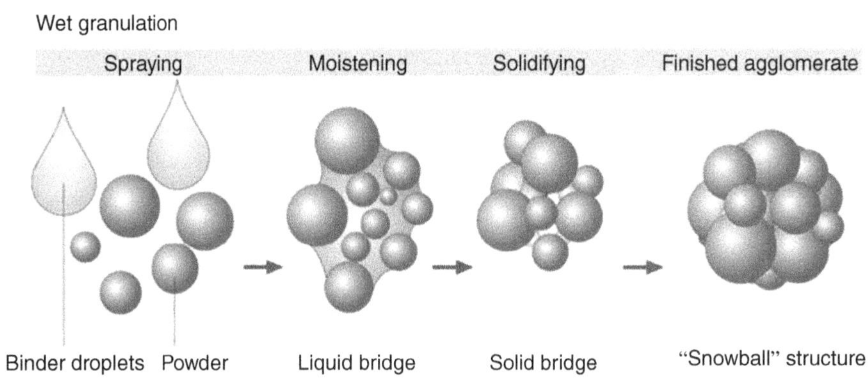

Figure 10.13 *F*-Stages of granule formation during wet granulation. *Source*: https://www.saintytec.com/wet-granulation-the-ultimate-guide-for-beginners-and-professionals.

much more rapidly as a result of more available freebase [21]. The nonlinear relationship of nitrosation kinetics to solution pH is well documented with the rate being the highest at pH of 3.4. However, what is less understood is the role of pH within solid products where the moisture is present at low levels (e.g. <5 wt.%). Unbound moisture in the interstitial spaces between particles may activate nitrite to nitrous acid and carry forward to react with a nearby secondary amine. However, this remains an area of study to further understand influence on formation risk in solid drug products.

In response to the recent emergence of *N*-Nitrosamine risk in drug products, numerous pharmaceutical companies are experimentally studying formation in model formulation systems. The aim of this work is to determine the possible extent of *N*-Nitrosamine formation, in comparison to the theoretical maximum discussed earlier, and gain understanding of the key risk factors which should drive prioritized focus in risk assessments. One such recent study evaluated reactivity of 4-phenylpiperidine HCl salt (4-PhP HCl) in solution and oral solid dosage (OSD) formulations (Figure 10.14). 4-PhP HCl was chosen as it is a simple molecule with one reactive site for nitrosation and a p*K*a of 10.2 for the conjugated base.

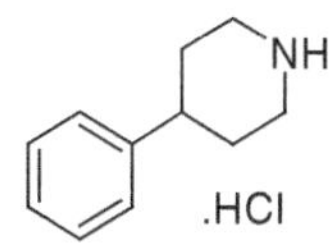

Figure 10.14 Structure of 4-phenylpiperidine HCl salt.

4-PhP HCl was demonstrated to be reactive to nitrosation as 73% of 0.1 M converted to *N*-nitroso-4-phenylpiperidine after 24 hours in aqueous media at pH = 3 with 0.2 M nitrite present. This demonstrated that 4-PhP HCl is a suitable model amine to study nitrosation propensity in OSD formulations given the presence of the 4-PhP HCl and potential nitrosating sources (e.g. nitrite from excipients). A study was conducted in OSD formulations containing common excipients including microcrystalline cellulose (MCC), lactose, silicon dioxide, sodium starch glycolate, and magnesium stearate. Four formulations were processed at 10 wt.% of 4-PhP HCl by the following means: (i) Dry blend, lubrication with magnesium stearate, and compression to 100 mg tablet weight referred to as direct compression, (ii) Wet granulation, tray drying, sieve, lubrication with magnesium stearate, and compression to 100 mg tablet weight and referred to as wet granulation.

The dry and wet formulations were processed using native excipients from the suppliers as well as two batches containing a 1000 ppm spike of $NaNO_2$ (nitrite) to look at a worst-case scenario. The spike was accomplished by dissolving the nitrite in water and granulating it onto MCC. This was then dried and mixed into the formulation as a portion of the MCC to hit a target of 1000 ppm added nitrite. The formulations containing native excipients were estimated to have a nitrite concentration of 0.9 ppm based on the nitrite content documented in Table 10.2.

The formulation with nitrite spike was equilibrated to a water activity <10%RH. The formulations without the nitrite spike were equilibrated to a water activity of ~40%RH. After equilibration, vials containing formulations with the nitrite spike were placed in an oven at 70 °C and stressed for one week. The samples containing formulations without the nitrite spike were placed in a 40 °C chamber for four weeks. The samples were analyzed for *N*-Nitroso-4-phenylpiperidine right after manufacture and at their respective stability pulls.

The results in Table 10.3 show that *N*-Nitrosamine can form in OSD formulations even right after manufacture. However, the concentration of *N*-Nitrosamine was very low in both direct compression (i.e. dry formulation) and wet formulations without the nitrite spike. The formulations with the spike were characterized shortly after manufacture and showed significant levels of *N*-Nitrosamine which is likely due to the high concentration of nitrite and higher probability to find the 4-PhP HCl particles to react. In addition, significant growth occurred for both of the spiked formulations after one week of stressing. It is important to reiterate that both of these formulations

Table 10.3 *N*-Nitroso-4-phenylpiperidine in tablets from the direct compression and wet granulation formulations w/ and w/o a 1000 ppm $NaNO_2$ (nitrite) spike.

Formulation	Theoretical max nitrosamine (ng/tablet)	Theoretical max nitrosamine (PPb)	*N*-Nitroso-4-PhP (ppb) @ $t = 0$	*N*-Nitroso-4-PhP in OSD stability (ppb)	Reactivity factor (%)
Direct compression	371	3709	11	24	0.6
Direct compression (w/$NaNO_2$ spike)	413 365	4 133 650	2482	21 025	0.5
Wet granulation	371	3709	20	611	16.5
Wet granulation (w/$NaNO_2$ spike)	413 365	4 133 650	20 099	329 623	6.0

Summary of theoretical maximum values, measured values after production ($t = 0$), measured values after four weeks of stressing at 40 °C/40%RH (no spike), measured values after one week at 70 °C/desiccated (spike), and calculated extent of reactivity.

represent an unrealistic drug product formulation given the nitrite present is about three orders of magnitude higher than that expected in a typical drug product formulation.

After four weeks at 40 °C/40%RH, very little growth of *N*-Nitrosamine was observed for the dry formulation without the spike which suggests that without a suitable level of unbound water in the tablet, the lack of mobility of nitrite and amine hinders nitrosation reactivity. The wet granulation formulation did show some significant growth. 4-PhP HCl is very water soluble increasing the propensity for interaction between 4-PhP HCl and nitrite in the water phase.

In addition, it is important to note that physicochemical characterization studies on the 4-phenyl piperidine and wet granulation samples showed that the crystal structure was lost during wet processing and likely a large degree of amorphous form was created. It is well known that amorphous material can exhibit higher solid-state chemical reactivity [29]. To examine the extent of reactivity, we will define the reactivity factor % as observed *N*-Nitrosamine on stability divided by the theoretical maximum if all nitrite reacted. This gives a method to evaluate how much *N*-Nitrosamine we would expect to form in a similar formulation. It is interesting to note that independent of the nitrite level present, the direct compression (dry) formulation shows ≤1% reactivity. In contrast, about 8–16.5% reactivity is observed in the wet granulation. The smaller reactivity factor exists for the wet granulation formulation with nitrite spike. At first, this may be counterintuitive given the larger concentration of nitrite present to react and clearly larger amount of *N*-Nitrosamine formed. However, a significantly higher theoretical maximum value also exists because of the higher nitrite amount which brings down the ratio. This work shows that wet processed formulations pose higher risk than dry systems for *N*-Nitrosamine formation. Additionally, it shows that *N*-Nitrosamine formation is significantly hindered in solid dosage formulations given the low extent of reactivity.

An additional wet granulation of identical composition discussed previously was made following the same process details but at 0.1 wt.% 4-phenylpiperidine HCl. The intent of this work was to explore the risk of reactive impurities that could come along with the drug substance like DMA. Samples were equilibrated at 40 °C/75%RH and stressed for 12 weeks. The amount of *N*-Nitroso-4-phenylpiperidine observed after manufacture ($t = 0$) and after 12 weeks stressed stability at 40 °C/75%RH was 21 and 53 ppb, respectively. This represents very little growth even for a formulation where the physical stability of the crystal was compromised leading to a more reactive system as observed in the wet granulation formulation containing 10 wt.% 4-PhP HCl. This work shows that process solvent impurities, controlled by ICH Q3A limits [24], and/or degradants of similar concentrations in the drug substance are of lower risk for *N*-Nitrosamine formation in the drug product. Degradants formed in the drug product may be of lower risk to nitrosate themselves in OSD product given they are locked into a matrix of limited mobility compared with impurities in the drug substance. Drug substance impurities may be better homogenized in the drug product processing and further activated for reaction from stresses in the drug product process itself.

The pharmaceutical industry is engaged in active research trying to better understand *N*-Nitrosamine contamination and formation risk in drug products during manufacture and over shelf life. This section aims to cover what is known about contamination and formation risks in drug products as well as relay some considerations to help frame the *N*-Nitrosamine risk that could be present. However, given the as yet immature nature of this research area, much more remains to be discovered to inform the pharmaceutical scientist of how to avoid *N*-Nitrosamine risk during development of new products. The following factors should be given further consideration to study the impact toward *N*-Nitrosamine formation risk in drug product:

- Nitrite level in solid drug product (e.g. expected levels of nitrite).
- More specific process stressors in each given drug product manufacturing unit operation (e.g. type of wet granulation).
- Role pH plays in drug product processing and final products.
- Reactive impurities in drug substance (e.g. solvent impurities).
- Reactive degradants in drug product.
- Influence of particle morphology and form of reactants in the drug product (e.g. molecular distribution vs particles on order of micron size).
- Temperature and humidity influence on product shelf life.
- Reactive amine in crystalline vs amorphous phase state.

The information gained so far suggests that the main risk of contamination of *N*-Nitrosamines in drug product comes from the drug substance. The research and understanding demonstrated to date clearly shows that *N*-Nitrosamine formation in drug product is possible which warrants further research to build an understanding of risk factors and their impact such that a control strategy can be developed to eliminate the risk in new drug products. Although possible, nitrosation in solid drug products is significantly hindered compared with what is observed for reactive environments in solution. Additionally, reactive amines or precursors in the drug substance at or below 1000 ppm that carry forward to the drug product present low risk of *N*-Nitrosamine formation. The research of *N*-Nitrosamine risk in drug products is evolving but the approaches discussed here give an initial framework that can be useful in conducting risk assessments for existing and new drug products. A more detailed manuscript is under development and expected to be published in first half of 2021. This will aim to elaborate on the mechanisms of formation, sources of contamination, broaden understanding of the risk factors and how to carry out quantitative risk assessments in drug products.

10.5.4 Container Closure Systems

Within the revised EMA Question and Answer document, specific reference is made to the risk of *N*-Nitrosamines associated with the use of lidding foils in blister packs; this relates to the use of nitrocellulose-based materials. Such foils are relatively complex as shown in Figure 10.15.

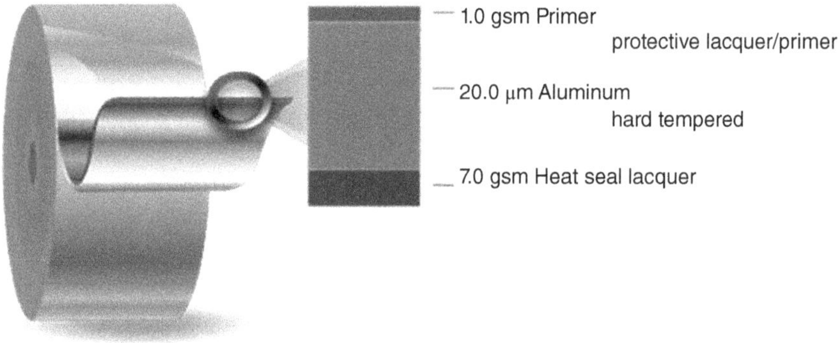

Figure 10.15 Illustration of a typical structure of a lidding foil and its associated layers.

Figure 10.16 Photograph of a lidding procedure.

Nitrocellulose may be used in the primer, the over-lacquer and even in the inks. A further factor is residual amines within the inks. The levels of small secondary amines present within inks can vary due to a number of factors including the color of the inks.

The process of lidding is shown in Figure 10.16.

Nitrocellulose is commonly used in blister lidding foils as a print primer and over-lacquer.

The Nitrosamine is believed to be generated during the printing process; the nitrocellulose degrades to give a series of NOx species that react with secondary amines in the inks to yield *N*-Nitrosamines. Amines in ink may be part of color pigments but are mainly trace impurities in the ink and are hence considered to be "Non-Intentionally Added Substances" (NIAS). Key factors to consider are extent of coverage, colors (e.g. reds, yellows are believed to be higher risk) and location (inner or outer surface).

Heat from sealing then volatilizes the Nitrosamine in the lidding foil, the subsequent vapor is captured in the unsealed blister pocket during sealing. In general, the risk is considered low as observed levels of Nitrosamines, when formed, have been very low and significantly below an ADI for the patient.

Given the risk of generating *N*-Nitrosamines, an evaluation of the blistering process, particular risks, and risk mitigation factors (e.g. ventilation) should be considered. Current understanding indicates the risk to be low, a few nanograms per cavity; this is believed to be due to a number of factors, these include:

1) Very low volatilization rates of NDMA and NDEA from common blister lidding and the short time of applied heat during sealing.
2) Convoluted pathways for vapors to migrate upstream into open blister wells.
3) Depletion of any volatilized nitrosamines from vapor cloud due to room air changeovers.

Despite this, it is suggested that screening of different nitrocellulose containing lidding foil types and inks (potentially containing residual amount of amines as NIAS) is performed as part of the overall *N*-Nitrosamine product risk assessment.

Should there remain a concern, perhaps for a multidose regimen (>4 tablets/day for example), then it may be prudent to consider removing the risk entirely, i.e. moving to nitrocellulose-free materials. This change can be implemented without prior regulatory approval ("do & tell"), as the lidding foil is not in contact with the product and the change is not expected to affect Quality & Stability, provided the aluminum layer and heat seal lacquer are unchanged (i.e. water vapor transmission rate, oxygen transmission rate; tightness unchanged).

Alternative mitigation solutions, such as local extract during blistering operation, would also reduce the risk.

Another option to consider is to seek to move to printing ink free of vulnerable amines. However, as secondary amines are NIAS, it is difficult to consistently avoid this risk without testing or robust certification.

10.5.5 Elastomeric Components

Nitrosamine formation in elastomers has a long history with initial concerns centered on elastomers and natural rubber (latex) and its use in babies' bottle nipples and dummies/pacifiers [31, 32]. In the 1980s, the link was made between elastomers compounded with accelerators or stabilizers derived from alkylamines and the formation of nitrosamines which were then found in artificial saliva extracts and hence the concern they might be dosed to babies using the elastomers [33].

Examples of large range of elastomer accelerators which hold a nitrosamine risk are shown in Table 10.4, listing the substance(s) as its corresponding nitrosamine. The complex nature of the curing process for elastomers allows for the presence of both secondary amines and the "nitrosating agent" NOx (NO^+, N_2O_3, N_2O_4, etc.) to form and thus create nitrosamines at the trace level.

Surveys were conducted by several European and international bodies including in the United States by the FDA. The nature of the nitrosamine detected was dependent on the nature of the compounding ingredients in use. FDA introduced into the federal register in 1984 [34] a requirement for no more than 60 ppb of nitrosamine in these products based on its extraction method employing methylene chloride (dichloromethane). In Europe, they adopted an artificial saliva-based method and lower limits (Germany 10 ppb, Netherlands 1 ppb); each of these were typically at or around the detection limits of the methods employed. The general method of detection was a chemiluminescence-based detector which gave high sensitivity and selectivity for nitrosamines when combined with gas chromatography [35].

During the mid-1990s, concern was turned to metered dose inhalers (MDIs). MDIs were first developed in the 1950s, with the Riker company (later acquired by 3 M Pharmaceuticals) in the United States introducing a formulation based on Freon 12 and Freon 114 with 35% ethanol. Ventolin, an MDI-delivering albuterol, was first commercialized in 1968. In the 1970s and 1980s,

Table 10.4 Nitrosamine sources from elastomeric cure accelerators.

Name of substance used as cure system	Chemical class(es)	Likely nitrosamine
Zinc dibutyl dithiocarbamate, Nickel bis(dibutyldithiocarbamate)	Dithiocarbamate	NDBA (*N*-nitrosodibutylamine)
Zinc dimethyldithiocarbamate, Sodium dimethyldithiocarbamate, Bismuth dimethyldithiocarbamate, Copper dimethyl dithiocarbamate Tetramethylthiuram disulfide Tetramethylthiuram monosulfide	Dithiocarbamate, Thiuram	NDMA (*N*-nitrosodimethyamine)
Zinc diethyl dithiocarbamate, Tellurium Diethyldithiocarbamate Tetraethylthiuram disulfide	Dithiocarbamate, Thiuram	NDEA (*N*-Nitrosodiethylamine)
Zinc-*N*-pentamethylene dithiocarbamate Dipentamethylenethiuram	Dithiocarbamate, Thiuram	NPIP (*N*-nitrosopiperidine)
Zinc pyrroline dithiobarbamate	Dithiocarbamate	NPYR (*N*-nitrosopyrrolidine)
Monopholine disulfide(4,4′-dithiodimorpholine) 2-Morpholinodithiobenzothiazole 2-Morpholinothiobenzothiazole *N*-Oxydiethylenethiocarbamyl-*N*′-oxydiethylenesulfenamide	Sulphur donor, Sulfenamide	NMOR (*N*-nitrosomorpholine)

MDIs continued to grow in popularity with the introduction of several important MDI products. In 1972, Allen & Hanburys commercialized the first beclomethasone dipropionate MDI (Becotide®) in Europe. Later, in 1982, beclomethasone dipropionate MDIs were marketed in the United States by Schering Corporation (Beclovent) and Glaxo Wellcome (Vanceril). Boehringer Ingelheim introduced Alupent (metaproterenol sulfate) in 1973. Two albuterol MDIs were introduced in 1981, Proventil by Schering Corporation and Ventolin by Glaxo Wellcome. Aerobid (flunisolide) was introduced by Forest in 1984. Atrovent (ipratropium bromide) was introduced by Boehringer Ingelheim in 1986 [36, 37].

These all contained metering valves with elastomeric components and these formulations were a potential source of nitrosamine in the same way as the baby products, additionally the propellant used aided in the leaching of substances (including nitrosamine) in the drug products formulation.

The concerns raised lead to introduction of methods and specifications very similar to those introduced for baby products using the same extraction and detection systems. Once again, the specification introduced was close to limits of detection of the methodology.

In the early 2000s, the original CFC propellants were replaced by HFA-based propellants due to concerns around these propellants effect on ozone layer as laid out in Montreal protocol, Decision XII/2 [38]. Pharmaceutical companies had the opportunity to replace their elastomers with elastomers which eliminated direct nitrosamines formation due to the selection of different compounding ingredients which did not produce nitrosamines and generally aimed to reduce leachables from the elastomeric parts in comparison to historically used components. Additionally, the change in propellant was not compatible with the original elastomer formulations. However, the regulators (in particular, FDA) were still insistent that methods and specifications were required to monitor and control nitrosamines citing the potential for nitrosamines to be present in production environment for the pharmaceutical elastomers, and thus they required methods for nitrosamines and corresponding acceptance criteria based upon the limits of quantitation of the methods. Thus, the original methods were once again employed, and acceptance criteria and specification were set at the limits of detection for each nitrosamine as defined by the original methods together with a total where they remain in place. Due to well-controlled manufacture within custom pharmaceutical production lines for the production of elastomeric components, the risk from nitrosamines is very low as contamination with nitrosamine-producing substances has been removed through process controls and careful choice on formulation ingredients all designed to keep general leachables low, including nitrosamines.

10.5.6 Nitrosamine Impurities in Biologics

As highlighted above, when the CHMP issued the final report relating to Article 5(3) [39], all biological medicinal products were brought within scope. This section considers the potential risk of nitrosamine impurities within biological medicinal products. It covers the different nature of products that loosely fall under the banner of Biologics as well as specific factors that may present a theoretical, if not actual, risk. The risk of nitrosamine impurities being introduced into biological products is discussed in three parts:

1) active substance
2) excipients
3) primary and secondary packaging/labeling

The term "biological medicinal products" is understood in accordance with Directive 2001/83/EC as a product containing a biological substance as the active component, where (with noted exceptions e.g. certain antibiotics) a biological substance is extracted from a biological source. Within this broad definition, it is anticipated that this will include recombinant protein expressed in a variety of systems, vaccines, advanced therapeutic medicinal products, etc. This section seeks to highlight how the major categories of biological products for a given class or modality could be assessed on the extent of risk defined.

It should be noted that the CHMP BWP itself concluded that there is only a very low risk of nitrosamines being present as impurities in biological medicinal products. It was though concluded that biological products containing chemically synthesized fragments, where risk factors similar to chemically synthesized active substances exist, were of more risk. It was also noted that biologicals packaged in blister packs containing nitrocellulose would also be at risk. It should be noted that there are few if any biologics packaged in such a manner.

10.5.6.1 Active Substance

Aside from Biologics incorporating chemical modification, e.g. antibody drug conjugates, ADCs, biological medicinal products are unlikely to involve a significant nitrosating agent present within the manufacturing process; furthermore, the conditions employed in their manufacture are suboptimal in terms of conditions for nitrosation (time, temperature, pH, nitrosating agent concentration). Downstream purification is also a significant factor. This typically involves chromatography steps and extensive dialysis to clear small-molecule process-related impurities. Also in terms of formulated product, such biologics are often simply solution products within purified water for injection or as lyophilized solids for reconstitution on administration. Excipients, if employed, are

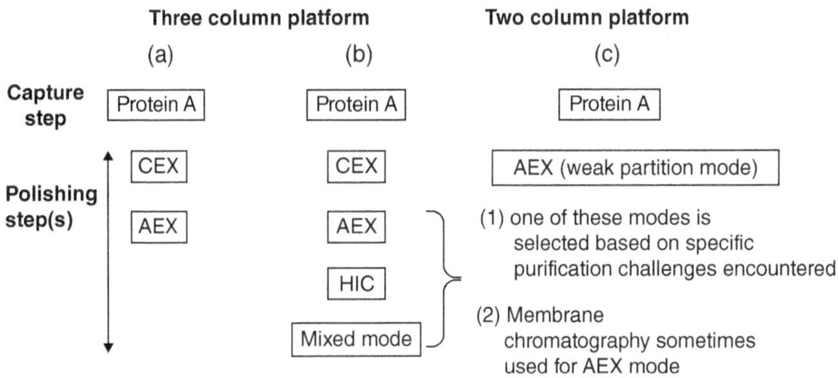

Figure 10.17 Illustration of common downstream process options.

limited. A generic mAb process is shown in Figure 10.17. A typical process downstream involves initial capture of the mAb by Protein A chromatography. This in then followed by a series of polishing steps, possible options are described.

The purging of small molecules was presented by Gong et al. [40]

Other key factors include the size and nature of the protein concerned. Proteins theoretically provide a potential source of amines (e.g. tryptophanyl, histidyl, prolyl residues in polypeptide) that could generate *N*-Nitrosamines through reaction with nitrosating agents (for example, nitrites in water). Even then only the molecule's outer, solvent accessible amino-acid side chains would be expected to be available for any nitrosating reaction (Figure 10.18). Although it is secondary or tertiary amines that have potential to form mutagenic derivatives, primary amines and thiols may also react with nitrosating agents. However, amino acids with primary amines (lysine), or thiols (cysteine) and the N-terminal amine group form unstable diazonium intermediates that rapidly

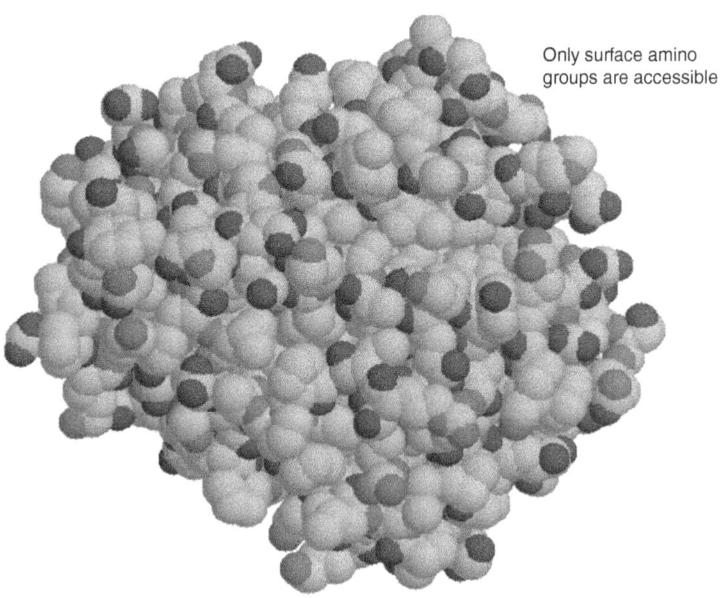

Figure 10.18 Image of a protein showing accessible surface amino acids.

decay to the hydroxyl form and nitrogen. Therefore, primary amine groups on protein or polypeptide may be considered as a scavenger of nitrosating agents [41]. This scavenger role may be enhanced by the reversible reaction of a secondary amine and NO^+ from nitrous acid to form a nitrosamine under appropriate conditions.

Another factor is steric hindrance. Large molecules containing any trace nitrosamine moieties cannot be activated to generate a potent mutagenic entity by the cellular mechanisms that activate small molecules. To form a potent mutagen, nitrosamines require metabolic activation by oxidation to form an α-hydroxynitrosamine that rapidly rearranges to a diazohydoxide form that can then itself generate a carbocation that can alkylate DNA [42]. Large protein molecules with a nitrosamine group would be sterically unfavorable substrates for CYP binding and activation in which the N–N needs to be in proximity to the haem group of CYP [43, 44]. Furthermore, mutagenicity of most *N*-Nitrosamines has been shown to decrease significantly as the size exceeds 12 to 14 carbons [45].

Even in the case of the theoretical risk in relation to secondary amines, those amino acids that are "vulnerable" are heteroaromatic secondary amines, and as described in more detail in Chapter 7 do not contain sp3 α hydrogens and thus cannot undergo metabolic activation to generate the alkyl diazonium intermediate and ultimate alkyl diazonium species even if access to the CYP enzyme cavity was possible.

10.5.6.2 The Water Used in Formulation Is Depleted in Nitrosating Agents

Products for parenteral use are generally formulated as aqueous solutions or lyophilized solids for reconstitution into water. Both require the use of Sterile Water for Injection (sWFI). sWFI is typically produced using distillation, reverse osmosis, and then either ultrafiltration or deionization (Guideline on the quality of water for pharmaceutical use, EMA/CHMP/CVMP/QWP/496873/2018). sWFI thus contains very low levels of nitrites that constitute the potential nitrosating agents in water. PhEur limit for nitrates for both Purified water (Ph Eur 0008) and WFI (Ph Eur 0169) is 0.2 ppm. In reality, as illustrated by Ashworth et al. [21], the levels are often significantly lower than the standards specified. Given this, combined with suboptimal conditions employed, the risk of introduction of *N*-Nitrosamines arising from the presence of nitrites in sWFI is negligible and can be discounted.

Based on the points highlighted above, it should be possible for non-chemically modified Biologics to be documented in a single, generic risk assessment for that specific type or class of biological medicinal product.

10.5.6.3 Bioconjugated or Chemically Modified Products

It is in many ways surprising that the scope of the guideline was revised given that Human medicinal products that contain a conjugated chemically synthesized API component such as Antigen-Drug Combination (ADC) products, PEGylated bioconjugates, were already within the scope of Article 5(3) [16, 39]. This would of course have focused on the synthetic aspect of such therapeutic modalities; however, as examined above, any risk is ultimately likely to relate to this aspect and not to the Biologic moiety within it.

Examining this through the context of an ADC, the risk assessment now needs to look at the drug-linker synthesis (drug intermediate), the recombinant protein production (drug intermediate) in addition to drug substance manufacture (conjugation of the drug intermediates) and drug product manufacture (filling into the primary container closure system and secondary packaging/labeling).

As highlighted by Gong et al. [40], the risk posed by small process-related impurities is negligible; this is based on the significant difference in molecular weights between small-molecule impurities such as *N*-Nitrosamines (if present) and the ADC, the conjugation potential of the small-molecule impurities, and the typical dosing concentrations and dosing schedule. The result is that exposure to small impurities in ADCs is so low as to often pose little or no significant safety risk. Indeed, for these drug products, the small-molecule impurities such as *N*-Nitrosamines can likely be assessed using a risk-based approach that is consistent with ICH Q3A [24] guidelines.

10.5.6.4 Excipients

In general, excipients used to formulate biological medicinal products should not be assessed any differently to excipients used for API products, including assessment of risk from the manufacture of the excipients. However, the excipients used for biologicals are worthy of some further reflection.

Biological products are usually stored as refrigerated liquid (in aqueous solution or suspension), "frozen liquid" or lyophilized at pH >5 – conditions unfavorable for nitrosamine formation and with the water as the sole source of trace levels of nitrite.

Vaccines may include adjuvants in their composition, commonly aluminum salts. Some less commonly used excipients for vaccines may also be of biological origin, including albumin. Many "traditional" vaccines also contain host cell components (protein, DNA, etc.).

Novel excipients or adjuvants containing chemically synthesized compounds should be considered within the scope of Article 5(3) [16, 39] and assessed accordingly.

The majority of current, biological medicinal products are formulated using excipients (or adjuvants) that are not susceptible to nitrosation and present no risk of nitrosamine impurity formation; this includes excipients (or adjuvants) with primary amine groups. Excipients or adjuvants that do contain vulnerable amine groups (e.g. histidine, proline), with potential for nitrosation, would require additional consideration and may leverage published literature to aid the evaluation.

L-histidine is a relatively common excipient in the formulation of biological medicinal products, used in low concentrations (e.g. 10 mM) as a buffering agent. While nitrosation of L-histidine is possible, only one derivative, (1 nitroso-1H-imidazol-4-yl) acetohydroxamic acid (NIAH), has been shown to be mutagenic. However, NIAH cannot exhibit the same metabolic pathway via a diazonium ion and thus is not part of the cohort of concern [46, 47]. Furthermore, NIAH is formed by the action of multiple equivalents of nitrosating agent which under conditions of negligible nitrosating agent content (from WFI) is considered highly unlikely.

L-proline: While nitrosation of the secondary amine of L-proline is possible, any nitroso-proline has been shown not to be carcinogenic as demonstrated in animal studies when L-proline and nitrite are co-ingested [48–50].

L-arginine: L-arginine is a common excipient in the formulation of biological medicinal products to reduce protein aggregation and enhance thermal stability [51] While L-arginine has no secondary or tertiary amine, nitrosation of the guanidino group of L-arginine is possible and occurs as part of endogenous cell metabolism to generate NO. However, the resulting derivative is not a nitrosamine but a nitrosourea form [30]. Studies indicate little to no carcinogenicity in animal co-fed arginine and nitrite compared with nitrite alone [52] and weakly mutagenic in an AMES test using one strain of salmonella [53]. Furthermore, the L-arginine primary amine group may also be nitrosated and hence act as a scavenger of low levels of nitrosating agent. It is concluded that the nitrosation products from L-arginine are not within the Cohort of Concern and should be considered in terms of ICH M7.

10.6 Regulatory Guidance Pursuant to *N*-Nitrosamines and its Implications

10.6.1 Article 31 Process and Outcomes

10.6.1.1 Article 31 Request

Issued in July 2018 [5], this laid out a series of specific actions required for the MAH of any Sartan. These are summarized below and specific sections reproduced in order to give a clear overview.

1) Obligatory risk assessments to be performed for manufacturing processes of Sartan drug substances in order to evaluate the theoretical risk of *N*-Nitrosamine formation and contamination.
2) Modifying manufacturing processes of Sartan manufacture, where necessary, to minimize contamination as much as possible.
3) Immediate implementation of analytical controls to detect and control *N*-Nitrosamine impurities (to defined very low levels) in the API (or intermediate, if justified).

The report then described requirements in detail including timelines (Table 10.5).

Table 10.5 Timescales for Article 31 process.

Conditions to the MAH	Due date
The MAH must ensure that the manufacturing processes of the drug substances used for their drug products are reviewed for the potential risk of formation of *N*-Nitrosamines and changed as necessary to minimize nitrosamine contamination as much as possible.	Within two years after Commission Decision
For all *N*-Nitrosamines, the MAH must ensure a control strategy is in place in drug substance batches used for their drug products.	At the time of Commission Decision
For *N*-Nitrosodimethylamine (NDMA) and *N*-Nitrosodiethylamine (NDEA), the MAH must introduce the following specifications for the drug substance: 1) Limits for NDMA and NDEA outlined below should be implemented for a transitional period of two years:	At the time of Commission Decision

Drug substance[a]	Max. daily dose (mg)	NDEA Limit in ng/day	NDEA Limit in ppm in API	NDMA Limit in ng/day	NDMA Limit in ppm in API
Valsartan	320	26.5	0.082	96.0	0.300
Losartan	150	26.5	0.177	96.0	0.640
Olmesartan	40	26.5	0.663	96.0	2.400
Irbesartan	300	26.5	0.088	96.0	0.320
Candesartan	32	26.5	0.820	96.0	3.000

[a] These limits are not applicable for batches where more than one of the above *N*-Nitrosamines has been identified simultaneously; such batches should be rejected.

2) After the transitional period of two years, a limit for NDMA and NDEA of maximum 0.03 ppm should be implemented.	Within two years after Commission Decision

10.6.2 Sartans Lessons Learnt Report

In November 2019, a meeting was held between regulatory agencies and various stakeholders including the Pharmaceutical Industry. The intent of the meeting was to conduct a review into the issues surrounding Sartans to understand what lessons could be learnt. In June 2020, the final report was issued [54]. This comprehensive report made a series of recommendations, 40 in total relating to areas including:

- Guideline revision
- ICH Guidelines including GMP
- Communications
- International cooperation
- Data solutions
- Training
- Other aspects

In addition to these recommendations, the bulk of the report focused on the technical background, of specific note are:

- *Prevention, chemistry, and root causes* – this covers the bulk of the technical document and provides a very comprehensive overview of the specific issues observed and risk factors.
- *GMP and inspections* – this recognizing the importance of GMP is a number of incidents particularly in terms of recycling of contaminated solvents.

At the time of publication, there were a number of concerns:

- While the report provided a lot of information about the scientific root causes of nitrosamine potential formation, it did not examine the systems failures, whether in relation to manufacturers of the Sartans or in terms of regulatory oversight of the manufacturing process, nor the associated Certificate of Suitability (CEP) process or Active Substance Master File (ASMF) review. Is it key to resolution of such matters to recognize that a technical investigation is only one aspect of effective control. Other areas such as cGMP, overall quality systems, etc., are also vital.
- In terms of the recommendations while many are understandable and reasonable, the practicalities of their implementation were and remain a significant concern.
- A particular concern was the timing of implementation of the recommendations, which preceded the completion of the product risk assessments and were also ahead of critical research into the proposed risk factors. That in seeking adoption at an early stage would lead to disproportionate or nonscientific expectations that simply added complexity rather than guarantee quality assurance of either medicines supplied to patients, or support/facilitate continued innovation of medicines. This reality was clearly evident at the time, in particular in relation to new applications, many submitted in advance of this report and indeed the Article 5(3) process with the net result that compliance with criteria not evident at time of filing were requested, resulting in major objections especially in Europe.
- Another major concern was the benefit of hindsight approach taken in the report. It specifically states that "Despite available guidance, the potential for *N*-Nitrosamine impurities in Sartans was not recognised during the development, manufacture and evaluation of medicines subsequently found to contain them." This statement of concern essentially ignores the fact that neither ICH M7 nor any of the earlier documents addressing mutagenic impurities were in place at the time of the development (or marketing license approval) of the Sartans. It is important to emphasize that since the beginnings of the development of specific guidance on GTIs/mutagenic impurities, there has been a paradigm shift in the risk evaluation of

manufacturing processes by industry (and in their assessment by regulators) such that an issue like that which affected certain Sartans would now be highly unlikely to occur.

- Reflecting in more detail on specific sections:

10.6.2.1 Reflection on the Initial Section of the EMA Report

Few would argue that the Summary and Recommendations pages contain some important suggestions, but concern was and still is at the time of writing that many have been established prematurely and in isolation by the EU regulatory group. What is not clear was whether or not discussions had been inclusive of the wider global regulatory community. Also not clear and apparent within the report was the awareness of the progression of scientific understanding relating to *N*-Nitrosamine risk factors, this being described in detail later in this chapter. As a consequence of these concerns both regulatory and technical, with this report and Article 5(3) [39], there was a general concern over consistent implementation and regulatory interpretation. That the matters highlighted had not been addressed globally, specifically at an ICH level; this being the most effective way of ensuring harmonized guidance and its consistent implementation.

10.6.3 Article 5(3) Report

The finalized Article 5(3) report was published by the EMA and CHMP in June 2020. [39] This report seeks to examine all aspects of both safety and quality related regarding the presence and risk associated with *N*-Nitrosamines within Medicines.

The report itself states that the areas covered include:

10.6.3.1 Quality

10.6.3.1.1 Root Causes for the Presence of N-Nitrosamines and Proposed Measures to Mitigate Them.

This includes:

- Environment, food, and drinking water.
- Chemistry – highlighting the Valsartan incident and also other examples, specifically Pioglitazone.
- Impact of water quality, the risk of generation of disinfection by-product of chloramination (by the reaction of monochloramine with dimethylamine).
- Theoretically possible root causes for *N*-Nitrosamines in pharmaceutical products linked with solvents, reagents, and catalysts. This talks specifically about the risk of secondary amine impurities in amide solvents, e.g. DMF, NMP.
- Confirmed root causes – this focuses on the formation of *N*-Nitrosamines within Sartans, formed as a result of the chemistry employed in the generation of the tetrazole ring, specifically the use of sodium azide and the subsequent use of sodium nitrite to quench it.
  Contained within this section is a useful guide to the pathways associated with the formation of the common synthetic *N*-Nitrosamines (Figure 10.19).
- Other cases, historical, including Ranitidine, linked to the molecule itself and potential NDMA formation on stability and aminophenazone, formation of NDMA resulting from hydrolysis and subsequent nitrosation, Figure 10.20.
- Other sources mentioned include excipients and primary packaging. In both cases the potential risk is described without detail as to the extent of the risk and clear identification of the specific risk factors and their control. Unfortunately, the report was issued with minimal consultation with industry and before outcome of risk assessments was complete. In other words, before a thorough and complete understanding of the science underpinning these risks was established. This is examined in detail in Section 10.4.1.

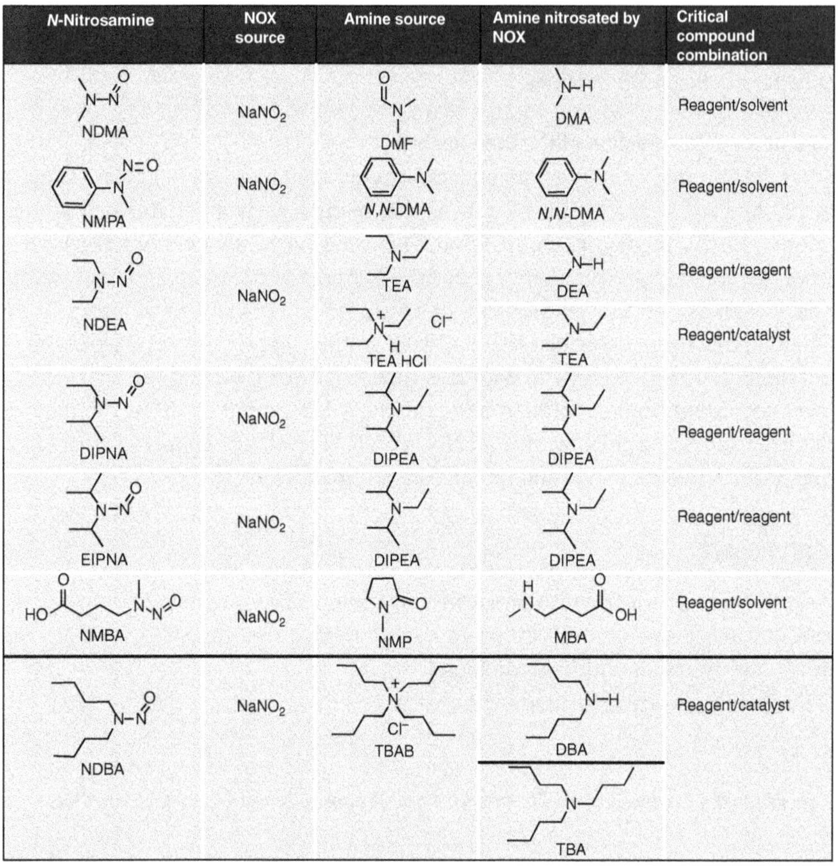

Figure 10.19 Reproduced from Article 5(3) report.

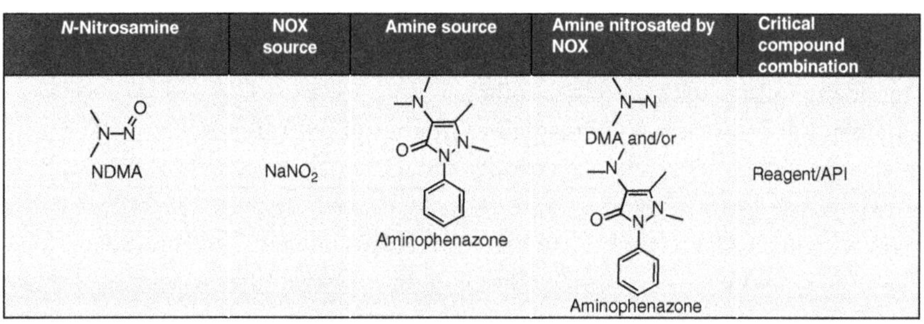

Figure 10.20 Reproduced from Article 5(3) report.

10.6.3.2 Consideration for Analytical Method Development to Identify and Quantify *N*-Nitrosamines in Drug Substances and Medicinal Products

Key Points:

- Techniques – This highlights the potential to measure *N*-Nitrosamines through use of either non-specific techniques or specifically with a clear preference of a combination of separative techniques and sensitive detection, i.e. LC/GC combined with mass spectrometry.
- Risk of erroneous results – this highlights risks such as in situ generation during analysis, sample contamination during sample preparation and misassignment – particularly DMF being misassigned as NDMA.

- Sample preparation.
- Current OMCL methods.
- Sensitivity.

The analysis of *N*-Nitrosamines is examined in detail in Chapter 12 of this book.

10.6.3.3 Safety

10.6.3.3.1 Considerations for Calculating Risk for Exposed Patients in Case of Detection of N-Nitrosamines in Medicinal Product(s)

- Background exposure – this section within the report provides a detailed overview of potential levels of exposure to *N*-Nitrosamines. These include processed foods, personal care products, and tobacco. It looks in depth at food in terms of both dietary exposure and also levels of *N*-Nitrosamines in drinking water. Of particular interest is the level of *N*-Nitrosamines observed in human urine, this being a key indicator of the extent of exposure both exogenous and endogenous. Specifically in terms of NDMA, studies indicate that the total NDMA exposure is very likely larger than ~1 μg/day for at least a part of the population. It is difficult not to reflect on this and form a view as whether the risk posed by pharmaceuticals is being addressed in the right context but that debate will likely remain active for some period of time and is beyond the intended scope of this work.
- Mutagenicity and Carcinogenicity of N-Nitrosamines. This highlights that most *N*-Nitrosamines are both mutagenic and carcinogenic but also notes that potencies vary significantly. This is an important point and highlights the need when considering the risk associated with *N*-Nitrosamines to examine this in the context of small dialkyl *N*-Nitrosamines such as NDMA and NDEA and also in larger *N*-Nitrosamines such as those of secondary amine drugs; this point is reflected on further in the Section 10.6.3.4 below relating to the recommendations made with the Article 5(3) report. The mutagenicity and carcinogenicity of *N*-Nitrosamines is also discussed in depth in Chapters 7 and 8 of this book.

Within this section, the mechanism of carcinogenicity is discussed, this being related to the generation of diazonium ion in vivo and this is illustrated in Figure 10.21.

N-Nitrosamines need to be activated metabolically to form different diazonium ions (e.g. methyl diazonium, ethyl diazonium, etc.). Alkydiazonium ions are precursors of reactive electrophilic carbenium ions, which directly react with DNA thereby forming stable adducts mainly with nitrogen and oxygen of guanine, cytosine, and thymidine.

Also examined in the context of potency is the fate of such adducts and the capacity of the body to repair DNA damage; different adducts are repaired by different cellular repair mechanisms with different capacity, velocity, and accuracy.

The report, in particular, highlights that the mutagenic and carcinogenic potential of *N*-Nitrosamines correlates with the presence of α-hydrogens, that their presence is necessary in order for metabolic activation by α-hydroxylation which subsequently leads to formation of alkyl diazonium

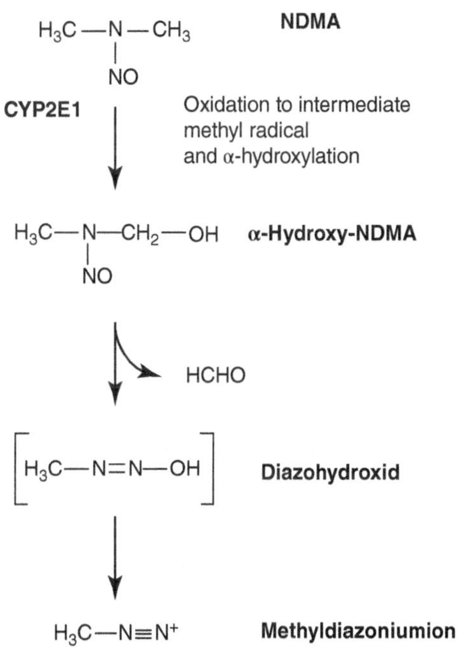

Figure 10.21 Metabolic activation of NDMA to generate the electrophilic methyl diazonium ion.

ions. Other factors such as stability of diazonium ions and steric factors are described. All of these factors again highlight that a "one size fits all" approach certainly to establishment of limits may be inappropriate, again this is examined in Chapters 7 and 8.

10.6.3.3.2 N-Nitrosamine Carcinogenicity in Animals

Presented within the Article 5(3) [39] report is a table of values derived from the Carcinogenicity Potency database. The values reported relate to a cancer incidence rate of 50%, i.e. TD50 values. These values are then linearly extrapolated back to a 1 in 100 000 risk, consistent with the TTC used in ICH M7 [1]. This is widely recognized as a very conservative approach effectively ignoring effective DNA damage repair mechanisms.

This table is reproduced here for reference (Table 10.6).

The report then discusses in some detail the variability of such data, depending on the number of dose groups within the studies conducted. Again the reader is referred to Chapter 8 for a more comprehensive assessment of *N*-Nitrosamine carcinogenicity data.

10.6.3.3.3 Use of in vitro Mutagenicity Data for Carcinogenicity Potency Ranking of N-Nitrosamines

This section looks at arguably one of the most contentious aspects of *N*-Nitrosamines. ICH M7 [1] makes clear that the primary concern is in relation to mutagenic carcinogens. Given that the mechanism of action associated at least primarily with *N*-Nitrosamines is mutagenic, then it may be expected that the principles defined within ICH M7 would apply, i.e. that to investigate potential carcinogenicity the primary test would be a bacterial reverse mutation assay, Ames test.

In the case of N Nitrosamines this has been challenged, the report states the following:

- *N*-Nitrosamines need to be activated metabolically and the artificial rat liver S9 mix used for simulation of metabolism in in vitro assays only provides limited metabolic competence,
- all standard Ames strains are alkyl transferase proficient and effectively repair alkylated guanine caused by small alkyl-*N*-Nitrosamines.

It also describes work conducted using alternative strains, by Wagner et al. [55] using the alkyltransferase-deficient Salmonella strain YG7108, which is specifically sensitive for *N*-Nitrosamine mutagenicity.

Separate from the Article 5(3) report, concerns over the use of DMSO have been reported also the method used, i.e. the need for preincubation being described and also the use of Hamster S9 as opposed to Rat S9 [56–59].

These concerns have called into question the validity of the Ames test as a suitable test for examining the risk posed by *N*-Nitrosamines, even to suggestions that alternatives such as transgenic rodent studies are needed.

This section also provides a summary of exogeneous exposure, reiterating that evidence from analysis of NDMA in urine shows levels to be 1–2 μg/day. No conclusion of relative risk is made in terms of exposure to ng levels potentially present in some pharmaceuticals.

What it does though is provide a useful overview of factors for consideration when assessing carcinogenic potency.

- The ability of the *N*-Nitrosamine to be metabolically activated.
- The metabolic competence and capacity of the tissue to form diazonium/carbenium ions.
- The nature and stability of the diazonium/carbenium ion and the DNA-adducts formed.

Table 10.6 Safety Limits for commonly observed M Nitrosamines – part 1

Agent	Abbreviation	IARC Group	TD_{50} [mg/kg/day] harmonic mean rat, CPDB	TD_{50} [mg/kg/day] most relevant study, sensitive species (tissue), CPDB	TD_{50} [mg/kg/day]other species, CPDB	Mutagenicity
Nitroso-*N*-methyl-*N*-(2-phenyl)ethylamine	NMPEA		0.00998 male only	0.00788, rat (ugi), Lijinsky et al 1982		Ames test positive (CPBD)
N-Nitrosodiethylamine	DENA, NDEA	2A	0.026	0.05, rat (liv), Peto et al 1991b; 0.026, rat (eso), Lijinsky et al 1981	0.00725, cynomolgus; 0.012 bush babies; 0.054, rhesus (harmonic means)	Ames test positive (CPBD)
N-Nitrosomethylethylamine	NMEA	28	0.053 (1 dose group)			Ames test positive (CPDB)
N-Nitrosodimethylamine	DMN, NDMA	2A	0.096	0.04 rat (liv), Peto et al 1991b; 0.06, rat (liv), Lijinsky et al 1984	0.189, mouse (harmonic mean)	Ames test positive (CPDB)
N-Nitrosonornicotine	NNN	1	0.096 (1 dose group)		10.8, hamster (harmonic mean)	Ames test positive, Padma et al 1989
4-(*N*-Nitrosomethylamino)-1-(3-pyridyl)-1-butanone	NNK	1	0.0999	0.182, rat (lun), Rivenson et al 1988		
N-Nitrosomorpholine	NMOR	2B	0.109	0.127, rat (liv), Lijinsky et al 1988	3.57, hamster (harmonic mean)	Ames test positive (CPDB)
N-nitrosomethylaniline	NMA, NMPA		0.142 (2 dose groups)		0.034 rat, Schmahl et al 1976	Positive in the hisG428 Salmonella strain TA104
N-Nitrosodi-n-propylamine	NDPA	2B	0.186 (1 dose group)		0.012 rhesus (liv)	Ames test positive (CPDB)
Nitrosodibutylamine	NDBA	2B	0.691 (1 dose group)		1.09 mouse (liv)	Ames test positive (CPDB)
N-nitrosopyrrolidine	NPVR	2B	0.799		1.7 rat (liv), Gray et al; 2.43 rat (liv), Berger et al 1987 0.697 mouse; (harmonic mean)	Ames test positive (CPDB)
N-Methyl-*N'*-nitro-*N*-nitrosoguanidine	MNNG	2A	0.803	0.284 rat (pyl), Zaidi et al 1993	2.03 mouse (harmonic mean)	Ames test positive (CPDB)
4-methyl)(nitroso)amino) butanoc acid	NMBA		0.982 (1 dose group)			AMES test negative (CPDB) Ames test positive, Inami et al 2013
N-Nitrosopiperidine	NPIP	2B	1.43	1.31 rat (eso), Gray et al 1991	1.3 mouse (harmonic mean)	Ames test positive (CPDB)
N-Nitrosodiethanolamine	NDELA	2B	3.17	0.19 rat (liv) Lijinski et al 1985		Ames positive (CPDB)
N,N-diisopropylethyl-*N*-ethylamine	DIPNA		none		positive male only no TD_{50} calculated	Ames test negative, Kameswar et al 1979
N-nitrosodiphenylamine	NDPhA	3	167 (2 dose groups)		mouse, no positive	Ames test negative (CPDB)

Abbreviations: CPDB, carcinogenic potency database; eso, oesophagus; liv, liver; lun, lung; pyl, pylorus; ugi, upper gastrointestinal tract.

- The capacity, velocity, and accuracy of the different cellular repair mechanisms responsible for the repair of the different DNA-adducts in tissues.
- Susceptibility (metabolic and proliferative) of the tissues exposed.

Critically, it looks to focus primarily on these *N*-Nitrosamines containing an α-hydrogen that can be metabolically activated as potentially mutagenic and carcinogenic to humans. This is important as other *N*-Nitrosamines such as heteroaromatic *N*-Nitrosamines of, for example, Histidine and Tryptophan are reported to be Ames positive but contain no α-hydrogen. In such instances, the profile of the Ames test is significantly different, histidine and Tryptophan *N*-Nitrosamines are positive without S9 activation, thus they do not undergo the same mechanism as dialkyl *N*-Nitrosamines that generate the highly reactive and carcinogenic alkyl diazonium ion.

Unfortunately, no specific carcinogenicity data exist for either of these two heteroaromatic *N*-Nitrosamines; however, given the different mechanism, it seems reasonable to exclude these from the specific cohort of concern associated with dialkyl *N*-Nitrosamines possessing the capacity to generate an alkyl diazonium ion.

Arguably the most contentious point made in the entire document is in relation to *N*-Nitrosamines where robust TD50 values are not available. In such cases, the report recommends using a class-specific threshold of theoretical concern (TTC) of 18 ng/day as the default option with the possibility to justify a higher limit based on the structure–activity relationship (SAR) approach described in ICH M7(R1). It also states that the class-specific AI of 18 ng/day for nitrosamines was determined using a novel methodology not widely used previously in the industry. The specific approach used to derive a class-specific TD50 was to use the TD50 data of all nitrosamines listed in Lhasa carcinogenicity potency database (LCDB) and use of the lower 5th percentile. This TD50 was then used to calculate the excess risk which would in theory not be exceeded with 95% probability by any nitrosamine.

Allied to this there is discussion around the possible presence of more than one *N*-Nitrosamine within a specific medicine and also multiple *N*-Nitrosamines across multiple medicines, although in the case of the latter it is difficult to see how this can be addressed. These scenarios, considered additive by EMA, raise the possible need to sum risks. In the specific case where more than one *N*-Nitrosamine occurs in manufacture, it suggests that it may be acceptable to limit the sum of *N*-Nitrosamines to the limit of the most potent one found.

In a further departure from ICH M7 [1], the concept of limits based on less than lifetime is not recommended for *N*-Nitrosamines, it being stated that LTL approach could lead to high acute nitrosamine intake, especially with medicines given at high doses and for a short period of time and that this could lead to overload of repair mechanisms. Unfortunately, no data are provided as evidence of this risk.

Taking all of the factors described above, a default limit of 18 ng/day, additive affects, and prohibition of use of the LTL approach lead to a very conservative, some might say restrictive, approach that could have a significant impact.

In the final part of the section addressing safety risk, the report provides:

- Comparison of means by which limits may be established covering analytical capability, i.e. sensitivity, ALARP, limits based on ICH M7 [1] methodology factoring in LTL and poly pharmacy, concluding that the correct approach is to use ICH M7 methodology and lifetime limits.
- Also discussed is the potential for future evaluation of *N*-Nitrosamine-related risk. This challenges the value of further carcinogenicity studies based on time, cost, proposing instead the

using of transgenic mutation studies; see Chapter 8 for further discussion on safety testing of mutagenic impurities.

- Also discussed is the potential for epidemiological studies, the scale of this though is potentially huge and it is difficult to see how such studies would ever distinguish risk posed by low-level exposure to *N*-Nitrosamines in pharmaceuticals to that of exogenous and endogenous exposure, especially as most pharmaceuticals will not contain *N*-Nitrosamines at levels of concern.

Chronology

Also included within the report is a detailed chronology of the expert opinion provided by both the EMA Quality Working Party and Safety Working Party. This is an interesting insight into the evolution of views of the different parties based on a series of predefined questions. It is beyond the scope of the chapter to examine these and to comment around the thinking of both SWP and QWP, but it is nevertheless interesting to read this in full to obtain a holistic understanding of the thought process and debate that lead to the final recommendations of the report.

10.6.3.4 Conclusions

This section summarizes the conclusions drawn in relation to all of the areas covered by the report, these are summarized in Table 10.7. Included within this table are comments/reflections of the authors.

10.6.4 EMA Question and Answer Document [6]

First issued to accompany the Article 5(3) request [16], this has gone through two significant iterations, the first on 25 March 25, one day before the original deadline. Within this, the EMA announced the postponement of the deadline, extending this to 1 October, an extension of six months, the primary reason being the challenge of completing risk assessments due to the Covid-19 pandemic. To accompany this, several revisions were made to the EMA Q and A document [6]. In relation to the original Questions and their associated answers, the changes of most significant were:

Text added to address new marketing applications in Europe; this made clear that new applications must address the risk of *N*-Nitrosamines within the submission itself, that without doing so could delay approval. This correlates with the experience of applicants where many received major objections during regulatory questions ahead of approval due to the absence of an assessment of *N*-Nitrosamine risk. This is somewhat unfortunate as these submissions clearly predated clarification of requirements.

This specifically states that:

- *At the submission stage: – For the risk evaluation, Applicants are required to follow the principles stated in "step 1 risk evaluation" of the Information on nitrosamines for marketing authorisation holders and to submit the a risk evaluation documentation as part of their MAA.*
- *If at this stage, a risk of presence of nitrosamines in the medicinal product is already identified, the applicants are required to provide the risk assessment outlining the impact on the benefit/risk balance of the product and a risk mitigation strategy. Applicants should also submit confirmatory testing plans or confirmatory testing data as mentioned in the step 2 of the "Information on nitrosamines for marketing authorisation holders."*

Table 10.7 Conclusions drawn in relation to all areas covered by the Article 5(3) report (with comments and reflections of the authors).

Key area	Conclusion	Comment
Root cause of the presence of *N*-Nitrosamines	**Identified causes were:**	
	• Use of sodium nitrite ($NaNO_2$), or other nitrosating agents in the presence of secondary or tertiary amines or quaternary ammonium salts, or in combination with reagents, solvents, and catalysts, which are susceptible to degradation to secondary or tertiary amines. • Use of contaminated raw or recovered materials, e.g. solvents, reagents, and catalysts (GMP issue, API).	This correlates well with the root causes identified by industry and many of these have been observed, primarily in the context of the Sartans investigations.
	• Use of nitrosamine-contaminated starting materials or intermediates (API). • Cross contaminations (related to GMP) due to different processes run on the same line and due to operator-related errors such as inadequate phase separations (API). • Degradation processes of starting materials, intermediates, and drug substances, including those induced by inherent reactivity in combination with carryover of sodium nitrite ($NaNO_2$), or other nitrosating agents. • Contamination from blister packaging materials.	The challenge is to define a risk assessment process to address the evaluation of the extent of each of these factors. While some of these relate specifically to the chemistry employed and the nature of specific drug substance, others relate to the very basics of good manufacturing practice (GMP).
		A risk assessment process is described above in earlier sections.
	In addition to the identified root causes, a series of theoretical risks were also defined, these being: • Additional GMP issues may include cross contamination during medicinal product manufacture, e.g. due to contaminated solvents or process equipment. • Formulation in general: nitrites from excipients could react with amines in APIs, or low molecular weight amine impurities. The experts were therefore in favor of testing of excipients for nitrite. Since excipients are generally the greatest component of medicinal products, high amounts of nitrite could be present as a reaction partner. Contamination of excipients with nitrosamines was also seen as a theoretical possibility. • Storage conditions of APIs (e.g. impact of container).	While apparently reasonable in their definition and rationale, a number of these theoretical risks present a substantive practical challenge from the perspective of understanding the extent of the risk. This is particularly true in the context of the risk associated with formulation and the potential formation of *N*-Nitrosamines due to the presence of traces of nitrites in certain excipients.
	• Packaging composition (e.g. use of nitrocellulose beyond blister packaging). • Additional degradation pathways: subsequent degradation of a nitrosated API or nitrosated impurity to smaller nitrosamines. • Storage conditions after packaging. • Water quality: nitrosamines in treated water as a result of the use of chloramine (or chlorine which can form chloramines with any amines present) and further reaction to nitrosamines. Other oxidants (e.g. ozone) can lead to NOx formation which could then react with amines to generate nitrosamines.	At the time, the Article 5(3) report was finalized and many of these factors were still under investigation.

• Possible reactions of volatile low molecular weight amines occurring in the manufacturing process (solvents, raw materials, in combination with nitrosating agent). Their volatility means they could potentially carry over during e.g. distillation processes. • Structure inherent to the API molecule or to intermediates in its synthesis, including presence or generation of amines susceptible to nitrosation in the manufacturing process. • Reactions of quaternary amines in addition to secondary and tertiary amines (e.g. use of Tetrabutylammonium bromide (TBAB) which could give rise to *N*-nitrosodibutylamine (NDBA) although the lack of an available electron lone pair makes this an unlikely mechanism, nevertheless, NDBA has been found in some instances. Nitrosamine formation seems in general however more likely to be a result of tertiary and secondary amine impurities. Also, dimethylacetamide was mentioned as a reaction partner with nitrosating agents. • Nitroalkanes are also known nitrosating agents. Other potential root causes might include emissions from vulcanization processes (rubber, also when it is in contact e.g. with product) and carbon capture technology.	
Biologics: • The BWP concluded that there is only a very low risk of nitrosamines being present as impurities in biological medicinal products. At higher risk would be biological products containing chemically synthesized fragments, where risk factors similar to chemically synthesized active substances should be considered, or biologicals packaged in blister packs containing nitrocellulose. Consideration should be given to extending the risk evaluation to classes of biological product (see Section 4.3) using processes where nitrosating reagents are deliberately added. The CHMP agrees with the BWP advice and considers that a risk evaluation/risk assessment for biological medicinal products should be performed taking into consideration the abovementioned risk factors.	It is interesting that it was felt necessary to effectively extend the scope to biologics, particularly as those biologics containing synthetic fragments, e.g. antibody drug conjugates were already considered to be in scope. The risk of other biologics, e.g. proteins and antibodies is described above. However, in general, consistent with the statement in the report itself the risk is very low. It does somewhat ask the question – was this necessary?

(*Continued*)

Table 10.7 (Continued)

Key area	Conclusion	Comment
Analytical methods for *N*-Nitrosamines	• The conclusions specifically discuss the publication on 30 March 2020, of three analytical methods for quantifying *N*-Nitrosamines in Pharmeuropa 32.2, 2.4.36, which are stated there as method A (LC-MS), method B (GC-MS, and method C (GC-MS). • The conclusion makes the following statement: – The combination of these three methods allows to analyze the following *N*-Nitrosamines: *N*-nitrosodimethylamine (NDMA, methods ABC); *N*-Nitrosodiethylamine (NDEA, methods ABC); *N*-nitroso-dibutylamine (NDBA, method C); *N*-nitroso-*N*-methyl-4-aminoburyric acid (NMBA, method A); *N*-nitroso-diisopropylamine (NDiPA or DIPNA, methods AC); *N*-nitroso-ethyl-isopropylamine (NEiPA or EIPNA, methods AC), and *N*-nitroso-dipropylamine (NDPA, method C) in Sartan-containing products and is considered suitable for additional APIs and finished products.	The analysis of *N*-Nitrosamines is discussed in detail in Chapter 12. This directly challenges the assertion that the methods are suitable for other, additional APIs and finished products. This in practice is likely to be incorrect; such analysis as indeed noted elsewhere in the Article 5(3) report [39] is complex and often needs specific attention to factors such as sample preparation and also avoidance of artifacts that may generate misleading results. Another major challenge is the insistence that LoDs must be ≤10% of the safety limit and that testing must be performed on the drug product. Given the complexity associated with the effect of the matrix associated with the drug product and a default safety limit of 18 ng/day this will push LoDs to below 1 ppb.
Setting Limits for *N*-Nitrosamines in human medicinal products	This section again revisits some of the debate over whether to adopt an avoidance/ALARP approach as taken for Sartans, concluding that: • The CHMP considers that setting limits for individual nitrosamines in human medicinal products based on ICH M7 [1] principles for substances of the "cohort of concern" and calculated considering a lifetime daily exposure is recommended as the best option after careful consideration of patient safety and regulatory practical requirements. This also comments on how to address more than one *N*-Nitrosamine: • The sum of all nitrosamine-specific risk levels should not exceed a total risk level of 1 in 100 000. This being achieved by controlling the sum of all nitrosamines to the limit of the most potent one.	It is interesting to reflect on each of these conclusions. While stating that the approach is consistent with ICH M7, it is clear in key areas it is not. Most notably the rejection of the less than lifetime approach described in ICH M7.

- *In case applicants have not submitted a risk evaluation and, if applicable, confirmatory testing plans with their MAA, these should be submitted during the marketing authorisation review procedure.*

In addition, other changes included:

Question 5 – When should MAHs report to competent authorities?

This stated that – The step 1 response template should already be used if a risk has been identified for the API. In this case, the MAH is advised to proceed directly with step 2 confirmatory testing of the finished product. This is concern for a number of reasons: first, if the risk relates to the synthesis, then it makes scientific sense to focus testing on the API; moreover, the analysis is far simpler when testing the API as there is far less issue with the sample matrix, as identified in Chapter 12, matrix effects and sample preparation can have a major impact on the suitability or otherwise of an analytical method. As examined above in the evaluation of the Article 5(3) final report, the requirement to test drug product remains in place despite concerns.

Another key area relates to question 16 – What limits will apply for nitrosamines in medicinal products based on lifetime and less than lifetime use? This stated that:

For any new cases of nitrosamine detection in a medicinal product, the MAH should apply, whilst waiting for the outcome of the CHMP Art 5(3) procedure, interim limits calculated for a lifetime treatment and based on a maximum daily dose of the medicine. These interim limits (ILs) have been defined for NDMA and NDEA impurities in the Article 31 referral assessment report. Furthermore, for NMBA, NDBA, DIPNA and EIPNA, additional interim limits calculated by the Safety Working Party (SWP) and agreed by the CHMP and CMDh are summarized in Table 10.8.

It also stated that:

Where the interim limit is exceeded for medicinal products with a limited treatment period or intermittent treatment (e.g. once a week), higher daily exposures may be used as an adjusted interim limit. The approach described in the ICH M7 guideline as the Less Than Lifetime (LTL) approach can be used to calculate adjusted interim limits for impurities present in medicinal products given for LTL and these are described in Table 10.9.

This would clearly seem to state that a LTL approach was permissible, this would of course be entirely consistent with the principles outlined in ICH M7 [1]. This was contradicted less than four months later with the publication of the Article 5(3) report.

Table 10.8 Nitrosamines and their interim limits.

Nitrosamine	Interim limit (ng/day)
NDMA, NMBA	96
NDEA, NDBA, DIPNA, EIPNA	26.5

Table 10.9 Adjusted interim limits for impurities present in medicinal products.

Duration	1 day to 1 month	1 month to 1 year	1 year to 10 years	10 years to lifetime
Daily intake	80×IL	13.3×IL	6.7×IL	IL

10.6.4.1 Further Revision of the EMA Question and Answer Document

One consequence of both the Sartans lessons learnt report and in particular the Article 5(3) report [39] was that the question and answer document revised in March 2020 was no longer reflective of the EMA position, thus in August 2020, a third version add ref of the document was issued. It is not the intent to reproduce this but to reflect on the most significant changes and their implications (Table 10.10)

Yet again, although associated with a welcome extension to the deadline in Europe as a result of the publication of the Article 5(3) report and revision of the associated Q&A document, industry faced significant and new implementation challenges.

10.6.5 FDA Guideline

Published in September 2020, the FDA guideline rather as expected largely mirrors the position defined in the Article 5(3) report and the associated revised Q&A document. There are though areas that do differ, some general, others specific. These are outlined below including reflections on their potential impact:

10.6.5.1 Introduction and Background

Contained within sections I and II, like the EMA Article 5(3) report, the FDA guideline provides an introductory commentary highlighting the sequence of events relating to unfolding of issues relating to *N*-Nitrosamines. It also provides details of those *N*-Nitrosamines observed to date and general root causes. There is nothing novel in this but it does indicate that quaternary amines are a risk as well as secondary and tertiary amines. In reality, there is little substantive evidence to support this. Within the root causes section, it highlights sources of amines as well as issues relating to cross contamination, recovered solvents, quenching of azides and other sources of risk including process water. Again, little additional insight is provided but this a useful concise overview. Some elements of this are though concerning:

- It states that Tertiary amines, such as triethylamine, have been shown to contain low levels of other secondary amines (such as dipropylamine and isopropyl ethylamine) – *what does this mean? That Triethylamine contains these? If so, what is the evidence for this?*
- Nitrosamine contamination has occurred when fresh solvents (ortho-xylene, toluene, and methylene chloride) were contaminated during shipment from vendors (e.g. during transfer between storage vessels). *Again, where is the evidence of this? As written, this would seem to imply the need for routine testing of fresh solvents.* Especially when combined with this statement:
- Secondary or tertiary amines have been reported as impurities in some raw materials (see details in section II.B.2 in this guidance) and in fresh solvents such as toluene.

10.6.5.2 Recommendations

The main body of guideline is taken up with recommendations; it begins with a general recommendation that manufacturers should prioritize evaluation of APIs and drug products based on factors such as maximum daily dose, duration of treatment, therapeutic indication, and number of patients treated. *Are these really the right prioritization factors? Surely risk of generation of an N-Nitrosamine should be the primary determining factor?*

Table 10.10 Most significant changes and their implications of the question and answer document (version August 2020).

Question/Answer	Comment
Q2: Scope – Biologics Following the conclusion of the review under Article 5(3), the CHMP considered that there is also a risk of the presence of nitrosamines in biological medicinal products, in particular for the biological medicines with the following risk factors: • Biologicals containing chemically synthesized fragments, where risk factors similar to chemically synthesized active substances are present; • Biologicals using processes where nitrosating reagents are deliberately added; • Biologicals packaged in certain primary packaging material, such as blister packs containing nitrocellulose.	This is surprising in that the risk associated with biologicals containing chemically synthesized fragments was already effectively in scope. The most obvious example being antibody drug conjugates (ADCs). The risk associated with ADCs is discussed below. In terms of wider risks, these are again discussed above but few if any Biologics use blister packs and the generic risk is low given factors such as: • Molecular size, Monoclonal Antibodies having masses 150 kDa, far too large for even product-related *N*-Nitrosamines if they are formed to be mutagenic • Processing – purification likely to remove certainly any small molecular weight process-related *N*-Nitrosamines or Nitrosating agents
Q3: Modified Timelines – Submission of step 1 outcome For product containing chemically synthesized APIs, the step 1 risk evaluation should be concluded and reported at the latest by 31 March 2021. For product containing biological APIs, step 1 risk evaluation should be concluded and reported at the latest by 1 July 2021.	This now aligned with the position stated in the conclusions of the Article 5(3) report
Q8: Confirmatory tests – risks of unrepresentative result Given the trace levels of nitrosamines to be measured, the following technical aspects should be considered when developing analytical methods: • Interference caused by the presence of trace amounts of nitrosamines in testing materials utilized (e.g. water, airborne sources, plastics products, and rubber/elastomeric products); • Contamination during sample preparation (avoiding cross contaminations from gloves, membranes, solvents, etc.) which could lead to false-positive results; • In situ formation of nitrosamines during analysis; • Use of accurate mass techniques is required (MS/MS or high-resolution accurate mass systems) in order to overcome interference in the identification of the specific peak of a certain nitrosamine (e.g. false positives have been observed from DMF co-eluting with NDMA).	This is useful in that it recognizes the complexity of the challenge of analysis

(Continued)

Table 10.10 (Continued)

Question/Answer	Comment

Q9: Analytical method requirements:

Focused on Sensitivity – this states the following:

If quantitative testing is performed as a routine control, the LoQ should be ≤ of the acceptable limit based on the relevant acceptable intake (AI) for the respective nitrosamine impurity;

- If quantitative testing is performed to justify skip testing, the LoQ of the analytical procedure employed should be ≤30% of the acceptable limit based on the AI;
- **If quantitative testing is performed to justify omission of specification, the LoQ of the analytical method employed should be ≤10% of the acceptable limit based on the AI;**
- Exceptions are anticipated for medicinal products used at high daily doses (AI may be below technical feasibility of the method), or in case more than one nitrosamine is anticipated or identified in a given medicinal product.

Comment: The most significant of these is that highlighted in bold. Given the insistence that testing be focused on the Drug product as opposed to the Drug substance, the requirement to demonstrate levels at ≤10% of AI. For high-dose products where the generic limit of 18 ng/day is applied to an *N*-Nitrosamine (where there is no specific safety data), this is likely to require LOQs in the picogram range. Given the complexity associated with the drug product matrix (see Chapter 12), this is believed by many to be at the limits of, if not beyond, technical feasibility. This is caveated with the exceptions comment on high daily dose but who will decide when this can be employed?

Q10: Which limits apply for nitrosamines in medicinal products?

The following limits are defined for a series of common *N*-Nitrosamines – reproduced from the EMA Q&A document

***N*-Nitrosamine (CAS number)**	**ng/day**[a]
NDMA[b] (62-75-9)	96.0
NDEA[b] (55-18-5)	26.5
EIPNA[c] (16339-04-1)	26.5
DIPNA[c] (601-77-4)	26.5
NMBA[c] (61445-55-4)	96.0
MeNP[c] (16339-07-4)	26.5
NDBA[c] (924-16-3)	26.5
NMPA[b] (614-00-6)	34.3

Also stated is that:

- A class-specific TTC for nitrosamines of 18 ng/day (derived from the Lhasa carcinogenic potency database) can be used as default option.

It also states that:

- an approach based on SAR considerations to derive an acceptable intake limit is acceptable, if appropriately justified.

  Also stated is:

  For determining limits in the case of presence of more than one nitrosamine, two approaches are considered acceptable in order not to exceed the acceptable risk level of 1:100 000 as outlined in the ICH M7(R1) [1] guideline:

1) The total daily intake of all identified *N*-Nitrosamines not to exceed the AI of the most potent *N*-Nitrosamine identified, or
2) Total risk level calculated for all identified *N*-Nitrosamines not to exceed 1 in 100 000.

Comment: The adoption of 18 ng/day as a default limit is a concern. Generation of limits for *N*-Nitrosamines is covered in detail in Chapter 6.

While the potential use of SAR and read across is welcome, it is anticipated that there may be practical challenges in seeking to do this.

[a] Limit calculated on the basis of harmonic mean TD50 derived from carcinogenic potency database (CPDB).
[b] Limit derived using structure-activity-relationship (SAR)/read-across approach.
[c] The conversion to a specification limit in ppm for a particular medicinal product is calculated by dividing the respective above limit (ng) by the maximum daily dose (mg) of a given product as reflected in the SmPC.

10.6.5.3 Acceptable Intakes (section III.A)

In comparison to the highly detailed section within the EMA Article 5(3) report, this more limited and vanilla in nature. This includes specific recommendations on limits for individual *N*-Nitrosamines that mirror those in the EMA Q and A document:

Nitrosamine	AI Limit (ng/day)[a,b]
NDMA	96
NDEA	26.5
NMBA	96
NMPA	26.5
NIPEA	26.5
NDIPA	26.5

[a] The AI limit is a daily exposure to a compound sush as NDMA, NMBA, NMPA, NIPEA, or NDIPA that approximates a 1:100 000 cancer risk after 70 years of exposure. Appendix B includes a description of the AI derivation for NDMA, which is an example of how FDA applied ICH M7 (R1) to set a limit.
[b] The conversion of AI limit into ppm varies by product and is calculated based on a drug's maximum daily dose (MDD) as reflected in the drug label (ppm = AI (ng)/MDD (mg)).

However, while aligned in terms of those specific *N*-Nitrosamines for which there are effective data, there are areas that remain ambiguous:

- **Limits as they related to the clinical phase.** It seems that in line with the EMA position, lifetime limits will be applied to commercial products but it is unclear if the principles of LTL apply, as industry believe they should, during the in clinical phase.
- Unlike the EMA guideline where a default limit in the absence of specific safety of 18 ng/day is applied, the FDA position is less clear, the guideline simply stating that *"If nitrosamines without published AI limits are found in drug products, manufacturers should use the approach outlined in ICH M7(R1) to determine the risk associated with the nitrosamine and contact the Agency about the acceptability of any proposed limit."* This may be considered as a positive offering the potential at least for some flexibility.
- **Multiple *N*-Nitrosamines** The guidance states that the limits are applicable only if a drug product contains a single nitrosamine. If more than one nitrosamine impurity is detected, the limit for total level is capped at 26.5 ng/day (the AI for the most potent nitrosamines). Further stating that the manufacturer should contact the Agency for evaluation if this is exceeded. The requirement for the multiple nitrosamines to not exceed the sum total of 26.5 ng/day is not aligned with the ICH M7 principles for multiple impurities.
- **ICH S9** The document states that M7 principles apply but unlike the EMA guidance it is not clear if this includes key points such as S9 considerations for products which are known to be cytotoxic or for advanced cancer indications.
- **Safety testing** – noticeable by its absence is the fact that the FDA guideline is silent on what testing is needed to eliminate safety concerns. Clearly, ICH M7 is predicated on the conduct of an Ames test. It is though unclear whether or not the FDA supports this view or requires additional testing, e.g. Transgenic studies.

10.6.5.4 Quality/Chemistry and Controls

10.6.5.4.1 Section III (B) Recommendations to API Manufacturers

Overall, there are elements of this section that align well with the concepts of a risk-based approach described in both ICH M7 and ICHQ9. However, there is a concern that the FDA remains focused on testing, even when the risk is negligible, in a manner that is not aligned with ICH principles. Also of concern is the concept within this that somehow the synthesis can be designed in such a way as to eliminate even theoretical risks, among several these include:

- recommendations to "Use bases other than secondary, tertiary, or quaternary amines (when possible) if ROS conditions may form nitrosamines."
- Replacing nitrites with other quenching agents for azide decomposition processes. It is not even known if this is possible and that if possible such procedures are safe at typical manufacturing scale.

Within Chapter 3, the Candesartan case studies clearly illustrate that control rather than avoidance is a perfectly adequate means by which the risk of *N*-Nitrosamines can be eliminated. Hazard and risk are not the same thing.

Other specific areas of concern relate to the following comments:

- "Given existing uncertainties regarding nitrosamine impurities and their presence in drugs, for APIs with an impurity detected above the LOQ or at-risk APIs, testing of each batch on release should be conducted." This makes no statement about the level detected and the relationship between this and the acceptable limit. Contrast that with the EMA guideline where a requirement of <10% permissible limit is stated, below this no testing is required.

10.6.5.4.2 Recommendation for Drug Product Manufacturers

Again, there are real concerns over specific statements within this section, these include:

- *"If a risk of nitrosamines in a drug product is identified, confirmatory testing of batches should be conducted using sensitive and appropriately validated methods. If a nitrosamine impurity is detected, manufacturers should investigate the root cause and implement changes in the manufacturing process to mitigate or reduce nitrosamine impurities."* Why if levels are << levels defined as safe is a root cause investigation automatically triggered?
- *"Drug product manufacturers must test representative samples of <u>all</u> incoming components, including lots of at-risk API, prior to use, as required under 21 CFR 211.84 . . . (At-risk APIs include APIs with secondary, tertiary, and quaternary amine functional groups. They also include any API with an ROS using at-risk materials) . . . drug product manufacturers should continue testing API lots until they have verified that the API supplier can consistently manufacture API without unacceptable levels of nitrosamine."* There appears little thought around the practicalities of this, again this reflects a philosophy allied almost entirely to a testing to address hazard as opposed to testing where an actual risk is identified, philosophy.
- Another concerning aspect is the following statement:

"Under section 501 of the Food, Drug, and Cosmetic Act (FD&C Act), a drug that is not manufactured, processed, packed, or held in conformity with CGMP to ensure that the drug meets certain quality and purity standards is <u>considered adulterated</u>. FDA may exercise regulatory discretion when warranted to prevent or mitigate a shortage of a drug." It is perhaps surprising the word adulterated is chosen here given the clear definition of adulteration that this is a deliberate act.

What is ultimately clear with both this guidance and that of the EMA is need for harmonization and alignment under ICH M7.

10.7 Way Forward

To have any chance of addressing the concerns surrounding *N*-Nitrosamines needs industry and regulators to work in partnership to ensure the real risks are identified, targeted, and removed. What is clear is that with the requirement to report detectable levels and agencies themselves reporting methods capable of limits of detection of 5 ppb or lower, we are at the vanguard of what is experimentally achievable.

The start point should be ICH M7 [1], which already provides the necessary framework to conduct a risk-based approach to the assessment. Without proper focus, there is a real risk that in targeting every scenario that time is lost in identifying where the real problems lie. Like all such issues the answer is in the science and it is in everyone's interest to ensure that the science in this area is properly defined.

One approach would be to define a systematic process for such as risk assessment, see Figure 10.3.

A critical first step is to ascertain whether or not a Nitrosating agent is used or is present within the synthesis employed. Nitrosating conditions include diazotization reactions and the use of the following reagents sodium nitrite, nitrous acid, nitric oxide and hydroxylamine, nitrosyl halides, dinitrogen trioxide, and dinitrogen tetroxide. Nitration reactions also pose a risk due to the potential presence of nitrous acid formed by the reduction of nitric acid. This evaluation should include the use of all chemicals within a process including those during a quench (such as azide or similar).

Consider potential sources of secondary amines such as NMP, DMF, and diisopropylethylamine. Secondary amines can be present in reagents and solvents as impurities, degradants, or an integral part of API or intermediate structures.

For example, amide solvents can degrade to secondary amines which are known sources of nitrosamines (such as DMF, NMP, or *N*,*N*-dimethylacetamide). For example, *N*-methyl-4-aminobutyric acid can form from the degradation of NMP. Tertiary amines including common bases are already implicated in *N*-Nitrosamine formation (i.e. triethylamine, diisopropylethylamine). Other less common bases are sometimes used in manufacturing processes, for example, *N*-methyl morpholine, tributylamine, diisopropylethylamine, trimethylamine, dimethylamine, diethylamine, or dibutyl amine. Other sources of secondary and tertiary amines include impurities in or degradants of quaternary ammonium salts such as tetrabutylammonium bromide (TBAB) or even in primary amines such as monomethylamine. This list of sources is not exhaustive as many other amine reagents (including alkylated anilines such as *N*-methylaniline), catalysts, or solvents can be used to mediate a range of synthetic transformations. Other reagents containing amine functionality should be considered for the potential risk of *N*-Nitrosamine formation.

Critical in any such evaluation is proximity of any secondary amine to a nitrosating agent. The most significant risk, as demonstrated by Valsartan, is where a nitrosating agent and a secondary amine are present in the same process step. There also a risk of course of carryover between process stages and it is important in any risk assessment to consider this. Also, critical, even where a risk of *N*-Nitrosamines is present, is proximity of the stage of formation to the final API, as multiple process steps may result in purging of any trace *N*-Nitrosamines formed; however, this of course needs to be thoroughly assessed and not simply assumed.

As described above, the risk of extrinsic contamination from recycled solvents may also needs to be considered. In doing so, it may be useful to determine the exact nature of any such process, distinguishing between recycled solvents and fresh solvents.

A recycled solvent is typically a solvent used exclusively in the same step of a synthesis from which it was recovered. Recycled solvents are recovered by distillation carried out in a reactor as part of the standard manufacturing method.

Distilled solvent is a solvent derived from a distillation carried out in a solvent distillation plant dedicated to solvent recovery by means of a multistage distillation column. The solvent loaded into the rectification distillation process can originate from multiple steps.

In addition, it is critical further scientific understanding of risks associated with formulated products and packaging materials is needed and a common understanding of such risks and how to address them is gained.

Ultimately, to address this issue in an appropriate manner in the timescales defined by Authorities is a huge undertaking and one where a combination of science and risk-based approaches is critical.

Another important factor is harmonization of guidelines, and the relationship to ICH M7 [1]. It is interesting to look at the relationship of this issue and examine its alignment to the risk assessment principles outlined in ICH M7. ICH M7 defines the scope of risk assessment expected of the supplier of a medicine. The guidance states that a risk assessment should include an assessment of potential impurities associated with the drug substance manufacturing process; this including starting materials, reagents, and intermediates in the route of synthesis from the starting material to the drug substance. It also states that the risk of impurity carryover into the drug substance should be assessed for impurities that are present in starting materials and intermediates, and impurities that are reasonably expected by-products in the route of synthesis of the drug substance.

Therefore, it may be argued that rigorous evaluation of the valsartan manufacturing process, according to ICH M7, should have identified the risk of *N*-Nitrosamine formation. The issue is therefore one associated with the comprehensiveness of the risk assessment as opposed to a flaw in the guideline.

Critical in understanding the context of this issue with risk assessment is the evaluation of manufacturing by-products.

In the case of the issues with Sartans. Is the formation of *N*-Nitrosamines a reasonably expected by-product that should have been identified and controlled? With hindsight, the formation of nitrosamines is a predictable by-product in the manufacture of certain Sartans, but the formation needs several peripheral process elements to come together for the formation of the potential impurity and thus one might be able to understand how the potential for nitrosamine formation could initially have been missed.

Now that this potential is clear and fully understood, the issue can be addressed and effectively managed. This will ultimately be true of the other risk factors and thus it is hoped the ICH M7 guideline and guidance relating to *N*-Nitrosamines will be aligned and based on science.

Acknowledgments

The work and findings discussed within this book chapter have been culminated from a strong collaboration among numerous pharmaceutical companies, raw material suppliers, and through health authorities. The authors would like to specifically call out the main contributors toward the experimental design, data generation, and thoughtful interpretation of the meaning. The following contributors from Merck & Co., Inc., Kenilworth, NJ, USA are, in no particular order: Andrew

Gmitter, Daniel Lee, Karen Thompson, Kausik Nanda, Tony Hudgens, Jinjian Zheng, Sanjay Mudur, Sandy Sandbeck, Tan Truong, John Carboy, Steven Wang. From AstraZeneca: Dawn Sievwright, Mark Harrison, Ed Bush, Heather Akehurst, Paula Tomlin.

References

1. ICH M7. Assessment and control of DNA reactive (mutagenic) impurities in pharmaceuticals to limit potential carcinogenic risk. https://www.ich.org/fileadmin/Public_Web_Site/ICH_Products/Guidelines/Multidisciplinary/M7/M7_R1_Addendum_Step_4_2017_0331.pdf (accessed September 2020).
2. EMA Press Release. 5 July 2018. https://www.ema.europa.eu/en/news/ema-reviewing-medicines-containing-valsartan-zhejiang-huahai-following-detection-impurity-some (accessed September 2020).
3. FDA News Release. https://www.fda.gov/news-events/press-announcements/fda-provides-update-its-ongoing-investigation-valsartan-products-and-reports-finding-additional (accessed September 2020).
4. FDA News Release. https://www.fda.gov/drugs/drug-safety-and-availability/fda-updates-and-press-announcements-angiotensin-ii-receptor-blocker-arb-recalls-valsartan-losartan#5cc3b83d3b22e (accessed September 2020).
5. CHMP. List of questions to be addressed by the API manufacturers for valsartan-containing medicinal products. Referral under Article 31 of Directive 2001/83/EC 16 July 2018 EMA/CHMP/466974/2018.
6. Questions and answers for marketing authorisation holders/applicants on the CHMP Opinion for the Article 5(3) of Regulation (EC) No 726/2004 referral on nitrosamine impurities in human medicinal products. 03 August 2020 EMA/409815/2020.
7. Assessment Report. Referral under Article 31 of Directive 2001/83/EC angiotensin-II-receptor antagonists (sartans) containing a tetrazole group Procedure no: EMEA/H/A-31/1471. 14 February 2019 EMA/217823/2019.
8. Update on nitrosamine impurities: EMA continues to work to prevent impurities in medicines. 26 April 2019 EMA/241020/2019.
9. https://www.fda.gov/news-events/press-announcements/statement-alerting-patients-and-health-care-professionals-ndma-found-samples-ranitidine (accessed September 2020)
10. EMA to review ranitidine medicines following detection of NDMA, 13 September 2019 EMA/503622/2019. https://www.ema.europa.eu/en/documents/press-release/ema-review-ranitidine-medicines-following-detection-ndma_en.pdf (accessed September 2020).
11. https://www.valisure.com/wp-content/uploads/Valisure-Ranitidine-FDA-Citizen-Petition-v4.12.pdf (accessed September 2020).
12. King, F.J., Searle, A.D., and Urquhart, M.W. (2020). Ranitidine—investigations into the root cause for the presence of N-Nitroso-N,N-dimethylamine in ranitidine hydrochloride drug substances and associated drug products. *Org. Process Res. Dev.* 24 (12): 2915–2926. https://doi.org/10.1021/acs.oprd.0c00462.
13. https://www.fda.gov/drugs/drug-safety-and-availability/laboratory-tests-ranitidine (accessed September 2020).
14. https://www.fda.gov/news-events/press-announcements/fda-requests-removal-all-ranitidine-products-zantac-market (accessed September 2020).

15 EMA to provide guidance on avoiding nitrosamines in human medicines, 13 September 2019 EMA/500009/2019.

16 Article 5(3) EMA advises companies on steps to take to avoid nitrosamines in human medicines, 26 September 2019 EMA/511347/2019.

17 ICH Q9 ICH guideline on quality risk management. September 2015 EMA/CHMP/ICH/24235/2006.

18 Information to Marketing Authorization Holders (MAHs) of Human Pharmaceutical Products Regarding Nitrosamine Impurities. 2 October 2019, 19-119258–557.

19 Potential contaminants with Nitrosamines: Call for risk evaluation. 15 November 2019. Published Swiss Medic Website.

20 Christopher J. Borths*, Michael Burns, Timothy Curran, and Nathan D. Ide Organic Process Research & Development 2021, 25, 8, 1788–1801 (Review) Subscribed Access Publication Date (Web):July 18, 2021.

21 Ashworth, I.W., Dirat, O., Teasdale, A., and Whiting, M. (2020). Potential for the formation of *N*-nitrosamines during the manufacture of active pharmaceutical ingredients: an assessment of the risk posed by trace nitrite in water. *Org. Process Res. Dev.* 24 (9): 1629–1646.

22 Lim, S., Lee, W. et al. (2016). N-nitrosodimethylamine (NDMA) formation during ozonation of N,N-dimethylhydrazine compounds: reaction kinetics, mechanisms, and implications for NDMA formation control. *Water Res.* 105: 119–128.

23 ICH Q10 – Guideline on pharmaceutical quality system. September 2010 EMA/CHMP/ICH/214732/2007.

24 ICH Q3A. Impurities in new Drug Substances. October 2006 CPMP/ICH/2737/99.

25 Comins, D.L. and Joseph, S.P. (2001). *N,N*-Dimethylformamide. In: *N,N-Dimethylformamide. Encyclopaedia of Reagents for Organic Synthesis.* Wiley https://doi.org/10.1002/047084289x.rd335. ISBN 9780470842898.

26 Wu, Y., Levons, J., Narang, A.S. et al. (2011). Reactive impurities in excipients: profiling, identification and mitigation of drug-excipient incompatibility. *AAPS PharmSciTech* **12** (4).

27 Ashadullah, A.K.M. and Kishimoto, N. (2016). Applicability of a reverse osmosis (RO) membrane to the treatment of wastewater contaminated with nitrite, nitrate, iron and manganese. *Intl. J. Water Wastewater Treatment* **2** (1): 1–9.

28 https://www.usp.org/sites/default/files/usp/document/harmonization/excipients/2011-02-25-povidone.pdf (accessed September 2020).

29 Byrn, S.R., Xu, W., and Newman, A.W. (2001). Chemical reactivity in solid-state pharmaceuticals: formulation implications. *Adv. Drug Deliv. Rev.* **48**: 115–136.

30 Mirvish, S.S. (1975). Formation of *N*-nitroso compounds: chemistry, kinetics and in-vivo occurrence. *Toxicol. Appl. Pharmacol.* **31**: 325–351.

31 Ireland, C.B., Hytrek, F.P., and Lasoskl, B.A. (1980). Aqueous extraction of *N*-nitrosamines from elastomers. *Am. Ind. Hyg. Assoc. J.* **41** (12): 895–900.

32 Sen, N.P., Seaman, S., Clarkson, S. et al. (1984). Volatile *N*-nitrosamines in baby bottle rubber nipples and pacifiers. Analysis, occurrence and migration. *IARC Sci. Publ.* 57: 51–57.

33 56988, Children's pacifiers containing Nitrosamines enforcement policy. (1983). 57014 Baby bottle nipples, rubber; action levels for total volatile *N*-nitrosamines: compliance policy guide. *Federal Register* **48**, 56927–57100.

34 Food and Drug Administration. (1984). Compliance Policy Guide CPG Sec. 500.450 Volatile *N*-nitrosamines in rubber baby bottle nipples.

35 Bottles, D.C., Havery, D.C., and Fazio, T. (1982). Estimation of volatile *N*-nitrosamines in rubber nipples for babies. *Food Chem. Toxicol.* **20** (6): 939–944.

36 (2017). *J. Aerosol Med. Pulm. Drug Deliv.* **30** (1): 20–41. https://doi.org/10.1089/jamp.2016.1297.

37 Stein, S.W., Sheth, P., Hodson, P.D., and Myrdal, P.B. (2014). Advances in metered dose inhaler technology: hardware development. *AAPS PharmSciTech* **15** (2): 326–338.

38 Montreal Protocol on Substances that Deplete the Ozone Layer, Decision XII/2. https://ozone.unep.org/treaties/montreal-protocol/meetings/twelfth-meeting-parties/decisions/decision-xii2-measures?source=decisions_by_issue&args%5B0%5D=456&parent=3014&nextParent=3013 (accessed September 2020).

39 Procedure under Article 5(3) of Regulation EC (No) 726/2004 Nitrosamine impurities in human medicinal products. Procedure number: EMEA/H/A-5(3)/1490. 25 June 2020 EMA/369136/2020.

40 Gong, H.H., Ihle, N., Jones, M.T. et al. (2018). Control strategy for small molecule impurities in antibody-drug conjugates. *AAPS PharmSciTech* https://doi.org/10.1208/s12249-017-0943-6.

41 Kato, T. and Kikugawa, K. (1992). Proteins and amino acids as scavengers of nitrite: inhibitory effect on the formation of nitrosodimethylamine and diazoquinone. *Food Chem. Toxicol.* **30** (7): 617–626.

42 López-Rodríguez, R., McManus, J.A., Murphy, N.S. et al. (2020). Pathways for *N* nitroso compound formation: secondary amines and beyond. *Org. Process Res. Dev.* **24**: 1558–1585.

43 Benkaidali, L., Andre, F., Moroy, G. et al. (2019). Four major channels detected in the Cytochrome P450 3A4: a step toward understanding its multispecificity. *Int. J. Mol. Sci.* **20**: 987–1009.

44 Urban, P., Lautier, T., Pompon, D., and Truan, G. (2018). Ligand access channels in cytochrome P450 enzymes: a review. *Int. J. Mol. Sci.* **19**: 1617–1638.

45 Guttenplan, J.B. (1987). *N*-Nitrosamines: bacterial mutagenesis and in vitro metabolism. *Mutat. Res.* **186**: 81–134.

46 Danno, G.-I., Kanazawa, K., Toda, M. et al. (1993). A mutagen from histidine reacted with nitrite. *J. Agric. Food Chem.* **41**: 1090–1093.

47 Kushida, H., Fujita, K.-I., Suzuki, A. et al. (2000). Metabolic activation of *N*-alkylnitrosamines in genetically engineered *Salmonella typhimurium* expressing CYP2E1 or CYP2A6 together with human NADPH-cytochrome P450 reductase. *Carcinogenesis* **21**: 1227–1232.

48 Ohshima, H., Mahon, G.A.T., Wahrendorf, J., and Bartsch, H. (1983). Kinetic model for predicting carcinogenic effects caused by endogenous nitrosation. *Cancer Res.* **43**: 5072–5076.

49 Garcia, H. and Lijinsky, W. (1973). Studies of the tumorigenic effect in feeding of nitrosamino acids and of low doses of amines and nitrite to rats. *Zeitschrift fur Krebsforschung und Klinische Onkologie* **79**: 141–144.

50 Lhasa Carcinogenicity database. https://carcdb.lhasalimited.org/carcdb-frontend (accessed September 2020).

51 Ohtake, S., Kita, Y., and Arakawa, T. (2011). Interactions of formulation excipients with proteins in solution and in the dried state. *Adv. Drug Deliv. Rev.* https://doi.org/10.1016/j.addr.2011.06.011.

52 Hsieh, J. and Dunn, A. (2016). *Evidence on the carcinogenicity of nitrite in combination with amines or amides.* Office of Environmental Health Hazard Assessment https://oehha.ca.gov/media/downloads/crnr/nitritehid08262016 (accessed September 2020).

53 Endo, H., Takahashi, K., and Aoyagi, H. (1974). Screening of compounds structurally and functionally related to *N*-methyl-*N′*-nitro-*N*-nitrosoguanidine, a gastric carcinogen. *Gann* 65: 45–54.

54 Lessons learnt from presence of *N*-nitrosamine impurities in sartan medicines. 23 June 2020 EMA/526934/2019.

55 Wagner, E.D., Hsu, K.M., Lagunas, A. et al. (2012). Comparative genotoxicity of nitrosamine drinking water disinfection by products in Salmonella and mammalian cells. *Mutat. Res.* 741 (1-2): 109–115.

56 Prival, M.J., King, V.D., and Sheldon, A.T. Jr. (1979). The mutagenicity of dialkyl nitrosamines in the salmonella plate assay. *Environ. Mutagen.* **1**: 95.
57 Prival, M.J. and Mitchell, V.D. (1981). Influence of microsomal and cytosolic fractions from rat, mouse, and hamster liver on the mutagenicity of dimethyl nitrosamine in the Salmonella plate incorporation assay. *Cancer Res.* **41**: 4361.
58 Lijinsky, W.T. and Andrews, A.W. (1983). The superiority of hamster liver microsomal fraction for activating nitrosamines to mutagens in *Salmonella typhimurium*. *Mutat. Res.* **111**: 135–144.
59 Control of Nitrosamine Impurities in Human Drugs Guidance for Industry. https://www.regulations.gov/docket?D=FDA-2020-D-1530 (accessed September 2020).

11

Conditions Potentially Leading to the Formation of Mutagenic Impurities

Lucie Lovelle[1], Andrew Teasdale[2], Ian Ashworth[2], Adrian Clarke[1], and Alan Steven[3]

[1] *Novartis Pharma AG, Technical Research and Development, Chemical and Analytical Development, Basel, Switzerland*
[2] *Chemical Development, Pharmaceutical Technology and Development, Operations, AstraZeneca, Macclesfield, UK*
[3] *CatSci Ltd, CBTC2, Cardiff, UK*

Following the valsartan contamination with nitrosamines [1], the potential of a similar incident happening in other products became a recurrent concern among process chemists. The formation of mutagenic side products – arising from reactions between process reagents (as opposed to active pharmaceutical ingredient [API]-related substances) has therefore attracted growing attention from the industry. Convinced that only a finite number of those reactions exist, and to mitigate the risk of further API contamination, a list of reactions potentially leading to mutagenic side products has been gathered. This approach is based on the structural alerts for mutagenicity [2] (Figure 11.1) and describes the most common formation pathways. This work should not be considered as a mutagenic assessment, and the use of the specific tools recommended by health authorities should still be applied to the specific structure; however, it should prove extremely valuable in understanding the risk posed by such interactions. This work is based on the expert knowledge combined with extensive literature searches. The focus is on the most common alerts relevant to the pharmaceutical industry, leaving aside very specific cases rarely encountered in our industry[1]. Given the nature of this work, it cannot be considered exhaustive, and it is well plausible that other combinations of reagents that lead to the formation of mutagenic side products may be identified. However, this comprehensive list aims to alert process chemists on the potential formation of mutagenic impurities (MIs). It does not in any case mean that they actually form, merely the risk exists. Process conditions such as concentration, solvent, temperature, pH, time, phase separation, etc. are still critical parameters to take into consideration when assessing the effective formation of those impurities. Alongside, it is also important to consider their degradation and potential depletion [3–6]. The comprehensive overview detailed case studies on nitrosamines, ester sulfonates, and nitro compounds are also presented in the following sections of this chapter.

Among the MIs of greatest interest is the so-called cohort of concern (COC) mentioned in the International Council of Harmonisation (ICH) M7(R1) guideline [7], which comprises aflatoxin-like, *N*-nitroso, and alkyl-azoxy compounds. Those compounds were found to have such high mutagenic potency that the use of the standard Threshold of Toxicological Concern (TTC)

1 Propiolactones, for instance, were omitted from this study.

Mutagenic Impurities: Strategies for Identification and Control, First Edition. Edited by Andrew Teasdale.
© 2022 John Wiley & Sons, Inc. Published 2022 by John Wiley & Sons, Inc.

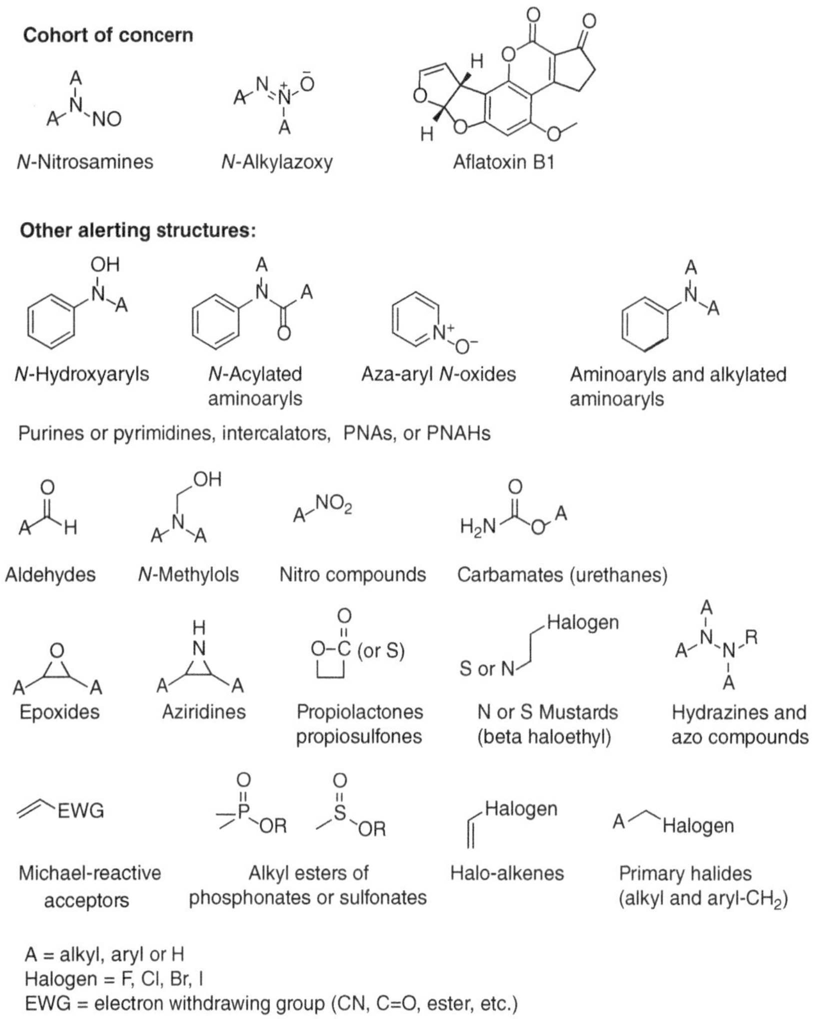

Figure 11.1 Structural alerts for mutagenicity.

approach is not justified[2] and thus, limits must be assessed and established on a specific basis. The recent recalls prompted health authorities to issue general guidance and recommendations on mitigating and preventing the presence of nitrosamines in human medicinal products [1, 8, 9]. Pharmaceutical companies were requested to avoid as much as possible the use of conditions possibly generating nitrosamines and to modify the synthesis routes accordingly. Regulatory authorities around the world, in a dialog with cross-industry working groups, have been cooperating to align requirements in this fast-changing environment.

Although not formally part of the COC, dioxins also have a very low acceptable intake and are often pooled with the COC compounds as their potential presence can lead to the need to modify the design of the synthetic route.

2 As defined in the ICH M7 guideline (see ref. [7]), the TTC concept was developed to provide a limit for unstudied potential carcinogens. It is based on the linear extrapolation from the dose giving a 50% tumor incidence (TD50) to a 1 in 106 incidence rate. For compounds with therapeutic benefits, the TTC was fixed at 1.5 μg/day, corresponding to a potential increased cancer risk of 1 in 100 000.

11.1 Problematic Reagent Combinations per Structural Alert

In the following tables, combination of chemicals potentially leading to the formation of MIs is highlighted. The entries listed highlight that the combination of compounds may lead to the generation of MIs. They listed in a conservative way that simply highlights the potential transformations that should be considered during risk assessments. In each instance the detailed process conditions should be carefully evaluated to assess whether the formation of the MIs is real and to what level and/or to which extent they are purged. In order to support chemists in this exercise, each section is complemented by relevant literature references and remarks, regarding the formation, the reactivity, and/or the purge of these potential contaminants.

11.1.1 *N*-Nitroso Compounds (COC)

11.1.1.1 Amines and Nitrosating Agents [10]

The most common pathway for the formation of nitrosamine requires an amine and a nitrosating agent, such as for example N_2O_3, most often formed from the presence of nitrites under acidic conditions [11]. However, other reagents such as nitric acid and also NaN_3 [12] can also be source of nitrosating agents (NO_2^-). The quality of water used for workup in presence of amine or amine residues also needs to be considered. The presence of nitrates or nitrites in the process water as well as residues from disinfection agents can be problematic[3] [13–15]. Indeed, *N*-nitroso compounds can also be formed via other pathways for instance via oxidative conditions such as chloramination or ozonolysis [16]. In addition, some reaction components can act as catalyst. One example is formaldehyde, which increases the kinetics of nitrosation of secondary amines under neutral and basic conditions [17, 18]. For a detailed study on the mechanism and processing factors affecting the formation of *N*-nitrosamines, see Section 11.3.

Primary amines are generally not considered as a source of nitrosamine as the monoalkylnitrosamines are unstable and usually yield the corresponding alcohols and olefins [19]. Their presence can even help in mitigating the formation of more stable nitrosamines [20]. As a side note, amines are possible contaminants – degradation and hydrolysis products of many solvents and reagents. During risk assessment therefore, amines that were not intentionally introduced in the reaction mixture, but likely to be present, should also be considered.

First component	Second component	By-products of concern
N,N-Dialkylamines, *e.g.* dimethylamine, diethylamine, Di-isopropylamine (DIPA), pyrrolidine, etc.	NO_2^-	*N*-Nitrosodialkylamines, *e.g. N*-nitrosodimethylamine (NDMA) [11], *N*-nitrosodiethylamine (NDEA), *N*-nitrosodiisopropylamine (DIPNA), *N*-nitrosopyrrolidine, etc.
N,N,N-Trialkylamine, *e.g.* Et_3N, N,N-Diisopropylethylamine (DIPEA)	NO_2^-	*N*-Nitrosodialkylamines [21], *e.g.* NDEA, DIPNA, and *N*-nitrosoethylisopropylamine (EIPNA)
N,O-Dialkylhydroxylamines, *e.g. N,O*-dimethylhydroxylamine	NO_2^-	*N*-Nitroso-*N,O*-dialkylhydroxylamines [22, 23]
N-Alkylanilines	NO_2^-	Alkyl nitrosoanilines
N,N-Dialkylanilines	NO_2^-, tBuONO	*N*-Nitroso-*N*-alkylanilines [24, 25]

3 Chloramines and nitrites are present in drinking water. Water quality should be taken into account during risk assessment.

First component	Second component	By-products of concern
Tertiary amine oxides e.g. Et_3NO	NO_2^-	*N*-Nitrosodialkylamines [26], e.g. NDEA
N,N,N-Dialkylchloramine	NO_2^-	*N*-Nitrosodialkylamines [27]
N,N-Dialkylamine	$NOBF_4$	*N*-Nitrosodialkylamine
N,N-Dialkylamine	NOCl	*N*-Nitrosodialkylamine
N,N-Dialkylamine	$CHBrNO_2$, CCl_3NO_2, $C(NO_2)_4$	*N*-Nitrosodialkylamines

11.1.1.2 Amine Derivatives and Nitrosating Agents

First component	Second component	By-products of concern
N-Alkylamide	NO_2^-	*N*-Nitrosoalkylamide [28]
Dimethylformamide (DMF), Dimethylacetamide (DMAc)	NO_2^-	NDMA [29]
N-Methylpyrrolidone	NO_2^-	4-(Methyl(nitroso)amino)butanoic acid [30]
Weinreb amide	NO_2^-	*N*-Nitroso-*N*-Methoxy-*N*-methylamine[a]
N,N-Dialkylcarbamoyl chloride	$(Ag)NO_3$	*N*-Nitrosodialkylamine [31]
N,N-Dialkylsulfamoyl chloride	$(Ag)NO_3$	*N*-Nitrosodialkylamine [32]
N-Alkylureas	NO_2^-	*N*-Nitrosourea [33, 34] (*e.g.* *N*-nitroso-*N*-methylurea: NMU) and potentially nitrosamines
N-Alkylcarbamates	NO_2^-	*N*-Nitrosocarbamate [28]
Acetylated (alkyl)anilines	NO_3^-, NO_2^-	*N*-Nitroso-*N*-alkylanilines [25]
Guanidines	NO_2^-	*N*-Nitrosoguanidines [35], especially methylnitronitroso-guanidine
Amidines	NO_2^-	*N*-Nitrosamines, *N*-nitrosoamides [36]
Cyanamides	NO_2^-	*N*-Nitrosocyanamide [37]

[a] N-Nitroso-N-methoxy-N-methylamine is mutagenic (and a prophage inducer) but is not carcinogenic. See: [38]

11.1.1.3 Other

First component	Second component	Third component	By-products of concern
Phenol	NO_2^-	–	*p*-Nitrosophenol [39] (quinone oxime)
N-Nitroguanidine	Reducing agent	–	Nitrosoguanidines [40]
Hydrazine	Alkylating agent	Oxidant	*N*-Alkylnitrosamines, alkyl-azoxy [41]
N-Alkylhydrazone	Ozonolysis		*N*-Alkylnitrosamines [42]
Hydrazide	Oxidant		*N*-Alkylnitrosamines [43]
Amines, anilines	Dichloroamines	Oxidant	*N*-Alkylnitrosamines, alkyl-azoxy [44, 45]
N,N-Dialkylamine	CH_3NO_2	Oxidant	*N*-Nitrosodialkylamines[a]
RMgX	NOCl, NO		*N*-Nitrosohydroxylamine [46]
Aryl boronic acid	Alkyl nitrite	Cu	*N*-Arylnitrosamines [47]
N-Nitramine	Reducing agent		*N*-Alkylnitrosamines [48]

[a] Various sets of conditions reported. For an example, see: [49].

Diazald (*N*-methyl-*N*-nitroso-*p*-toluenesulfonamide) and other precursors of diazomethane are nitrosamines. Therefore, it is possible that the use of diazomethane in a synthesis introduces some nitrosamines as a reagent impurity.

Nitrosamide, nitrosocyanamides, and nitrosoguanidines can be destroyed in basic conditions. Nitrosamides also decompose quantitatively in acidic conditions[4]. Their carryover across a multistep synthesis is therefore unlikely. They are, nonetheless, potential contaminants when formed in the final steps of an API synthesis.

Finally, it is worthy to note that the principal mechanism of action of dialkylnitrosamines is proposed to go via the alpha- (in limited cases beta-) hydroxylation by the CYP2A6 and 2E1 enzyme families. The hydroxylated compound can then eliminate to form diazomethane. The alkyl carbocation is released and can then alkylate deoxyribonucleic acid (DNA). Therefore, when the alpha-hydroxylation is not feasible, the mutagenic activity is reduced. For instance, *N*-nitrosamines bearing only aryl groups are much weaker carcinogens, most likely through a different mechanism (see Chapter 7) [50, 51].

11.1.2 Alkyl-azoxy Compounds (COC)

11.1.2.1 Reduction [52–54]

First component	Second component	Third component	By-products of concern
R-NO_2	H_2	Metal catalyst	Azo, azoxy, hydroxylamines
R-NO_2	AcOH/Ac_2O	Zn	Azoxy compound
R-NO_2	Alcohol	Tl, Mg	Azo, azoxy compound
R-NO_2	KOH	EtOH	Azoxy compound
R-NO_2	NaOH	Glucose	Azoxy compound
Oxime	Reductant		Azoxy compound
Nitroso	Hydroxylamine		Azoxy compound
Dimeric nitroso	H_2	Palladium	Azoxy compound

11.1.2.2 Oxidation

First component	Second component	By-products of concern
Hydrazine	Oxidant	Azoxy compound [55]
1,2-Dialkylhydrazine	Oxidant	Azoxy compound [56]
Azo compounds	Oxidant	Azoxy compound [57]
Hydrazone	Oxidant	Azoxy compound
Hydroxylamine	Oxidant	Azoxy compound

11.1.2.3 Others

First component	Second component	Third component	By-products of concern
Diazenolate	Alkyl halide		Azoxy compound
Dimeric nitroso	Hydroxylamine		Azoxy compound
Nitroso	*N,N*-Dihaloamine	KOH, CuCl	Azoxy compound

4 IARC Scientific Publication no. 3, 81–86.

Azoxy compounds are 1,3-dipoles that could undergo 1,3 dipolar cycloaddition with double bonds. It is possible to leverage from this reactivity to purge them.

11.1.3 Other N-O Compounds

11.1.3.1 Reduction of Nitro Groups

First component	Second component	Third component	By-products of concern
R-NO_2	H_2	Catalyst	Azo, azoxy, hydroxylamines[a]
R-NO_2	Alcohol	Ti, Mg	Azo, azoxy compound
Ar-NO_2	H_2	Catalyst	Azo, azoxy, hydroxylamines, anilines

[a] For an example of the use of QbD to assess the risk of GTI contamination from a nitro reduction step, see: [58].

For a detailed study on the formation, fate, and purge of impurities arising from the hydrogenation of nitroarenes to anilines, see Section 11.4.

11.1.3.2 Oxidation of Amines and Hydroxylamines

First component	Second component	By-products of concern
Primary amine	Peracids, peroxides, and other oxidants (=Oxidant)	Nitroalkyl, hydroxylamine
Secondary amine	Oxidant	Hydroxylamine
Tertiary amine	Oxidant	*N*-Oxides
Pyridine and other heterocyclic compounds	Oxidant	*N*-Oxides
Oxime	Oxidant	Nitro compound
ArNHOH	Diethyl azodicarboxylate (DEAD)	Nitroso aromatics

11.1.4 Nitration

First component	Second component	By-products of concern
Chloroacetic acid	$NaNO_2$	Nitromethane
Haloalkane	$NaNO_2$	Nitroalkane
Electron-rich aromatics	HNO_3	Nitroaromatics

11.1.5 Other N-N Compounds [59, 60]

First component	Second component	By-products of concern
Hydrazine	–	Hydrazine
Hydrazine	Alkylating agent	Alkyl hydrazines

First component	Second component	By-products of concern
Alkyl ketone and aldehyde	Hydrazine	Alkyl hydrazones
Aromatic ketone and aldehyde	Hydrazine	Aryl hydrazones, benzidine
Esters	Hydrazine	Acyl hydrazines
Amines	Haloimides, X_2, hypochlorite	Alkyl and aryl hydrazines
HOBt		Hydrazine [61]

11.1.6 Aflatoxin-like Compounds [62] (COC)

Aflatoxin-like compounds are part of the COC of ICH M7. However, they are mostly synthesized by fungi and, unless the API structure is related, they are unlikely to be formed in standard chemical processes. They should, nonetheless, be kept in mind in case of biological processes.

11.1.7 Dioxin-like Compounds (Including Polychlorinated Biphenyls = PCBs) [63]

Polychlorinated dibenzodioxins (PCDDs), polychlorinated dibenzofurans (PCDF), and polychlorinated biphenyls (PCB) are a group of persistent lipophilic organic compounds that are significant environmental pollutants. The toxicity of fluorinated, brominated, and iodinated dioxins is unknown. Due to the very low levels to be detected and controlled (often ppb levels), precursors leading to problematic dioxins should be avoided by design and alternate synthetic routes may be required.

First component	Second component	Third component	By-products of concern
Chlorohalophenyl	Transition metal		PCBs, dioxin-like
Halophenol	–	Heat	Dioxin-like
Phenol	Chlorinating agent		Dioxin-like
Phenol	Halogen salt	Oxidant (e.g. O_2)	Dioxin-like
Aryl halide	Oxidant (e.g. O_2)		Dioxin-like
Catechol	Halophenol		Dioxin-like
Haloboronic esters/acids	(Transition metal)	Oxidant (e.g. O_2)	Dioxin-like
Haloboronic esters/acids	Halophenol	Oxidant (e.g. O_2)	Dioxin-like

11.1.8 Alkyl and Acyl Halides

11.1.8.1 $ROH + HCl \rightarrow RCl + H_2O$

The reaction between alcohol and halogenated acids is often well known from process chemists and considered during salt formation reaction. However, alcohols are potential contaminants of various solvents and reagents and care should be taken that the alcohols that were not intentionally introduced in the reaction are considered as well.

As a side note, the *t*BuCl formed during the deprotection of Boc-protected amines is now widely accepted as nonmutagenic and should be controlled according to ICH Q3A only [64].

First component	Second component	By-products of concern
R-OH (e.g. MeOH, EtOH, PrOH, *i*PrOH)	HCl, HBr	R-X (e.g. MeCl, EtCl, PrCl, *i*PrCl, MeBr, EtBr, PrBr, *i*PrBr [65]
Pinacol	HCl, $SOCl_2$, SO_2Cl_2	3-Chloro-2,3-dimethyl-butan-2-ol; 2,3-dichloro-2,3-dimethyl-butane; 3-chloro-2,3-dimethylbut-1-ene [66]
R-OH	$SOCl_2$, $POCl_3$, PCl_5 or PCl_3	Alkyl halides, alkyl phosphonates, and sulfonates
R-OH	MsCl or TsCl	Alkyl halides and alkyl sulfonates

11.1.8.2 Ether Opening with Halides

First component	Second component	By-products of concern
Tetrahydrofuran (THF)	SiX_4, AlX_3, HX, NaX	4-Halobutan-1-ol; 1,4-dihalobutane; and higher molecular mass compounds [67, 68]
Me-THF	SiX_4, AlX_3, HX, NaX	4-Halopentan-1-ol; 1,4-dihalopentane; and higher molecular mass compounds [69]
THF	$SOCl_2$, SO_2Cl_2	2-Chloro-THF, Bis-chlorobutyl ether [70]
MeTHF	AcCl and HCl	4-Chloropentanol [71]
ArOMe	HX, BX_3 or AlX_3	MeX [72]
Phenol ether	HX, BX_3 or AlX_3	Alkyl halides [72]
Bis-alkylethers	$AlCl_3$	Alkyl halides [73]
Et_2O	Cl_2	1-Chloro-1-ethoxyethane [74]
Ethyl acetate	TCCA[a]	1-Chloroethyl acetate [75]
Alkyl carboxylates	Haloimides, X_2, $SOCl_2$, etc.	1-Haloalkyl carboxylates

[a] TCCA: trichloroisocyanuric acid.

11.1.9 Methyl Sulfoxides and Pummerer Rearrangement

The Pummerer rearrangement is especially something to consider when choosing the solvent for an Ames test submission[5]. Indeed, dimethylsulfoxide (DMSO) is often a solvent of choice for Ames test; however, DMSO is not inert to all functionalities and, for instance, the reaction of DMSO with acyl chlorides led to numerous false positive Ames test due to the formation of the mutagenic chloromethyl (methyl)thioether (see Chapter 5).

First component	Second component	By-products of concern
DMSO or $RSOCH_3$	Acyl, carbamoyl or cyanuric chloride, triphosgene, etc.	Chloromethyl (methyl)thioethers
DMSO or $RSOCH_3$	Sulfonyl or sulfuryl chlorides	Chloromethyl (methyl) thioethers [77]

5 For differences in Ames test outcome depending on solvent, see: [76].

First component	Second component	By-products of concern
DMSO or $RSOCH_3$	N-Bromosuccinimide and N-chlorosuccinimide	Halomethyl (methyl)thioethers
DMSO or $RSOCH_3$	POX_3, PX_3 or other $R_2P(O)X$	Halomethyl (methyl)thioethers
DMSO or $RSOCH_3$	HX [78], R_3SiX [79] or BX_3 [80]	Halomethyl (methyl)thioethers
DMSO or $RSOCH_3$	CX_4 [81] or C_2X_6	Halomethyl (methyl)thioethers

11.1.10 Acyl Chlorides Formation [82]

First component	Second component	By-products of concern
DMF	$SOCl_2$, $SOBr_2$, $POCl_3$ or $POBr_3$	Dimethylcarbamoyl chloride (DMCC) or bromide [83]
DMAc	$SOCl_2$, $SOBr_2$, $POCl_3$ or $POBr_3$	DMCC + MeCl

11.1.11 Halogenation of Unsaturated Compounds

First component	Second component	By-products of concern
Formaldehyde	HCl [84], HSO_3Cl [85], $POCl_3$ [86], chlorides [87], etc.	Bis(chloromethyl)ether[a]
Double bonds	HX or X_2	Alkyl halides
Triple bonds	HX or X_2	Alkyl halides, halo alkene

[a] BCME is an ICH M7 class 1 compound with a lifetime acceptable intake (AI) of 4 ng/day.

11.1.12 Ammonium Salts (Hofmann Elimination)

First component	Second component	By-products of concern
NR_4^+	X^-	Alkyl halides [88]
DMTMM[a]	Organic solvents	MeCl [89]

[a] DMTMM: 4-(4,6-dimethoxy-1,3,5-triazin-2-yl)-4-methyl-morpholinium chloride.

11.1.12.1 Alkyl Sulfonates [90]

Whether they are used as catalysts or in stoichiometric amounts, alkyl sulfonates are a potential risk in API synthesis [91].

Applications of alkyl sulfonates in API synthesis	
API salt-forming agent	
Good leaving group	Etherification
	Hydroxyl to halogen, to sulfur, or to amine
	Amine or amide to nitrile
	Isocyanate to amine

Applications of alkyl sulfonates in API synthesis	
Cyclization reactions	Formation of aziridines, oxazoline, pyrrolidine, lactone, or oxirane
	Cyclodehydration
Protecting group	
Protecting group removal	
Mitsunobu rearrangement	
Double bond migration	
Enamine-amine reduction	
Sulfonamide formation	
Esterification	
Resolution of enantiomers	

First component	Second component	By-products of concern
MsCl, TsCl, EsCl, BesCl, TfCl, or NsCl, etc.	Alcohols: hydroxy groups, solvents, impurities	Alkyl sulfonate
H_2SO_4	Alcohols	$ROSO_3H$
Sulfonic acids or salts (e.g. *p*TsOH, MsOH, TfOH)	Alcohols	Alkyl sulfonate [92, 93]
Sulfonyl anhydride	Alcohols	Alkyl sulfonate
Camphorsulfonate salts	Alcohols	Alkyl camphorsulfonate

In the Viracept™ case (contamination with ethyl methanesulfonate, EMS), the root cause identified was the use of ethanol for the cleaning and the absence of drying of the head tank after the EtOH wash, which was used for the dosing of the methanesulfonic acid (MSA). This led to a contamination of the drug with significant levels of EMS [94]. As a general consideration, when using alcohols in combination with sulfonyl halides, both alkyl sulfonates and alkyl halides can be expected. However, it is relatively easy to mitigate the formation of those MIs with the process parameter. Indeed, the effect of pH, temperature, water, or the addition conditions (order of addition, time, stirring speed, etc.) drastically affect not only their kinetics of formation but also their hydrolysis [92]. For a detailed study on the mechanism and processing parameters affecting the formation of sulfonate esters, see Section 11.5.

11.1.13 Epoxides and Aziridines [95–97]

Aziridines and epoxides are used as pesticides for their particular toxicities. Although aziridines have some specific properties, they usually react in the same way as other powerful alkylating agents. Epoxides and aziridines can be used as such but are also common reaction intermediates, degradation products, or side products. The undesired formation of three-membered rings is most likely due to ring closure side reactions. Aziridines and epoxides can be degraded by addition of aqueous acids to form the corresponding aminoalcohol, prone to be soluble in water. Kinetics of degradation highly depends on substitution.

First component	Second component	By-products of concern
Olefin	Oxidant	Epoxides
Ketone/aldehyde	α-Haloester	Epoxides
Halohydrin	(Base)	Epoxides
Allyl chloride	HOCl	Epoxides
Glycerol, glycol, diols	HCl/HBr	Epoxides
Glycol, sugar	Heat	Glycidol
β-Aminoalcohol	Dehydrator, activator	Azirines
β-Haloamines		Aziridines
α,β-Dihalocompounds	Primary amine	Aziridines
Hydrazinium	β-Proton	Aziridines
Oxime	Grignard reagent, $LiAlH_4$	Aziridines

11.2 Miscellaneous

11.2.1 B and P Based Compounds

First component	Second component	By-products of concern
Boronates	H_2O	Boronic acids [98]
Phosphonic acid	Alcohols	Alkylphosphonates
$POCl_3$, PCl_5, PCl_3	Alcohols	Alkylphosphonates, alkyl chlorides

Boronic acids and esters are mainly encountered when performing Suzuki or borylation reactions. In these circumstances, the hydrolysis of the boronic ester during reaction/workup should always be considered. In addition, in the case of a borylation, bispinacolato diboron is a known *in vivo* mutagen and its purge should be studied. Alkylphosphonates are mainly encountered when performing a Wittig olefination. Their degradation by aqueous basic treatment can be used to deplete the potential MI, but the kinetics of hydrolysis should then be carefully studied to ensure the depletion.

11.2.2 Formation of *N*-Methylol

First component	Second component	By-products of concern
Nitro compounds	(para) Formaldehyde	*N*-Methylol
Amines (1°, 2°, and 3°)	(para) Formaldehyde	*N*-Methylol

11.2.3 Acetamide

Acetamide is not mutagenic, but it is a known carcinogen [64]. Acetamide might be present as an impurity in relatively common acetamide derivative building blocks (trifluoroacetamide, *N*--bromoacetamide, 2-bromoacetamide, etc.)

First component	Second component	By-products of concern
MeCN	Acidic or basic conditions	Acetamide [99]
MeCN	Ac_2O	Acetamide
N-Bromoacetamide	Alkenes	Acetamide
2-Bromoacetamide	Reductant	Acetamide
(Activated) Acetic acid	Ammonia	Acetamide

11.2.4 Quinones and Quinone Derivatives

First component	Second component	By-products of concern
Phenols or anilines	Strong oxidant ($NaOCl$, $KClO_3$, $KMnO_4$, etc.)	Benzoquinone
Anilides	Strong oxidant	

11.2.5 Anilines [100]

First component	Second component	By-products of concern
Protected aniline	Hydrolytic conditions	Aniline
$ArNO_2$	Reductant	Azo, azoxy, hydroxylamines, aniline, benzidines
Arylamides	$Br_2/NaOH$	Aniline via Hofmann rearrangement
Arylazides	Heat	Aniline via Curtius rearrangement
Arylazides	Phosphine	Aniline via Staudinger reduction
Arylhalides	Amines + Metal	Aniline via cross-coupling
1,2-Diphenylhydrazine	Acids	Benzidine (5,5-sigmatropic rearrangement)
4,4′-Disubstitutedbiphenyl (see rows above for substituents and conditions at risk)		Benzidine

Anilines and benzidines are usually easily depleted via aqueous acidic washes. Some anilines can also be depleted via azeotropic distillation with water[6].

6 Azeotropic mixture of water and aniline boils at 98.5 °C. Based on vapor pressures, 0.06 mol of aniline are distilled off per mol of water. (see ref. [100]).

11.2.6 Michael Acceptors

First component	Second component	By-products of concern
Beta haloacid or ester	Base	Unsaturated acid or ester
Activated acrylic acid	Alcohol	Alkyl acrylate
Activated acrylic acid	Amines	Acrylic amides

11.2.7 Others

First component	Second component	By-products of concern
Alcohol	Oxidant	Aldehyde
Tetrahydropyran (THP)-protected compound	Brønsted or Lewis acid	Dihydropyran (DHP)

11.3 Mechanism and Processing Factors Affecting the Formation of *N*-nitrosamines

11.3.1 Introduction

The finding that a number of the sartans (Chapter 10) and several other pharmaceutical products were contaminated with dialkyl *N*-nitrosamines has led to significant regulatory scrutiny of the risk that nitrosamines could be introduced into medicinal products as trace contaminants during the synthesis of APIs [8, 101]. An understanding of the formation of dialkyl *N*-nitrosamines is therefore of use to scientists involved in the development of API and drug product manufacturing processes.

The formation of *N*-nitroso compounds has been the subject of a number of thorough reviews, which provide a basis for considering the risks of nitrosamines forming [10, 11, 102–106]. The specific risks of nitrosamines forming due to trace nitrite in the water used during API synthesis have recently been addressed [15].

From the perspective of scientists involved in the development of medicinal products, there are two different scenarios that need to be covered in any risk assessment process. First, there is the risk posed by the use of nitrosating conditions during the synthesis of the API if traces of a compound that can give rise to a nitrosamine are present. Second, there is the risk that nitrosatable compounds used within the API synthesis encounter traces of nitrosating agents. The low exposure limits in place [9] for nitrosamines in medicinal products mean that trace impurities can give rise to levels of nitrosamines that could lead to a medicinal product being contaminated with nitrosamines at levels that would require process change and/or analytical controls.

11.3.2 Mechanisms of Amine Nitrosation

11.3.2.1 Nitrosation of Secondary Amines

The nitrosation of secondary amines was the subject of much study following the finding that dialkyl *N*-nitrosamines are carcinogenic [107]. At a general level, the reaction fits into the wider landscape of diazotization chemistry, as nitrosamines are formed as unstable intermediates when primary amines are nitrosated (Figure 11.2). The first step in the reaction is the reaction between the neutral unprotonated amine and a nitrosating agent (NO^+ or a carrier thereof). In the case of secondary amines, the nitrosamine formed by this nitrosation step is incapable of tautomerization to give a diazohydroxide intermediate and secondary nitrosamines are therefore stable.

11.3.2.2 Aqueous Nitrosation

In water nitrosation is usually carried out by species derived from nitrous acid [104, 108], a weak acid with a pK_a of 3.15 [109], which is not a nitrosating agent in its own right. Under strongly acidic conditions (>4 M H^+), nitrous acid is converted to the nitrosonium ion (NO^+) [110]. At lower acidities (<4 M H^+), high concentrations of nitrous acid give rise to dinitrogen trioxide, N_2O_3 (Figure 11.3), while at lower nitrous acid concentrations, the nitrous acidium ion, $H_2NO_2^+$, or traces of NO^+ generated through the protonation of nitrous acid are the dominant nitrosating species. Mechanistic investigations have so far failed to differentiate between these kinetically equivalent pathways. Addition of nonbasic nucleophiles such as chloride [111], bromide [112], and thiocyanate [113] leads to the formation of nitrosyl nucleophile adducts such as the known nitrosating agent nitrosyl chloride, ClNO (Figure 11.4).

The kinetics of nitrosation by N_2O_3, ClNO (and other nitrosyl nucleophile adducts), and the nitrous acidium ion have been studied, and the observed rate laws are given (Eqs. 11.1–11.3). As these contain terms for the concentrations of the free unprotonated amine, nitrous acid, and sometimes H^+, the rates of reaction are strongly dependent upon the solution pH. A thorough discussion of the dependence of the rates of reaction upon the nitrous acid (nitrite) concentration and solution pH has been published [15] and shows that the nature of the dominant nitrosating species depends upon the conditions. This is illustrated in Figure 11.5, which compares the contributions of the different pathways to the initial rate of nitrosation of diethylamine (1 mM) by aqueous nitrite (1 mM) in the presence of chloride (1 M) as a function of pH. In this example the reaction is dominated by N_2O_3 over most of the pH range and shows a peak in the initial rate around the pK_a of

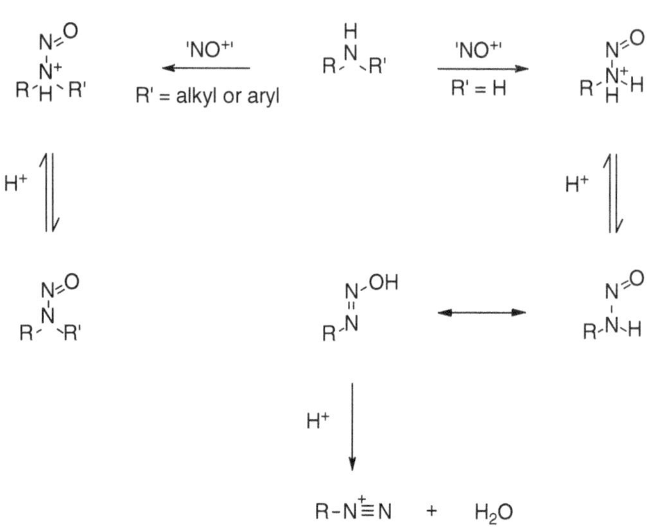

Figure 11.2 Comparison of primary and secondary amine nitrosation.

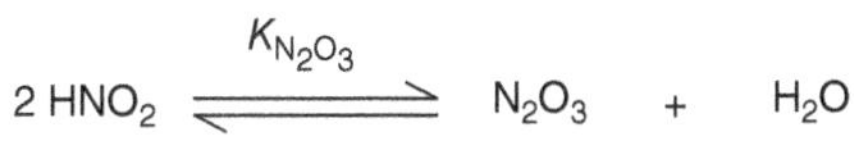

Figure 11.3 Formation of dinitrogen trioxide, N_2O_3.

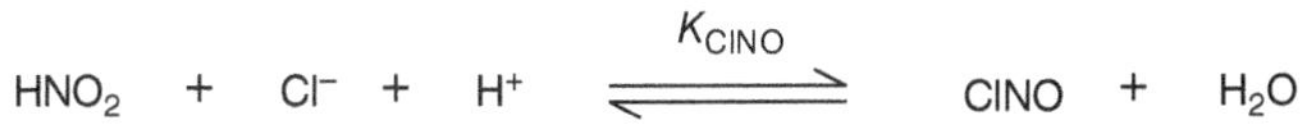

Figure 11.4 Formation of nitrosyl chloride, ClNO.

nitrous acid. At high pH the rate falls due to the concentrations of nitrous acid and free H^+ decreasing.

$$\text{Rate } N_2O_3 = k_{N_2O_3} K_{N_2O_3} [R_2NH][HNO_2]^2 \tag{11.1}$$

$$\text{Rate ClNO} = k_{ClNO} K_{ClNO} [R_2NH][HNO_2][H^+][Cl^-] \tag{11.2}$$

$$\text{Rate } k_{H_2NO_2^+} [R_2NH][HNO_2][H^+] \tag{11.3}$$

Kinetic studies of the nitrosation of a range of secondary amines by N_2O_3 [114] and BrNO [115] have determined that rate constants for the nitrosation reactions vary little with the structure and pK_a of the amine. These observations coupled with the low enthalpy of activation observed for the nitrosation of secondary amines by N_2O_3 [114] have been taken as evidence of secondary amine nitrosation being an encounter-controlled process. Therefore, it follows that less-basic amines will nitrosate faster than more basic amines at a particular pH due to more of the less-basic amine being in the reactive free base form. For example, morpholine (pK_a 8.49) [116] will react 270 times faster than diethylamine (pK_a 10.92) [117].

Figure 11.5 The pH initial rate profiles for the nitrosation of Et_2NH (0.001 M) by aqueous nitrite (0.001 M) in the presence of added chloride (1.0 M) at 25 °C. Calculated using the models published by Ashworth et al. [15]

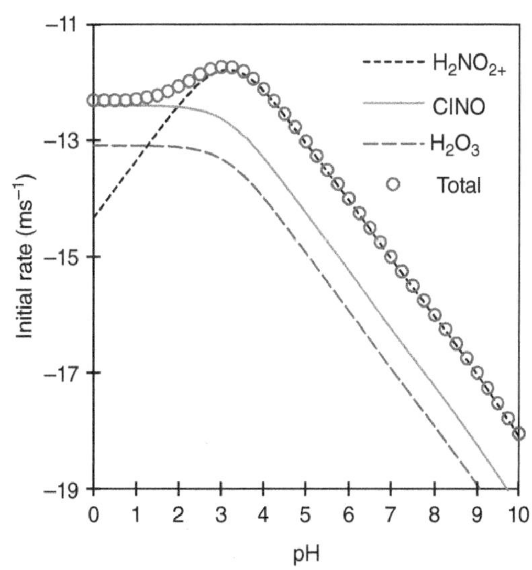

The fact that the rate of amine nitrosation falls with pH means that the rate of nitrosation is usually insignificant at pH 7 and above for basic secondary amines (pK_a > 9.5) [15]. However, there is one important exception to this rule. It has been observed that secondary amines can undergo nitrosation by nitrite in the presence of formaldehyde at high pH [17]. Subsequent studies [18, 118, 119] have shown that other relatively unhindered aldehydes such as benzaldehyde can also catalyze the reaction and support the proposed mechanistic rationale (Figure 11.6) of the amine reacting with the aldehyde to form an iminium ion that then reacts with nitrite to form the nitrosamine.

11.3.2.3 Nitrosation in Organic Solvents

The kinetics and mechanism of the reaction between secondary amines and nitrosating agents have been less thoroughly investigated than their aqueous nitrosation chemistry. A thorough review of synthetic approaches to amine nitrosation has recently been complied by López-Rodríguez et al. [10] A number of the approaches to carrying out nitrosations under mild conditions use a nitrite source and an organic acid under biphasic conditions and are likely to generate a nitrous acid derived nitrosating agent, which partitions into the immiscible organic solvent where nitrosation occurs.

Studies of nitrosation in organic solvents and water by gaseous N_2O_3, dinitrogen tetroxide N_2O_4 [120], and ClNO have shown the reaction, which is postulated to proceed through the unprotonated amine, to be rapid. Stored solutions of secondary amines should therefore be protected from air as N_2O_3 and N_2O_4 are components of NO_x and may give rise to traces of nitrosamines.

One of the most widely used classes of organic soluble nitrosating agents are alkyl nitrites. These are effective nitrosating agents in organic solvents and under aqueous conditions [121]. In organic solvents the amine reacts directly with the weakly electrophilic nitrogen of the alkyl nitrite in a reaction that is postulated to occur via a zwitterionic intermediate (Figure 11.7). Under acidic aqueous conditions, alkyl nitrites undergo rapid hydrolysis to liberate nitrous acid, which is transformed into a nitrosating agent. Under basic aqueous conditions, nitrosation is competitive with hydrolysis and exhibits simple second order kinetics. The fact that alcohols will react reversibly with nitrosating agents [104] means that alkyl nitrites can be formed as side products when synthetic nitrosating conditions are used and can lead to the carryover of nitrosating potential into the downstream process steps.

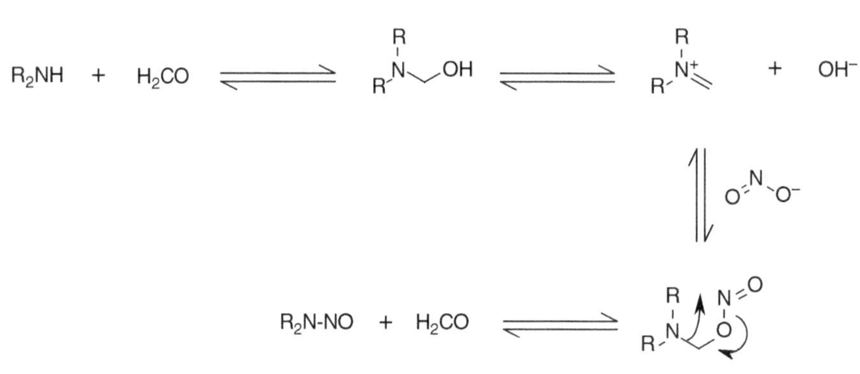

Figure 11.6 Mechanistic scheme for the formaldehyde-catalyzed nitrosation of secondary amines by nitrite.

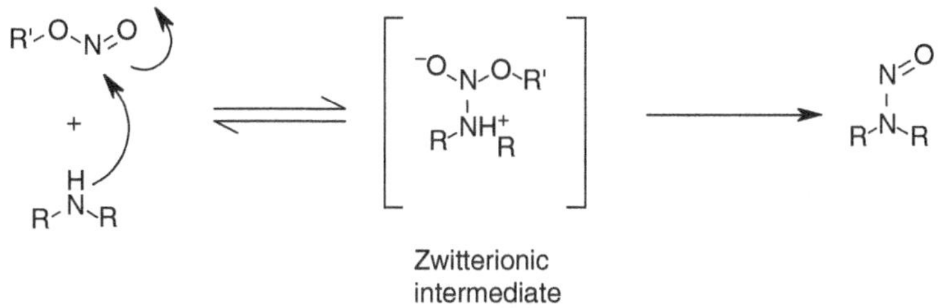

Figure 11.7 Mechanistic scheme for the nitrosation of a secondary amine by an alkyl nitrite in an organic solvent.

11.3.3 Nitrosation of Tertiary Amines

Simple tertiary alkylamines can nitrosate to form nitrosamines [122]. The reaction is much slower than secondary amine nitrosation, and the mechanistic proposal [123] (Figure 11.8) of reversible formation of a nitroso ammonium ion followed by elimination of a nitroxyl (NOH) to give an iminium ion has been validated kinetically [124]. Hydrolysis to a secondary amine and a carbonyl compound is finally followed by nitrosation to give a nitrosamine. Unsymmetrical tertiary amines can give rise to multiple nitrosamines depending upon which alkyl group is lost. For tertiary amines carrying a methyl or benzyl group, there is the possibility that crossover may occur onto the nitrite-mediated manifold seen in formaldehyde catalyzed nitrosation (see Section 11.3.2.3).

A small number of more structurally complex tertiary amines have been identified that rapidly nitrosate to liberate a low molecular weight nitrosamine. An early example of this was the analgesic aminopyrine that liberates NDMA directly upon nitrosation [125, 126]. Many of the complex tertiary amines that have been found to display this behavior contain functional groups within the molecule that are capable of driving the liberation of a nitrosamine directly from the initially formed nitroso ammonium ion. The liberation of nitroso cyclohexylmethyl amine from the mucolytic bromohexine (Figure 11.9) [127] exemplifies this pathway. The recent identification of NDMA in medicinal products containing ranitidine [128] (Chapter 10) may in part stem from this mechanism of tertiary amine nitrosation.

11.3.3.1 Nitrosation of Quaternary Amines

The nitrosation of a limited number of quaternary ammonium ions has been reported [129]. These were proposed to occur via an initial nucleophile-mediated demethylation reaction to liberate a

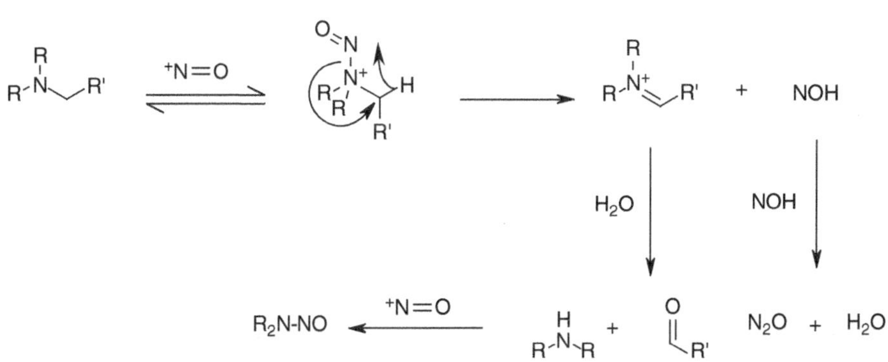

Figure 11.8 Mechanistic scheme for the dealkylative nitrosation of simple tertiary amines.

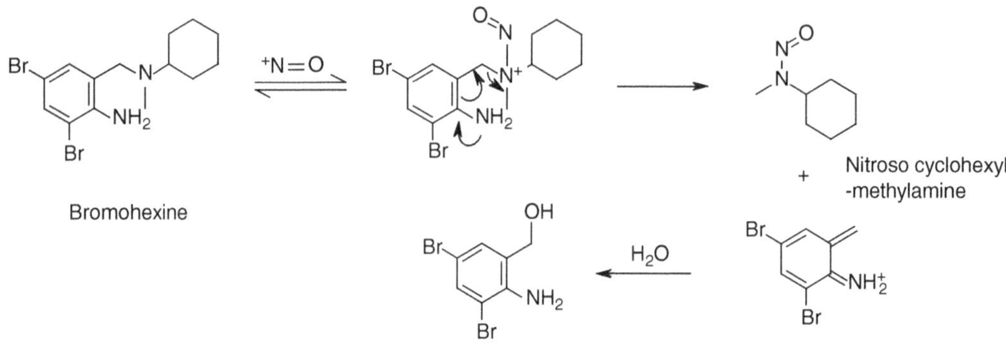

Figure 11.9 Mechanistic scheme for the formation of nitroso cyclohexyl methylamine from bromohexine.

tertiary amine, which then underwent nitrosation (Figure 11.10). Dealkylation of quaternary ammonium ions possessing β-hydrogen atoms via Hoffmann elimination [130] is a known problem in phase transfer catalysis when operating under basic conditions [131]. Nucleophilic substitution-based degradation of quaternary ammonium ion based phase transfer catalysts containing methyl [132] or benzyl groups [133] has also been reported.

Given that quaternary ammonium ions nitrosate following their degradation to tertiary amines, it follows that they will not generate nitrosamines any faster than the intermediate tertiary amine undergoes nitrosation.

11.3.3.2 Nitrosation of Amine Oxides

Very few studies of the nitrosation of amine oxides have been reported. Nitrosation of tribenzylamine *N*-oxide has been reported to give low levels of dibenzyl nitrosamine alongside benzaldehyde as a coproduct [123]. Trimethylamine *N*-oxide has been shown to give similar yields of NDMA to trimethylamine under relatively forcing nitrosation conditions [134]. The mechanism has not been proven, but it has been proposed that nitrosation occurs on the oxygen of the *N*-oxide followed by the elimination of nitrous acid to yield an iminium ion that hydrolyzes to give a secondary amine and an aldehyde (Figure 11.11).

11.3.4 Sources of Nitrosating Agents

Some materials that may not be expected to have nitrosating potential are considered in this section.

11.3.4.1 Process Water

The potential for traces of nitrite present in the water used in the synthesis of APIs to generate significant levels of nitrosamines has been evaluated [15]. To meet pharmacopoeial requirements, the minimum standard of water that should be used in API synthesis is potable water [135], which can contain no more than 3 mg/l nitrite (World Health Organization [WHO] guideline). While

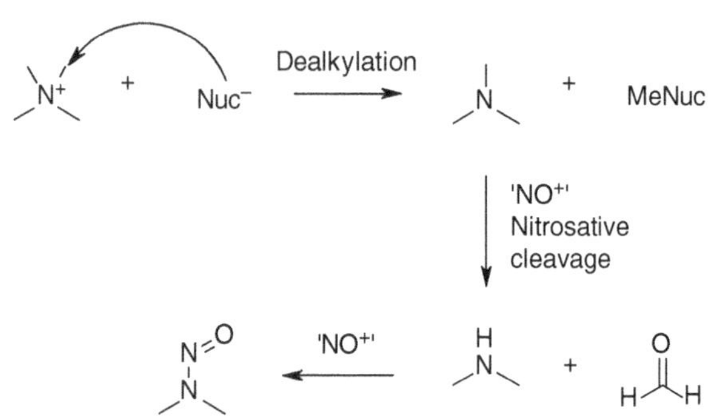

Figure 11.10 Proposed mechanistic scheme for the formation of NDMA from tetramethylammonium chloride.

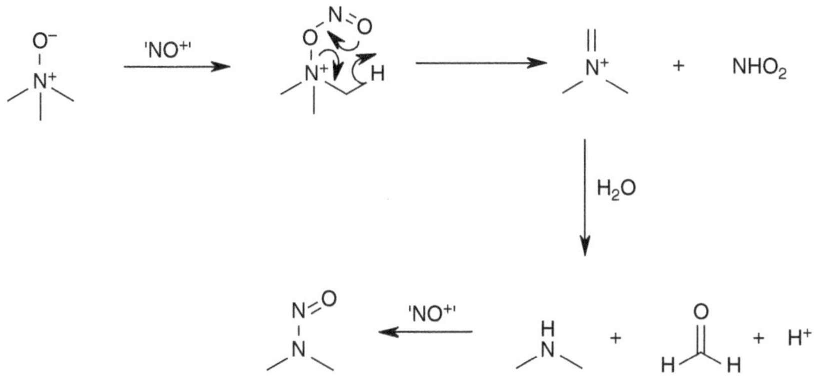

Figure 11.11 Proposed mechanistic scheme for the formation of NDMA from trimethylamine *N*-oxide.

this value may seem low, it is sufficient to potentially give rise to levels of nitrosamines that could be a concern under some processing conditions. In Europe the guideline nitrite content of potable water is lower at 0.5 mg/l (European Commission (EC) directive), and typical levels are often at the limit of quantification of the analytical methods employed. For example, routine analysis of water in Macclesfield, UK, reports a maximum nitrite content of <0.0115 mg/l (https://www.unitedutilities.com/help-and-support/your-water-supply/drinking-water-quality/water-quality-search-results/ for SK10 2NA, accessed 04 September 2020, range <0.0016 to <0.0115 mg/l, average < 0.0066 mg/l). The risks of basic secondary amines ($pK_a > 9.5$) giving rise to significant levels of nitrosamines through contact with water containing such low levels of nitrite are insignificant under most processing conditions [10]. This highlights the importance of understanding the actual quality of the water used when conducting assessments of the risk of nitrosation in processes that use secondary amines, as the only safe assumption in the absence of data is the WHO guideline limit.

Potable water is specified as the source water for the production of higher pharmacopoeial grades of water [135], and purified water and water for injection should therefore contain even lower levels of nitrite than the source water and should therefore under most circumstances not represent a risk.

11.3.4.2 Nitric Acid

Nitric acid is a reasonably strong oxidizing agent and can be reduced to nitrous acid by contact with most metals. Concentrated solutions of nitric acid are also prone to photochemical degradation to give NO_2, which is in equilibrium with a nitrosating agent in the form of its dimer, dinitrogen tetroxide N_2O_4 (Figure 11.12) [136, 137]. Solutions of nitric acid are therefore often capable of nitrosating secondary amines or tertiary amines, which means that nitration reactions pose two risks. First, the may nitrosate vulnerable amines present in the reactants, and second, the nitrated products may be contaminated with traces of nitrosating agents.

11.3.4.3 Atmospheric Sources

A series of equilibria link the oxides of nitrogen in the gas phase. The mixture of nitric oxide (NO) and nitrogen dioxide present (NO_2) in the atmosphere are commonly referred to as NO*x*. Both are

$$2\,HNO_3 \xrightarrow{h\nu} H_2O + 1/2\,O_2 + 2\,NO_2 \rightleftharpoons N_2O_4$$

Figure 11.12 Photochemical degradation of nitric acid.

radicals and combine to form the nitrosating agents, dinitrogen trioxide, and dinitrogen tetroxide (Figure 11.13) in low levels [138]. Exposure of secondary amines to atmospheric NOx therefore has the potential to lead to low levels of nitrosation occurring over prolonged periods of time. This is a potential reason for the observation that many samples of secondary amines are contaminated with the corresponding nitrosamine [139].

$$NO + 1/2\ O_2 \rightleftharpoons NO_2$$

$$NO + NO_2 \rightleftharpoons N_2O_3$$

$$2\ NO_2 \rightleftharpoons N_2O_4$$

Figure 11.13 Equilibria between nitrogen oxides.

11.3.4.4 Excipients Used in Drug Product Manufacture

It has been reported that a range of excipients commonly used in drug product formulations can contain trace levels of nitrite [140]. While the observation is correct, more recent studies carried out under the umbrella of the Innovation and Quality (IQ) consortium have shown that typical levels of nitrite present in excipients are significantly lower[7]. There is therefore a real risk that vulnerable amine APIs and potentially secondary amine impurities could undergo nitrosation to form low levels of nitrosamines during drug product processing and storage [8].

11.3.4.5 Nitrocellulose

The interaction between nitrocellulose primers and printing inks containing amines to form NDMA and NDEA during the printing of lidding foils has also been highlighted [8]. The high temperatures involved in sealing blister packs when using contaminated lidding foils are believed to have resulted in low levels of volatilized nitrosamine being trapped within the sealed blisters.

11.3.4.6 Nitrosating Agent Scavengers

Having discussed some of the potential sources of trace nitrosating agents that may be relevant during API synthesis, it is appropriate to discuss how these risks and the risk introduced by the deliberate use of a nitrosating agent may be mitigated. One significant facet of the chemistry of nitrous acid that should not be neglected is its well-known instability with respect to decomposition to form nitric oxide (Figure 11.14) [138, 141]. This behavior has been used to remove traces of nitrosating agents from the product of a nitration through a reduced pressure distillation of water in the synthesis of the herbicide Trifluralin [142].

The principal method, however, of removing residual nitrosating species is to use a nitrous acid scavenger. This works by two primary mechanisms: they are either reducing agents that reduce nitrous acid to nitric oxide or compounds that nitrosate to give stable non-nitrosating products. The relative reactivities of a range of nitrous acid scavengers including among others azide, sulfamic acid, hydroxylamine, and urea (Figure 11.15) have been compared [143]. Urea is notable because, while not very reactive, it is effective under the strongly acidic conditions often employed for nitrations. Primary amines will also remove nitrosating agents as they nitrosate irreversibly to form an alkyl diazonium species.

$$3\ HNO_2 \longrightarrow 2\ NO + HNO_3 + H_2O$$

Figure 11.14 Decomposition of nitrous acid to nitric oxide.

---

7 At the time of writing work was taking place under the umbrella of the IQ Consortium's Short Duration Working Group on Nitrosamines to investigate the levels of nitrite to be found in a wide range of excipients. The results of these investigations were being built into a database by Lhasa Ltd.

$$HNO_2 + HN_3 \longrightarrow N_3NO + H_2O \longrightarrow N_2 + N_2O$$

$$HNO_2 + NH_2SO_3H \longrightarrow N_2 + H_2O + H_2SO_4$$

$$HNO_2 + NH_2OH \longrightarrow N_2O + 2\,H_2O$$

$$2\,HNO_2 + CO(NH_2)_2 \longrightarrow 2\,N_2 + 3\,H_2O + CO_2$$

Figure 11.15 Nitrous acid scavenging reactions of hydrazoic acid, sulfamic acid, hydroxylamine, and urea.

11.3.4.7 Removal of Nitrosamines

The removal of nitrosamines may be affected through chemical and physical processes. In general, the low reactivity of dialkyl nitrosamines means that specific conditions are required for their effective removal. One noteworthy method stems from the fact that nitrosation is reversible under strongly acidic conditions meaning that it is possible to transfer the nitroso group to another nitrosatable substrate [104]. The reactivity of nitrosamines is the subject of a recent review [144].

11.4 Formation, Fate, and Purge of Impurities Arising from the Hydrogenation of Nitroarenes to Anilines

The reduction of nitroarenes to anilines provides valuable intermediates for the manufacture of many agrochemicals, pharmaceuticals, dyes, and pigments [145]. The transformation has been subject to a number of reviews [146, 147]. Hydrogenative conditions are particularly significant: 40–50% of hydrogenation reactions performed in fine chemicals and pharmaceutical manufacturing in 2001 were estimated as involving the reduction of nitro groups [148]. In contrast to other nitroarene reduction methods (notably Fe/HCl [149], H_2S, or NaSH) [150][8], the stoichiometric by-product of hydrogenation using a precious metal catalyst is simply water. The opportunity to simply filter away a heterogeneous catalyst at the end of the reaction adds to the environmental credentials of these processes. A drawback, though one that is not peculiar to the use of metal-catalyzed hydrogenation conditions, is that the hydroxylamine intermediate and other reaction by-products are potentially mutagenic. If not controlled, the potential presence of these materials in the aniline product poses a health hazard to patients, as well as the operators and technicians involved in the manufacturing process.

11.4.1 Primary Reaction Mechanism

Historically, mechanistic thinking around the reduction of a nitroarene using a supported noble metal catalyst has used a sequence that routes via aryl nitroso and aryl hydroxylamine intermediates. This was first proposed by Haber following his electrochemical study of the reduction of nitrobenzene [152]. Figure 11.16 (including red arm) shows how such a sequence is possible by the sequential transfer of dissociated hydrogen atoms. The so-called Haber mechanism, with its use of

8 Other than heterogeneous hydrogenation, other approaches to nitro reduction include catalytic transfer hydrogenation, hydride transfer and metal dissolving reductions. [151].

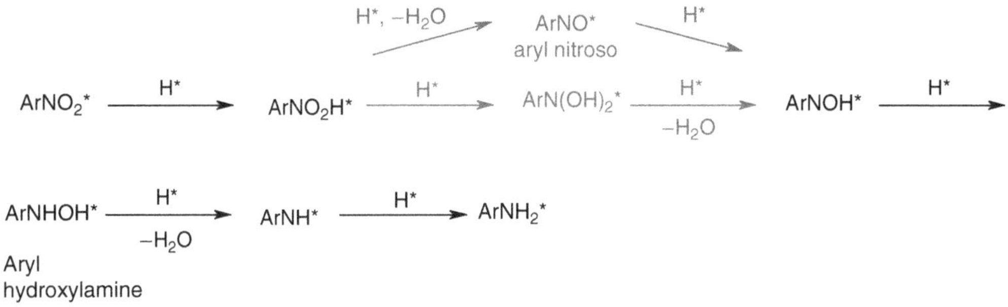

Figure 11.16 Primary mechanistic options for nitroarene reduction using a noble metal catalyst and dissociated hydrogen.

the aryl nitroso species, lost some credence when independently prepared nitrosobenzene was shown to react more slowly in the hydrogenation than nitrobenzene using Pd-C [153, 154]. This implied that the reduction of nitrobenzene did not proceed via the nitroso intermediate. Studies using an Au/TiO_2-supported catalyst were also shown to be inconsistent with an aryl nitroso intermediate [155]. In the case of supported palladium and platinum catalysts, a credible alternative to the aryl nitroso route for the main pathway uses the alternative series of dissociated hydrogen additions shown in Figure 11.16 (including green arm) [156]. Suffice to say, in the case of supported noble metal catalysts, the importance of an aryl nitroso intermediate in the main pathway is still an area of continued debate and is complicated by the impact of the mode (parallel or vertical) of adsorption of the aryl moiety to the metal surface [157]. The nitroso is still thought to be important where the catalyst is non-noble: [158] its comparative oxophilicity facilitating dissociation of the N—O bond at the expense of an early hydrogen transfer step [159–161].

11.4.2 Mass and Heat Transfer Effects

Above a certain point, further increases in the loading of a supported catalyst does not increase the reaction rate in a linear fashion. Under these circumstances, the chemical reaction of dissolved hydrogen with the nitroarene is replaced by mass transfer considerations as being rate-limiting. Movement between all the phases present, namely, hydrogen gas, reaction solution, the catalyst, and potentially undissolved substrate, creates multiple potential sources of mass transfer resistance. The concentration gradients concerned for a fully soluble substrate are shown in Figure 11.17. When changing scale, it is advisable to strive to match mass transfer coefficients in order to reproduce the original reaction rate and selectivity [162]. Simply trying to match geometry, agitation rates, or other mixing characteristics, in isolation, will inevitably alter the mass transfer resistance. Depending on the catalyst loading, this could lead to a change in the reaction outcome.

For a batch process where the substrate is present in solution at the outset of the reaction, mass transfer to the liquid phase concerns itself solely with the passage of hydrogen gas into this phase. The rate of gas–liquid mass transfer is affected by the difference between the actual and saturation solution concentrations of hydrogen, with the latter being a function of hydrogen headspace pressure. How quickly a saturation concentration could be achieved (though not what this saturation concentration is[9]) is encapsulated by the mass transfer coefficient (k_La). This is affected by the

9 The maximum concentration of hydrogen in the liquid phase is determined by the pressure, temperature and the solvent choice.

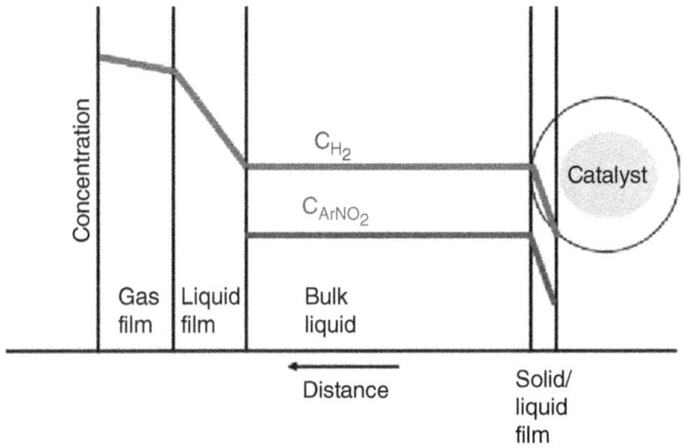

Figure 11.17 Concentration gradients (not to scale) associated with mass transfer resistance in nitroarene hydrogenation on a heterogeneous catalyst.

agitator and reactor geometries, vessel fill, surface area between the two phases (which varies with agitation), and the physical properties of the fluid. For fast chemical processes, the rate with which the hydrogen passes into the solution may not be large enough to saturate the reaction mixture and hence the catalyst, limiting the overall rate. Partitioning of the hydroxylamine intermediate via the condensation pathway (*vide infra*) is reported to be more likely when the reaction mixture is starved of hydrogen [163][10].

Once in solution, bulk diffusion of the dissolved hydrogen and the substrate to the boundary layer at the surface of the catalyst is required. Either transport process is unlikely to be an issue for reasonable agitation rates. The material that has made it to the boundary layer around the catalyst particles needs to diffuse across this film ahead of reaction on the catalyst's surface. Where an eggshell catalyst (where the metal is restricted to the surface of the particle) is not being used, the nitroarene and hydrogen will need to make it into the catalyst pores. Adsorption can either take place at a pore entrance ahead of diffusion along the internal surface of the pore or be preceded by diffusion into the catalyst pores.

The chemical reaction of the dissociated hydrogen atoms with the adsorbed nitroarene and intermediates on the catalyst surface has been discussed previously (Figure 11.16). This reactivity uses Langmuir–Hinshelwood–Hougen–Watson behavior, in which the nitroarene and hydrogen are adsorbed to the catalyst surface ahead of reaction, with the hydrogen dissociating in the adsorption process [164]. The presence of hydrogen but not the nitroarene in the rate expression when using a platinum or palladium catalyst is consistent with comparatively low catalyst coverage of hydrogen versus the nitroarene. The hydrogen is thought to occupy catalyst sites not occupied by the nitroarene or reaction intermediates. Before moving on, it should be pointed out that there are isolated reports of the hydrogen reacting without first adsorbing onto the chemical surface (Eley–Rideal behavior) [165, 166], with their paucity perhaps due to the ease with which hydrogen adsorbs on and dissociates at platinum or palladium surfaces. Once the aniline product has formed, it desorbs from the catalyst surface to create a vacant site for fresh material. The aniline then back diffuses out of the pores and across the film layer into the bulk liquid.

Mass transfer aside, the high exothermicity of nitroarene hydrogenations (c. −500 kJ/mol) means heat transfer effects also need to be considered. The associated adiabatic temperature rise,

10 This can be addressed by increasing the impeller's agitation rate, addressing any interruption to the hydrogen supply or lowering the catalyst loading to bring the reaction back under kinetic control.

if uncontrolled, creates a safety issue as well as disturbing competing reaction rates responsible for the reaction's impurity profile. For the reactor jacket to control the exotherm, it may be necessary to operate under continuous or semibatch conditions[11,12].

11.4.3 Condensation Chemistry

Of the intermediates shown in Figure 11.16, only the starting nitroarene, hydroxylamine, and aniline product are typically detected in solution using online Raman and infrared (IR) spectroscopic techniques [167], reaction calorimetry [168], and classical offline analysis [154, 169][13]. Other materials are also typically present, however, on account of the ability of the hydroxylamine intermediate to disproportionate [170]. This disproportionation generates the aniline and the aryl nitroso[14]. The aryl nitroso and residual hydroxylamine species can condense with one another with a selectivity that is increased by the medium being alkaline [171] or the presence of DMSO [172, 173]. As previously discussed, while there is not a consensus around the mechanism of nitrobenzene reduction, some studies clearly support the reduction of any nitrosobenzene to aniline via predominantly the condensation route [154][15]. The condensation affords an azoxy species (small amounts of which can confer intense color to a mixture), which is then reduced further to form a hydrazo species[16] and then the aniline. This forms a third means of forming the aniline on top of hydroxylamine reduction and disproportionation. In a synthesis of the mammalian target of rapamycin complex (mTORC) inhibitor AZD3147, the amounts of materials at the end of the reduction of nitro compound **1** to aniline **2** are shown in Figure 11.18 [17]. Azoxy compound **3**, azo compound **4**, and hydrazine **5** were thought to arise from the disproportionation of hydroxylamine **6**.

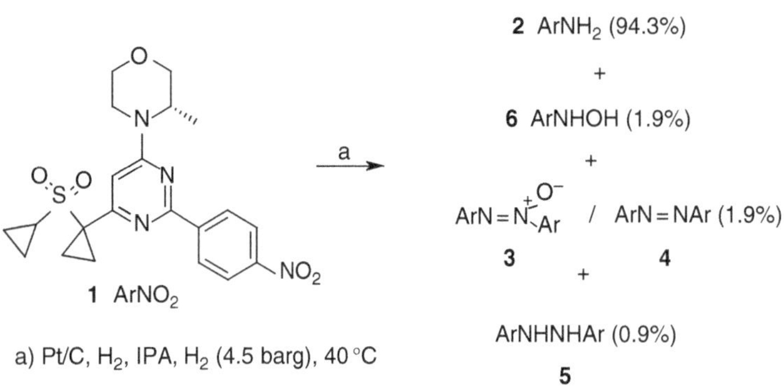

Figure 11.18 Condensation chemistry impurities at the end of a nitro reduction used to make AZD3147.

---

11 Alternative heat transfer solutions related to the reactor choice and setup have included fitting a traditional slurry tank reactor with a pump around loop with a heat exchanger, or the use of a falling film reactor.
12 Alternative means of controlling the exotherm are to lower the catalyst charge or agitation rate.
13 The detection of very low concentrations of the aryl nitroso intermediate is consistent with its absence from the primary reaction, though its strong adsorption to the metal surface and a rapid onward reduction to the hydroxylamine would also explain these concentrations.
14 This is a highly exothermic transformation, but can not be controlled by shutting off the hydrogen supply as it does not rely on hydrogen.
15 Whilst not well documented in the literature, pharmaceutical development exercises in documenting the mechanisms by which impurities could form, should note reports of aryl nitroso compounds dimerising under thermal conditions to form the diaryl azoxy [174].
16 Accumulated hydrazine can also rearrange to a p,p′-bianiline in an acid-catalysed process.
17 AstraZeneca unpublished results.

The reduction of hydrazo dimer to the aniline is comparatively slow compared to the reduction of the azoxy and azo compounds, and this can result in accumulation of the hydrazo dimer. The effect of ring substitution, hydrogen pressure, catalyst choice, and DMSO on levels of this dimer have all been studied [173]. The hydrazo dimer can be easily oxidized to the azo dimer upon exposure to atmospheric oxygen leading to the observation of artificially high levels of the azo compound where offline analysis has been used [175][18]. An ever growing capacity throughout manufacturing facilities for leveraging inline measurements avoids the artifacts associated with offline analysis. In this regard, ReactIR offers one such means of monitoring levels of the hydroxylamine intermediate so as to inform the decision about when to stop the hydrogenation of a nitroarene [177]. The principal chemical processes associated with the reduction of a nitroarene on a supported noble metal catalyst are summarized in Figure 11.19.

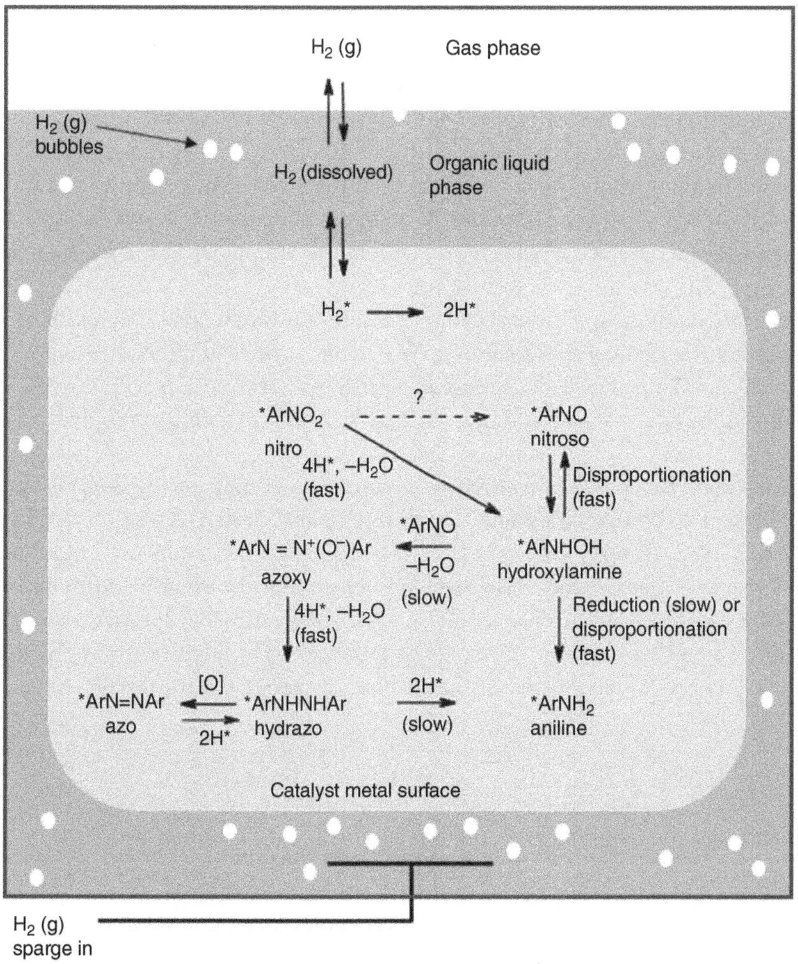

Figure 11.19 Mechanistic representation of $ArNO_2$ reduction. The * represents species adsorbed/bound on the metal surface.

18 For completeness when capturing failure modes, it should be noted that the azo dimer can also form from the acid-catalysed condensation of the aryl nitroso intermediate with the aniline product, though it is quickly reduced to the hydrazo under the reaction conditions [176].

While not reported as a by-product in Figure 11.19, when under acidic conditions, accumulated hydroxyaniline is susceptible to participation in a Bamberger rearrangement. This is exemplified by the chemistry in Figure 11.20 where the presence of hydrogen chloride results in the contamination of aniline **7** with hydroxyaniline **8** [177][19].

11.4.4 Factors Affecting Aryl Hydroxylamine Accumulation

Having covered the chemical and physical processes operating, the accumulation and control of the principal potential mutagenic liability of a nitroarene reduction, the aryl hydroxylamine intermediate, will now be examined in more detail.

When using a noble metal catalyst, the formation of ArNH* from the adsorbed hydroxylamine involves the cleavage of an N—O bond and is often cited as being rate-limiting [179]. The risk of significant hydroxylamine accumulation is greater where a platinum, rather than a palladium, catalyst has been used [180] and where the arene is electron deficient [180, 181]. Steric effects may also play a part for *ortho* substituents [181].

Hydroxylamine accumulation can often be inferred if the reaction rate has slowed significantly after two-thirds of the theoretical hydrogen amount has been used [147]. This is illustrated with the hydrogenations of 3,4-dichloronitrobenzene and *p*-chloronitrobenzene using a platinum catalyst. Two-thirds of the way through their consumption, the amount of hydroxylamine was adjudged to be 35% in the case of the more electron-deficient 3,4-dichloro compound and 1.9% for the 4-monochloro [182].

Lowering the temperature (discouraging the aryl hydroxylamine's onward reaction versus higher temperatures) or increasing the hydrogen pressure will also encourage accumulation. Where a catalyst has been used for a prolonged period, or reused, its surface characteristics will move further and further away from those of virgin catalyst, reducing its ability to facilitate the hydroxylamine reduction.

The presence of particular components of a reaction mixture also encourages hydroxylamine accumulation. When using Pt/C, these include small amounts of DMSO [180, 183–187], phosphorus compounds [182, 185, 188], nitrogenous bases [172, 184, 188], and the latter two factors in combination [188]. Inorganic impurities have also been implicated in hydroxylamine formation. This point was tragically illustrated with a reduction of 3,4-dichloronitrobenzene (*vide supra*) where an imperfect upstream phase cut contaminated the hydrogenation mixture with a nitrite salt [181, 182, 189][20] [190]. This retarded the hydrogenation forcing recourse to the use of a higher

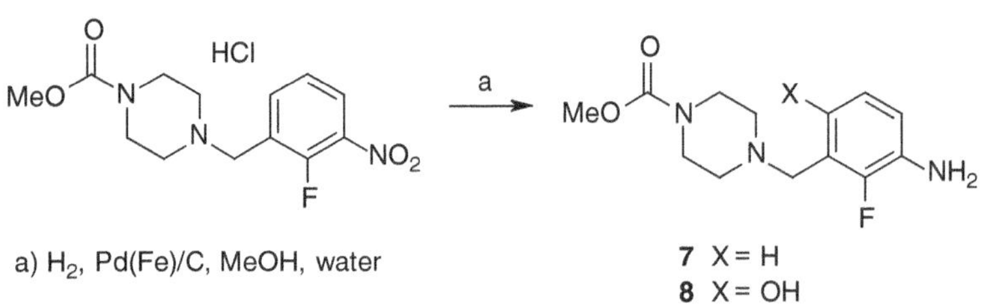

Figure 11.20 Contamination of aniline with a hydroxyaniline arising from a Bamberger rearrangement.

19 The hydrogenation of nitrobenzene in 15% sulfuric acid using a Pt-C catalyst is reported to give 83.3% *p*-aminophenol [178].

20 The hydroxylamine was normally present at about 5%, but in the case of the explosion, had increased to about 35% due to a slow reaction attributed to nitrite contamination of the nitro starting material.

reaction temperature. The accumulated hydroxylamine intermediate disproportionated in an exothermic fashion, resulting in an explosion.

While the present discussion is concerned with understanding the formation, fate, and purge of the hydroxylamine intermediate, among other by-products, the reaction can be biased such that the hydroxylamine, rather than the aniline, is the major product. This has been achieved through the use of pyridine as a solvent [191] or through catalyst choice [192–194]. As an example of the latter, the use of a sulfided platinum catalyst for the reduction of 2-iodo-5-nitropyridine resulted in 66 Liquid chromatography (LC) area percent (LCAP) of the hydroxylamine [195].

11.4.5 Aryl Hydroxylamine Control

11.4.5.1 Use of Cocatalysts

The disproportionation of the hydroxylamine to the nitroso and aniline is accelerated by using vanadium-containing additives during catalyst preparation (Figure 11.21) [196–199]. This can act as an important control against hydroxylamine accumulation. In the case of the hydrogenation of *N*-cyclohexyl-*N*-methyl-2-nitrobenzenesulfonamide on Pd/C at 120 °C, the yield of the corresponding arylhydroxylamine drops from 41% to <1% using a vanadium salt (Figure 11.22) [200][21].

Figure 11.21 Vanadium-catalyzed disproportionation of hydroxylamine.

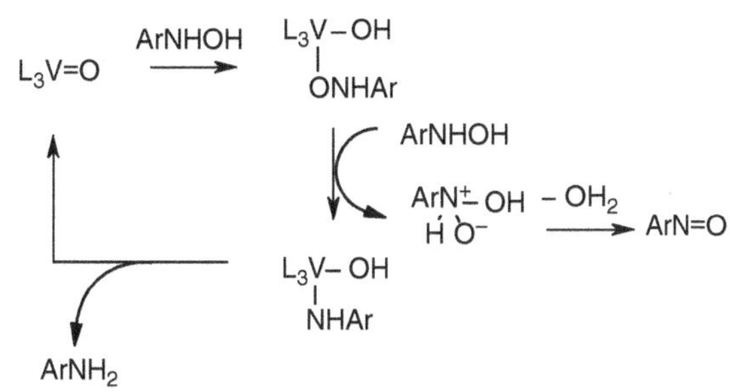

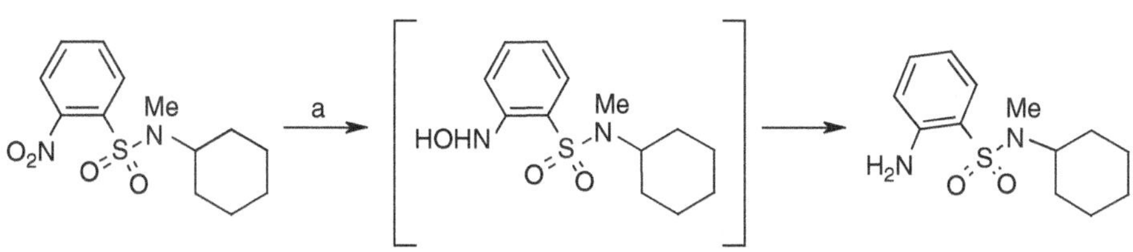

a) Pd-C, H_2, MeOH

Additive	Hydroxylamine(%)
None	41
NH_4VO_3	<1
V	<1

Figure 11.22 Effect of vanadium cocatalyst on outcome of *N*-cyclohexyl-*N*-methyl-2-nitrobenzene sulfonamide hydrogenation.

21 A steel autoclave used for a reduction with the vanadium salt additive showed a temporary "memory effect", even after it had been carefully cleaned. A further three runs in its absence were required for hydroxylamine levels to return to the levels seen before the additive was used.

Another example can be found in the reduction of nitro compound **9** to aniline **10** as part of the synthesis of the AstraZeneca tyrosine kinase inhibitor AZD8931 (Figure 11.23) [189]. An ammonium vanadate cocatalyst helped to force the disproportionation of hydroxylamine **11**, reducing remaining levels available to react intramolecularly to form amide by-product **12**. This amide was the major product of the reaction in the absence of the cocatalyst.

Iron(II) chloride has also been used as an alternative modifier that suppresses hydroxylamine accumulation [201]. The iron(II) has been linked to improved selectivity, whereas the increased activity, with specifically the chloride counterion, is suggestive of an electrochemical reaction mechanism. Modifiers apart, standalone catalysts consisting of gold nanoparticles supported on titanium dioxide or iron(III) oxide [202, 203] or a Pt-Sb/TiO_2 [204] are also reported to suppress aryl hydroxylamine formation.

11.4.5.2 Physical Adsorption

Where the chemical consumption of the hydroxylamine during the reaction is insufficient, a physical process for its removal from the aniline product can be considered. The hydroxylamine formed during a nitro reduction as part of a synthesis of ZD2315, an mammalian target of rapamycin complex (MHC) II antagonist, was removed by adsorbing the more basic aniline product onto an acidic ion exchange resin [189]. This left the residual hydroxylamine intermediate in solution.

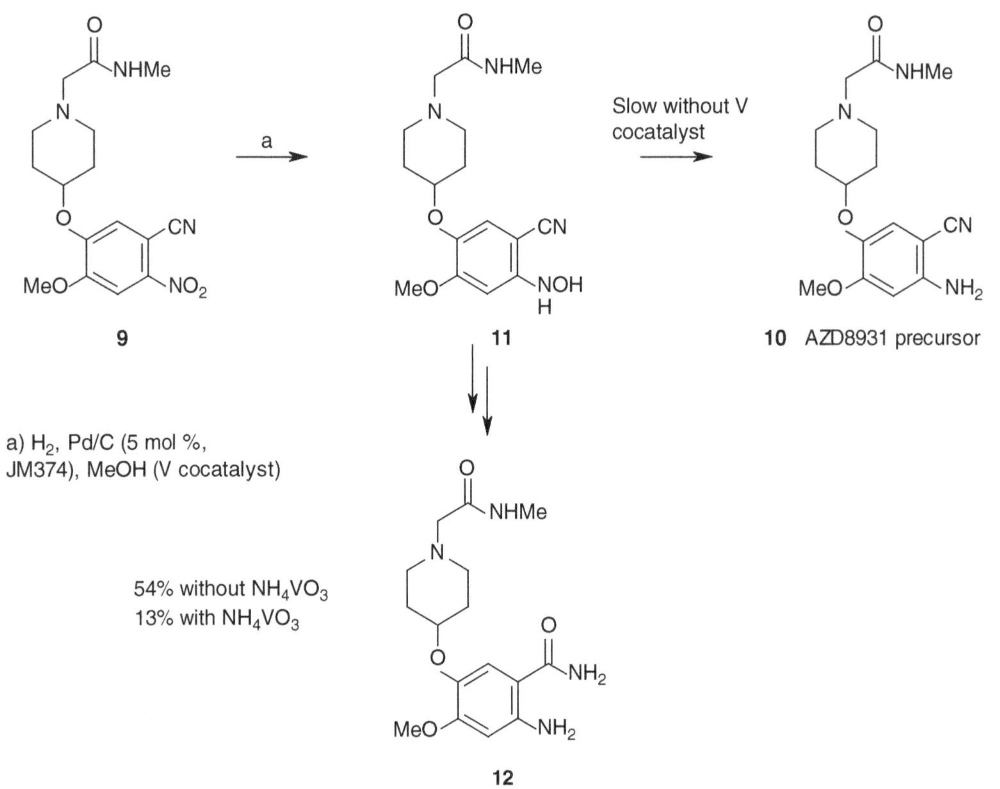

Figure 11.23 Effect of vanadium cocatalyst on nitro reduction used in AZD8931 synthesis.

11.4.5.3 Kinetic Understanding Around Formation and Consumption

Quantification of MI levels through the development of analytical methods requires significant investment of resources due to the ppm-level sensitivity required. The preparation and isolation of these materials to create analytical standards for offline analysis is complicated by the inherent reactivity of some MIs. Where a purge factor calculation is being used (ICH M7, Option 4), in lieu of analytical testing, the reactivity purge factor component can be accurately evaluated by establishing the half-life of the impurity [205]. This in turn requires knowledge of the rate constants and energies of activation for steps where MIs are being generated and consumed.

As part of the late-stage development of the heterogeneous catalytic hydrogenation of an unspecified dinitro compound (O_2N–core–NO_2), AbbVie scientists acquired such understanding [206]. To allow for the detection of the intermediates in meaningful amounts, the envisaged hydrogenation process was perturbed by lowering the reaction temperature, hydrogen pressure, catalyst loading, and by using more solvent. The observed concentration-time course data was congruous with the predictions arising from one of the possible sequence of elementary steps, though only after building in inhibition behavior displayed by the growing levels of dianiline product. Rate constants were extracted for the elementary steps, allowing extrapolative predictions around the consumption of the starting material and intermediates within the operating space envisaged for the commercial manufacture. To provide assurance that the reaction outcome would still be dominated by intrinsic kinetics in the envisaged operating space, the hydrogen-liquid mass transfer coefficient was measured in the commercial equipment using the final batch volume of the reaction solvent. This produced a good fit between observed and predicted data (Figure 11.24). Robustness was built into the control strategy by extending the mass transfer hold time, and the reaction kinetics control hold time, well beyond the time predicted to lower levels of potentially mutagenic intermediates to the target level of 1 ppm.

11.4.5.4 Holistic Control of Impurity Profile

While this discussion has focused on impurities arising from the reduction of the nitroarene moiety in the substrate molecule, it should be recognized that the reaction conditions will frequently lead to side reactions elsewhere in the molecule [147]. Reaction optimization will strive to deliver a reaction outcome that takes into consideration the likely formation and downstream purge of all

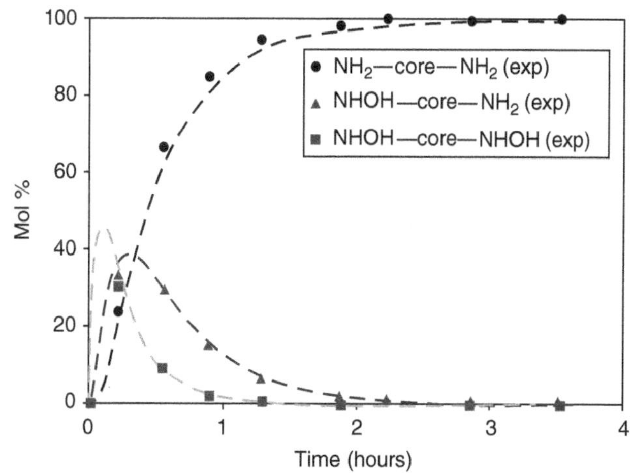

Figure 11.24 Concentration profiles of "HOHN-core-NHOH," "HOHN-core-NH_2," and "H_2N-core-NH_2" species. Dashed lines represent values calculated by the kinetic model. (Copied with permission [206].)

impurities, nitro reduction-related or otherwise. A nitroarene substrate will often have unsaturated functionality or C_{Ar}–Hal functionality that could be reduced. The limited purge of deshalo impurities often observed during crystallization means alternative control strategies are required for their removal. The risk of hydrodehalogenation depends on the halogen type (I, Br, Cl) and its position on the aromatic ring. The susceptibility increases when the halogen is on the same ring as the nascent aniline, due to the increased electron richness of this ring once the nitro has been reduced.

The selection of conditions for the reduction of nitro compound **13** to aniline **14** as part of studies toward the synthesis of vismodegib provides an example of managing competing concerns with respect to a reaction's impurities (Figure 11.25) [207]. An initial screen revealed that while low pressure conditions protected against pyridine reduction (impurity **15**) or dechlorination (impurity **16**), they led to the contamination of the aniline with hydroxylamine **17** or the daughter materials arising from its onward condensation. Persisting with platinum and optimizing around the cocatalyst, catalyst loading and temperature allowed aniline of >99.9 LCAP to be obtained. This material contained less than 0.12 LCAP of hydroxylamine **17**, a level that provided assurances around API quality as a result of accompanying spike and purge studies.

The partial and undesired loss of chlorine during the reduction of nitroarene **18** to aniline **19** as part of the synthesis of the antiplatelet compound ticagrelor provides an example where dehalogenation was addressed by catalyst modifier selection (Figure 11.26). Hypophosphorus acid was first used to modify the platinum-carbon catalyst [189, 201], to address hydroxylamine accumulation and sluggish hydrogen consumption at the end of the reaction. Using $VO(acac)_2$ as a further catalyst modifier, and increasing the reaction temperature, accelerated the disproportionation of the hydroxylamine **20** to the nitroso **21** and aniline **19**. This allowed the combined levels of nitroarene **18**, aryl nitroso **21**, and hydroxylamine **20** to meet the end of reaction criterion of Not more than (NMT) 0.1 LCAP, without any significant impact on levels of dechlorinated impurities. It was later found that replacing the vanadium additive with molybdenum (in the form of ammonium orthomolybdate) [208] led to an even cleaner reaction profile and was more economical.

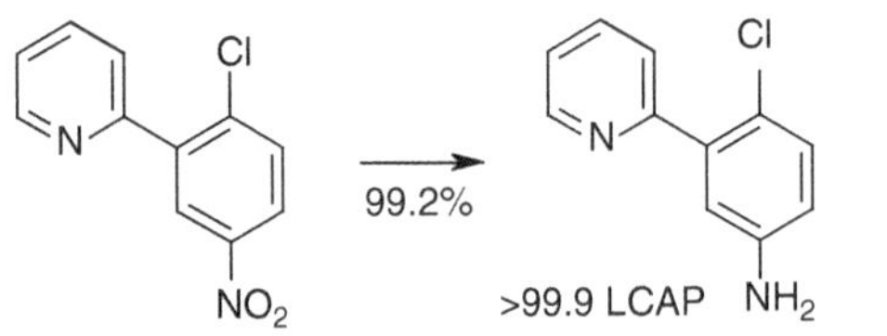

13 **14**

Optimized conditions: Pt(1%)/V(2%)-C (cat.), MeOH, AcOH, H_2 (5 bar), 50–55 °C

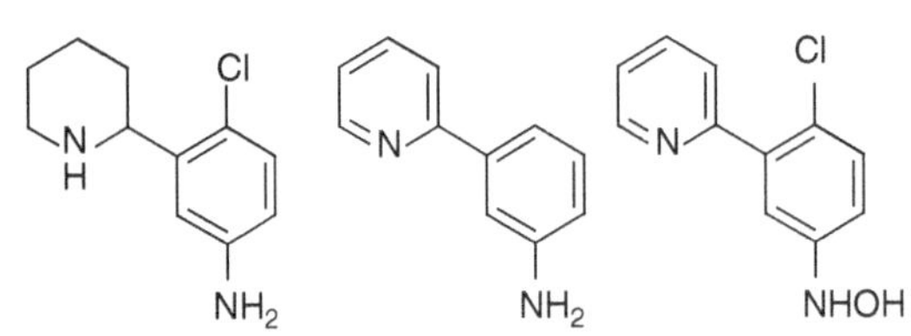

15 **16** **17**

Figure 11.25 Nitro reduction during synthesis of vismodegib.

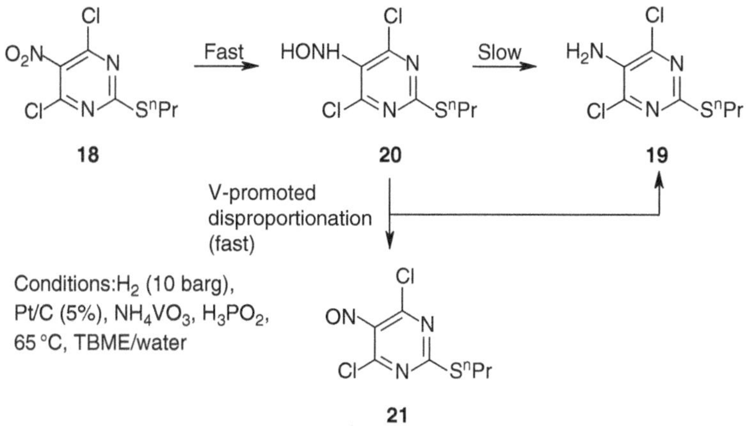

Figure 11.26 Nitro reduction during synthesis of ticagrelor.

In a further example of catalyst modification, the use of a commercially available sulfided platinum catalyst has shown utility where the C_{Ar}–Cl and C_{Ar}–NO_2 groups are both on a pyridine ring [195]. Its high chemoselectivity is likely due to the occupation of the most active sites on the platinum by the sulfur. Amgen scientists used a low loading of this catalyst enroute to a B-raf kinase inhibitor (Figure 11.27).

Even when clearly signposted catalyst modifiers are not present, nominally similar supported catalysts can produce significantly different impurity profiles. This is illustrated by a GSK study of the reduction of 2-bromo-4-chloro-nitrobenzene where key impurities proved to be the diaryl hydrazine and desbromo impurities [209]. Medium- or high-throughput catalyst screening, enabled by automation, allows the performance of nominally similar Pd-C (e.g. all labeled "5% Pd-C") and Pt-C (e.g. all labeled "5% Pt-C") catalysts to be differentiated. In the case of the High Throughput Chemistry and Chemical Catalysis group at GSK, the screen chosen depends on whether or not the nitroarene bears a halide or a hydrogen-sensitive group like a nitrile, carbonyl, olefin, *O*-benzyl, *N*-benzyl, or benzyloxycarbonyl[22].

11.4.6 Controlling Residual Nitroarene

While this report has hitherto concentrated on the hydroxylamine intermediate, the nitroarene starting material, while typically readily consumed under the reaction conditions, is also a potential mutagen. In Figure 11.28, the aniline product was a retigabine-like drug substance and any unreacted nitroarene starting material was a critical quality attribute (CQA) that needed to be controlled to low levels [211]. When using a vanadium-doped catalyst, reaction profiling enabled by *in situ* UV-vis

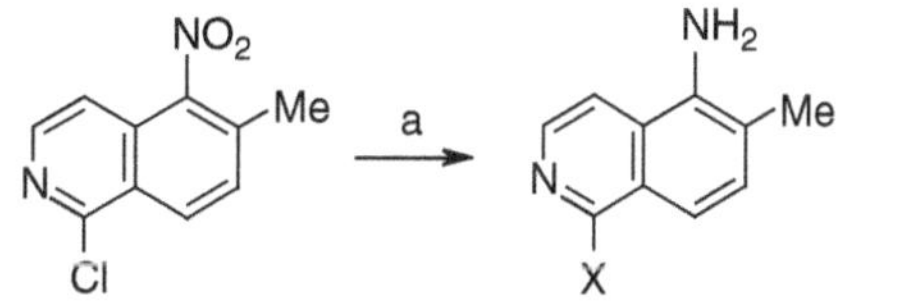

Figure 11.27 Use of a sulfided platinum catalyst.

22 In another illustration of the opportunity for judicious catalyst choice, nitro reductions in syntheses of the active pharmaceutical ingredients clofazimine and vismodegib suppressed dechlorination using Raney cobalt [210].

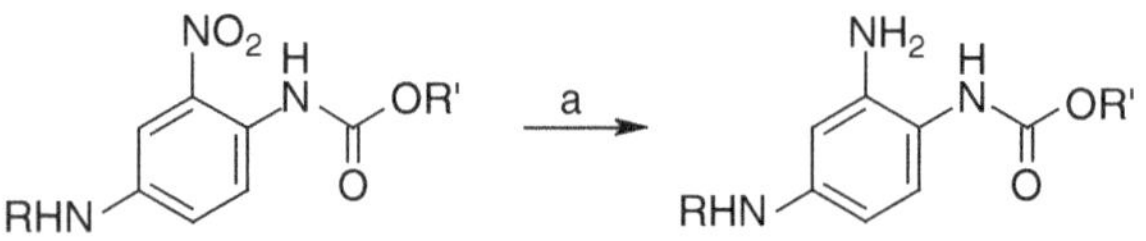

Figure 11.28 Reduction of a retigabine-like drug substance.

a) H_2, 1% Pt + 2% V/C, EtOH

and IR absorbance data supported the initial reduction of the nitro group as being the slowest elementary mechanism step.

While behavior at low hydrogen partial pressures was complex, when precautions were taken to ensure that the reaction was always under kinetic control, a single model describing substrate adsorption to a catalyst active site followed by surface reaction could be fitted to the observed data. This allowed levels of the nitroarene input to be predicted, at any scale, given knowledge of the initial substrate concentration, reaction temperature, catalyst concentration, and hydrogen partial pressure. Given the need for the model to be leveraged over the course of the remainder of the lifecycle of the product and in a commercial manufacture setting, it was futureproofed through testing with different batches of input material. This revealed the need for its modification with a term for catalyst deactivation by measured but uncontrolled levels of sodium ions in the input[23]. With this modification in hand, Eq. 11.4) was found to describe the reaction time taken for a particular change in substrate concentration. Ten thousand simulations were performed *in silico*, and a Bayesian statistical approach was used to account for model error when evaluating the reaction time that would consume 99% of the input with a 95% confidence interval. While the outcome of any discussions with health authorities is not disclosed, the authors express confidence that the kinetic understanding supporting the design-space-type region proposed circumvents the need to verify it at a commercial scale. While not explicitly mentioned in this case, process design to remove residual nitroarene at the end of a reaction may in general be able to take advantage of differing behavior displayed by it and a basic aniline product, when using an acidic aqueous phase and a water-immiscible organic phase.

$$t = t_0 + \frac{\ln\frac{[\mathrm{Sub}]_0}{[\mathrm{Sub}]} + K_A\left([\mathrm{Sub}]_0 - [\mathrm{Sub}]\right)}{k_B K_A e^{-\frac{E_a}{RT}}[\mathrm{C_{tot}}](1 - \mathrm{a}[\mathrm{Na}])p_{H_2}} \tag{11.4}$$

where K_A = equilibrium constant for adsorption
k_B = reaction rate constant
E_a = activation energy
a = fraction of catalyst deactivation
R = gas constant
T = temperature
[Sub] = substrate concentration
C_{tot} = total catalyst concentration
p_{H_2} = hydrogen partial pressure
t = time

23 The working hypothesis was that a sodium salt was adsorbed onto the surface of the catalyst, blocking active sites, or otherwise prevented the initial reduction of some of the active sites.

11.4.7 Specific Considerations of Alkyl Nitro Reductions

The present review has focused on the reduction of aryl nitro groups to anilines. The importance of this approach is enabled by the pharmaceutical scientist's ability to nitrate arenes, or to functionalize nitroarenes, when establishing a route to a drug substance. The reduction of alkyl nitro groups, often introduced through the use of nitromethane as a reactant, is much less common in pharmaceutical development and manufacture. This is evidenced by the limited appearance of alkyl nitro reductions in scaleup literature [212]. While all potential mechanisms of mutagenicity for azoxy compounds are not clear, alkyl-azoxy materials alone can form alkyl carbocations with extremely high carcinogenic potency [213]. As a result, they are included in ICH M7's COC [7]. The limited use of the reduction of alkyl nitro groups in late stage pharmaceutical development means case studies documenting the associated control strategies are still to make it into the public domain.

11.4.8 Closing Comments on Hydrogenation of Nitroarenes to Anilines

The operational simplicity and low waste streams associated with the metal-catalyzed reduction of a nitroarene are undoubtedly attractive when an aniline features as an intermediate in an API synthesis. However, the stepwise nature of these reactions brings the need to understand the formation, fate, and purge of the intervening materials, as well as the purge of unreacted nitroarene. Building this understanding is complicated by ambiguity around the relative importance of pathways leading to the aryl nitroso intermediate but is increasingly being enabled by the application of *in situ* process analytical technology.

The most popular means of controlling levels of the aryl hydroxylamine intermediate itself (as well as the materials arising from its onward condensation, disproportionation, and rearrangement) uses a cocatalyst to accelerate its disproportionation. It is important to know the rate of hydrogen delivery to and the ability to remove heat from the solution if the impurity profile is to be reliably maintained when changing scale or equipment. With this knowledge, the catalyst charge can be selected to maintain a kinetically controlled regime. If a pharmaceutical development scientist can predict levels of potential MIs at the end of a nitroarene reduction, confidence is built into the associated purge factors. In turn, this potentially reduces the need for, or criticality of, costly and cumbersome analytical testing of the impurities when manufacturing control strategies are being evaluated and assessed.

11.5 Mechanism and Processing Parameters Affecting the Formation of Sulfonate Esters – Summary of the PQRI Studies

11.5.1 Introduction

Within the variety of classes of MIs until *N*-nitrosamines none have come under more scrutiny than sulfonate esters, the theoretical product of a reaction between a sulfonic acid and an alcohol. Indeed, concerns over sulfonic acids and the potential to generate sulfonate esters when in contact with alcoholic solvents predate the advent of the ICH M7 guideline. The first clear reference to such concerns came with the publication within PharmEuropa in 2000 of a short article that drew attention to the potential risk of formation of sulfonate esters as a result of a combination of sulfonic acids in alcoholic solution as part of a salt formation process [214]. At the time of publication, this was merely a call for "further information," it being an attempt to prompt the generation of data to understand better the extent of any of the risks. This publication is now seen by many as one of the triggers eventually leading to the original European Medicines Evaluation agency (EMEA) guideline. Thus, in many ways the concern over sulfonate esters was itself the trigger for the drive for specific control over the broader range of MIs.

The specific reason for the concerns relating to sulfonate esters stems from the properties of the sulfonate group and its labile nature i.e. the fact that it can readily be displaced by a nucleophile, resulting in the alkylation of the nucleophile concerned. This of course includes biological nucleophiles such as DNA. Indeed, studies conducted by Glowienke et al. [90] demonstrate that many such sulfonate esters are mutagenic.

During the period of time between 2000 and mid-2006, when the EMEA guideline was itself under development, sulfonate esters came under ever increasing scrutiny. This led to calls either to avoid the use of alcoholic solvents when handling sulfonic acids in salt formation processes or, even more extreme, to eliminate their use altogether i.e. not to develop sulfonic acids as counterions at all.

The potential loss of sulfonic acids as counterions was a major concern [215]. Doing so would have a significant impact. It would eliminate an entire class of pharmaceutically useful counterions, leading to the potential development of suboptimal API forms e.g. hygroscopic. It could also impact the synthesis of the API through restricting the use of reagents such as mesylates/tosylates. The frustration was compounded by the fact that both industry and regulators alike poorly understood the extent of the risk and even the mechanism involved in the formation of sulfonate esters.

Matters to some extent came to a head during a Drug Information Association (DIA) meeting held in Bethesda, Maryland, in late 2005. During one of the sessions at this meeting, a debate took place between Food and Drug Administration (FDA) officials and members of the audience from the pharmaceutical industry. FDA queried as to why was industry still using sulfonic acids as counterions. The response for a number of participants was to look to defend their use on the basis that many had examined the resultant APIs e.g. mesylates and tosylates, and no one had found predicted esters such as EMS at a level of concern in isolated materials. The debate continued after the session with the FDA showing interest in the scientific understanding behind these results. What was clear was that although ester formation was controlled, exactly how was not fully understood. This directly set up the challenge for industry to develop a detailed understanding of the factors associated with the formation of sulfonate esters, and the work described in this chapter was born.

The Product Quality Research Institute (PQRI) is a nonprofit consortium of organizations involving the pharmaceutical industry and FDA's Center for Drug Evaluation and Research (CDER) and Health Canada. Its mission is work together to generate and share relevant information that advances drug product quality and development. Given the nature of the problem with sulfonate esters and the FDA's interest, PQRI was seen as the ideal vehicle through which to conduct research into this area.

The keen interest in this initiative resulted in the formation of a multidisciplinary team, with representatives from a significant number of major pharmaceutical companies. Individuals were selected to ensure the team had the requisite skill set in the critical areas of analytical and physical organic chemistry/kinetics. Another critical decision taken at the onset of the project was to involve the Research Institute of Chromatography (RIC). The RIC, led by Professor Pat Sandra, is renowned for its expertise in the field of trace analysis and thus was able to make a telling contribution from a technical perspective. Furthermore, by taking this approach, the project had dedicated resource that was a critical factor in conducting the work in the desired timeframe.

Soon after the work was initiated, the importance of the research became even clearer. In May–June 2007, it was widely reported in both the scientific and general press [216] that patients taking Viracept (nelfinavir mesylate), an antiviral marketed by Roche, had complained of the tablets possessing a strong odor and of adverse reactions such as nausea. It subsequently became clear that the tablets were contaminated with the sulfonate ester EMS, with levels of up to 2300 ppm eventually being reported. This led to the temporary withdrawal of Viracept and also further heightened concerns over sulfonic acid salts in general [217, 218].

11.5.2 Reaction Mechanism

Sulfonic acids are widely used as counterions for APIs [219], to modify and manipulate the physical properties of the API to develop an appropriate physical form. Low molecular weight alcohols such as methanol, ethanol, and isopropanol are widely used as solvents in processes used to produce such salts and thus the risk is that the sulfonic acid can react with the alcohol during the manufacturing process, yielding the corresponding sulfonate ester.

The solvolytic behavior of sulfonate esters has been studied in great detail, particularly with respect to both kinetics of hydrolysis and the products from sulfonates of increasingly complex alcohols. Analytical methodologies based on spectroscopy or conductance used for some early kinetic studies assumed the solvolysis proceeded to completion, thereby implying no mechanism to form sulfonate esters in alcoholic solutions. Little consideration had been given to assessing any levels of residual ester, which might have characterized propensity for the reverse reaction – the formation of esters. A *post-facto* justification of the assumption of "complete" solvolysis can be found in the work described by Teasdale et al. [92], as the low levels of ester reported would not been sufficient to have introduced noticeable errors in kinetic measurements.

At the advent of these studies into the potential reaction between sulfonic acids and alcohols, little was known about the mechanism of formation of sulfonic acid with only one identified reference in the literature [220]. Within this paper, Snodin briefly commented on the possible mechanism and the low likelihood of reaction; however, there was no experimental data to provide proof of the postulated mechanism. That no data is available is unsurprising as in synthetic terms the reaction is of very little value. This is because there are far more effective means of generating sulfonate esters, usually employing the corresponding sulfonyl chloride.

As a result of the paucity of data, it was decided to look to study the reaction mechanism in order to be able to understand the factors that impact the reaction, ultimately to be able to manipulate these to enable effective control during salt formation processes.

Ahead of the study, two mechanistic pathways were postulated, as shown in Figure 11.29, using the example of the reaction between MSA and methanol.

Pathway 1 involves protonation of the alcohol by the sulfonic acid and subsequent nucleophilic attack of the resultant sulfonate anion on the protonated alcohol, yielding the corresponding sulfonate ester. Pathway 2 is analogous to the $A_{AC}2$ mechanism associated with acid-catalyzed esterification of carboxylic acids.

In order to elucidate the mechanism, a simple experiment was designed, using ^{18}O-labeled methanol. Were the mechanism to follow pathway 1, then it would be expected that all of the ^{18}O label would be present in the water molecule. The reverse is true for pathway 2 where the label would expect to reside in the subsequent methyl methanesulfonate (MMS) molecule generated by the reaction.

Solutions of ^{18}O-methanol/MSA and unlabeled methanol/MSA were heated at reflux for a short period of time (two hours), and then the resultant solutions were analyzed by direct injection gas

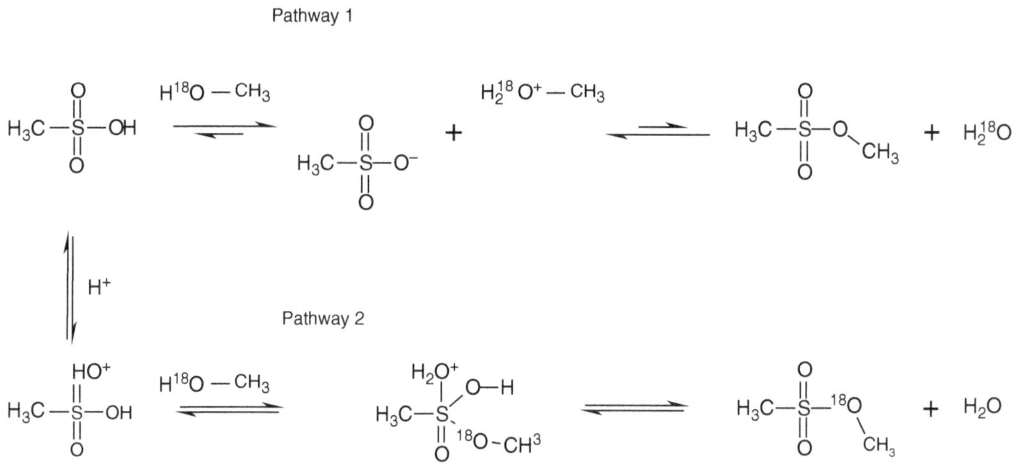

Figure 11.29 Potential mechanistic pathways for the formation of sulfonate esters.

chromatography-mass spectrometry (GC-MS). Within the resultant chromatograms, peaks corresponding to excess methanol, water, and MMS were detected, the latter being positively identified based on its mass spectrum.

The level of conversion observed was low. Examining the mass spectrum of the MMS formed using isotope-labeled methanol showed only one molecular ion *m*/*z* 110. This was identical to that obtained from the corresponding reaction involving the unlabeled methanol. This clearly demonstrated that the ^{18}O was not incorporated into the MMS molecule, supporting pathway 1. Further evidence was provided through the extraction of ion chromatograms at *m*/*z* 18 (water) and 20 ($H_2{}^{18}O$). When these were overlaid, a peak at *m*/*z* 20 was only observed in the reaction involving the labeled methanol confirming that the isotope label was incorporated into the water molecule. These findings were critical not only in clearly demonstrating the essential features of the mechanism but also in proving that sulfonic acids are not analogous with their carboxylic acid counterparts as might have been postulated.

Another key observation is the proton dependency of the reaction. As will be illustrated later, this underpins all of the results seen in the kinetic studies.

In addition to pathways 1 and 2 describing reversible reactions, the well-established (*vide supra*) solvolysis is a third competing reaction limiting net ester formation. This is shown in Figure 11.30 for the consumption of MMS. In the experiment diethyl ether was detected, eluting just before methanol in the GC analysis. This was confirmed by analysis of the mass spectrum. In the case of the reaction involving the labeled methanol, the peak had shifted by two mass units showing that the ^{18}O label had been incorporated into the ether.

This method was eventually selected as the basis for the studies described in the chapter.

However, the original method was focused on the determination of trace levels of sulfonate ester in API. In contrast the present studies were focused on the formation of sulfonate ester in

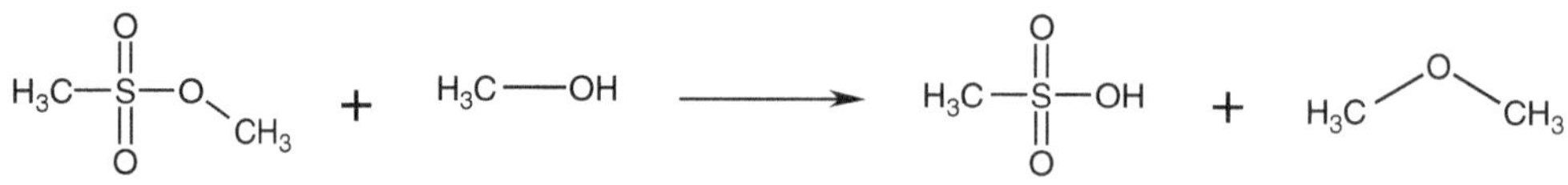

Figure 11.30 Solvolysis of sulfonate esters.

concentrated reaction mixtures. Thus, minor modifications (changes to the levels of derivatizing agent and NaOH added) were made to the method to ensure it demonstrated the requisite precision and reproducibility over the wide linear dynamic range required. One further modification was the incorporation of a second internal standard (IS) pentafluoroanisole (PFA). This IS, which has a chemical structure similar to the derivatized solutes, (methyl ether instead of ethyl sulfide) was used to monitor instrument performance. A large deviation observed on the peak area of PFA would indicate an error in IS addition (liquid handling) and/or in static headspace analysis.

This derivatization-headspace GC-MS method was ultimately fully automated using a robotic system and applied to the analysis of multiple sulfonic acid/alcohol systems. A schematic representation of the system is provided in Figure 11.31.

It should be noted that rather than performing one single reaction from which aliquots could be withdrawn, a series of identical reactions were established, these being placed in individual vials within the sample tray. Full details of the method are described in the paper published by Jacq et al. [221].

11.5.3 Experimental Results

11.5.3.1 Experimental Results from Study of the Ethyl Methanesulfonate (EMS) System

11.5.3.1.1 EMS Formation – Effect of Temperature

The effect of temperature on the level of EMS formed is graphically illustrated in Figure 11.32. This represents the molar conversion of sulfonate anion to ester in solution. Even at elevated temperature, 70 °C, less than 0.4% conversion was observed after 20 hours. After 80 hours, reaction was approaching a pseudo-equilibrium molar conversion at c. 0.65%. This again illustrates the point made earlier that from a synthetic perspective this is an extremely ineffective way to synthesize sulfonate esters. i.e. it is a chemically useless reaction.

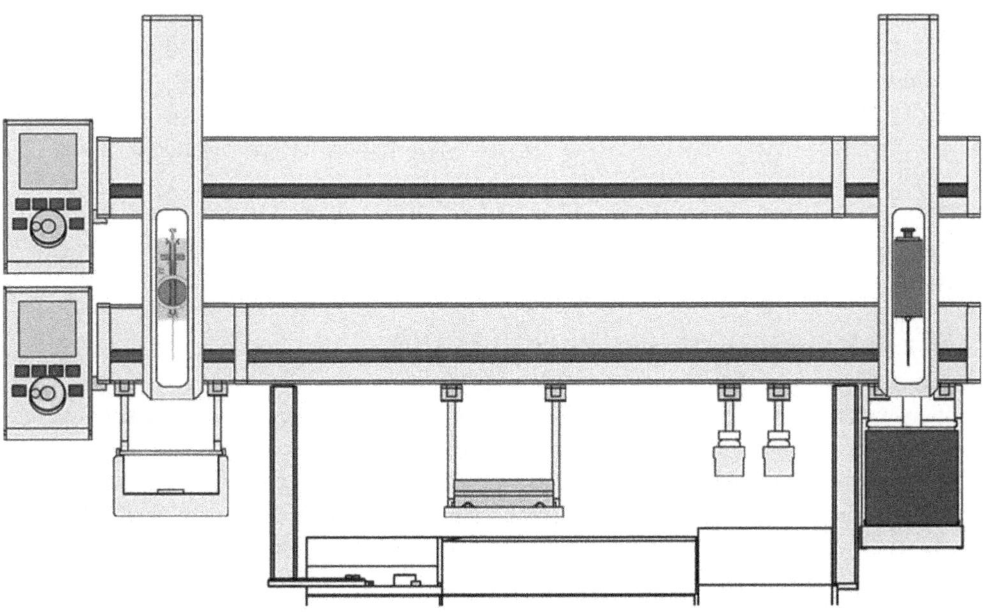

Figure 11.31 Schematic representation of the instrument used for the conduct of studies.

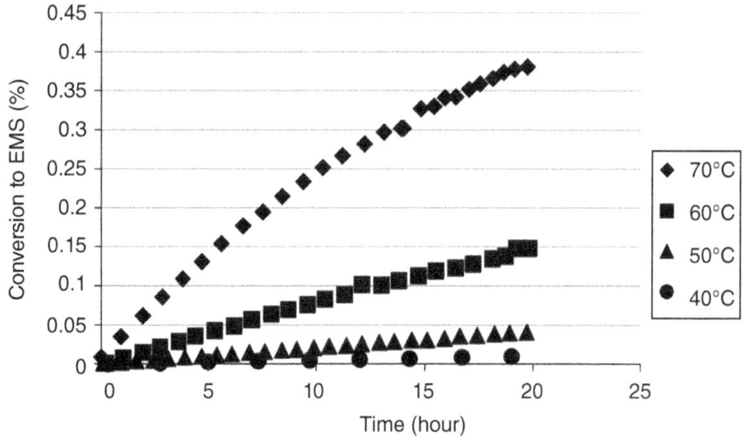

Figure 11.32 EMS formation under anhydrous conditions and the effect of temperature.

From this it can be clearly seen that controlling temperature can minimize the level of ester formed. It also illustrates the time taken to reach equilibrium, showing that minimizing the residence time of sulfonic acid in alcoholic solution can dramatically reduce the level of ester formed.

It is important to recognize this study, focused on the simple binary (acid + alcohol) system, effectively demonstrates the upper limit that may be formed in solution. In API manufacture, isolation processes would be expected to significantly reduce the levels, the majority remaining in the mother liquors. Thus, the level seen in solution represents an upper limit in terms of potential contamination, and by comparing this to the level in isolated material, it is possible to determine the efficiency of the salt formation process itself to purge any sulfonate ester present.

11.5.3.1.2 EMS Formation – Effect of Water

Reaction profiles in the presence of added water are shown in Figure 11.33. The water content in each reaction mixture was measured by Karl–Fisher titration at the start of the reaction and expressed as %w/w. The presence of even moderate water at levels of 5 %w/w reduced the levels of EMS by a factor of approximately one-third, to below 1000 ppm molar conversion at 70 °C after a reaction time of 18 hours. This can be rationalized both by the stronger solvation of protons by water compared with ethanol reducing the forward reaction and by the enhanced rate of hydrolysis of EMS compared to ethanolysis.

Where high levels of water are present in the system, the level of conversion is very effectively suppressed.

11.5.3.1.3 EMS Formation – The Impact of the Presence of Base

Low but appreciable sulfonate ester formation, as illustrated, does occur under strongly acidic conditions when combined with high temperature. However, salt formations using sulfonic acid counterions often employ either stoichiometric amounts or at most small excesses of acid.

Experiments were therefore designed to have the sulfonate anion present at a comparable concentration to the binary system (acid + alcohol only) but with reduced proton availability. The formation of EMS under conditions of lower acidity was tested using the weak base 2,6-lutidine as a surrogate API. No measurable rate of EMS formation was observed (above background noise) when a slight excess of 2,6-lutidine was used, the "reaction" being studied over a time period in excess of 12 hours at 70 °C (see Figure 11.34).

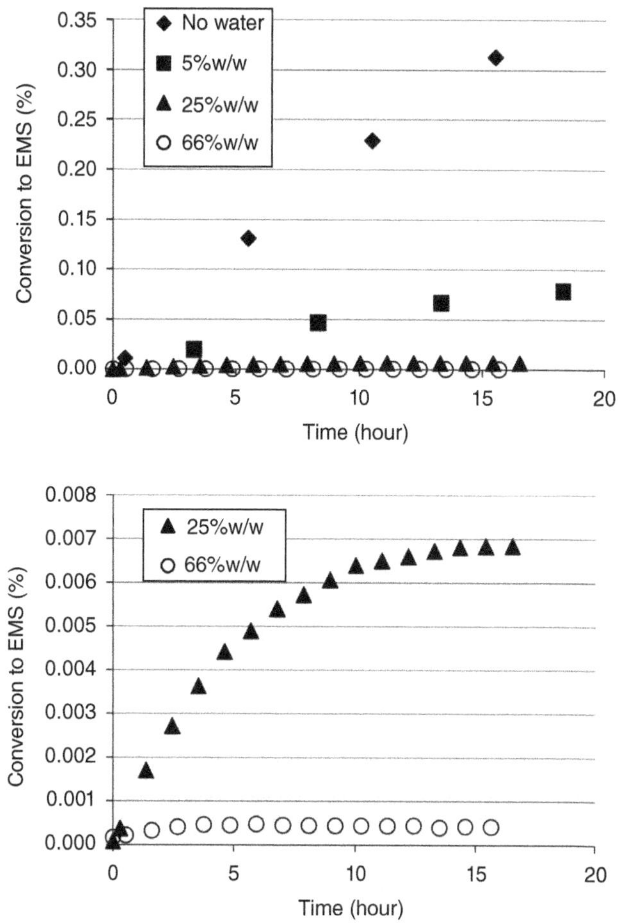

Figure 11.33 EMS formation – the effect of water on conversion; including expansion of plots for 25 and 66% w/w reaction systems.

When a 2% excess of MSA was present, a reaction did occur, however, at a very slow rate (approximately 0.004% conversion after 12 hours at 70 °C).

The lack of any reaction when an excess of base is present is entirely consistent with the reaction mechanism and the fact that the reaction is proton mediated. Without the transfer of the proton to the alcohol no reaction can occur. The proton will react overwhelmingly with the strongest base present, in this case 2,6-lutidine, in preference to ethanol. This also illustrates that the conjugate acid of 2,6-lutidine is not strong enough to protonate ethanol.

This was tested further by conducting an experiment where a small excess of lutidine was used to which was added a 10% excess of concentrated phosphoric acid, thus neutralizing the excess base. Even under these conditions, no reaction was observed, illustrating that an acid strength exceeding that of phosphoric acid is required in order for sulfonate ester formation to occur. The same experiment was conducted for all three alcohols studied – methanol, ethanol, and isopropanol; no reaction was observed in any of the systems.

11.5.3.2 Other Methanesulfonic Acid Systems

11.5.3.2.1 Experimental Results from Study of the MMS System

The above set of experiments was repeated studying the reaction between MSA and methanol. The results were consistent with those obtained for EMS. Figure 11.35 shows the results obtained for

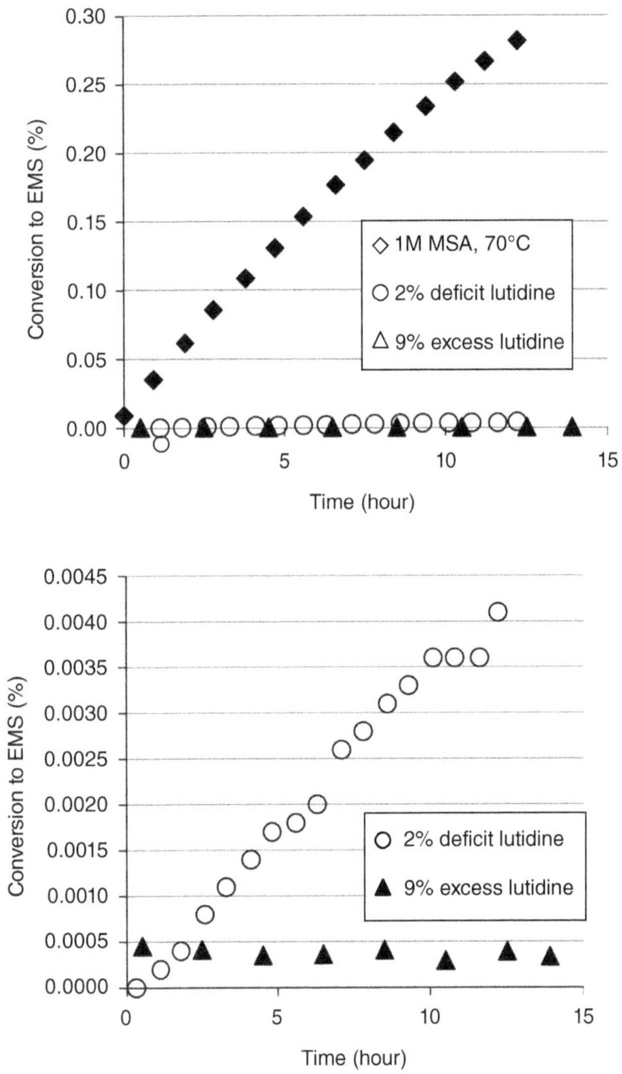

Figure 11.34 Effect of added base on EMS formation.

MMS, the plot showing the impact of both temperature and water content on the level of conversion.

Most critically of all, no reaction was seen in the presence of a slight molar excess of 2,6-lutidine.

11.5.3.3 Experimental Results from Study of the Isopropyl Methanesulfonate (IMS) System

In terms of reaction profile, the results were very similar to those seen for the EMS and IMS systems (Figure 11.36). The level of conversion under anhydrous/high temperature conditions was higher at around 1% conversion in comparison to EMS/MMS where levels were less than 0.5% over the same time period.

Again, critically, no reaction was seen in the presence of a slight molar excess of lutidine.

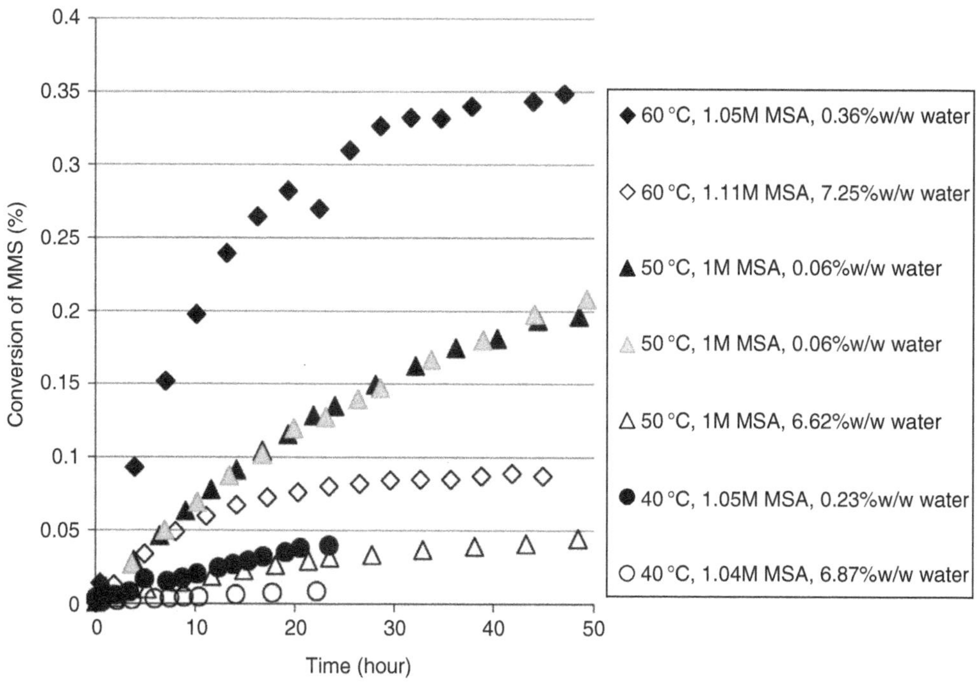

Figure 11.35 Combined plot showing the impact of both temperature and water content on conversion rates.

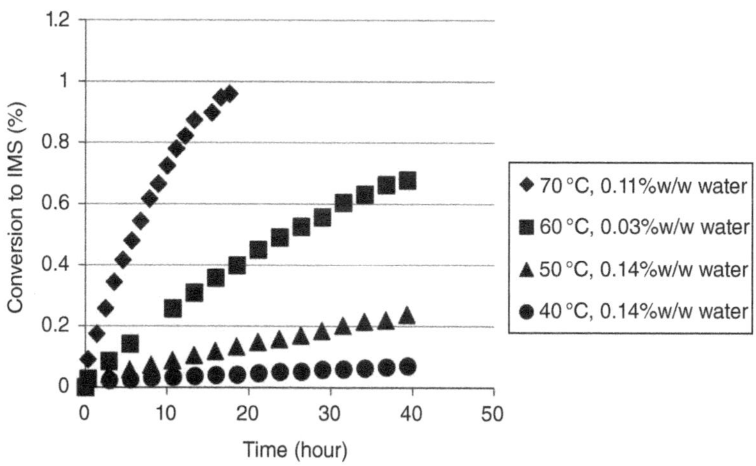

Figure 11.36 IMS formation under anhydrous conditions and the effect of temperature.

11.5.4 Experimental Results from Study of Toluenesulfonic (Tosic) Acid Systems

As well as studying the reaction between MSA and alcohols an identical series of studies was performed for toluenesulfonic (tosic) acid, another common salt counterion. Studying this system also allowed us to determine what differences there were, if any, between an alkyl and aryl sulfonic acid, in particular whether there were any significant differences in terms of proton dependence and extent of reaction.

11.5.4.1 Experimental Results from Study of the Ethyl Tosylate (ETS) System

The reaction profile for formation of ETS was directly comparable to that of the equivalent alkyl system (MSA + ethanol). Figure 11.37 shows the impact of temperature and water on conversion rates. It is important to note that the studies described were performed using toluenesulfonic acid monohydrate; this was used as the anhydrous form is not commercially available and therefore any production of a tosylate salt would involve the use of the monohydrate.

A further critical result was the observation that for all of the tosylate systems studied (methanol, ethanol, and isopropanol) no observable reaction was observed where an excess of lutidine was present. This clearly demonstrates that the mechanism of reaction between sulfonic acid and alcohol (methanol, ethanol, and isopropanol) is common to both MSA and tosic acid. It would therefore seem reasonable to postulate that the reaction mechanism elucidated through a study of MMS using isotope-labeled ^{18}O methanol is common to all sulfonic acid–alcohol reactions.

The data collected allows a direct comparison to be made between the EMS and ETS systems. This is illustrated in Figure 11.38. Visual comparison shows no substantive difference between representative alkyl- and aryl-sulfonic acids.

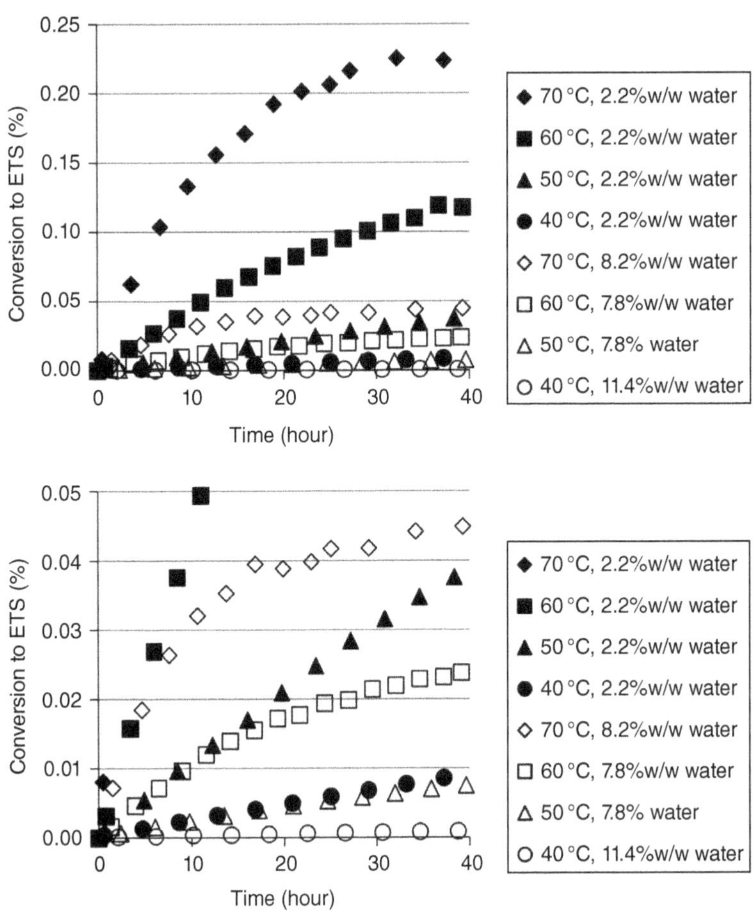

Figure 11.37 Combined plot showing the impact of both temperature and water content on conversion rates.

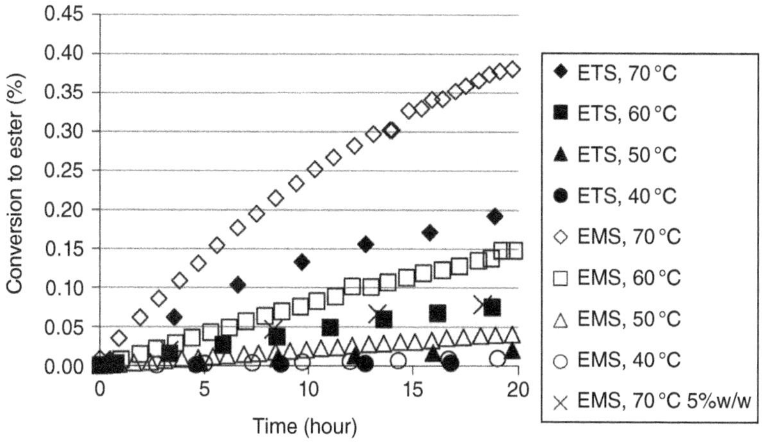

Figure 11.38 Comparison between conversions of MSA and TSA to ETS and EMS in ethanol.

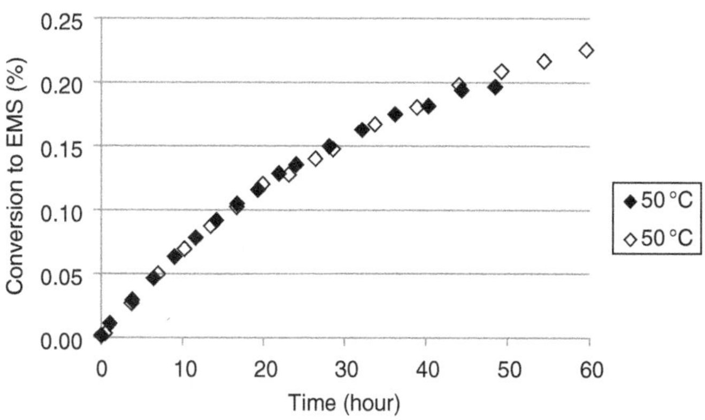

Figure 11.39 Comparison of data from two separate determinations of the conversion to MMS under anhydrous conditions at 50 °C.

11.5.4.2 Kinetic Modeling

At the onset of the work described within this chapter, the quality of the data collected through the experimental studies was recognized as critical were we to be able to achieve the stated aim of being able to generate a kinetic model of the system. As can be seen from the data illustrated above the quality of the data is very high, which is clearly demonstrated by the smoothness of the reaction profiles. Indeed, the method used for the analysis, when validated, was found to achieve relative standard deviations of <5% across the full data range. To illustrate this point further, Figure 11.39 shows an overlay of two entirely separate experiments performed for MMS formation. As can be seen, they show excellent agreement.

Thus, with the assurance of the quality of data it was possible to kinetically model each of the systems to determine rate constants and activation energies. Fitting data for the formation of sulfonate ester in "anhydrous" alcohols (no base present, Figure 11.40) required a kinetic model, which was first order in sulfonic acid. Plots of fractional conversion to ester showed (initial) rates independent of concentration of sulfonic acid.

The kinetic modeling was performed using Dynochem®. Rate constants for hydrolysis and alcoholysis were refined by including data from solvolysis reactions.

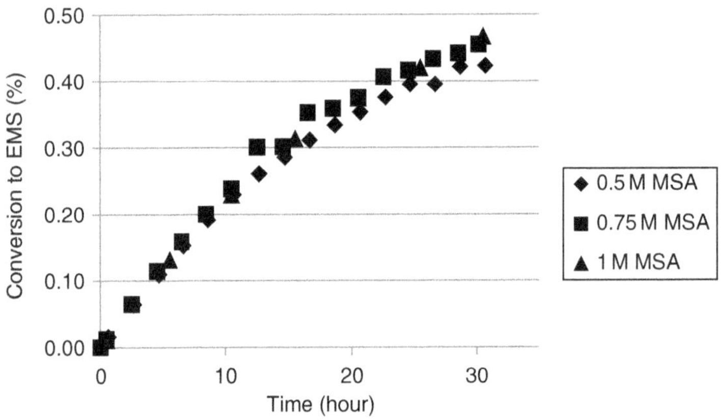

Figure 11.40 Conversion of MSA to EMS as a function of MSA concentration in "anhydrous" ethanol.

Table 11.1 illustrates the full kinetic data derived from the mesylate systems studied. Included are rate constants for formation, alcoholysis, and hydrolysis at reference temperatures along with activation energies.

Using these data it becomes possible to predict sulfonate ester formation as a function of temperature.

Under anhydrous conditions the observed kinetics, first order in terms of sulfonic acid, can be interpreted as a reaction of ion pairs (a "single" reacting species). However, this model does not describe the measured data in aqueous alcoholic systems. Plots of fractional conversion to ester show formation rates are dependent on concentration of sulfonic acid, indicating non-first-order kinetics (Figure 11.41).

Reactions of separated sulfonate anions and (solvated) protons might be expected to demonstrate second-order kinetics. The hydration of proton would favor ion separation in aqueous systems. However, clean second-order kinetics were not observed, even in solvent systems with high water content (where conversion to ester was very low), making good estimates of the second-order rate constant difficult to attain. Both mechanisms are operational in the presence of water.

Table 11.1 Measured kinetic and thermodynamic constants involving sulfonate ester formation and hydrolysis.

Sulfonate ester	Forward rate constant (s)	Activation energy (kJ/mol)	Hydrolysis rate constant[a] (l/mol-s)	Alcohol	Alcoholysis rate constant[a] (l/mol-s)	Activation energy (kJ/mol)[d]
MMS	7.10*E*−08 (60 °C)	115[b]	3.03*E*−06	Methanol	8.50*E*−07	95
EMS	7.90*E*−08 (70 °C)	114[c]	4.80*E*−06	Ethanol	6.0*E*−07	85
IMS	2.26*E*−07 (70 °C)	123[c]	1.09*E*−5	Isopropanol	1.03*E*−06	105

a Rate constants were measured at the corresponding reference study temperature.
b Forward rate constants were measured at 60, 40, and 30 °C and the activation energy calculated using Dynochem.
c Forward rate constants were measured at 70, 60, 50, and 40 °C and the activation energy calculated using Dynochem.
d Estimate calculated from difference in equilibrium value projected at various temperatures.

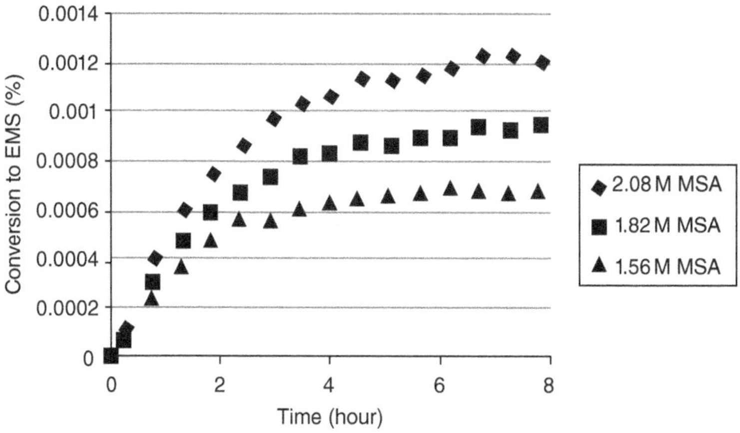

Figure 11.41 Conversion of MSA to EMS as a function of MSA concentration in 60%w/w water in ethanol.

Fitting multiple data sets covering a range of water contents did not lead to satisfactory estimates of the second-order rate constant, as the partition between ion pairs and separated ions proved not to be a simple function of water content. Both kinetic models explain the requirement for acid to be present for the formation of sulfonate esters.

The conclusion from the available data was that the first-order model appears sound and represents a "worst-case" model for predicting ester formation in systems where there is an excess of sulfonic acid present. In the presence of water, models are not yet sufficiently refined to be able to make accurate predictions, but the data sets provide a series of "reference points," representing specific water contents, against which potential processes may be assessed.

11.5.4.3 Key Learnings and Their Implications for Process Design

What is clear from these studies is that under certain specific ranges of conditions low levels of sulfonate esters can form. In particular, conditions favorable to the low-level formation of sulfonate ester include the presence of sulfonic acids under essentially anhydrous conditions in alcoholic solvents at elevated temperature, held there for prolonged periods. For example, levels of up to 0.3 mol% EMS are formed after MSA is held in solution in ethanol for 24 hours at 70 °C.

However, as would be expected of any kinetic process, the reaction rate can be modified through adjusting the reaction temperature thus immediately offering a relatively simple way of affecting the rate of ester formation. Close examination of Figure 11.5 shows that the conversion can be reduced to well below 0.1% simply by reducing the temperature to below 40 °C. This provides clear guidance to a process chemist to run the desired chemistry at the lowest practical temperature to minimize the ester formation.

Another key parameter affecting the extent of reaction is the level of water present. The presence of water has a twofold impact. First, formation rates are reduced due to the competition for protons between water and the alcohol molecules, reducing the number of alcoholic molecules protonated. Second, the accumulation of sulfonate ester is reduced by the back-reaction i.e. hydrolysis. This, as demonstrated by the data in Table 11.1, is a significantly faster reaction than the analogous solvolytic pathway involving solvolysis through reaction between the sulfonate ester and alcohol. Close examination of Figure 11.7 shows the profound impact high water levels can have – under highly aqueous conditions – in excess of 50%w/w water, virtually no reaction occurs, at least in the timescales typically associated with a salt formation process. This again is a very simple precautionary step that can be taken should the process concerned be amenable to such high water levels without having a negative impact e.g. poor yields or a reduction in quality i.e. higher levels of other impurities. Perhaps the simplest example of how this could be achieved is through the use of an MSA solution in water as opposed to neat MSA. A 70%w/w aqueous solution of MSA is commercially available.

By far the most significant factor though in terms of control is the manipulation of the level of acidity. Where a stoichiometric amount of base was present no observable reaction was seen. This was the same within all of the systems studied. This observation is entirely consistent with the mechanism of the reaction where it was elucidated that the first step in the reaction involved protonation of the alcohol. Where base is present, there is competition for the proton between the base and the alcohol; however, given the relative basicity of the base in comparison to alcohols, the proton almost exclusively resides on the base. Use of a stoichiometric amount of base effectively removes the proton source that drives sulfonate ester formation. This is further explained by considering the relative rates associated with this proton transfer compared to the rates associated with sulfonate ester formation; proton transfer being many orders of magnitude faster.

All of the studies described here were carried out in solution; however, another important factor in control of sulfonate esters is their solubility. Levels of sulfonate ester are very likely to be substantially lower in isolated material when the solubility of such species is considered, sulfonate esters effectively being freely soluble in alcohols. Certainly, the levels of ester formed are well below any practical limit in solubility terms. Thus, even in scenarios whereby esters are formed these can be easily removed by effective/efficient de-liquoring during workup, thus providing yet more effective protection from the risk of sulfonate ester contamination of APIs.

11.5.4.4 Processing Rules

Taking the key points from these studies, it is possible to devise a set of very simple rules that, where applied, mean that sulfonic acids can be used to form salts without risk of sulfonate ester contamination. These are:

Step 1: Use a stoichiometric amount or slight excess of free-base API when forming the salt. Provided this does not significantly impact the yield.

Step 2: If it is necessary to use a substoichiometric amount of base, try to introduce water into the process to inhibit sulfonate ester formation. This can be introduced for example in the case of MSA through the use of an aqueous solution, a 70%w/w solution in water being commercially available.

Step 3: The greatest risk of sulfonate ester formation relates to where sulfonic acids are mixed with alcohols, in the absence of base. In terms of salt formation, this is most likely to occur where a solution of the acid needs to be pre-prepared. Where this is necessary, try to use a partially aqueous solvent system, use the minimum temperature, and keep hold times to a minimum. Applying such controls should restrict levels of ester formed to a level that is inconsequential.

11.5.5 What About Viracept™?

A common challenge to this work is the above question and the assertion that the events associated with Viracept conflict with these studies. The issues associated with Viracept were the result of prolonged contact between trace amounts of ethanol and MSA inadvertently present together within a head tank. Roche have disclosed the details of their investigation into the contamination [94]. This occurred due to a good manufacturing practice (GMP) failure. During the routine maintenance/cleaning of the plant, cleaning with ethanol was carried out. However, the tank was not dried, resulting in a small amount of ethanol remaining in the head tank. The tank was then charged with MSA and over a period of time, several months, significant amounts of EMS were formed that, when introduced into the salt formation process, ultimately contaminated the isolated salt, levels of up to 2300 ppm being observed. Closer examination of this chain of events shows that rather than conflicting with the findings described above they are in fact entirely consistent with them. Reflecting on the mechanism of sulfonate ester formation, as illustrated this is proton mediated. In the Viracept incident rather than the typical scenario whereby acidity is limited, in that case there was an unlimited source of acidity, MSA being effectively the solvent. Thus, it is entirely understandable that appreciable levels of EMS formed over the time period involved, the only limiting factor being that the "reaction" was held at room temperature.

11.5.6 What About Other Sources of Sulfonate Esters?

Another factor to take into consideration is the quality of raw materials i.e. that of the sulfonic acid. The sulfonic acid itself can be contaminated with sulfonate esters or other reactive species such as sulfonyl chlorides that can readily form sulfonate esters on contact with alcohols. The level and nature of these is dependant upon the manufacturing process used. This is of particular significance in relation to MSA.

Traditionally, MSA has been manufactured through a two-step process involving the oxidative chlorination of methyl sulfide (methyl mercaptan) followed by hydrolysis of the resultant sulfonyl chloride (Figure 11.42).

The resultant MSA can often contain several thousand ppm of residual methane sulfonyl chloride that if introduced into a salt formation involving an alcoholic solvent will rapidly react with the alcohol to form a sulfonate ester. MSA produced via this route can also often contain both MMS and EMS, at variable, but appreciable, levels.

An alternative process has been devised to manufacture MSA, this involving a two-step process. However, in this instance the starting material is methanol.

The resultant MSA generated from this alternative process is significantly cleaner with virtually no sulfonyl chloride or EMS/MMS being present. MSA manufactured via this process is also available as an aqueous (70% w/w) solution further eliminating any risk from a sulfonate ester perspective.

Historical process

$$CH_3SH + 3\,Cl_2 + 2\,H_2O \longrightarrow CH_3SO_2Cl + 5\,HCl$$

$$CH_3SO_2Cl + H_2O \longrightarrow CH_3SO_3H + HCl$$

Alternative MSA manufacturing process

$$H_2 + 2\,CH_3OH + 2\,S \longrightarrow CH_3\text{-}S\text{-}S\text{-}CH_3 + 2\,H_2O$$

$$CH_3\text{-}S\text{-}S\text{-}CH_3 + 5/2\,O_2 + H_2O \xrightarrow{\text{Catalyst}} 2\,CH_3\text{-}SO_3H$$

Figure 11.42 Manufacture of MSA.

11.5.7 Potential for Ester Formation in the Solid Phase

A question that is often asked is: What is the potential risk of formation of sulfonate esters either during formulation processes such as wet granulation of a sulfonic acid salt in alcohols or within the solid dosage form on storage?

Although no formal solid-phase studies were performed as part of the PQRI research, many of the parties involved had themselves studied sulfonate ester formation both during formulation and subsequently on stability. No ester formation had ever been observed.

A recent paper relating to the chemical side of the Viracept incident [13] looked in detail at the level of EMS present in film-coated tablets and how this changed on storage. What they observed was that rather than any additional sulfonate ester forming on stability the exact opposite occurred i.e. the level was seen to reduce as a result of hydrolysis. This was seen under all storage conditions studied and is illustrated in Figure 11.43.

Using data from these studies, Roche reported hydrolysis rates of 0.3%/day at 25 °C within Viracept 250 mg film-coated tablets.

These results and those informally reported by other organizations are not surprising when the postulated reaction is scrutinized. For any reaction to occur, there first has to be a proton source to initiate the reaction between the sulfonic acid and any alcohol. This would require the salt to disassociate. This in itself is very unlikely, as any such salt would have been selected on the basis of among several parameters, its stability. Furthermore, even if this was to occur, the kinetics of any reaction between sulfonic acid and residual levels of any alcohol present in the solid formulation (tablet) are likely to be extremely slow. Given the concentrations involved and the relatively low temperatures, levels formed are therefore likely to be miniscule. Far more likely as demonstrated by the Roche data is that any pre-formed ester present in formulated product will degrade through hydrolysis either as a result of water present directly in the formulation or through moisture within the air picked up on storage.

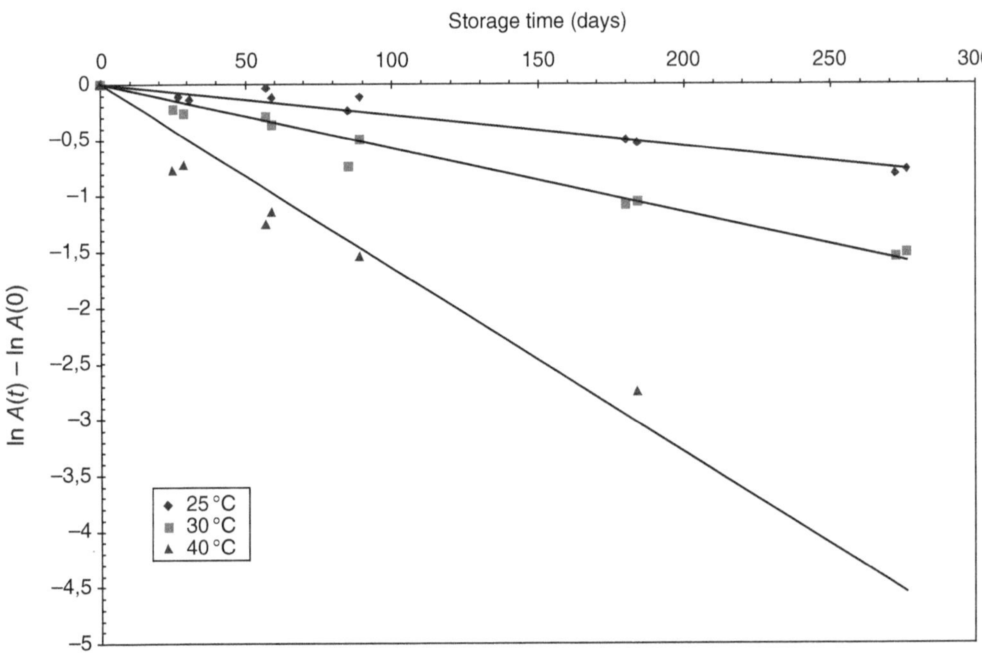

Figure 11.43 Logarithm of EMS concentration versus time at different storage temperatures.

11.5.8 Conclusions

At the onset of this work many organizations had, through analysis of isolated sulfonic acid salts, demonstrated rudimentary control over levels of sulfonate esters. However, this was based on little or no understanding of the extent or nature of the reaction between sulfonic acids and alcohols.

These studies have delivered a clear understanding both of the reaction mechanism itself and of the key parameters that affect the kinetics of the reaction. This has made it possible to manipulate these key parameters in order to control sulfonate ester formation to such an extent as to render the risk it poses in terms of contamination of sulfonic acid salt inconsequential.

References

1 Teasdale, A. and Popkin, M. (2019). Regulatory highlights. *Org. Process Res. Dev.* 23: 1292–1297.
2 Robinson, D.I. (2010). Control of genotoxic impurities in active pharmaceutical ingredients: a review and perspective. *Org. Process Res. Dev.* 14: 946–959.
3 Teasdale, A., Elder, D., Chang, S.-J. et al. (2013). Risk assessment of mutagenic impurities in new chemical entities: strategies to demonstrate control. *Org. Process Res. Dev.* 17: 221–230.
4 Szekely, G., Amores de Sousa, M.C., Gil, M. et al. (2015). Genotoxic impurities in pharmaceutical manufacturing: sources, regulations, and mitigation. *Chem. Rev.* 2015 (115): 8182–8229.
5 Elder, D.P. and Teasdale, A. (2015). Is avoidance of genotoxic intermediates/impurities tenable for complex, multistep syntheses? *Org. Process Res. Dev.* 19: 1437–1446.
6 Lapanja, N., Zupančič, B., Toplak, R. et al. (2015). A generic industry approach to demonstrate efficient purification of potential mutagenic impurities in the synthesis of drug substances. *Org. Process Res. Dev.* 19: 1524–1530.
7 ICH Harmonised Guideline. (2018). Assessment and control of DNA reaction (mutagenic) impurities in pharmaceuticals to limit potential carcinogenic risk: "M7(R1)", final version, March 2018.
8 European Medicines Agency (2020). Assessment Report; Procedure under Article 5(3) of Regulation EC (No) 726/2004: Nitrosamine Impurities in Human Medicinal Products; EMA/369136/2020.
9 European Medicines Agency (2020). Questions and Answers for Marketing Authorisation Holders/ Applicants on the CHMP Opinion for the Article 5(3) of Regulation (EC) No 726/2004 Referral on Nitrosamine Impurities in Human Medicinal Products; EMA/409815/2020.
10 López-Rodríguez, R., McManus, J.A., Murphy, N.S., and Burns, M.J. (2020). Pathways for *N*-Nitroso compound formation: secondary amines and beyond. *Org. Process Res. Dev.* 24 (9): 1558–1585. and references therein.
11 Mirvish, S.S. (1975). Formation of *N*-nitroso compounds: chemistry, kinetics, and in vivo occurrence. *Toxicol. Appl. Pharmacol.* 31: 325–351.
12 Bräse, S.; Mende, M.; Jobelius, H. H. and Scharf, H.-D. (2015) Hydrazoic acid and Azides Ullman's encyclopedia of industrial chemistry.
13 Schreiber, I.M. and Mitch, W.A. (2006). Nitrosamine formation pathway revisited: the importance of chloramine speciation and dissolved oxygen. *Environ. Sci. Technol.* 40 (19): 6007–6014.
14 Iitsuka, M., Hoshino, T., Kato, T., and Kikugawa, K. (1985). Formation of nitrosamines in organic solvents and aqueous systems containing organic solvents. *Chem. Pharm. Bull.* 33: 2516–2521.
15 Ashworth, I.W., Dirat, O., Teasdale, A., and Whiting, M. (2020). Potential for the formation of *N*-nitrosamines during the manufacture of active pharmaceutical ingredients: an assessment of the risk posed by trace nitrite in water. *Org. Process Res. Dev.* 24 (9): 1629–1646. and references therein.

16 Sgroi, M. (2018). *N*-nitrosodimethylamine (NDMA) and its precursors in water and wastewater: a review on formation and removal. *Chemosphere* 191: 685–703.
17 Keeper, L.K. and Koller, P.P. (1973). *N*-nitrosation by nitrite ion in neutral and basic medium. *Science* 181: 1245.
18 Casado, J., Castro, A., Lopez Quintela, M.A., and Vazquez Tato, J. (1981). Kinetic studies on the formation of *N*-nitroso compounds. *Zeitschrift fur physikalische Chemie* 127: 179–192.
19 (a) Glaser, H., Möller, F., Pieper, G., Schröter, R., Spielberger, G., and Söll, H. (1958). Vol 11/1: *Nitrosation of Secondary Amines and Denitrosation: Houben-Weyl: Methoden der Organischen Chemie*, 4e, 957–959. Stuttgart: Thieme; (b) Freytag, H., Möller, F., Pieper, G., and Söll, H. (1958). Vol. 11/2: *Nitrosation of Secondary Amines and Denitrosation: Houben-Weyl: Methoden der Organischen Chemie*, 4e, 171–181. Stuttgart: Thieme.
20 Loeppky, R.N., Yen, T., Bao, Y.T. et al. (1994). Blocking nitrosamine formation - understanding the chemistry of rapid nitrosation. *Nitrosam. Rel. N-Nitroso Comp.* 28: 52–65.
21 Ender, F. and Čeh, L. (1971). Conditions and chemical reaction mechanisms by which nitrosamines may be formed in biological products with reference to their possible occurrence in food products. *Eur. Food Res. Technol. (Z. Lebensmittel-Untersuch. u. -Forsch.)* 145 (3): 133–142.
22 Bhat, J.I., Clegg, W., and Maskill, H. (2000). *N*-Nitroso-*N,O*-dialkylhydroxylamines: preparation, structure, and mechanism of the hydronium ion catalysed solvolytic nitrous oxide extrusion reaction. *J. Chem. Soc. Perkin Transactions* 2 (7): 1435–1446.
23 Chaudhary, P., Gupta, S., Muniyappan, N. et al. (2016). An efficient synthesis of: *N*-nitrosamines under solvent, metal and acid free conditions using tert-butyl nitrite. *Green Chem.* 18 (8): 2323–2330.
24 Teuten, E.L. and Loeppky, R.N. (2005). The mechanistic origin of regiochemical changes in the nitrosative *N*-dealkylation of *N,N*-dialkyl aromatic amines. *Org. Biomol. Chem.* 3: 1097.
25 Qiu, S., Guo, C., Wang, M. et al. (2018). Mild dealkylative *N*-Nitrosation of *N,N*-dialkylaniline derivatives for convenient preparation of photo-triggered and photo-calibrated NO donors. *Org. Chem. Front.* 5: 3206–3209.
26 Lijinsky, W., Conrad, E., and Van der Bogart, R.J. (1973). Feeding studies of nitrilotriacetic acid and derivatives in rats. *Natl. Cancer Inst.* 49: 1239.
27 Nakajima, M., Warner, J.C., and Anselme, J.-P. (1984). *N*-nitrosamines from the reaction of *N*-chlorodialkylamines with sodium nitrite. *J. Chem. Soc. Chem. Commun.* 7: 451–452.
28 White, E.H. (1955). The chemistry of the *N*-alkyl-*N*-nitrosoamides. I. Methods of preparation. *J. Am. Chem. Soc.* 77 (22): 6008–6010.
29 Mitch, W.A. and Sedlak, D.L. (2004). Characterization and fate of *N*-nitrosodimethylamine precursors in municipal wastewater treatment plants. *Environ. Sci. Technol.* 38: 1445–1454.
30 Gil, R., Casado, J., and Izquierdo, C. (1997). Structural effects on the *N*-nitrosation of amino acids. *Int. J. Chem. Kinet.* 29: 495–504.
31 Norris, W.P. (1959). The reaction of dialkylcarbamyl chlorides with silver nitrate (1). *J. Am. Chem. Soc.* 81 (13): 3346–3350.
32 Warner, J.C., Nakajima, M., and Anselme, J.-.P. (1984). *N*-Nitrosamines from the reaction of sulfamoyl chlorides with sodium nitrite. *Bull. Soc. Chim. Belges* 93: 919–920.
33 Sueyoshi, S. and Tanno, M. (1985). Preparation and properties of novel trisubstituted *N*-nitrosoureas which decompose to afford triazenes and nitro compounds under mild conditions. *Chem. Pharm. Bull.* 33 (2): 488–496.
34 von Brüning, G. (1889). Ueber das Methylhydrazin. *Justus Liebigs Ann. Chem.* 253: 5–14.
35 Kamiya, S.A. (1972). Convenient synthesis of 1-(2-halogenoethyl)-3-nitro-1-nitrosoguanidines. *Chem. Pharm. Bull.(Tokyo)* 20 (11): 2497–2500.
36 Loeppky, R.N. and Yu, H. (2004). Amidine nitrosation. *J. Org. Chem.* 69 (9): 3015–3024.

37 Jian, R.-C., Tsai, C., Hsu, L.-C., and Chen, H.-L. (2010). Theoretical study on reaction mechanisms and kinetics of cyanomidyl radical with NO. *J. Phys. Chem. A* 114 (13): 4655–4663.

38 Lijinsky, W. (1992). *Chemistry and Biology of N-Nitroso Compounds*. Cambridge University Press.

39 Rubio, M.A., Lissi, E., Olivera, N. et al. (2014). Reactions of p-substituted phenols with nitrous acid in aqueous solution. *Int. J. Chem. Kinet.* 46: 143–150.

40 Lieber, E. and Smith, G.B.L. (1935). Reduction of nitroguanidine. IV. Preparation of nitrosoguanidine by catalytic hydrogenation. *J. Am. Chem. Soc.* 57 (12): 2479–2480.

41 Mitch, W.A. and Sedlak, D.L. (2002). Formation of *N*-nitrosodimethylamine (NDMA) from dimethylamine during chlorination. *Environ. Sci. Technol.* 36: 588–595.

42 Enders, D., Eichenauer, H., Baus, U. et al. (1984). Asymmetric syntheses via metalated chiral hydrazones: overall enantioselective α-alkylation of acyclic ketones. *Tetrahedron* 40 (8): 1345–1359.

43 Harrington, G.W., Eshraghi, J., Pylypiw, H.M. Jr. et al. (1986). Formation of nitrosamines by oxidation. *Cancer Lett.* 32: 187–191.

44 Huang, M.E., Huang, S., and McCurry, D.L. (2018). Re-examining the role of dichloramine in high-yield *N*-nitrosodimethylamine formation from *N,N*-dimethyl-α-arylamines. *Environ. Sci. Technol. Lett.* 5 (3): 154–159.

45 Cicchi, S., Corsi, M., and Goti, A. (1999). Inexpensive and environmentally friendly oxidation of hydroxylamines to nitrones with bleach. *J. Org. Chem.* 64: 7243.

46 Müller, E. and Metzger, H. (1956). Über Nitrosoverbindungen, V. Mitteil.: Einwirkung von Nitrosylchlorid Oder Stickstoffmonoxyd Auf Cyclohexylmagnesiumchlorid Und Auf Benzylmagnesiumchlorid. *Chem. Ber.* 89 (2): 396–406.

47 Levitskiy, O.A. and Magdesieva, T.V. (2019). Amination of aryl boronic acids with alkylnitrites: a convenient complement to Cu-promoted reductive amination. *Org. Lett.* 21 (24): 10028–10032.

48 Polivin, Y.N., Karakhanov, R.A., Kelarev, V.I. et al. (1993). Reaction of methylmagnesium iodide with dibutylcyanamide and dibutylnitrosamine. *Russ. Chem. Bull.* 42 (12): 1991–1994.

49 a) Sakai, N., Sasaki, M., and Ogiwara, Y. (2015). Copper(II)-catalyzed oxidative *N*-nitrosation of secondary and tertiary amines with nitromethane under an oxygen atmosphere. *Chem. Commun.* 51 (58): 11638–11641; b) Potturi, H.K., Gurung, R.K., and Hou, Y. (2012). Nitromethane with IBX/TBAF as a nitrosating agent: synthesis of nitrosamines from secondary or tertiary amines under mild conditions. *J. Org. Chem.* 77 (1): 626–631.

50 Crebelli, R., Falcone, E., Aquilina, G., and Carere, A. (1984). in vitro mutagenicity of rubber chemicals and their nitrosation products. *Toxicol. Lett.* 23: 307–313.

51 Challis, B.C. and Osborne, M.R. (1973). The chemistry of nitroso-compounds. Part VI. Direct and indirect transnitrosation reactions of *N*-Nitrosodiphenylamine. *J. Chem. Soc. Perkin Trans II*: 1526–1533.

52 Kabalka, G.W. (1991). 2.1.3 Reduction of aromatic nitro and nitroso compounds to azo and azoxy compounds, comprehensive organic synthesis.

53 Gilchrist, T.L. (1991). 6.1.5.3 Oxidation of azo to azoxy compounds and 2.2.2.1.1 azo and azoxy to hydrazo groups: comprehensive organic synthesis.

54 Boysen, M.M.K. (2010). Product Class 5: Aliphatic Azoxy Compounds (Aliphatic Diazene Oxides) Science of Synthesis, 41: Category 5, Compounds with One Saturated Carbon Heteroatom Bond, 449.

55 Browne, A.W. and Overman, O.R. (1916). Die angebliche Bedeutung der salpetrigen Säure bei der Bildung von Stickstoffwasserstoffsäure durch Oxydation von Hydrazin. *Z. Anorg. Allg. Chem.* 94: 217–239.

56 Knorr, L. and Weidel, A. (1909). Hydrazo-phenylmethyl aus Phenyl-pyrazol. *Ber. Dtsch. Chem. Ges.* 42: 3523–3529.

57 Azoxy Compound (2019). Azoxy compounds contain in their molecule a bond similar to amine N-oxides, and their formula can be described by the following two structures (a coordinate covalent bond is indicated as "→"):(9.6.18). https://www.sciencedirect.com/topics/chemistry/azoxy-compound (accessed March 2020).

58 Looker, A.R., Ryan, M.P., Neubert-Langille, B.J., and Naji, R. (2010). Risk assessment of potentially mutagenic impurities within the framework of quality by design. *Org. Process Res. Dev.* 14: 1032–1036.

59 Elder, D.P., Snodin, D., and Teasdale, A. (2011). Control and analysis of hydrazine, hydrazides and hydrazones – mutagenic impurities in active pharmaceutical ingredients (APIs) and drug products. *J. Pharm. Biomed. Anal.* 54: 900–910.

60 Kimball, R.F. (1977). The mutagenicity of hydrazine and some of its derivatives. *Mut. Res. Rev. Gen. Toxicol.* 39: 111–126.

61 Treitler, D.S., Marriott, A.S., Chadwick, J., and Quirk, E. (2019). Mutagenic impurities in 1-hydroxybenzotriazole (HOBt). *Org. Process Res. Dev.* 23: 2562–2566.

62 Wogan, G.N. and Paglialunga, S. (1974). Carcinogenicity of synthetic aflatoxin M1 in rats. *Food Cosm. Toxicol.* 12: 381.

63 a) Kulkarni, P.S., Crespo, J.G., and Afonso, C.A.M. (2008). Dioxins sources and current remediation technologies – a review. *Environ. Int.* 34: 139–153; b) Canady, R., Crump, K., Feeley, M. et al. (2002). *Safety Evaluation of Certain Food Additives and Contaminants: Polychlorinated Dibenzodioxins, Polychlorinated Dibenzofurans, and Coplanar Polychlorinated Biphenyls*, WHO Food Additives Series: 48. World Health Organization.

64 Bercu, J.P., Galloway, S., Teasdale, A. et al. (2018). Potential impurities in drug substances: compound-specific toxicology limits for 20 synthetic reagents and by-products, and a class-specific toxicology limit for alkyl bromides. *Regul. Toxicol. Pharmacol.* 94: 172–182.

65 Yang, Q., Haney, B.P., Vaux, A. et al. (2009). Controlling the genotoxins ethyl chloride and methyl chloride formed during the preparation of amine hydrochloride salts from solutions of ethanol and methanol. *Org. Process Res. Dev.* 13: 786–791.

66 Nicholas, P.P. (1979). Hexakis[μ-(ethyl chlorocarbamato-*N,O*)]-μ_4-oxo-tetrazinc and its thermolysis in simple olefins. *J. Org. Chem.* 13: 2126–2130.

67 Shuikin, N.I. and Belsky, I.F. (1956). Opening of the tetrahydrofuran ring under the action of aluminum halides *Russ. Chem. Bull.* 5: 761–762.

68 Roy, R.N. and Sen, B. (1968). Thermodynamic studies of hydrochloric acid in tetrahydrofuran-water mixtures. *J. Chem. Eng. Data* 13: 79–82.

69 Amouroux (1987). Ouverture Regioselective De Tetrahydrofurannes Fonctionnalises Par Le Systeme Chlorure De Trialkylsilane/iodure De Sodium. *Bull. Soc. Chim. Fr.* 3: 505.

70 De Buyck, L. and De Pooter, H. (1992). On the mechanism of selective chlorination of ethers with sulfuryl chloride in the dark. *Bull Soc. Chim. Belg.* 101 (9): 807–815.

71 Peterson, P.E. and Kamat, R.J. (1969). Solvents of low nucleophilicity. XII. Triple-bond participation in the acetolysis, formolysis, and trifluoroacetolysis of 6-heptyn-2-yl tosylate and 6-octyn-2-yl tosylate. *J. Am. Chem. Soc.* 91 (16): 4521–4527.

72 Wuts, P.G.M. and Greene, T.W. (2006). *Greene's Protective Groups in Organic Synthesis*, 4e. Wiley.

73 Menzel, W. (1942). Über die Einwirkung von Aluminiumhalogeniden auf aliphatische Äther, I. Mitteil.: Über Aluminium-oxy-halogenide. *Chem. Ber.* 75 (9): 1055–1061.

74 Fritsch, P. and Schumacher, W. (1894). II. Ueber die Chlorirung des Aethers. *Justus Liebigs Ann. Chem.* 279: 301–310.

75 Gaspa, S., Carraro, M., Pisano, L. et al. (2019). Trichloroisocyanuric acid: a versatile and efficient chlorinating and oxidizing reagent. *Eur. J. Org. Chem.* 22: 3544–3552.

76 Amberg, A., Harvey, J.S., Czich, A. et al. (2015). Do carboxylic/sulfonic acid halides really present a mutagenic and carcinogenic risk as impurities in final drug products? *Org. Process Res. Dev.* 19: 1495–1506.

77 Richtzenhain, H. and Alfredsson, B. (1953). Über die Chlorierung des Dimethylsulfids. *Chem. Ber.* 86: 142–148.

78 Lange, H.G. and Shreeve, J.M. (1985). On the reactions of dimethylsulfoxide with acyl fluorides: Pummerer Rearrangements and formation of monofluoromethylesters. *J. Fluorine Chem.* 28: 219–227.

79 Mohanazadeh, F., Momeni, A.R., and Ranjbar, Y. (1994). Deoxygenation of sulfoxides with silica chloride. *Tetrahedron Lett.* 35 (33): 6127–6128.

80 Lappert, M.F. and Smith, J.K. (1961). 633. Reactions of sulphoxides with some Group III and IV halides. *J. Chem. Soc.*: 3224–3230.

81 Tanemura, K., Suzuki, T., Nishida, Y., and Horaguchi, T. (2010). Chlorination of various substrates in subcritical carbon tetrachloride. *Tetrahedron* 66 (15): 2881–2888.

82 Kimball, R.F. (1977). The mutagenicity of hydrazine and some of its derivatives. *Mutat. Res. Rev. Gen. Toxicol.* 39: 111–126.

83 Stare, M., Laniewski, K., Westermark, A. et al. (2009). Investigation on the formation and hydrolysis of *N,N'*-Dimethylcarbamoyl Chloride (DMCC) in Vilsmeier reactions using GC/MS as the analytical detection method. *Org. Process Res. Dev.* 13 (5): 857–862.

84 Frankel, L.S., McCallum, K.S., and Collier, L. (1974). Formation of bis(chloromethyl) ether from formaldehyde and hydrogen chloride. *Environ. Sci. Technol.* 8 (4): 356–359.

85 Stephen, H., Short, W.F., and Gladding, G. (1920). The introduction of the chloromethyl group into the aromatic nucleus. *J. Chem. Soc Trans.* 117: 510–527.

86 Backes, M. (1933). Action de l'hoxychlorure de phosphore sur quelques aldéhydes. *Comptes Rendus Hebdomadaires des Seances de l'Academie des Sciences* 196: 1675.

87 Tou, J.C. and Kallos, G.K. (1976). Possible formation of bis(chloromethyl) ether from the reactions of formaldehyde and chloride ion. *Anal. Chem.* 48 (7): 958–963.

88 Lawson, A.T. and Collie, N. (1888). XLVII.-The action of heat on the salts of tetramethylammonium. *J. Chem. Soc. Trans.* 53: 624–636.

89 Raw, S.A. (2009). An improved process for the synthesis of DMTMM-based coupling reagents. *Tetrahedron Lett.* 50: 946–948.

90 Glowienke, S., Allmendinger, T., Muller, L. et al. (2005). Structure/activity considerations and in vitro approaches to assess the mutagenicity of 19 methane, benzene- and toluenesulfonic acid esters. *Mutat. Res.* 581: 23–34.

91 Szekely, G., Amores de Sousa, M.C., Gil, M. et al. (2015). Genotoxic impurities in pharmaceutical manufacturing: sources, regulations, and mitigation. *Chem. Rev.* 115: 8182–8229.

92 Teasdale, A., Eyley, S., Delaney, E. et al. (2009). Mechanism and processing parameters affecting the formation of methyl methanesulfonate from methanol and methanesulfonic acid: an illustrative example for sulfonate ester impurity formation. *Org. Process Res. Dev.* 13: 429–433.

93 Teasdale, A., Eyley, S., Delaney, E. et al. (2010). A detailed study of sulfonate ester formation and solvolysis reaction rates and application toward establishing sulfonate ester control in pharmaceutical manufacturing processes. *Org. Process Res. Dev.* 14: 999–1007.

94 Gerber, C. and Toelle, H.-G. (2009). What happened: the chemistry side of the incident with EMS contamination in Viracept tablets. *Toxicol. Lett.* 190: 248–253.

95 Padwa, A. and Murphree, S. (2006). Epoxides and aziridines - a mini review. *ARKIVOC* iii: 6–33.

96 Elder, D.P. (2010). Analytics for epoxides. *J. Pharm. Biomed. Anal.* 51: 1015–1023.

97 Sweeney, J.B. (2002). Aziridines: epoxides' ugly cousins? *Chem. Soc. Rev.* 31: 247–258.

98 Hansen, M.M., Jolly, R.A., and Linder, R.J. (2015). Boronic acids and derivatives: probing the structure-activity relationships for mutagenicity. *Org. Process Res. Dev.* 19: 1507–1516.

99 Bodrikov, I.V., Michurin, A.N., Vasyanina, G.I. et al. (1980). Reaction of nitrile with 100% sulfuric acid. Influence of the structural effects of nitriles on the direction of transformation, the character of complex formation and the form of the kinetic equation. *J. Org. Chem. USSR (English Translation)* 16 (12): 2123–2133. [Zhurnal Organicheskoi Khimii, 16, (12), 2492–2503].

100 a) Kahl, T., Schröder, K.W., Lawrence, F.R. et al. (2011). Aniline. In: *Ullmann's Encyclopedia of Industrial Chemistry*. https://doi.org/10.1002/14356007.a02_303.pub2. b) Schwenecke, H. and Mayer, D. (2000). Benzidine and benzidine derivatives. In: *Ullmann's Encyclopedia of Industrial Chemistry*. https://doi.org/10.1002/14356007.a03_539.

101 U.S. Department of Health and Human Services Food and Drug Administration. (2020). Control of Nitrosamine Impurities in Human Drugs: Guidance for Industry.

102 Turney, T.A. and Wright, G.A. (1959). Nitrous acid and nitrosation. *Chem. Rev. 59*: 497–513.

103 Challis, B.C. and Challis, J.A. (1982). *N-nitrosamines and N-nitrosoimines*. In: *Supplement F. The Chemistry of Amino, Nitroso and Nitro Compounds and their Derivatives, Part 2* (ed. S. Patai), 1151–1223. Chichester: Wiley.

104 Williams, D.L.H. (1988). *Nitrosation*. Cambridge: Cambridge University Press, 1–35 &77–112.

105 Williams, D.L.H. (2004). *Nitrosation Reactions and the Chemistry of Nitric Oxide*, 1–56. Amsterdam: Elsevier.

106 Dekant, W., Rastogi, S., Sanner, T., et al. (2012). *SCCS (Scientific Committee on Consumer Safety), Opinion on Nitrosamines and Secondary Amines in Cosmetic Products*, SCCS/1458/11 European Commission, 27 March 2012.

107 Magee, P.N. and Barnes, J.M. (1956). The production of malignant primary hepatic tumours in the rat by feeding dimethylnitrosamine. *Br. J. Cancer 10*: 114–122.

108 Ridd, J.H. (1961). Nitrosation, diazotisation, and deamination. *Q. Rev. Chem. Soc. 15*: 418–441.

109 Tummavuori, J. and Lumme, P. (1968). Protolysis of nitrous acid in aqueous sodium nitrate and sodium nitrite solutions at different temperatures. *Acta Chem. Scand. 22*: 2003–2011.

110 Singer, K. and Vamplew, P.A. (1956). Oxidation by nitrous and nitric acid. Part IV. Spectroscopic investigation of the equilibrium between NO^+ and nitrous acid in aqueous perchloric acid. *J. Chem. Soc.*: 3971–3974.

111 Schmid, H. and Hallaba, E. (1956). Kinetik und Mechanismus der Diazotierung. *X. Monatsh. Chem. 87*: 560–573.

112 Schmid, H. and Fouad, M.G. (1957). Kinetik und Mechanismus der Diazotierung XII. *Monatsh. Chem. 88*: 631–638.

113 Fan, T.-Y. and Tannenbaum, S.R. (1973). Factors influencing the rate of formation of nitrosomorpholine from morpholine and nitrite. Acceleration by thiocyanate and other anions. *J. Agric. Food Chem.* 21: 237–240.

114 Casado, J., Castro, A., Leis, J.R. et al. (1983). Kinetic studies on the formation of *N*-nitroso compounds VI. The reactivity of N_2O_3 as a nitrosating agent. *Monatsh. Chem. 114*: 639–646.

115 Castro, A., Leis, J.R., and Pena, M.E. (1986). The efficiency of nitrosyl bromide in nitrosating aliphatic amines. *J. Chem. Res. Synop.*: 216–217.

116 Christensen, J.J., Izatt, R.M., Wrathall, D.P., and Hansen, L.D. (1969). Thermodynamics of proton ionization in dilute aqueous solution. Part XI. pK, $\Delta H°$ and $\Delta S°$ values for proton ionization from protonated amines at 25 °C. *J. Chem. Soc. A*: 1212–1223.

117 Ilcheva, L. and Bjerrum, J. (1976). Metal ammine formation in solution. XVII. Stability constants of Copper(II) methylamine complexes obtained from solubility measurements with gerhardtite, $Cu(OH)_{1.5}(NO_3)_{0.5}$. *Acta Chem. Scand. A 30*: 343–350.

118 Archer, M.C., Tannenbaum, S.R., and Wishnok, J.S. (1976). Nitrosamine formation in the presence of carbonyl compounds. In: *Environmental N-Nitroso Compounds, Analysis and*

Formation, IRAC Science Publication No. 14 (eds. E.A. Walker, P. Bogovski and L. Griciute), 141–146. Lyon: International Agency for Research on Cancer.

119 Casado, J., Mosquera, M., Paz, L.C. et al. (1984). Nitrite ion as a nitrosating reagent. Nitrosation of morpholine and diethylamine in the presence of formaldehyde. *J. Chem. Soc. Perkin Trans. 2*: 1963–1966.

120 Challis, B.C. and Kyrtopoloulos, S.A. (1979). The chemistry of nitroso-compounds. Part 11. Nitrosation of amines by the two-phase interaction of amines in solution with gaseous oxides of nitrogen. *J. Chem. Soc. Perkin Trans. 1*: 299–304.

121 Iglesias, E. and Casado, J. (2002). Mechanisms of hydrolysis and nitrosation reactions of alkyl nitrites in various media. *Int. Rev. Phys. Chem. 21*: 37–74.

122 Smith, P.A.S. and Pars, H.G. (1959). Nitrosative cleavage of *N′,N′*-dialkylhydrazides and tertiary amines. *J. Org. Chem. 24*: 1325–1332.

123 Smith, P.A.S. and Loeppky, R.N. (1967). Nitrosative cleavage of tertiary amines. *J. Am. Chem. Soc. 89*: 1147–1157.

124 Gowenlock, B.G., Hutchison, R.J., Little, J., and Pfab, J. (1979). Nitrosative dealkylation of some symmetrical tertiary amines. *J. Chem. Soc. Perkin Trans. 2*: 1110–1114.

125 Lijinsky, W. and Greenblatt, M. (1972). Carcinogen dimethylnitrosamine produced *in vivo* from nitrite and aminopyrine. *Nat. New Biol. 236*: 177–178.

126 Mirvish, S.S., Gold, B., Eagen, M., and Arnold, S. (1974). Kinetics of the nitrosation of aminopyrine to give dimethylnitrosamine. *Z. Krebsforsch. Klin. Onkol. 82*: 259–268.

127 Conrad, J., Schlemmer, K., and Eisenbrand, G. (1990). Nitrosation of bromohexine. *Drug Dev. Eval. 16*: 181–197.

128 European Medicines Agency (2020). Suspension of Ranitidine Medicines in the EU; EMA/231394/2020 Rev.1.

129 Fiddler, W., Pensabene, J.W., Doerr, R.C., and Wasserman, A.E. (1972). Formation of *N*-nitrosodimethylamine from naturally occurring quaternary ammonium compounds and tertiary amines. *Nature 236*: 307–307.

130 Landini, D., Maia, A., and Rampoldi, A. (1986). The stability of quaternary onium salts under phase-transfer conditions in the presence of aqueous alkaline solutions. *J. Org. Chem. 51*: 3187–3191.

131 Starks, C.M., Liotta, C.M., and Halpern, M. (1994). *Phase-Transfer Catalysis: Fundamental, Applications & Industrial Perspectives*, 288–292. Chapman & Hall.

132 Dou, H.J.M., Gallo, R., Hassanaly, P., and Metzger, J. (1977). Behaviour and stability of catalysts in Bi-and triphase transfer catalysis. *J. Org. Chem. 42*: 4275–4276.

133 Kametani, T., Kigasawa, K., Hiiragi, M. et al. (1969). Novel debenzylation of quaternary ammonium salts with thiophenol. *Tetrahedron Lett. 10*: 635–638.

134 Lijinsky, W., Keefer, L., Conrad, E., and Van de Bogart, R. (1972). Nitrosation of tertiary amines and some biological implications. *J. Natl. Cancer Inst. 49*: 1239–1249.

135 European Medicines Agency (2002). *Note for Guidance on Quality of Water for Pharmaceutical Use*, EMEA/CPMP/QWP/158/01 Revision. *Chapter 1231, Water for Pharmaceutical Purposes*, USP. http://www.uspbpep.com/usp29/v29240/usp29nf24s0_c1231.html (accessed 3 April 2020).

136 Cotton, F.A. and Wilkinson, G. (1988). *Advanced Inorganic Chemistry*, 112–113. New York: Wiley.

137 Addison, C.C. (1980). Dinitrogen tetroxide, nitric acid, and their mixtures as media for inorganic reactions. *Chem. Rev. 80*: 21–39.

138 Stedman, G. (1979). Reaction mechanisms of inorganic nitrogen compounds. *Adv. Inorg. Chem. Radiochem. 22*: 113–170.

139 Spiegelhalder, B., Eisenbrand, G., and Preussmann, R. (1978). Contamination of amines with *N*-nitrosamines. *Angew. Chem. Int. Ed. Engl. 17*: 367–368.

140 Wu, Y., Levons, J., Narang, A.S. et al. (2011). Reactive impurities in excipients: profiling, identification and mitigation of drug-excipient incompatability. *AAPS PharmSciTech 12*: 1248–1263.

141 Lewis, G.N. and Edgar, A. (1911). The equilibrium between nitric acid, nitrous acid and nitric oxide. *J. Am. Chem. Soc. 33*: 292–299.

142 Habig, K., Baessler, K., Schulz, L., and Schutte, H. (1982). Method for Removing Nitrosation Agent(s) From a Nitrated Aromatic Compound, US4338473.

143 Fitzpatrick, J., Meyer, T.A., O'Neil, M.E., and Williams, D.L.H. (1984). Comparison of the reactivity of nine nitrous acid scavengers. *J. Chem. Soc. Perkin Trans. 2*: 927–932.

144 Borths, C.J., Murns, M., Curran, T., Ide, N.D. Nitrosamine Reactivity: A Survey of Reactions and Purge Processes. *Org. Process Res. Dev.* 2021, 1788–1801.

145 Blaser, H.U., Siegrist, U., and Steiner, H. (2001). *Ullmann's Encyclopedia of Industrial Chemistry* (eds. R.A. Sheldon and H. van Bekkum), 389–406. New York: Wiley.

146 Blaser, H.U., Siegrist, U., Steiner, H., and Studer, M. (2001). *Aromatic Nitro Compounds*, 389–406. Wiley-VCH Verlag GmbH.Dixon, D.J. and Morejon, O.P. (2004). *Recent Developments in the Reduction of Nitro and Nitroso Compounds*, 479–492. Elsevier B.V.Song, J., Huang, Z.-F., Pan, L. et al. (2018). Review on selective hydrogenation of nitroarene by catalytic, photocatalytic and electrocatalytic reactions. *Appl Catal B 227*: 386–408.Orlandi, M., Brenna, D., Harms, R. et al. (2018). Recent developments in the reduction of aromatic and aliphatic nitro compounds to amines. *Org. Process Res. Dev. 22*: 430–445.

147 Blaser, H.-U., Steiner, H., and Studer, M. (2009). Selective catalytic hydrogenation of functionalized nitroarenes: an update. *ChemCatChem 1*: 210–221.

148 Auer, E., Berweiler, M., Gross, M. et al. (2001). *Catalysis of Organic Reactions*, vol. 82 (ed. M.E. Ford), 293–300. New York: Marcel Dekker, Inc.

149 Béchamp, A. (1854). De l'action des protosels de fer sur la nitronaphtaline et la nitrobenzine. nouvelle méthode de formation des bases organiques artificielles de Zinin. *Annales de Chimie (Paris) 42*: 186–196.

150 Kadam, H.K. and Tilve, S.G. (2015). Advancement in methodologies for reduction of nitroarenes. *RSC Adv. 5*: 83391–83407.

151 Orlandi, M., Brenna, D., Harms, R. et al. (2018). Recent developments in the reduction of aromatic and aliphatic nitro compounds to amines. *Org. Process Res. Dev. 22*: 430–445.

152 Haber, F. (1898). Gradual electrolytic reduction of nitrobenzene with limited cathode potential. *Zeit. Elektrochem. 4*: 506–513.

153 Gelder, E.A. (2005). The hydrogenation of nitrobenzene over metal catalysts. University of Glasgow Thesis.

154 Gelder, E.A., Jackson, S.D., and Lok, C.M. (2005). The hydrogenation of nitrobenzene to aniline: a new mechanism. *ChemComm.*: 522–524.

155 Corma, A., Concepción, P., and Serna, P.A. (2007). Different reaction pathway for the reduction of aromatic nitro compounds on gold catalysts. *Angew. Chem. Int. Ed. 46*: 7266–7269.

156 Sheng, T., Qi, Y.-J., Lin, X. et al. (2016). Insights into the mechanism of nitrobenzene reduction to aniline over Pt catalyst and the significance of the adsorption of phenyl group on kinetics. *Chem. Eng. J. (Amsterdam, Neth.) 293*: 337–344.

157 Zhang, L., Shao, Z.-J., Cao, X.-M., and Hu, P. (2018). Insights into different products of nitrosobenzene and nitrobenzene hydrogenation on Pd(111) under realistic reaction conditions. *J. Phys. Chem. C 122*: 20337–20350.

158 Formenti, D., Ferretti, F., Scharnagl, F.K., and Beller, M. (2019). Reduction of nitro compounds using 3d-non-noble metal catalysts. *Chem. Rev. 119*: 2611–2680.

159 Millán, R., Liu, L., Boronat, M., and Corma, A.A. (2018). New molecular pathway allows the chemoselective reduction of nitroaromatics on non-noble metal catalysts. *J. Catal. 364*: 19–30.

160 Burge, H.D., Collins, D.J., and Davis, B.H. (1980). Intermediates in the Raney nickel catalyzed hydrogenation of nitrobenzene to aniline. *Ind. Eng. Chem. Prod. Res. Dev. 19*: 389–391.

161 Mahata, A., Rai, R.K., Choudhuri, I. et al. (2014). Direct vs. indirect pathway for nitrobenzene reduction reaction on a Ni catalyst surface: a density functional study. *Phys. Chem. Chem. Phys. 16*: 26365–26374.

162 Machado, R.M. (2020). Fundamentals of mass transfer and kinetics for the hydrogenation of nitrobenzene to aniline. https://www.mt.com/dam/mt_ext_files/Editorial/Generic/3/Hydrogenation-Nitrobenzene-Aniline-Mass-Transfer_Editorial-Generic_1204310241270_files/fundamentals_of_masstransferandkineticshydrogenation.pdf (accessed 18 June 2007).

163 Britest (2018). Knowledge and Understanding DOmainS article on Hydrogenation. https://www.britest.co.uk/ (accessed 26 April 2020).

164 Yang, K.H. and Hougen, O.A. (1950). Determination of mechanism of catalyzed gaseous reactions. *Chem. Eng. Prog. 46*: 146–157.

165 Holler, V., Wegricht, D., Yuranov, I. et al. (2000). Three-phase nitrobenzene hydrogenation over supported glass fiber catalysts: reaction kinetics study. *Chem. Eng. Technol. 23*: 251–255.

166 Figueras, F. and Coq, B. (2001). Hydrogenation and hydrogenolysis of nitro-, nitroso-, azo-, azoxy- and other nitrogen-containing compounds on palladium. *J. Mol. Catal. A: Chem. 173*: 223–230.

167 Dyson, R.M., Hazenkamp, M., Kaufmann, K. et al. (2000). Modern tools for reaction monitoring: hard and soft modeling of "non-ideal", on-line acquired spectra. *J. Chemometr. 14*: 737–750.

168 Visentin, F., Puxty, G., Kut, O.M., and Hungerbuehler, K. (2006). Study of the hydrogenation of selected nitro compounds by simultaneous measurements of calorimetric, FT-IR, and gas-uptake signals. *Ind. Eng. Chem. Res. 45*: 4544–4553.

169 Gelder, E.A., Jackson, S.D., and Lok, C.M. (2007). *Catalysis of Organic Reactions* (ed. S.R. Schmidt), 167–176. Boca Raton, FL: CRC Press LLC.

170 Makaryan, I.A. and Savchenko, V.I. (1993). *N*-Arylhydroxylamines transformation in the presence of heterogeneous catalysts. *Stud. Surf. Sci. Catal. 75*: 2439–2442.

171 Augustine, R.L. (1995). *Heterogeneous Catalysis for the Synthetic Chemist*, 473–510. New York: Marcel Dekker, Inc.

172 Coe, C. G. and Brockington, J. W. (1988): In *Catalysis of Organic Reactions*; Rylander, P. N. G. H.; Augustine, R. L., Ed. Marcel Dekker: New York; Vol. 33, pp 101–119.

173 Karwa, S.L. and Rajadhyaksha, R.A. (1988). Selective catalytic hydrogenation of nitrobenzene to hydrazobenzene. *Ind. Eng. Chem. Res. 27*: 21–24.

174 Chen, Y.-F., Chen, J., Lin, L.-J., and Chuang, G.J. (2017). Synthesis of azoxybenzenes by reductive dimerization of nitrosobenzene. *J. Org. Chem. 82*: 11626–11630.

175 Cole, K.P., Johnson, M.D., Laurila, M.E., and Stout, J.R. (2017). An automated repeating batch with catalyst recycle approach to nitro group hydrogenolysis. *React. Chem. Eng. 2*: 288–294.

176 Ogata, Y. and Takagi, Y. (1958). Kinetics of the condensation of anilines with nitrosobenzenes to form azobenzenes. *J. Am. Chem. Soc. 80*: 3591–3595.

177 Caille, S., Allgeier, A.M., Bernard, C. et al. (2019). Development of a factory process for omecamtiv mecarbil, a novel cardiac myosin activator. *Org. Process Res. Dev. 23*: 1558–1567.

178 Lee, L.T., Chen, M.H., and Yao, C.N. (1989). Preparation of *p*-aminophenol by hydrogenation of nitrobenzene in aqueous acid media. US4885389A, December 5.

179 Zhang, L., Jiang, J., Shi, W. et al. (2015). Insights into the hydrogenation mechanism of nitrobenzene to aniline on Pd_3/Pt(111): a density functional theory study. *RSC Adv. 5*: 34319–34326.

180 Karwa, S.L. and Rajadhyaksha, R.A. (1987). Selective catalytic hydrogenation of nitrobenzene to phenyl hydroxylamine. *Ind. Eng. Chem. Res. 26*: 1746–1750.

181 Freifelder, M. (1971). *Practical Catalytic Hydrogenation: Techniques and Applications*, 177–182. New York, NY: Wiley-InterScience.

182 Kosak, J.R. (1988). *Catalysis of Organic Reactions* (eds. P.N. Rylander, H. Greenfield and R.L. Augustine), 135–147. New York: Marcel Dekker.

183 Rylander, P.N., Karpenko, I.M., and Pond, G.R. (1970). Selectivity of hydrogenation over platinum metal catalysts: nitroaromatics. *Ann. N. Y. Acad. Sci. 172*: 266–275.

184 Rylander, P.N. (1985). *Hydrogenation Methods*, 106–107. New York, NY: Academic Press.

185 Stratz, A.M. (1984). *Catalysis of Organic Reactions* (ed. J.R. Kosak), 335. New York: Marcel Dekker.

186 Takenaka, Y., Kiyosu, T., Choi, J.-C. et al. (2009). Selective synthesis of *N*-aryl hydroxylamines by the hydrogenation of nitroaromatics using supported platinum catalysts. *Green Chem. 11*: 1385–1390.

187 Rylander, P.N., Karpenko, I.M. and Pond, G.R. (1971). Selective catalytic hydrogenation of nitro aromatics to the corresponding N-arylhydroxyamines. DE2118369A, October 28.

188 Sharma, A.H. and Hope, P. (1992). Process for the preparation of a hydroxylamine. US 5166435A, November 24.

189 Tong, W.R., Seagrave, R.L., and Wiederhorn, R. (1977). 3,4-Dichloroaniline autoclave incident. *Loss Prev. 11*: 71–75.

190 Rains, R.K., Lambers, E.A., and Genetti, R.A. (1996). *Catalysis of Organic Reactions*, vol. 68 (eds. E. Russell and J. Malz), 43–52. New York: Marcel Dekker.

191 Le Ludec, J. (1974). The hydrogenation of 2,5-dichloronitrobenzene over Pt/C in pyridine gave the corresponding hydroxylamine in 80% yield. Chlorinated phenylhydroxylamines DE 19742455887, November 26.

192 Oxley, P.W., Adger, B.M., Sasse, M.J., and Forth, M.A. (1989). *N*-acetyl-*N*-phenylhydroxylamine via catalytic transfer hydrogenation of nitrobenzene using hydrazine and rhodium on carbon. *Organic Synth. 67*: 187–192.

193 Savchenko, V.I., Dorokhov, V.G., Tsareva, R.S. et al. (1978). Hydrogenation of aromatic nitro compounds in the presence of an iridium catalyst. *Izv. Akad. Nauk SSSR, Ser. Khim* 27: 2062–2065.

194 Taya, K. (1966). Hydrogenating activity of an iridium catalyst. *Chem. Commun. (London)*: 464–465.

195 Kasparian, A.J., Savarin, C., Allgeier, A.M., and Walker, S.D. (2011). Selective catalytic hydrogenation of nitro groups in the presence of activated heteroaryl halides. *J. Org. Chem. 76*: 9841–9844.

196 Studer, M., Neto, S., and Blaser, H.U. (2000). Modulating the hydroxylamine accumulation in the hydrogenation of substituted nitroarenes using vanadium-promoted RNi catalysts. *Top. Catal. 13*: 205–212.

197 Studer, M. and Baumeister, P. (1996). Catalytic hydrogenation of aromatic nitro compounds. WO9636597A1, November 21.

198 Blaser, H.U. and Studer, M. (1999). The role of catalysis for the clean production of fine chemicals. *Appl. Catal., A 189*: 191–204.

199 Moebus, K., Wolf, D., Benischke, H. et al. (2010). Hydrogenation of aromatic nitrogroups with precious metal powder catalysts: influence of modifier on selectivity and activity. *Top. Catal. 53*: 1126–1131.

200 Baumeister, P., Blaser, H.U., and Studer, M. (1997). Strong reduction of hydroxylamine accumulation in the catalytic hydrogenation of nitroarenes by vanadium promoters. *Catal. Lett. 49*: 219–222.

201 Siegrist, U., Baumeister, P., Blaser, H.-U., and Studer, M. (1998). *Catalysis of Organic Reactions*, vol. 75 (ed. F.E. Herkes), 207–219. New York: Marcel Dekker, Inc.

202 Boronat, M., Concepción, P., Corma, A. et al. (2007). A molecular mechanism for the chemoselective hydrogenation of substituted nitroaromatics with nanoparticles of gold on TiO_2 catalysts: a cooperative effect between gold and the support. *J. Am. Chem. Soc. 129*: 16230–16237.

203 Corma, A. and Serna, P. (2006). Chemoselective hydrogenation of nitro compounds with supported gold catalysts. *Science 313*: 332–334.

204 Meng, X., Lan, M., Zhong, F. and Chen, Y. (2006). Method for preparing aromatic amine compound through selective hydrogenation of aromatic nitro compound. CN106146232A, November 23.

205 Betori, R.C., Kallemeyn, J.M., and Welch, D.S. (2015). A kinetics-based approach for the assignment of reactivity purge factors. *Org. Process Res. Dev. 19*: 1517–1523.

206 Verma, A.A., Richter, S., Kotecki, B., and Diwan, M. (2019). Chapter 43 Process Modeling Applications Toward Enabling Development and Scale-Up: Chemical Reactions. *Chemical Engineering in the Pharmaceutical Industry: Active Pharmaceutical Ingredients* (eds. D.J. Ende and M.T. Ende). Wiley.

207 Angelaud, R., Reynolds, M., Venkatramani, C. et al. (2016). Manufacturing development and genotoxic impurity control strategy of the hedgehog pathway inhibitor vismodegib. *Org. Process Res. Dev. 20*: 1509–1519.

208 Quittman, W., Belser, T.P., and Aufdenblatten, R.N. (2011). Catalytic hydrogenation of nitro compounds. US0071290A1, March 24.

209 Mennen, S.M., Alhambra, C., Allen, C.L. et al. (2019). The evolution of high-throughput experimentation in pharmaceutical development and perspectives on the future. *Org. Process Res. Dev. 23*: 1213–1242.

210 Loos, P., Alex, H., Hassfeld, J. et al. (2016). Selective hydrogenation of halogenated nitroaromatics to haloanilines in batch and flow. *Org. Process Res. Dev. 20*: 452–464.

211 Crump, B.R., Goss, C., Lovelace, T. et al. (2013). Influence of reaction parameters on the first principles reaction rate modeling of a platinum and vanadium catalyzed nitro reduction. *Org. Process Res. Dev. 17*: 1277–1286.

212 a) Deshpande, M.N., Cain, M.H., Patel, S.R. et al. (1998). Scalable process for the novel antidepressant ABT-200. *Org. Process Res. Dev. 2*: 351–356. b) Wen, L., Tang, F., Ge, C. et al. (2012). Practical large-scale preparation of (*R*)-Rolipram using chiral nickel catalyst. *Synth. Communicat. 42*: 3288–3295. c) Moccia, M., Cortigiani, M., Monasterolo, C. et al. (2015). Development and scale-up of an organocatalytic enantioselective process to manufacture (S)-Pregabalin. *Org. Process Res. Dev. 19*: 1274–1281. d) Carangio, A., Edwards, L.J., Fernandez-Puertas, E. et al. (2020). Evaluation of sponge metal catalysts in a trickle bed reactor for the continuous hydrogenation of an aliphatic nitro intermediate. *Org. Process Res. Dev. 24*: 1909–1915.

213 Galloway, S.M., Vijayaraj Reddy, M., McGettigan, K. et al. (2013). Potentially mutagenic impurities: analysis of structural classes and carcinogenic potencies of chemical intermediates in pharmaceutical syntheses supports alternative methods to the default TTC for calculating safe levels of impurities. *Regul. Toxicol. Pharmacol. 66*: 326–335.

214 European Directorate for the Quality of Medicines and Healthcare (2000). Enquiry: alkyl mesilate (methane sulfonate) impurities in Mesilate salts. *PharmEuropa* 12: 27.

215 Elder, D.P. and Snodin, D.J. (2009). Drug substances presented as sulfonic acid salts: overview of utility, safety and regulation. *Pharm. Pharmacol.* 61: 269–278.

216 EMEA Press Release (2007). European Medicines Agency agrees on action plan following the recall of Viracept and recommends suspension of the Marketing Authorisation. Doc ref. EMEA/275367/2007.

217 Swissmedic Department for Control of the Medicinal Products Market (2007). Mesylate Ester Type Impurities Contained in Medicinal Products.

218 Coordination Group for Mutual Recognition-Human committee (CMDh) (2008). Request to Assess the Risk of Occurrence of Contamination with Mesilate Esters and Other Related Compounds in Pharmaceutical, EMEA/CMDh/98694/2008, London.

219 Elder, D.P., Delaney, E., Teasdale, A. et al. (2010). The utility of sulfonate salts in drug. *Dev. J. Pharm. Sci.* 99 (7): 2948–2961.

220 Snodin, D.J. (2006). Residues of Genotoxic alkyl mesylates in mesylate salt drug substances: real or imaginary problems? *Regul. Toxicol. Pharmacol.* 45: 7–90.

221 Jacq, K., Teasdale, A., Delaney, E. et al. (2008). Development and validation of an automated static headspace gas chromotography-mass spectrometry (SHS-GC-MS) method for monitoring the formation of ethyl methane sulfonate from ethanol and methane sulfonic acid. *J. Pharm. Biomed. Anal.* 48 (5): 1339.

12

Strategic Approaches to the Chromatographic Analysis of Mutagenic Impurities

Frank David[1], Gerd Vanhoenacker[1], Koen Sandra[1], Pat Sandra[1], Tony Bristow[2], and Mark Harrison[2]

[1] *RIC-group, Kortrijk, Belgium*
[2] *Chemical Development, Pharmaceutical Technology and Development Operations, AstraZeneca, Macclesfield, UK*

12.1 Introduction

Chromatographic analytical techniques have gained a prominent position in research and quality control within the pharmaceutical industry. For many years, capillary gas chromatography (CGC) and high performance liquid chromatography (HPLC) have been widely used to monitor the quality of starting materials, intermediates, drug products, and formulations. For the determination of potentially mutagenic impurities (PMIs) and mutagenic impurities (MIs), standard QC methods are, however, often not adequate, because of the low maximum allowable limits. Based on the threshold of toxicological concern (TTC) concept (1.5 μg/day for lifetime exposure [1]) [2], methods for PMI/MI analysis in pharmaceutical products often need to be able to quantify the target analytes at the trace level, often at or below 1 ppm (1 μg/g), which is typically 500 times lower than for classical impurity analysis in pharmaceutical quality control (at 0.05% level). Indeed for certain classes of MIs, such as *N*-nitrosamines, even low ppb (ng/g) concentration levels should be monitored, requiring highly sensitive analytical methods.

Since the inception of concerns relating to MIs, at the beginning of the millennium, analytical methods have been developed for PMI/MI impurity analysis which in many cases are "compound & matrix" specific, e.g. focused on a specific mutagenic analyte present in a specific drug product. Developing methods on an individual basis (one PMI/MI in one matrix) is, however, hugely time and resource consuming for pharmaceutical companies [3]. Therefore, a program was established to develop a systematic approach to method development, one that could be applied to multiple impurities within a MI class and across a wide selection of matrices. Due to the extremely wide ranges of polarities and volatilities of the possible target analytes, in combination with a large range of matrices (APIs and intermediates) often with significantly different physicochemical characteristics (water soluble, ionic, polar, apolar, basic, acidic,...), this is a challenging task. No single method, not even employing the most advanced chromatographic and mass spectrometric instrumentation, can cover all PMI/MIs in all matrices. The project therefore aimed to develop a strategic approach based on knowledge of the properties of the analyte and matrix. The work resulted in a method selection chart (decision tree) that can be used to guide analysts through the selection of the most appropriate method to apply to a specific PMI/MI analysis [4]. Methods were developed either using gas chromatography (GC) or liquid

Mutagenic Impurities: Strategies for Identification and Control, First Edition. Edited by Andrew Teasdale.
© 2022 John Wiley & Sons, Inc. Published 2022 by John Wiley & Sons, Inc.

chromatography (LC), both in combination with a single quadrupole mass spectrometer as a detector, since these techniques are commonly available in a routine pharma QC environment.

In addition, several sample preparation methods were tested on different classes of PMI/MI in a selection of APIs with different physicochemical properties (polarity, functionalities,…). Attention was specifically focused on robust sample preparation methods that could be automated, and online coupled to GC or LC. The aim being to minimize manual handling that could require a high degree of expertise and thus lead to potential variability within the resultant method, especially were this is to be transferred between laboratories.

The result of this research is the decision tree chart presented in Figure 12.1. Briefly, this chart starts from the question: Is the PMI/MI amenable to analysis by GC or not? If yes, the next question is whether the PMI/MI has sufficient vapor pressure to be present in the headspace phase of a concentrated solution of the API (in water, DMSO, or other low volatile solvent). If yes, static headspace (SHS), solid-phase micro-extraction (SPME) or dynamic headspace (DHS) can be used. Since most API are not volatile, headspace methods can therefore be considered as "first-to-try," since it allows contamination of the analytical system (GC inlet, column, detector) by the matrix to be avoided. The choice between SHS, SPME, and DHS depends on the volatility of the analytes and the desired sensitivity. This is discussed in detail in the section relating to the analysis of alkyl- and aryl halides.

In some cases, analytes do not have enough vapor pressure for headspace analysis and *in-situ* derivatization can be applied to generate a volatile derivative of the PMI/MI that can be analyzed by SHS, SPME, or DHS. This will be illustrated by the analysis of sulfonates and *N*-mustards. If the solute is GC amenable (stable and eluting from the GC column at moderate temperatures, e.g. <320 °C), but not volatile enough for headspace techniques (even after derivatization), a direct injection of a concentrated API solution in GC can be used. Hereby, attention should be paid to the volatility and thermal stability of the analytes and the matrix (API). Since the matrix is also introduced, the API itself, other API impurities, and/or API decomposition products (e.g. formed in the hot GC inlet) can interfere with the target PMIs/MIs or can influence the system performance (contamination). In this respect, state-of-the-art GC techniques such as back-flushing and two-dimensional GC (2D-GC, GC-GC) can be very useful, as illustrated by the analysis of Michael acceptors and haloalcohols.

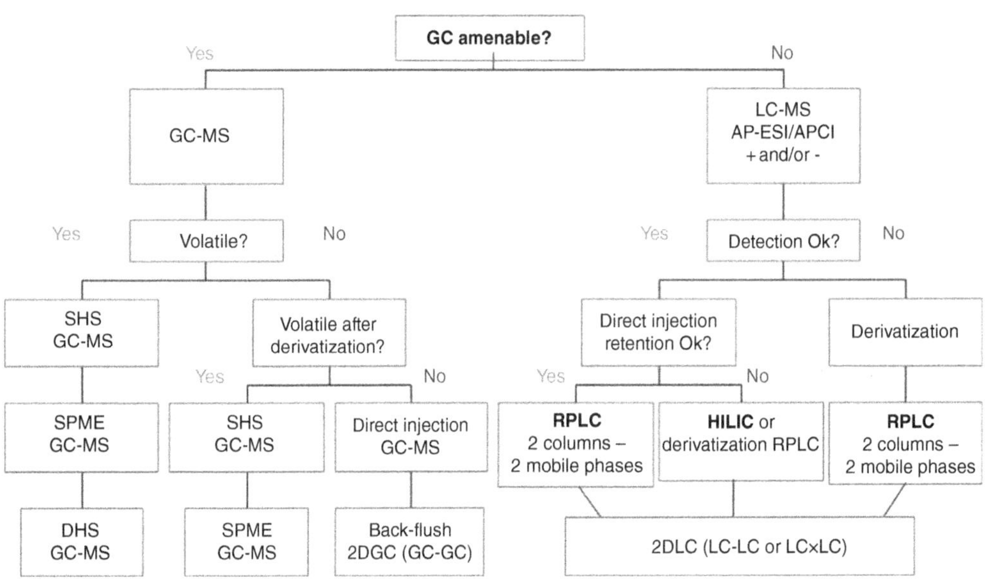

Figure 12.1 Method selection chart for PMI/MI determination in API.

In the studies described, several PMIs/MIs were found amenable to GC and because of the availability of robust and low-cost GC-MS systems, we recommend that the GC-MS option should be evaluated first. If the analytes are not amenable to GC, LC-MS is used (Figure 12.1, right-hand side pathway). In selecting LC-MS, first a selection of the ionization mode is needed. Both atmospheric pressure electrospray ionization (AP-ESI or ESI) and atmospheric pressure chemical ionization (APCI), either operating in the positive ion or negative ion detection mode, can be evaluated in terms of sensitivity and selectivity. Both modes are rather complementary. One mode can, for instance, result in a higher absolute response, but combined with lower selectivity and/or higher background, the overall result might be that the other ionization mode performs better for a given application. Flow injection of solutions of the PMI/MI and MS acquisition in alternating positive and negative ion scan mode and the use of a multimode ESI/APCI source can hereby be useful. Obviously, also other LC detectors such as Ultraviolet-Diode Array Detection (UV-DAD) or fluorescence detection (FLD) can be applied if sufficient sensitivity and selectivity can be obtained. Recent developments in UV-DAD systems and flow cells have significantly increased sensitivity, making this solution attractive for UV-detectable PMI/MI [5].

Once the detection mode is selected, direct injection of a concentrated API solution, followed by reversed-phase HPLC (RP-LC) is then the next step. Obviously, the drug product (main solute) should be well separated from the PMI/MI. Evaluation of two RPLC columns and different mobile-phase compositions can be helpful here. If the analyte is poorly retained in RPLC mode, hydrophobic interaction liquid chromatography (HILIC) or pre-column derivatization (into more hydrophobic derivatives) can be used as alternative, as illustrated by the analysis of aziridines, aryl amines, and aminopyridines. All these LC modes can be combined with MS, using either ESI or APCI. Although MS is often considered as "universal" detection, some target analytes can give very low or nonselective response, in both ESI and APCI ionization modes. In these cases, derivatization of the target compound(s) to enhance detectability prior to LC-MS can be useful, as will be illustrated for the analysis of hydrazines and aldehydes. Also for alkyl halides and epoxides, the use of chemical derivatization, eventually in combination with coordination ion-spray mass spectrometry has been demonstrated [6]. Especially for neutral molecules, ionization efficiency in atmospheric pressure ESI can be poor and derivatization in a more polar, highly ionizable or permanently charged derivative can boost detectability. Derivatization in HPLC is thus very useful, both to increase retention and to enhance detectability [7].

This flow chart, already presented in 2009, has proven its value in developing new methods and is still valid today [8].

An alternative flow chart was presented by Liu and Kord in 2013 [9]. The flow chart proposed by Liu and Kord is given in Figure 12.2 and shows a similar approach compared with Figure 12.1, but starting with a first question on stability of matrix and solute. If stable, volatile compounds can be analyzed by GC, thereby avoiding issues of matrix interference. HPLC is the alternative for nonvolatile solutes. In that flow chart, also the options of other detectors such as flame ionization detection (FID) or electron capture detection (ECD) for GC and UV/DAD or element-specific detection for HPLC are included. Special attention is paid to problems that can be encountered with non-stable analytes and matrices. Stabilization by derivatization of analytes is thereby suggested. This paper also introduces the concept of matrix deactivation. This differs from conventional chemical derivatization where only the specific analyte is chemically transformed, to one where possible reactive interfering species within the sample matrix are also deactivated [10].

In the following section of this chapter, some general information is included on method development and method validation. Next, instrumental configurations that have been applied in the author's laboratories are described. Finally, more detailed information is given on potential methods for various classes of PMIs/MIs.

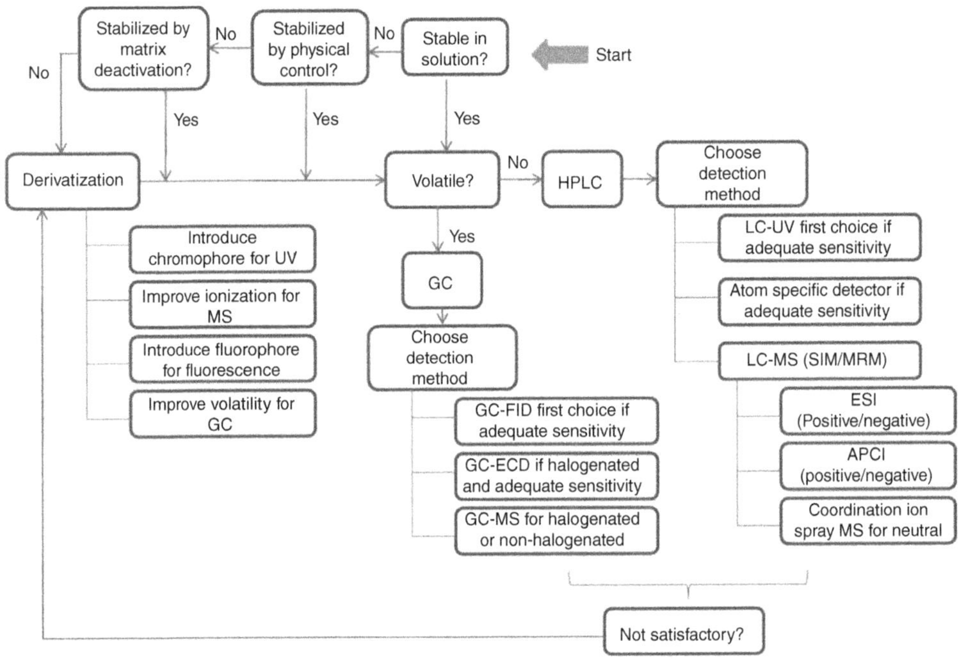

Figure 12.2 Alternative method selection chart for designing methods for PMI/MI determination. *Source:* Reproduced from Liu and Kord [9].

12.2 Method Development and Validation

In our project to create a set of "generic" methods, method development was performed for sets of selected PMI/MI analytes that are representative for organic compound classes with certain structural alert functionalities [11]. The selected target compounds as well as the representative APIs used as "matrix" were all commercially available. The API used as matrices covered different physicochemical properties and included promethazine (apolar, base), carbamazepine (medium polar, neutral), ampicillin (polar, zwitterion), Vitamin C (polar, acidic), penicillin V (polar, acidic), ephedrine (medium polar, base), acetyl salicylic acid (Aspirin, medium polar, acidic), bromhexine (medium polar, base), guaifenesin (polar, neutral), and doxylamine (medium polar, base).

After selection of a possible route in the method selection chart presented in Figure 12.1 and based on prior experiences in our laboratories, method development work typically started with the evaluation of the sensitivity that can be reached for the target solutes using a certain sample introduction, separation, and detection method. After this detectability check, attention was paid to the chromatographic separation of the target solutes and the matrices. This is the most crucial step, since co-elution of solutes with API influences detectability, even if triple quadrupole or high-resolution MS systems are used. Presence of bulk API in the GC inlet and column can destroy system performance. In LC-MS, co-elution of analyte and matrix in the ionization source leads to ionization suppression. Therefore, it is of utmost importance that either a selective sample preparation is used, avoiding API introduction, or that chromatographic resolution is obtained before detection.

Several groups have applied quality by design (QbD) and design of experiment (DoE) approaches in method development and optimization. In the first instance, these approaches can help to determine the required LOD/LOQ, based on TTC and API dosage [8]. During method development,

DoE is useful to optimize methods in terms of chromatographic resolution, specificity, and sensitivity. This is, for instance, illustrated by Székezy et al. for 4-dimethylaminopyridine [12], and for 1,3-di-isopropylurea [13]. LC parameters such as flow, gradient, and injection volume, and MS parameters such as cone voltage and MS/MS collision energy, were optimized. Grigori et al. [14] used a Box–Behnken design to optimize the LC mobile-phase composition for the resolution of three nitroaromatic PGIs from meropenem API. This allowed to divert the API to waste via a switching valve, before the bulk of the matrix could enter the MS ionization source. The same QbD approach was also used for the determination of two PMIs in rabeprazole [15]. Optimized LC parameters included mobile-phase composition, pH, and buffer salt concentration. Schmidtsdorff and Schmidt [16] applied QbD for the development of a Supercritical Fluid Chromatography (SFC) method for the determination of nitrosamines and other sartan-related impurities. First, critical quality attributes were selected. These included peak resolution, peak height and symmetry, signal-to-noise, and retention. The critical process parameters that were optimized included stationary phase, modifier and additives to mobile phase, column temperature, gradient slope of the mobile phase, type and flow rate of the makeup solvent (for SFC-MS/MS coupling), and mass spectrometric instrument parameters. To verify the effects and the interactions of all these parameters, data evaluation using a statistical approach was applied. The final method allowed sensitive determination of nine *N*-nitrosamines at sub-ppm level in Valsartan and Losartan.

After method development, method validation is required. This is based on ICH Q2 [17]. Trace analysis methods can either be validated as "quantitative methods" or as "limit tests" [3]. While the first requires the most intensive validation (with demonstration of accuracy, repeatability, reproducibility, linearity, sensitivity, specificity, robustness), the latter can be used as a faster approach if no PMI/MI is expected at a requested LOD/LOQ and if the method only aims at demonstrating that the target solute is below a threshold value. This can be achieved by performing tests on blank samples and on samples spiked at the LOD/LOQ level, typically 10–30% of the toxicological threshold (maximum residue level).

12.3 Analytical Equipment for Mutagenic Impurity Analysis

For the development of platform-based methods for the determination of PMIs/MIs in pharmaceutical products, two main instrumental platforms were selected: a gas chromatograph and a liquid chromatograph, both in combination with a single quadrupole mass spectrometer (SQ) as detector. These configurations are commonly available in a routine QC environment. Obviously, other detectors, such as FID or ECD for GC and UV/DAD or FLD for HPLC, can also be used if sufficient sensitivity and specificity can be obtained.

More recently, a clear trend has been observed toward the use of high-end mass spectrometers in mutagenic impurity analysis [18–20]. The methods described in this chapter can of course also be applied using triple quadrupole (QQQ) instruments using multiple reaction-monitoring (MRM) or using ion trap (IT, Orbitrap) and time-of-flight (TOF) mass spectrometers. These systems generally provide higher sensitivity and specificity compared with single quadrupole systems, but are less routinely available, have a higher investment cost, and can lead to a "sensitivity overkill" [8]. Only in some specific cases, the use of more sensitive MS system can be required, as demonstrated for the analysis of *N*-nitrosamines.

The generic methods described for the specific chemical classes were developed on two platforms, both consisting of a dedicated sample preparation unit/autosampler, a dedicated chromatographic system, and a SQ MS. Since the aim of the project was to develop methods that can be applied to various classes of solutes and matrices, the systems were configured for maximum flexibility and "multi-tasking." To this, different sample introduction options were available and multidimensional and column switching options were included.

GC-MS methods were developed and validated on Agilent 6890/7890 GCs combined with a 5975/5977 MSDs (Agilent Technologies, Wilmington, DE, USA). A typical GC configuration for PMI/MI analysis is presented in Figure 12.3. The GC was equipped with a split/splitless inlet and a programmed vaporizing inlet (CIS 4 PTV, Gerstel GmbH, Mulheim, Germany). Injection was performed using a multipurpose sampler (MPS Robotic, Gerstel), which allows liquid injection, SHS, and SPME. A Robotic PRO configuration as shown in Figure 12.3a allows automated sequences of sample preparation steps, such as dilution, reagent, and IS addition. This is followed by liquid, headspace, or SPME injection. This is possible since the newest versions of the autosampler allow automatic exchange of syringe modules (Figure 12.3b). Different vial trays (2 ml, 20 ml vials, reagents, rinse solvents) and thermostated agitators can also be installed. The system could also be equipped with thermal desorption (TDU) and DHS (Gerstel). The GC was also equipped with capillary flow technology (CFT) options, such as a purged Tee, allowing column back-flush and/or a CFT Deans switch, allowing to direct the flow from the GC column to a first detector or to a second column, eventually in a separate low thermal mass oven (LTM, Agilent Technologies) for further separation of heart-cut fractions. Carrier gas flows through the columns and restrictors are controlled by electronic pressure control (EPC) modules. The installation of CFT devices in a GC oven is illustrated in Figure 12.3c.

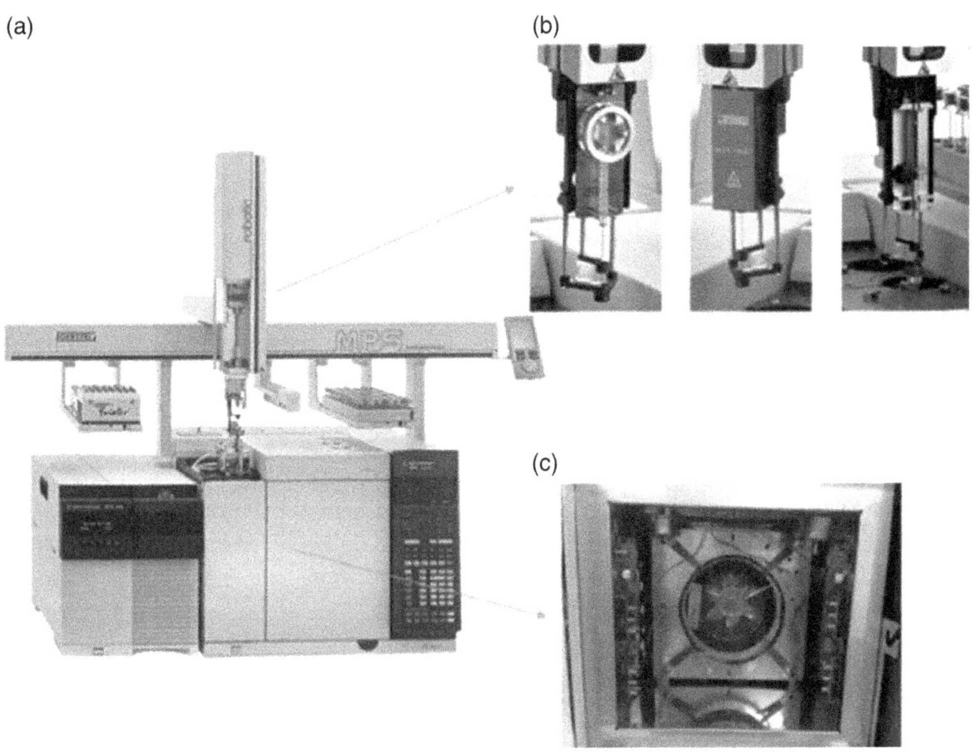

Figure 12.3 GC-MS configuration: (a) GC with SSL and PTV inlet, thermal desorption (TDU) option and MPS Robotic autosampler; (b) Syringe options for liquid, SHS, and SPME; (c) GC oven with CFT options.

LC-MS methods were developed on Agilent 1200 LC systems combined with a single quadrupole LC-MSDs equipped with an electrospray ionization (ESI) or APCI source (Agilent Technologies, Walbronn, Germany). A typical LC-MS configuration is shown in Figure 12.4. The HPLC system consisted of a binary pump (with solvent selection valve), an automated liquid sampler, a thermostated column oven equipped with a column switching valve, and a diode array detector (optional). A six-port/two-position valve allows unattended selection and switching between column 1 and column 2 as shown in Figure 12.5 (top). Automated sample introduction can either be done using a standard liquid sampler or a MPS robot (as in Figure 12.3a, but equipped with a liquid injection valve).

The MSD is typically used in SIM mode using either positive or negative ion detection. During the elution of the API, the LC column effluent could be diverted to waste, in order not to contaminate the ionization chamber.

An interesting option is the addition of a second HPLC system that can be coupled to the main HPLC system via a multidimensional switching valve, as shown in Figure 12.5 (bottom). This valve is mounted after the first HPLC system (after UV/DAD detector) and can be used both in comprehensive LC (LC × LC) mode and in heart-cut LC-LC mode. The latter technique is of particular interest for impurity and PMI/MI analysis. After a first dimension separation (^{1}D pump – ^{1}D column – ^{1}D detector), the elution window that contains the suspected impurity can be selectively transferred to a second dimension separation (^{2}D pump – ^{2}D column – ^{2}D detector and MS) for additional separation and interference-free identification and quantification. This setup can also be used to combine LC-UV methods utilizing nonvolatile buffers with MS. A fraction containing an impurity eluting under non-MS compatible conditions in the first dimension column is thereby transferred to a short ^{2}D column for separation of the impurity from the nonvolatile buffer before MS detection.

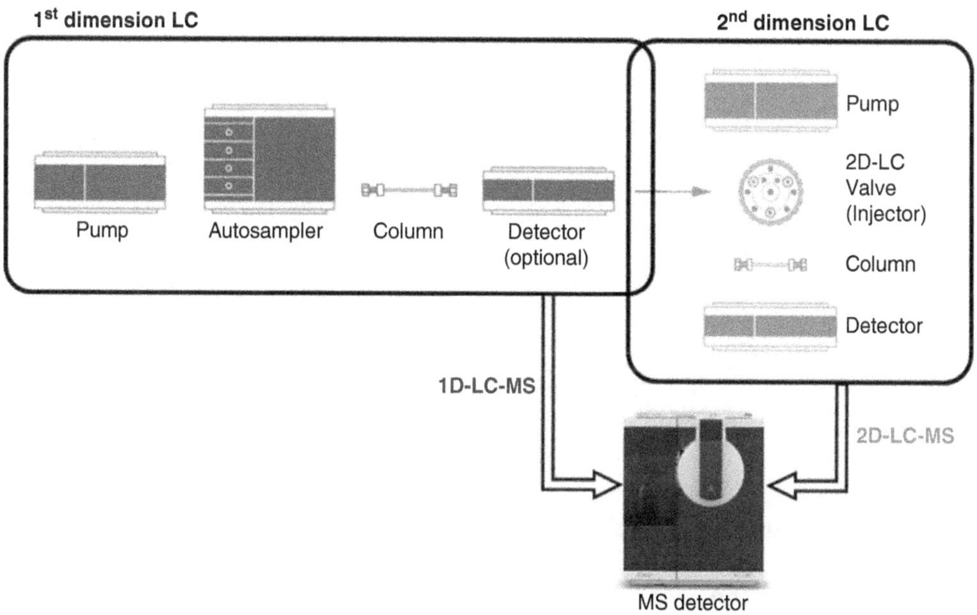

Figure 12.4 Typical LC-MS configuration, with additional LC for 2D-LC operation.

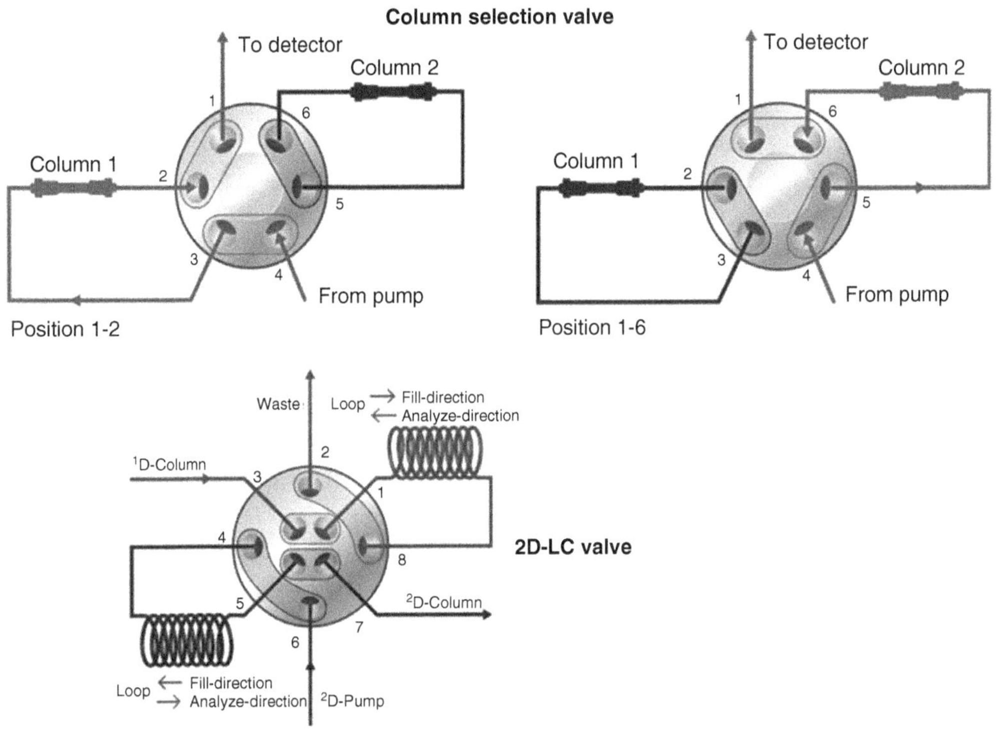

Figure 12.5 Valve systems for column selection (top) and 2D-LC (bottom).

The instrumental configurations described above are only given as examples of robust and flexible systems used in the author's laboratories. Equivalent systems can also be applied.

Besides GC and HPLC, also other separation techniques have been used in the determination of PMIs/MIs in API. These techniques include SFC [16], ion chromatography (IC) [21–23], capillary electrophoresis (CE) [24–27], and GC and LC hyphenated to inductively coupled plasma mass spectrometry (ICP-MS) [28, 29]. However, compared with GC and HPLC, these techniques are only used for a very limited number of applications, mostly in an R&D environment.

Another trend is the application of direct (real-time) techniques that do not include a chromatographic separation. These include direct-MS and NMR. The potential of NMR is described in a separate chapter. Direct MS techniques such as selected ion flow tube mass spectrometry (SIFT-MS) [30] or TDU in combination with atmospheric pressure ionization MS [31, 32] have been tested for some specific applications, but cannot be considered as "generic" methods and are more applicable for high-throughput screening in a development phase of a specific API or for continuous manufacturing control [8].

12.4 Alkyl Halides and Aryl Halides

12.4.1 Method Selection

Alkyl halides are used as alkylating agents in synthesis. Analytical methods for this class of PMIs/MIs were reviewed by Elder et al. [33] and include both GC and LC methods. A typical list of possible target analytes is given in Table 12.1. Most of these target analytes are stable, amenable to

Table 12.1 Typical alkyl halide and aryl halide analytes.

					SHS	SPME
Peak	**Analytes**	**RT (min)**	**Quant ion**	**Qual ion**	**LOD (ppm)**	**LOD (ppm)**
1	Chloromethane	4.21	50	52	0.42	ND
2	Vinyl chloride	4.49	62	64	0.65	ND
3	Bromomethane	5.04	94	96	0.31	1.05
4	Vinyl bromide	5.62	106	108	0.09	0.10
5	1-Chloropropane	6.30	63	42, 78	0.11	0.61
6	Iodomethane	6.71	142	127	0.13	0.01
7	2-Chloropropane	7.23	63	42, 78	0.08	0.07
8	E-1,2-dichloroethene	7.65	61	96	0.02	0.03
9	2-Bromopropane	8.13	43	122, 124	0.06	0.57
10	Z-1,2-dichloroethene	8.66	61	96	0.03	0.03
11	2-Chloroacrylonitrile	9.04	87	52	0.08	0.01
12	1-Chloro-2-methylpropene	9.11	55	90	0.02	0.04
13	1-Bromopropane	9.24	122	43, 124	0.11	0.19
14	2-Iodopropane	10.56	127	70	0.64	0.68
15	1-Bromo-2-methylpropene	11.34	55	134	0.17	0.03
16	1-Iodopropane	11.75	127	43, 170	0.13	0.13
17	E-1,2-dibromoethene	12.38	186	105	0.11	0.01
18	Z-1,2-dibromoethene	13.21	186	105	0.12	0.01
19	2-Iodoethanol	14.20	172	127	ND	<1[a]
20	3-Bromo-2-methylacrylonitrile (cis)	15.11	66	145	0.29	0.02
21	3-Bromo-2-methylacrylonitrile (trans)	16.45	66	145	2.24	0.20
22	4-Fluorobenzyl chloride	19.04	109	144	0.25	0.01
23	Benzyl chloride	19.04	91	126	0.38	0.01
24	4-Fluorobenzyl bromide	20.83	109	83	2.50	0.27
25	Benzyl bromide	20.86	91	170	ND	2.70
26	4-Methylbenzyl chloride	21.28	105	140	1.83	0.02
27	4-Methylbenzyl bromide	22.98	105	184	ND	3.70
28	4-Chlorobutylether	26.11	91	55	1.15	0.01

RT: retention time (min) on DB-VRX column using the conditions listed in text, Quant ion: ion used for quantification, Qual ion: qualifier ions, LOD: limit of detection ($S/N = 3$) for, respectively, SHS and SPME.
[a] Bad peak shape.

GC, and are also volatile enough to allow headspace analysis. Therefore, SHS sampling in combination with GC-MS detection is recommended as the first-to-try method.

SHS is often available in pharmaceutical QC labs since the same methodology is used for residual solvent analysis. While residual solvents are normally analyzed using FID, the selection of mass spectrometric detection operating in selected ion monitoring (SIM) mode results in enhanced selectivity and sensitivity needed for trace analysis. As state-of-the-art single quadrupole MS systems allow simultaneous scan and SIM operation, screening of unknowns is also included, making SHS-GC-MS a generic approach for volatile impurities.

Sample preparation typically includes the dissolution of the sample (API, formulation,…) in water or a nonvolatile solvent, such as dimethyl sulphoxide (DMSO), dimethyl acetamide (DMAC), or dimethyl imidazolidinone (DMI). SHS is performed at temperatures between 70 and 120 °C, depending on the solvent.

An interesting approach described by Ho et al. [34] is the use of ionic liquids as solvent for headspace analysis. Due to their very low vapor pressure, ionic liquids can be used for SHS analysis at high temperatures (100–200 °C), broadening the applicability to less volatile solutes. For alkyl halides and nitroaromatics, sensitivities down to low ppb levels were obtained using ECD detection.

In case SHS sampling does not provide sufficient sensitivity, headspace sampling in combination with enrichment can be applied. This enrichment can be obtained by SPME or DHS sampling. The applicability of headspace-SPME for alkyl halides depends on the affinity of the solutes for the fiber versus the solution. Both the equilibrium between solution and air (headspace) and between headspace and fiber determine the enrichment factor. The applicability of SPME for alkyl halides is demonstrated below.

In DHS sampling, the headspace of the sample, placed in, for instance, a 20 ml vial, is purged with a controlled flow of inert gas (usually carrier gas). The gas is sent through a trap containing an adsorbent such as Tenax or charcoal. The trapped compounds can subsequently be desorbed using dedicated TDU equipment [35]. Since DHS sampling aims at exhaustive extraction of the target solutes, a gain in sensitivity in the order of a factor 10–100 versus SHS can be obtained and therefore DHS can be considered as the sample preparation method offering the highest sensitivity for volatile solutes.

Alternatively, LC-MS approaches have also been applied for the analysis of organohalides. Since the compounds are typically apolar, ESI efficiency is rather poor and therefore either chemical derivatization or coordination ion-spray MS have been applied, as demonstrated by Bai et al. [6] for bis(2-chloroethyl)ether and 3-chloro-1,2-propanediol.

In general, we consider the SHS-GC-MS approach as the most generic method for alkyl halides and aryl halides, and typical SHS-GC-MS conditions that are used in our laboratory are described below.

12.4.2 Typical Conditions Used for Alkyl- and Aryl Halide Analysis by SHS-GC-MS and SPME-GC-MS

The following conditions were applied in a generic method for the determination of the alkyl- and aryl halides listed in Table 12.1. Analyses were performed on a GC equipped with split/splitless inlet and combined with a single quadrupole MSD.

12.4.2.1 Sample Preparation

- The API (50 mg) was dissolved in 2 ml DMSO/water (1:1) in a 20 ml headspace vial. The solvent(s) can be adapted according to API solubility (first dissolved in 1 ml DMSO, then add 1 ml water or vice versa). Alternatively other high boiling solvent can be used (e.g. dimethylacetamide, ionic liquids [34],…).
- For calibration, a set of target compounds was prepared in methanol (at a concentration of 50 ng/μl). Aliquots (0.5–10 μl) of this solution were spiked in (blank) API solution. Resulting spiked concentrations are in the order of 0.5–10 ppm (μg/g API). An internal standard, such as fluorobenzene, 3-fluorotoluene, or a deuterated solute, can be added. Such internal standards are utilized where there is a need to enhance the precision of the analysis.

12.4.2.2 GC-MS Parameters

- SHS was performed at 80 °C equilibration temperature during 15 minutes equilibration time (while shaking). 1 ml headspace gas was injected in split mode (split ratio 1/10, inlet temperature: 250 °C).
- Headspace SPME was performed using a 75 μm/85 μm Carboxen/PDMS fiber at 80 °C equilibration and extraction temperature during 20 minutes.
  - GC-MS conditions were identical for SHS-GC-MS and SPME-GC-MS:
  - Column: 60 m × 0.25 mm i.d. × 1.4 μm d_f DB-VRX (Agilent Technologies)
  - Carrier gas: helium, constant flow (1.5 ml/minute)
  - Oven: 40 °C – 2 minutes – 10 °C/minute – 250 °C – 4 minutes
  - Detection: MS in simultaneous SIM/SCAN mode
    - Scan range: 29–350 *m/z*
    - Selected ions (SIM): see Table 12.1
    - Solvent delay: 3.5 minutes

12.4.3 Typical Results Obtained for Alkyl- and Aryl Halide Analysis by SHS-GC-MS and SPME-GC-MS

A typical chromatogram (SIM trace) obtained for an API (promethazine) spiked with 28 analytes (Table 12.1) at 5 ppm (5 μg/g API) level is shown in Figure 12.6.

From the chromatogram, it is clear that most analytes are well detected. Their retention times are included in Table 12.1. Good linearity ($r^2 > 0.99$) was obtained in a range from 0.5 to 10 ppm (μg/g API). Peak area repeatability was better than 10% RSD at 5 ppm level, except for chloromethane (RSD ≈ 20%). The limit of detection was below 0.5 ppm for most analytes, as shown in Table 12.1. It is clear from these results that very volatile analytes, such as vinyl chloride, can be analyzed with excellent sensitivity using these SHS-GC-MS conditions. It can also be observed that the relative response of the late eluting, less volatile compounds drop significantly, as can be expected from their lower vapor pressure. 4-Methylbenzylbromide (solute 27) could not be

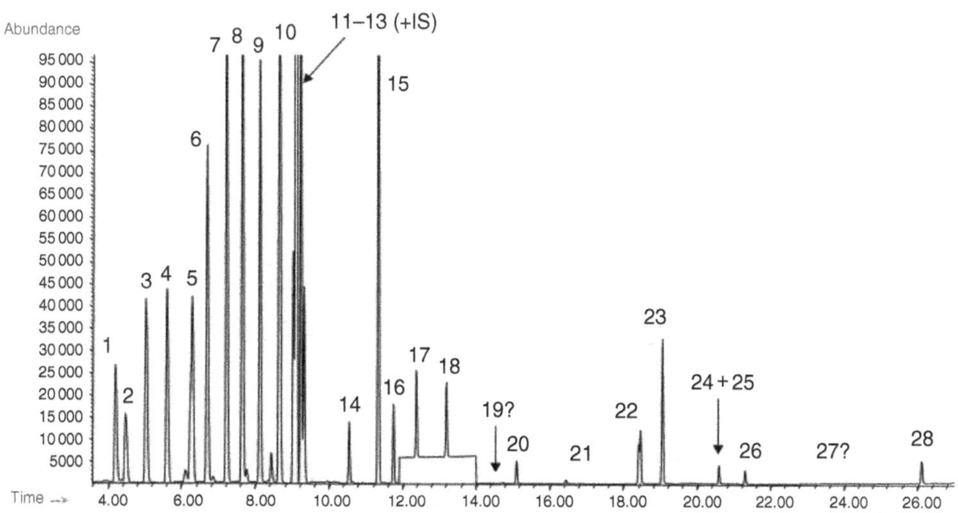

Figure 12.6 Analysis of organohalides in API (promethazine) by SHS-GC-MS (SIM). Sample spiked at 5 μg/g API. SHS at 80 °C. Analytes: see Table 12.1.

detected at this level and a more polar solute, such as iodo ethanol (solute 19), was not detected with the SHS-GC-MS method. Also, other haloalcohols, such as 2-chloroethanol, 2-bromoethanol, and 2-(2-chloroethoxy)ethanol, were tested, but these more polar analytes could also not be detected by SHS-GC-MS.

Based on these results, it is clear that SHS-GC-MS can be utilized as a general screening method applicable to the analysis of a wide range of alkyl- and aryl halides.

In order to increase the sensitivity for some analytes, headspace solid-phase micro-extraction (HS-SPME) was also evaluated. A Carboxen/PDMS fiber was used since this fiber is recommended for the analysis of VOCs. The SIM chromatograms for the analysis of promethazine, spiked at 0.5 ppm level, obtained by, respectively, SPME and SHS are compared in Figure 12.7. The LODs for the 28 analytes obtained by SPME are compared with SHS in Table 12.1. The late eluting analytes, with favorable PDMS-air partitioning coefficients ($K_{PDMS/air}$), are efficiently concentrated in the fiber and enriched, resulting in very low detection limits. This is obvious for the analytes eluting after 12 minutes (peaks 16–28). As shown in Table 12.1, the gain in sensitivity can be as high as a factor 10 (as illustrated by dibromo ethene, 4-fluorobenzylchloride, and 4-chlorobutylether, for instance). 4-Methylbenzylbromide (peak 27), not detected by SHS-GC-MS, can now be detected. Also, iodo-ethanol (peak 19) could be detected, but repeatability was not good, and another method is therefore recommended (see Section 12.9). In general, SPME is complementary to SHS and for analytes typically eluting after toluene on an apolar column (log $K_{PDMS/air} > 3$), SPME leads to higher sensitivity. It should, however, be noted that for the very volatile analytes (e.g. chloromethane (1), vinyl chloride (2)), SHS is superior to SPME since no enrichment on the fiber is obtained for these analytes.

It can be concluded that for the analysis of most alkyl halides, SHS followed by GC-MS operated in SIM mode can be employed and this should be the start point when looking to determine a suitable method. SHS-GC-MS was also successfully applied for online monitoring of methyl iodine purging during the synthesis of ephedrine [36]. In case the sensitivity obtained by SHS-GC-MS is not sufficient for a given application, SPME-GC-MS can be employed for volatile analytes with log $K_{PDMS/air} > 3$. Also, DHS can be applied if ultimate sensitivity is required (or if very limited sample amount is available).

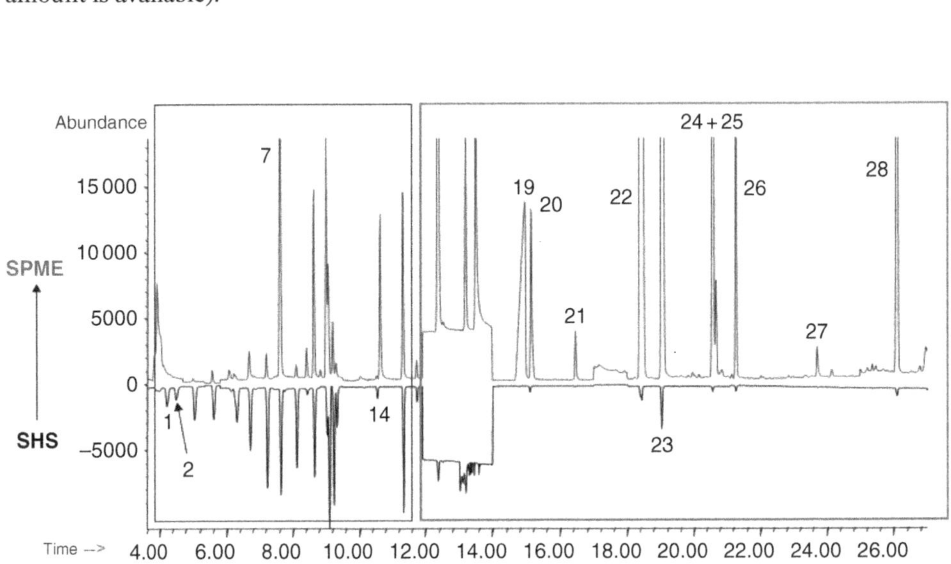

Figure 12.7 Analysis of organohalides in API (promethazine) by headspace-SPME-GC-MS (SIM) (top) and SHS-GC-MS (SIM) (bottom, inversed). Sample spiked at 0.5 µg/g. SPME at 80 °C. Analytes: see Table 12.1.

12.5 Sulfonates

12.5.1 Method Selection

Methyl-, ethyl-, and isopropyl esters from methanesulfonic acid (mesylates), benzenesulfonic acid (besylates), and tolyl sulfonic esters (tosylates) are probably the best known class of MIs. These esters can potentially be formed from volatile alcohols (used as solvent) and sulphonic acids (used to produce API-salts). The formation reaction is given in Figure 12.8a. The most important sulfonate and sulfate esters are listed in Table 12.2.

The analysis of sulfonate esters is not straightforward. Methods were reviewed by Elder et al. [37] and include both GC- and LC-based methods. Sulfonate esters cannot be analyzed by headspace techniques since the vapor pressure is too low. Direct analysis of alkyl- and aryl sulfonates by GC

Figure 12.8 (a) Reaction of sulfonic acid with alcohol and formation of sulfonate; (b) derivatization reaction according to Alzaga et al. [41].

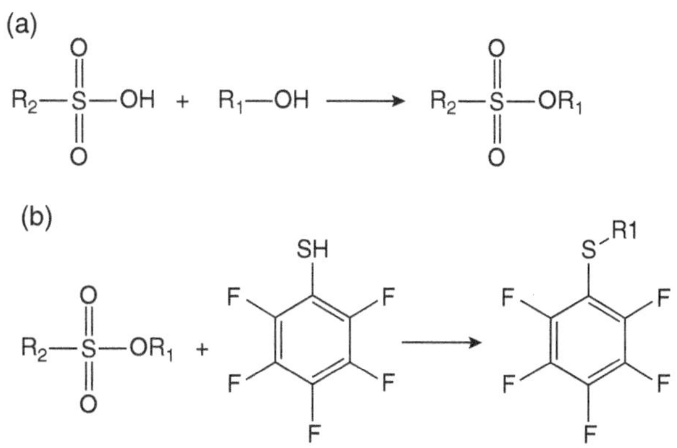

Table 12.2 Sulfonate and sulfate ester target compounds.

				Synthesis	
Name	**Abbreviation**	**R_1**	**R_2**	**Alcohol**	**Acid**
Dimethyl sulfate	DMS	CH_3	OCH_3	Methanol	Sulfuric acid
Diethyl sulfate	DES	C_2H_5	OC_2H_5	Ethanol	Sulfuric acid
Diisopropyl sulfate	DIS	C_3H_7	OC_3H_7	Isopropanol	Sulfuric acid
Methyl methane sulfonate	MMS	CH_3	CH_3	Methanol	Methane sulfonic acid
Ethyl methane sulfonate	EMS	C_2H_5	CH_3	Ethanol	Methane sulfonic acid
Isopropyl methane sulfonate	IMS	C_3H_7	CH_3	Isopropanol	Methane sulfonic acid
Methyl benzene sulfonate	MBS	CH_3	C_6H_5	Methanol	Benzene sulfonic acid
Ethyl benzene sulfonate	EBS	C_2H_5	C_6H_5	Ethanol	Benzene sulfonic acid
Isopropyl benzene sulfonate	IBS	C_3H_7	C_6H_5	Isopropanol	Benzene sulfonic acid
Methyl p-toluene sulfonate	MpTS	CH_3	C_7H_8	Methanol	p. Toluene sulfonic acid
Ethyl p-toluene sulfonate	EpTS	C_2H_5	C_7H_8	Ethanol	p. Toluene sulfonic acid
Isopropyl p-toluene sulfonate	IpTS	C_3H_7	C_7H_8	Isopropanol	p. Toluene sulfonic acid

or HPLC has been described [38–40], but is difficult and prone to artifacts such as possible formation of the MIs by decomposition of the API-salts (e.g. in the GC inlet = false positives), hydrolysis of the sulfonates (e.g. in aqueous media), and/or API interference.

Therefore, the use of a GC-MS method, involving derivatization was developed as a generic approach for sulfonate analysis. An elegant derivatization method was described by Alzaga et al. [41] using in-situ derivatization in combination with SHS-GC-MS. In this method, the sulfonates are derivatized (and thereby also stabilized) by reaction with pentafluorothiophenol (Figure 12.8b). The methyl-, ethyl-, or isopropyl- derivatives are stable and volatile and can be analyzed by SHS-GC-MS. Isotope labeled analytes, prepared in-house from deuterated alcohols reacting with sulphonic acids or sulphonyl chlorides, are used as internal standards. This method, including sampling from a reaction vial, derivatization, reaction quenching, and online SHS-GC-MS, was completely automated on a dual rail autosampler (Gerstel) and applied in a project from the Product Quality Research Institute (PQRI) subgroup to assess the likelihood of formation of ethyl methane sulfonic acid (EMS) from methane sulfonic acid and ethanol under various reaction conditions and to follow the reaction kinetics of the esterification reaction [42].

It should be noted that using this methodology, differentiation between sulfonates is only made based on differences in the R_1-group (Figure 12.8b). This means that no differentiation can be made between ethyl methane sulfonate (EMS), ethyl benzene sulfonate (EBS), and ethyl para-toluene sulfonate, since all three result in the same reaction product (ethyl-PFTP). Differentiation can only be made between the different mesylates: methyl methane sulfonate (MMS), EMS, and isopropyl methane sulfonate (IMS). In pharmaceutical QC, this is sufficient since in most cases it is known if mesylates, besylates, tosylates, or sulphates are potentially formed.

If confirmation analysis is required or speciation is needed, additional GC-MS or LC-MS analysis can be used.

Wollein and Schramek [38] described a GC-MS method for the determination of mesylates and besylates using liquid extraction with hexane. The method was, however, only applied to a lactose/microcrystalline cellulose matrix, and some mesylate salts of API, insoluble in hexane.

Direct injection of liquid extracts was also used in combination with GC-MS/MS [39] or LC-MS/MS [40]. For LC-MS/MS of the underivatized sulfonates, APCI ionization in negative ion mode was found to be superior to ESI in terms of ionization efficiency and sensitivities down to low ppb level were reported.

Another interesting approach was described by An et al. [43]. In that approach, the sulfonates were converted to the ammonium salts. The positively charged ions can be detected very well using ESI ionization and are suitable for LC separation using HILIC. The less polar APIs are not strongly retained under HILIC conditions, while the sulfonate-ammonium salts are well separated.

Also, alternative derivatization methods in combination with GC-MS/MS [44] or LC-MS/MS [45] were reported. These methods can be considered as complementary to the method used in our laboratory, applying derivatization in combination with SHS-GC-MS. Below, typically analytical conditions and results obtained using the derivatization SHS-GC-MS method are given. Also, an example is included demonstrating the use of programmed temperature vaporizing (PTV) injection in combination with GC-MS for direct sulfonate analysis.

12.5.2 Typical Conditions Used for Sulfonate Analysis by Derivatization SHS-GC-MS

The following conditions were applied in a generic method for the determination of sulfonates. Analyses were performed on a GC equipped with split/splitless inlet and combined with a single quadrupole MSD and a Gerstel MPS autosampler.

12.5.2.1 Sample Preparation

- API (50 mg) was dissolved in 4 ml of a dimethylacetamide/water or DMSO/water mixture (1:1, v/v) in a 20 ml HS vial. (DMAC gives a lower background than DMSO if sulfates should be monitored, DMSO gives slightly higher sensitivity.)
- Internal standard was added (typically 10 µl of a 5 ng/µl solution, see below). For spiking experiments, the sulfonate esters are also added from an acetone solution.
- 100 µl of the derivatization solution, containing 6.4 mg/ml pentafluorothiophenol in 1 M NaOH was added through the septum of the vial. Derivatization was performed for 15 minutes at 105 °C (headspace equilibration time and temperature).

12.5.2.2 Synthesis of Deuterated Internal Standards

- A deuterated internal standard was synthesized for each solute from the corresponding acid and deuterated alcohol (see Table 12.2).
- For each IS, 100 µl deuterated alcohol (e.g. d6-ethanol) and 0.188 mmol acid (e.g. methane sulfonic acid) were mixed in 5 ml reagent tubes, heated for two hours at 100 °C, then cooled down at room temperature and diluted in 5 ml acetone. These solutions were stored at 4 °C (stock solution). The exact concentration and purity of these solutions were measured using GC-FID and GC-MS against a non-deuterated standard.

12.5.2.3 GC-MS Parameters

- SHS: 15 minutes equilibration time, 105 °C equilibration temperature
- GC-MS
  - Injection: 1 ml, 1:10 split ratio, 250 °C
  - Column: 60 m × 0.25 mm i.d. × 1.4 µm d_f DB-VRX (Agilent)
  - Carrier gas: helium, constant flow (2.4 ml/minute)
  - Oven: 60 °C – 1 minute – 10 °C/minute – 200 °C – 30 °C/minute – 250 °C – 1.33 minutes
  - Detection: MS in SIM mode
    - Selected ions (SIM):
      10.50–13.27 minutes: 199, 200, 214, 217
      13.28–14.00 minutes: 200, 201, 228, 233
      14.00–18.00 minutes: 200, 201, 242, 249
    - Dwell times: 100 ms
    - Solvent delay: 10.5 minutes

Quantification was performed using the following ions:

- *m/z* 214 for methylthiopentafluorobenzene (Methyl-PFTP)
- *m/z* 217 for deuterated methylthiopentafluorobenzene (Methyl-PFTP-d_3)
- *m/z* 228 for ethylthiopentafluorobenzene (Ethyl-PFTP)
- *m/z* 233 for deuterated ethylthiopentafluorobenzene (Ethyl-PFTP-d_5)
- *m/z* 242 for isopropylthiopentafluorobenzene (Isopropyl-PFTP)
- *m/z* 249 for deuterated isopropylthiopentafluorobenzene (Isopropyl-PFTP-d_7)

12.5.3 Typical Results Obtained Using Derivatization – SHS – GC-MS

An example of a chromatogram obtained by the derivatization – SHS-GC-MS method for the analysis of ampicillin spiked at 1 ppm level with methyl methanesulfonate (MMS) and ethyl methanesulfonate (EMS) is shown in Figure 12.9. Detection was done in SIM mode at *m/z* 214 (MMS-derivative), 217 (d3-MMS-derivative), 228 (EMS-derivative), and 233 (d5-EMS-derivative). The target analytes and their respective internal standards can easily be detected.

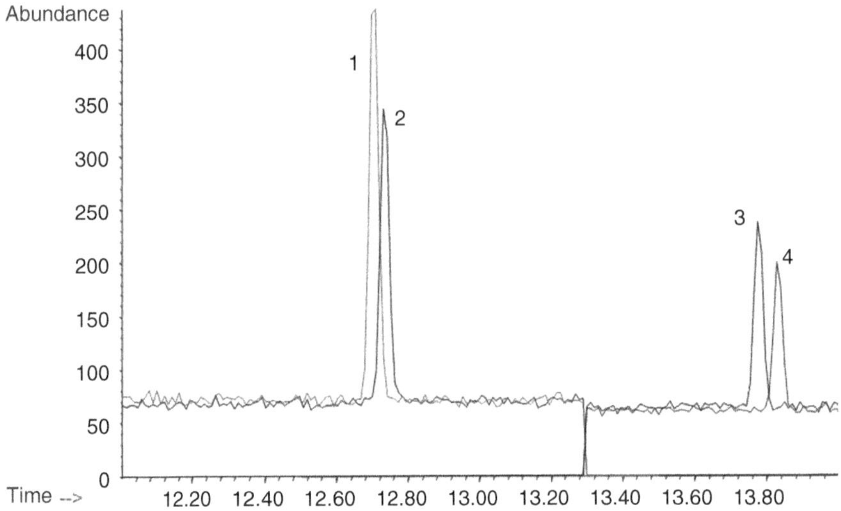

Figure 12.9 Analysis of sulfonate esters in API (ampicillin) by derivatization – SHS-GC-MS (SIM). Sample spiked at 1 μg/g. Derivatization with pentafluorothiophenol. Analytes: 1. d3-MMS, 2. MMS, 3. d5-EMS, 4. EMS.

The method was validated (Table 12.3). The linearity, measured in a concentration range from 0.2 to 10 ppm (adding 10, 25, 50, 100, 250, and 500 ng/50 mg API), was excellent. The LODs ($S/N = 3$) were below 0.5 ppm, except for IMS (isopropyl-PFTP derivative) which showed a slightly lower sensitivity (LOD = 3 ppm). RSDs were measured at 1 ppm level (at 5 ppm for IMS) and were below 10%.

The full automation of the method and the resulting repeatability allowed us to use this method to be applied for kinetic studies on the formation of sulphonate esters from alcohol/acid mixtures under different reaction conditions (pH, temperature, water content) [42].

12.5.4 Confirmation Analysis by PTV-GC-MS

As described above, the generic SHS-GC-MS method does not allow differentiating between mesylates, besylates, and tosylates. The derivatization reaction results in methyl-, ethyl-, or isopropyl pentafluorothiophenol derivatives.

In case of positive detection, a confirmation analysis might be required. Depending on the API and possible sulphonate, either GC or LC methods can be selected [37].

An example of a GC-MS approach that can be used is described below. Direct liquid injection of a concentrated API solution was done in a GC-MS system equipped with a PTV inlet and an automated liner exchange option (ALEX, Gerstel GmbH). This system allows automated exchange of the PTV inlet liner after a single or a few injections, avoiding cross contamination and reducing contamination of the analytical column and detector. The PTV exchangeable liner was filled with polydimethylsiloxane (PDMS) particles that retain the API and avoids the introduction of the drug product in the GC. As illustration, a 50 mg/ml solution of promethazine (API) was spiked with 0.5 μg (10 ppm) sulfonates (EMS, IMS, and EpTS) in chloroform. From this solution, 1 μl was injected in splitless mode. The PTV injector was heated from 20 to 220 °C at 12 °C/s. Separation was done on a 30 m×0.25 mm ID×0.25 μm HP-5MS column, using a temperature program from 40 °C (1 minute) at 8 °C/minute to 280 °C (2 minutes). Carrier gas was helium at a constant flow of 1.6 ml/minute. MS detection was done in scan/SIM mode.

The chromatogram is shown in Figure 12.10a. EMS, IMS, and EpTS are clearly detected. The relatively low end temperature of the PTV (220 °C) avoids the introduction of the bulk of the API,

Table 12.3 Validation results for sulfonate ester analysis by derivatization – SHS-GC-MS.

		LOD	Linearity	RSD (%)
Type	Solute	ppm (50 mg API)	r^2 (0.2–10 ppm)	(@ 1 ppm)
Sulfates	*DMS*	0.04	0.999	9
	DES	0.08	0.999	8
Mesylates	*MMS*	0.11	0.998	6
	EMS	0.26	0.999	9
	IMS	3.2	0.999	9
Besylates	*MBS*	0.06	0.998	3
Tosylates	*MpTS*	0.08	0.998	3
	EpTS	0.11	0.999	7

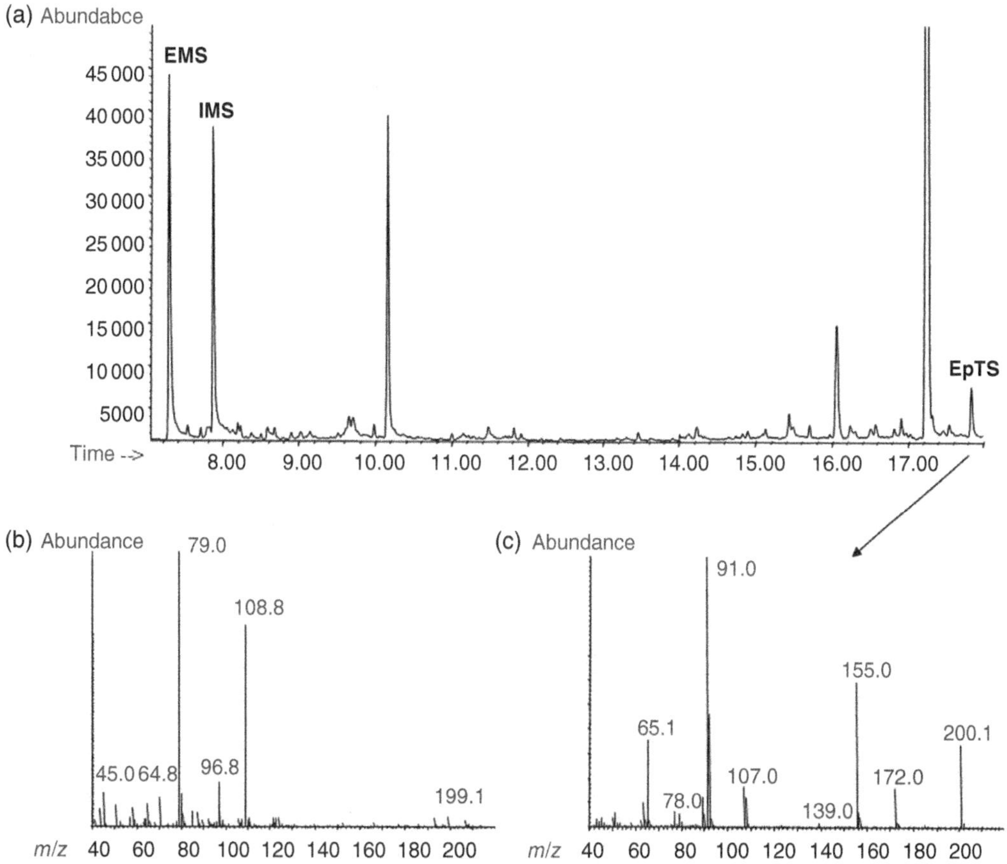

Figure 12.10 Analysis of sulfonate esters in API (promethazine) by direct injection-GC-MS (scan) using an automated liner exchange system. (a) TIC chromatogram, (b) mass spectrum of EMS, (c) mass spectrum of EpTS.

while the target analytes are quantitatively transferred in the column. As illustration, the obtained mass spectra for EMS and EpTS are given in Figure 12.10b and c, respectively.

With this method, the presence of sulfonates can be confirmed, and differentiation can be made between mesylates (EMS) and tosylates (EpTS).

12.6 *S*- and *N*-mustards

12.6.1 Method Selection

Mustards are β-halogenated dialkylsulfides (*S*-mustards) or β-halogenated amines (*N*-mustards). These compounds are also used as alkylating reagents in chemical synthesis. Typical representatives of this class of PMI/MIs are given in Table 12.4.

Some of these analytes, such as 2-chloroethyl methyl sulfide and 2-chloroethyl ethyl sulfide, are volatile and can be analyzed by SHS GC-MS using the method described for halides (Section 12.4).

N-mustards (or "mustard-like" analytes), with a primary or secondary amine functionality, on the other hand, do not have sufficient vapor pressure to be analyzed by headspace techniques. For these analytes, a method based on derivatization followed by SHS or SPME was developed. A derivatization reaction that can be performed in-situ in aqueous media was selected. The same method could also be applied in nonaqueous solvent systems.

Successful analysis of the target *N*-mustard analytes could be achieved by in-situ derivatization using ethyl chloroformate. This derivatization is also used for the GC analysis of amino acids. Primary amine-groups ($-NH_2$) are derivatized to carbamates ($-NH-CO-OEt$). The resulting compounds are more volatile and can be enriched on an SPME fiber. Details of this derivatization – SPME-GC-MS method are described below.

Alternatively, LC-MS(/MS) can be applied for the analysis of *N*-mustards. Shackman [46] described the determination of 2-chloroethylamine, together with aziridine, using HILIC – mass spectrometry. Separation was done on a Waters BEH HILIC column (50 mm × 2.1 mm, 1.7 μm particle size) using 0.1 M ammonium formate/acetonitrile (5/95) mobile phase and single quadrupole MS detection with ESI+ ionization and operated in SIM mode.

Derivatization in combination with LC-MS was applied by Sun et al. [10] for the analysis of bis(2-chloroethyl)amine. Acidified acetonitrile was used for matrix stabilization and the *N*-mustard was derivatized to the quaternary ammonium salt using dimethylamine. Analysis was performed by HILIC (50 mm × 2.1 mm × 3 μm Atlantis HILIC Silica) in combination with ESI-MS.

Table 12.4 Selected *S*- and *N*-mustard target analytes.

		CAS n°	RT[a] (min)	Quant ion	Qual ion
S-mustards					
2-Chloroethyl methyl sulfide	ClEtSMe	542-81-4	12.10	61	110, 112
2-Chloroethyl ethyl sulfide	ClEtSEt	693-07-2	13.50	75	124, 126
2-Chloroethyl phenyl sulfide	ClEtSPhe	6276-54-6	22.28	123	172, 174
N-mustards					
2,2,2-Trifluoroethylamine[a]	CF_3EtNH_2	870-24-6	11.05	144	106, 126
2-Chloroethylamine[a]	$ClEtNH_2$	821-48-7	17.27	102	63
3-Chloropropylamine (IS)[a]	$ClPropNH_2$	5535-49-9	19.71	102	165
bis-(2-Chloroethyl)amine[a]	Bis ClEtNH	753-90-2	22.42	164	56, 92

[a] As derivatives.
RT = Retention time (minutes) obtained on DB-VRX column.

12.6.2 Typical Analytical Conditions for the Analysis of *N*-mustards by Derivatization – SPME-GC-MS

The *S*-mustards 2-chloroethyl methyl sulfide, 2-chloroethyl ethyl sulfide, and 2-chloroethyl phenyl sulfide can be analyzed using the SHS – GC-MS method described for alkyl- and aryl halides (retention times: see Table 12.4). The LODs were lower than 1 ppm for a 50 mg API in 2 ml solvent solution.

The following conditions were applied for the determination of *N*-mustards using derivatization – SPME-GC-MS. Analyses were performed on a GC equipped with split/splitless inlet and combined with a single quadrupole MSD and a Gerstel MPS autosampler.

12.6.2.1 Sample Preparation

- The API (100 mg) was dissolved in 2 ml of a solvent mixture containing water/ethanol/pyridine (2:1:1, v/v). Depending on the API solubility, the order of solvent addition can be changed.
- Internal standard (for instance: 3-chloropropylamine, as hydrochloride) was added.
- 50 µl ethyl chloroformate was added and the derivatization is performed at room temperature during 15 minutes in an ultrasonic bath.

12.6.2.1.1 GC-MS Conditions

- SPME: PDMS fiber (100 µm), 2 minutes incubation and 10 minutes extraction time at 80 °C
- Injection: split/splitless inlet in split (1/10) mode, 250 °C
- Column: 60 m × 0.25 mm i.d. × 1.4 µm d_f DB-VRX (Agilent Technologies)
- Carrier gas: helium, constant pressure (235.9 kPa))
- Oven: 50 °C – 1 minute – 10 °C/minute – 250 °C (4 minutes); run time: 20 minutes
- Detection: MS in simultaneous SIM/SCAN mode
  - Scan range: 29–350 m/z
  - Selected ions (SIM): see Table 12.4
  - Solvent delay: 3.5 minutes

12.6.3 Typical Results for *N*-mustards by Derivatization – SPME-GC-MS

The chromatogram (EIC of SIM) obtained for an API sample (doxylamine) spiked at 0.5 ppm level with *N*-mustard target solutes is shown in Figure 12.11. The derivatives of trifluoro-ethylamine (CF_3EtNH_2) (ion 144), 2-chloroethylamine ($Cl\text{-}EtNH_2$) (ion 102), 3-chloropropylamine (IS) (ion 102), and bis(2-chloroethyl)amine (bisClEtNH) (ion 164) can be detected. The linearity was tested in the range from 0.5 to 10 ppm and the correlation coefficients were in the order of 0.995–0.999. The RSDs measured for APIs spiked at 1 ppm level were in the order of 13–15%. The LODs, measured using the lowest spike level and corresponding to a signal-to-noise of 3, were 0.2 ppm for trifluoro-ethylamine (as derivative) and 2-chloroethylamine (as derivative), 0.1 ppm for 3-chloropropylamine (IS, as derivative), and 0.5 ppm for bis(2-chloroethyl)amine (as derivative). The response for this later compound is lower. In order to obtain higher sensitivity for this solute, derivatization followed by direct injection can be considered, as could the employment of alternative detection techniques, e.g. triple quadrupole MS.

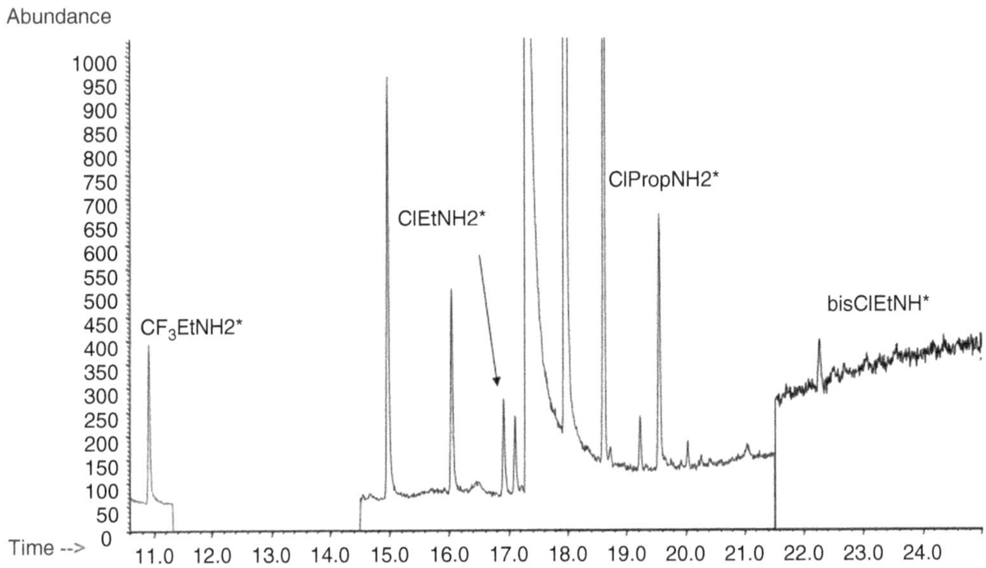

Figure 12.11 Analysis of *N*-mustards in API (doxylamine) by derivatization – SPME-GC-MS. Sample spiked at 0.5 μg/g. Derivatization with ethyl chloroformate. Analytes: acyl-derivatives of trifluoro-ethylamine (CF_3EtNH_2), 2-chloroethylamine (Cl-$EtNH_2$), 3-chloropropylamine (IS), and bis(2-chloroethyl)amine (bisCl$EtNH_2$).

12.7 Michael Reaction Acceptors

12.7.1 Method Selection

Michael reaction acceptors are reagents that have an α,β-unsaturation next to an electron-withdrawing functionality (such as a carbonyl or nitrile). Representative target analytes of this class of PMI/MIs are listed in Table 12.5. SHS – GC-MS was initially tested. Several solutes, including acrylonitrile, methacrolein, methyl acrylate, and ethyl acrylate, could be measured in API matrices at very low concentration using the generic method that is also applicable to organohalides (Section 12.4). Less volatile Michael reactive acceptors were not detected by SHS-GC-MS and were analyzed using direct liquid injection of an API solution or extract. Since a concentrated API solution is injected in the GC-MS, this can potentially contaminate the system and therefore back-flushing or Deans switching can be applied. The principle of back-flushing and its application to the analysis of nonvolatile Michael reaction acceptors are described in this section. The application of Deans switching is discussed in detail in Section 12.9.

For the analysis of Michael reaction acceptors, also LC methods are described. Some Michael reaction acceptors are strong UV absorbers and can be detected at trace levels using UV/DAD. For others, MS or MS/MS are used. One interesting approach was the use of glutathione as selective derivatization reagent for the determination of 2-pyranone [47].

12.7.2 Typical Analytical Conditions for Michael Reaction Acceptors

The following conditions were applied for the determination of Michael reaction acceptors using SHS-GC-MS or liquid injection GC-MS. Analyses were performed on a GC equipped with a PTV inlet, which was used for both headspace gas injection and liquid injection. For each mode, a dedicated liner

Table 12.5 Selected Michael reaction acceptor target analytes.

Peak	Solute	CAS Nr	Method	RT (min)	Target ion	Qualifiers ions
1	Acrylonitrile	107-13-1	SHS	6.19	53	52
2	Methacrolein	78-85-3	SHS	6.99	70	39, 41
3	Methyl acrylate	96-33-3	SHS	7.85	55	85
4	2-Chloroacrylonitrile	920-37-6	SHS	7.96	52	87
5	Crotononitrile (Z)	4786-20-3	SHS	8.70	67	39, 41
6	Ethyl acrylate	140-88-5	SHS	9.73	55	99
7	cis-2-Pentenenitrile	25899-50-7	SHS	10.75	54	81
8	4-Methyl-3-pentene-2-one	141-79-7	SHS	12.25	83	55, 98
9	Methyl tiglate	6622-76-0	SHS	13.57	83	55, 114
10	2-Cyclohexen-1-one	930-68-7	LIQ	7.44	68	96
11	Methyl-3-aminocrotonate	14205-39-1	LIQ	8.83	84	42, 115
12	Cinnamonitrile	1885-38-7	LIQ	13.07	129	102
13	3-Ethoxy-2-cyclohexenone	5323-87-5	LIQ	13.13	68	140

Methods: SHS: static headspace, LIQ: liquid injection.

was used. Sampling introduction was done with a Gerstel MPS autosampler with SHS sampling and liquid injection options, and detection was done using a single quadrupole MSD.

12.7.2.1 Sample Preparation

- For SHS analysis, the API (50 mg) was dissolved in 2 ml of a solvent mixture containing water/DMSO (1:1, v/v) in a 20 ml headspace vial. Depending on the API solubility, the order of solvent addition can be changed. The target analytes were dissolved in acetone for spiking in the API solution. Typically, 5 μl of a 10 ng/μl solution was added to obtain a 1 ppm spike level in the API.
- For the nonvolatile analytes, a 50 mg/ml solution was made in dichloromethane or pyridine). Injection of 1 μl in splitless mode corresponds to the introduction of 50 μg API and a 1 ppm PMI/MI level corresponds to the injection of 50 picogram PMI/MI on-column.

12.7.2.2 Parameters for SHS-GC-MS

- SHS equilibration was performed at 80 °C during 15 minutes. 1 ml headspace was injected in split mode (split ratio 1/10) in a PTV inlet with a Tenax TA packed liner.
- PTV inlet temperature program: 20 °C (0.1 minute) – 720 °C/minute – 250 °C, CO_2 cooling
  GC-MS conditions:
  - Column: 60 m × 0.25 mm i.d. × 1.4 μm d_f DB-VRX (Agilent Technologies)
  - Carrier gas: helium, constant pressure (200 kPa, with QuickSwap at 28 kPa)
  - Oven: 60 °C (0.5 minute) – 8 °C/minute – 100 °C (3 minutes) – 30 °C/minute – 250 °C (7.5 minutes)
  - Detection: MS in simultaneous SIM/SCAN mode
    - Scan range: 29–350 m/z
    - Selected ions (SIM): see Table 12.5, dwell times: 100 ms
    - Solvent delay: 3.5 minutes

12.7.2.3 Parameters for Liquid Injection and GC-MS with Back-flush

- Injection was performed in a PTV inlet with an empty, baffled liner. 1 μl was injected in splitless mode.
- PTV inlet temperature program: 60 °C (0.1 minute) – 720 °C/minute – 250 °C
- GC-MS conditions:
  - Column: 20 m × 0.18 mm i.d. × 1 μm d_f DB-VRX (Agilent Technologies)
  - Carrier gas: helium, constant flow 1.0 ml/minute (with post-column CFT at 28 kPa)
  - Oven: 60 °C (1 minute) – 10 °C/minute – 250 °C (10 minutes)
  - Detection: MS in simultaneous SIM/SCAN mode
    - Scan range: 45–350 *m/z*
    - Selected ions (SIM): see Table 12.5, dwell times: 100 ms
    - Solvent delay: 5 minutes
  - Post-run back-flush at 20 minutes:
    - MS is switched off
    - Inlet pressure: 1 kPa
    - Pressure at post-column CFT: 300 kPa
    - Back-flush time: 10 minutes

12.7.3 Typical Results Obtained for Trace Analysis of Michael Reaction Acceptors

12.7.3.1 SHS with PTV

SHS was first used in combination with a classical split/splitless inlet. For the selected solutes, better results were obtained using a PTV inlet at 20 °C (and using a Tenax packed liner). This injection technique resulted in peak focusing before introduction in the column. Good peak shapes were obtained for the volatile solutes.

The SHS-GC-MS analysis is illustrated by the analysis of a promethazine sample spiked with 1 ppm of the target analytes. The obtained chromatograms (extracted ion chromatograms (EIC) from SIM acquisition mode) are shown in Figure 12.12. Solutes 1–9 in Table 12.5 can easily be detected (compound 5 not shown).

For these volatile solutes, the method was validated and excellent linearity ($r^2 > 0.99$ in the range from 0.2 to 5 ppm), repeatability (RSD < 5% at 1 ppm), and sensitivity (LOD < 0.1 ppm) were obtained as summarized in Table 12.6.

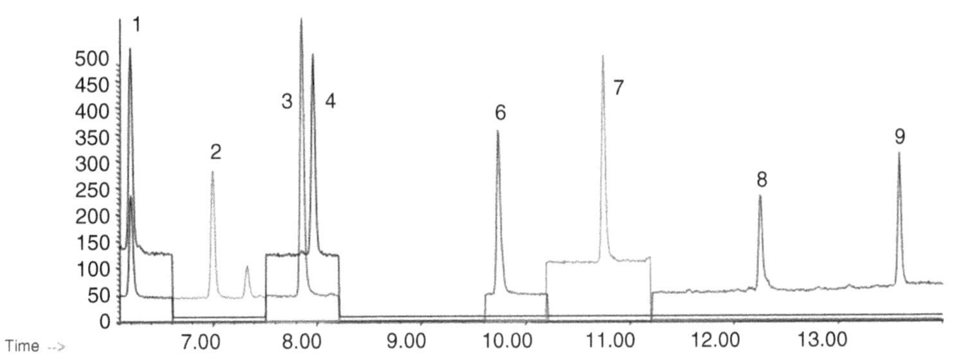

Figure 12.12 Analysis of Michael reaction acceptors in promethazine spiked at 1 ppm level by SHS-GC-MS (SIM). Peaks and extracted ions: see Tables 12.5 and 12.6.

Table 12.6 Validation results for Michael reaction acceptor analysis by SHS-GC-MS and liquid injection GC-MS.

Peak nr	Solute	Method	RSD (%)	r^2	LOD
1	Acrylonitrile	SHS	3.6	0.999	0.02
2	Methacrolein	SHS	2.9	0.997	0.02
3	Methyl acrylate	SHS	1.4	0.996	0.02
4	2-Chloroacrylonitrile	SHS	3.9	0.994	0.02
5	Crotononitrile (Z)	SHS	3.4	0.994	0.04
6	Ethyl acrylate	SHS	3.4	0.996	0.02
7	cis-2-Pentenenitrile	SHS	3.7	0.999	0.02
8	4-Methyl-3-pentene-2-one	SHS	2.5	0.994	0.05
9	Methyl tiglate	SHS	3.2	0.993	0.05
10	2-Cyclohexen-1-one	LIQ	7.6	0.995	0.02
11	Methyl-3-aminocrotonate	LIQ	3.7	0.974	0.30
12	Cinnamonitrile	LIQ	12.5	0.997	0.03
	Cinnamonitrile	*Liq – 2D*	*1.6*	*nm*	*0.01*
13	3-Ethoxy-2-cyclohexenone	LIQ	8.2	0.999	0.30
	3-Ethoxy-2-cyclohexenone	*Liq – 2D*	*1.5*	*nm*	*0.05*

nm: not measured; SHS: static headspace GC-MS method; Liq: liquid injection of 5% solution, GC-MS and back-flush; Liq – 2D: liquid injection of 5% solution, two-dimensional GC-MS (see Sections 12.8 and 12.9).

12.7.3.2 Liquid Injection GC-MS

The less volatile target analytes (2-cyclohexen-1-one, methyl-3-aminocrotonate, 3-ethoxy-2-cyclohexenone, and cinnamonitrile, compounds 10–13 in Table 12.5) are also amenable to GC but are not volatile enough for headspace analysis. For these compounds, direct injection of a concentrated (5%) solution of the API was used. Hereby, the API and/or its impurities and degradation compounds are also introduced in the GC column. To eliminate these, column back-flush was used. To this, the GC was equipped with a post-column back-flush option. This consisted of a CFT Tee-piece at which the pressure at the outlet of the analytical column can be controlled. Connection to the MS is done with a piece of deactivated fused silica.

The principle of column back-flush is illustrated in Figure 12.13. In normal GC mode, a high pressure on the inlet and a low pressure at the post-column CFT results in a forward capillary column flow rate of 1 ml/minute. Typically, the outlet pressure is set at 4 psi (= 28 kPa), which results, in combination with a 17 cm × 110 μm capillary in the transfer line at 260 °C, with a flow rate of about 2 ml/minute to the MS.

After elution of the solutes of interest, the inlet pressure is decreased to a minimum, just enough to flush the inlet, and the pressure at the outlet (transfer line) is raised. The MS is switched off (too high flow). In this way, the column outlet pressure is higher than the inlet pressure and the capillary column flow is reversed. Higher molecular weight solutes (such as the API) are back-flushed and exit the GC via the split vent. Back-flush time needs to be optimized according to final GC oven temperature and column choice [48].

A direct injection GC-MS method with this back-flush option was applied for the less volatile Michael reaction acceptors. A sample of carbamazepine spiked at 1 ppm level with the 4 solutes

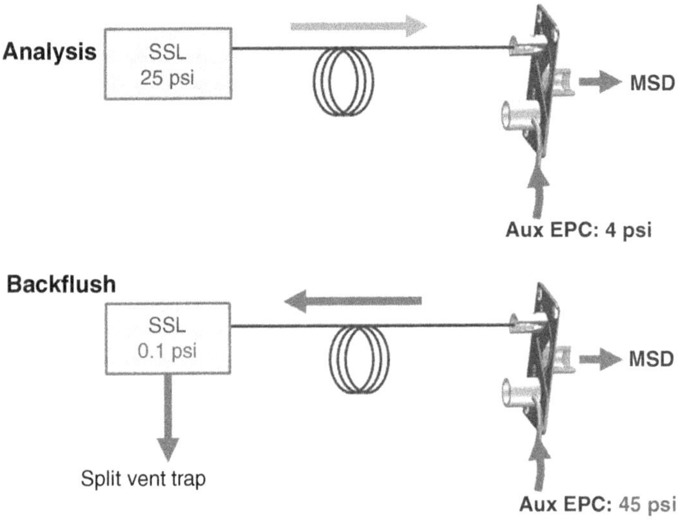

Figure 12.13 Principle of capillary column back-flush. *Source:* Agilent Technologies

(compounds 10–13, Table 12.5) was analyzed. The chromatogram obtained in scan mode without back-flush is shown in Figure 12.14a. The solutes could not be detected (present at 50 pg level), but the peaks for carbamazepine and its degradation compound (iminostilbene) overload the chromatogram. Next, the same analysis was performed with a back-flush initiated at 20 minutes (Figure 12.14b). A blank run (solvent injection) was performed afterward. API (carbamazepine) and iminostilbene were completely removed, only some minor peaks (solvent impurities) were detected in the blank run (Figure 12.14c).

The EIC from the SIM acquisition are shown in Figure 12.14d. The target solutes (2-cyclohexen-1-one, methyl-3-aminocrotonate, 3-ethoxy-2-cyclohexenone, and cinnamonitrile) can easily be detected. Under these GC conditions, cinnamonitrile and 3-ethoxy-2-cyclohexanone co-elute, but they can be deconvoluted based on their mass spectrum and selective ions. The method was validated, and the results are included in Table 12.6. RSDs at 1 ppm level were between 5 and 15% and LODs were lower than 0.5 ppm.

Target analytes 12 and 13 were also analyzed by using two-dimensional heart-cutting GC-MS [49]. Details on the analytical method are included in Section 12.9. The validation results obtained by 2D-GC-MS are included in Table 12.6 and show that repeatability, measured on a carbamazepine sample spiked at 1 ppm level, was better than 2% ($n = 6$) and that the LOD was lower than 0.1 ppm. This better sensitivity is mainly due to the removal of background by the two-dimensional approach. Moreover, the 2D-GC approach also resulted in complete separation of 3-ethoxy-2-cyclohexenone and cinnamonitrile (co-eluting in Figure 12.14) [49].

12.8 Epoxides

12.8.1 Method Selection

Epoxides form a very heterogeneous group of PMI/MIs and include both volatile and nonvolatile, polar and apolar solutes. Methods for trace analysis of epoxides and hydroperoxides in API, drug products, and in herbal products were reviewed by Elder et al. [50]. Most of the referenced papers

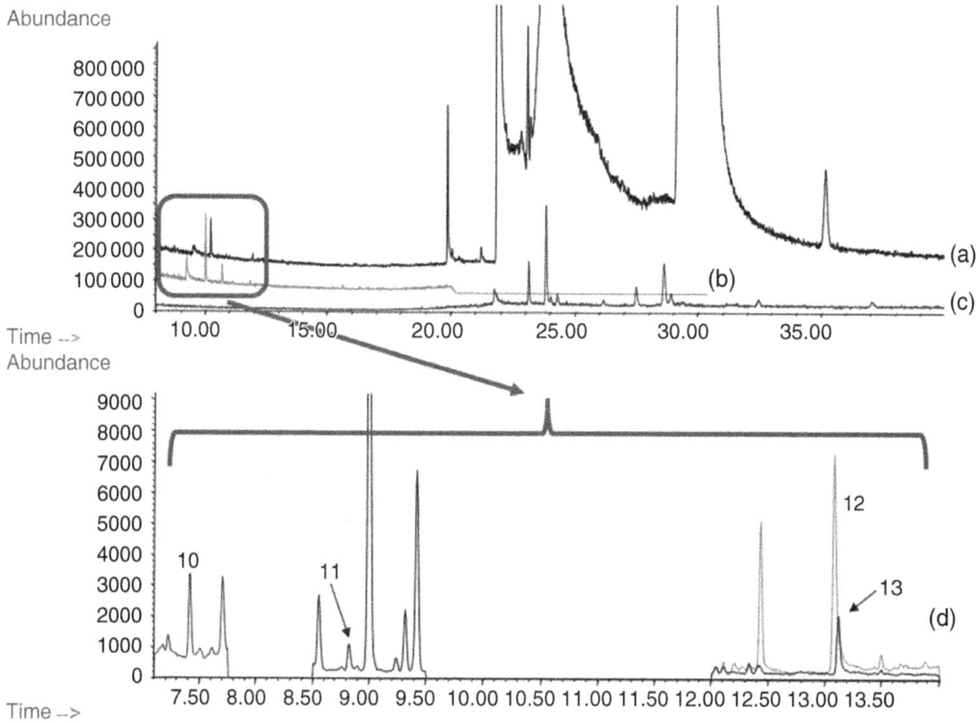

Figure 12.14 Analysis of carbamazepine using liquid injection GC-MS with back-flush. (a) No back-flush, (b) Back-flush at 20 minutes, (c) Blank run after back-flush, (d) EICs for Michael reaction acceptors spiked at 5 ppm level in carbamazepine (with back-flush). Peaks and extracted ions: see Tables 12.5 and 12.6.

included specific methods based on HPLC, and only few generic HPLC or GC methods were found. This can be explained by the fact that several applications described the analysis of Chinese herbal medicines, potentially containing epoxides or hydroperoxides as oxidation products of natural constituents and for these high molecular weight solutes, HPLC is the method of choice.

Short chain aliphatic epoxides, such as ethylene oxide, propylene oxide, epoxy butane, styrene oxide, etc., on the other hand, are commonly used in synthetic processes and these epoxides can be analyzed by GC. In the analysis of organic molecules containing an epoxide function, special attention should, however, be paid to their thermal stability and reactivity. This is the biggest challenge in epoxide analysis. If a GC method is applied, the selection of the GC inlet and inlet parameters should be optimized to avoid degradation of solutes in the inlet [51–53]. Klick [51, 52] demonstrated that hot splitless injection resulted in substantial solute degradation in the GC inlet, while much better results were obtained by cool on-column injection. For sample extracts also containing nonvolatile material, cool on-column injection is not an option, and PTV injection in cold split or splitless mode offers an interesting alternative. More recently, interesting LC-MS approaches are described. Bai et al. [6] applied a derivatization reaction with dimethylamine to obtain a 2-hydroxy-dimethylamine derivative from a volatile aryl epoxide. The derivative is polar and highly ionizable with ESI, resulting in excellent sensitivity.

In our project, the same approach was used for trace analysis of epoxides as for the Michael reaction acceptors. First, SHS – GC-MS was tested for the representative epoxide solutes listed in Table 12.7. For the analytes that could not be detected, liquid injection – GC-MS in combination with back-flush or Deans switch 2D-GC was applied.

Table 12.7 Selected epoxide target analytes.

Name	CAS n°	RT (min)	Ions	Method
2-Methyloxirane = propylene oxide	75-56-9	6.19	43, 58	SHS
1,2-Epoxybutane = ethyl oxirane	106-88-7	7.87	71, 41, 57	SHS
2,3-Epoxy-2-methylbutane	5076-19-7	8.44	58, 41, 43	SHS
2-Propyloxirane = 1,2-epoxypentane	1003-14-1	10.03	71, 41	SHS
2-Oxiranylmethanol = glycidol	556-52-5	10.23	44, 43	**LIQ**
Epichlorohydrin	106-89-8	10.44	57, 27	SHS
Glycidyl isopropyl ether	4016-14-2	13.42	43, 59	SHS
Methyl 2-methylglycidate	58653-97-7	13.52	43, 57, 84	SHS
Cyclohexene oxide	286-20-4	13.96	83, 41, 54	SHS
Exo-2,3-epoxynorbornane	3146-39-2	15.68	81, 39, 54	SHS
Glycidyl acrylate	106-90-1	15.92	55, 27	SHS
2-(4-Fluorophenyl)oxirane	18511-62-1	17.33	137, 119	SHS
Styrene oxide	96-09-3	17.37	119, 91	SHS
1-Phenylpropylene oxide (S,S)	4518-66-5	17.97	90, 105, 133	SHS
(2,3-Epoxypropyl)benzene	4436-24-2	18.52	91, 134, 105	SHS
Ethyl 3-methyl-3-phenylglycidate I	77-83-8	21.62	132, 104, 205	**LIQ**
Ethyl 3-methyl-3-phenylglycidate II	77-83-8	23.05	132, 104, 205	**LIQ**
2-(4-Nitrophenyl)oxirane (R)	78038-43-4	23.40	89, 118, 148	**LIQ**

SHS: static headspace GC-MS method; Liq: liquid injection of 5% solution, GC-MS and back-flush (see Section 12.7) or two-dimensional GC-MS (see Section 12.9).

12.8.2 Typical Analytical Conditions for the Analysis of Volatile Epoxides by SHS-GC-MS

Volatile epoxides were analyzed on an Agilent 6890GC in combination with a 5975MSD. The GC was equipped with a Gerstel PTV inlet and automated injection was performed using a Gerstel MPS autosampler.

12.8.2.1 Sample Preparation

- The API (50 mg) was dissolved in 2 ml of a solvent mixture containing water/DMSO, (1:1, v/v) in a 20 ml headspace vial. Depending on the API solubility, the order of solvent addition can be changed. The target analytes were dissolved in acetone for spiking in the API solution. Typically, 5 μl of a 10 ng/μl solution was added to obtain a 1 ppm spike level in the API.

12.8.2.2 SHS-GC-MS Parameters

- SHS was performed at 80 °C during 15 minutes. 1 ml headspace was injected in split mode (split ratio 1/10) in a PTV inlet with a Tenax TA packed liner.
- PTV inlet temperature program: 50 °C (0.1 minute) – 720 °C/minute – 250 °C, CO_2 cooling
- GC-MS conditions:
  - Column: 60 m × 0.25 mm i.d. × 1.4 μm D_f DB-VRX (Agilent Technologies)
  - Carrier gas: helium, 1.2 ml/minute constant flow (with Quick Swap at 28 kPa)

- Oven: 50 °C (0.1 minute) – 8 °C/minute – 150 °C – 25 °C/minute – 250 °C (7.5 minutes)
- Detection: MS in simultaneous SIM/SCAN mode
  - Scan range: 27–350 m/z
  - Selected ions (SIM): see Table 12.7, dwell times: 100 ms
  - Solvent delay: 3.5 minutes

If the solutes could not be detected with the SHS method (at levels below 10 ppm), they were analyzed by a direct liquid injection method, as described in Sections 12.7 and 12.9. Trace analysis of glycidol using derivatization and liquid injection 2DGC-MS is included in Section 12.9.

12.8.3 Typical Results Obtained for Volatile Epoxides Using SHS-GC-MS

As with Michael reaction acceptors, better peak shape and sensitivity were obtained using a PTV inlet with Tenax packed liner for the SHS injection. Optimum initial PTV temperature was 50 °C.

Most of the solutes listed in Table 12.7 could be detected at LODs below 1 ppm using the SHS-GC-MS method. The retention times obtained using the above mentioned conditions are included in the table. For these solutes, the SHS-GC-MS method could also be validated.

Three compounds of this list were not detected: glycidol, ethyl-3-methyl-3-phenylglycidate (both isomers), and 2-(4-nitrophenyl)oxirane. Although these solutes are GC amenable, as indicated by their retention times measured by liquid injection, their vapor pressure is too low for SHS analysis. Ethyl-3-methyl-3-phenylglycidate (both isomers) and 2-(4-nitrophenyl)oxirane should either be analyzed by liquid injection and GC-MS, eventually applying back-flush or 2D-GC or by LC-MS. Glycidol, on the other hand, elutes relatively early, but the presence of the hydroxyl-function makes this analyte less volatile for headspace analysis and more difficult to analyze by GC. This compound could be analyzed together with haloalcohols (see Section 12.9).

12.9 Haloalcohols

12.9.1 Method Selection

During the method development for alkyl halides (Section 12.4), it was observed that some more polar halogenated compounds, such as 2-iodoethanol, could not be extracted from an API solution using headspace techniques and/or that poor peak shape and low repeatability and linearity were obtained (see compound 19 in Figures 12.5 and 12.6). Also, analysis of other haloalcohols using SHS-GC-MS and SPME-GC-MS was evaluated, but due to their low vapor pressure and high polarity, these compounds could not be measured at trace levels in API solution using these headspace techniques. A derivatization method that could be performed in-situ (in a concentrated API solution) and combined with headspace analysis (SHS or HS-SPME) was not readily available.

Haloalcohols, such as chloro-, bromo-, iodo-ethanol, -propanol, or -butanol, are mostly GC amenable, and analysis by GC using liquid injection could thus be considered. For trace level analysis, however, better repeatability, sensitivity, and robustness, are obtained if the hydroxyl-function is derivatized. This is easily done by silylation (formation of trimethylsilyl-ethers), but silylation reactions are normally performed in nonaqueous media and removal of residual reagent (or derivatization by-products) can be difficult. Silylation followed by direct liquid injection and GC-MS analysis was found to work well for different haloalcohols, but, since the derivatized solutes are quite volatile, interferences from the residual reagent and derivatization by-products were observed. Moreover, in contrast to HS techniques, a complex mixture is introduced in the analytical system containing derivatized solutes, excess reagent, reaction by-products, and (derivatized) API or more abundant API impurities (or degradation compounds). Solvent(s), reagent(s), and matrices can thus interfere with PMI/MI determination and, moreover, contaminate the analytical system (GC column, MS detector). For this reason, heart-cutting two-dimensional GC (2D-GC-MS) was evaluated as an alternative [49]. The analytical setup for this approach is shown in Figure 12.15. A concentrated solution of the API is injected (in split/splitless inlet) on an apolar first dimension column (column 1). The fraction containing the solvent and excess derivatization reagent is diverted to a monitor FID detector by a makeup flow added after the first column. The fraction containing the PMI/MI analytes is heart-cut to a second dimension column (column 2), placed in a separate low thermal mass column oven module (LTM, Agilent Technologies). After the heart-cut, the remaining fraction eluting from the first column is also diverted to the monitor FID or can be back-flushed (see Section 12.7). The target analytes are further separated from other potentially interfering impurities on the second dimension column, and detected by MS (scan/SIM mode). The main analytes (API, derivatization reagents, solvent, etc.) are not introduced in the second column, avoiding overloading and contamination of this column and of the MS source. The use of the low thermal mass oven enables an independent

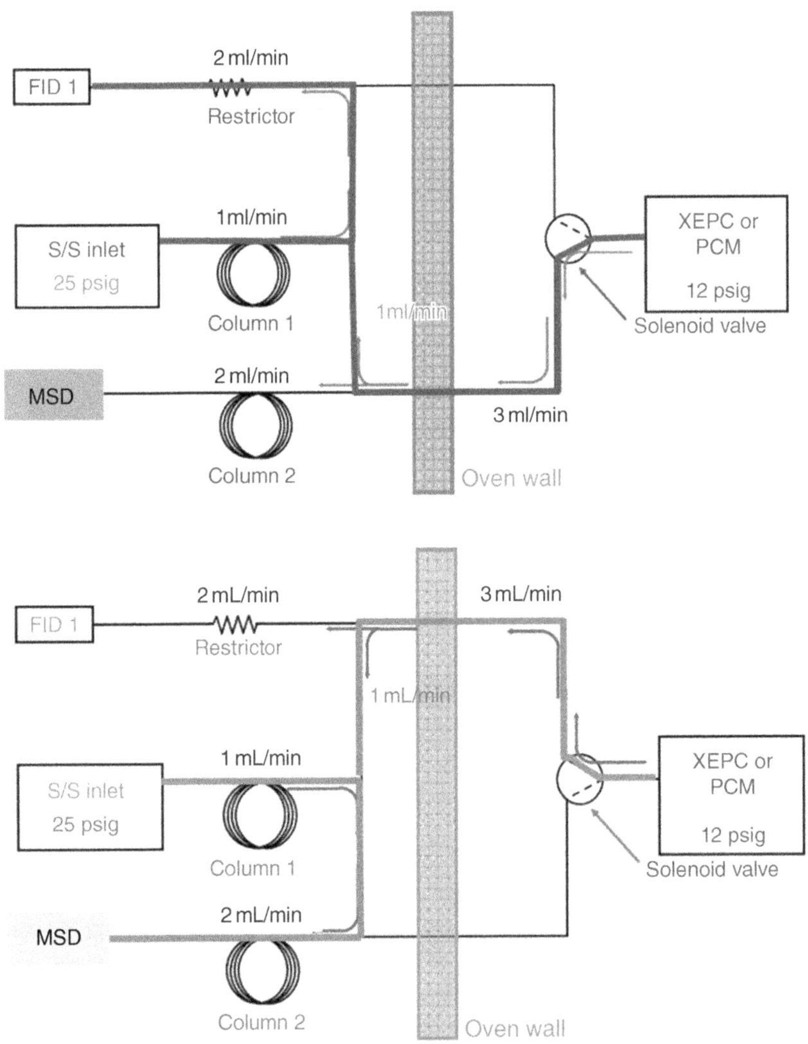

Figure 12.15 Deans switch 2D-GC-MS setup for the analysis of PMI/MI in API by direct injection. (a) column 1 effluent diverted to monitor FID, (b) heart-cut fraction to column 2. *Source:* Agilent Technologies, USA

temperature control of the second dimension column and more flexibility in method optimization. If not available, a configuration with both columns in the GC oven and with a cold trap at the inlet of the second column can be used as alternative [54].

A 2DGC-MS method was applied to several halo-alcohol analytes (2-chloroethanol, 2-bromoethanol, 2-iodoethanol, 4-chloro-1-butanol, 2-(2-chloroethoxy)ethanol, and 11-bromo-1-undecanol) and to glycidol, an epoxide that could not be analyzed by the methods described in Section 12.8. The same method can also be applied for less volatile epoxides and Michael reaction acceptors [49]. Method details are described below. The selected target analytes are given in Table 12.8.

Alternative methods for the analysis of haloalcohols include derivatization combined with LC-MS/MS [6] and LC-ICP-MS [29].

Analytical conditions for trace analysis of haloalcohols by derivatization and liquid injection – 2DGC-MS

The analysis of haloalcohols, after derivatization by silylation, was performed on an Agilent 7890 GC equipped with a split/splitless inlet, a FID detector (monitor detector), a CFT Deans switch (with an auxiliary EPC), and a LTM oven (for temperature control of the second dimension column). Detection was done using an Agilent 5977 MSD. Automated sample preparation and injection was performed using a Gerstel MPS autosampler.

12.9.2 Analytical Conditions for Trace Analysis of Halo-alcohols by Derivatization and Liquid Injection - 2DGC-MS

12.9.2.1 Sample Preparation

- The API was dissolved in dry pyridine at 50 mg/ml. From this solution, 100 μl (=5 mg API) was placed in a 2 ml vial and spiked with 5 μl of a 1 ng/μl PMI/MI test mixture (corresponding to 1 ppm PMI/MI in API).
- Then, 100 μl BSTFA was added. The vial was heated at 70 °C for 30 minutes.
- After cooling, 500 μl dichloromethane and 500 μl water were added. The mixture was vortexed and injection (1 μl in splitless mode) was performed from the lower organic layer. This sample preparation was fully automated on an MPS2 autosampler.

Table 12.8 Figures of merit for 2D-GC-MS analysis of haloalcohols and glycidol.

			RSD (%)			LOD[c]
Haloalcohols[a]	Monitored ions	r^{2b}	1 ppm	5 ppm	10 ppm	ppm
Glycidol	59, 101	0.999	5.4	4.8	7.0	0.34
2-Chloroethanol	93, 137, 73	0.998	7.5	8.2	9.1	0.15
2-Bromoethanol	137, 139, 181, 183	0.999	2.5	3.1	3.0	0.17
2-Iodoethanol	185, 229	1.000	1.2	2.6	3.2	0.06
4-Chloro-1-butanol	93, 123, 165	1.000	2.2	2.4	2.8	0.10
2-(2-Chloroethoxy)ethanol	73, 93, 137, 181	0.998	9.8	7.1	1.5	0.16
11-Bromo-1-undecanol	97, 169, 83, 75	0.999	13.0	10.6	8.2	0.38

[a] As TMS-derivatives.
[b] r^2 from 6 replicates at 1, 5, and 10 ppm.
[c] LOD in ppm calculated for $S/N = 3$.

12.9.2.2 2D-GC-MS Parameters

- Injection: 1 μl in splitless mode, 250 °C
- Column 1: 30 m × 0.25 mm i.d. × 0.25 μm D_f HP-5MS (in GC oven)
- Column 2: 30 m × 0.25 mm i.d. × 0.25 μm D_f DB-17 (in LTM oven)
- Carrier gas: helium, 170 kPa at inlet (flow column 1 = 1 ml/minute at 50 °C), 120 kPa at Deans switch (flow column 2 = 2 ml/minute at 50 °C)
- GC oven temperature: 50 °C (1 minute) – 20 °C/minute – 300 °C (20 minutes)
- LTM oven temperature: 50 °C (11 minutes) – 10 °C/minute – 140 °C (4 minutes) – 25 °C/minute – 300 °C (3.1 minutes)
- Monitor detector: FID, 300 °C, 30 ml/minute H_2, 400 l/minute air
  Main detection: MS in simultaneous SIM/SCAN mode
  - Scan range: 45–350 m/z
  - Selected ions (SIM) (100 ms dwell times), see Table 12.8

12.9.3 Typical Results for Analysis of Halo-alcohols by Derivatization and Liquid Injection - 2DGC-MS

The application of Deans switching is illustrated by the analysis of selected haloalcohols in carbamazepine. In first instance, the first dimension retention times of the derivatized target compounds are determined by injection of a concentrated solution (1–10 ng/μl) using the system in standby mode (Figure 12.15a). Heart-cut windows are determined for each solute and are typically set at retention time ±0.1 minute. For closely eluting target compounds, a single larger heart-cut window can be used.

Next, the derivatized sample is analyzed using a method with time programmed heart-cut. The GC-FID profile obtained for a sample of carbamazepine spiked at 1 ppm level with the selected target compounds after the first separation on the apolar HP-5MS column is shown in Figure 12.16. The PMIs (present at 50 picogram each) cannot be detected in the FID chromatogram, while

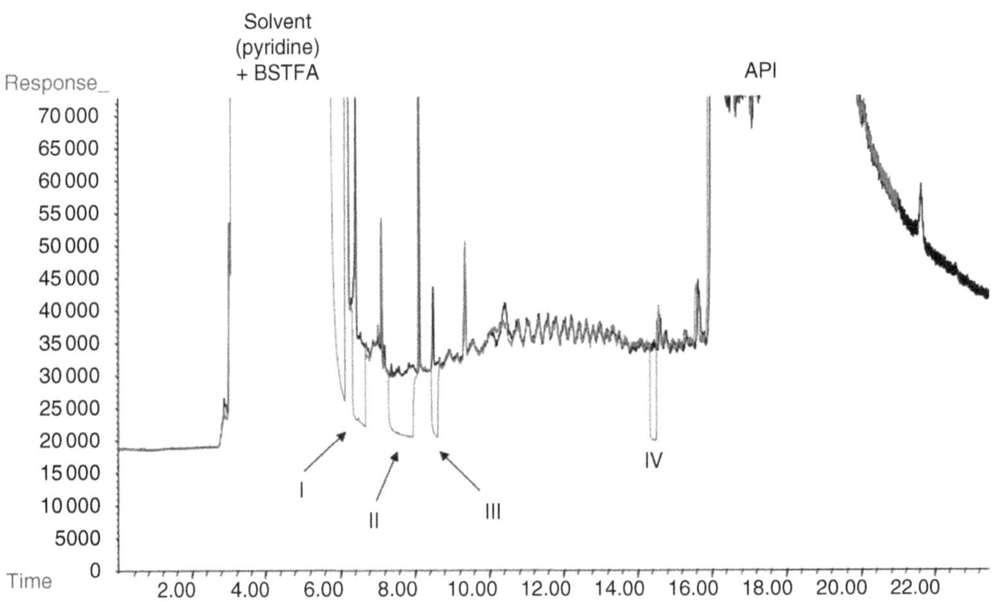

Figure 12.16 FID monitor detector trace from first dimension separation of the analysis of haloalcohols and glycidol in API (carbamazepine) by 2D-GC-MS. Selected heart-cut windows are indicated by I, II, III, IV.

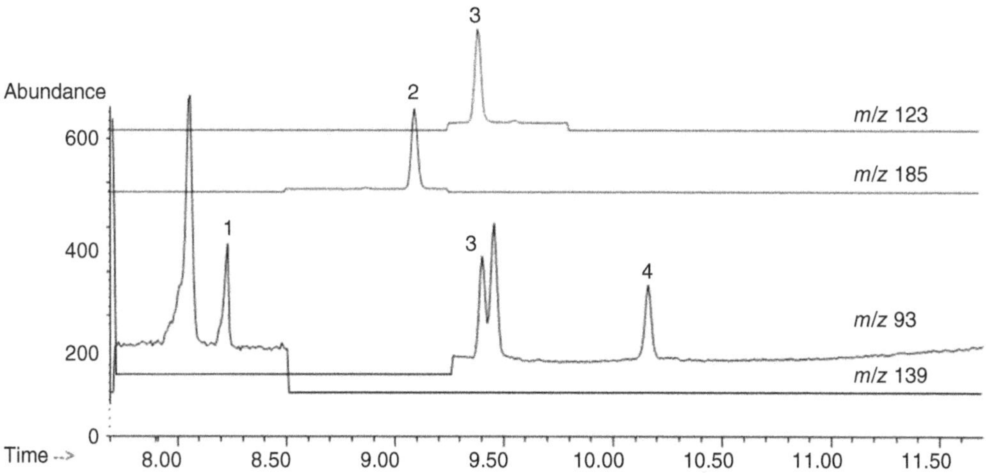

Figure 12.17 Extracted ion chromatogram from second dimension GC-MS analysis of haloalcohols in API (carbamazepine). Peaks: 2-bromo-ethanol (peak 1), 2-iodo-ethanol (peak 2), 4-chloro-1-butanol (peak 3), 2-(2-chloroethoxy)-ethanol (peak 4) (as TMS ethers).

solvent, residual silylating agent, and API (also derivatized) overload the column. The target analytes eluting between 8 and 14 minutes were heart-cut using small heart-cut windows around the peak elution times. These heart-cuts are "visible" as signal drops in Figure 12.15 (I, II, III, IV).

The heart-cut fractions are further analyzed on the second dimension column. The EIC (from SIM acquisition) in Figure 12.17 show the detection of 2-bromo-ethanol (*m/z* 139), 2-iodo-ethanol (*m/z* 185), 4-chloro-1-butanol (*m/z* 123), and 2-(2-chloroethoxy)-ethanol (*m/z* 93) free from interferences.

The method was validated [49] and the results are summarized in Table 12.8. The linearity was very good ($r^2 > 0.99$) in the range from 1 to 10 ppm, the RSDs are below 10%, except for the late eluting bromo-undecanol (RSD: 10–15%). The LODs were below 0.5 ppm, which is also better than obtained by direct liquid injection and one-dimensional GC-MS. Also, glycidol, an epoxide that could not be analyzed by the methods described in Section 12.8, was analyzed with this derivatization – 2D-GCMS method with excellent linearity, repeatability, and LOD below 1 ppm. The 2D-GC approach is thus very useful for problem cases where no headspace sampling is possible. Since only the selected target compounds are introduced in the analytical second dimension column and MS, this most critical part of the GC-MS is kept clean. Two-dimensional heart-cut GC can thus be considered as a very selective sample preparation and sample introduction technique, applicable in challenging determinations of GC amenable solutes [54].

12.10 Aziridines

12.10.1 Method Selection

Aziridines form a broad group of volatile and nonvolatile solutes that have an ethyleneimine functionality in common. Aziridine (ethyleneimine, C_2H_5N, CAS 151-56-4) itself can be analyzed using derivatization-SPME, similar to the method described for *N*-mustards. Zapata et al. [55] developed an automated analytical method for the screening of aziridine and 2-chloroethylamine in pharmaceutical active principles, based on derivatization-SPME of the amines using a PDMS/DVB fiber that was first saturated with 2,3,4,5,6-pentafluorobenzoyl chloride derivatization reagent. The amines are extracted from the headspace of the API samples, dissolved or suspended in an alkaline solution, and, after in-fiber derivatization, the PFB-derivatives are thermally desorbed in a GC inlet and analyzed by GC-MS (with negative ion chemical ionization). Alternatively, Shackman described a HILIC-MS method for the non-derivatized aziridine and chloroethylamine [46]. The HILIC approach using a 50 mm × 2.1 mm ID × 1.7 μm dp Waters BEH HILIC column showed excellent sensitivity (low ppb level) and high throughput (<5 minutes).

For nonvolatile and polar solutes containing an aziridine functionality, LC-MS is the method of choice. With the aim to develop generic methods that are applicable to a wide range of solutes and matrices, a versatile HPLC configuration was set-up in our laboratory that enabled both reversed-phase liquid chromatography (RPLC) and HILIC methods with fast switching between the modes. This allowed to develop in parallel several HPLC methods based on two columns, each with (minimum) two mobile-phase combinations, resulting in 4 methods that can be applied in an automated sequence during method selection/optimization. The instrument diagram is shown in Figure 12.4. With this configuration, scouting runs can be made to select the optimal column/phase combination for a given set of PMIs/MIs in a given API that minimize possible interference of matrix with the target solute(s) and avoid ionization suppression in MS detection. Either two RPLC columns (e.g. C18 + phenyl) or a RPLC and a HILIC column can be combined. HILIC is very complementary to RPLC, especially for more polar analytes that elute after the (less polar) API. This is illustrated for the set of aziridine solutes listed in Table 12.9.

12.10.2 Typical Analytical Conditions for RPLC-MS and HILIC-MS Analysis of Aziridines

The analysis of aziridines using both LC modes was performed on an Agilent 1200 series HPLC, consisting of an autosampler, a binary pump, a column oven compartment with column selection valve (see Figure 12.5a), a UV/DAD detector, and a 6100 MSD.

12.10.2.1 Sample Preparation

- Samples were dissolved or dispersed in acetonitrile (for HILIC) or in water/acetonitrile 90/10 (v/v, for RPLC) to obtain a concentration of 100 mg/ml. The mixture was placed in an ultrasonic bath for five minutes to complete extraction and dissolution of target analytes. Some samples were not completely dissolved, but the selected aziridines were easily solubilized and extracted. The sample was centrifuged and/or filtered if necessary.
- Spiked sample solutions can be prepared from a standard solution of the aziridines. To obtain a 1 ppm spiked sample, 100 ng (10 μl of 10 ng/μl solution) was added to the sample at a concentration of 100 mg/ml.

Table 12.9 Selected aziridine target analytes.

Peak	Name	CAS Nr	MW	LC mode	SIM ion
1	1-(2-Cyanoethyl)aziridine	1072-66-8	96	HILIC	97.2
2	1-Isobutyrylaziridine	20286-12-8	113	HILIC	114.2
3	cis-2,3-diphenyl-1-propylaziridine	314062-46-9	237	RPLC	238.1
4	2-Aziridin-1-yl-1-(4-nitrophenyl)-ethanol	21719-28-8	208	RPLC	209.1

12.10.2.2 RPLC-MS Method Parameters

- Column: Alltima HP C18 column, 15 cm × 3 mm ID × 3 μm, (Grace, Belgium)
- Column temperature: 30 °C
- Injection: 50 μl
- Mobile phase A: 0.1% ammonium acetate + 0.1% acetic acid in water
- Mobile phase B: acetonitrile.
- RPLC gradient: 10% B to 100% B at 10 minutes (5 minutes hold)
- Flow rate: 0.5 ml/minute
- MSD:
  - ESI, positive ion detection mode
  - SIM acquisition (selected ions: see Table 12.9)
  - Drying gas 350 °C, 12 l/minute, 40 psig pressure, cap V: 3.6 kV

12.10.2.3 HILIC-MS Method Parameters

- Column: Prevail Silica HILIC column, 15 cm × 4.6 mm ID × 3 μm (Grace)
- Column temperature: 20 °C
- Injection: 50 μl
- Mobile phase A: 0.1% ammonium acetate + 0.1% acetic acid in water
- Mobile phase B: acetonitrile
- HILIC gradient: 100% B (2 minutes hold) to 50% B at 15 minutes (1 minute hold)
- Flow rate: 1 ml/minute
- MSD:
  - ESI, positive ion detection mode
  - SIM acquisition, ions: see Table 12.9
  - Drying gas 320 °C, 12 l/minute, 45 psig pressure, cap V: 2.8 kV

Note that for both methods, the same solvents (A and B) are used. This allows automatic switching between methods in a sequence if a column selection valve is installed (Figure 12.5a).

For the analytes listed in Table 12.9 and the tested APIs, these conditions resulted in sufficient selectivity. It might be needed to adapt the gradients for other analytes and matrices.

12.10.3 Typical Results Obtained for Aziridine Analysis Using RPLC and HILIC

The chromatograms (EIC from SIM) for a Vitamin C sample, dissolved at 100 mg/ml in acetonitrile/water (9/1), and spiked at 1 ppm level with the 4 analytes, are given in Figure 12.18. Target compounds 1 and 2 (most polar, lowest mass) are not retained on RPLC and are measured by the HILIC method. Excellent peak shapes are obtained.

Target analytes 3 and 4 show more retention in RPLC and can be separated from the API using the RPLC conditions described above. No interferences from the API, or other impurities (present at high level) were observed.

For all analytes, excellent signal-to-noise ratios are obtained. LODs were well below 0.1 ppm. Only compound 4 showed lower response and some peak broadening (using the RPLC method). This is probably due to the acid/base characteristics of this analyte.

The RPLC and HILIC methods could be validated. The linearity was good ($r^2 > 0.99$) in the range from 0.1 to 20 ppm for the RPLC method and $r^2 > 0.97$ in the range from 0.1 to 2 ppm for the HILIC method.

This example clearly illustrates the complementary nature of RPLC and HILIC for PMI/MI analysis. For other solutes and API combinations, methods need to be adapted, but the instrumental configuration described above allows this in a flexible manner.

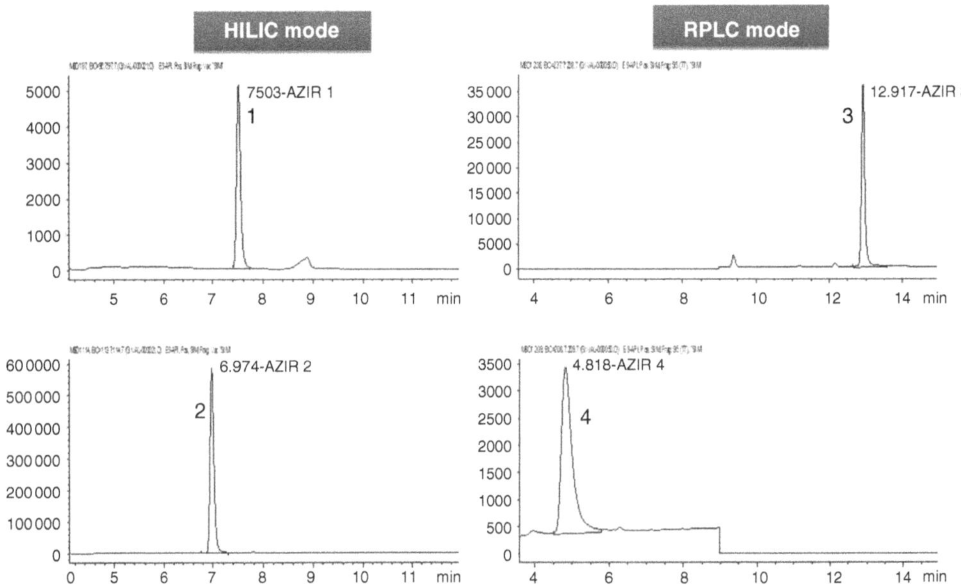

Figure 12.18 Analysis of aziridines in Vitamin C spiked at 1 ppm using HILIC and RPLC methods, depending on solute polarity. Peaks: see Table 12.9.

12.11 Arylamines and Amino Pyridines

12.11.1 Method Selection

Arylamines and aminopyridines are often used as starting materials in organic synthesis and an important class of PMI/MIs. While some of these compounds can be analyzed by GC, the primary method used for their trace analysis is clearly HPLC. This is based on the fact that arylamines and aminopyridines are quite polar and often not volatile enough for headspace analysis. Moreover, they often possess good UV absorbance characteristics, making them detectable at low levels by LC-UV. Developments in LC-UV/DAD technology enabled the analysis of some of these compounds in API matrices at sub-ppm level [5]. Obviously, the use of LC-MS or LC-MS/MS offers enhanced selectivity and sensitivity and is therefore recommended [12, 56, 57]. For some specific cases, such as for the analysis of piperazine, pre-column derivatization with dansyl chloride, followed by LC-MS analysis was used [58].

In our project, we developed a generic method for arylamines and aminopyridines based on reversed-phase HPLC [56]. A selection of 14 target analytes was analyzed without and with derivatization, each applying two RPLC methods. The selected target compounds are listed in Table 12.10. For the analytes with low retention in RPLC (mostly those with primary amine functions and no alkyl-group), pre-column derivatization with hexyl chloroformate to form a $-NH-CO-O-C_6H_{13}$ derivative improved detectability (M+128 = higher mass ion) and increased retention in RPLC due to the hydrophobicity of the hexyl-group. Two columns (C18 and phenyl) were used and for each column, different gradients (combinations of solvents A1/B1, A1/B2, A2/B1, and A2/B2) could be applied to optimize the selectivity of the separation between API and the PMIs/MIs. Detection was done using a single quadrupole mass spectrometer in the SIM mode. Like the approach for the aziridines, all methods could be run on the same HPLC configuration that allowed fast switching between methods. Method details and typical results are described below.

Table 12.10 Selected target analytes for arylamine and aminopyridine analysis.

Peak	Analyte	CAS No	Retention[a]	MW	MW der[b]
1	1-Phenylpiperazine	92-54-6	Yes	162	290
2	*N,N*-dimethyl-m-toluidine	121-72-2	Yes	135	–
3	5-Amino-2-chloropyridine	5350-93-6	Yes	128	256
4	4-Aminopyridine	504-24-5	No	94	222
5	4-Amino-2-methylpyridine	18437-58-6	No	108	236
6	3-Aminobenzonitrile	2237-30-1	Yes	118	246
7	2-Aminophenol	95-55-6	No	109	237
8	5-Amino-2-methylpyridine	3430-14-6	No	108	236
9	4-Methyl acetanilide	103-89-9	Yes	149	-
10	3-Aminopyridine	462-08-8	No	94	222
11	5-Fluoro-2-methyl aniline	367-29-3	Yes	125	253
12	N-ethyl anthranylic acid	89-50-0	Yes	165	293
13	5-Aminoindole	5192-03-0	No	132	260
14	Aniline	62-53-3	No	93	221

[a] Retention under RPLC conditions: yes = well retained (retention factor $k > 3$).
[b] MW after hexyl chloroformate derivatization.

Obviously, triple quadrupole MS in MRM mode can also be used, further increasing sensitivity and selectivity, and allowing to increase speed of analysis [57].

It should be noted that also nitro-aromatic compounds can typically be analyzed with the same methods as described here. However, nitro-aromatics give significantly lower response in ESI ionization. Therefore, reduction of the NO_2-functionality into a $-NH_2$ group has been used [59].

12.11.2 Typical Analytical Conditions for Arylamines and Aminopyridines by RPLC-MSD

The analysis of arylamines and aminopyridines was performed on an Agilent 1290 Infinity LC, consisting of an autosampler, a binary pump, a thermostated column compartment with column selection valve (see Figure 12.5a), a UV/DAD detector, and a 6100 MSD or 6460 triple quadrupole MS.

12.11.2.1 Sample Preparation

- The API was solubilized at 100 mg/ml concentration in acetonitrile. Internal standard and spiking solutions (for instance, 10 μl of 10 ng/μl PMI/MI) were added to this sample. The injected concentration of the PMI/MI corresponds to 0.1 ng/μl. Solubilization was accelerated by using ultrasonic treatment during 5 minutes. An aliquot of this sample was filtered and analyzed directly (= non-derivatized PMI/MI method).
- From the same solution, 1 ml was added to 9 ml borate solution (15 mM sodium tetraborate, adjusted to pH 9.5 with sodium hydroxide 1N) in a 20 ml vial. Next, 0.5 ml hexachloro formate solution (2% in acetonitrile) was added. Reaction was performed at room temperature during 30 minutes. Finally, 0.2 ml of phosphoric acid solution (85%) was added to stop the reaction and neutralize the mixture. This mixture was analyzed for the derivatized PMIs/MIs.

12.11.2.2 HPLC-MS Parameters

- Column A: Zorbax Eclipse Plus C18, 150 mm L×3.0 mm ID, 3.5 μm particle size (Agilent Technologies)
- Column B: Zorbax Eclipse XDB Phenyl, 150 mm L×3.0 mm ID, 3.5 μm particle size (Agilent Technologies)
- Column temperature: 30 °C
- Injection volume: 5 μl (underivatized sample) – 50 μl (derivatized sample)
- Mobile phase: A1 = 0.05% v/v formic acid in water, B1 = acetonitrile
- (A2: 10 mM ammonium formate in water, B2: methanol)
- Gradient 1: from 5% B1 to 100% B1 at 10 minutes (4 minutes hold)
- (other gradients: see [56])
- Flow: 0.5 ml/minute
- MSD:
  - ESI, positive ion detection mode
  - Nitrogen drying gas: 350 °C, 12 l/minute,
  - Nebulizer pressure: 35 psig
  - Capillary voltage: 4 kV
  - SIM acquisition (selected ions: see Table 12.11)

The different injection volumes compensate for the tenfold dilution during the derivatization step. For the non-derivatized and derivatized sample, the same quantity of API is injected.

Table 12.11 Retention times, selected SIM ions, and recovery for arylamines and aminopyridines in ephedrine.

		Non-derivatized			Derivatized		
Peak	**Analyte**	**RT (min)**	**Ion**	**Rec**[a]	**RT (min)**	**Ion**	**Rec**[a]
1	1-Phenylpiperazine	5.54	163	51.4[b]			
2	*N*,*N*-dimethyl-m-toluidine	5.85	136	90.9			
3	5-Amino-2-chloropyridine	6.15	129	80.4	*11.25*	*257*	
4	4-Aminopyridine				7.02	223	94.2
5	4-Amino-2-methylpyridine				7.19	237	67.2
6	3-Aminobenzonitrile	7.29	119	81.4			
7	2-Aminophenol				7.34	238	94.1
8	5-Amino-2-methylpyridine				7.34	237	93.1
9	4-Methyl acetanilide	7.77	150	76.5			
10	3-Aminopyridine				7.88	223	34.1[c]
11	5-Fluoro-2-methyl aniline	8.39	126	78.9			
12	*N*-ethyl anthranylic acid	8.80	166	86.7			
13	5-Aminoindole				10.80	261	65.1
14	Aniline				11.61	222	99.4

RT (min): retention time on Eclipse Plus C18 column, using conditions described above.
[a] Recovery (accuracy) measured as response relative to external standard.
[b] Recovery increased to 80.8% on Zorbax Eclipse XLB Phenyl column using an ammonium formate/methanol gradient [56].
[c] Recovery increased to 103.5% on Zorbax Eclipse XLB Phenyl column using an ammonium formate/methanol gradient [56].

12.11.3 Typical Results for Arylamines and Aminopyridines by RPLC-MSD

The chromatograms (EIC) obtained for a non-derivatized and derivatized sample of ephedrine spiked at 1 ppm level with the target analytes are shown in Figures 12.19 and 12.20, respectively. The peak identification, retention times, and measured ions (SIM) are listed in Table 12.11. The 7 target analytes that have sufficient retention using standard RPLC conditions (C18 column, gradient 1) are well detected. The 7 target analytes that initially showed low retention in RPLC, are better retained, and more easily detected in the API after derivatization. Some solutes (such as 5-amino-2-chloropyridine) can be detected in both analyses.

The responses for the compounds were compared with the responses for the injection of the same concentration of target analytes in a 0.1 ng/μl standard solution without API (same injected amount of PMI/MI) and the obtained recoveries (%, = (area target analyte in API/area in standard solution)*100) are also given in Table 12.11.

For the analysis of the non-derivatized sample, good recoveries (between 75 and 100%) are obtained, except for 1-phenylpiperazine (51%). For this solute, the analysis on column B (phenyl phase) resulted in better recovery [56].

For the analysis of the derivatized sample, recoveries were between 65 and 100%, except for 3-aminopyridine. Also, for this solute, better recovery was obtained using the phenyl column [56].

These lower recoveries are often due to ionization suppression by co-eluting API or other impurities (present at higher concentration). This was also observed in a more recent work applying HPLC in combination with a triple quadrupole MS detector operated in MRM mode [57]. The effect of ionization suppression was clearly observed during the analysis of 3-aminobenzonitrile in bupivacaine. The extracted MRM chromatogram (parent ion m/z 119 > daughter ion m/z 92) of a 3-aminobenzonitrile standard solution analyzed using a formic acid/acetonitrile gradient is shown in Figure 12.21a. The target solute elutes at 2.6 minutes. In the

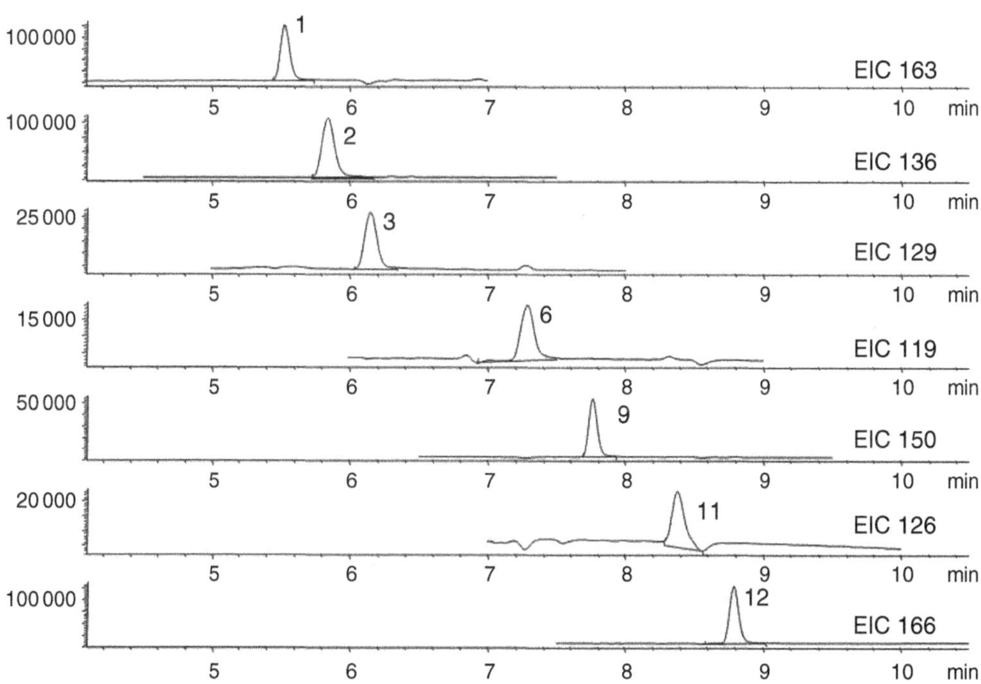

Figure 12.19 RPLC-MS analysis of non-derivatized arylamines and aminopyridines in ephedrine spiked at 1 ppm, Peaks: see Table 12.12.

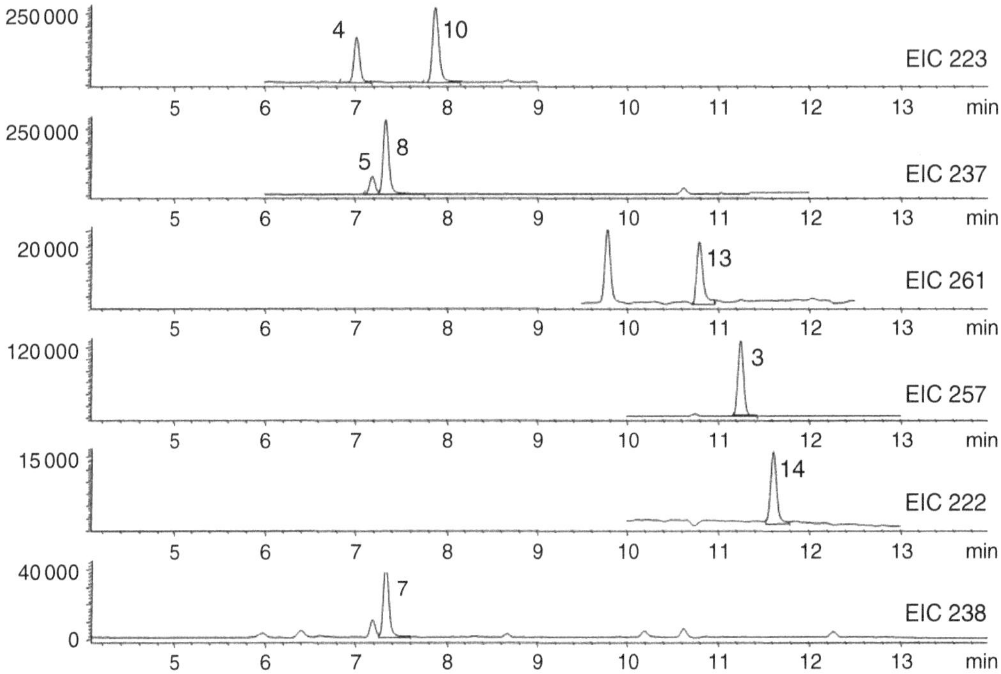

Figure 12.20 RPLC-MS analysis of derivatized arylamines and aminopyridines in ephedrine spiked at 1 ppm. Peaks: see Table 12.12.

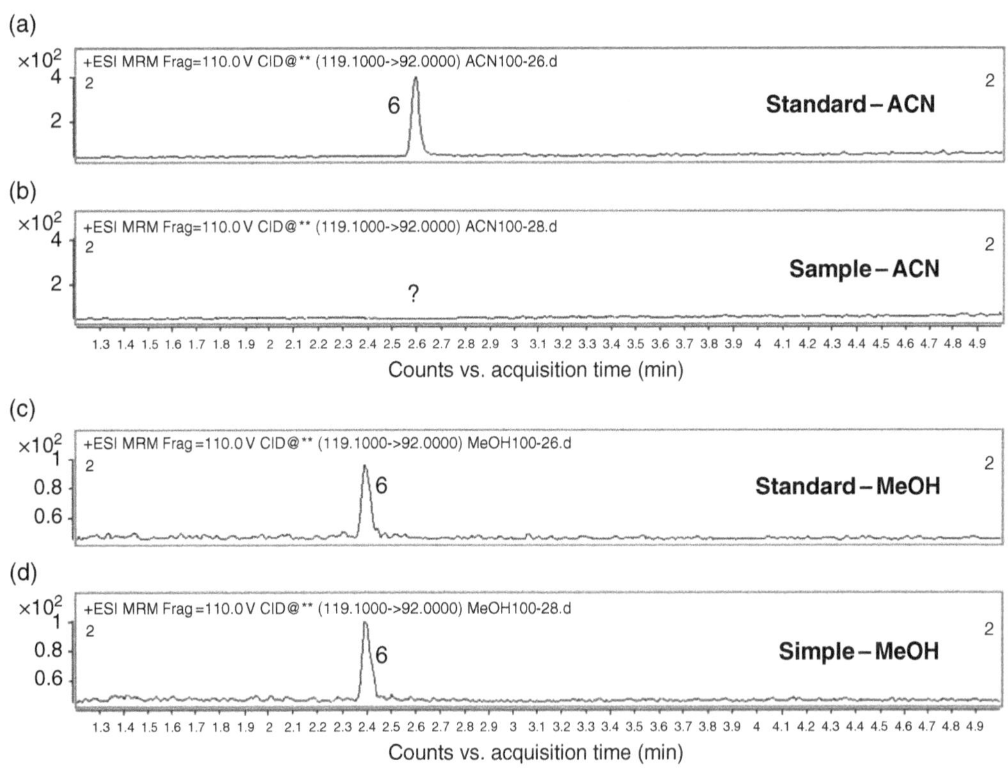

Figure 12.21 Analysis of 3-aminobenzonitrile in bupivacaine spiked at 1 ppm by RPLC-MS/MS without derivatization. (a) analysis of standard using default method (formic acid in water + acetonitrile gradient), (b) analysis of spiked API using default method (formic acid in water + acetonitrile gradient), (c) analysis of standard using modified method (formic acid in water + methanol gradient), (d) analysis of spiked API using modified method (formic acid in water + methanol gradient). *Source:* From Ref. [57].

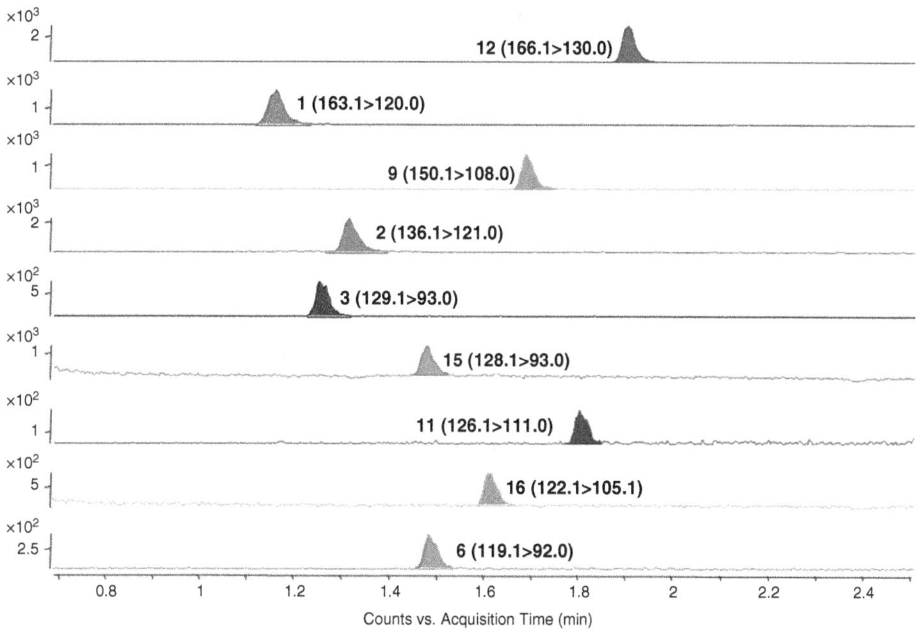

Figure 12.22 High-throughput analysis of non-derivatized arylamine and aminopyridine PMI/MIs in diclofenac at 0.1 ppm level using LC-MS/MS in MRM mode (peaks: see Table 12.12, peak 15: 4-chloroaniline, peak 16: 2,6-dimethylaniline, MRM transitions given between brackets, for analytical conditions, see Ref. [57]).

extracted MRM chromatogram obtained for a sample of bupivacaine spiked at the same level (1 ppm), and analyzed by the same method, no peak was detected for the target compound (Figure 12.21b). After changing to a formic acid/methanol gradient, 3-aminobenzonitrile could be detected both in standard (Figure 12.21c) and in API (Figure 12.21d) with similar response. UV data showed that in the first case the API co-eluted with the target compounds, while methanol resulted in separation of API and PMI. This example illustrates that selectivity tuning can be necessary with some combinations of PMIs/MIs and APIs, even using a highly sensitive and selective detection such as triple quadrupole MS. Once the selectivity of the method is optimized to avoid ionization suppression, speed of analysis can be increased using a LC-MS/MS system. This is illustrated in Figure 12.22, showing the detection of 9 underivatized PMI/MIs spiked at 0.1 ppm in diclofenac. Analysis time was under 3 minutes. In addition to the analytes retained under RPLC condition, this analysis also included two additional PMIs: 4-chloroaniline (peak 15) and 2,6-dimethylaniline (peak 16).

12.12 Hydrazines and Hydroxylamine

12.12.1 Method Selection

Hydrazine (NH_2-NH_2), methylated hydrazines, and related hydrazines are highly reactive bases with reducing properties and common synthesis intermediates or degradation products of pharmaceutical products. Analytical methods for trace analysis of these compounds in API were reviewed by Elder et al. [60] and include HPLC, GC (after derivatization), IC, thin layer chromatography, and CE.

Analysis of hydrazine and the related hydrazines is challenging due to their volatility, high polarity, low molecular weight, and absence of chromophore. Direct analysis by GC is difficult due to too low vapor pressure for headspace analysis and system contamination by the matrix if liquid injection of a concentrated API solution is attempted. Sun et al. [61] described an in-situ derivatization headspace GC-MS method for the determination of hydrazine in API at low ppm level. Hydrazine was converted with acetone (or the deuterated analog acetone-d6) to acetone-azine (acetone-azine-d12). The derivatives could be analyzed by headspace-GC-MS and an LOD of 0.1 ppm was obtained in the presence of 10 mg API.

Alternatively, also methods based on derivatization, followed by HPLC analysis are described for trace analysis of hydrazine. Wang et al. [62] used 2-hydroxy-1-naphthaldehyde as derivatization reagent. Analysis was done using UV detection. More recently, Cui et al. [63] described a method using benzaldehyde derivatization followed by LC-MS analysis for trace analysis of hydrazine and aceto-hydrazine in API. Using a "matrix matching" calibration, involving the acidification of the derivatization reaction medium with benzoic acid, resulted in improved accuracy (recovery) compared with a standard comparison of spiked samples with standard solutions. Sensitivities down to 1 ppm were obtained.

Although less volatile solutes such as phenylhydrazine are better retained in LC analysis, pre-column derivatization is also used for these solutes to increase detection selectivity, especially if UV/DAD detection is applied [64].

For the selected hydrazines listed in Table 12.12, we used a simple in-situ derivatization using hexyl chloroformate. This derivatization converts the primary amine function (s) in N-acyl derivatives, increasing retention in RPLC analysis and detectability in MS (more selective higher masses). Using the appropriate autosamplers, derivatization and injection can be fully automated. Details of this method are described below.

A particular case is the analysis of hydroxylamine (NH_2-OH). This very polar, water soluble, and reactive solute, without chromophore, was not included in our study. Recent papers have described the analysis of hydroxylamine by HPLC-UV or HPLC-MS after derivatization with 9-fluorenylmethyl chloroformate (FMOC) [65] or with benzaldehyde [66].

12.12.2 Analytical Conditions for the Analysis of Hydrazines Using Derivatization and HPLC-MS

The solutes listed in Table 12.12 were analyzed on an Agilent 1200 series LC, consisting of an autosampler, a binary pump, a thermostated column compartment with column selection valve (see Figure 12.5a), a UV/DAD detector, and a 6100 MSD.

Table 12.12 Selected hydrazine target analytes.

Peak Nr	Name (CAS)	CAS	$MW_{original}$	$MW_{derivate}$
1	1-Methylhydrazine	60-34-4	46	174
2	Acetohydrazide	1068-57-1	74	202
3	2-Hydrazinoethanol	109-84-2	76	204
4	Phenylhydrazine	100-63-0	108	236

12.12.2.1 Sample Preparation

- APIs (Vitamin C, Penicillin V) were dissolved at 100 mg/ml in acetonitrile and spiked with 1-methylhydrazine, acetohydrazide, 2-hydroxyethanol, and phenylhydrazine at 1 ppm level.
- To the sample, 7 ml borate buffer (15 mM sodium tetraborate, pH 9.5 in 10% acetonitrile) was added, followed by 0.5 ml hexyl chloroformate solution (2% in acetonitrile). After reaction during 15 minutes at room temperature, the reaction was stopped by adding 0.5 ml phosphoric acid (17% in water) and injection was performed.

12.12.2.2 HPLC-MS Parameters

- Column: Zorbax Eclipse Plus C18, 150 mm L×3.0 mm ID, 3.5 μm particle size (Agilent Technologies)
- Column temperature: 35 °C
- Injection volume: 50 μl
- Mobile phase: A1 = 0.05% formic acid, B1 = acetonitrile
- Gradient: from 15% B1 to 100% B1 at 19 minutes
- Flow: 0.5 ml/minute
- MSD:
  - ESI, positive ion detection mode
  - Nitrogen drying gas: 340 °C, 12 l/minute,
  - Nebulizer pressure: 35 psig
  - Capillary voltage: 4 kV
  - SIM acquisition (selected ions: $[M+H]^+$ ions at *m/z* 175, 203, 205 and 237)
- Alternative method for phenylhydrazine in Penicillin V:
  - Column B: Zorbax Eclipse XDB Phenyl, 150 mm L×3.0 mm ID, 3.5 μm particle size (Agilent Technologies)
  - Mobile phase: A2= 10 mM ammonium acetate, B1=acetonitrile
  - Gradient: from 10% B1 to 100% B1 at 19 minutes
  - Flow: 0.5 ml/minute

12.12.3 Typical Results Obtained for Hydrazines Using Derivatization LC-MS

The EIC (from SIM acquisition) obtained for the analysis of a Vitamin C sample spiked at 1 ppm level with the four selected analytes is shown in Figure 12.23.

All four analytes are detected at this level. For analyte 2 (aceto-hydrazine), confirmation analysis by another method is advised since it is not completely resolved from other impurities.

The repeatability of this derivatization method was better than 10% RSD and good linearity ($r^2 >$ 0.99) was obtained in a range from 0.25 to 5 ppm.

Another example is shown in Figure 12.24. A Penicillin V sample, also spiked at 1 ppm level was analyzed by the default method (C18 column, formic acid/acetonitrile). The EIC for phenylhydrazine (analyte 4) is shown in Figure 12.24a. The peak elutes just after another peak, also detected at *m/z* 237 and is largely suppressed (recovery < 10%). The same sample was reinjected using a Phenyl column and ammonium formate (A2)/acetonitrile (B1) gradient. The EIC at *m/z* 237 is shown in Figure 12.24b. The analyte elutes at 10.75 minutes, while the interference is shifted to a longer retention time. Phenylhydrazine can be better detected and recovery was > 90%.

This example shows again that selectivity tuning might be needed. Using a HPLC configuration equipped with 2 columns and 4 solvents, this can be done in an automated sequence.

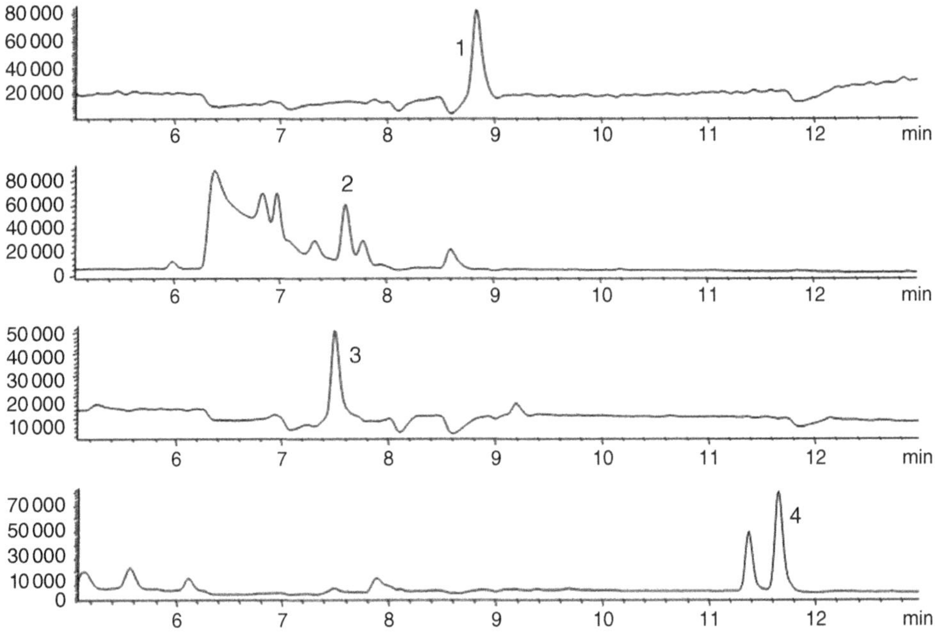

Figure 12.23 Analysis of hydrazines in Vitamin C spiked at 1 ppm level by derivatization with hexyl chloroformate, followed by RPLC-MS. Peak identification: see Table 12.12.

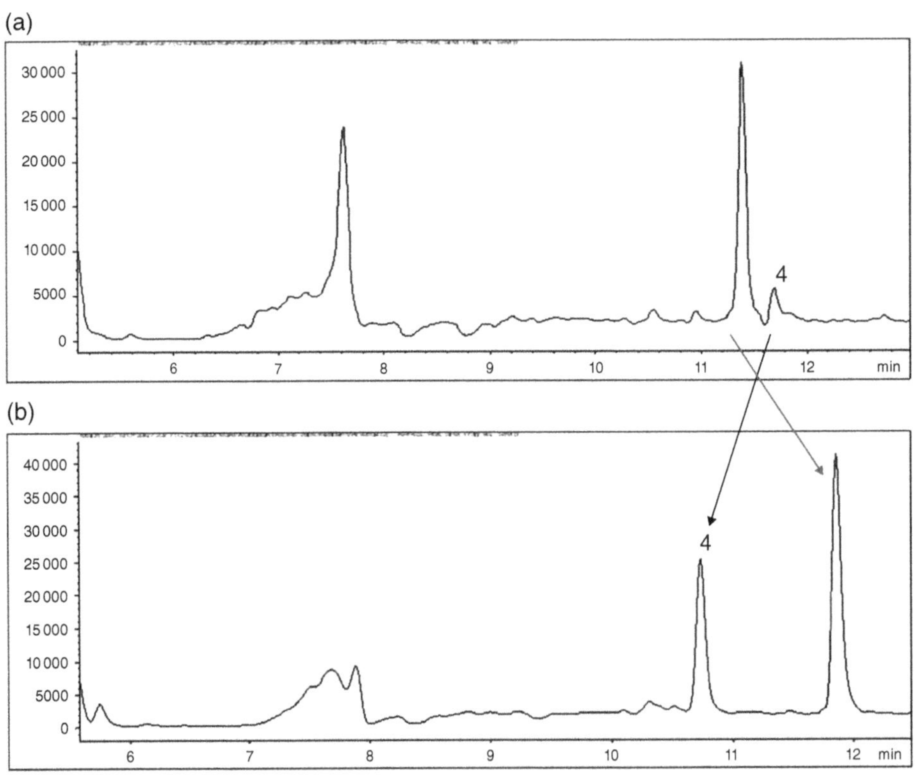

Figure 12.24 Analysis of phenylhydrazine in Penicillin V spiked at 1 ppm level by derivatization with hexyl chloroformate, followed by RPLC-MS. (a) default method (C18 column, formic acid in water + acetonitrile gradient), (b) modified method (Phenyl column, ammonium acetate + acetonitrile gradient).

12.13 Aldehydes and Ketones

12.13.1 Method Selection

Aldehydes and ketones can be analyzed either by GC or with LC. The main challenge in trace analysis of low molecular weight (C1–C5) aldehydes is detection. Formaldehyde, for instance, elutes very fast in GC and no specific ions are available for sensitive and selective detection by MS, neither with GC-MS nor LC-MS. Therefore, derivatization of aldehydes and ketones is applied to enhance detectability, both in GC and LC methods. For GC analysis, O-2,3,4,5,6-pentafluorobenzylhydroxylamine (PFBHA) is the mostly used derivatization reagent since it can be applied in-situ (in an aqueous solution) and the PFBHA-oxime derivatives are volatile making them amenable to headspace sampling and GC-MS analysis [67]. For HPLC analysis, both with UV/DAD or MS detection, 2,4-dinitrophenylhydrazine (DNPH) is the derivatization reagent of choice. The DNPH-derivatives of carbonyl containing compounds can be detected at 360 nm by UV/DAD and/or by MS.

In our study, the pre-column derivatization using DNPH was applied for analytes with carbonyl functionality, since this method is applicable to a wide range of aldehydes and ketones, including very polar solutes such as acrolein, and less volatile analytes such as 4-hydroxybenzaldehyde. Isotopically labeled aldehydes were used as internal standards to improve the repeatability and accuracy of the method.

The selected target analytes and some commercially available isotope labeled internal standards are listed in Table 12.13.

12.13.2 Typical Analytical Conditions for Analysis of Aldehydes and Ketones by DNPH Derivatization, Followed by LC-MS Analysis

The aldehydes and ketones were analyzed on an Agilent 1200 series LC, consisting of an autosampler, a binary pump, a thermostated column compartment with column selection valve (see Figure 12.5a), a UV/DAD detector, and a 6100 MSD. Pre-column derivatization with DNPH was performed in the autosampler.

Table 12.13 Selected aldehyde target analytes.

Peak	Name (CAS)	CAS No	$MW_{original}$	$MW_{derivate}$
1	Formaldehyde	50-00-0	30	210
2	Propionaldehyde	123-38-6	58	238
3	Hexanal	66-25-1	100	280
4	Nonanal	124-19-1	142	322
5	4-Hydroxybenzaldehyde	123-08-0	122	302
6	4-Cyanobenzaldehyde	105-07-7	131	311
7	2,3-Dimethylbenzaldehyde	5779-93-1	134	314
8	3,4-Dimethylbenzaldehyde	5973-71-7	134	314
9	4-Hydroxy-3-methylbenzaldehyde	15174-69-3	136	316
IS1	Formaldehyde-13C,d2		33	213
IS2	Propionaldehyde-2,2-d2		60	240
IS3	4-Hydroxybenzaldehyde-2,3,5,6-d4		126	306

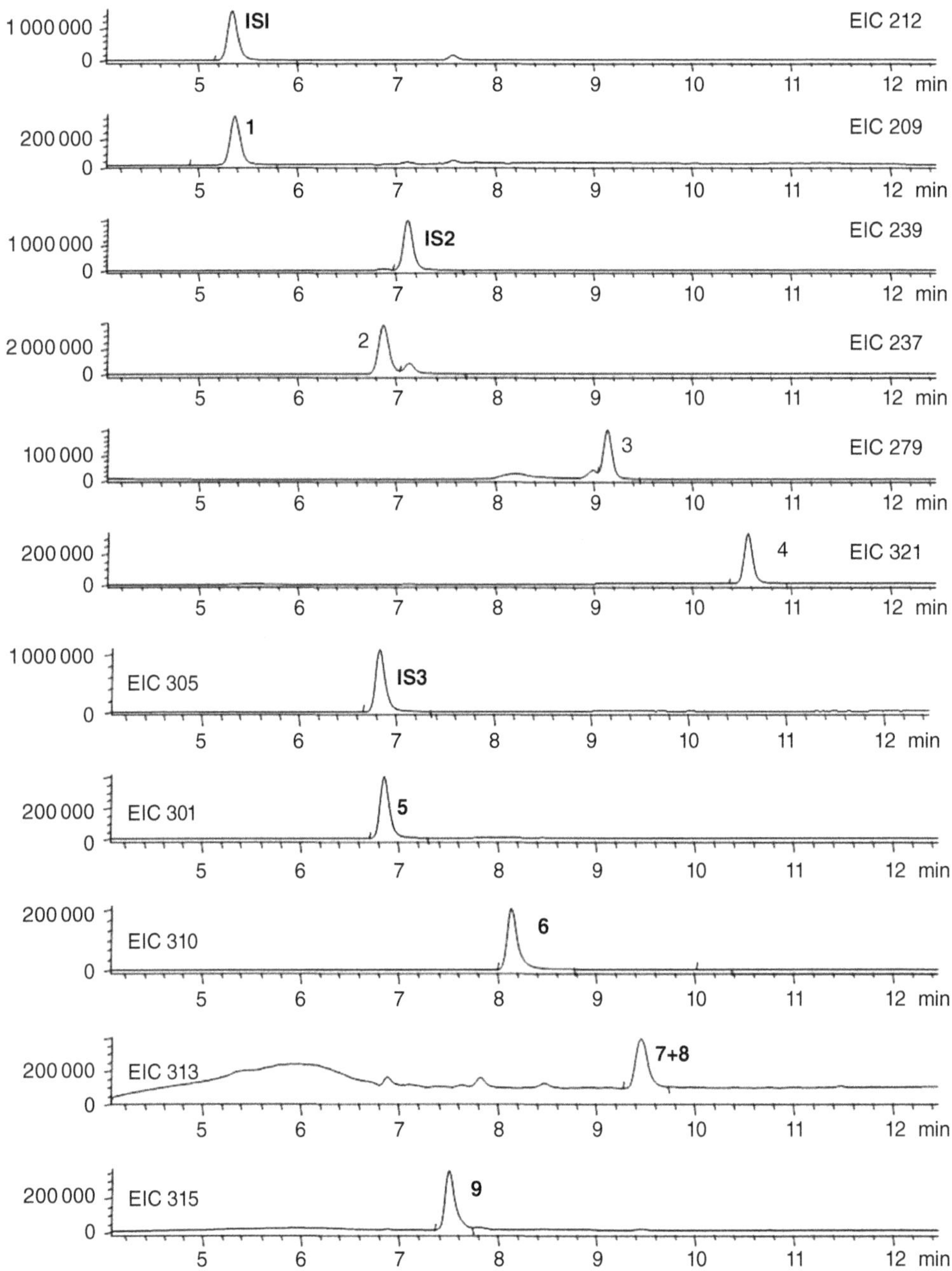

Figure 12.25 Analysis of aldehydes by DNPH derivatization – LC-MS, Peaks: see Tables 12.13 and 12.14.

12.13.2.1 Sample Preparation

- The sample was dissolved in a mixture of methanol/acetonitrile/water (2/1/1) at 50 mg/ml. Internal standard was added at 0.5 ppm level.
- Derivatization was automated in the Agilent 1100/1200 series ALS. From the sample solution, 4 μl was drawn in the sampler, followed by 1 μl derivatization reagent solution (see below – injector program).

12.13.2.2 Derivatization Reagent Solution

- 500 mg DNPH is dissolved in 10 ml methanol. 1 ml conc sulphuric acid is added and the solution is diluted to 50 ml with methanol (resulting DNPH concentration: 1%).

12.13.2.3 HPLC-MS Parameters

- Column: Zorbax Stablebond Phenyl column, 150 mm L × 3.0 mm ID, 3.5 µm particle size (Agilent Technologies)
- Column temperature: 40 °C
- Injection volume: 5 µl (with injector program)
  - Draw 4 µl sample
  - Add 1 µl DNPH reagent
  - Mix
  - Wait 4 minutes
  - Mix
  - Inject
- Mobile phase: A = 10 mM ammonium acetate in water, B = acetonitrile
- Gradient: from 36% B to 100% B at 12.5 minutes, 3.5 minutes hold
- Flow: 0.65 ml/minute
- MSD:
  - ESI, negative ion detection mode
  - Nitrogen drying gas: 330 °C, 11 l/minute
  - Nebulizer pressure:50 psig
  - Capillary voltage: 3.5 kV
  - SIM acquisition (selected ions: $[M-H]^-$ ions were monitored, see Table 12.14)

Table 12.14 Retention times, SIM ions, and recovery for aldehydes spiked in Vitamin C and Ephedrine.

Peak	Name (CAS)	RT (min)	Ion	Recovery % (Vit C)	Recovery % (ephedrine)
1	Formaldehyde	5.37	209	[a]	110
2	Propionaldehyde	7.14	237	83	96
3	Hexanal	9.14	279	98	102
4	Nonanal	10.57	321	105	101
5	4-Hydroxybenzaldehyde	6.86	301	106	98
6	4-Cyanobenzaldehyde	8.14	310	121	108
7	2,3-Dimethylbenzaldehyde	9.46	313	124	103
8	3,4-Dimethylbenzaldehyde	9.46	313	(co-elutes with 7)	
9	4-OH-3-methylbenzaldehyde	7.51	315	18	97
IS1	d2,C13-Formaldehyde	5.35	212	–	–
IS2	d2-Propionaldehyde	7.12	239	–	–
IS3	d4-Hydroxybenzaldehyde	6.84	305	–	–

[a] Trace was present in sample at 10 ppm level.

12.13.3 Typical Results Obtained for Aldehyde Analysis by DNPH Derivatization – LC-MS

Figure 12.25 shows the analysis of a 0.1 ng/μl standard solution of selected aliphatic and aromatic selected aldehyde analytes. This concentration corresponds to 2 ppm relative to the API (50 mg sample). The retention times measured for the solutes using the generic RPLC method described above are given in Table 12.14. All compounds, including the internal standards, are easily detected. The most volatile aldehyde (formaldehyde) can also be detected at this level without API interference.

The method was validated using several APIs. Aliphatic aldehydes are calibrated using IS1 and IS2, aromatic aldehydes using IS3. In general, good linearity ($r^2 > 0.99$) was observed in a concentration range from 0.5 to 10 ppm. Repeatability (RSD) was better than 10% at 1 ppm level. The recoveries for the analytes (accuracy) were determined in Vitamin C and ephedrine (spiked at 1 ppm) versus a standard solution and are listed in Table 12.14. In most cases, recoveries were between 80 and 120%, except for 4-hydroxy-3-methylbenzaldehyde in Vitamin C. In this case, ion suppression is observed, and the method requires modification to alter chromatographic selectivity.

12.14 Nitrosamines

12.14.1 Method Selection

Nitrosamines have been present in the chemical literature for over a hundred years. In 1956, after an incident of accidental poisoning, Barnes and Magee [68] discovered nitrosodimethylamine (NDMA) caused liver tumors in rats. This discovery prompted scientists globally to investigate nitrosamines for carcinogenicity and mutagenicity. Nitrosamines are widely prevalent in the environment, as their precursors, secondary amines and nitrosating agents, occur commonly and the reaction is quite facile. This potential background contamination presents a challenge to trace analysis, where ppb levels must be reported. Therefore, great care must be taken to avoid false positive identification.

Nitrosamines in pharmaceuticals are present as a by-product of secondary, and to a lesser extent tertiary amines, reacting with nitrosating agents such as sodium nitrite under favorable e.g. acidic conditions. The recent regulatory focus on nitrosamines began with the sartans, and specifically Valsartan, where excess sodium azide used to generate the tetrazole functionality is quenched with sodium nitrite. In the case of Valsartan, the issue appears to have arisen from a change to the route of synthesis and the use of DMF as the solvent, potentially introducing dimethylamine [69]. However, the scope has widened to other alkyl *N*-nitrosamines (Table 12.15) primarily due to intrinsic and extrinsic (e.g. solvent recycling and facility reuse) sources of other nitrosamines. This is examined in detail in Chapter 10.

Prior to the Valsartan contamination issue, a limited number of methods for analysis of *N*-nitrosamines in pharmaceuticals had been published. These methods largely used GC coupled with a niche thermal energy analyzer (TEA) for detection. Dawson and Lawrence [70] applied GC-TEA to a number of drug products available in the Canadian market, for NDMA and NDEA content. NDMA was determined at levels up to 12 ppb, with NDEA found at levels up to 0.6 ppb. Taylor et al. [71] applied GC-TEA for the determination of NDMA and NDEA in antihistamines and cough syrups. Castegnaro et al. [72] also applied GC-TEA to quantify the levels of NDMA and NDEA in various formulations containing aminopyridine, disulfiram, or oxytetracycline. Severin [73] investigated an alternative approach, applying trace enrichment HPLC for the determination of *N*-nitrosohexamethylenimine in tolazamide bulk drug and pharmaceutical dosage forms, reporting accurate quantitation of the analyte down to 1 ppb.

Table 12.15 *N*-nitrosamine target compounds.

Analyte		CAS number	MW
NDMA	*N*-nitrosodimethylamine	62-75-9	74.05
NDEA	*N*-nitrosodiethylamine	55-18-5	102.08
NDEIPA	*N*-nitrosoethylisopropylamine	16339-04-1	116.10
NDIPA	*N*-nitrosodiisopropylamine	601-77-4	130.11
NMBA	*N*-nitroso-*N*-methyl-4-aminobutyric acid	61445-55-4	146.07
NDBA	*N*-nitrosodibutylamine	924-16-3	158.15
NDPA	*N*-nitrosodipropylamine	621-64-7	130.11

The ever-growing importance of the area is reflected in a number of detailed publications that have reviewed the analytical methodology developed for *N*-nitrosamines of pharmaceutical interest. These include the publications in 2019 and 2020 by Parr and Joseph [69], Sorgel et al. [74, 75], and Shaik et al. [76] describing LC and GC methods hyphenated with high-performance MS. The publications reflect the requirement for enhanced analytical technology, methodology, and scientist skills, driven by the trace analysis LOQs, LODs, and the specificity required to minimize risk of false positives. A list of typical analytes associated with Sartans is given in Table 12.15, with other *N*-nitrosamines also hypothesized to be present [77]. For the smaller alkyl *N*-Nitrosamines in Sartans, the Pham Eur. Monograph [78] mandates testing to 30 ppb LOQ. The monograph also details validated GC and LC-MS and MS/MS methods suitable for analysis of all seven nitrosamines in the various Sartan matrices. The FDA require that "non detected" must be reported. This must be equivalent with the best sensitivity currently published in an Official Medicines Control Laboratory (OMCL) or FDA method, which is presently in the range of 1–5 ppb. An OMCL is a European laboratory, independent from pharmaceutical companies, that supports regulatory authorities by controlling the quality of medicinal products for human or veterinary use. There are now more than 20 official methods published covering the analysis of trace *N*-Nitrosamines for Sartan, Ranitidine, and Metformin drug substance and drug product matrices [79].

The high sensitivity and specificity required for these analytes creates a degree of technical challenge and significantly increases the risk of false positives from external contamination. This risk of contamination has been strikingly illustrated by the publication by Yang et al. in 2020 [80]. A private testing laboratory reported in a Citizen Petition (CP) to FDA that 16 of 38 metformin drug products they had tested contained NDMA levels above the allowable intake (AI) of 96 ng/day. In response, the FDA carried out testing on the same set of 38 samples with orthogonal procedures, reporting amounts over the AI in only 8 of the 38 products and generally observed lower values than reported by the private testing laboratory. A subsequent investigation identified the cause of the discrepancy to be the presence of *N*,*N*-dimethylformamide (DMF) which had interfered with NDMA measurements in the analysis by the private laboratory. Specifically, an isotopic peak associated with the allowed DMF impurity was shown to impact the private laboratory's method because insufficient mass accuracy was used. This example clearly illustrates the necessity for applying highly selective and/or high-resolution MS detection to assure the specificity of the analytical method. This investigation also illustrated the importance of orthogonal methods.

All analytes in Table 12.15, except *N*-nitroso-*N*-methyl-4-aminobutyric acid (NMBA) are amenable to GC analysis and volatile enough for headspace analysis. NMBA requires LC for direct analysis. *N*-

Nitrosamines are thermally stable but are subject to photolysis [81], which can be a supplementary means of confirming identity of a peak suspected of being an *N*-nitrosamine. As previously discussed, SHS coupled with GC-MS is widely available. A SIM MS method is described below for all key *N*-nitrosamines (except NMBA). It is important to recognize that the elevated temperatures used in SHS incubation have the potential to produce artifactual levels of nitrosamines when a vulnerable amine and a nitrosating agent are present. This may be mitigated by minimizing the incubation temperature or using ascorbic acid or sulphamic acid to inhibit the nitrosation reaction [82]. SHS-GC or liquid extraction and direct injection HPLC in combination with MS/MS (using a triple quadrupole system) or with a high resolution accurate mass (HRAM) MS system (QTOF, Orbitrap) offer great specificity, confidence in assignment of identity, and potentially better signal to noise. In addition, to enhance sensitivity and to potentially reduce matrix interference solid-phase matrix extraction (SPME) has been evaluated [83].

In order to analyze NMBA, LC methodology coupled with MS/MS or HRAMS MS/MS is required [84]. LC-MS methods also provide an orthogonal approach to confirm positive results from GC analysis and a means to overcome matrix effects, and vice versa. As a further alternative, Schmidtsdorff and Schmidt [16] proposed SFC in combination with MS for the detection of nitrosamines and other sartan-related impurities in active pharmaceutical ingredients. SFC allowed a broad spectrum of nonpolar and very polar impurities can be separated and analyzed in under 20 minutes with validation for limit testing according to ICH Q2(R1) and fulfilling default thresholds of EMA and FDA.

12.14.2 Sample preparation for SHS-GC-MS Analysis (according to ref [85])

12.14.2.1 SHS-GC-MS Analysis [85] Sample Preparation

500 mg of sample (API, or drug product containing 500 mg API) was weighed into a 20 ml amber headspace vial and 5 ml DMSO was added to solubilize the sample. It is recommended that the diluent contains a deuterated internal standard at an appropriate concentration (e.g. 40 ppb NDMA D6) to improve the accuracy, precision, and reliability of the analysis. The solvent can be changed based on API or drug product solubility.

Note that DMF as a solvent carries a risk of generating false positive due results to traces of secondary amines reacting with nitrosation agents. Experiments to explore artifactual formation of *N*-nitrosamines can be carried out by spiking with $NaNO_2$ to determine the potential for formation. Sulfamic acid and ascorbic acid may be used to control nitrosation [70, 86, 87]. The physical conditions of the matrix (e.g. acidity, counterions, etc.) are important and certain chemicals and compounds drive nitrosation (e.g. formaldehyde and chloride) [88, 89].

To calibrate the method, a set of calibration standards was prepared in diluent. A typical calibration range is from 0.4 to 8 ng/ml (4–80 ppb with respect to API).

12.14.2.2 GC-MS (HRAM-MS) Conditions

Analyses were performed on a Thermo Scientific Exactive™ GC Orbitrap™ GC-MS system using the following parameters. SHS analysis was done on a valve-type SHS sampler, equipped with a 1 ml sample loop.

- SHS:
- Incubation for 15 minutes at 120 °C
- 1 ml injected, 1:5 split at 220 °C
- GC:
- Column: Trace GOLD TG-WAXMS B, 30 m × 0.25 μm ID × 0.5 mm df
- Temperature program: 40 °C (0.5 minute), 20 °C/minute to 160 °C then 10 °C/minute to 210 °C, hold for 1 minute

Table 12.16 Selected ions for HR-MS detection of *N*-nitrosamines using GC-EI-MS.

Compound	Mass (*m/z*)	Start (min)	End (min)
NDMA	74.048 01	6.00	6.65
NDEA	102.079 31	6.70	7.20
NEIPA	116.094 96	7.00	7.40

- MS
  - SIM/Scan, scan range 70–300
  - Resolving power 30 000 (FWHM measured at *m/z* 200)
  - Automatic gain control (AGC) 5E4
  - Isolation window 20
  - MSX counts 4
  - EI ionization at 70 eV
  - Ion source at 250 °C, transfer line 220 °C.
  - Selected ions, see Table 12.16
- Solvent delay 6.00 minutes

12.14.2.3 UHPLC-MS Analysis

In the case of the less volatile nitrosamines, LC-based methods are necessary. NMBA and the nitrosamines formed from the API or intermediates in the synthesis of the API are not suitable for GC analysis. In addition, having an orthogonal method of analysis is recommended for verification of results and investigation of analytical artifacts. The small alkyl nitrosamines, and especially NDMA, are, however, characterized by limited retention by the standard OMCL LC methods. The Kinetex Biphenyl column (Phenomenex) described here offered suitable retention with an optimal matrix suppression profiling for typical Metformin API, especially with respect to the elution window for NDMA. The column demonstrated an added advantage through its ability to resolve DMF and NDMA responses, eliminating one of the false positive risks noted earlier [80].

12.14.2.4 Sample Preparation for Hydrophilic Samples (e.g. Metformin)

- 100 mg of sample was weighed into a suitably sized Eppendorf tube.
- 0.5 ml deionized water containing suitable quantities of internal standard was added (e.g. 4 ng/ml NDMA-d6 and NDEA).
- The sample was vortexed for 10 seconds.
- Liquid–liquid extraction was performed with 0.5 ml dichloromethane (trace grade) applying vortex for 10 seconds.
- The mixture was centrifuged for five minutes at 13 000 rpm and 200 μl DCM phase was transferred into a clean Eppendorf tube.
- 50 μl of water was added and DCM was carefully evaporated using an automated shaker at 1200 rpm at 30 °C until all DCM has been removed.
- The residue was reconstituted with 150 μl deionized water, vortexed for 10 seconds, and transferred to a clean HPLC vial, containing a low-volume insert.
- Note: this sample preparation method involving liquid–liquid extraction with DCM gives very poor recovery for NMBA.

12.14.2.5 Sample Preparation for Hydrophobic Matrices

- 100 mg sample of API (or drug product containing 100 mg of API) was weighed into an Eppendorf tube.
- 1 ml diluent (95% deionized water with 5% Methanol (v/v)) was added. The diluent contained an appropriate concentration of a suitable deuterated *N*-nitrosamine internal standard (e.g. 4 ng/ml NDMA-D6 and NDEA-D10).
- The solution was vortex mixed, then agitated on a knee mixer for 30 minutes.
- After agitation, the sample was centrifuged at 3000 rpm for five minutes.
- The supernatant was transferred to a fresh Eppendorf tube and centrifuged for a further three minutes at 3000 rpm.
- After five minutes, the clear supernatant is placed in a clean HPLC vial (ensure the vial and cap provide a clean blank).
- Note: In this sample preparation method, solvent selection and API/solvent ratio might require modification for individual drug product matrices.

Both sample preparations are improved by being automated from the particular perspective of limiting the risk of contaminating samples when manipulating samples. This is compounded by the requirement to have a lowest standard at 0.5 ng/ml for an effective 5 ppb (5 ng/g) LOD with respect to a 100 mg/ml sample preparation.

12.14.2.6 UHPLC Conditions

- Kinetex Biphenyl, 100 mm × 2.1 mm ID, 2.6 μm (Phenomenex)
- Eluent A: deionized water with 0.1% trace grade formic acid; eluent B: methanol (Pestinal or equivalent) with 0.1% trace grade formic acid
- Gradient: 0 minute: 100% A, 2.5 minutes: 100% A, 4 minutes: 50% A, 8 minutes: 5% A, 11 minutes: 5% A, 11.1 minutes: 100% A – total run time 15 minutes.
- Flow rate: 0.35 ml/minute
- Column temperature: 40 °C
- Injection volume: 20 μl
- Metformin API – divert to waste 0.00–1.70 minutes. Other APIs and drug products will require assessment of matrix and/or interference elution.

Note: The column and UHPC-MS may need additional cleaning procedure if sensitivity is seen to decrease during repeated sample analysis.

12.14.2.7 HRAM-MS and MS/MS Conditions

- Ionization:
  - ESI positive (ESI negative for NMBA) on Thermo Scientific Exactive™ LC Orbitrap™ GC-MS
  - APCI positive on Sciex Triple Quadrupole (QqQ) MS
- HRAM-MS acquisition mode in SIM or MRM acquisition with triple quadrupole.
- Ion selection: see Table 12.17

12.14.3 Typical Results Obtained for Volatile *N*-nitrosamines Using SHS-GC-MS

Typical total ion chromatograms (TIC) and EIC obtained for the analysis of *N*-nitrosamines using SHS-GC-HRMS are shown in Figure 12.26. The chromatograms obtained for blank solvent (DMSO), solvent spiked at 10 ppb level, blank Valsartan, and Valsartan spiked at 10 ppb level are compared.

Table 12.17 Selected ions for LC-HRAM MS or LC-MS/MS analysis of *N*-nitrosamines.

Compound	Mass (*m/z*) Orbitrap	Mass (*m/z*) Quant transition QqQ	Mass (*m/z*) Qual transition QqQ
NDMA	74.048 01	75.0–75.0	75.0–43.0
NDMA-D6	81.0930	81.1–46.0	N/A
NMBA	145.0619	147.1–87.1	147.1–117.3
NDEA	102.079 31	103.0–74.8	103.0–47.1
NDEA-D10	113.1494	113.0–81.1	N/A
NEIPA	116.094 96	117.1–75.1	117.1–43.0
NDIPA	131.1179	131.0–89.1	131.0–43.0

Headspace sampling removed the complexity of the matrix in a simple and generic fashion giving excellent signal to noise as seen in the EIC for NDMA, NDEA, and NEIPA in Figure 12.26. The LOD was well below the FDA requirement for all analytes at 5 ppb. Excellent mass accuracy (<2 ppm) was also obtained for the three analytes. This aligns with SANTE guidance [90] resulting in high confidence of identity in the context of trace quantitation. The risk of artifactual generation during the headspace heating phase has been observed when temperatures are too high and held for an excessive length of time [91]. This must be investigated during development and quenching of nitrosation may possibly remediate this effect.

In the case of the larger, less volatile *N*-nitrosodibutylamine, liquid injection GC-MS or GC-MS/MS is an approach that can be used, and several methods have been published by the OMCL groups [74]. Typical sample preparation involves liquid–liquid extraction of the analytes using a suitable solvent [92]. The shorter exposure to high temperature in liquid injection, compared SHS-GC-MS, may minimize artifactual formation of *N*-nitrosamines. However, artifactual generation of *N*-nitrosamines during sample preparation or during injection of certain drug product extracts has been observed (private communication) and this should always be considered in method development. It is possible that adding a nitrosation quenching reagent, as described above for SHS-GC-MS, could also control this effect.

12.14.4 Typical Results Obtained for *N*-nitrosamines Using LC-MS

The UHPLC method has been proven validated on an UHPLC-HRAM Orbitrap MS system and on a Sciex Triple Quadrupole UHPLC-MS/MS system. Using dichloromethane liquid–liquid extraction method, an LOD of 5 ppb was comfortably achieved on both systems. This is illustrated in Figure 12.27 by the extracted MRM chromatograms for four *N*-nitrosamines spiked into generic Metformin API at 15 ng/g level. NDMA-d6 and NDEA-d10 were added as isotope labeled internal standards. Analysis was performed on a SCIEX triple quadrupole MS in MRM mode.

The only drawback with the DCM extract is the poor recovery obtained for NMBA. Alternative aqueous methanol sample preparations [93] can be used to increase recovery, but this method must be tailored to individual matrices. This is illustrated in Figure 12.28, showing the EIC for NDMA, NMBA, NDEA, NEIPA, and NDIPA (with ISTDs NDMA-d6 and NDEA-d10) spiked at 10 ng/g in Metformin API obtained by LC-HRAM MS on an Orbitrap MS system using the water/methanol extraction method.

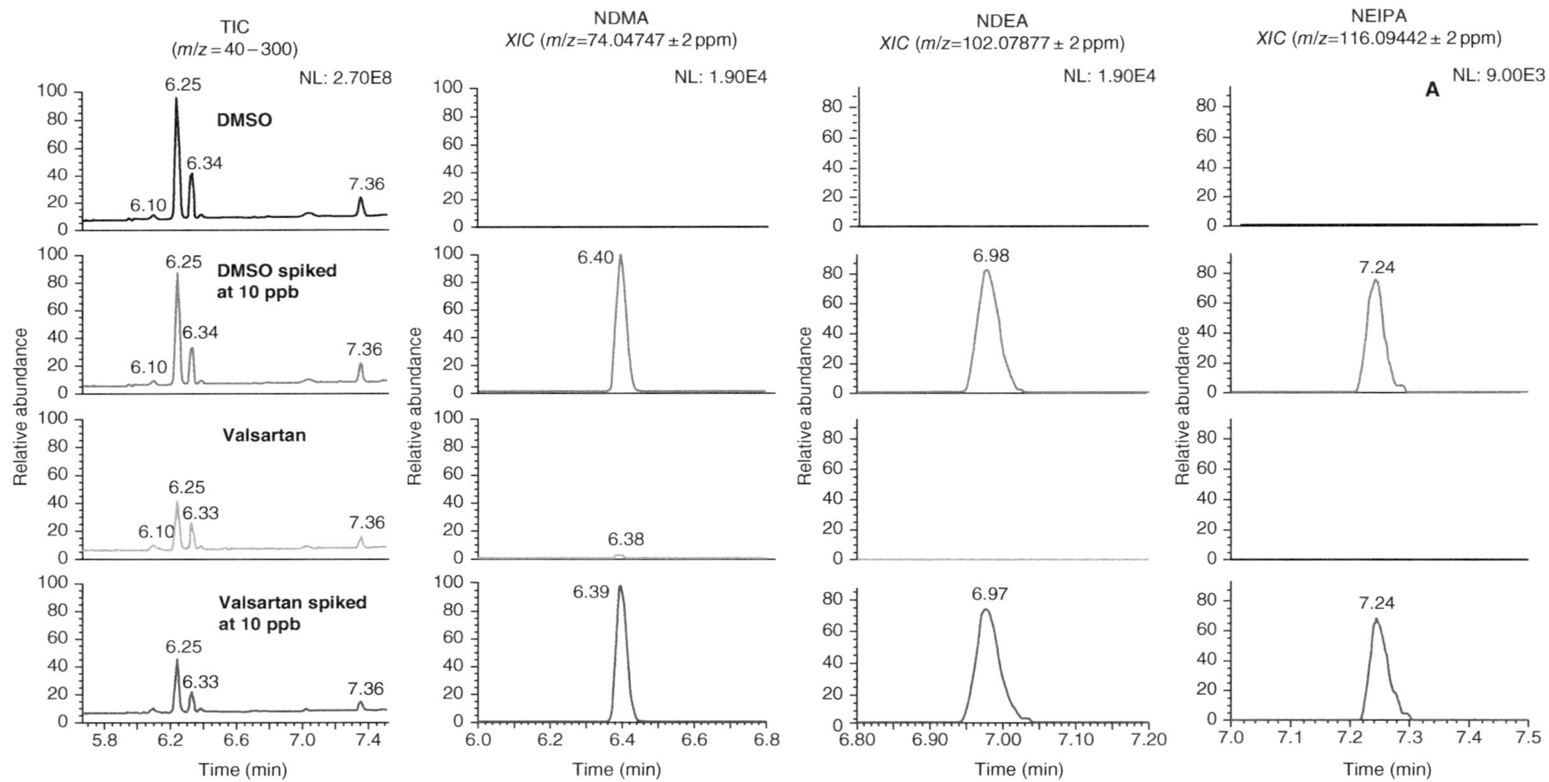

Figure 12.26 TIC and EIC chromatograms for blank solvent (DMSO), spiked solvent, Valsartan blank, and spiked Valsartan. *Source:* Copyright Thermo Scientific.

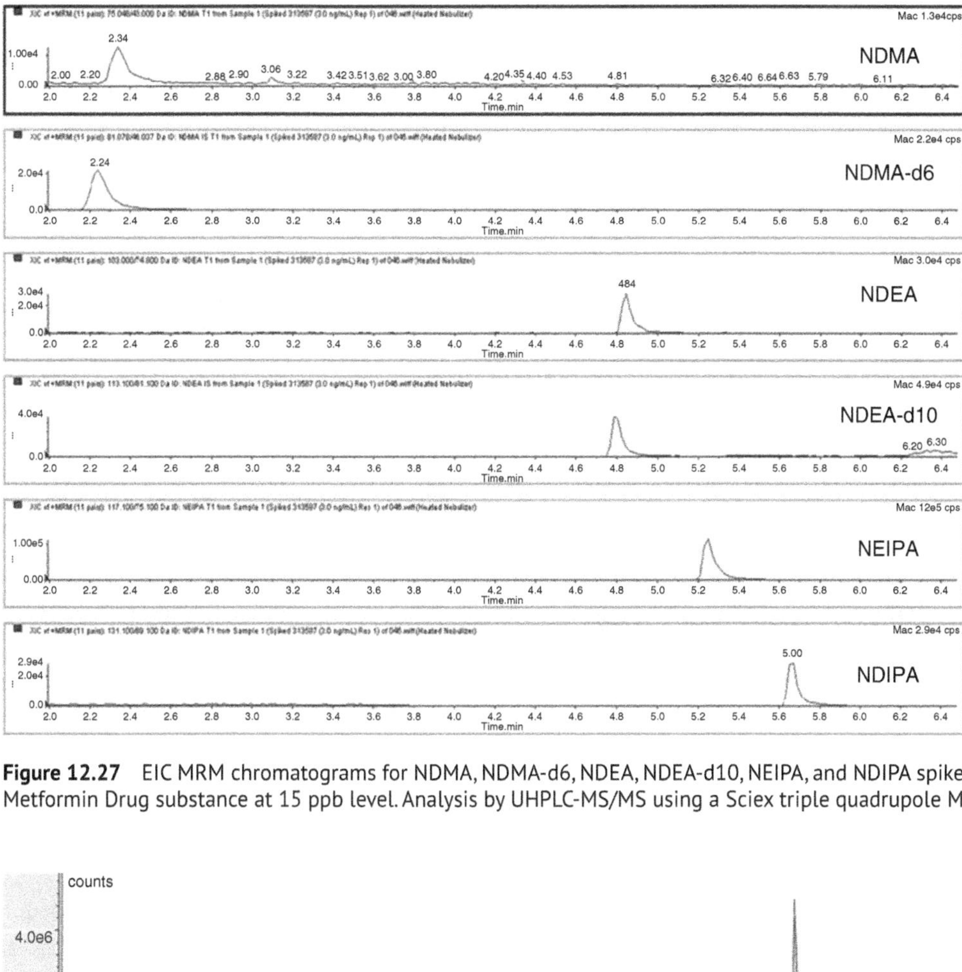

Figure 12.27 EIC MRM chromatograms for NDMA, NDMA-d6, NDEA, NDEA-d10, NEIPA, and NDIPA spiked in Metformin Drug substance at 15 ppb level. Analysis by UHPLC-MS/MS using a Sciex triple quadrupole MS.

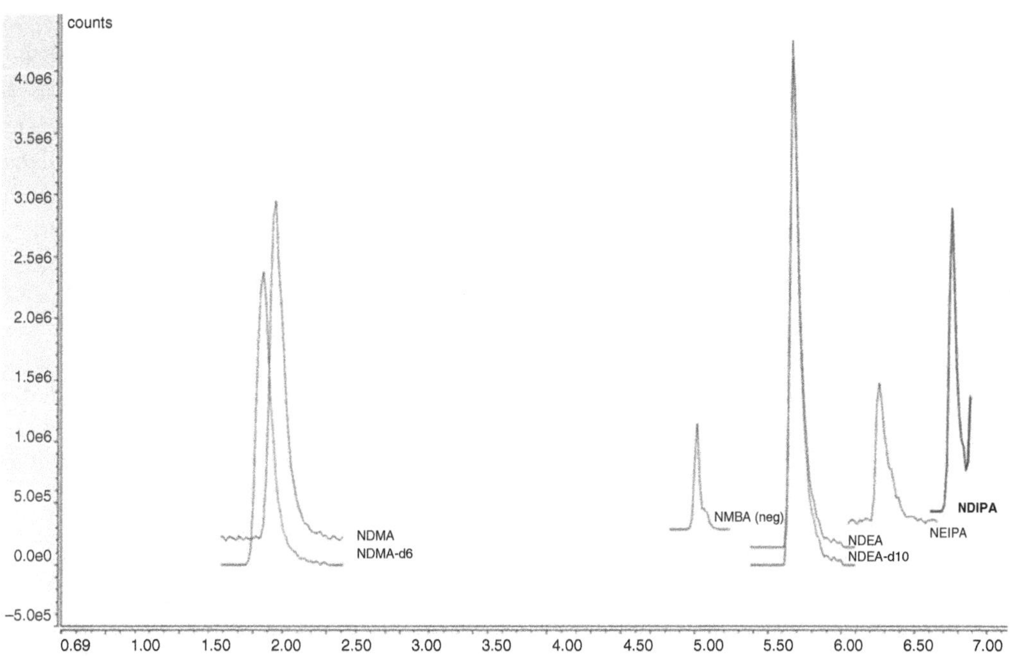

Figure 12.28 EIC chromatograms for NDMA, NDMA-d6, NMBA, NDEA, NDEA-d10, NEIPA, and NDIPA spiked in Metformin Drug substance at 10 ppb level. Analysis by UHPLC-HRAM MS using a Thermo Fisher Orbitrap LC-MS system.

For Metformin API matrix, good recovery for all analytes was achieved (all within 70–130% at 15 ppb) with excellent repeatability (<5% RSD achieved at the 15 ppb LOQ for all analytes). The standards and samples were stable in amber glassware and vials for 24 hours. The aqueous-based preparation dissolves a significant amount of Metformin and similarly for Ranitidine API or their various drug product formulations.

As an alternative approach, the published FDA official methods for *N*-nitrosamine analysis in Metformin [94] and Ranitidine [95] use a methanol extraction. This extract is cleaner than an aqueous organic dissolution for ranitidine and metformin, and provides good recoveries for NMBA, as well as for the *N*-Nitrosamines listed in Table 12.15.

12.15 Nontarget Analysis of PMI/MIs

The methods described above can be considered as generic methods targeting on specific classes of PMI/MIs. Recently, there has been a trend toward untargeted analysis, whereby impurities in API are analyzed by a screening method. Certainly, the availability of very sensitive HRAM systems, such as Q-TOF or Orbitrap mass spectrometers, in combination with ultrahigh performance liquid chromatography (UHPLC), have made such untargeted screening of solutes at low levels possible [18]. Based on accurate masses detected in API (or stressed API) solutions, molecular formula of previously unknown impurities or degradants can be generated, and molecular structures can be postulated. This can lead to the discovery of additional impurities that contain structural alert functionalities.

In addition to the developments in HRAM MS systems, also the developments in multidimensional separations should be mentioned here. Especially, comprehensive HPLC (LC×LC) can be considered as a powerful tool to enable the detection of "hidden" unknown impurities. This technique has been successfully applied in our laboratory for impurity analysis in API.

The potential of LC×LC in impurity detection is here illustrated by the analysis of unknown impurities in metoclopramide. In the classical generic impurity screening method, applying one-dimensional RPLC-DAD/MS, the separation was performed on a Bonus-RP column (150 mm × 2.1 mm ID, 1.8 μm) using a 0.2% ammonium acetate/methanol gradient. The DAD profile obtained for a sample spiked with known impurities and for a metoclopramide sample stressed at 40 °C is shown in Figure 12.29 (left). Four spiked impurities eluting after the main peak (API) are clearly visible in the upper chromatogram. These impurities are not detected in the stressed sample. A general concern in impurity analysis is, however, the possible co-elution of minor impurities with the API. To verify this, LC×LC was used. A dual HPLC configuration (Figure 12.4) equipped with a LC×LC switching valve (Figure 12.5, bottom) was used. The first dimension separation was the same RPLC method using a Bonus-RP column, operated at a low flow rate (80 μl/minute). The second dimension separation was also performed by RPLC, but using Zorbax Eclipse Plus C18 column (50 mm × 3 mm ID, 3.5 μm), operated at high flow rate (2.8 ml/minute) with a very fast (10.8 seconds) formic acid/acetonitrile gradient. The 40 μl collection loops were switched every 18 seconds (modulation time).

The obtained 2D plots for the spiked metoclopramide and stressed metoclopramide are given in Figure 12.29 (right). The four impurities eluting after the main peak are again detected and are slightly shifted to higher retention compared with the API in the second dimension separation. For analytes a and b, partly co-eluting in the first dimension separation, complete resolution is obtained in the second dimension. The most interesting observation is, however, the detection of a new impurity (marked with arrow, eluting just above the main API in second dimension separation). This impurity is completely co-eluting with the API in the first dimension separation but could now be detected and identified as an impurity with a structure alert functionality (structure not disclosed). This example demonstrates the potential of LC×LC-DAD/MS for untargeted impurity screening.

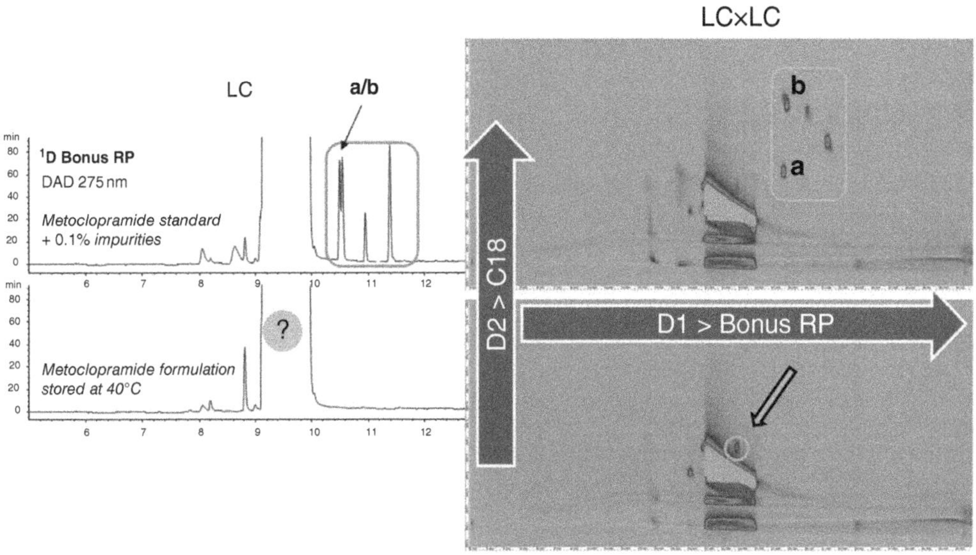

Figure 12.29 Untargeted analysis of unknown impurities in metoclopramide by 1D-LC (left) and LC × LC analysis (right). Conditions: see text. Unknown PMI detected in 2D indicated by arrow.

This untargeted screening is also applied in extractable and leachable studies, whereby the potential contamination of API or drug formulations by packaging material constituents is evaluated [8, 96]. In such studies, a whole range of analytical methods are applied, including headspace GC-MS techniques (for volatiles), and liquid extraction followed by GC-MS or LC-MS for semi-volatiles and nonvolatiles. In fact, the method selection charts presented in Figures 12.1 and 12.2 can also be used as guidelines for extractable and leachable studies.

12.16 Conclusions

This chapter describes a number of platform methods that can be applied for the determination of specific classes of PMI/MI solutes in various API at the trace (low ppm) level. Method selection was based on a previously presented method selection chart [4], still valid today and given in Figure 12.1. This method selection chart has proven to be a very effective tool that can be employed in the systematic selection of methods to be employed in the determination of PMIs/MIs in APIs.

The methods presented in the different sections should be considered a "generic," not optimized for a given PMI/MI in a specific matrix. However, we believe that each described method is a good starting point for further method development and a good guideline for solutes similar to the selected target compounds.

In general, SHS sampling in combination with GC-MS is a good starting point for all solutes with sufficient vapor pressure or that can be volatilized after derivatization. SPME and DHS can be used to further increase sensitivity, if needed.

Solutes with lower vapor pressure, but GC amenable, can be analyzed by direct liquid injection (eventually after derivatization) in combination with GC-MS. Heart-cutting 2D-GC is hereby useful to avoid solute interference and MS source contamination.

Nonvolatile compounds are analyzed by LC-MS. An automated selectable column approach using two reversed-phase LC or a RP-LC/HILIC approach was used for aziridines, arylamines and aminopyridines. Analytes with low retention in RP-LC can also be analyzed after pre-column derivatization. Derivatization is also useful to increase detectability in MS, as demonstrated by the analysis of aldehydes and hydrazines.

Since the development of these generic methods, new developments in GC, HPLC, and mostly MS instrumentation, have offered additional options for further increasing sensitivity, selectivity, and speed of analysis, as illustrated for the analysis of arylamines using UHPLC-MS/MS [57]. HRAM analyzers, eventually in combination with multidimensional LC, will further increase impurity coverage. These trends should, however, be evaluated in terms of real needs, fit-for-purpose performance, rather than for "fishing expeditions" [8].

Acknowledgements

The authors would like to thank all the individuals and technical groups within AstraZeneca who contributed to the setup and running of this project

The HRAMS-GCMS data were kindly provided by Thermo Fisher Scientific, UK.

References

1 ICH M7 (R1). Assessment and control of DNA reactive (mutagenic) impurities in pharmaceuticals to limit potential carcinogenic risk. Current Step 4 version dated 31 March 2017.
2 Muller, L., Mauthe, R.J., Riley, C.M. et al. (2006). *Regul. Toxicol. Pharmacol.* 44: 198.
3 Liu, D.Q., Sun, M., and Korda, A.S. (2010). *J. Pharm. Biomed. Anal.* 51: 999–1014.
4 David, F., Jacq, K., Vanhoenacker, G. et al. (2009). *LC.GC Europe* 11 (11): 552.
5 Vanhoenacker, G., David, F., and Sandra, P. (2011). Sensitivity enhancement for potential genotoxic impurity determination using the Agilent 1290 Infinity LC System and a 60 mm Agilent Max-Light Cartridge Cell. Agilent Technologies Publication Number 5990-7443EN.
6 Bai, L., Sun, M., An, J. et al. (2010). *J. Chromatogr. A* 1217: 302–306.
7 Raman, N.V.V.S.S., Prasad, A.V.S.S., and Reddy, K.R. (2014). *J. Pharm. Biomed. Anal.* 89: 276–281.
8 Teasdale, A. and Elder, D.P. (2018). *TrAC Trends Anal. Chem.* 101: 66–84.
9 Liu, D.Q. and Kord, A.S. (2013). *TrAC Trends Anal. Chem.* 49: 108–117.
10 Sun, M., Bai, L., Terfloth, J. et al. (2010). *J. Pharm. Biomed. Anal.* 52: 30–36.
11 Information on structural alert functionalities. http://ecb.jrc.ec.europa.eu/documents/QSAR/EUR_23844_EN.pdf (October 2020).
12 Székely, G., Henriques, B., Gil, M. et al. (2012). *J. Pharm. Biomed. Anal.* 70: 251–258.
13 Székely, G., Henriques, B., Gil, M., and Alvarez, C. (2014). *Drug Test. Anal.* 6: 898–908.
14 Grigori, K., Loukas, Y.L., Malenović, A. et al. (2017). *J. Pharm. Biomed. Anal.* 145: 307–314.
15 Iliou, K., Malenović, A., Loukas, Y.L., and Dotsikas, Y. (2018). *J. Pharm. Biomed. Anal.* 149: 410–418.
16 Schmidtsdorff, S., Schmidt, A.H., and Pharm, J. (2019). *Biomed. Anal.* 174: 151–160.
17 ICH Topic Q 2 (R1), Validation of Analytical Procedures: Text and Methodology (CPMP/ICH/381/95). https://www.ema.europa.eu/en/documents/scientific-guideline/ich-q-2-r1-validation-analytical-procedures-text-methodology-step-5_en.pdf (October 2020).
18 Singh, S., Handa, T., Narayanam, M. et al. (2012). *J. Pharm. Biomed. Anal.* 69: 148–173.
19 Baldwin, S., Bristow, T., Ray, A. et al. (2016). *Rapid Commun. Mass Spectrom.* 30: 873–880.
20 Ramachandra, B. (2017). *Crit. Rev. Anal. Chem.* 47: 24–36.
21 Hadda, P.R., Nesterenko, P.N., and Buchberger, W. (2008). *J. Chromatogr. A* 1184: 456–473.

22 Karthikeyan, K., Shanmugasundaram, P., Ramadhas, R., and Chandrasekara, P.K. (2011). *J. Pharm. Biomed. Anal.* 54: 203–207.

23 Prasanna, S.J., Sharma, H.K., Mukkanti, K. et al. (2009). *J. Pharm. Biomed. Anal.* 50: 1065–1069.

24 Altria, K. and Elder, D.P. (2004). *J. Chromatogr. A* 1023: 1–14.

25 Zhu, Q. and Scriba, G.K.E. (2018). *J. Pharm. Biomed. Anal.* 147: 425–438.

26 Khan, M., Kumar, M.S., Jayasree, S.K. et al. (2012). *J. Pharm. Biomed. Anal.* 58: 27–33.

27 Van Wijk, A.M., Niederländer, H.A.G., Van Ogten, M.D., and De Jong, G.J. (2015). *Anal. Chim. Acta* 874: 75–83.

28 Patel, I., Venkatramani, C.J., Stumpf, A. et al. (2017). *Org. Process Res. Dev.* 21: 182–186.

29 Harigaya, K., Yamada, H., Yaku, K. et al. (2014). *Anal. Sci.* 30: 377–382.

30 Langford, V.S., Graves, I., and McEwan, M.J. (2014). *Rapid Commun. Mass Spectrom.* 28: 10.

31 Devenport, N.A., Sealey, L.C., Alruways, F.H. et al. (2013). *Anal. Chem.* 85: 6224–6227.

32 Smith, R.W., Reynolds, J.C., Lee, S.-L., and Creaser, C.S. (2013). *Anal. Methods* 5: 3799–3802.

33 Elder, D.P., Lipczynski, A.M., and Teasdale, A. (2008). *J. Pharm. Biomed. Anal.* 48: 497–507.

34 Ho, T.D., Yehl, P.M., Chetwyn, N.P. et al. (2014). *J. Chromatogr. A* 1361: 217–228.

35 DHS brochure Gerstel GmbH. http://www.gerstel.com/en/dynamic-headspace.htm (June 2020).

36 David, F., Alegre, M.R., Vanhoenacker, G., and Sandra, P. (2013). *LCGC*: 31–34.

37 Elder, D.P., Teasdale, A., and Lipczynski, A.M. (2008). *J. Pharm. Biomed. Anal.* 46: 1–8.

38 Wollein, U. and Schramek, N. (2012). *Eur. J. Pharm. Sci.* 45: 201–204.

39 Liu, Z., Fan, H., Zhou, Y. et al. (2019). *J. Pharm. Biomed. Anal.* 168: 23–29.

40 Guo, T., Shi, Y., Zheng, L. et al. (2014). *J. Chromatogr. A* 1355: 73–79.

41 Alzaga, R., Ryan, R.W., Taylor-Worth, K. et al. (2007). *J. Pharm. Biomed. Anal.* 45: 472.

42 Jacq, K., Delaney, E., Teasdale, A. et al. (2008). *J. Pharm. Biomed. Anal.* 48: 1339.

43 An, J., Sun, M., Bai, L. et al. (2008). *J. Pharm. Biomed. Anal.* 48: 1006–1010.

44 Liu, X.-W., Zhang, W.-P., Han, H.-Y. et al. (2018). *J. Pharm. Biomed. Anal.* 155: 104–108.

45 Grinberg, N., Albu, F., Fandrick, K. et al. (2013). *J. Pharm. Biomed. Anal.* 75: 1–6.

46 Shackman, J.G. (2019). *Methods X* 6: 2176–2180.

47 Chadwick, J., Kelly, A., and Quirk, E. (2017). *Anal. Methods* 9: 1603–1610.

48 Klee, M.S. (2009). *J. Sep. Sci.* 32: 88–98.

49 David, F., Jacq, K., Sandra, P. et al. (2010). *Anal. Bioanal. Chem.* 396: 1291–1300.

50 Elder, D.P., Snodin, D., and Teasdale, A. (2010). *J. Pharm. Biomed. Anal.* 51: 1015–1023.

51 Klick, S. (1995). *J. Pharm. Biomed. Anal.* 13: 563–566.

52 Klick, S. (1995). *J. Chromatogr. A* 689: 69–76.

53 Chen, L., Zhang, W., and Hu, S. (2017). *J. Pharm. Biomed. Anal.* 146: 103–108.

54 David, F. (2019). *Hyphenations of Capillary Chromatography with Mass Spectrometry*, 1ste (eds. P. Tranchida and L. Mondello). Elsevier. ISBN: ISBN: 9780128096383.

55 Zapata, J., Temprado, J., Mateo-Vivaracho, L., and Ferreira, V. (2011). *J. Pharm. Biomed. Anal.* 55: 458–465.

56 Vanhoenacker, G., Dumont, E., David, F. et al. (2009). *J. Chromatogr. A* 1216: 3563–3570.

57 Vanhoenacker, G., David, F., and Sandra, P. (June 2010). *Agilent Technologies Application Note 5990-5732EN*. Agilent.

58 Dong, S., Yan, Z., and Yang, H. (2016). *Anal. Sci.* 32: 1333–1338.

59 Wu, X., Zhu, L., Visky, D. et al. (2014). *Anal. Methods* 6: 7277–7284.

60 Elder, D.P., Snodin, D., and Teasdale, A. (2011). *J. Pharm. Biomed. Anal.* 54: 900–910.

61 Sun, M., Bai, L., and Lui, D.Q. (2008). *J. Pharm. Biomed. Anal.* 49: 529–533.

62 Wang, J., Yang, S., and Zhang, K. (2016). *J. Pharm. Biomed. Anal.* 126: 141–147.

63 Cui, L., Jiang, K., Liu, D.Q., and Facchine, K.L. (2016). *J. Chromatogr. A* 1462: 73–79.

64 Li, J., Nadine Joelle, E.N., Yang, Q. et al. (2019). *Anal. Methods* 11: 6146–6152.

65 Song, M., Wu, S., Lu, P.-B. et al. (2016). *Anal. Methods* 8: 8352–8361.

66 Kumar, T., Xavier, N., and Ramya, M. (2019). *J. Chromatogr. Sci.* 57: 63–70.

67 Li, Z., Jacobus, L.K., Wuelfing, W.P. et al. (2006). *J. Chromatogr. A* 1104: 1–10.

68 Magee, P.N. and Barnes, J.M. (1956). *Br. J. Cancer* 10: 114–122.

69 Parr, M.K. and Joseph, J.F. (2019). *J. Pharm. Biomed. Anal.* 164: 536–549.

70 Dawson, B.A. and Lawrence, R.C. (1987). *J. Assoc. Off. Anal. Chem.* 70: 554–556.

71 Taylor, P., Braddock, P., and Carter, D. (1980). *IARC Sci. Publ.* 31: 575–587.

72 Castegnaro, M., Pignatelli, B., and Walker, E.A. (1981). *Food Cosmet. Toxicol.* 19: 489–491.

73 Severin, G. (1987). *J. Chromatogr.* 386: 57–63.

74 Sörgel, F., Kinzig, M., Abdel-Tawab, M. et al. (2019). *J. Pharm. Biomed. Anal.* 172: 395–405.

75 Scherf-Clavel, O., Kinzig, M., Besa, A. et al. (2019). *J. Pharm. Biomed. Anal.* 172: 278–284.

76 Shaik, K.M., Sarmah, B., Suresh Wadekar, G., and Kumar, P. (2020). *Crit. Rev. Anal. Chem.*: 1–19.

77 EMA. (2020). *Article 5.3 Opinion Nitrosamine impurities in human medicinal products.*

78 European Pharmacopoeia Nitrosamine monograph. Ph. Eur., 2019. 10th edition.

79 Council of Europe, E. Ad-hoc projects of the OMCL Network. (2019). Methods for determination of nitrosamines in sartans. EDQM and European Pharmacopoeia:State-of-the-art Science for Tomorrow's Medicines, Poster Session [Web]. https://www.edqm.eu/en/ad-hoc-projects-omcl-network (accessed August 2019).

80 Yang, J., Andres Marzan, T., Ye, W. et al. (2020). *AAPS J.* 22: 89.

81 Standardization, E.c.f. (2016). Safety of toys – Part 12: *N*-Nitrosamines and *N*-nitrosatable substances, 30 October 2016.

82 Combet, E., Paterson, S., Iijima, K. et al. (2007). *Gut* 56: 1678–1684.

83 Roasa, J., Liu, H., and Shao, S. (2019). *Food Addit. Contam. Part A Chem. Anal. Control Expo. Risk Assess* 36: 396–404.

84 Wierer, M., Wanko, R., and Bertrand, M. (2019). OMCL Network contribution to regulatory action with respect to the "sartan crisis". https://www.edqm.eu/sites/default/files/medias/fichiers/Events/edqm_european_pharmacopoeia_state-of-the-art_science_for_tomorrows_medicines_-_poster_session.pdf (June 2020).

85 SHS-GC-MS method was kindly provided by Thermo Fisher Scientific, United Kingdom.

86 Granger, J. and Sigman, D.M. (2009). *Rapid Commun. Mass Spectrom.* 23: 3753–3762.

87 Tanaka, K., Chung, K.C., Hayatsu, H., and Kada, T. (1976). *Food Cosmet. Toxicol.* 16: 209–215.

88 Lv, C.-L., Liu, Y.D., and Zhong, R.-G. (2009). *J. Phys. Chem.* 113: 713–718.

89 Mirvish, S.S. and National Cancer, J. (1970). *Institute* 44: 633–639.

90 (EURL), E.R.l.f.r.o.p. (2020). Method validation & quality control procedures for pesticide residues analysis in food & feed. SANTE/12682/2019.

91 Lim, H.-H., Oh, Y.-S., and Shin, H.-S. (2020). *J. Pharm. Biomed. Anal.* 189: 113460.

92 Abraham, M.H., Al-Hussaini, A.J.M., and Environmental, J. (2002). *Monitoring* 4: 743–746.

93 LGL, O. de_lgl_losartan_method_parameters_nmba_lcms. 2019. https://www.edqm.eu/sites/default/files/de_lgl_losartan_method_parameters_nmba_lcms.pdf (July 2020).

94 FDA, FY20-106-DPA-S_Liquid Chromatography-Electrospray Ionization-High Resolution Mass Spectrometry (LC-ESI-HRMS). (2020). Method for the Determination of Nitrosamine Impurities in Metformin Drug Substance and Drug Product, p. 11.

95 FDA, FY19-177-DPA-S_Liquid Chromatography-High Resolution Mass Spectrometry (LC-HRMS). (2019). Method for the Determination of NDMA in Ranitidine Drug Substance and Drug Product.

96 Scherer, N., Marcsekova, K., Posset, T., and Winter, G. (2019). *Int. J. Pharm.* 555: 404–419.

13

Analysis of Mutagenic Impurities by Nuclear Magnetic Resonance (NMR) Spectroscopy

Andrew R. Phillips[1] *and Stephen Coombes*[2]

[1] *Pharmaceutical Sciences, R&D Biopharmaceuticals, AstraZeneca, Macclesfield, UK*
[2] *Chemical Development, Pharmaceutical Technology and Development, Operations, AstraZeneca, Macclesfield, UK*

13.1 Introduction to NMR

Nuclear magnetic resonance (NMR) is a familiar technique in the sphere of structure elucidation. In combination with mass spectrometry (MS), the information present in an NMR spectrum can allow the full determination of unknown structures. More recently, NMR is playing an increasingly key role within the pharmaceutical industry for quantitative analysis.

NMR is an inherently insensitive technique and therefore not the natural choice for trace analysis. However, if sufficient sensitivity can be achieved through the correct experimental setup or the development of new technology, NMR becomes a natural choice for mutagenic impurity (MI) analysis because of its structural specificity and quantitative nature.

The aim of this chapter is to describe the theory required to understand the advantages and challenges of MI analysis by NMR and present a number of salient examples.

For a more general background to NMR, the book *High-Resolution NMR Techniques in Organic Chemistry* by T. D. W. Claridge provides an excellent introduction [1]. A wealth of other useful background information can be found in References [2–5].

13.2 Why Is NMR an Insensitive Technique?

13.2.1 Nuclear Spin

The nuclei of some atoms have an intrinsic property called nuclear spin angular momentum, which is characterized by the quantum number, I. This quantum number takes half integer values. Nuclei with $I = 0$ are NMR silent, such as carbon-12. Nuclei with $I > 1/2$ are NMR active, but they often display "exotic" NMR properties that result in effects such as very broad lines. Therefore, nuclei that are suitable for trace analysis have $I = 1/2$, for example ^{1}H and ^{19}F.

For a spin-1/2 nucleus, there are two possible quantum states, denoted α and β, that in a magnetic field have different energies. The energy difference, ΔE, increases with increasing magnetic field strength:

$$\Delta E = h\nu = \frac{h\gamma B_0}{2\pi}$$

Mutagenic Impurities: Strategies for Identification and Control, First Edition. Edited by Andrew Teasdale.
© 2022 John Wiley & Sons, Inc. Published 2022 by John Wiley & Sons, Inc.

where ν is the Larmor frequency in s^{-1}, γ is the gyromagnetic ratio in rad/s/T and B_0 is the magnetic field the nucleus experiences in T.

13.2.2 Boltzmann Distribution

What is crucial here is that for a collection of similar nuclei the total signal depends on the population difference between the α and β states – this is given by the Boltzmann distribution:

$$\frac{N_\alpha}{N_\beta} = e^{\Delta E/k_B T}$$

where $N\alpha, \beta$ are the number of nuclei in that state, kB is the Boltzmann constant, and T is the temperature.

At the field strengths that are used in NMR at room temperature this corresponds to a very small population difference – about 1 part in 10 000 for protons. This explains why NMR is such an insensitive technique.

The sensitivity depends on the nuclei used, both because of the different gyromagnetic ratio and the natural abundance. Table 13.1 lists selected spin-1/2 nuclei and shows very clearly that the relative sensitivities of ^{1}H and ^{19}F are significantly higher than other spin-1/2 nuclei, and hence these are the only nuclei that are suitable for trace analysis.

13.3 How Could NMR Be Used for Trace Analysis?

To understand how NMR could be used for trace analysis it is first important to explain a little about how an NMR spectrum is generated.

13.3.1 Generating an NMR Spectrum

The easiest way to understand how an NMR spectrum is generated is to think of the population difference between two spin states giving rise to a bulk magnetization vector, M_0, that behaves according to classical mechanics. This means, if perturbed, the vector will precess like a gyroscope around any applied field.

For example, if a radiofrequency (RF) pulse of the correct frequency is applied, it will cause the magnetization to precess. A correct choice of amplitude and length for this pulse results in rotation onto the x–y plane. The vector then precesses around the static field at a specific frequency, referred to as the Larmor frequency, causing an induced voltage in the RF coil surrounding the NMR sample. Over time the system will return to equilibrium resulting in a decay of this voltage – this is termed relaxation.

Table 13.1 Sensitivity of common spin-1/2 nuclei.

Nuclei	Natural abundance (%)	Gyromagnetic ratio/10^6 rad/s/T	NMR frequency υ (MHz)	Relative sensitivity
^{1}H	99.98	267.5	400.0	1.0
^{13}C	1.11	67.3	100.6	1.76×10^{-4}
^{15}N	0.37	−27.1	40.5	3.85×10^{-6}
^{19}F	100.00	251.8	376.3	0.83
^{29}Si	4.7	−53.2	79.5	3.69×10^{-4}
^{31}P	100.00	108.4	161.9	6.63×10^{-2}

Frequencies are for a 400 MHz (9.4 T) magnet.

Traditionally, this signal is called the free induction decay (FID), an amplitude time signal. Fourier transformation converts this into an amplitude frequency signal – with a signal appearing at the Larmor frequency. The whole process is illustrated in Figure 13.1.

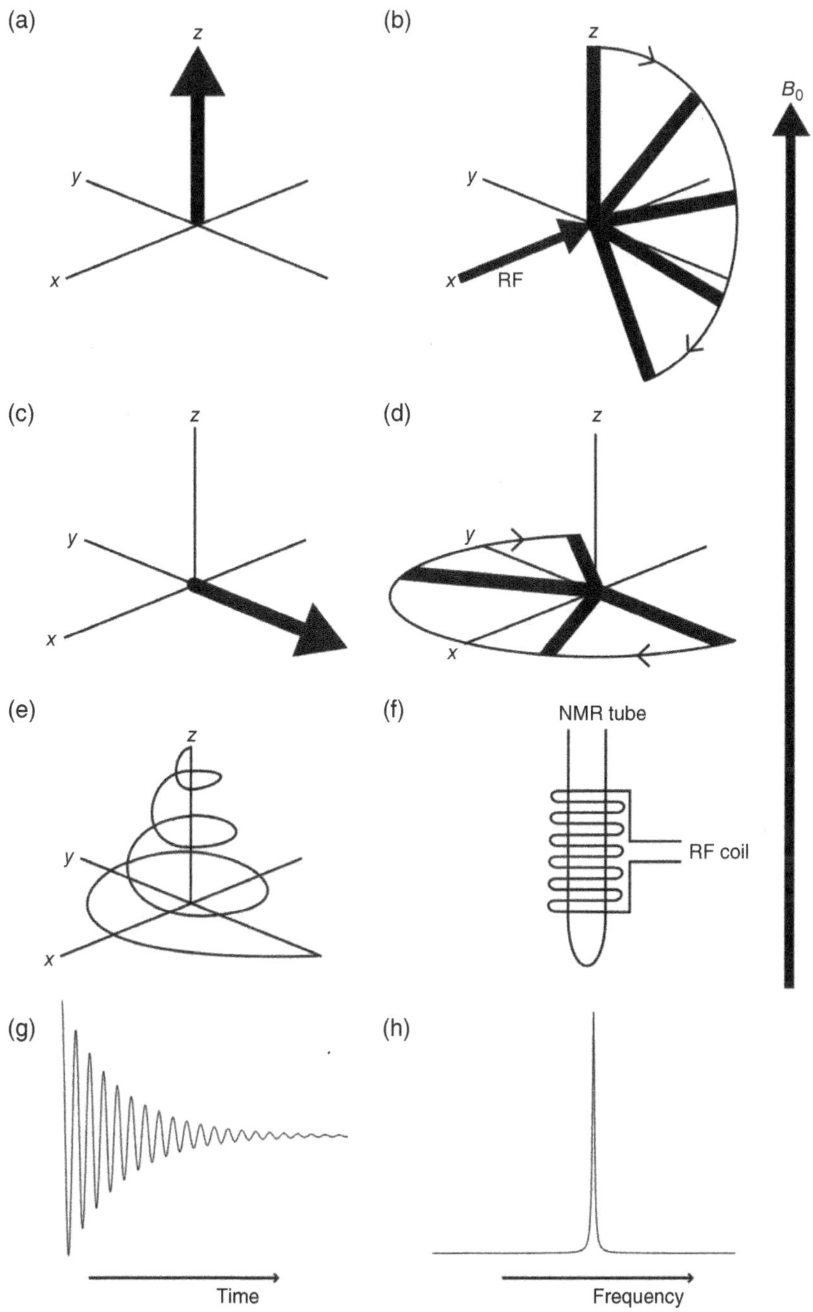

Figure 13.1 Diagrammatic representation of how an NMR spectrum is generated. (a) The bulk magnetization vector, M_0, at equilibrium. (b) Precession due to the applied RF pulse. (c) After an RF pulse has been applied for the correct length to time to rotate the magnetization by 90°. (d) Precession at the Larmor frequency around the static field, B_0. (e) Relaxation of magnetization back to equilibrium. (f) The RF coil that detects the precessing magnetization, which results in (g) the FID. Finally, Fourier transformation gives a spectrum shown in (h).

Two features of acquiring a spectrum that become important when thinking about MI analysis are firstly the ability to change how the RF pulse is applied (see Section 13.1.4.8.1) and secondly understanding the relaxation in the system to ensure the experiments are quantitative (see Section 13.1.3.5).

13.3.2 Chemical Shift

If all ^{1}H nuclei had the same Larmor frequency, NMR would not be a very useful technique. However, the beauty of NMR is that the local chemical environment alters the exact magnetic field a particular nucleus experiences. As these differences are tiny, it is conventional to express frequency in terms of a chemical shift, δ, in parts per million (ppm) defined as:

$$\delta = \frac{\nu - \nu_{\text{ref}}}{\nu_0}$$

where ν is the frequency of the nucleus in question compared to a reference, ν_{ref}, and ν_0 is the operating frequency of the NMR instrument. Table 13.2 lists some examples – for more detailed information see *Spectroscopic Methods in Organic Chemistry* by Williams and Fleming [6].

Table 13.2 Some examples of ^{1}H and ^{19}F chemical shifts.

Functional group	Range of ^{1}H chemical shifts (ppm)	Example ^{19}F chemical shifts (ppm)
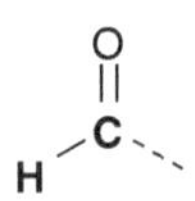	9.5–10.0	—
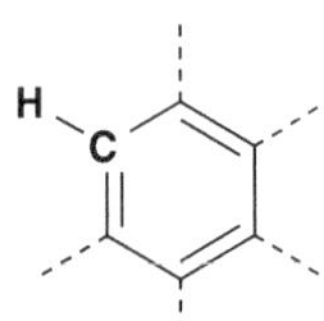	6.5–9.0	—
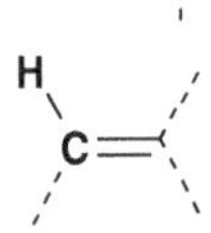	4.5–6.5	—
-O**CH2**CH_2CH_3	3.5–4.0	—
-OCH_2**CH2**CH_3	1.0–2.0	—
-OCH_2CH_2**CH3**	0.5–1.0	—
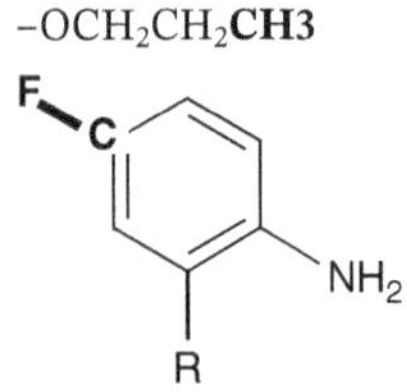	—	~−112
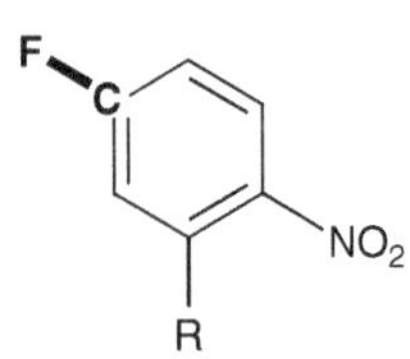	—	~−105
R–N(CH_3)CH_2CH_2–**F**	—	~−217
R–N(CH_2OH)CH_2CH_2–**F**	—	~−219

In favorable circumstances it is this chemical shift difference between atoms than can give resolution between a substrate and a potential MI.

For example, it can be seen that a common structural alert, an aldehyde functionality, has a distinct chemical shift in comparison to most other functional groups. It is therefore very likely when trying to detect an aldehyde proton in the presence of a substrate that it will be resolved. Signal resolution is also one of the reasons why ^{19}F NMR is an attractive nucleus for MI analysis given the high signal dispersion typically seen. Most common fluorine signals are observed over a spectral width covering approximately 300 ppm (compared to ~12 ppm for ^{1}H NMR), and the fluorine nucleus is exquisitely sensitive to changes in chemical structure, even when these changes are remote (>5–6 bonds away) from the fluorine atom. This means in most cases, impurities containing fluorine atoms will be resolved from the parent molecule in the resultant ^{19}F NMR spectrum.

There is also a dependence on the NMR solvent used. With modern spectrometers, any solvent can be used for NMR and there is no longer the requirement for the NMR solvent to be deuterated, although this can still be beneficial. It is conceivable that even if no resolution between two atoms is seen in one solvent the choice of a different solvent may achieve sufficient resolution.

The chemical shift of atoms near a possible protonation site will also change depending on the protonation state. This effect can be utilized to achieve resolution between nuclei that may not be resolved originally. For example, in compounds (**1**) and (**2**) the two protons H_A and H_B have approximately the same chemical shift in d_6-DMSO (dimethyl sulfoxide) and hence are not resolved from each other. The addition of a base deprotonates the NH_2 group, causing a change in the chemical shift of H_B. H_B is now resolved from H_A.

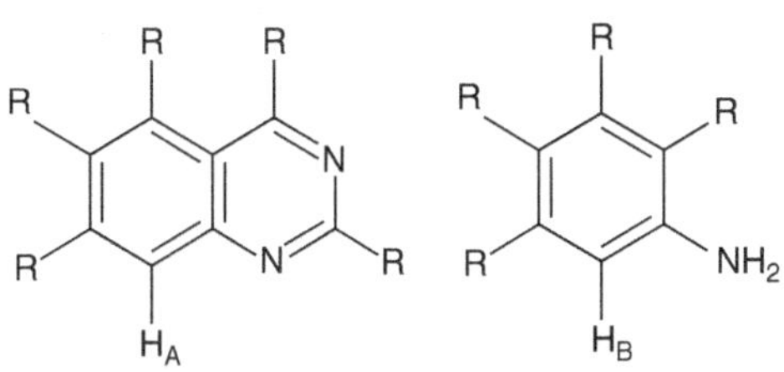

(1) (2)

13.3.3 Scalar Coupling

The chemical shift of a given nucleus is not only affected by the local chemical environment, but the exact frequency also depends on the spin-state of other nearby nuclei. For example, if you consider two nearby protons A and B as in Figure 13.2, the frequency of proton A is different depending on whether proton B is in spin-state α or β. The result of this is that the signal from proton A appears as a doublet (the signal is split into two signals of approximately equal intensity). The effect, called scalar coupling, continues for all nearby protons, often resulting in a complex multiplet structure.

Scalar coupling is extremely useful for structure elucidation, because it contains a wealth of molecular connectivity information. However, the coupling can cause a problem for trace analysis as it not only reduces sensitivity but can also result in less resolution between signals as shown in Figure 13.3.

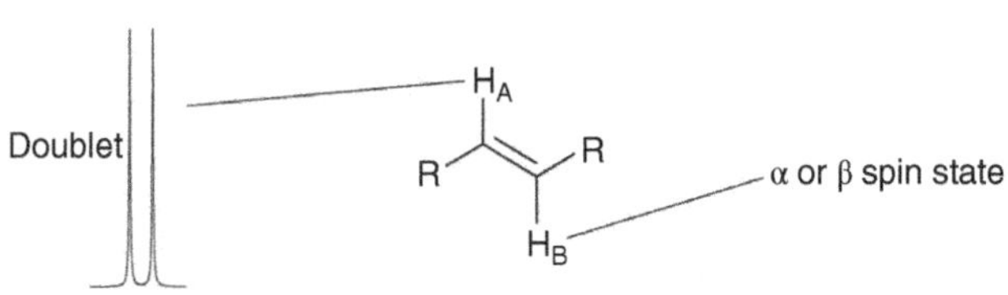

Figure 13.2 The different spin-states of H_B causes the signal from H_A to split into a doublet.

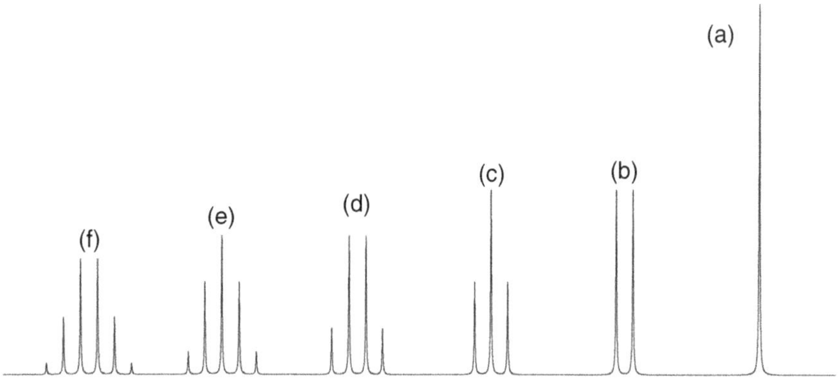

Figure 13.3 The effect of coupling on a single nucleus (a) not coupled and coupled to (b) 1, (c) 2, (d) 3, (e) 4, and (f) 5 other equivalent nuclei. As the degree of coupling increases, the signal intensity is reduced, leading to less sensitivity and the overall width of the signal increases, leading to an increase in the likelihood of signal overlap.

13.3.4 The Quantitative Nature of NMR

Despite the chemical shift and the scalar coupling dramatically changing the nature of the NMR spectrum, neither affects the total intensity that arises from each nucleus. This means that the intensity, as long as relaxation is taken into account, is a direct reflection of the total number of nuclei present. This is the basis of assay measurement. It is possible to calculate the purity of a substrate by the addition of a known purity standard and the measurement of the integral ratio between the purity standard and the substrate in question, using the formula:

$$\text{Assay}\%\text{w}/\text{w} = \frac{I_s \times MW_s \times P_p \times N_p \times W_p}{I_p \times MW_p \times N_s \times W_s}$$

where Ip,s are the respective integrals of the internal purity standard (p) and substrate (s), MWp,s are the molecular weights (MWs), Np,s are the number of protons in the signals integrated, Wp,s the weights added, and Pp the percentage purity of the standard.

In Figure 13.4, integration of the internal purity standard, tetrachloronitrobenzene (TCNB), relative to the six protons from the two CH_3 groups in atenolol enables an assay value for atenolol of 99.2% to be calculated.

In the same way, the ratio between a substrate and a residual solvent or a trace level impurity can be used to calculate the amount present without the need for an internal standard. The following equation can be used to calculate the % w/w of any impurity relative to the main component:

$$\text{Imp}\%\text{w}/\text{w} = 100 \times \frac{I_i \times MW_i \times P_s \times N_s}{I_s \times MW_s \times N_i}$$

where Ii,s are the respective integrals of the impurity (i) and substrate (s), MWi,s are the MWs, Ni,s the number of protons in the signals integrated, and Ps the purity. For trace impurities it is more convenient to convert to ppm (do not confuse with the NMR chemical shift):

$$\text{Imp}(\text{ppm}) = \text{Imp}\%\text{w}/\text{w} \times 10000$$

Many examples of quantitative NMR applied to pharmaceutical analysis have been published [7].

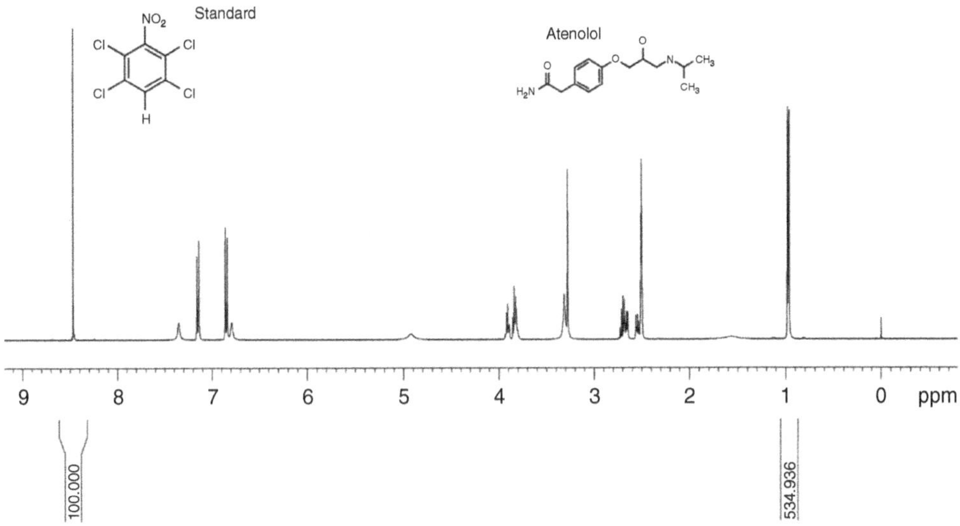

Figure 13.4 Quantitative ^{1}H spectra of atenolol (12.36 mg, MW 266.3 g/mol) in d_6-DMSO recorded at 400 MHz; 13.50 mg TCNB (purity 99.8%, MW 260.9 g/mol) has been added as internal standard.

13.3.5 Relaxation

In NMR there are two important relaxation phenomena termed longitudinal and transverse relaxation. Longitudinal relaxation can be understood as the exponential recovery of magnetization along the *z*-axis (*Mz*) following an RF pulse as shown in Figure 13.1e. Mathematically this is given at a time *t* as:

$$M_z = M_0\left(1 - e^{-t/T_1}\right)$$

where T_1 is the time constant in seconds. Figure 13.5 shows this decay and as can be seen, 95% of magnetization has returned to *z* after 3× T_1, while 99.3% has returned after 5× T_1.

For small- to medium-sized organic molecules, proton T_1 values range from 0.4 to 10 seconds. Generally, the smaller the molecule the longer the T_1, and it must be remembered that different nuclei in a molecule will have different T_1 values. Measurement of T_1 values can be done using the inversion recovery sequence – see Claridge [1] for practical details.

Therefore, it is important that any quantitative NMR experiment ensures that following a 90° pulse, sufficient time (5 × T_1) is left to allow full relaxation before the experiment is repeated (for signal averaging). Otherwise the differential relaxation across the spectrum will result in the integrals not being a direct reflection of the number of that nuclei present in the sample.

It should be noted that a fully quantitative spectrum is obtained at the expensive of sensitivity (per unit time) – see Section 13.1.4.5 for more details.

Transverse relaxation can be described as the "fanning out" of magnetization in the *x*–*y* plane. The time constant, T_2, for this relaxation manifests itself in the linewidth of signals. However, there is also a contribution to the linewidth from magnetic field inhomogeneity ($T_{2(\Delta B_0)}$). Traditionally, a combined time constant T_2^*, which is related to the actual observed linewidth, is defined as:

$$\frac{1}{T_2^*} = \frac{1}{T_2} + \frac{1}{T_{2(\Delta B_0)}}$$

In small- to medium-sized organic molecules in normal low-viscosity NMR solvents, the greatest contribution to the linewidth comes from this inhomogeneity term. In terms of quantification,

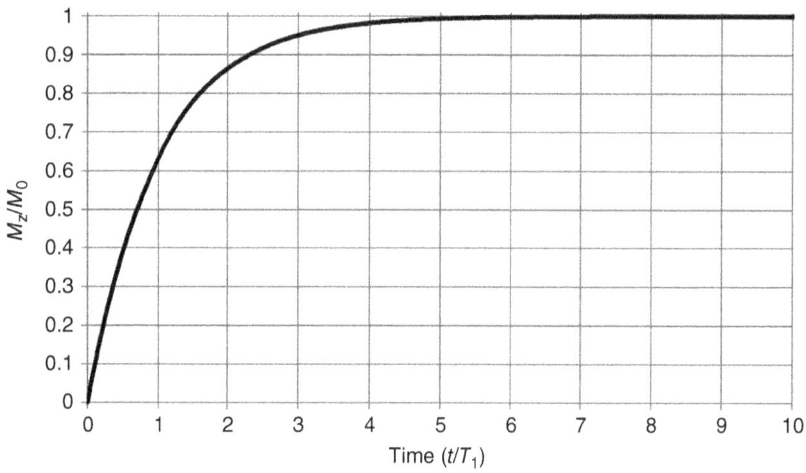

Figure 13.5 The return to equilibrium magnetization is governed by the relaxation time T_1 and is greater than 99% complete after a period of $5 \times T_1$.

because in general T_2 is similar to, but never bigger than T_1, it is not significant when ensuring that experiments are quantitative.

13.3.6 Summary

In summary, there a number of reasons why NMR could be a great tool for MI analysis:

- Chemical shift differences and the ability to manipulate them to give resolution between substrates and MIs.
- The selectivity offered by the ability to analyze different NMR active nuclei (e.g. ^{1}H or ^{19}F).
- The NMR spectrum is quantitative. As long as the correct precautions are taken, it is possible to directly calculate impurity levels without need for any calibration.
- Sample preparation is straightforward, and often little method development is required. More often than not, it is possible to determine if resolution can be achieved using reference data for each of the components; therefore, the substrate simply needs to be dissolved in the NMR solvent and the sample is ready for analysis.

As a result of these factors, NMR can be considered as a true orthogonal technique to liquid chromatography/gas chromatography (LC/GC) for MI analysis. However, this is of little value unless we can overcome the sensitivity issues. This is the topic of the next section.

13.4 What Can Be Done to Maximize Sensitivity?

We have already discussed the inherent insensitivity of NMR. This section explores the different factors that determine the sensitivity of a given experiment.

The signal-to-noise (S:N) ratio of a given signal depends on many factors and for our purposes can conveniently be represented by:

$$\frac{S}{N} \propto P \cdot C \cdot MW_i^{-1} \cdot NS^{1/2} \cdot A \cdot \gamma^{5/2} \cdot N \cdot T_2^*$$

where P is a “performance factor” for the NMR system used, C is the substrate concentration, MWi is the MW ratio of the MI, NS is the number of scans, A is the natural abundance of the nuclei, γ is the gyromagnetic ratio, N is the number of atoms contributing to a signal, and T_2^* is the apparent transverse relaxation time constant. Additionally, the sensitivity is affected by signal multiplicity as was illustrated in Figure 13.3.

Each of these factors will be considered in turn along with the impact of signal resolution.

13.4.1 System Performance

The performance factor, P, of a given NMR system depends on a number of features, primarily the magnetic field strength, B_0, and the probe performance for a given nuclei.

13.4.1.1 Field Strength

In terms of field strength:

$$\frac{S}{N} \propto B_0^{3/2}$$

This means, as shown in Figure 13.6, increasing the field strength does give a significant sensitivity enhancement (for example, going from 400 to 600 MHz gives a theoretical increase of ×1.8). However, as field strength increases, magnet costs also increase dramatically.

13.4.2 Probe Performance

Probe performance depends on a number of factors including probe design and diameter, which are discussed in the following sections.

13.4.2.1 Probe Design

Most NMR probes are designed to allow RF pulses to be applied to two or more nuclei of different frequencies at the same time. This allows the acquisition of a wide range of different experiments. For example, acquisition of one nuclei while the other is being decoupled (such as the common ^{13}C experiment with ^1H decoupling) or advanced two-dimensional experiments.

These experiments are achieved by the presence of two or more separate coils surrounding the sample. The ability of the coil to detect an NMR signal depends on many factors such as choice of material and geometry. However, one generally accepted factor is the "filling factor," or how "close" the coil is to the sample (Figure 13.7).

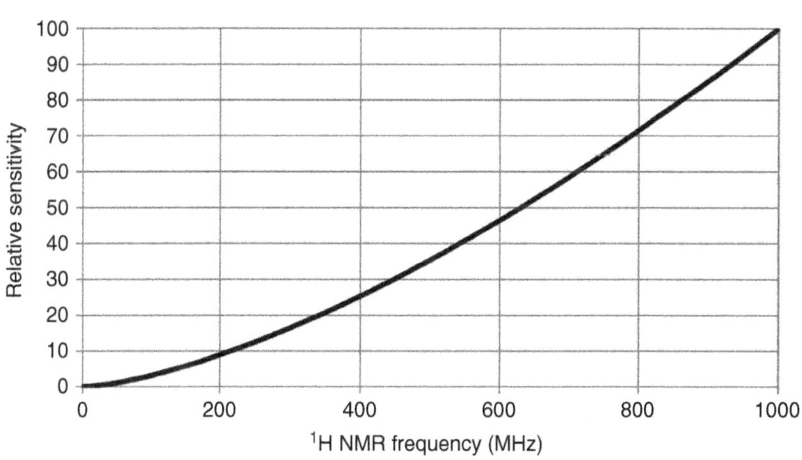

Figure 13.6 Relative sensitivity of the NMR experiment as a function of magnetic field strength.

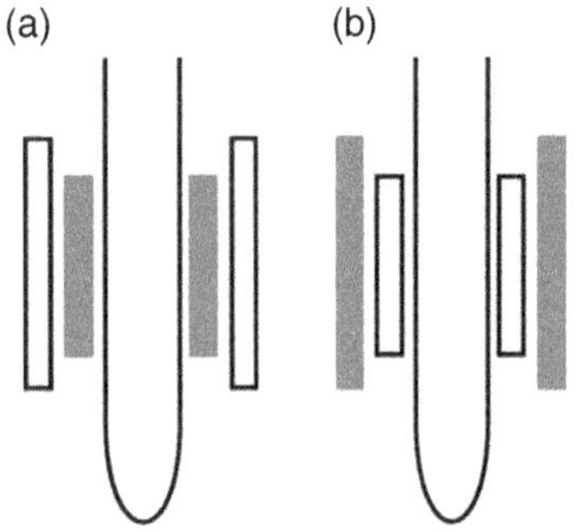

Figure 13.7 (a) Traditional and (b) inverse probe designs. The ^{1}H coil is shown in white and the X coil in gray.

The traditional design places the coil for the less sensitive nuclei (referred to as the X channel) nearest the sample. With the correct electronic setup (and dependent on the exact probe design), this coil is able to observe a wide range of frequencies, i.e. ^{13}C, ^{15}N, ^{31}P, and ^{19}F. Proton observation is then performed with the ^{1}H coil outside this X coil.

In comparison, the so-called inverse design places the proton coil nearer the sample. Clearly, to optimize sensitivity for the nuclei used for MI analysis, it is important to select a probe that is optimized: i.e. a traditional design for ^{19}F and an inverse design for ^{1}H (although some modern probe designs now allow for the ^{1}H coil to also be tunable to ^{19}F with only a small loss in sensitivity).

13.4.2.2 Probe Diameter

The second important factor when choosing an appropriate probe is the sample size. Probes are available that will allow the use of NMR tubes with an outside diameter of between 1 and 10 mm (or larger) as shown in Figure 13.8. Clearly, the larger the tube size the larger the active volume within the coil and hence a greater mass of material.

There are two ways to understand the sensitivity these different probe sizes allow. The first is to consider a fixed mass of material that is dissolved in an appropriate amount of NMR solvent. This method is of crucial importance when looking at mass-limited samples, for example, suppose an impurity has actually been isolated using chromatography and let us say 50 μg of material is available. In this case dissolving the material in a small amount of solvent and using a probe designed for a smaller NMR tube results in better mass sensitivity.

Second, as is more often the case for MI analysis, we are not usually sample limited. Generally, a large amount of substrate is available and to maximize sensitivity we need to maximize the total amount of material within the coil (i.e. by increasing the concentration). Therefore, what is relevant is the concentration sensitivity of the probe. The best way to measure this is to consider the $S:N$ obtained from the classic measure of instrument sensitivity – 0.1% ethyl benzene in $CDCl_3$. In this case the sensitivity will be proportional to the amount of material actually within the coil and hence the sample volume. This of course means the larger the probe diameter the better the concentration sensitivity. Table 13.3 lists the concentration sensitivity of a number of probe designs at different field strengths – the importance of correct probe design is clear.

13.4.2.3 Cryogenically Cooled Probes

Table 13.3 makes reference to cryogenic probes that give significantly enhanced sensitivity, but what are these?

In a "normal" probe, the RF coils are at room temperature and it is possible to represent the noise generated by the electronic hardware by the equation:

$$\frac{S}{N} \propto \frac{1}{\sqrt{T_C R_C + T_a\left(R_C + R_s\right) + T_S R_S}}$$

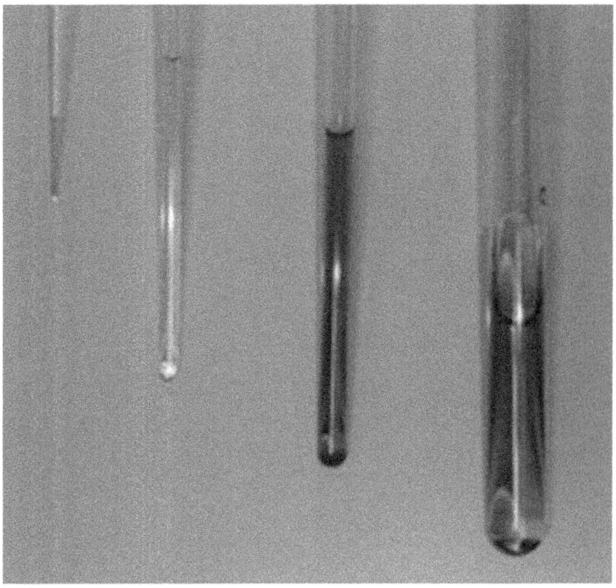

Figure 13.8 A variety of NMR tube diameters: 1, 3, 5, and 10 mm.

Table 13.3 Concentration sensitivity of some example probe designs at different field strengths.

Field (MHz)	Probe type	0.1% Ethyl benzene sensitivity
400	Traditional 5 mm	220 : 1
500	Traditional 5 mm	330 : 1
500	*Inverse* 5 mm	900 : 1
600	*Inverse* cryogenic probe 5 mm	6000 : 1

where T_C and R_C represent the temperature and resistance of the coil, T_a the effective noise temperature of the preamplifier, T_S the sample temperature, and R_S the resistance generated in the coil by the sample [8]. Although complex, this equation simply means that a decrease in the temperature of the coil results in a reduction of the noise from the preamplifier and thus a corresponding increased $S:N$.

Despite the engineering challenges of having an RF coil at a very low temperature while keeping the sample at room temperature, cryogenic probes that cool the RF coils and preamplifier down to 25 K are now relatively common in many research labs.

As can be seen in Table 13.3, these types of probes result in at least a fourfold increase in sensitivity (the additional increase seen arises from the increase in field strength). A cryoprobe installation is shown in Figure 13.9.

13.4.3 Substrate Concentration

As has already been stated, NMR reflects the actual number of nuclei within a sample. This means that the sensitivity increases linearly with concentration, and therefore increasing sample concentration is the easiest and cheapest option to ensure sufficient sensitivity. In MI analysis, the aim is to detect and measure the concentration of an impurity relative to the concentration of the substrate. For example, if you need to detect an MI, which is present at 1 ppm of the concentration of your substrate, and you only dissolve 1 mg of substrate in 1 ml of solvent, the NMR systems need to be capable of detecting 1 ng of MI (a tough task, even for ultrahigh field NMR spectrometers). In

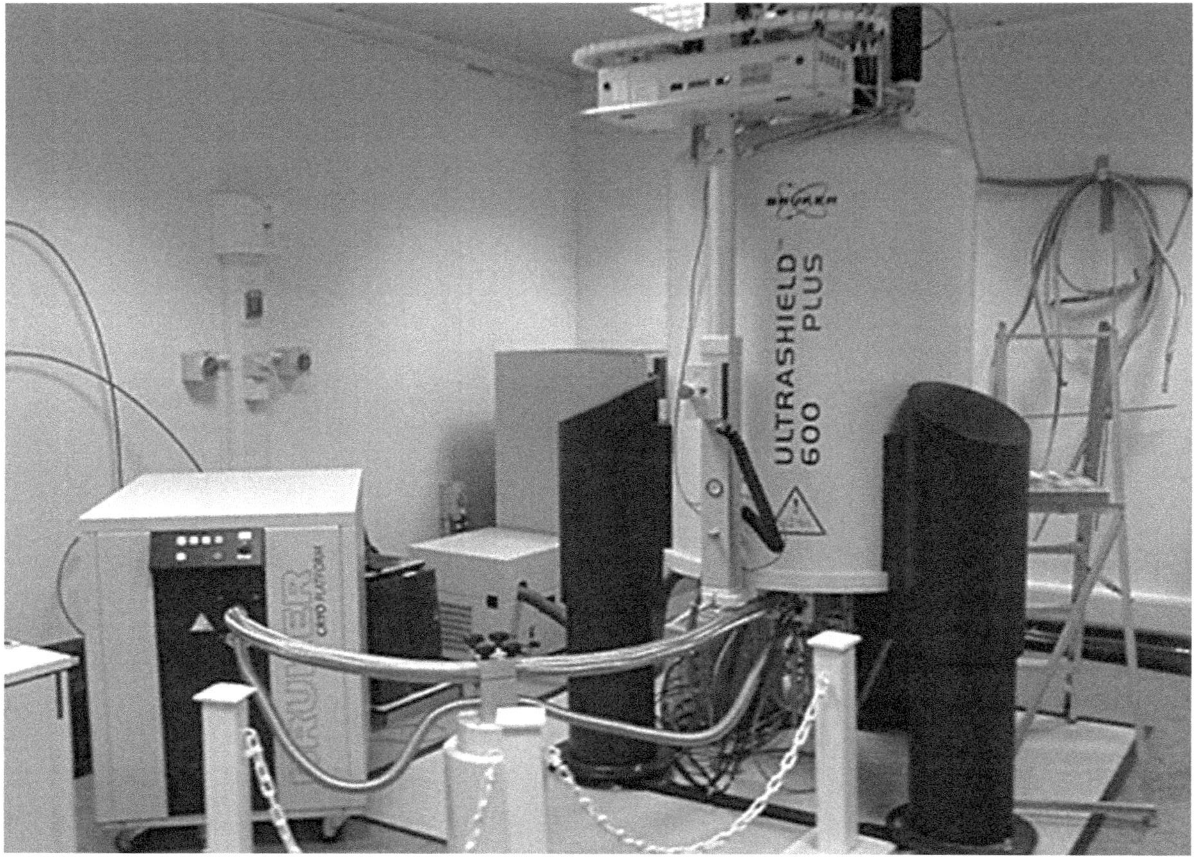

Figure 13.9 A cryoprobe installation on 600 MHz magnet. The unit on the left generates cold helium gas, which is transferred into the probe by the top connecting pipe. The bottom connecting pipe provides the vacuum, which isolates the sample from the cold coils.

contrast, if you dissolve 100 mg of substrate in 1 ml of solvent, the NMR only needs to be capable of detecting 100 ng of MI, which is much more realistic.

The substrate concentration that can be achieved in solution will therefore have a direct correlation with the levels of MI detection that are likely to be achieved.

Generally, the higher concentration the better, i.e. there is no limit as long as the material stays in solution for the period of the experiment, does not cause stability problems, or causes significant spectral broadening due to the high viscosity of the solution.

So, can NMR detect at these levels? Focusing on two amounts, 100 ng (1 ppm at 100 mg/ml) and 2 μg (20 ppm at 100 mg/ml), and using the $S:N$ ratios in Table 13.3, it is possible to estimate levels of detection (assuming a molecular weight identical to ethyl benzene).

As can be seen in Table 13.4, it is conceivable that even on a "normal" probe, 20 ppm could be detected, whereas 1 ppm might be difficult unless circumstances are favorable (assuming a typical limit of detection [LOD] of 3 : 1 and limit of quantification [LOQ] of 10 : 1). However, on a cryoprobe system, it appears theoretically possible to detect even 1 ppm. At this point it should be stated that this is a theoretical calculation – practically there are problems that will be discussed in a later section on dynamic range.

13.4.4 Molecular Weight Ratio

Sensitivity is proportional to the MW of the MI. Table 13.4 estimated whether there was enough sensitivity in an NMR experiment assuming that the MW was identical to ethyl benzene. The determination of the level of an MI in ppm is calculated on a weight to weight basis, so for example, 1 ppm of an MI at 100 mg/ml substrate concentration will always correspond to 100 ng of MI (irrespective of its MW). If the MI had a MW 10 times that of ethyl benzene, there would be 10 times less in terms of the number of moles of the MI in the 100 ng (compared to the number of moles of ethyl benzene present in 100 ng). NMR is a molar detection technique, and therefore a 10-fold reduction in the number of moles present would result in the sensitivity being 10 times less (compared to ethyl benzene).

Therefore, it will always be easier to detect small MW MIs by NMR in comparison to MIs that are perhaps analogs of bigger MW drug substances.

13.4.5 Acquisition Time and Signal Averaging

As noise is random, it increases in size more slowly than the signal, leading to a square root relationship between $S:N$ and the number of times the experiment is repeated, *NS*. As can be seen in Figure 13.10, this means to double the $S:N$ ratio from 0.5 to 1 takes four times the number of scans (25 vs. 100 scans) – and hence four times the time.

Table 13.4 Signal to noise in one hour assuming it is possible to pulse every five seconds – corresponding to 720 scans in one hour and hence a $\sqrt{720} = 26.8$ times increase in $S:N$ ratio.

Probe type	$S:N$ of 0.1% (1 mg) ethyl benzene in one scan	$S:N$ in one hour of acquisition	$S:N$ of 1 ppm	$S:N$ of 20 ppm
400 MHz 5 mm	220 : 1	5903 : 1	0.6 : 1	12 : 1
500 MHz inverse 5 mm	900 : 1	24 120 : 1	2.4 : 1	48 : 1
600 MHz inverse cryogenic 5 mm	6000 : 1	160 800 : 1	16 : 1	320 : 1

$S:N$ ratio at 1 and 20 ppm calculated assuming peak shape, MW, and multiplicity is identical to ethyl benzene.

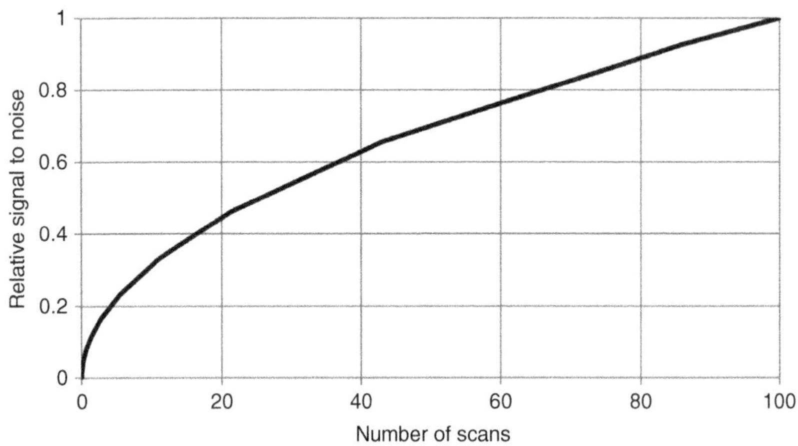

Figure 13.10 The $S:N$ as a function of the number of scans.

Therefore, although running the sample of interest for more time to increase the $S:N$ has great benefit, it only goes so far. For example, if you need to run the sample for 12 hours to detect 10 ppm – to get the same $S:N$ on 5 ppm would take $12\times4 = 2$ days – very quickly the time required can become very long. It is therefore much more beneficial to optimize the system in terms of correct probe/instrument selection and sample concentration rather than relying on simply running the sample for longer and longer. In addition, the issue of dynamic range means there comes a point when running the sample for longer will no longer make a difference.

The other time optimization factor that we can change is of course how long is left between each scan to allow for complete relaxation. As already stated, if we wish our experiment to be totally quantitative, we must leave $5\times T_1$ between each scan. Therefore, based on our approximate measurements of the T_1 values for both the substrate signals of interest and our MI, we have a minimum repetition time between scans. This of course can have a dramatic impact on the sensitivity we will be able to obtain, for example, consider two systems, A and B. In system A, the longest relaxing species has a T_1 of 500 ms, so in order to be fully quantitative our repetition between scans is 2.5 seconds and therefore in 1 hour of acquisition we are able to acquire 1440 scans. In system B, the longest relaxing species has a T_1 of 4 seconds (not unreasonable for a small MI molecule), leading to a repetition time between scans of 20 seconds, and so in 1 hour we are only able to acquire 180 scans (eight times fewer scans than in system A). Taking into account the square root factor, this corresponds to 2.8 times less $S:N$ in system B than in system A, over the same time.

However, there is something we can do when all other factors such as concentration or choice of instrument have been exhausted. Although leaving $5\times T_1$ gives quantitative spectra, it essentially *wastes time* – if pulses were applied more quickly, more sensitivity could be achieved in the same time but at the expense of quantitation. The use of a reduced flip angle, for example a 30° pulse, would only require a repetition delay of $3\times T_1$ for the experiment to be quantitative; however, using a reduced flip angle leads to a reduced signal intensity (and hence loss of sensitivity) and as such is undesirable for low-level signal detection.

For a given repetition time, T_R, and a given T_1, it is possible to calculate a pulse angle (i.e. not a 90° pulse) that optimizes sensitivity – this is called the Ernst Angle, α, which over time maximizes the signal intensity [3].

$$\cos\alpha = \exp\left(-\frac{T_R}{T_1}\right)$$

The strategy to maximize the sensitivity is to have a repetition time as fast as possible and then optimize the pulse angle based on this time and the relaxation time of the MI. How short *TR* can be is governed by T_2^* and hence how long the FID takes to decay – if the FID is not acquired for long enough, the signal is truncated, which leads to unwanted artifacts. It is also important to optimize the pulse angle based on the MI relaxation time, as of course this is the signal for which the maximum sensitivity is required.

The amplitude of each signal, ε is given by:

$$\varepsilon = \frac{1-\exp\left(-\frac{T_R}{T_1}\right)}{1-\exp\left(-\frac{T_R}{T_1}\right)\cos\alpha}\sin\alpha$$

For a given pulse angle and T_R, it is possible to calculate the signal intensity expected from each signal of interest in the spectrum (with different T_1 values) and hence the expected error in any quantitative measurement. Of course, this error is only a guide and so if this approach is used, a calibration curve must be generated.

For example, in a particular MI problem when ^{19}F NMR was used, it was shown that at a concentration of 400 mg/ml substrate it was possible to detect the impurity down to the required level of 8 ppm. A sample spiked with 1000 ppm of MI was used to measure T_1 values. As relaxation times depend on the sample conditions, it is important to measure them in as representative sample as possible. In this case, while it was possible that the relaxation time of the MI was different when present at 8 or 1000 ppm, it was clearly not feasible to perform the T_1 measurement on an 8 ppm sample. It was, however, very important to measure the MI relaxation time in the presence of the substrate.

In this case using the inversion recovery experiment, the T_1 of the substrate was estimated at 0.6 seconds and the MI at 1.0 seconds. Therefore, to ensure the experiment was quantitative, five seconds were left between scans. Using this repetition time, however, it was shown that 17 hours would be required to achieve the required $S{:}N$ ratio at 8 ppm – this was deemed too long. It was therefore decided to optimize the experiment time at the expense of quantification. Looking at the FID (see Figure 13.11) it was possible to reduce the repetition time down to 1.3 seconds without truncating the signal – therefore using the MI relaxation time, the optimum pulse angle of 74° was used. In addition, by calculating the amplitude expected from the substrate and MI signals, it was shown that the expected error in quantification was about 16% (as the MI is relaxing more slowly than the substrate, its signal intensity will be less than if the experiment were fully quantitative).

It is also possible to calculate the theoretical sensitivity enhancement. As can be seen in Figure 13.12, at the Ernst angle for a single scan a repetition time of 1.3 seconds results in only 75% of the signal. However, in a fixed time it is possible to acquire nearly 4 times as many scans, taking into account the square root factor giving a final enhancement of 1.5 times, which equates to a 2.2 times reduction in time. This was borne out experimentally, as using this setup there was sufficient sensitivity to detect the MI signal at a level of 8 ppm with an acquisition time of approximately six hours (as shown in Figure 13.11).

13.4.6 Number of Protons and Linewidth

NMR signals in a spectrum have areas that are proportional to the number of nuclei, *N*, present. However, each signal is not of equal height. Theoretically, NMR lines have a Lorentzian lineshape given by the following equation:

$$S(\omega) = \frac{R}{R^2 + (\omega - \Omega)^2}$$

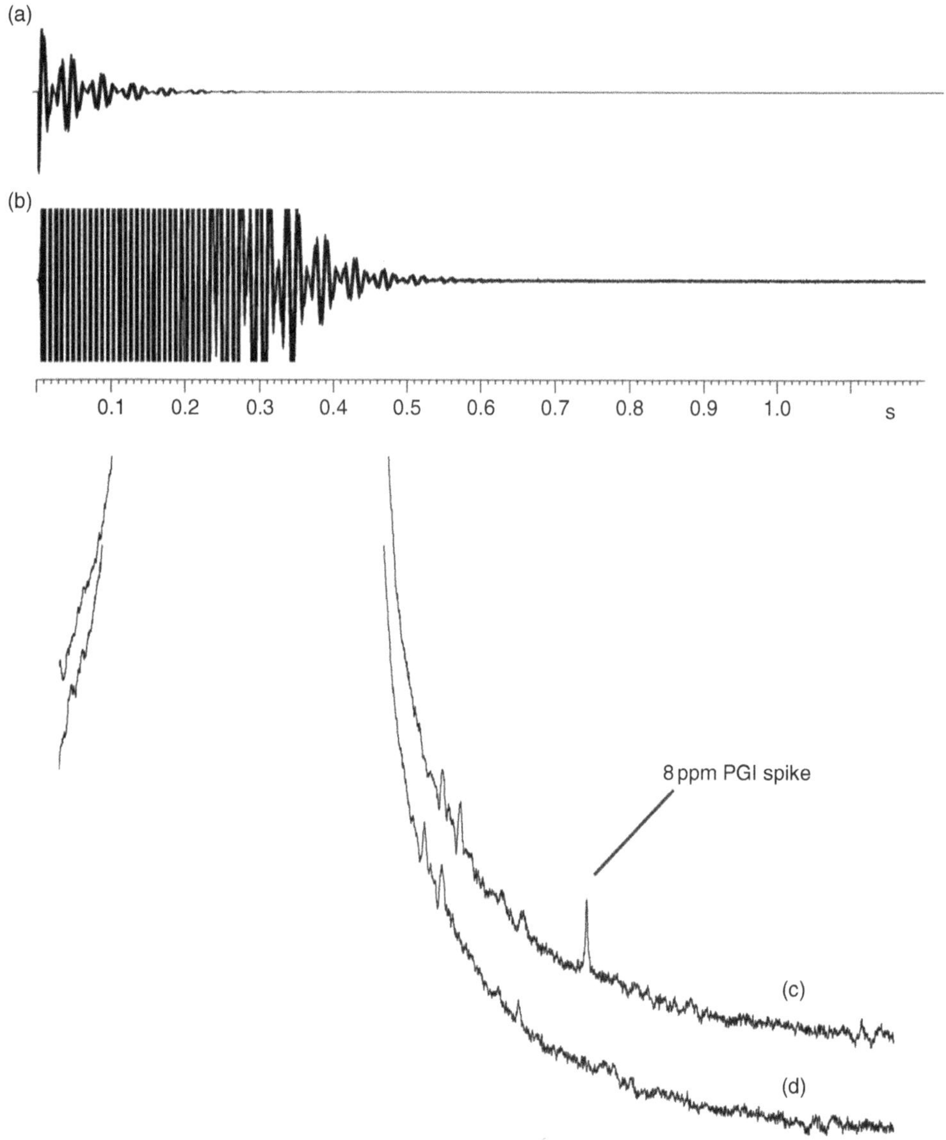

Figure 13.11 400 mg/ml substrate (a) FID of ^{19}F NMR spectrum with ^{1}H decoupling and (b) same FID expanded vertically. As can be seen, 1.3 seconds is more than enough time to allow the signal to disappear. Fourier-transformed spectra showing the region where the MI signal appears, (c) with and (d) without 8 ppm MI spike. Signal from substrate is on the left disappearing upward. Data was acquired as per the text with a 74° pulse angle and 1.3 seconds repetition time in 6 hours.

where $S(\omega)$ is the signal height, ω is the signal frequency, Ω is the spectral offset, and R is defined as $1/T_2^*$. From this equation it is easy to see (by setting $\omega = \Omega$, i.e. on resonance) that the signal height is $1/R = T_2^*$, and hence sensitivity is directly proportional to the relaxation, T_2^*.

Figure 13.13 shows the effect of increasing linewidth on the intensity and sensitivity of signals. Therefore, in terms of MI analysis, the linewidth of any MI signal will have a direct impact on the levels of detection that are achievable. This should be remembered when optimizing experiments in terms of concentration, solvent choice, and temperature; all of which can have an effect on linewidth.

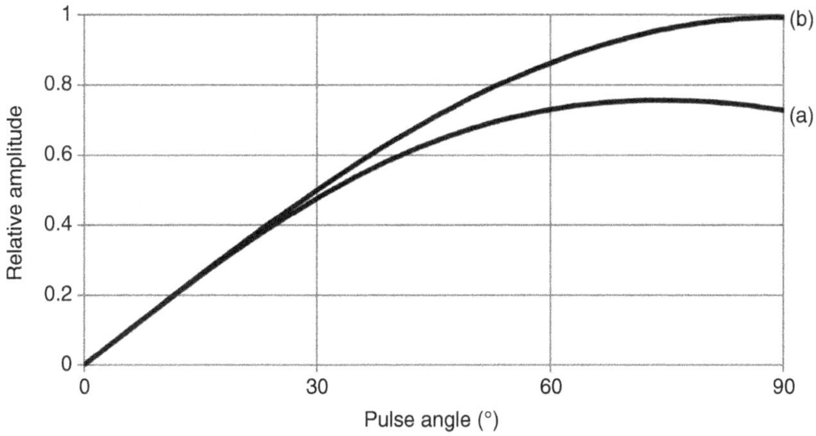

Figure 13.12 Relative amplitude as a function of pulse angle for a fixed relaxation time of 1 seconds and a repetition time of (a) 1.3 seconds and (b) 5 seconds (quantitative conditions). Although the maximum amplitude is less when the repetition time is only 1.3 seconds, in a fixed time, more scans will have been acquired leading to an overall 1.5 times enhancement in sensitivity.

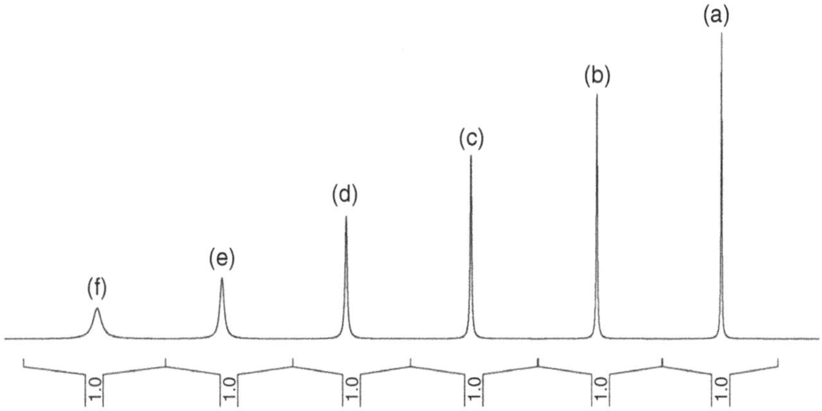

Figure 13.13 The effect of linewidth on sensitivity. All signals from a single proton (a)–(f) have the same peak area but increasing linewidth that reduces the signal intensity and leads to a reduction in sensitivity.

13.4.7 Resolution

We have now addressed all the terms in the original equation leaving just the resolution between signals to consider. In this context, the linewidth of substrate signals and the frequency difference between an MI signal and substrate will directly influence the sensitivity that can be achieved. This is because NMR Lorentzian lines have long tails that at ppm levels can easily obscure very small signals. This is clearly seen in Figure 13.14 as, for example, an MI signal present at a level of 1 ppm with a chemical shift difference of 0.9 ppm from the substrate signal is easily detected and resolved, whereas a signal with a chemical shift difference of 0.2 ppm is only just detected and certainly could not be integrated. However, at a level of 20 ppm, a signal with a chemical shift difference of 0.2 ppm is easily detected.

13.4.8 Dynamic Range

So far, all the discussion has been around the theoretical ability of NMR to detect the very low mass amounts of MIs that are present in solution. It has been argued that although a large number of factors have an impact, NMR does possess enough sensitivity to make such analysis a practical

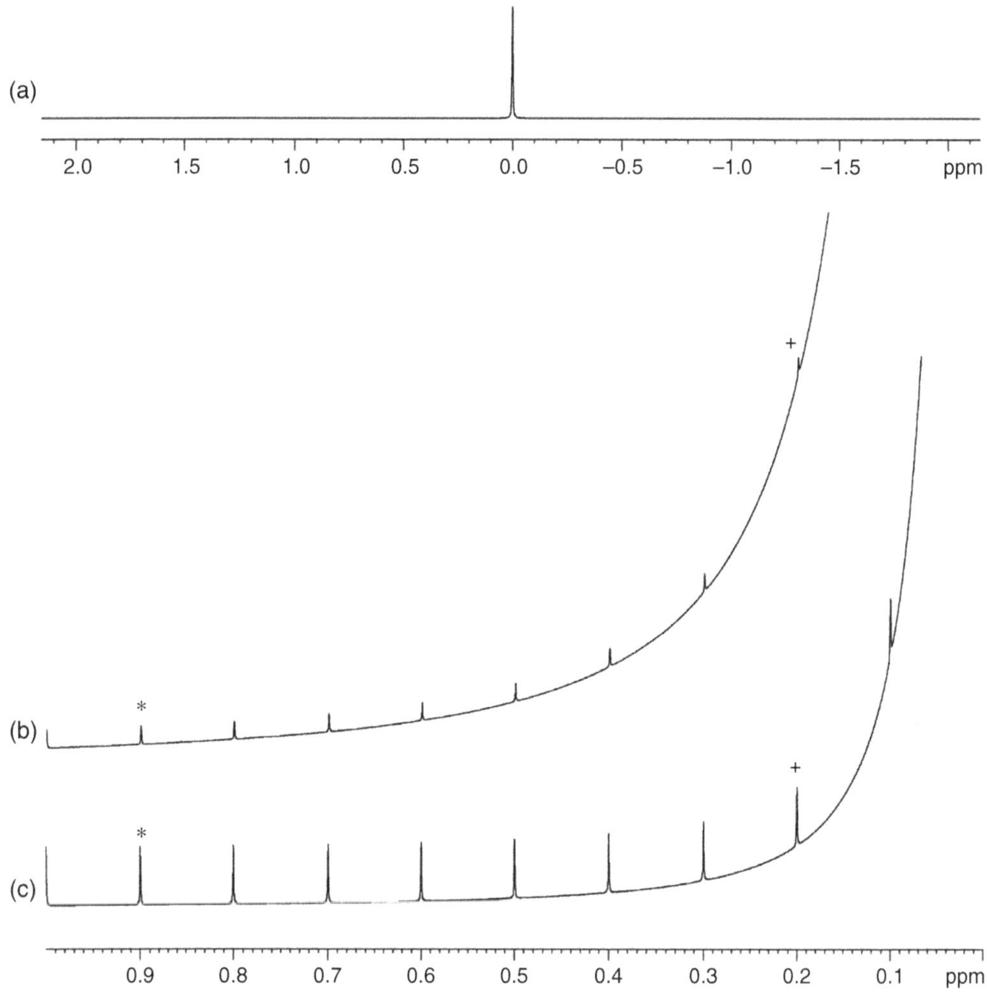

Figure 13.14 (a) Simulated spectrum (with no noise) of a substrate signal at δ_H 0 ppm with a series of MI signals at 0.1 ppm intervals from 1 ppm down to 0 ppm and zoomed spectra of (b) 1 ppm concentration of MI and (c) 20 ppm concentration of MI (with respect to the substrate signal). Signals at a chemical shift of 0.9 ppm (marked by *) are easily detected in both cases, but at a chemical shift of 0.2 ppm (marked by +), the 1 ppm MI signal is virtually obscured by the tail of the huge substrate signal.

reality. However, what we have yet to consider is the impact of the substrate – does the fact that you are trying to detect a signal that could be one million times smaller than most signals in the NMR spectrum cause a problem? To answer this question we have to consider whether the NMR instrumentation has sufficient dynamic range to detect these small signals.

When the FID is detected, it is digitized by an analog to digital converter (ADC). The performance of this ADC determines both the frequency and amplitude range that can be detected. Traditionally, ADCs have 16 bits, one is reserved for the sign of the signal leaving a dynamic range of $2^{15} - 1 = 32\,767$ to represent the values.

Assuming the receiver gain of the instrument is adjusted in such a way that it exactly fills the ADC (with a value of 32 767), the smallest signal that can be detected has a value of 1, anything smaller than this will not be detected in a single scan. Therefore, if we are looking at ppm levels, this dynamic range can prevent detection of the small signals, even if the sensitivity is theoretically high enough.

The situation is improved by the fact that in reality smaller signals will be detected as they can "ride" on the noise and with signal averaging will gradually sum to give signals. Additionally, the most recent ADCs available in spectrometers have up to 22 bits. This corresponds to $2^{21} - 1 = 2\,097\,151$, which is clearly significantly more favorable.

Figure 13.15 shows the dramatic difference between an older instrument with a 16 bit digitizer and a new instrument with a 22 bit digitizer when detecting very small signals in the presence of a very large water signal. Spectra (b) and (d) show little difference (approximately 10 000 difference in intensity between the water signal and the impurity signal), whereas there is a huge difference in the sensitivity between spectra (c) and (e) when the intensity difference is nearer 100 000.

An additional problem, that of distortion, can be encountered particularly with the more sensitive cryoprobe systems. If the largest signal in the system is so large that it cannot fit into the ADC range, even after adjustment of the receiver gain, this results in severe distortion of the spectra as shown in Figure 13.16.

Cryoprobes are traditionally used for mass-limited samples and are not designed to look at high concentration samples. The use of a 16 bit digitizer for samples at 100 mg/ml or higher concentration will almost certainly cause this distortion, which will prevent any MI detection.

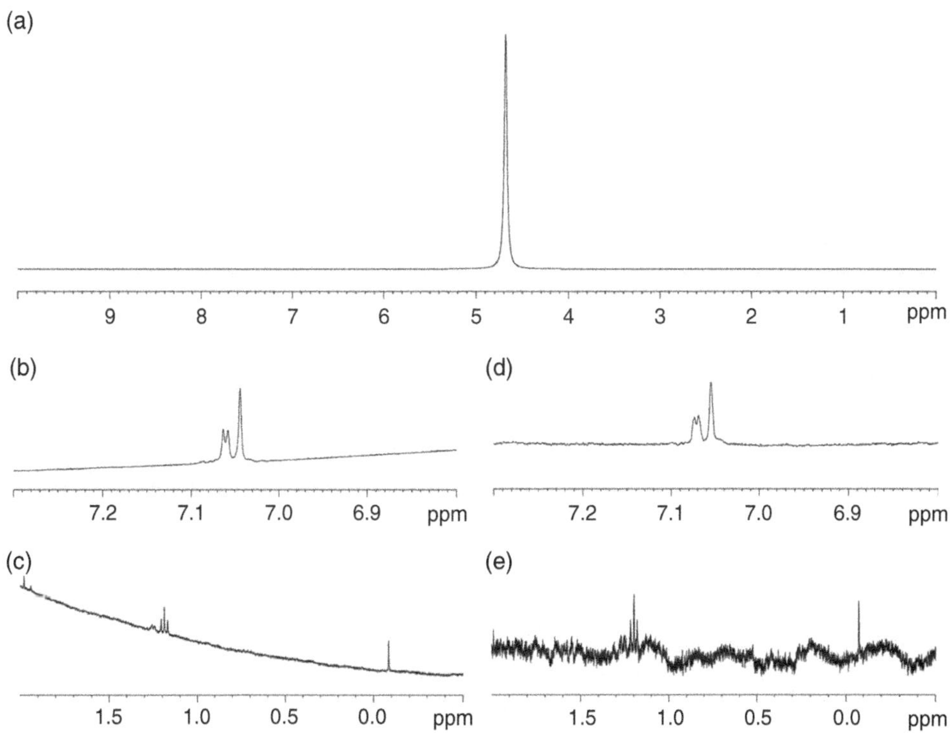

Figure 13.15 ^1H spectra of H_2O (spiked with 5% D_2O for a lock) with impurities present. (a) Full expansion, (b and c) acquired on a new generation 400 MHz spectrometer with 22 bit ADC, and (d and e) acquired on an older 400 MHz spectrometer with 16 bit ADC. In both cases the sample and all experimental conditions were identical. In spectra (b and d) the intensity difference between the impurity and the water signal is approximately 10 000, whereas in (c and e) it is nearer 100 000.

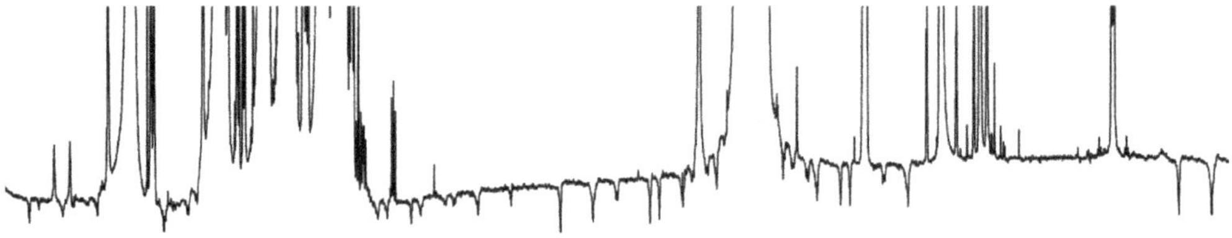

Figure 13.16 Example of the distortion caused by receiver overload. The form of the distortion can vary significantly depending on the sample concentration and the instrumentation used.

13.4.8.1 Selective Excitation

These dynamic range and receiver overload problems are very common in the field of biological NMR when looking at proteins dissolved in water. Typically, in these samples there is a very intense water signal and a large number of very small signals from the protein. To overcome this, there are a large number of so called "solvent suppression" strategies that remove the water signal from the spectra, hence allowing an increase in receiver gain and so detection of the protein signals [9–11]. Figure 13.17 shows an example of a spectrum acquired with simple presaturation water suppression.

Unfortunately, this approach is not useful for MI analysis because the substrate is not a single signal, rather a large number of very intense signals across the entire spectrum.

The solution is therefore to take the opposite approach – instead of trying to suppress the large signals, the experiment is performed in such a way as to only "excite" the small MI signal using selective pulses.

A normal 90° pulse often described as a hard pulse can take 10 μs (t_{90}) to rotate the magnetization from the z-axis onto the x–y plane. This rotation can be translated into a frequency:

$$\frac{\omega_1}{2\pi} = \frac{1}{4t_{90}}$$

If t_{90} is 10 μs then $\omega_1 = 157 \times 10^3$ rad/s. In addition, one can choose at which chemical shift or frequency in the spectrum to apply the pulse. The distance between a signal and this chosen frequency is referred to as the spectral offset, Ω.

The general condition is that if $\omega_1 \gg |\Omega|$ then the full spectrum will receive a complete 90° pulse. For example, in a proton spectrum, signals usually appear between 0 and 10 ppm, so if the RF pulse is applied at 5 ppm, the maximum offset, Ω, is $5 \times 400 = 2000$ Hz (or 13×10^3 rad/s) at a field strength of 400 MHz. It is clear then that ω_1 is significantly larger than any possible offset, and hence all signals will be excited.

If you significantly reduce the power of the RF pulse, and so increase the length of time it takes to rotate the magnetization, the width of excitation will decrease. Figure 13.18 shows the excitation profile for a 10 μs, 100 μs, and 10 ms 90° pulse. If a 10 ms pulse was applied, only a small region of the spectrum would be excited, this is termed a selective or soft pulse.

Using this selective pulse, a region where an MI signal is known to appear can be excited fully while the more intense regions are not, removing the problem that the intense signals cause.

The wiggly nature of the excitation profile, as illustrated in Figure 13.18, can cause a problem as some signals may still be excited to a small degree, even if they are significantly far away from the central excitation region. Given that these could be the very intense signals, they could still be large enough to cause a problem.

13.4.8.2 Shaped Pulses

It is possible to describe the normal RF pulse, whether applied in a *hard* or *soft* manner as a rectangular pulse, i.e. the RF is turned on for a given period and then turned off. Mathematically is it possible

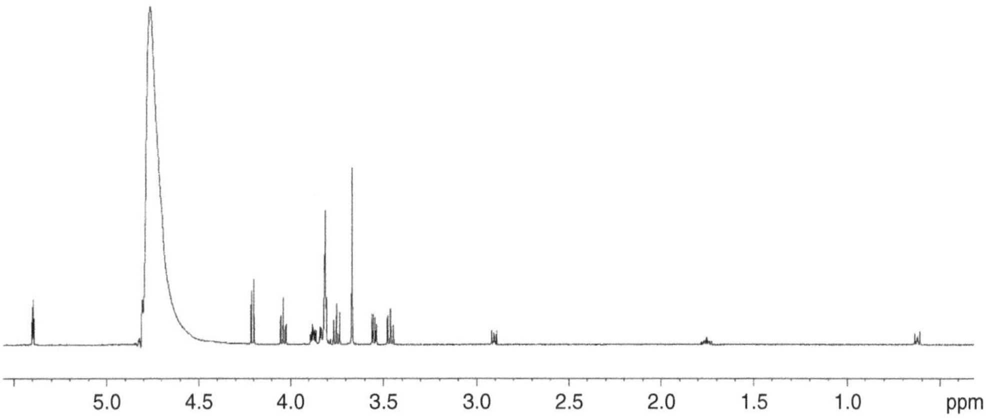

Figure 13.17 2 mM sucrose in 90% H_2O. Without water suppression the water signal would be huge but using simple presaturation, the size of the water signal is reduced allowing the detection of the sucrose signals.

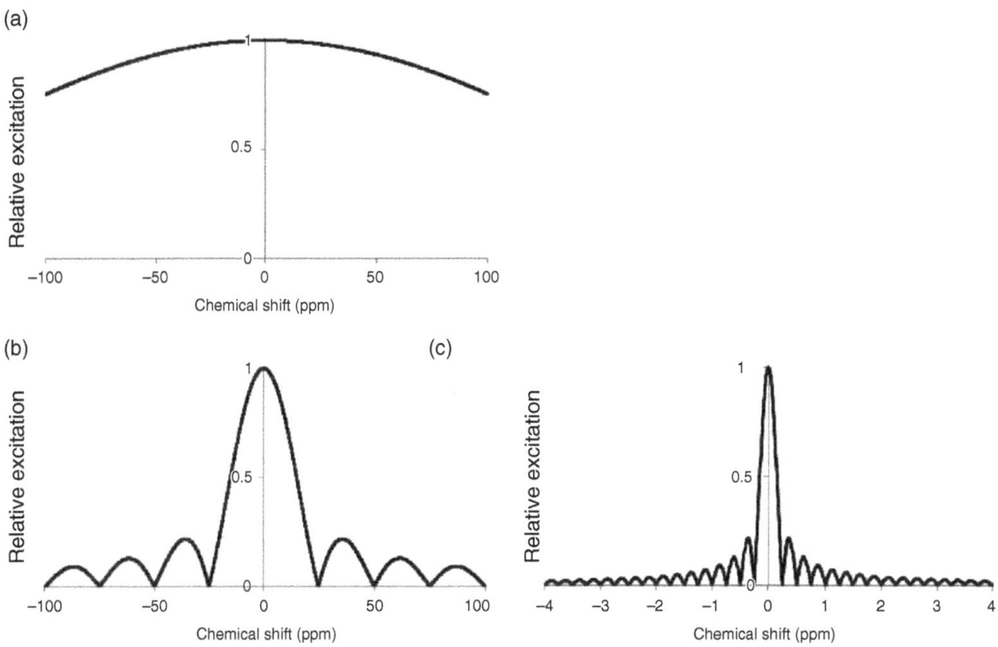

Figure 13.18 Excitation profiles for rectangular 90° pulses of (a) 10 μs, (b) 100 μs, and (c) 10 ms duration. Both the 10 and 100 μs pulses easily excite the full spectral width of a proton spectrum (maximum 10–15 ppm), whereas a 10 ms pulse will only excite a very small region around where the pulse is applied.

to calculate the shape of the excitation profile by Fourier transformation of the shape of the RF pulse? The Fourier transform of a rectangle is a sinc function – hence the wiggly nature. However, the Fourier Transform of a Gaussian shape is another Gaussian shape. If the RF power is therefore turned on and off in a Gaussian profile, the resulting excitation will be much flatter outside the main region [12].

This principle can be extended much further to a whole a range of different shapes that have different properties and that can be used in a wide range of applications. For example, Urquhart et al. [13] have shown how selective excitation (using the *selzg* pulse sequence with a 90° Eburp2 shaped pulse) can be utilized to detect trace levels of impurities down to 1 μg/g.

The generation and use of different-shaped pulses is beyond the scope of this chapter – further details can be found in references [1, 14, 15] – for the case studies described later in this chapter, a Gaussian pulse has been applied and has proved to be more than adequate.

Figure 13.19a shows how the selective Gaussian pulse is applied in practice on a cryoprobe system – by excitation of a small signal, present in this case at approximately 4.8 ppm.

Although residual artifacts of the intense sample signals appear, their intensity is significantly reduced, removing receiver distortion and allowing a dramatic increase in the instrument receiver gain (the receiver gain is essentially the level of signal amplification and is automatically set by the spectrometer based on the most intense signal in the spectrum). The result is being able to fully utilize the theoretical sensitivity of the instrument to detect the low-level signal (by ignoring the intense signals).

From our experience, this type of experiment is beneficial over a normal pulse on a cryoprobe system or normal probe systems with older-type ADCs. For normal probe systems with new higher dynamic range ADCs, no benefit is seen.

13.4.8.3 Quantification Using Selective Pulses

One down side of this approach is that the selection of only the MI signal removes any reference signals in the spectrum (to integrate against), so how can one of the big advantages of NMR, quantification, be utilized?

The easy solution is to make use of the flexibility of shaped pulses. By modulating the amplitude and phase of the shaped pulse, it is possible to cause two regions of the spectrum to be excited (see

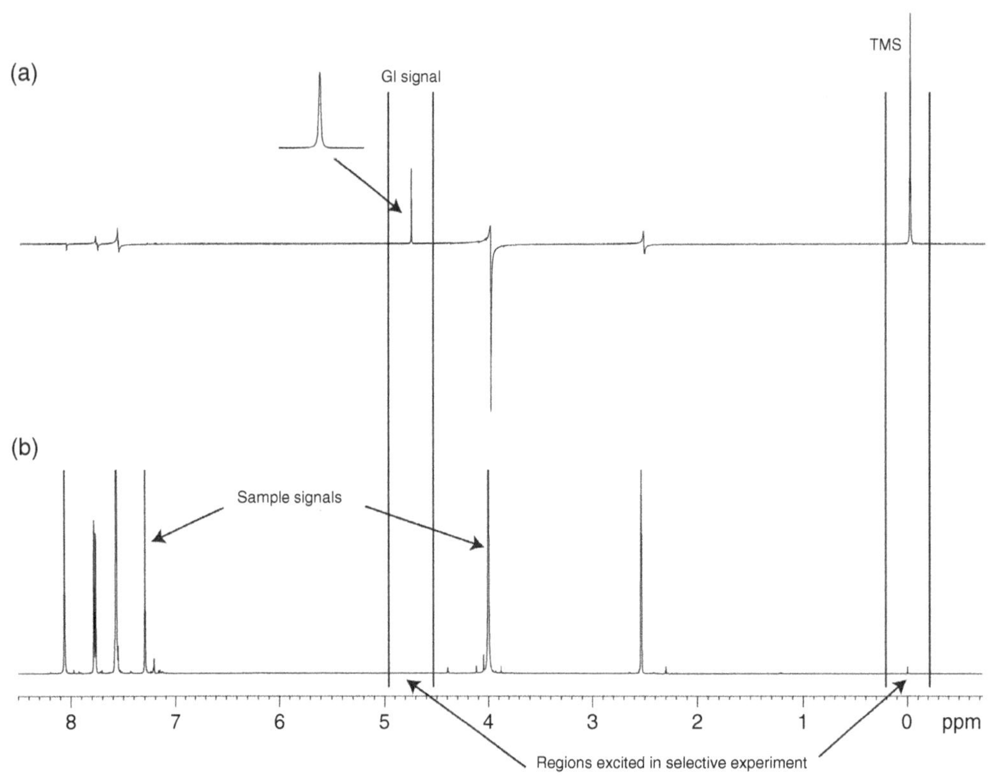

Figure 13.19 (a) The selective pulse experiment overlaid with (b) a normal companion proton.

Freeman [14] for more details). This then allows you to obtain an accurate integral ratio between the substrate and the MI signal by using a two-step approach described below.

First, a normal companion proton spectrum is acquired in which an intermediate signal can be integrated relative to the substrate – this allows a ratio between the substrate and the intermediate signal to be obtained. Second, a shaped pulse is created that selectively excites both the MI signal and the intermediate signal. The selective experiment is acquired using this shaped pulse, and the MI and intermediate signals are both integrated that establishes the integral ratio between the MI signal and the intermediate signal. Finally, the integral value of the intermediate signal (in the selective experiment) is set to the same value obtained in the companion proton spectrum, giving a direct ratio between the substrate and the MI signals.

The two main challenges here are to ensure that the relaxation conditions are the same in both experiments and that there is no distortion of the spectrum (due to receiver overload) in the companion proton experiment. This is easily achieved by significantly reducing the pulse angle of the hard pulse used for the companion proton experiment (sometimes as little as a 1° pulse is used).

It is possible to choose any signal present at an intermediate level to excite with the MI, but in our experience tetramethylsilane (TMS), the chemical shift reference, is an ideal choice as it is always resolved and present at an appropriate level. The whole process is summarized in Figure 13.19.

13.4.8.4 Excitation Sculpting

Although selective-pulse acquire experiments dramatically reduce the intensity of the strong signals (as shown in Figure 13.19), they still appear as dispersive signals (having both positive and negative components). These dispersive signals have long tails, and if there is not sufficient resolution between the MI signal and other signals, these tails can prevent accurate integration. As per any quantitative method, integration on a flat baseline is crucial for accurate results.

The solution is the use of a slightly more complicated NMR pulse sequence – called excitation sculpting [16, 17] – which actively removes unselected signals. This pulse sequence comes in two varieties incorporating either a single or double gradient echo.

There are positive and negative factors for each selective pulse sequence, with the main disadvantage of the excitation sculpting pulse sequences being a slight loss in the $S{:}N$ ratio. Pulse sequence selection is summarized in Table 13.5.

The difference between spectra obtained from the simple selective experiment (*selzg*) and the double echo excitation sculpting experiment can clearly be seen by comparison of Figures 13.26a and 13.21c.

13.4.9 Limit Tests

NMR is an inherently quantitative technique, and so it is possible to calculate levels without calibration. However, in many MI problems, all that is required is a limit test, i.e. is the MI present at the Threshold of Toxicological Concern (TTC) or not?

This type of test can easily be performed by NMR – by simple comparison of spectra acquired under identical parameters with and without a spike at the appropriate level. This approach has become the method of choice for NMR MI analysis – most commonly utilizing the double echo excitation sculpting pulse sequence. Running the NMR MI analysis as a limit test is advantageous for a number of reasons:

- Method development can be very rapid – all that is required is one detectable signal for the MI, which is resolved from all parent substrate signals.
- Reduced validation requirements for limit tests mean that validation is essentially complete once the method has been developed.

Table 13.5 Choice of selective experiment pulse sequence.

Name	Performance comments: Positive	Performance comments: Negative	Pulse sequence name (Bruker)
Selective ^{1}H	Highest signal-to-noise	Only suitable for well-resolved singlets	*selzg*
Excitation sculpting: single echo	Suitable for all signals of any multiplicity	Only possible to analyze a single signal due to phase errors with ≥2 signals	*selgpse*
Excitation sculpting: double echo	Best option for selecting >2 signals simultaneously	Lower $S:N$ than single echo	*seldpfgse*

- No signal integration is required, and simple comparison of spectra from a spiked and unspiked sample is all that is needed – this means that it is possible to use a limit test even if the MI signal is not baseline resolved from background signals (see Section 13.1.5.5).
- There is no need for the experiment to be run under fully quantitative conditions, meaning there is no requirement to wait 5× T_1 in between each scan that reduces the recycle time and speeds up data acquisition.

13.4.9.1 Method Development

Method development for an NMR limit test can be very quick, either to confirm that NMR is an appropriate technique to use (i.e. there is a resolved signal from the MI, which is resolved from the parent material) or equally to confirm that NMR cannot be used if no resolved signals from the MI can be observed. This "*yes or no*" decision can typically be achieved in a matter of minutes, which is why NMR should be the first technique that is considered for MI analysis.

Given below is the typical approach to be taken to determine if an NMR MI limit test method could be developed:

- Confirm it is possible to dissolve the substrate at high concentration (e.g. 100–200 mg/ml) and acquire the ^{1}H NMR spectrum (typically using a standard NMR system, i.e. a 400 MHz spectrometer with a conventional room temperature probe).
- Confirm the MI is soluble in the same solvent and prepare a sample at a "normal" concentration (e.g. 10 mg/ml) and acquire the ^{1}H NMR spectrum (note, these samples do not need to be prepared in deuterated solvents as most modern spectrometers are capable of acquiring NMR data *unlocked*, i.e. without the presence of a deuterium lock signal).
- Compare/overlay the NMR spectra of the substrate and the MI to determine if there is at least one signal for the MI that is resolved from the substrate and solvent signals.
- Prepare a high-level spiked sample of the MI in a "clean" substrate sample (i.e. a sample known to not contain any of the MI of interest), e.g. a 1000 ppm spike of the MI in a 100 mg/ml sample of substrate and acquire the ^{1}H NMR spectrum of the mixture to confirm there are no matrix affects and the signal of the MI is indeed resolved from all other signals in the sample. It should be possible to observe the 1000 ppm spiked signal using a standard 400 MHz NMR spectrometer and probe, assuming a suitable number of scans have been acquired (e.g. 64 or 128 scans).

If the signal from the MI is observed, this establishes specificity for the method and further method development is then required using a high sensitivity probe and spectrometer (e.g. 600 MHz spectrometer with an inverse cryoprobe) in order to be able to detect the MI signal down to typical TTC levels (e.g. 1–10 ppm). The 1000 ppm spiked sample can be used to set up the excitation sculpting pulse at the required frequencies for the MI signal and the TMS signal (which is typically used as a system suitability test signal). This parameter set is then saved, and lower level spiked samples are analyzed (e.g. 100, 10 ppm and TTC level) under exactly the same experimental conditions to confirm there are no matrix effects that change in the chemical shift of the MI signal – additional scans will be required for the lower level samples to ensure there is sufficient S:N. Finally, the substrate (blank) sample is analyzed under the same conditions (i.e. same number of scans as the TTC spiked sample), and the absence of a signal where the MI signal is known to resonate confirms that no MI is detected, with an LOD reported as the TTC level.

13.4.9.2 Validation

The International Council for Harmonisation of Technical Requirements for Pharmaceuticals for Human Use (ICH) validation requirements [18] for an impurities limit test are much reduced from those needed for a quantitative method (see Table 13.6).

Validation of a limit test only requires that specificity and LOD are established, both of which are addressed in the initial method development. Specificity is demonstrated through the selection of a resolved signal for the MI and the creation of the selective pulse, which will only irradiate signals at this frequency. LOD is established by running the TTC spiked sample with sufficient number of scans to be able to detect the MI signal at this level (i.e. with an S:N ratio of >3 : 1). This confirms that the NMR experiment is capable of detecting a signal at the TTC level, and so if no signal is observed (or if a smaller signal than the TTC level spike is observed by overlaying the spectra), then the result is "non-detected" with a quoted LOD of the TTC level.

13.4.9.3 Unresolved Signals

Unlike quantitative methods, a limit test requires no true measurement of the signal you are detecting, you just need to be able to show that you can detect the signal if it is present (at a specified level). This opens up the possibility to test for the presence of signals that are not fully resolved from all other signals – as there are no concerns with the effect of overlapping signals increasing the integral value of the analyte signal. This can be particularly useful in the analysis of crude

Table 13.6 ICH validation requirements.

Parameter	Identity	Impurities: Quantitative	Impurities: Limit test	Assay
Accuracy	–	+	–	+
Precision				
Repeatability	–	+	–	+
Intermediate	–	+	–	+
Specificity	+	+	+	+
LOD	–	[a]	+	–
LOQ	–	+	–	–
Linearity	–	+	–	+
Range	–	+	–	+

[a] May be required for some applications.

chemistry samples where there may be background signals from unrelated low-level materials in the same region as the MI signal. As long as the TTC spiked sample can be detected over and above any background signals, the method can be considered to be suitable for analysis. Section 13.1.5.5 demonstrates this perfectly, highlighting the fact you do not even need to be able to see all of the MI signal as long as you can detect some of it.

13.4.9.4 Rapid Analysis

The final advantage of the limit test approach is the ability to rapidly increase the $S:N$ ratio of the MI signal by acquiring multiple scans quickly, without the concern of incomplete signal relaxation and nonquantitative results. As a limit test approach requires the comparison of a standard sample of known analyte concentration versus your test sample, as long as each experiment is performed under the same conditions, it does not matter how much of the available signal you are detecting. This enables us to optimize the NMR experiment for speed, rather than quantitative accuracy, as it means there is no requirement to wait $5 \times T_1$ in between each scan. Instead, we can rapidly pulse the sample (for example, only using a 1 second inter-scan delay rather than a quantitative 10 or 20 seconds delay), meaning we can acquire more scans over a giving time period and increase the $S:N$. In practice this means that it is routinely possible (given a high substrate concentration) to establish NMR limit tests, which can detect single figure ppm levels of MIs in less than 10 minutes.

13.4.10 Expanded Use of MI NMR Methodology

The success of using NMR for low-level MI analysis has led to NMR being used for other low-level detection analysis, for example, to show ppm level control of critical (none MI) impurities during manufacturing processes and also supporting cleaning validation methods. In respect to the latter, in 2014 the Food and Drug Administration (FDA) introduced legislation stating that visual inspection of manufacturing equipment to confirm it was "clean" prior to continued use was no longer acceptable and data from validated sources were now required. The NMR limit test methodology discussed above that was developed for low-level MI analysis is an ideal tool for such analysis and has been deployed to support several cleaning validation campaigns (see Section 13.1.5.6).

13.4.11 Summary

In this section we have discussed the many factors that determine the sensitivity of the NMR experiment – and highlighted that perhaps the most important for trace analysis is obtaining sufficient solubility. In addition, the difficulties of using high sensitivity probes have been highlighted and a solution proposed by the use of selective pulses.

We are now in a position to consider some real examples to highlight how useful NMR can be in this area.

13.5 Case Studies

As has been shown, if sufficient solubility can be achieved, NMR does have enough sensitivity to detect at ppm levels. The following examples highlight the use of ^{1}H NMR using normal systems, cryoprobe systems coupled with selective excitation, and the use of ^{19}F NMR.

13.5.1 Case Study 1 – An Aldehyde Functionalized MI

During early Phase I development of compound (**4**), a synthetic intermediate functionalized with an aldehyde group (**3**) was identified as an MI. A fit-for-purpose method was therefore required to detect and, if present, quantify this impurity down to about 50 ppm.

By a simple comparison of the partial active pharmaceutical ingredient (API) structure (**4**) and the aldehyde (**3**), it was immediately obvious that the proton signal from the aldehyde would be likely to appear in a totally unoccupied region of the API NMR spectrum.

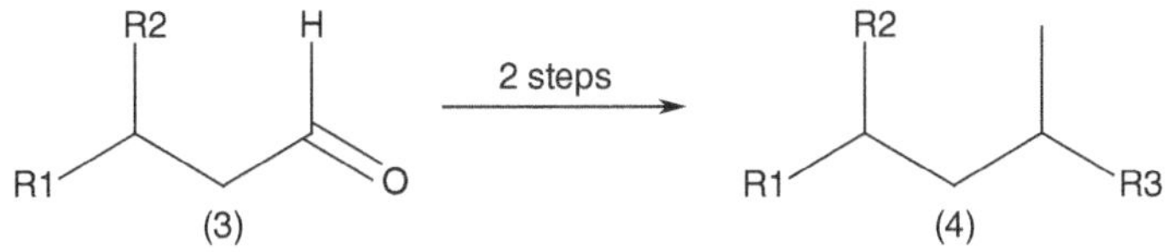

A solubility of 100 mg/ml of the API was easily achieved in d_6-DMSO, so a 500 ppm spike of the aldehyde impurity was added and the sample was run under normal proton conditions, ensuring the spectrum was quantitative. As the 500 ppm spike was clearly resolved, a further series of two duplicates of a "blank" sample along with 50/100/150 ppm spikes were successfully performed again under normal proton conditions on a 400 MHz instrument. Figure 13.20 summarizes the results of these experiments. They clearly demonstrate that low levels of the impurity are present in the API.

As a comparison, the same samples were run on a 600 MHz instrument with a cryoprobe, using a simple selective experiment with companion proton (Figure 13.21). There is an $S:N$ ratio increase along with a dramatic decrease in experiment time (3 hours to 10 minutes).

Simple integration of the aldehyde signal relative to a drug signal allowed quantification of the levels present as shown in Table 13.7. Excellent recoveries were also observed.

If lower detection limits were required, it would have been possible to acquire data for longer, and it was estimated that in three hours a level of detection <5 ppm could have been achieved.

All the work required for this project was performed in less than one day and provided all the answers that were required at that time.

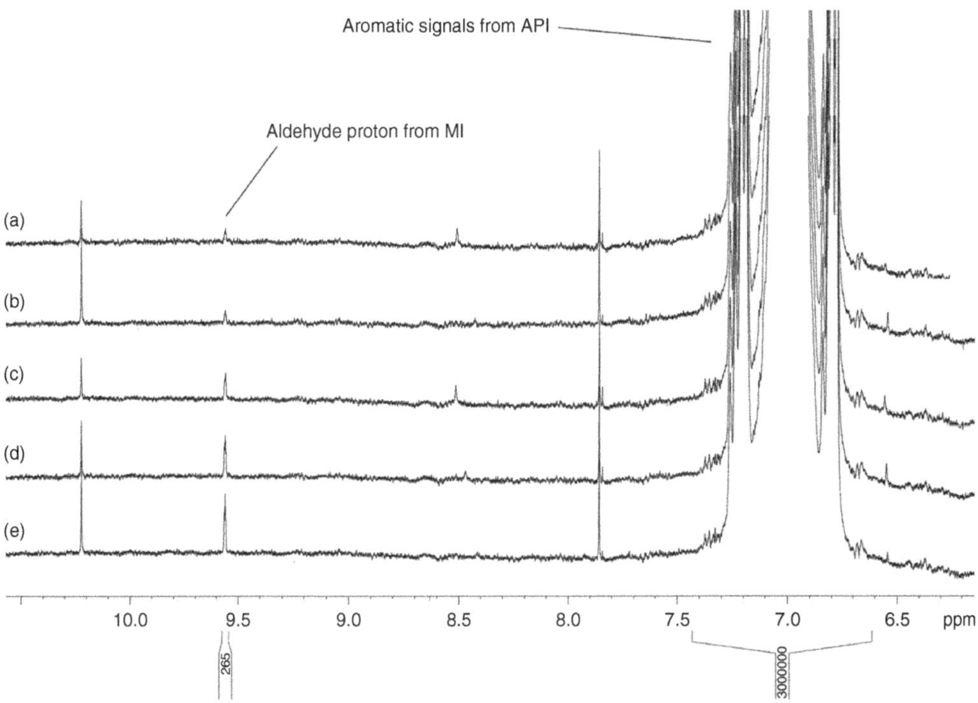

Figure 13.20 Proton spectra of 100 mg API (**4**) in d_6-DMSO (a and b) duplicates with no spike. Spiked with (c) 50 ppm aldehyde MI, (d) 100 ppm, and (e) 150 ppm. Spectra run at 400 MHz and 300 K over three hours of acquisition.

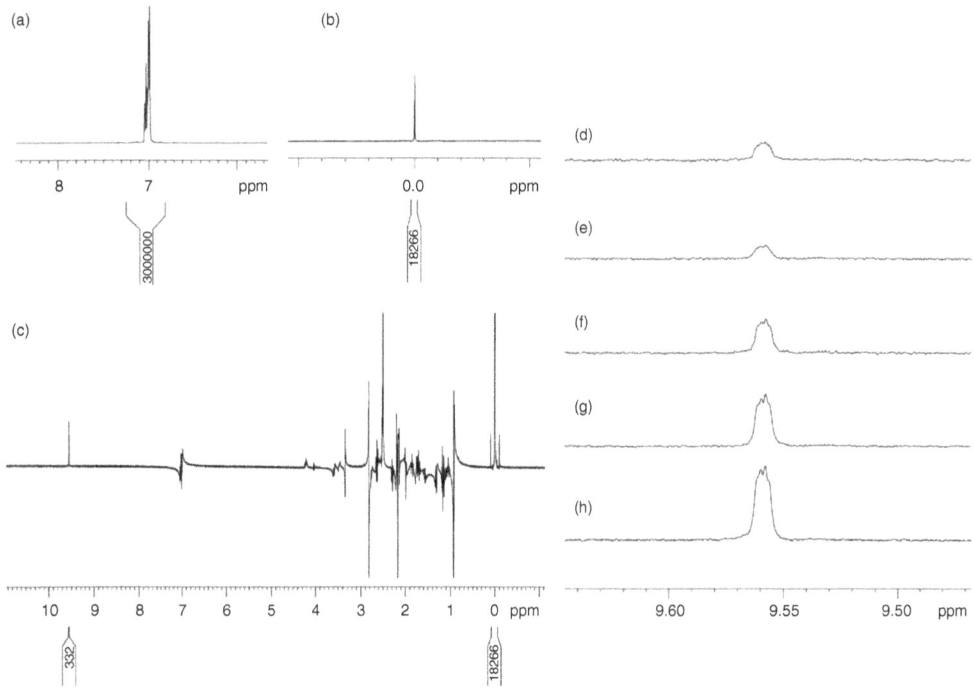

Figure 13.21 Proton spectra of 100 mg API (**4**) in d_6-DMSO (a and b) companion proton experiment with 150 ppm spike. (c) Selective pulse experiment with 150 ppm spike; (d and e) duplicates with no spikes. Spiked with (f) 50 ppm aldehyde MI, (g) 100 ppm, and (h) 150 ppm. Spectra run at 600 MHz and 300 K over 10 minutes of acquisition.

Table 13.7 Calculated ppm levels and recoveries from a selective experiment on a 600 MHz instrument.

	Level (ppm)	Recovery (%)
Duplicate 1	41	—
Duplicate 2	45	—
50 ppm spike	89	95
100 ppm spike	137	96
150 ppm spike	191	99

13.5.2 Case Study 2 – Use of ^{19}F NMR

Compound (**6**) is an intermediate in the development of a fluorinated API. At this intermediate stage of the synthesis, it was necessary to control the level of impurity (**5**), trifluoronitrobenzene (TFNB), down to 10 ppm.

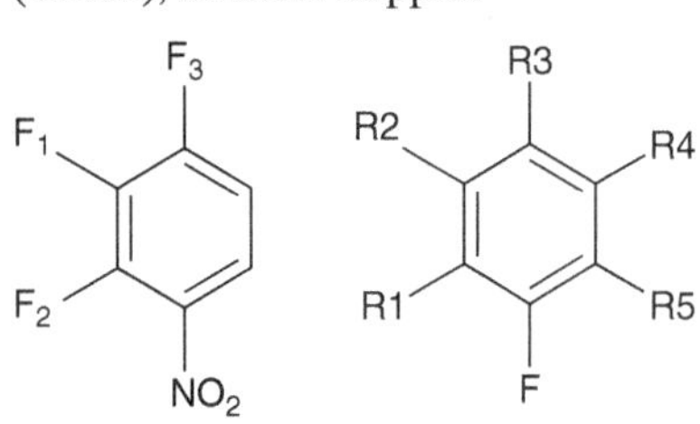

(5) (6)

It was noted that the solubility of intermediate (**6**) was extremely high (>3 g/ml in d_6-DMSO) so both ^{1}H and ^{19}F experiments could conceivably give the necessary sensitivity. Initially, ^{1}H experiments were attempted but although the MI signals were in a region of the spectrum distinct from signals from the intermediate, other impurities and poor resolution prevented detection below 50 ppm despite experimental optimization, as shown in Figure 13.22.

However, ^{19}F with ^{1}H decoupling experiments were more successful. Figure 13.23 shows that the three ^{19}F signals from (**5**) are well resolved from the fluorine signal of intermediate (**6**) at −147 ppm. F_1 at −156 ppm was chosen for further study as it had the simplest multiplet structure and hence would give the highest sensitivity.

To achieve the required sensitivity, the sample had to be prepared with 1 g of (**6**) in 300 µl d_6-DMSO. Due to the very high concentration and hence high viscosity, it was necessary to run the NMR

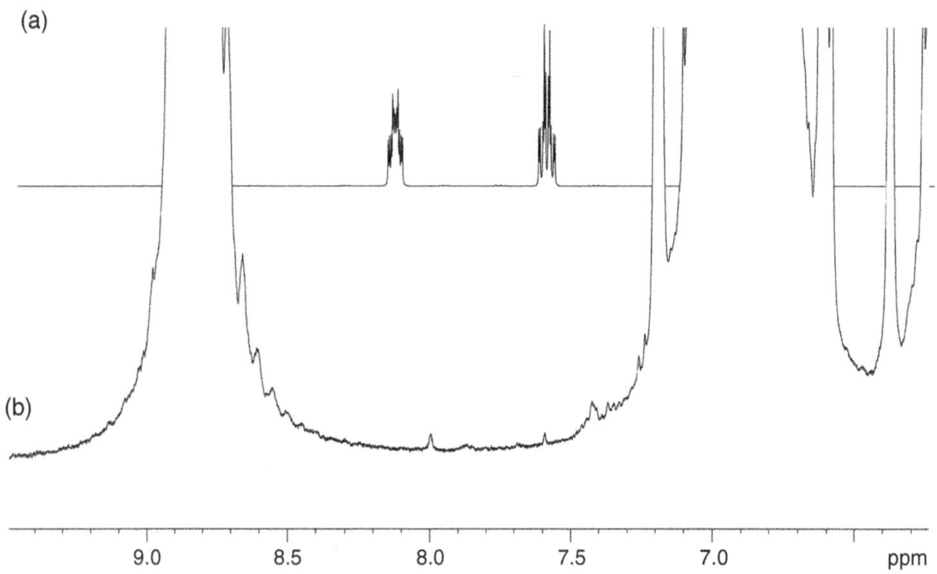

Figure 13.22 (a) ^{1}H NMR spectrum of TFNB and (b) 50 ppm spike of TFNB in 500 mg/ml intermediate (**6**).

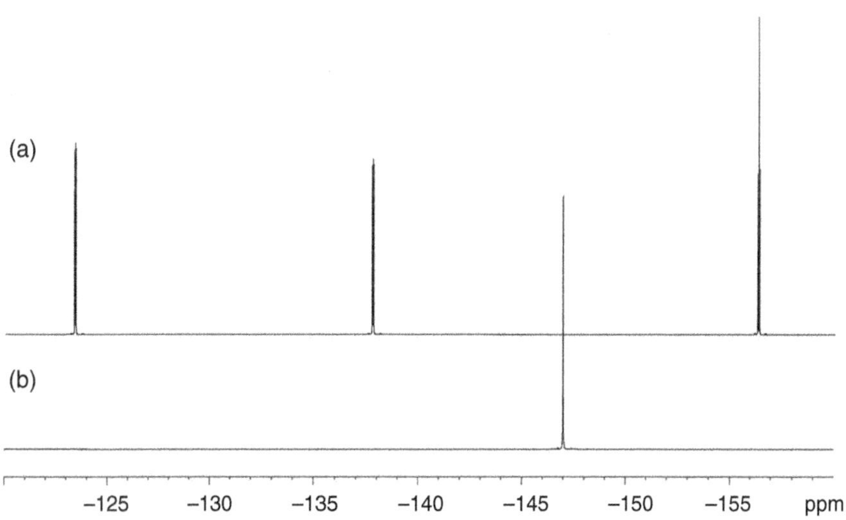

Figure 13.23 ^{19}F with ^{1}H decoupling spectra of (a) TFNB and (b) intermediate (**6**).

spectra at 70 °C to achieve sharp signals. Under these conditions, the required level of detection was achieved on a 400 MHz instrument in 3.5 hours of acquisition (see Figure 13.24).

Calculation of ppm levels was performed by simple integration, and recoveries from spiking experiments were comparable to those expected from other techniques for trace analysis such as LC-UV or GC. The data obtained are summarized in Table 13.8.

Using this method, a number of batches of material were screened and it was quickly shown that none contained the TFNB impurity.

13.5.3 Case Study 3 – Epoxide and Chlorohydrin MIs

In another compound, two MIs were of concern, an epoxide (**7**) and a chlorohydrin (**8**), which were early synthetic intermediates of the API. Initially, other techniques apart from NMR were proposed; GC was not considered an alternative due to likely matrix interference, so first to be tried was high-performance liquid chromatography (HPLC)-MS; however, it was quickly apparent that method development would not be straightforward.

Ion chromatography was then attempted, and in the absence of the API, the MIs were readily retained and detected in aqueous media. However, in the presence of the API, there were interfering signals in addition to concerns around the solubility of the API in water. This left NMR as the only real alternative.

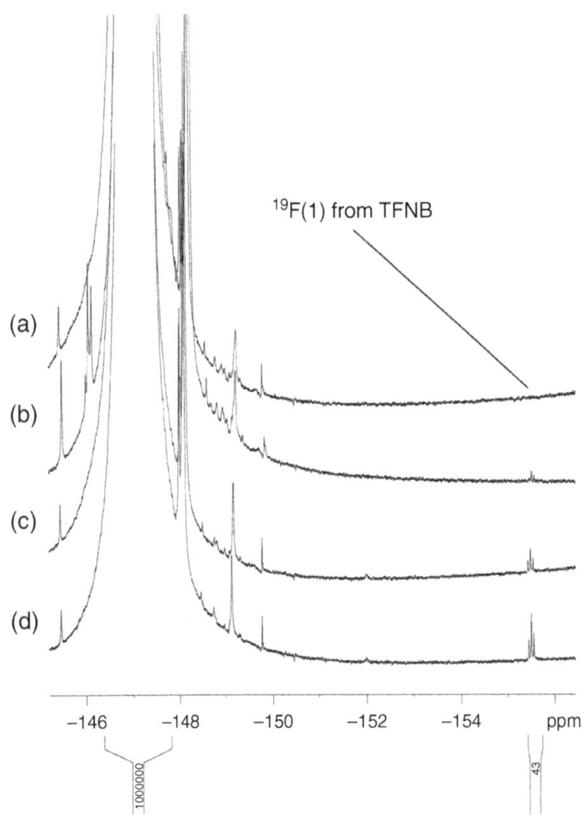

Figure 13.24 ^{19}F spectrum with ^{1}H decoupling of intermediate (**6**) at approximately 1 g in 300 ml *d6*-DMSO, recorded on a 400 MHz instrument in 3.5 hours at 70 °C (a) not spiked. Spiked with TFNB (b) 9 ppm, (c) 23 ppm, and (d) 47 ppm.

Table 13.8 TFNB recoveries.

	TFNB level (ppm)	Recovery (%)
Intermediate (**6**) no spike	Not detected	—
(**6**) + 9 ppm spike	10	111
(**6**) + 23 ppm spike	22	97
(**6**) + 47 ppm spike	45	96

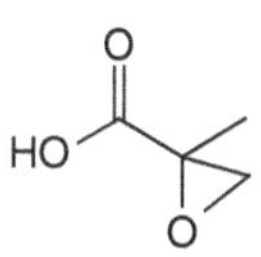

(7)

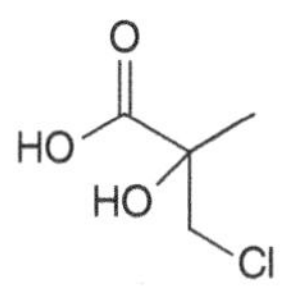

(8)

An initial comparison of ^{1}H spectra for the API, epoxide and chlorohydrin, quickly suggested that resolution could be achieved, and hence the NMR methodology was pursued. It was determined that the maximum API solubility that could be obtained was only approximately 60 mg/ml, and with the required level of detection of 10 ppm, a 600 MHz system with a cryoprobe was required to be utilized.

Spikes were performed of both epoxide (**7**) and chlorohydrin (**8**) at 1000, 100, 30, 10, and 3 ppm to confirm the method was specific and had sufficient sensitivity. To obtain maximum sensitivity, a companion proton and pair of selective experiments (one selecting the chlorohydrin and one the epoxide) were used.

As can be seen in Figure 13.25, both MIs were detected at the required level and the recoveries obtained were acceptable (Table 13.9).

A series of batches were screened, and it was shown that no batches contained the MIs. In this case, to achieve a greater level of confidence, each sample analyzed was subsequently spiked with 10 ppm epoxide and chlorohydrin. These samples were then rerun to prove 10 ppm could be detected.

13.5.4 Case Study 4 – Sulfonate Esters

Sulfonate esters are a very common class of MIs. Although they are easily detected using a range of other traditional techniques, it is also possible to detect them by NMR, as long as they are resolved from the substrate.

In this example, ethyl methanesulfonate (EMS) (**9**) and methyl methane sulfonate (MMS) (**10**) are detected in the presence of a certain API. Proton A in (**9**) is a quartet at 4.26 ppm and proton B in (**10**) is a singlet at 3.87 ppm in d_6-DMSO. Both appear in an unoccupied region of the spectrum of the API in question. However, in contrast to previous examples, excitation sculpting has been used rather than a simple selective pulse. As described in Section 13.1.4.8.4, this allows for the measurement of both impurities in the same experiment and the removal of artifacts from the intense signals, at the expense of only a small amount of sensitivity. This removal of artifacts is clearly seen when comparing Figure 13.26a with Figure 13.21c.

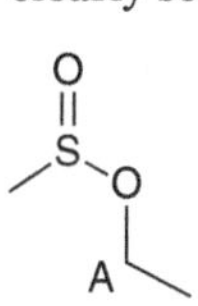

(9)

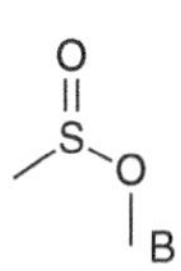

(10)

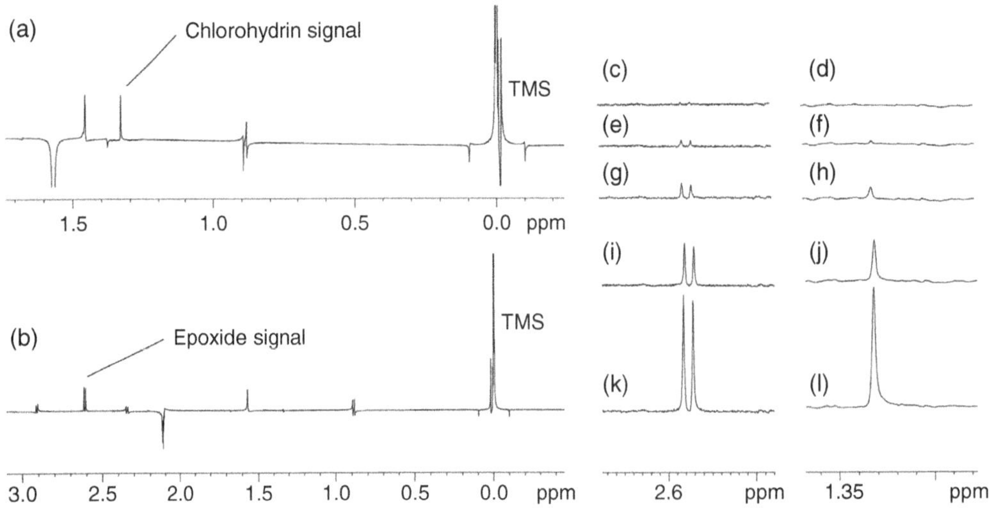

Figure 13.25 (a) Selective excitation experiment on spiked sample selecting chlorohydrin signal at 1.35 ppm and TMS at 0 ppm. (b) As (a) but selecting epoxide at 2.56 ppm and TMS. (c and d) Epoxide and chlorohydrin selective experiments, respectively, on sample with no spike. Samples with epoxide and chlorohydrin spikes (e and f) 3 ppm, (g and h) 10 ppm, (i and j) 30 ppm, and (k and l) 100 ppm. All spectra recorded at 600 MHz.

Table 13.9 Epoxide (**7**) and chlorohydrin (**8**) recoveries.

	Epoxide (7) level (ppm)	Recovery (%)	Chlorohydrin (8) level (ppm)	Recovery (%)
3 ppm spike	4.1	137	2.6	87
10 ppm spike	10.0	100	8.8	88
30 ppm spike	27.6	92	31.9	106
100 ppm spike	78.9	79	102.3	102

Figure 13.26b–g shows a series of spikes into the API present at a concentration of 100 mg/ml in d_6-DMSO. The NMR methodology is successful with a level of detection of approximately 4 ppm. This example highlights one difficulty with NMR; down at ppm levels of detection there are often a large number of other unknown impurity signals present, and if these happen to resonate at the same frequency as the signals of the MIs of interest, they can interfere. In this case while the signals C–E shown in Figure 13.26 (which are from unknown impurities) do not obscure the EMS and MMS signals, impurity E does make the integration of MMS (signal B) more difficult.

13.5.5 Case Study 5 – Limit Test for Poorly Resolved Signals

In the previous case studies, the signals from the MIs have been resolved from the signals of the parent compound; however, when detecting very low-level signals, resolution of the MI signals from the main component signals is not the only concern. For example, when supporting fate and purge experiments to see how well the chemistry tolerates being spiked with varying levels of MIs, the samples are often crude reaction liquors that may contain multiple low-level background materials that can result in a very noisy baseline. When this is the case, any method that requires a "clean" resolved signal to enable accurate integration to be achieved would not be possible, and this is where an NMR limit test approach can be utilized to maximum effect.

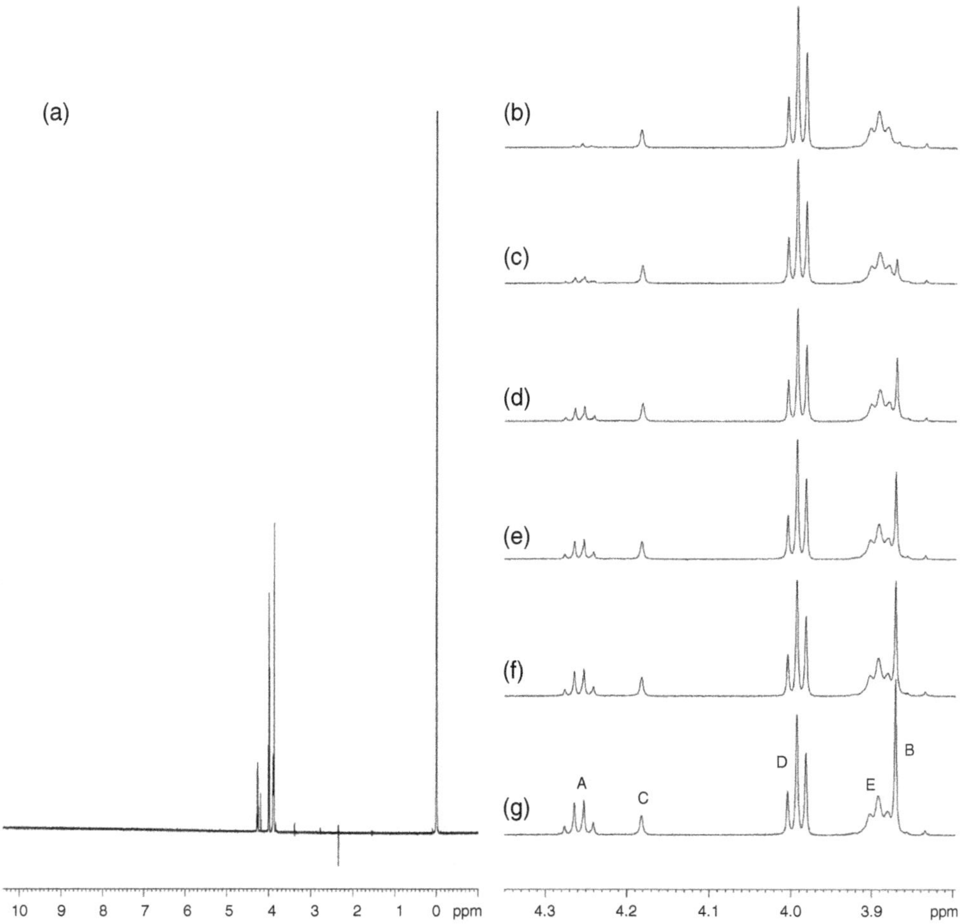

Figure 13.26 ^{1}H excitation sculpting with double gradient echo exciting the TMS signal at 0 ppm, the MMS singlet at 3.87 ppm (B), and the EMS quartet (A) at 4.26 ppm. Sample of API prepared at 100 mg/ml in d_6-DMSO (a) showing the full expansion of a 20 ppm spike of EMS and MMS. No other signals are present in the spectra apart from those around the excited regions. Expansion of the region around the two MI signals are shown: (b) no spike, and with spikes of EMS and MMS at (c) 4 ppm, (d) 8 ppm, (e)12 ppm, (f) 16 ppm, and (g) 20 ppm. C, D, and E are signals from unrelated impurities.

In this example, the two MIs of concern were isopropyl camsylate (**11**) and *n*-butyl camsylate (**12**), and the only signals that were resolved from the parent compound were the 4.92 ppm -O-**CH**- septet and 4.25 ppm -O-**CH**$_2$- multiplet signals of the alkyl chains from isopropyl camsylate and *n*-butyl camsylate, respectively.

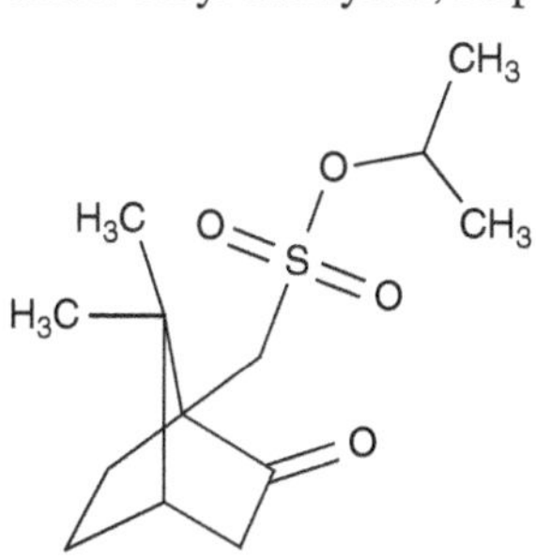

(11)

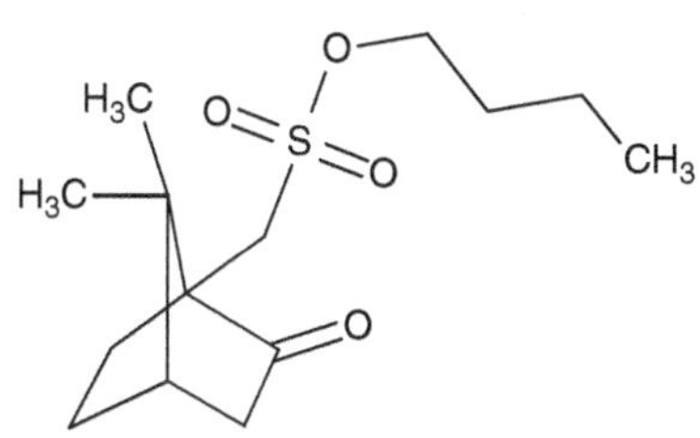

(12)

Again, the excitation sculpting with double gradient echo experiment was used to enable the detection of both MI signals in one NMR experiment while removing the intense signals of the API; however, as can be seen in Figure 13.27e, the background baseline in the region of interest is far from clean. That being said, for both the isopropyl camsylate signal at 4.92 ppm and the *n*-butyl camsylate signal at 4.25 ppm, by comparison of the blank API spectrum (e) and the 10 ppm spiked sample (d), you can clearly see the presence of the MI signals in the spiked spectrum, but none is detected in the blank spectrum despite the presence of overlapping signals. There is no way an accurate integral could be obtained for either of the MI signals, but by using the limit test approach we can confidently show we can detect the MIs down to 10 ppm, and that there is non-detected at this level in the API.

13.5.6 Case Study 6 – Using NMR MI Methodology for Cleaning Validation

Having established the robustness and sensitivity of NMR limit tests for low-level MI detection, the approach has also been applied to support cleaning validation activities. Cleaning validation is a good manufacturing practice (GMP) requirement to prevent contamination or cross-contamination of materials between different batches in a manufacturing campaign, and an NMR limit test approach has been validated for the analysis of acetic acid solutions used during the cleaning of equipment used in the production of the AstraZeneca oncology product Zoladex™. Zoladex is formulated as a depot injection of the decapeptide goserelin acetate (**13**) and a proprietary poly lactic-co-glycolic acid (PLGA) copolymer (**14**).

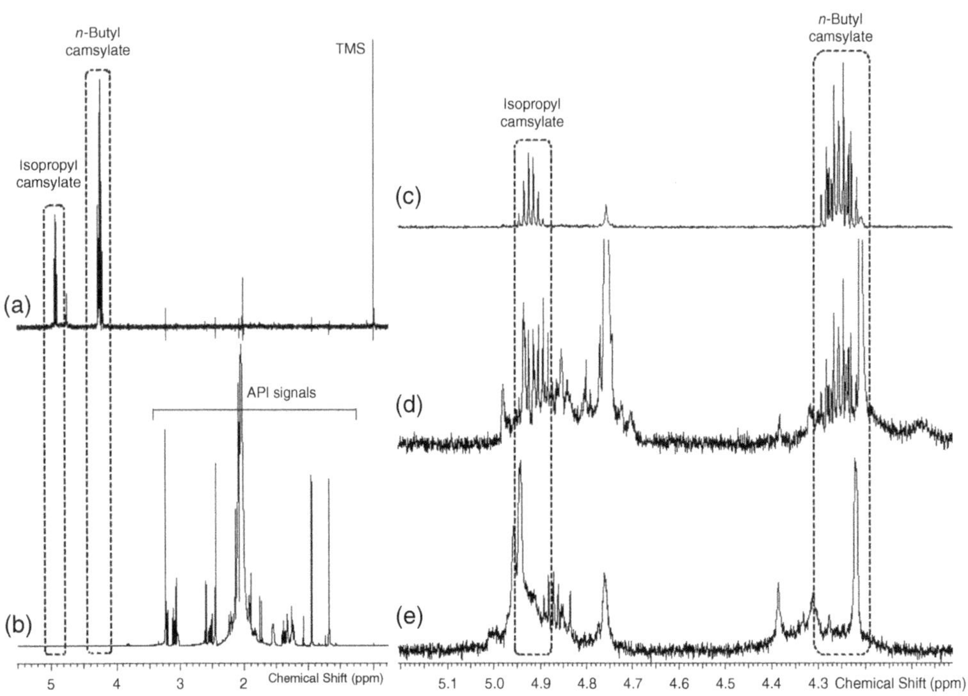

Figure 13.27 (a) ^1H excitation sculpting with double gradient echo of spiking solution containing 10 ppm isopropyl and 10 ppm *n*-butyl camsylate esters; (b) ^1H spectrum of API showing resolved region of interest between 4 and 5 ppm. Expansions of excitation sculpting with double gradient echo spectra are shown: (c) 10 ppm camsylate esters spiking solution, (d) API + 10 ppm camsylate ester spiking solution, and (e) blank API.

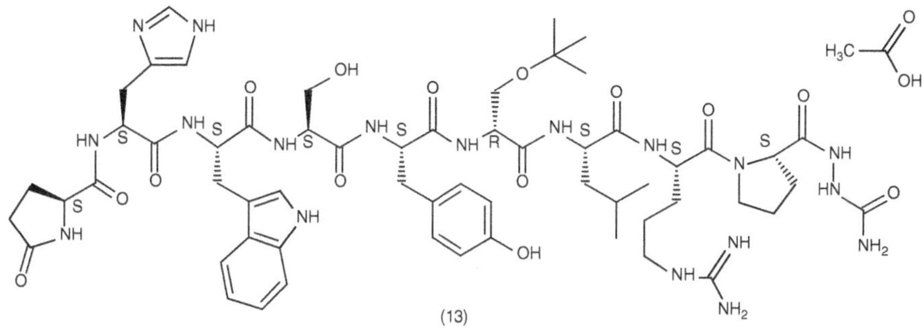

(13)

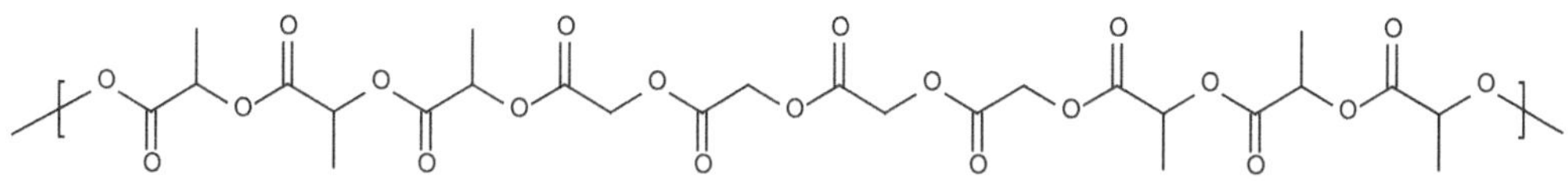

(14)

A double gradient echo excitation sculpting method was developed and validated, which could detect 10 ppm levels (i.e 10 μg/ml) of **(14)** in the cleaning solvent, acetic acid. There is a suitable signal for **(14)** in the ^{1}H NMR spectrum at 5.3 ppm, which is resolved from all of the signals of **(13)** and the methyl signal of the acetic acid solvent at approximately 2 ppm (note, the solutions to be analyzed are prepared in protic acetic acid, not deuterated acetic acid as would normally be used for routine NMR sample analysis). Having established specificity, the only remaining validation requirement for a limit test was to determine the LOD, which was achieved via a series of spiking experiments with decreasing concentrations of PLGA (Figure 13.28). Due to the intensity of the acetic acid solvent signal, a small proportion of the signal can be seen to "leak through" into the double gradient echo excitation sculpting spectra, Figure 13.28c–f, even so, the signal for PLGA can be detected down to a level of 0.5 ppm, confirming that the LOD was well below the specification level of 10 ppm. Once validated, the method was setup to run in full automation and has been used to analyze hundreds of cleaning validation solutions to date.

13.6 Conclusion

It has been demonstrated that NMR is a feasible technique for MI analysis that provides an excellent orthogonal technique to the more traditional methods.

It has a number of advantages over other techniques such as:

a) Inherent quantitative nature.
b) Ease and speed of sample preparation.
c) Ability to provide a quick answer as to whether NMR is a feasible technique.
d) Ability to detect more than one MI in the same analysis and conceivably combine with other measurements such as residual solvent analysis.

These advantages mean that NMR is a logical choice for problems particularly in the earlier stages of pharmaceutical development when fit-for-purpose answers are required in a timely fashion but can also be equally applied in late stage development or commercial settings where limit test approaches can be easily developed and validated. A suggested strategy is to consider NMR, investigating the spectra of both the substrate and MI independently and do an assessment of whether any MI signals are likely to be resolved. In addition, determine the likely solubility in a common NMR solvent, such as DMSO.

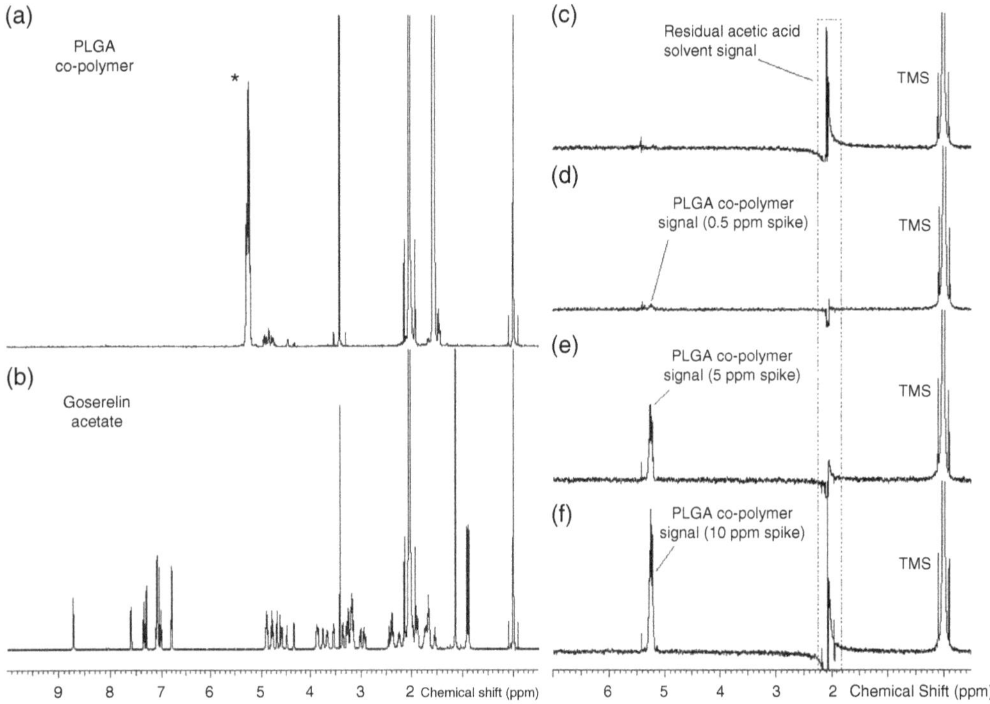

Figure 13.28 (a) ^{1}H NMR spectrum of PLGA copolymer highlighting resolved signal * at 5.3 ppm; (b) ^{1}H NMR spectrum of goserelin acetate. Expansions showing ^{1}H excitation sculpting with double gradient echo spectra exciting the regions at 0 ppm (TMS) and 5.3 ppm in (c) solvent blank, (d) 0.5 ppm PLGA spiked sample, (e) 5 ppm PLGA spiked sample, and (f) 10 ppm PLGA spiked sample.

If there is resolution and a high solubility can be obtained, performing a spike at a high level, such as 1000 ppm, will quickly give an indication as to the success of the method. If such a spike is successful, further work can continue at lower levels to determine whether NMR has sufficient sensitivity.

If the initial attempt fails, further experiments can be performed using different solvents or with the addition of additives to try and obtain suitable signal resolution and concentration. Alternatively, one can quickly move on and investigate the use of different techniques.

There are of course a number of disadvantages of NMR that must be considered:

a) Inherent insensitivity.
b) Expense and availability of some NMR equipment, particularly higher fields and cryoprobe systems.
c) If acceptable resolution cannot easily be achieved between the signals of the substrate and those of an MI, there is reduced scope for the amount of method development that can be done (compared to other techniques, such as chromatography). Method development options are typically limited to changing the solvent, the use of additives or utilizing higher magnetic fields to achieve greater signal dispersion.
d) It can be difficult to apply NMR to MI problems involving formulated materials due to the increased likelihood of signal overlap from excipients and the reduced ability to prepare samples of sufficiently high concentration for analysis.

In conclusion, NMR is a great orthogonal technique for the analysis of MIs and can be a very powerful tool that, for certain problems, can really speed up the tricky task that is trace analysis.

References

1 Claridge, T. (2016). *High Resolution NMR Technique in Organic Chemistry*, 3e. Cambridge: Wiley.
2 Cavanagh, J., Fairbrother, W.J., Palmer, A.G. III et al. (2006). *Protein NMR Spectroscopy: Principles and Practice*, 2e. San Diego, CA: Academic Press.
3 Ernst, R.R., Bodenhausen, G., and Wokaun, A. (1990). *Principles of Nuclear Magnetic Resonance in One and Two Dimensions*. Oxford: Clarendon Press.
4 Levitt, M.H. (2008). *Spin Dynamics*, 2e. Chichester, UK: WileyBlackwell.
5 Keeler, J. (2011). *Understanding NMR Spectroscopy*, 2e. Chichester, UK: Wiley.
6 Williams, D.H. and Fleming, I. (2007). *Spectroscopic Methods in Organic Chemistry*, 6e. Maidenhead, UK: McGraw-Hill Higher Education.
7 Wawer, I., Diehl, B., and Holzgrabe, U. (2008). *NMR Spectroscopy in Pharmaceutical Analysis*. Elsevier Science.
8 Kovacs, H., Moskau, D., and Spraul, M. (2005). Cryogenically cooled probes – a leap in NMR technology. *Progress in Nuclear Magnetic Resonance Spectroscopy* 46: 131–155.
9 Hore, P.J. (1989). Solvent suppression. *Methods in Enzymology* 176: 64–77.
10 Guéron, M., Plateau, P., and Decorps, M. (1991). Solvent signal suppression in NMR. *Progrsss in NMR Spectroscopy* 23: 135–209.
11 McKay, R.T. (2009). Recent advances in solvent suppression for solution NMR: a practical reference. *Annual Reports on NMR Spectroscopy* 66: 33–76.
12 Bauer, C., Freeman, R., Tom, F. et al. (1984). Gaussian pulses. *Journal of Magnetic Resonance* 58: 442–457.
13 Urquhart, M.W.J., Bardsley, B., Edwards, A.J. et al. (2018). Managing emerging mutagenicity risks: late stage mutagenic impurity control within the atovaquone second generation synthesis. *Regulatory Toxicology and Pharmacology* 99: 22–32.
14 Freeman, R. (1998). Shaped radiofrequency pulses in high resolution NMR. *Journal of Progress in Nuclear Magnetic Resonance Spectroscopy* 32: 59–106.
15 Vincent, S.J.F. and Zwahlen, C. (1997). Selective pulses in NMR. In: *Methods for Structure Elucidation by High-Resolution NMR* (eds. G. Batta, K.E. Kövér and C. Szántay Jr.), 1–17. Elsevier Science.
16 Stott, K., Stonehouse, J., Keeler, J. et al. (1995). Excitation sculpting in high-resolution nuclear magnetic resonance spectroscopy: application selective NOE experiments. *Journal of the American Chemical Society* 117: 4199–4200.
17 Hwang, T.-L. and Shaka, A.J. (1995). Water suppression that works. Excitation sculpting using arbitary waveforms and pulsed field gradients. *Journal of Magnetic Resoance. Series A* 112: 275–279.
18 ICH (1995). Eurpoean Medicines Agency. https://www.ema.europa.eu/en/documents/scientific-guideline/ich-q-2-r1-validation-analytical-procedures-text-methodology-step-5_en.pdf (accessed 5 November 2019).

14

Addressing the Complex Problem of Degradation-Derived Mutagenic Impurities in Drug Substances and Products

Steven W. Baertschi[1] *and Andrew Teasdale*[2]

[1] *Baertschi Consulting LLC, Carmel, IN, USA*
[2] *Chemical Development, Pharmaceutical Technology and Development, Operations, AstraZeneca, Macclesfield, UK*

14.1 Introduction

14.1.1 Background

The safety of potential or actual impurities in drug substances (DSs) and drug products (DPs) has been a significant concern within the pharmaceutical industry for many years, long before the landmark ICH Q3A and Q3B guidance on organic impurities, first adopted in 1995 and 1996, respectively [1, 2].

In more recent years, a focus on the potential for "unusually toxic" impurities, which was first discussed in the ICH Q3A and Q3B guidance, emerged. This eventually led to regulatory guidelines [3, 4] and ultimately the ICH M7 guideline [5], which was finalized in 2014 and followed by a revision (R1) in September of 2017. A fundamental requirement of M7(R1) is that all "Actual and potential impurities that are likely to arise during the synthesis and storage of a new drug substance, and during manufacturing and storage of a new drug product should be assessed" for mutagenicity. Thus, the need to predict, assess, and potentially control mutagenic degradation products came into sharp focus.

Strategies for assessing and controlling mutagenic impurities (MIs) arising from drug synthesis have received considerable attention in regulatory guidance [3, 4, 6] and scientific literature [7–11]. Much has also been written about the use of science-based mutagen risk assessments (MRAs) to evaluate drug substance quality [9, 12–20]. In these MRAs, the workflow includes evaluation of the starting materials and intermediates in tandem with the process chemistry to arrive at a complete view of potential process impurities and their possible "purgeability" in subsequent steps. Once this is completed, as outlined by ICH M7(R2), two or more (Q)SAR tools (e.g. DEREK and Sarah (Lhasa Ltd.), Leadscope (Leadscope Inc.), SciQSAR (Scimatics, Inc.), CaseUltra (MultiCASE, Inc.) should be used to assess whether these impurities have any structural alerts for mutagenicity. The two (Q)SAR tools should be complementary (expert- and statistical-based systems). The selection of chemical structures for (Q)SAR screening from the DS synthesis should be focused on the starting materials, intermediates, and reagents; additionally, major by-products and significant process-related impurities are usually discovered during the synthetic route development, and those should be considered for inclusion.

Mutagenic Impurities: Strategies for Identification and Control, First Edition. Edited by Andrew Teasdale.
© 2022 John Wiley & Sons, Inc. Published 2022 by John Wiley & Sons, Inc.

In contrast to control strategies for mutagenic process impurities, where there is a potential for elimination (e.g. by changing the synthetic route) or reducing ("purging") in downstream synthetic steps [21–23], there is no simple way to "remove downstream" or prevent degradation products that form (at potentially very low levels) during storage of the DS or DP [18, 24–26]. Consideration of the difficulties in assessing degradation products that may occur over the shelf life of the DS and DP is complicated by several factors, including:

- Both "potential" and "actual" degradation products should be assessed for mutagenic potential.
- The range of "potential" degradation products should be defined.
- Only a very small degree of degradation may be needed to produce a mutagenic degradation product above the acceptable thresholds.
- Experimentation to control degradation rates requires degradation rate projections over the shelf life of both the DS and DP.
- Detecting and quantifying low levels of mutagenic degradation products is particularly challenging due to the presence of the formulation matrix, requiring highly sensitive and selective separation and detection methodologies with good recovery.

14.2 Working Definitions

ICH M7(R1) defines "actual" and "potential" degradation products. *Actual* degradation products include those that are observed above the ICH Q3A (for DS) and Q3B (for DP) *reporting thresholds* over the shelf life, when stored at the proposed long-term storage conditions in the proposed packaging.

Potential degradation products are those that may be reasonably expected to form during long-term storage conditions, and include those that form above the ICH Q3A (DS) and Q3B (DP) identification thresholds during accelerated stability studies (e.g. 40 °C/75% RH) for 6 months, and confirmatory photostability studies as described in ICH Q1B, but are yet to be confirmed as forming under long-term storage conditions *in the primary packaging*.

Significant literature also exists [18, 24–29] that describes potential degradation products as those that form during stress testing studies as defined by ICH Q1A(R2) [30]. An excerpt of this definition is: "Stress testing. . .can help identify the likely degradation products, which can in turn help establish the degradation pathways and the intrinsic stability of the molecule and validate the stability indicating power of the analytical procedures used." Thus, potential degradation products can be reasonably defined to include those that form during stress testing studies in addition to those that form during accelerated stability studies and long-term stability studies not in the final packaging. Evidence has been presented that well-designed and executed stress testing studies (of both the drug substance and drug product) should produce all (or nearly all) probable degradation products [18].

A further class of degradation products can be considered: *theoretical* degradation products. Theoretical products are those that can be predicted to form under pharmaceutically relevant conditions based on chemical principles *in the absence of experimental data*. A computational tool called Zeneth™ is available to aid the predictions for such theoretical chemical degradation products [31]. Boyd and Sharp discussed various computational tools available to aid in predicting and investigating theoretical and actual degradation products and pathways [32]. The computational approach that has been studied the most in the context of predicting potential oxidative degradation pathways for pharmaceutical compounds is the susceptibility to radical-initiated oxidation (i.e. autoxidation). Such an approach involves calculation of bond dissociation enthalpies (BDEs) via hydrogen-atom abstraction, resulting in an unstable radical, leading to oxidative degradation. Such methods are fairly well developed and can provide predictive results for the sites in a drug molecule that are the most susceptible to radical-initiated oxidation [33, 34].

Finally, it is important to consider the definition of "identified" with respect to degradation products. As defined by ICH Q3A, an unidentified impurity is one "for which a structural characterization has not been achieved and that is defined solely by qualitative analytical properties (e.g., chromatographic retention time)." In contrast, an identified impurity is an "impurity for which structural characterization has been achieved." Although numerous structures may be proposed for potential and actual impurities/degradation products throughout the development of a compound, these structural proposals are often derived from nominal or accurate mass data, sometimes combined with a scientific rationale. For the purpose of a mutagenic risk assessment, however, it is important to have a high degree of confidence in the structure assignment. A strategy recommended in the literature [18] (and utilized by numerous pharmaceutical companies) is that tentative structures proposed from mass spectrometric data should be confirmed before being included in an MRA; confirmation of structure would involve support from additional evidence, e.g. unambiguous NMR data, independent synthesis, or a strong chemical rationale.

14.3 Challenges Associated with the Assessment of Risk Posed by (Potentially) Mutagenic Degradation Products

DSs that are small organic molecules typically possess a variety of functional groups, resulting in a wide array of potential inter- and intramolecular reactions, rearrangements, and other degradation pathways. Accurate prediction of potential degradation pathways is further complicated because of the relatively small amount of published information available regarding the degradation of pharmaceuticals, the length of time required to assess molecular stability, the significant effects of physical form, the formulation ingredients, and dosage strength on possible pathways. Consequently, degradation processes sometimes yield chemical structures that are not initially expected, predicted, or even seen under long-term storage conditions. Further complications result from the reality that degradation pathway knowledge associated with a particular compound often evolves (grows) during the development process. Such pathways are often not fully elucidated before the final formulation is established and are generally unknown prior to the start of clinical studies.

Although prediction of potential degradation pathways via in silico tools or knowledge-based approaches can help in the development of a mechanistic understanding of the possible degradation pathways, the overall strategy should be based on risk assessment. Potential degradation products should be determined to be either *relevant* and included in the MRA or *nonrelevant* and excluded from further consideration.

14.4 Risk Assessment Process for Mutagenic Degradants

14.4.1 Stability-Related MRA Process Overview

ICH M7(R1) states that "actual and potential degradation products likely to be present in the final drug substance or drug product and where the structure is known should be evaluated for mutagenic potential…." Thus, a critical aspect of mutagen risk assessment (MRA) for DS or DP degradation is the determination of degradation pathways and the structures of associated degradation products that are relevant to the manufacturing process and the proposed storage conditions in the proposed packaging. As mentioned in ICH M7, "knowledge of relevant degradation pathways can be used to help guide decisions on the selection of potential degradation products to be evaluated for mutagenicity, e.g., from degradation chemistry principles, relevant stress testing studies, and development stability studies."

Piecing together an overall strategy based on ICH M7(R1), the critical steps involve knowledge of potential degradation products, the relevance of those potential degradation products to patient exposure, and determination of the structures of those products determined to be potentially relevant (Figure 14.1). These relevant potential degradation products are then included in the MRA.

The overall strategy for the MRA of the potential degradation products where structures are known and have been selected for assessment is shown in Figure 14.2. As shown in the figure, the process starts with (Q)SAR analysis by two complementary in silico tools. From Section 6 of ICH M7(R1), a positive (Q)SAR result indicates that the degradation product is a class 1 or 2 (known mutagenic carcinogen or known mutagen) or class 3 (an alerting structure unrelated to the structure of the API).

14.4.2 Stress Studies

Well-designed stress testing studies, as recommended by ICH Q1A(R2) and discussed in more detail by others [24, 27, 28, 35], can yield a set of potential degradation products, whose chemical structures can be included in the MRA. Such products are termed "potential" since they may or may not form during ICH long-term and accelerated stability studies (see ICH Q1A(R2) for a definition of stress testing and ICH M7 Section 5.2).

14.4.3 Accelerated Stability Studies

Evaluation of ICH accelerated stability results (typically 40 °C/75% RH for six months for solid dosage forms, per ICH Q1A(R2)) may also reveal degradation products to be included in, and potentially

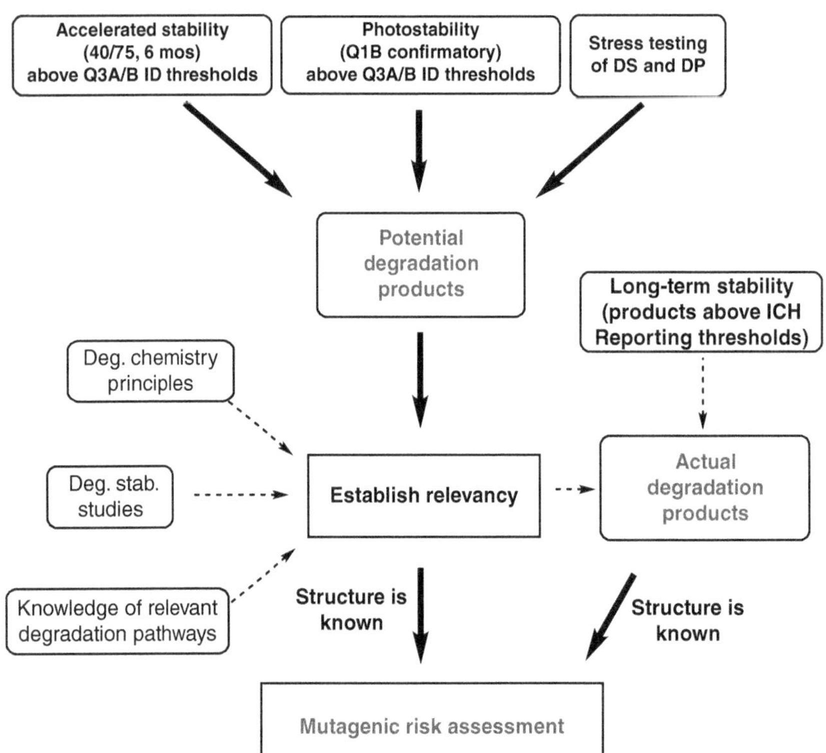

Figure 14.1 Critical aspects of determining potential degradation products to include in an MRA.

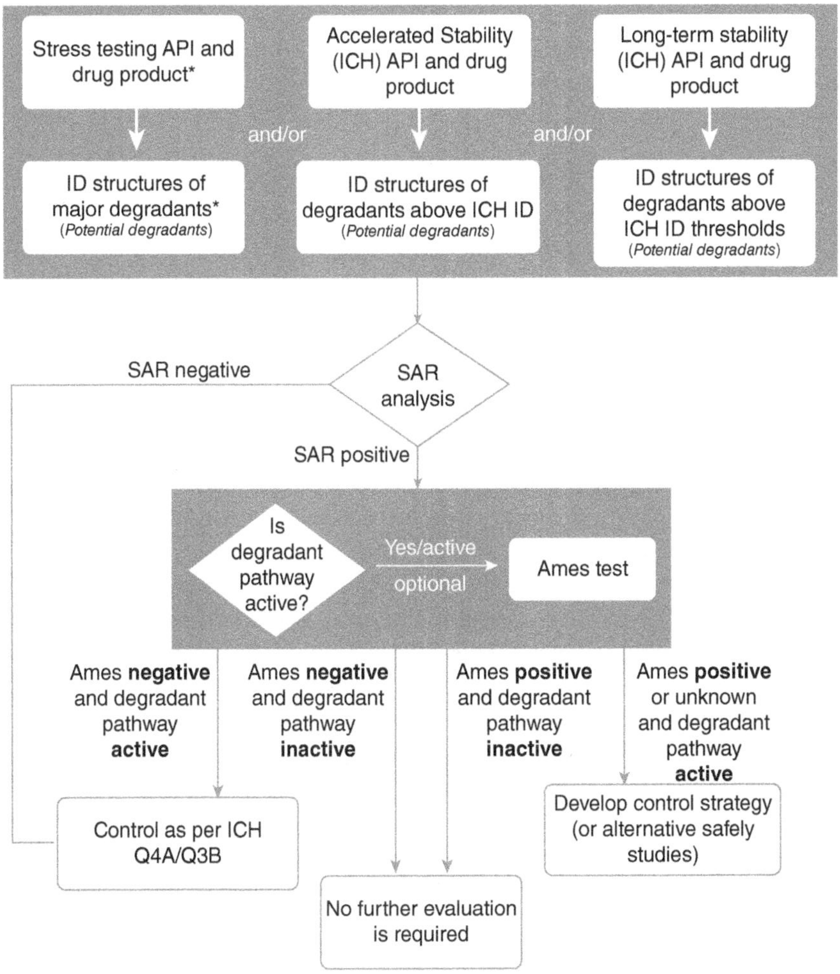

Figure 14.2 Proposed process flow for assessing degradation products in the DS and DP. (*See insert for color representation of this figure*).

which stress testing-derived potential degradation products to exclude from, the MRA. Such studies may encompass a range of stability conditions ranging from open storage to studies performed in the final packaged product. Per ICH M7(R1), those degradation products formed under accelerated conditions are also included in the set of "potential degradation products."

14.4.4 Long-term ICH Stability Studies

Degradation products formed (above the reporting thresholds) during the proposed long-term ICH stability storage conditions in the proposed primary and secondary packaging (per ICH Q1A(R2) and ICH M7(R1)) should also be included in the MRA and are included in the set of "actual degradation products" (see Section 1.2).

14.4.5 Deciding Which Products to Include in the MRA

As discussed above in this section, ICH M7(R1) indicates that the MRA should include potential and actual degradation products "likely to be present in the DS and DP and where the structure is known." As has been suggested in the literature, the MRA should focus on primary degradation pathways and their associated major degradation products, and that hypothetical or theoretical degradation products that are not observed *experimentally* would not need to be included.

14.4.6 In Silico Tools for the Prediction of Potential Degradation Products

Prior to conducting stress or other stability studies, in silico predictions are useful to help consider theoretical degradation pathways and potentially guide the strategy for analytical method development. There are relatively few commercial software programs designed specifically for predicting degradation pathways [24]. The commercially available predictive degradation software Zeneth (Lhasa Ltd) is currently the most well-documented software package available. Zeneth is a chemical degradation prediction application that utilizes a rules-based software platform with a growing number of chemical transformations in its knowledge base to predict possible degradation products from a given molecular structure, including potential interactions with excipients in the formulation [31, 36]. In cerebro, knowledge gleaned from chemical principles and expertise provides a means for interpreting the output from Zeneth predictions, aiding in the design of analytical methods to use for stress testing studies and in the interpretation of stress testing results.

In general, in silico tools tend to "over predict," i.e. predict more potential degradation products than are observed experimentally from stress testing studies (potential degradation products) or long-term stability studies (actual degradation products) [36]. Consequently, while hypothetical assessment of degradation products and pathways may provide potentially useful starting points for deciding which degradation products need to be considered in the MRA process, such theoretical assessments are not a substitute for stress testing studies.

It has been proposed that in silico prediction tools should not be used to initiate investigations (e.g. "fishing" or "hunting" exercises that evaluate numerous hypothetical structures whose significance has not been verified experimentally) [24]. This proposal is consistent with the ICH M7(R1) guidance that the MRA should focus on "reasonably expected" impurities.

14.5 Using Stress Testing to Select Degradation Products for Identification

As described above (see Section 14.2), degradation products produced by stress testing are "potential" degradation products, and potential degradation products that have been structurally identified should be included in the MRA. Thus, decisions around identification of degradation products from stress testing are impactful. Stress testing, also known as "forced degradation," is utilized in the pharmaceutical industry to learn more about the pharmaceutically relevant possible degradation pathways of the parent drug structure (by itself or when formulated in a drug product). Stress testing therefore becomes a critical filter for determining which *hypothetically predicted* degradation products to include in an MRA.

Due to the lack of detail in the ICH guidance documents, particularly ICH Q1A(R2), in relation to stress testing, the exact conditions of stress testing have been interpreted differently by individual companies leading to differences in approaches [28, 37, 38]. There is, however, a general industry consensus that only the "major" degradation products arising from these studies need to be evaluated [24, 28, 36, 38]. As described in ICH Q1A(R2), stress testing of the drug substance includes exposure of the drug substance to acid/base in solution, oxidative conditions, and solid-state exposure to photo-radiation and thermal/humidity stress (at least 10 °C above accelerated stability temperatures). Stress testing of the drug product typically involves exposure to elevated temperature and humidity, along with photo-stressing per ICH Q1B. The relevance of acid/base hydrolytic conditions and oxidative conditions is well recognized in the pharmaceutical industry. Hydrolytic and oxidative conditions are the two principle conditions that form the majority of relevant degradation products for most drugs [28, 38]. The acid/base conditions are designed to accelerate hydrolytic reactions via lowering of the Activation Energy (Ea) through acid or base catalysis. The oxidative conditions (e.g. via peroxides, radicals, or transition metals) are designed to significantly accelerate normal oxidative reactions. Since the degradation products that form under thermal/humidity stress are typically subsets of the products that form under these two conditions, the conditions are pharmaceutically relevant. In addition, such studies provide a wide range of potential degradation pathways at higher levels, in a shorter timeframe, than typically observed during solid-state stress studies; such studies can also be used for mechanistic investigations and insight. Overall, the conditions outlined here provide thorough coverage of the environmental conditions that the DS or DP may reasonably be exposed to and, hence, can be expected to cover all relevant degradation pathways.

Due to the different strategies that have been utilized in the industry, it has been suggested that any of the three following strategies can be considered for decisions related to identification of degradation products from stress testing. The choice of strategy can be dependent on a number of criteria, which may include the phase of development [39], disease indication, and knowledge of the primary degradation pathways of the molecular scaffold.

14.5.1 Approach 1: Criteria for Structure Identification After Observation in Accelerated and Long-term Stability Studies

The first strategy is the simplest to implement and uses forced degradation to help guide the accelerated and long-term stability studies. Partially degraded samples from forced degradation studies are used for analytical method development of the stability-indicating method(s). Forced degradation results in the generation of larger quantities of degradation products (both in number and amount), making degradation products easier to detect analytically. Degradation product structures are not *necessarily* elucidated unless they are also observed in accelerated or long-term ICH stability studies, as described in ICH M7.

14.5.2 Approach 2: Criteria for Structure Identification Through Use of an Algorithm in Stress Testing Studies

The second approach utilizes forced degradation studies to delineate potential degradation products and pathways, with structure elucidation focused on "major" degradation products, as outlined by the algorithm below. The overall systematic approach is based on work by Alsante et al. [28]. The algorithm involves defining a major degradation product as one that is present in partially degraded samples from forced degradation studies at levels greater than 10% of the total degradation *and* also >25% of the largest individual degradation product as illustrated in Figures 14.3 and 14.4. This strategy has been successfully applied [18] and has been shown to reliably focus on the major degradation pathways, creating a reasonable number of "potential" degradation products while comprehensively including "actual" degradation products (i.e. "actual" degradation products are a subset of "potential") [18]. Kleinman et al. [24] add further detail regarding the comparison of the two criteria.

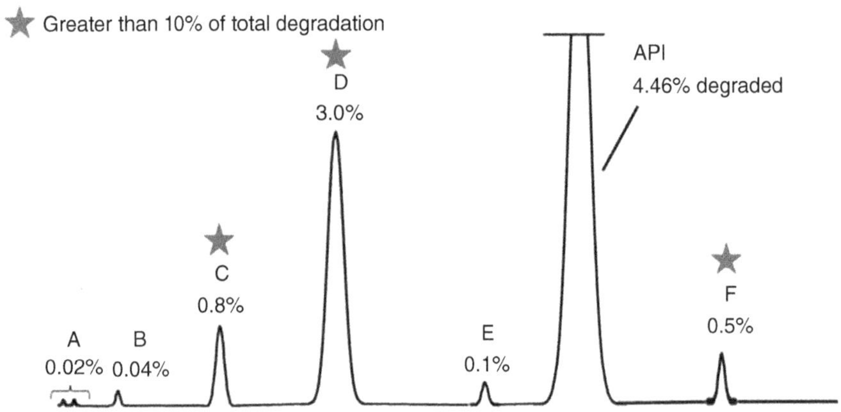

Figure 14.3 Illustration of criterion no. 1 in a sample chromatogram: Identify the degradation products only if they comprise at least 10% of the total degradation for appropriately stressed samples.

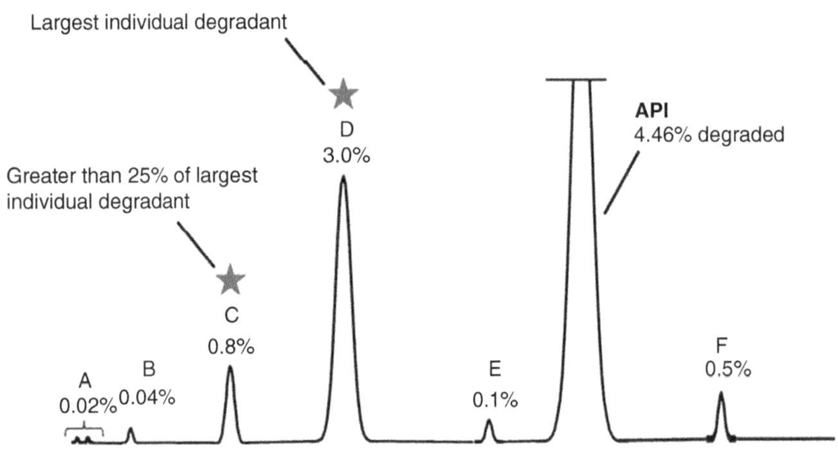

Figure 14.4 Illustration of combining criterion no. 1 with criterion no. 2 in a sample chromatogram: Identify additional peaks only if they are greater than 25% of the largest impurity (Table 14.1).

Table 14.1 Peaks are selected for identification based on meeting Criteria 1 (Identify largest impurity only if it comprises at least 10% of the total degradation for appropriately degraded samples) and Criteria 2 (Identify additional peaks only if they are greater than 25% of the largest impurity).

Peak	% Area or % weight	Criterion 1 assessment (rationale)	Criterion 2 assessment (rationale)	Selected for identification (meets criterion 1 and criterion 2)
A	0.02	No (below threshold)	No (below threshold)	No
B	0.04	No (below threshold)	No (below threshold)	No
C	0.8	Yes (meets criteria; 0.8%/4.46%)	Yes (above threshold)	Yes
D	3.0	Yes (largest impurity 3.0%/4.46%)	Yes (largest impurity)	Yes
E	0.1	No (below threshold)	No (below threshold)	No
API	95.54	API (4.46% degraded)	API	API
F	0.5	No (below threshold; 0.5%/4.46%)	No (below threshold)	No

In this case, only two degradation products are selected for identification from the forced degradation study.

14.5.3 Approach 3: Structure Identification Through Use of Kinetic Equivalence and Scaled ICH Q3B Thresholds

A third approach has also been developed based on the concept of "kinetic equivalence" (described below) and ICH Q3B thresholds. This strategy attempts to merge kinetics and ICH Q3B thresholds to define when to identify degradation products from forced degradation studies. The overall decision to identify is guided by thresholds that attempt to account for differences in the total amount of degradation generated in forced degradation vs the total amount of degradation expected in long-term stability studies.

14.5.3.1 Kinetic Equivalence

Generally, solid-state stress testing is aimed at achieving at least the same amount of "stress" (i.e. "kinetic equivalence") that would be experienced from storage at six months at 40 °C/75% relative humidity (RH) and at least 2 × ICH Q1B confirmatory photo-exposure, or in the range of 10–20% loss, whichever comes first [28, 35, 38]. Since relative humidity often plays a key role in the rate of degradation, the design of short-term studies to create a kinetic equivalence should take both temperature and RH into account [40, 41]. Two main options are available: (i) building the sensitivity to RH into the Arrhenius relationship (e.g. using the Accelerated Stability Assessment Program (ASAP) protocol) [40–42] or (ii) maintaining the same RH at the various temperatures used in an Arrhenius study. While both options are feasible, the latter approach offers a simpler conceptual approach and is the focus of the discussion here.

Accurate Arrhenius-based rate predictions require knowledge of the energy of activation (Ea) of degradation specific to the DS or DP being studied. If the Ea is not known, an Ea of 19.87 kcal/mol (83.14 kJ/mol), which is the default value used by the United State Pharmacopeia guidance for the calculation of the mean kinetic temperature [43], is recommended here as a reasonably conservative approach for use in a kinetic projection using the Arrhenius relationship. The Ea has a significant effect on the rate of degradation as a function of temperature, and this is illustrated in Table 14.2.

The different Ea's shown in the table, ranging from a very low 12–29.8 kcal/mol were included in the table for specific reasons. An Ea of 12 kcal/mol corresponds to the low end of documented Ea's for drug-like molecules, where as an Ea of 29.8 kcal/mol corresponds to an experimentally determined average Ea

Table 14.2 Rate of degradation (relative to 25 °C) assuming an Arrhenius kinetic relationship.

Temp (°C)	Relative rate[a] (Ea = 12 kcal/mol)	Relative rate[a] (Ea = 17 kcal/mol)	Relative rate[a] (Ea = 19.87 kcal/mol) [40]	Relative rate[a] (Ea = 25.8 kcal/mol)	Relative rate[a] (Ea = 29.8 kcal/mol)
25	1	1	1	1	1
30	1.4	1.6	1.7	2.1	2.3
40	2.6	4.0	5.0	8.1	11.2
50	4.8	9.2	13.5	29.2	49.3
60	8.4	20.4	34.1	97.7	198.9
70	14.3	43.2	81.9	304.8	739.8
80	23.6	86.6	187	891.2	2554.7

[a] The relative rate is meaningful only within the individual columns. Relative rates across rows should not be inferred.

from more than 100 solid-state drug studies involving c. 50 compounds [44] and unpublished work from Eli Lilly [24]. More recent data indicates that the average Ea for solid-state DPs is approximately 26–27 kcal/mol. As mentioned above, an Ea of 19.87 kcal/mol corresponds to the USP-recommended assumption for mean kinetic temperature calculations [43], which was based on the work by Kennon where Ea's were determined from evaluation of drug compounds *in solution* at different temperatures [45]. More recent work by MacFaul et al. [46] determined an average Ea of 23.6 kcal/mol, with a range of 11.9–47.2 kcal/mol for 166 drug-like compounds *in solution*. An Ea of 17 kcal/mol corresponds to the assumption that six months storage at 40 °C/75% RH is equivalent to two years at 25 °C/75% RH. An Ea of 25.8 kcal/mol corresponds to the "Joel Davis rule" [47], where three months storage at 40 °C/75% RH was assumed to be equivalent to two years at 25 °C/60% RH.

Table 14.3 illustrates the effect of Ea on the rate of degradation at different temperatures by calculating the number of days it would take to achieve a kinetic equivalence to six months at 40 °C/75% RH. Considering an Ea of 19.87 kcal/mol for the degradation of a drug molecule, a stress testing study at 70 °C with a length of 11 days would yield the kinetic equivalence of an accelerated stability study of 40 °C for 6 months; an Ea of 29.8 kcal/mol would reach kinetic equivalence in less than 3 days! These projections of course assume that the degradation pathways at 70 °C are the same as those at lower temperatures, with no phase changes over the range of temperatures studied; this has been discussed in more detail by Baertschi et al. [38].

14.5.3.2 Scaled ICH Q3B Thresholds

In alignment with ICH Q1A(R2) and with Q4 in the EMA Q&A on MIs [48], structure elucidation of degradation products that form during long-term stability studies is not required when the levels are below ICH identification thresholds. Thus, using ICH Q3B identification thresholds [2] for the drug product as a starting point, a scaled threshold for identification of degradation products observed during stress testing has been proposed [24]. These thresholds reflect the typically larger degradation levels observed during stress testing. Conservatively, assuming 2% as an acceptable amount of degradation in a drug product over the shelf life, a relationship to stress testing was proposed. Degradation of 5–10% in a stress testing experiment is 2.5–5 times higher than the 2% level at the end of shelf life. Thus, an identification threshold for degradation products arising during stress testing would be 2.5–5-fold higher than the corresponding thresholds for the DP under normal storage conditions (see Table 14.4).

Table 14.3 Number of days at specified temperatures calculated for a "kinetic equivalence" to six months storage at 40 °C for reactions with different energies of activation assuming Arrhenius kinetics.

Temp (°C)	Number of days (Ea = 12 kcal/mol)	Number of days (Ea = 15 kcal/mol)	Number of days (Ea = 19.87 kcal/mol)	Number of days (Ea = 25.8 kcal/mol)	Number of days (Ea = 29.8 kcal/mol)
25	482	615	912	1470	2040
30	345	405	524	718	888
40	182.5	182.5	182.5	182.5	182.5
50	100	86.5	67.8	50.5	41.4
60	57.2	42.8	26.8	15.1	10.3
70	33.7	22.1	11.1	4.8	2.8
80	20.5	11.8	4.9	1.7	0.8

Table 14.4 Proposed identification thresholds for major degradation products formed during stress testing based on scaled ICH Q3B thresholds for long-term storage.

Maximum daily dose (mg)	Identification (ID) threshold from ICH Q3B [1] (%)	ID threshold derived from ICH Q3B for stressed samples degraded 1–5% (%)	ID threshold derived from ICH Q3B for stressed samples degraded >5–10% (%)	ID threshold derived from ICH Q3B for stressed samples degraded >10–15% (%)	ID threshold derived from ICH Q3B for stressed samples degraded >15–20% (%)
>2000	0.10	0.25	0.5	0.75	1.0
>10–2000	0.2	0.5	1.0	1.5	2.0
>1–10	0.5	1.25	2.5	3.75	5.0
<1	1.0	2.5	5.0	7.5	10.0

14.6 Development Timeline Considerations

14.6.1 Drug Discovery Stage

The goal of stress testing or stability studies at this stage is primarily to determine whether or not a compound has sufficient stability for the desired routes of administration over the length of anticipated clinical studies. Such clinical studies are typically short in duration, limited in scope, and use analytical methodologies that are often generic (e.g. designed for high throughput, not specifically designed for the individual compound). Degradation products are typically viewed as "peaks in a chromatogram," not as structurally identified products. It may be prudent to evaluate the *theoretical* potential formation of mutagenic degradation products for particular structures/ scaffolds, since controlling degradation to the low levels required for MIs may be very difficult and could threaten the developability of the drug [25, 49].

It may also be prudent to consider the potential formation of mutagenic degradation products from acidic, nonenzymatic hydrolysis of the API in the stomach. Discovering such potential issues during development could lead to effective mitigation strategies (e.g. enteric-coating of the oral dosage form).

14.6.2 Preclinical to Phases 1/2

During Phase 1 clinical trials of up to 14 days, only known carcinogens and mutagens need to be flagged and limited to acceptable levels as described in ICH M7(R1). Other impurities, even those with mutagenicity-alerting structures, can be treated as non MIs because of the short duration of exposure. ICH M7(R1) acknowledges that not all impurities (including degradation products) are typically identified at this stage. Higher identification thresholds for actual impurities (including degradation products) are often employed, especially at the preclinical stage and can be 2× [18] or 3× [50] ICH Q3A/B guidelines. However, since ICH M7, which applies throughout clinical development, utilizes ICH Q1A and Q3B thresholds, some companies have moved away from using such modified identification thresholds, especially in the context of degradation products.

While stress testing studies are encouraged, but not specifically required, during this stage, the intent is to ensure stability and purity throughout the clinical trial(s) [51, 52]. While stability-indicating analytical methods, supported by stress testing studies, should be developed for the DS, the definitive identification of degradation products observed during stress studies is not expected. Nonetheless, such degradation pathway information can be very useful to the further development of the compound.

14.6.3 Phase 3 to New Drug Application (NDA) Regulatory Submission

Stress testing studies, with a full understanding of the "inherent stability of the drug substance, potential degradation pathways, and the capability and suitability of the proposed analytical procedures" [53] are expected to be completed by or during Phase 3, and certainly for the marketing application. ICH Q3A(R2) and Q3B(R2) reporting, identification, and qualification thresholds are typically applied at this stage of development for formal stability studies. It is notable that any degradation products (potential or actual) for which structures are known should be assessed for mutagenic potential, per ICH M7(R1).

14.6.4 Post-marketing/Line Extensions

After registration, changes to the DS or DP manufacturing process are often desired for cost reduction, quality or reliability increases, or environmental impact reduction. Manufacturing site and scale changes are also common. Risk-based guidance, such as ICH Q9 [54], can aid in assessing the significance of a process or formulation change that may require stability studies to be conducted to demonstrate that the proposed changes do not adversely impact the already established stability characteristics (i.e. degradation rate or profile) of the product.

Another important consideration during the lifecycle of a drug is the development of new dosage strengths, forms, formulations, and alternate routes of administration. Each new development will require new or modified stress testing and/or accelerated stability studies, as it cannot be assumed that degradation rates and pathways will remain the same as those in the original product. New or modified analytical methodologies may also be required, and therefore, new or revised accelerated stability studies will need to be performed as part of the stability-indicating method development process. New or modified analytical methodologies can also lead to the discovery of new impurities/degradation products (in line extensions and even in existing products) that were not detected with previous methods.

14.7 Developing Control Strategies for (Potential) Mutagenic Degradation Products

14.7.1 Determining Relevancy of Potential Degradation Products and Developing Control Strategies for Actual Degradation Products

Kleinman et al. [24] and Dow et al. [18] have provided systematic approaches for determining what degradation products to include in an MRA. The overall MRA process for degradation products is summarized in Figures 14.1 and 14.2. Potential degradation products are discovered via well-designed stress testing, photostability, and accelerated stability studies. Actual degradation products are defined via accelerated and long-term stability studies using stability-indicating analytical methods. Guided by chemistry understanding and identification thresholds described in this chapter, potential and actual degradation products are identified; once identified, they should be considered for inclusion in the MRA. Decisions about whether or not to include potential degradation products in the MRA can be guided by degradation chemistry principles, development stability studies, and knowledge of the pathways that are relevant to patient exposure. In this section, the use of accelerated and photostability studies along with degradation chemistry knowledge is discussed in the context of determining the relevancy of potential degradation products and developing appropriate control strategies.

14.7.2 Accelerated Stability (40 °C/75% RH Six months) or Kinetic Equivalent

In Section 8.4 of ICH M7(R1), [5] the control of degradation products is discussed: "For a potential degradation product that has been characterized as mutagenic, it is important to understand if the degradation pathway is relevant to the drug substance and drug product manufacturing processes and/or their proposed packaging and storage conditions." The guideline then describes the use of a "well-designed accelerated stability study (e.g., 40 °C/75% RH, six months) in the proposed packaging" as a method for determining the "relevance" of the degradation product. An option for shortening such a study using higher temperatures is mentioned, with the requirements being that the shorter study be "well-designed" and "kinetically equivalent"; these accelerated studies can be done "prior to initiating long term stability studies." A more detailed discussion of kinetic equivalence and how to conduct such studies is provided in section 2.3.3.1 of the manuscript by Kleinman et al. [24].

If the results of such studies indicate that the degradation product will form at "levels approaching the acceptable limit under the proposed packaging and storage conditions, then efforts to control formation. . .is expected," and confirmation of the control in primary stability studies is expected unless otherwise justified. Whether a specification would be required would depend on the results of the primary stability studies. Alternatively, the formulation, packaging, or storage condition could be changed and the determination of whether a pathway is active could be performed again.

The significant question here is what is meant by "levels approaching" the acceptable (Threshold of toxicological concern (TTC)-based) limit? Significant discussion of this question occurred during development of the M7 guideline, and since no consensus was reached, the vague wording was retained. In a grammatical sense, a level of >50% would constitute approaching; some have argued that for consistency with the Option 1 approach for process impurities, a 30% level could be adopted. What is meant by "approaching the acceptable limit" is important to consider since it will help define the sensitivity needed for analytical detection. Regardless, if the level of the product does not form during these relevancy studies at "levels approaching" the acceptable limit, the pathway can be deemed inactive and no further studies should be needed, although a conservative approach (as discussed by Dow et al. [18]) would be to confirm the result in primary stability studies.

14.7.3 Photostability Studies

As described in Sections 5 and 8 of ICH M7(R1), photodegradation products that are formed during ICH Q1B confirmatory photostability studies (directly exposed) are potential degradation products. Protection from light is an effective control strategy since primary and/or secondary packaging can easily be designed to block all UV and visible light transmission. This is in contrast to thermally induced excursions since, while packaging can be designed to protect from elevated humidity, packaging does not protect from temperature excursions during shipping, distribution, and storage. In addition, in-use photostability studies for I.V. or topical dosage forms may be critical since these formulations may be exposed to a significant amount of light during administration/patient use. A combination of in-use photo-exposure tests [55–57] and standard ICH Q1B confirmatory photostability tests can provide the understanding of photostability concerns and required packaging. The ICH Q1B confirmatory photostability test specifies a minimum exposure of 1.2 million lux·h for visible light and 200 W h/m^2 for UVA light; these exposures were intended to correspond to approximately three months of continuous exposure to ultraviolet (UV) and visible light without protective packaging in the pharmacy, warehouse, or at home [58]. Typical photostress testing conditions usually apply 2–5 times the specified ICH Q1B UV and visible dose, with twofold being the minimum recommended light exposure for stress studies [59]. If no degradation products are observed in the DS or DP after photo-stressing with or without packaging above the specified thresholds, then it is proposed that no further work is required. However, should a photodegradation product be observed above the levels discussed in the three approaches previously described, then it is suggested that identification of the photodegradation product be undertaken.

Photolytic degradation kinetics follow a linear relationship with the "light dose," i.e. there is a one-to-one relationship between the number of absorbed photons and the number of excited species created, and there is a direct correlation between the number of degradation products formed and the number of excited species [60]. If a photodegradation product is formed at a certain level after a specified amount of light exposure, it is likely to be observed at approximately half that level when the light exposure is cut in half. If light-protective packaging is impermeable to light, then the photodegradation product formation would be eliminated. The permeability of light through packaging can be tested by quantitatively measuring the light transmission, and the amount of protection can then be calculated. Combining the transmission results with expected patient in-use light exposure can be used to calculate a "light budget" [61] to keep photodegradation products below a required threshold.

14.7.4 Degradation Chemistry Knowledge

As discussed in ICH M7(R1), and summarized graphically in Figure 14.1, "knowledge of relevant degradation pathways can be used to *help guide decisions on the selection of potential degradation products to be evaluated for mutagenicity*, e.g., from degradation chemistry principles, relevant stress testing studies, and development stability studies" [italic added for emphasis]. Figure 14.1 captures these three guiding criteria as feeding into establishing "relevancy," a critical filter for bringing potential degradation products into the MRA. Just how to use the three guiding criteria is not explicitly clear from ICH M7(R1), so here we seek to provide thoughts around this critical decision point.

The concept of degradation chemistry principles is exemplified in Section 14.8.1, shown below, with the example of an actual degradation product that has an intermediate (predicted based on chemical principles) in the chemical degradation pathway with a mutagenic alerting moiety (i.e. a hydroperoxide). Further consideration of the predicted intermediate is described in the case study and in more detail in the publication by Raillard et al. [25] It was concluded that because no stress testing, mechanistic development stability studies, or ICH stability studies detected the hydroperoxide intermediate, the hydroperoxide was a transient intermediate, too unstable for patient exposure; therefore, no further MRA was needed.

A second hypothetical example is worth considering. Consider drug X, which during stress testing under basic conditions (0.1 N NaOH, 70 °C, one week) degrades to a major product with a mutagenic alerting structure, but does not form under *any* other stress conditions, including neutral pH conditions. Subsequent stress studies show that the product does not form appreciably unless the pH is ≥12. The synthetic route is evaluated and there is no exposure to such basic conditions during the synthesis. The DS is a solid powder and the DP is a solid-oral dosage tablet with a microenvironmental pH of c. 6. An argument could be made that this pathway is not relevant to the patient because the DS and DP will never be exposed to pH conditions above 8, which is four orders of magnitude away from pH 12. Based on this argument using data coupled with chemical principles and degradation chemistry knowledge, the degradation pathway is deemed as inactive, with no risk of exposure to the patient, and therefore no further mutagenic assessment is needed.

14.8 Risk Assessment Process Illustrated

The overall MRA process for degradation products is summarized in Figures 14.1 and 14.2. Potential degradation products are discovered via well-designed stress testing, photostability, and accelerated stability studies. Actual degradation products are discovered via accelerated and long-term stability studies using stability-indicating analytical methods. Guided by chemistry and identification thresholds described in this chapter, potential and actual degradation products are identified; once identified, they should be included in the MRA. Decisions about whether or not to include potential degradation products in the MRA can be guided by degradation chemistry principles, development stability studies, and knowledge of the pathways that are relevant to patient exposure. This process will be illustrated in four case studies: (i) Molecule A; (ii) Galunisertib; (iii) Naloxegol; and (iv) Selumetinib side chain.

14.8.1 Case Study #1: Molecule A

A case study was described by Kleinman et al. [24] and is reproduced here to help illustrate the overall risk assessment process. The degradation pathways of Molecule A were determined by stress testing, followed by accelerated and formal ICH stability studies (see Figure 14.5). The structures in the figure that are circled represent those degradation products that were detected above the various identification thresholds outlined for stress testing, accelerated, and long-term stability, and were thus structurally identified. The pathways shown in Figure 14.5 were deduced using degradation chemistry knowledge and principles. Thus, the scheme represents potential (products 1, 2, 3, and 4), theoretical (products 5, 7, and 8), and actual (product 6) degradation products.

Using the rationale described above in this chapter, degradation products 1–4 and 6 should be included in the MRA since they were observed in either the stress testing (potential), accelerated (potential), or long-term (actual) stability studies. Degradation products 7 and 8 are theoretical degradation products that were not observed during stress testing and therefore do not need to be included in the MRA. Degradation product 5 was proposed (based on chemical principles) to be a theoretical intermediate in the degradation pathway to 6. Product 5 was not observed during stress testing or any other stability studies above the recommended thresholds in (Table 14.1 or Table 14.4), or above the ICH reporting thresholds in any long-term stability studies; therefore, 5 may not need to be included in the MRA. Kleinman et al. point out, however, that further consideration may be warranted since the degradation pathway to 6 is active and intermediate 5 is a logical intermediate enroute to 6.

Degradation products 1–6 were evaluated using two complementary (Q)SAR tools, and two products were listed as having mutagenic potential (aniline 3 and hydroperoxide 5), being classified as ICH M7 Class 3 [7], having a different alert from the parent (which did not itself flag for mutagenicity). Based on these results, further consideration of 3 and 5 is warranted.

In the case of 3, further consideration would likely be running the Ames test to see if it is Ames positive. If negative, no further evaluation is needed. If it is positive, then a well-designed short-term

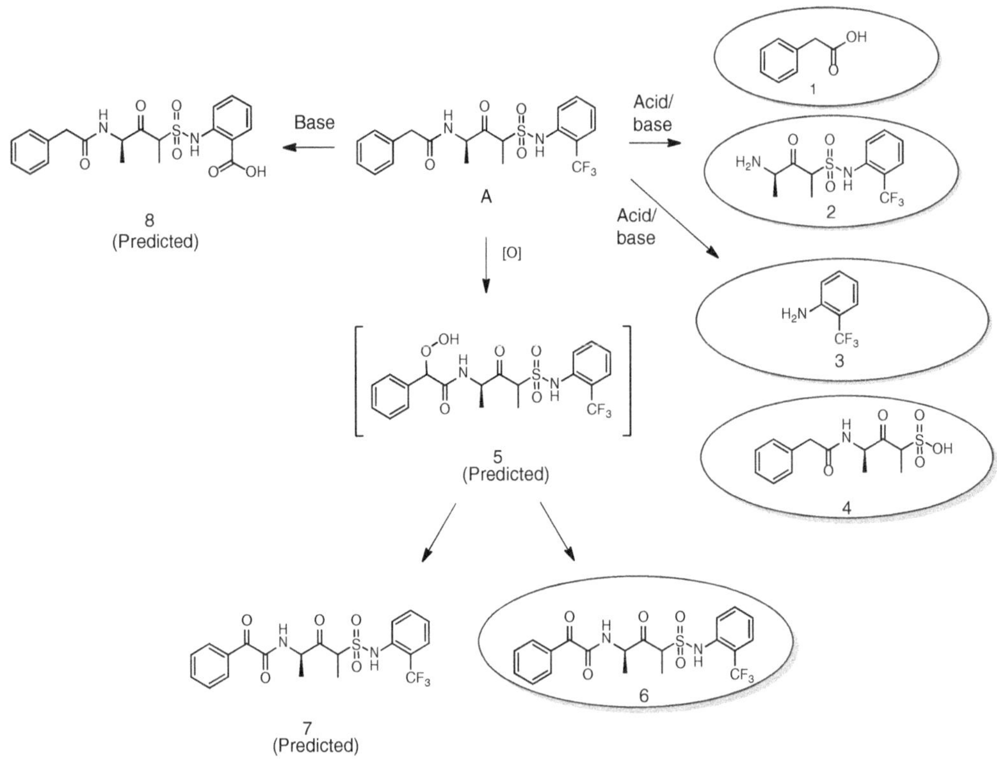

Figure 14.5 The degradation pathways of Molecule A in a solid-oral dosage form product. *Source:* From Kleinman et al. [24].

kinetic study should be conducted with a goal of exceeding the kinetic equivalence of 6 months at 40 °C/75% RH, if product 3 is formed in the DS or DP (in the proposed commercial packaging) at levels approaching the TTC, then further studies would be needed to demonstrate control to an acceptable level over the shelf life. Dow et al. [18] have described the studies recommended by Lilly to demonstrate acceptable control over the shelf life.

In the case of hydroperoxide 5, the decisions would be guided by degradation chemistry principles, development stability studies, and knowledge gained through stress testing and other investigations. As Kleinman et al. discussed, 5 was not observed in any stress testing or stability study, and consideration could be given as to whether the theoretical intermediate is a transient intermediate or is too unstable to allow analytical detection. In either case it would suggest that the compound is unstable and therefore the patient would not be expected to be physiologically exposed to hydroperoxide 5. The mutagenic impurity risk assessment (MRA) risk assessment conclusion would then be that no further evaluation is required. Kleinman points out that there are cases where a hydroperoxide is observed in stress studies or long-term studies [62–64], and therefore it should not simply be presumed that a given hydroperoxide will be too unstable to be relevant to patient exposure.

14.8.2 Case Study #2: Galunisertib

Galunisertib illustrates a case where an identified mutagenic impurity is both a potential process impurity and a potential degradation product [65]. DS stress testing studies performed during clinical development identified two degradation products resulting from peroxide-induced oxidation of a pyridine moiety and a quinoline moiety (to form two aromatic N-oxide structures, Figure 14.6). The aromatic N-oxides generated positive alerts for mutagenicity by in silico Q(SAR) mutagenicity assessments; subsequently, both tested positive in the Ames bacterial mutagenicity test. These compounds were also identified as potential process impurities that could theoretically form from reaction with hydrogen peroxide in the synthesis; hydrogen peroxide was used to promote hydrolysis of a cyano group to the corresponding carboxylate group in the final step of the synthetic route (Figure 14.7). A toxicology limit of not more than 166 ppm (w/w relative to galunisertib) for N-oxides 1 and 2 combined was assigned based on a clinical dose of 300 mg of galunisertib daily for two years, and a modified TTC approach appropriate for oncolytic agents [12], justifying a maximum tolerable dose of the N-oxides of 50 μg/day.

To eliminate the risk of N-oxide formation in the synthetic process, the reaction conditions were altered, and the use of hydrogen peroxide was eliminated. Whether or not the N-oxide *potential* degradation products were *actual* degradation products (i.e. did they form above reporting thresholds over the shelf life?) remained to be assessed.

An LC-MS method was developed and validated to provide the needed sensitivity to quantify the aromatic N-oxides to levels <10% of the limits (i.e. <8 ppm each, <16 ppm combined) in both the DS and the DP tablets. To evaluate whether the degradation pathway was relevant to DS storage, per Section 8 of ICH M7(R1), the DS was stored at accelerated conditions (40 °C/75% RH) for

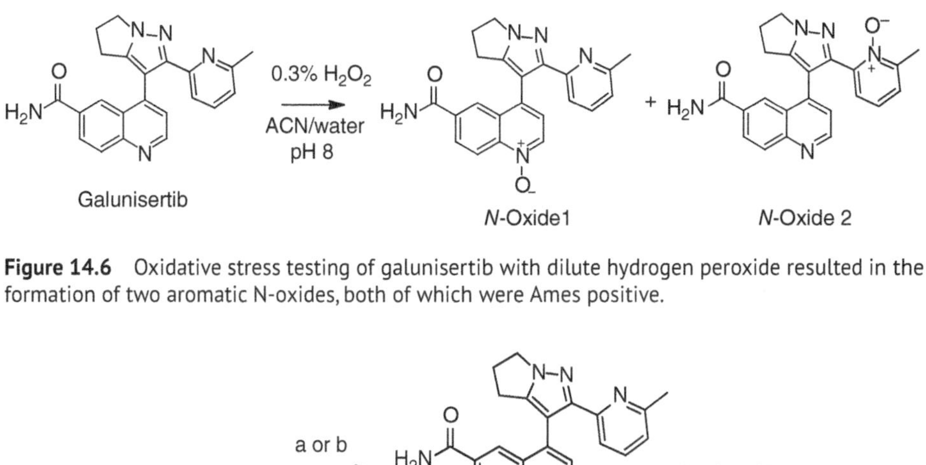

Figure 14.6 Oxidative stress testing of galunisertib with dilute hydrogen peroxide resulted in the formation of two aromatic N-oxides, both of which were Ames positive.

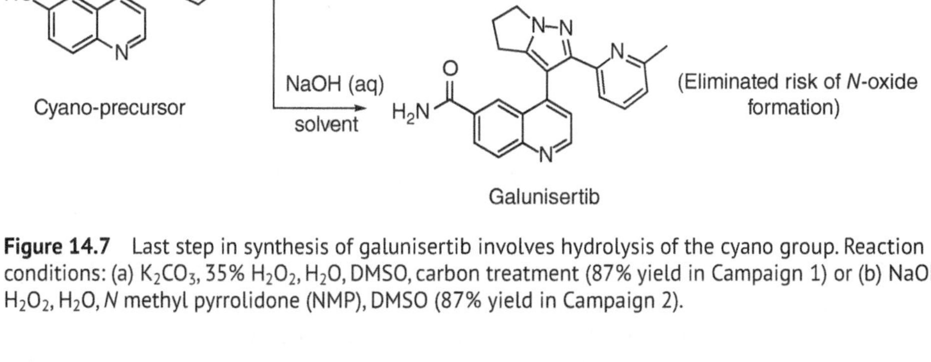

Figure 14.7 Last step in synthesis of galunisertib involves hydrolysis of the cyano group. Reaction conditions: (a) K_2CO_3, 35% H_2O_2, H_2O, DMSO, carbon treatment (87% yield in Campaign 1) or (b) NaOH, 35% H_2O_2, H_2O, *N* methyl pyrrolidone (NMP), DMSO (87% yield in Campaign 2).

6 months in the packaging used for clinical trial storage. LC-MS analysis revealed N-oxides 1 and 2 to be less than 8 ppm at the initial timepoint and throughout the six months storage. The same results were confirmed for long-term storage at 25 °C/60% RH for 36 mos.

In addition, tablets from two DP manufacturing processes were evaluated – high sheer wet granulation (HSWG) and roller compaction (RC). Tablets (10, 50, and 100 mg strengths) manufactured using HSWG were stored for 68 months at 25 °C/60% RH. The results from the LC-MS analysis indicated combined levels of the N-oxides to be less 16 ppm at all tablet strengths. Commercial development efforts were focused on the RC platform, with 80 and 150 mg tablet strengths. These (RC) tablets were exposed to 70 °C/75% RH for 30 days; following Section 2.3.3.1 of Kleinmaan et al. [24] and using Table 14.2, assuming a conservative 19.87 kcal/mol Ea, this 30 day stress is a kinetic equivalence to approximately 491 days (or 16.3 months) at 40 °C/75% RH (a significant excess to six months at 40 °C/75% RH). In addition, per ICH M7(R1), tablets were exposed to simulated sunlight significantly in excess to ICH Q1B confirmatory doses of UVA and Visible light to assess whether photodegradation pathways could lead to the N-oxides. In all cases, LC-MS analyses indicated combined levels of <16 ppm for the N-oxides. While ICH M7(R1) appears to indicate that these studies clearly show that the pathways to the N-oxides is not active in the DP, and therefore it could be argued that no further studies were needed to confirm this conclusion, a conservative approach was chosen and additional studies were performed on tablets stored at 40 °C/75% RH through 6 and 12 months on long-term storage (30 °C/65% RH). The same (<16 ppm) results were obtained, confirming that the degradation pathways are not active and no control strategy or specifications should be needed for the market. As pointed out by the authors [65] of this work, confirmation of these results in long-term primary stability studies should not be required based on the wording in ICH M7(R1).

14.8.3 Case Study #3: Naloxegol

A potentially mutagenic impurity was discovered within Naloxegol, due to the alerting structure of the mesylate group. The final steps in the synthesis of Naloxegol are shown in Figure 14.8.

The methanesulfonate impurity AZ13336989 is formed in the second to last step (below) due to the presence of a short chain impurity, AZ13378660, generated in the synthesis of Naloxegol (see Figure 14.9).

AZ13336989 became subject for Ames testing. As the phenanthrene-based chemical class is generally associated with weak *in vitro* genotoxicity, the Ames test of AZ13336989 was performed in parallel with the API; both of them contain the phenanthrene core structure.

While a weak bacterial mutagenicity of the NKTR-118 had been earlier detected in the Ames test, the compound was negative in the other regulatory genotoxicity tests, i.e. mouse lymphoma tk assay and in vivo micronucleus assay. Similar in vitro activity is seen for structurally similar compounds such as naloxone and other phenanthrene-based drugs, including nalbuphine, buprenorphine, and oxycodone. The overall weight of evidence supports the conclusion that phenanthrene-based drugs are not genotoxic in vivo, and do not represent a carcinogenic risk to patients.

Contradictory to the previously reported results, both Naloxegol and AZ13336989 returned strongly positive results in the Ames test. Subsequent Ames tests revealed large variations in mutagenic responses among various NKTR-118 batches, ranging from weakly to strongly positive.

In order to explain these unexpected findings, an extensive package of investigatory work was initiated.

The investigation also showed a clear correlation between sample storage conditions and positive Ames responses, and sample handling procedures therefore play a role in the variability in Ames responses. The degradation product glycidaldehyde, formed through oxidation of API, was identified

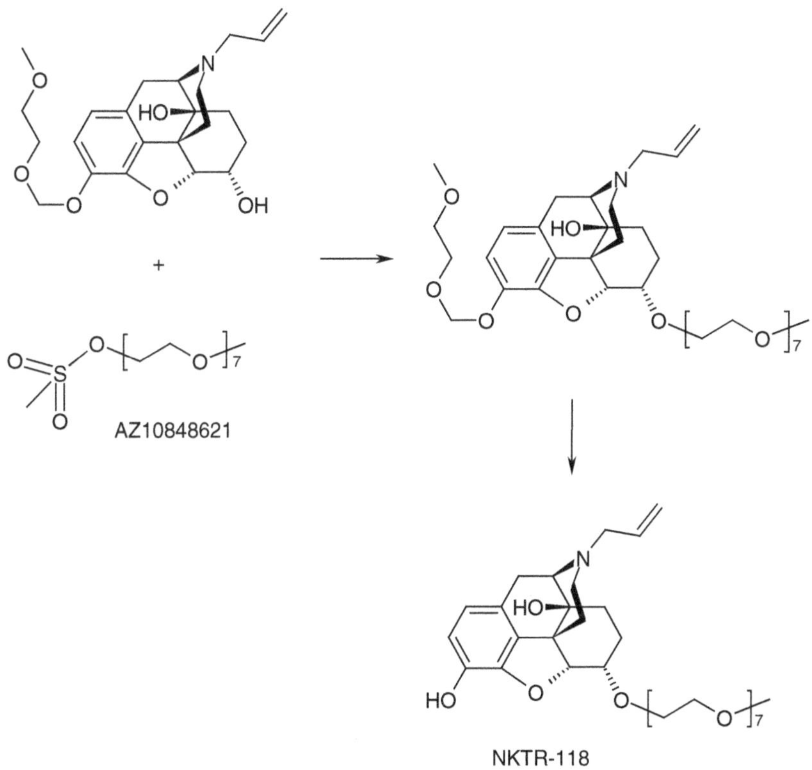

Figure 14.8 PEGylation of MEM-protected Naloxone.

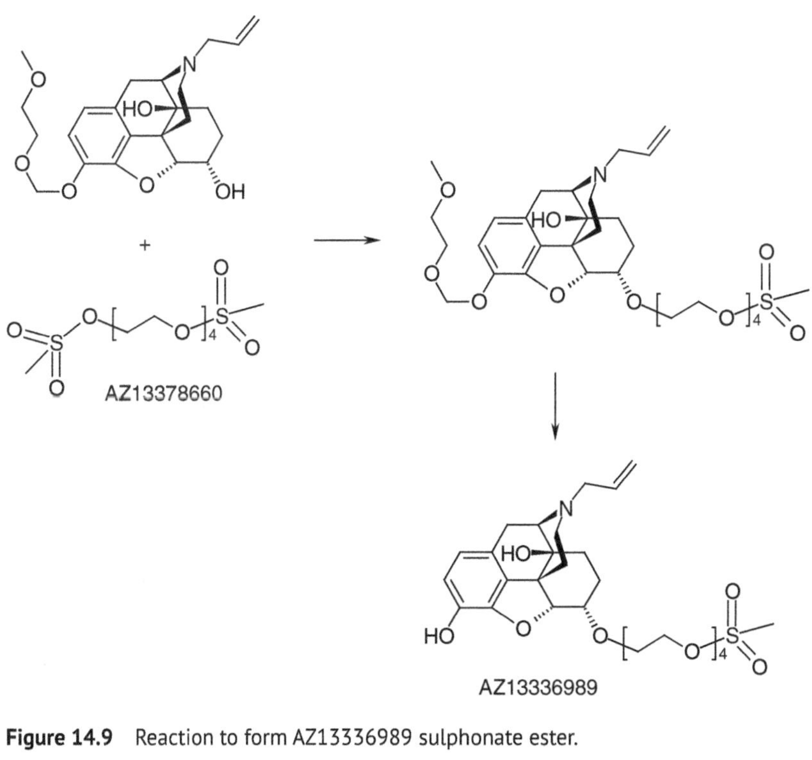

Figure 14.9 Reaction to form AZ13336989 sulphonate ester.

as the root cause to the Ames activity in the Naloxegol. Quantified levels of glycidaldehyde in degraded NKTR-118 correlated directly with the observed Ames response.

Glycidaldehyde is formed upon extended periods of exposure of the API to oxygen, through oxidation of the allylamine functional group (Figure 14.10). This is believed to be a radical-initiated autoxidation process and can be effectively controlled by storing the bulk API inerted under positive nitrogen pressure, and by the use of stabilizing ingredients in tablet formulation. In addition, and most importantly, the salt form was re-evaluated and the oxalate salt of NKTR-118 was shown negative in the Ames test; thus the issue was mitigated.

14.8.4 Case Study #4: Selumetinib Side Chain

Selumetinib drug substance can hydrolyze on storage to give two degradation products: (i) Selumetinib acid and (ii) Selumetinib side chain (Figure 14.11):

Twice the Selumetinib side chain had been Ames tested (Figure 14.12); however, the origin of the positive Ames response was not well understood. Nevertheless, the side chain degradation product was controlled to levels below the acceptable safety limit within the formulation.

It was decided to investigate this further by following ICH harmonized testing of genotoxicity (ICH S2(R1) [66]. ICH S2(R1) defines a standard battery for genotoxicity testing of pharmaceuticals consisting of an:

1) *in vitro* test for gene mutation in bacteria (e.g. Ames test) – Positive;
2) *in vitro* test of chromosomal damage with mammalian cells (e.g. mouse lymphoma assay (MLA) mouse lymphoma tk assay – no effect;
3) *in vivo* test for chromosomal damage (e.g. Rat micronucleus assay) – no effect.

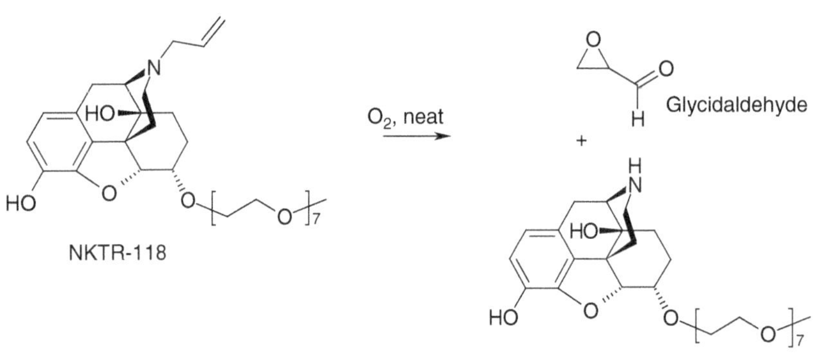

Figure 14.10 Degradation pathway of NKTR-118.

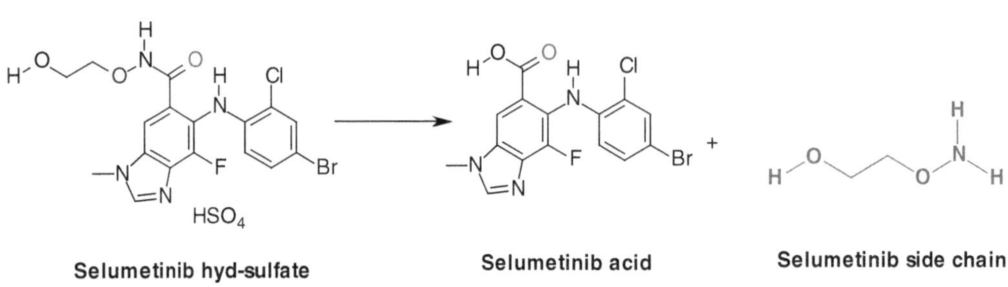

Figure 14.11 Degradation pathway of Selumetinib.

Structure

Adduct (salt)

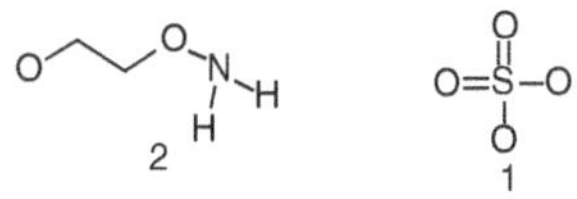

Responsible scientist CTL reported 2007

Results summary	TA1535 + S9 (1)	TA1535 + S9 (2)
Result	Positive	Positive
Maximum increase	2.4 times at 2500 μg/plate	2.7 times, at 2500 μg/plate
Lowest positive dose (μg/plate)	2500	2500
Highest scoreable dose (μg/plate)	5000	5000

Structure

Adduct (salt)

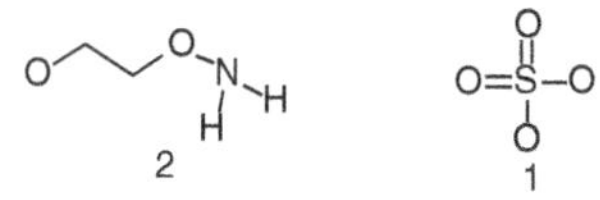

Responsible scientist Chris Mee

Results summary	TA1535 - S9	TA1535 + S9
Result	Positive	Positive
Maximum increase	3.1 times, at 1600 μg/plate	3.3 times, at 5000 μg/plate
Lowest positive dose (μg/plate)	1600	1600
Highest scoreable dose (μg/plate)	5000	5000

Figure 14.12 Ames test results for Selumetinib side chain.

It was noted that David Tweats et al. [67] had studied the mutagenicity of Fexinidazole analogues and found them to be mutagenic in the standard Salmonella/Ames test but non-mutagenic in Ames Salmonella strains lacking one or more nitroreductase(s) enzymes. Since mammalian cells do not contain nitroreductase, they concluded that these compounds were non-mutagenic in humans. Based on this, AZ11910553 was assayed for mutation in three histidine-requiring strains of *Salmonella typhimurium,* TA1535-functional nitroreductase enzyme, TA1535NR (deficient in "classical" nitroreductase enzyme), and YG7127 ("classical" nitroreductase gene deleted), both in the absence and in the presence of metabolic activation by an Aroclor 1254-induced rat liver post-mitochondrial fraction (S-9), in two separate experiments. All AZ11910553 treatments in this study were performed using formulations prepared in anhydrous analytical dimethyl sulphoxide (DMSO).

Mutation Experiment 1 involved the treatment of strains TA1535 and TA1535NR (deficient in "classical" nitroreductase enzyme). These were performed in the absence and in the presence of S-9, using final concentrations of AZ11910553 at 5, 16, 50, 160, 500, 1600, and 5000 μg/plate, plus vehicle and positive controls. Following these treatments, evidence of toxicity was observed at 5000 μg/plate in strains TA1535 and TA1535NR in the absence and presence of S-9.

Mutation Experiment 2 involved treatment of strain YG7127 (where the nitroreductase gene is deleted). Again, these experiments were performed in the absence and in the presence of S-9, using final concentrations of AZ11910553 at 5, 16, 50, 160, 500, 1600, and 5000 μg/plate, plus vehicle and positive controls. Following these treatments, evidence of toxicity was observed at 1600 and/or 5000 μg/plate in the absence and presence of S-9.

Vehicle and positive control treatments were included for all strains. The mean numbers of revertant colonies all fell within acceptable ranges for vehicle control treatments and were elevated by positive control treatments where applicable. The test article was completely soluble in the aqueous assay system at all concentrations tested.

Following AZ11910553 treatments of test strains TA1535 and TA1535NR increases in revertant numbers were observed ≥threefold the concurrent vehicle control at 1600 μg/plate in the absence of S-9 in strain TA1535 and at 1600 and 5000 μg/plate in the absence of S-9 and at 1600 μg/plate in the presence of S-9 in strain TA1535NR. These increases in revertants showed some evidence of a concentration relationship up to concentrations showing evidence of toxicity. These increases are therefore considered as evidence of AZ11910553 mutagenic activity in these strains.

A previous study had shown that strain TA1535NR (deficient in "classical" nitroreductase activity) has a small amount of nitroreductase activity remaining, indicating that there may be more than one nitroreductase pathway in this strain. Although there were increases in revertants observed in both strains TA1535 and TA1535NR, as TA1535NR has been shown to retain small levels of nitroreductase activity, therefore the increases in the number of revertants observed in this strain would not necessarily preclude nitroreductase involvement in the mechanism of action of the test article.

No increases in revertant numbers were observed that were ≥threefold the concurrent controls in strain YG7127 in the absence and presence of S-9 or in the presence of S-9 in strainTA1535. As there were no increases in the number of revertants ≥threefold the concurrent vehicle controls in strain YG7127 ("classical" nitroreductase gene deleted) in the absence and presence of S-9, this may indicate that there is some nitroreductase involvement in the mutagenic mechanism of the test article.

It was concluded that AZ11910553 induced mutation in two histidine-requiring strains of *Salmonella typhimurium* (TA1535 and TA1535NR) when tested under the conditions of this study. These conditions included treatments at concentrations up to 5000 μg/plate (the maximum recommended concentration according to current regulatory guidelines) in the absence (TA1535 and TA1535NR) and in the presence (TA1535NR) of a rat liver metabolic activation system (S-9).

AZ11910553 did not induce mutation in one histidine-requiring strain of *Salmonella typhimurium* (YG7127) when tested under the conditions of this study. These conditions included treatments at concentrations up to 5000 μg/plate in the absence and in the presence of S-9. The negative result observed with strain YG7127 with one of the nitroreductase genes deleted, may be indicative of nitroreductase involvement in the positive responses observed with strains TA1535 and TA1535NR.

Although controlled to appropriate levels, this nevertheless indicates that the mechanism may be bacteria specific.

14.9 Significance of the Risk of Forming Mutagenic Degradation Products

If drug degradation leads to the formation of a mutagenic product, only a very small amount of degradation may be necessary to produce levels above the acceptable thresholds (TTC) over the shelf life. Controlling such degradation processes to the low levels required to ensure patient safety can be very challenging. An understanding of the theoretical potential for degradation of a specific DS structure to a mutagenic product would be very helpful for (i) predicting the possibility of a particular DS to form mutagenic degradation products, and (ii) using mechanistic insight to aid in the development of control strategies.

14.9.1 Frequency of Alerting Structures in Degradation Products

Mining data found in a database of drug degradation products (containing at the time 322 unique parent drug structures and 1021 unique degradation products) [68], Raillard et al. [25] provided insight into the most common mutagenic alerting structures found in degradation products of typical drug molecules, along with the features in the drug molecules that can lead to degradation products with these alerting substructures. The alerting structures were determined using ToxTree (v1.51), an open-source software for predicting the toxic hazards of chemicals [69]. Using ToxTree predictions, the following results were obtained:

- 221 alerting structures among the 322 parent molecules (69%).
- 336 alerting structures among the 1021 degradation products (33%).
- 155 alerting structures (unique from the parent) among the 1021 degradation products (15%).

Using more sophisticated mutagenicity-predicting software (i.e. DEREK and MultiCASE), the 15% of unique alerting structures was reduced to c. 5–8% [70].

14.10 Degradation Reactions Leading to Alerting Structures in Degradation Products

Raillard et al. [25] then identified and classified the most common degradation reactions that formed alerting structures that were not present in the parent molecule (unique alerting structures). This analysis, summarized in Table 14.5, showed that eight functional groups accounted for almost 85% of the unique alerting structures (for the full results table, see Raillard et al. [25]). Further analysis revealed the functional groups in the parent drug molecule that led to the unique alerting structures via particular mechanisms. This further analysis is summarized in Tables 14.6, 14.7, 14.8, 14.9, and 14.10, and this information can be useful for identifying the functional groups in DS structures that could degrade to form potentially mutagenic degradation products.

Table 14.5 Reactions leading to the production of a mutagenicity alerting structure.

Structural alert and corresponding ToxTree™ alert number	Number of alerts in parent molecule structure	Number of alerts in degradation product structures	Number of unique alerts in degradation product structures
Aldehydes (SA 11)	2	40	34
α,β-unsaturated carbonyls (SA 10)	79	126	30
Primary aromatic amines, hydroxyl amines and its derived esters (SA 28)	73	93	23
Heterocyclic, polycyclic aromatic hydrocarbons (SA 19)	4	15	13
Epoxides and aziridines (SA 7)	9	17	12
Nitro aromatics (SA 27)	26	25	6
Aromatic ring N-oxides (SA 26)	0	6	6
Aliphatic halogens (SA 8)	12	12	6
Total	205	334	130
% of total alerts		93.8	83.9

Table 14.6 Degradation reactions leading to the production of a mutagenic alerting structure: aldehydes.

Functional group in parent leading to degradation	Mechanism (in parenthesis are listed the names of the corresponding drugs in the Pharma D3 database exhibiting this degradation pathway)
Imine 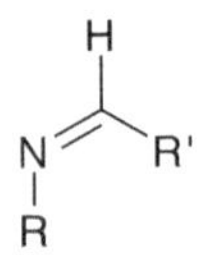	Hydrolysis (Nitrofurantoin) 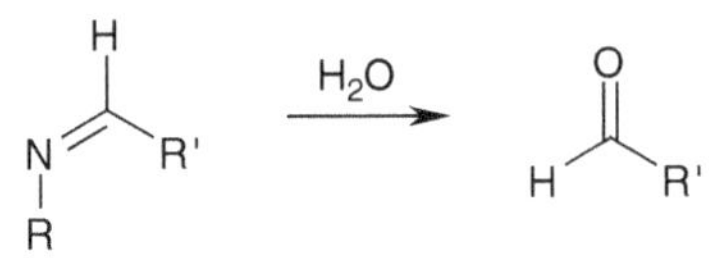
Benzylalcohols, Benzylamines, Benzylic carbons 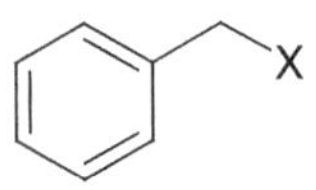 X = H, OR, NR1R2	Oxidation (Albuterol, Chloroamphenicol, Econazole, Ezlopitant, Ibuprofen, Losartan, Miconazole) 
Amino-cyclopropyl 	Photo-oxidation (Cryomazine) 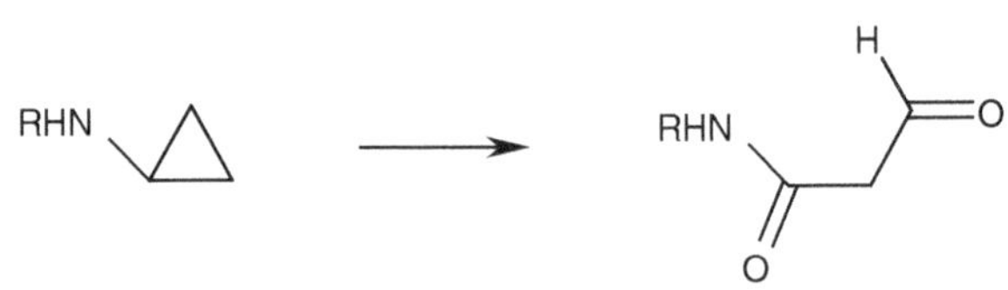
Aromatic acetic acid 	Oxidation + decarboxylation (Indomethacin) 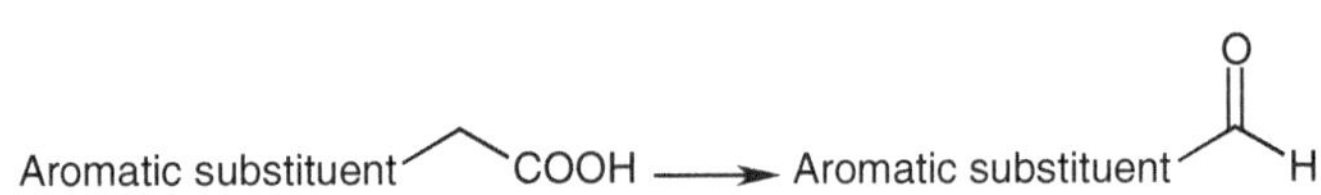
Ketone 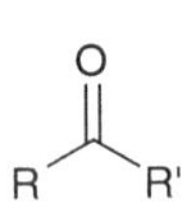	Norrish reaction (Methadone) 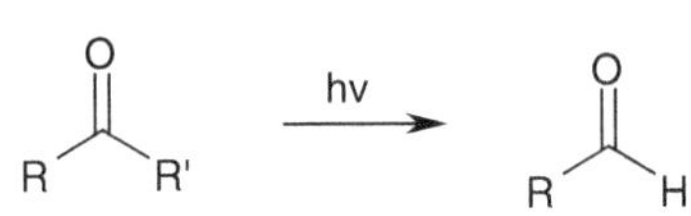
Amide 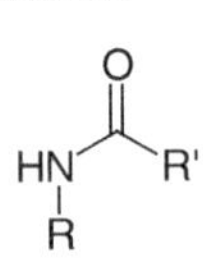 R' = alkyl	Photolysis (Flutamide) 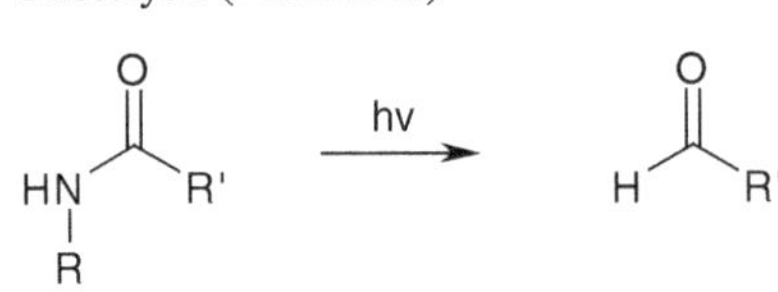

Table 14.7 Degradation reactions leading to the products of a mutagenic alerting structure: α,β-unsaturated carbonyls.

Functional group in parent leading to degradation	**Mechanism (in parenthesis are listed the names of the corresponding drugs in the Pharma D3 database exhibiting this degradation pathway)**
Allyl alcohol 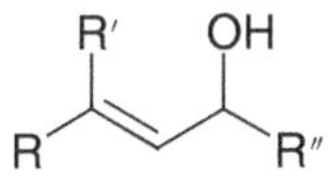	Oxidation (Docetaxel) 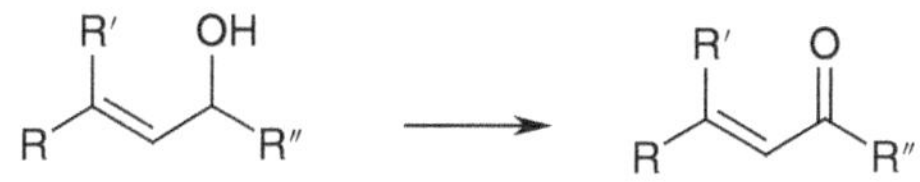
Quarternary ammonium group beta to carbonyl 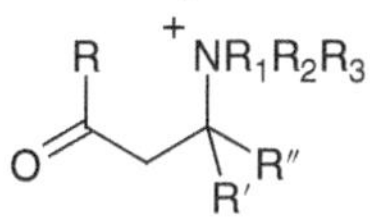	Hofmann elimination (Cisatracurium) 
Heteroatom β to carbonyl 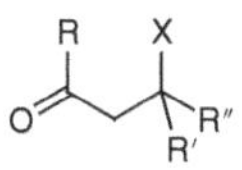	Substitution/elimination reaction induced with water (Dyclonine) 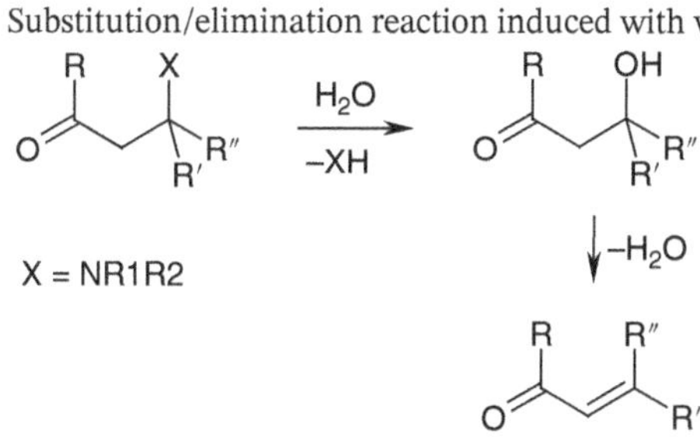
β-Hydroxycarbonyl 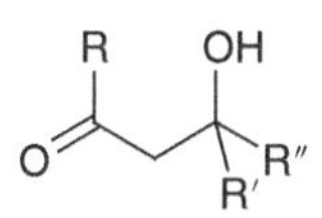	Elimination of water (Streptovitacine) 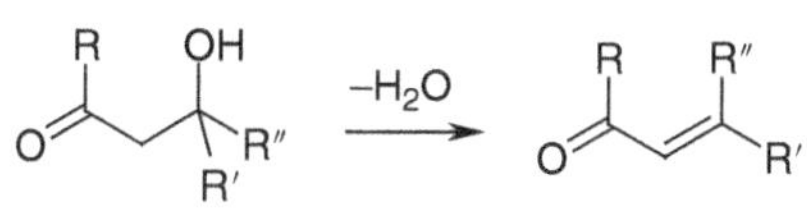
Cyclohexene 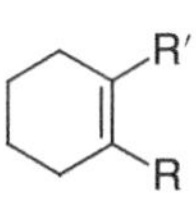	Oxidation α to double bond (Cyclobarbital) 
Phenol 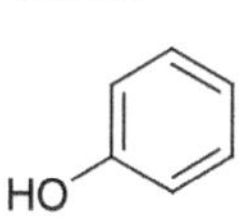	Oxidative dimerization to quinone (Propofol) 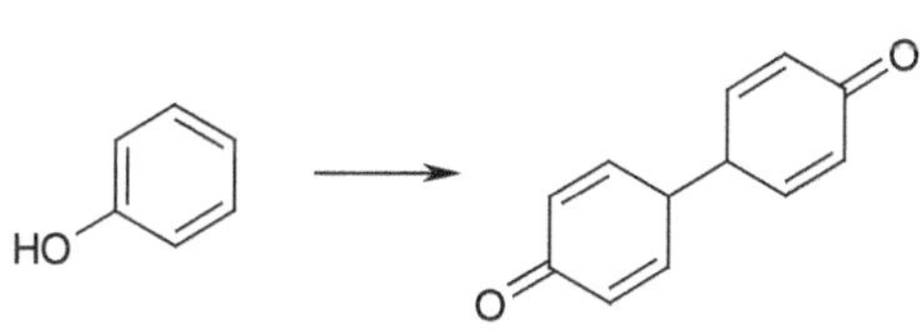
Catechol 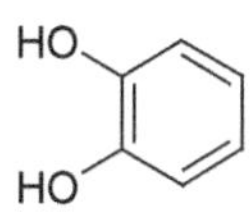	Oxidation to quinone (Adrenaline, Isoprenaline) 

Table 14.8 Degradation reactions leading to the production of a mutagenic alerting structure: primary aromatic amine, hydroxylamine, and its derived esters.

Functional group in parent leading to degradation	Mechanism (in parenthesis are listed the names of the corresponding drugs in the Pharma D3 database exhibiting this degradation pathway)
Aromatic N-acyl 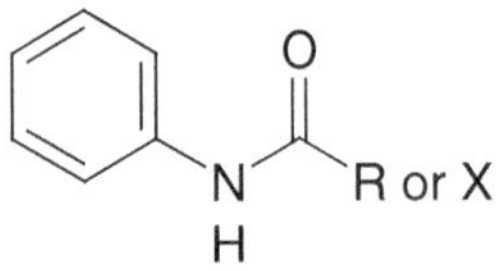 R = alkyl X = Heteroatom	Hydrolysis (Acebutolol, Acetaminophen, Bicalutamide, Cinalukast, Clanfenur, Imatinib) 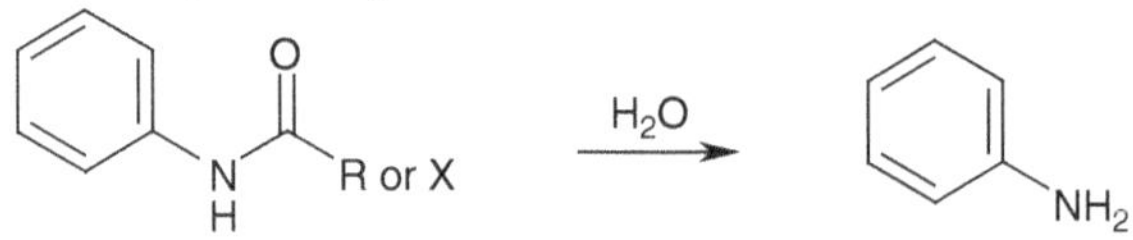
Phenyltriazene 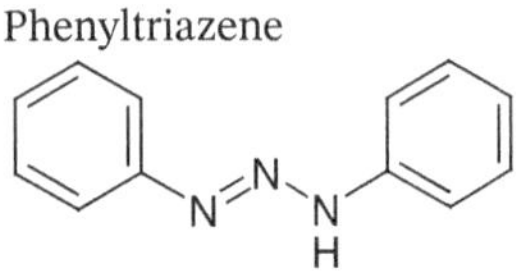	Hydrolysis (Diminazene) 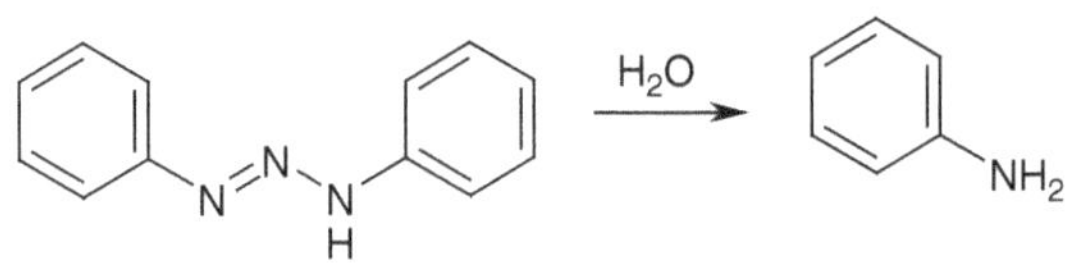
Benzodiazepine 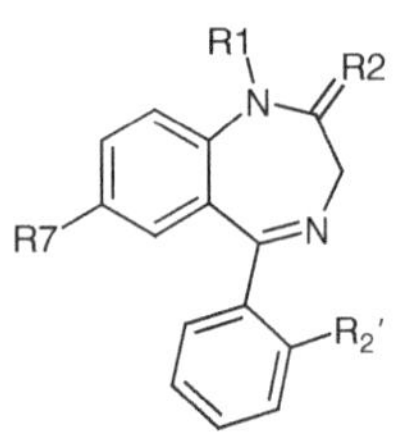 R1, R2, R2′, R7 refer to the Benzodiazepine substitution nomenclature	Photoinduced Hydrolysis (Midazolam) 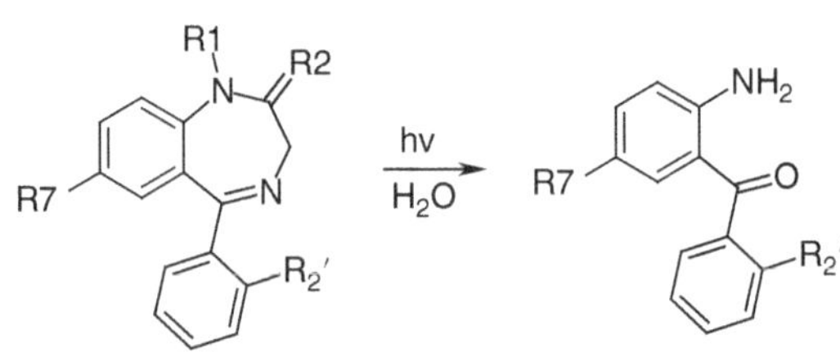

Table 14.9 Degradation reactions leading to the production of a mutagenic alerting structure: epoxides.

Functional group in parent leading to degradation	Mechanism (in parenthesis are listed the names of the corresponding drugs in the Pharma D3 database exhibiting this degradation pathway)
Double bond 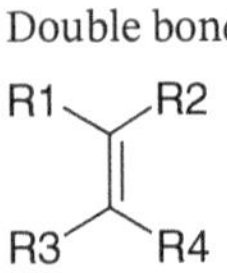	Oxidation (Indolizine, Menadione, Tanespimycin) 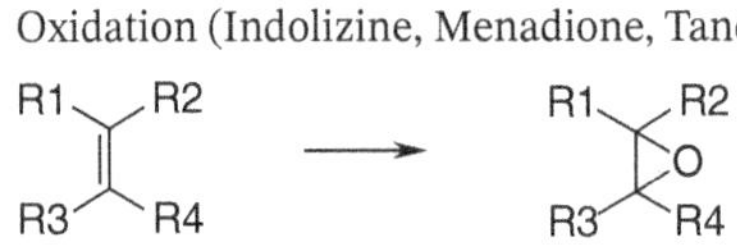
2-Hydroxy-1-chloroalkane 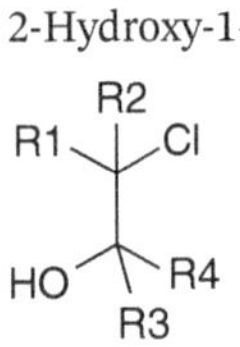	Cyclization via Nucleophilic substitution (Mometasone, Ornidazole) 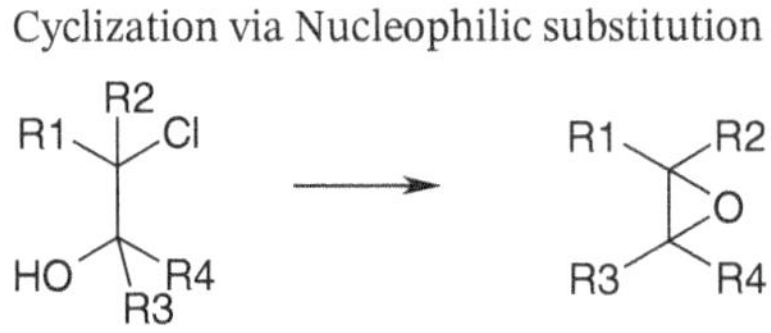

Table 14.10 Degradation reactions leading to the production of a mutagenic alerting structure.

Functional group in parent leading to degradation	Mechanism (in parenthesis are listed the names of the corresponding drugs in the Pharma D3 database exhibiting this degradation pathway)
Aromatic chloride in chloro-substituted diphenylamine	Photo-dehalogenation reaction, leading to cyclization (Diclofenac, Meclofenamic acid)
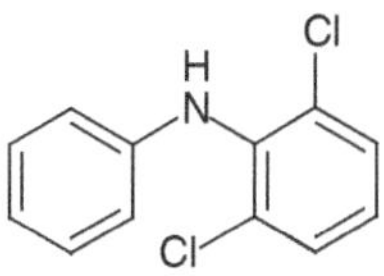	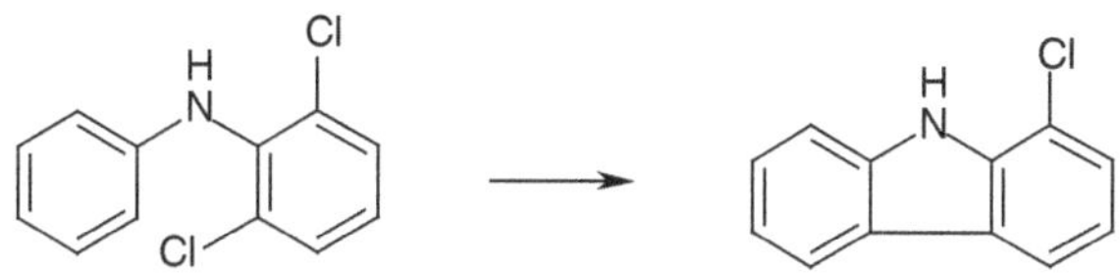

14.10.1 Frequency of Alerting Structures Giving Rise to Ames Positive Tests

Based on the work by Galloway et al. [71], about 54% of typical process impurities that are structurally alerting were found to be Ames positive. If you subtract the boronic acid derivatives (which were Ames positive about 82% of the time), the overall frequency drops to c. 52%. Unpublished data from Lilly has shown that Ames positive rates of degradation products closely parallel those of typical process impurities such as those discussed in the Galloway et al. study. Thus, we can expect that approximately 50% of the alerting degradation product structures will be Ames positive.

14.10.2 Mutagenic Degradation Products: Overall Predicted Frequency

As described by Dow et al. [18], Lilly internal benchmarking data from 15 DPs revealed that using DS and DP stress testing protocols described by Baertschi et al. [38] in conjunction with the Approach 2 threshold criteria shown above (Section 2.3.2), an average of 8.2 degradation products derived from stress testing will qualify as "major." This same benchmarking data show that approximately 36% of these major degradation products will form in stability studies at levels that will lead to assignment as "actual" degradation products.

Using these data, combined with the estimates that indicate that approximately 5–8% of degradation products will give rise to degradation products that have structural alerts unique to the products and the estimates that c. 50% of alerting structures may be Ames positive, we can calculate a rough expected frequency of "actual" degradation products that will be mutagenic for a typical DS.

Assuming 8.2 potential degradation products identified from stress testing:

8.2 × 0.36 (% of potential products that may be actual products) × 0.08 (% of identified deg products expected to be mutagenic alerting) × 0.5 (% of alerting structures expected to be Ames positive) = 0.118 × 100 = 11.8%.

Thus, from this limited data set, approx. 12% of new DSs can be predicted to form "actual" degradation products that are mutagenic.

14.11 N-Nitrosamines: Special Considerations

Issues associated with N-Nitrosamines are described in detail in Chapter 10. Initially associated with the chemistry of formation of Sartan drugs, N-Nitrosamines are potentially formed through reaction between secondary and tertiary amines, see Figure 14.13.

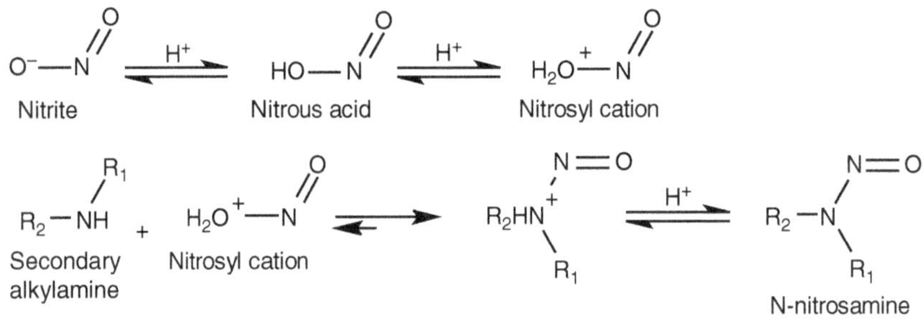

Figure 14.13 Mechanism of formation of N-Nitrosamines.

Primary amines are not a concern as they are unstable and immediately decompose.

While the risk is primarily associated with the synthetic process, in particular where a nitrosating agent is used in the synthesis as was the case in the revised Valsartan synthesis (changes made to the process placed traces of dimethylamine in the same processing step as sodium nitrite, introduced to quench highly reactive sodium azide, again this is described in Chapter 10). The risk of formation of N-Nitrosamines is also inherent, at least potentially, in drug product, this is examined in Chapter 10 but the risk is associated with the reaction of trace secondary amines (present as residues within the API as a result of the manufacturing process) and traces of nitrite in excipients. Work to understand and control such risks is at the time of writing still in progress. Of course, many drugs are themselves secondary amines and thus there is the risk of generation of the N-Nitrosamine of the actual API. However, while small molecular weight N-Nitrosamines such as N-Nitrosodimethylamine (NDMA) are potent carcinogens, larger more bulky API like N-Nitrosamines are typically less potent and in some cases non-mutagenic [72]. Again this is an area of rapid research.

Perhaps more worryingly, Ranitidine was reported [73–76] to contain unacceptably high levels of NDMA, resulting in its ultimate withdrawal. In the case of Ranitidine, the issue appears to be related to the presence of both a secondary amine pendant and subsequent reaction with a nitrite source (excipient related). Another theory is that the Nitro group within the molecule may lead to an intermolecular reaction, although in the absence of mechanistic evidence, the authors of this chapter are skeptical of this proposal. Recently, GSK published their investigation into the root cause [77]. These investigations indicated the presence of NDMA results from a slow degradation of the ranitidine molecule (Figure 14.14). Analysis using isotopically labeled ranitidine hydrochloride confirmed the formation of NDMA solely from an intermolecular reaction of ranitidine hydrochloride without involvement of impurities. Factors that influence the rate of degradation include heat, humidity, and the crystal morphology of ranitidine hydrochloride with the material exhibiting a columnar habit showing a slower rate of degradation.

What is clear is that critical investigations into such risks as described above need to be included in the design of experimental degradation studies.

14.11.1 Evaluation of Potential Formation of N-Nitrosamines in Drug Product

There are two requirements for the chemical formation of N-nitrosamine compounds: a "vulnerable" amine and a nitrosating reagent. A "vulnerable amine" is considered to be a compound containing a secondary or tertiary, present either within the API structure, the excipients in the formulation, or as an impurity or degradant. EFPIA proposed an overall drug product workflow, which is shown in Figure 14.15. International Pharmaceutical Excipients Council (IPEC) Europe also published a "Questionnaire for excipient nitrosamines risk evaluation" [78] that suggests that

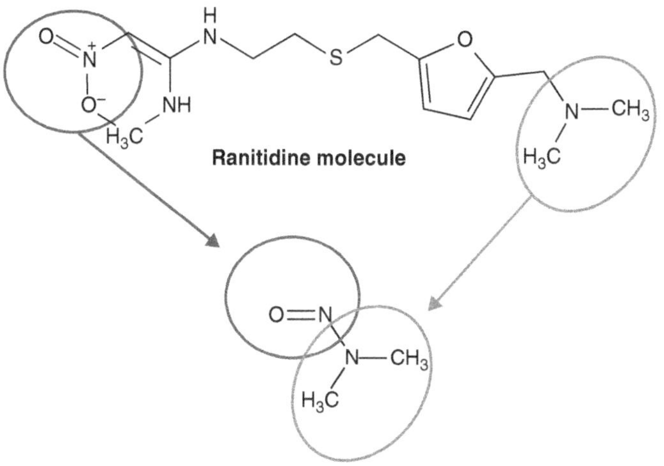

Figure 14.14 Potential formation of NDMA inter- or intramolecularly within ranitidine.

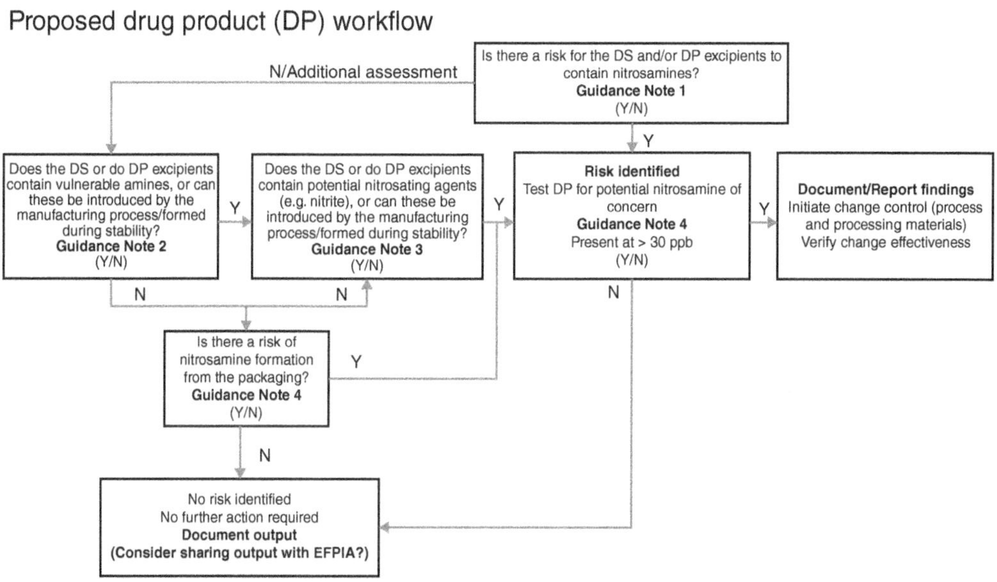

Figure 14.15 The proposed drug product workflow for assessing the risk of the presence of N-nitrosamines in formulated drug products (reproduced from ref [78]).

while secondary amines are of the greatest concern, all secondary, tertiary aliphatic, and aromatic amines should be considered in a risk assessment (of both the API and DP).

N-nitrosamines in the drug product could be present either as a result of:

1) presence of N-nitrosamines in the API;
2) presence of nitrosating reagents in the API (e.g. from contamination during the synthesis) reacting with the API, formulation ingredients, or API impurities or degradation products;
3) presence of nitrosating reagents introduced during drug product manufacturing, reacting with the API, formulation ingredients, or API impurities or degradation products;
4) presence of N-nitrosamines in the formulation ingredients/excipients or packaging reacting with the API, formulation ingredients, or API impurities or degradation products.

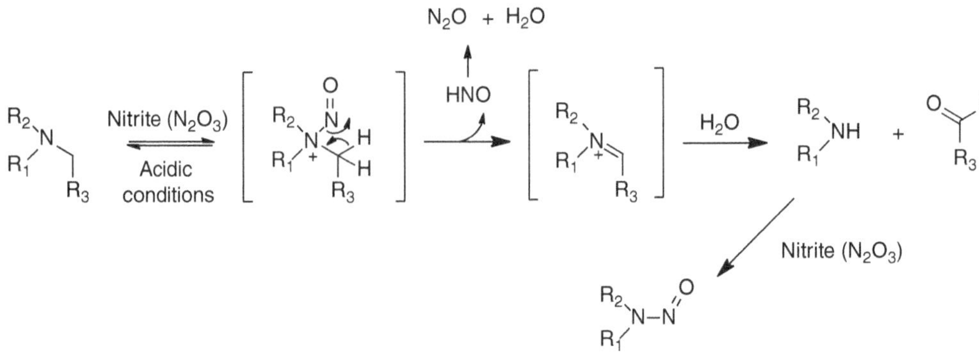

Figure 14.16 Pathway for N-nitrosation of tertiary amines from nitrites (more specifically, N_2O_3).

There is a potential for N-nitrosating species (specifically nitrites and nitrates) contamination in common excipients that has been discussed in the literature: "Nitrates and nitrites are common nitrosating impurities that can be found in most excipients at ppm levels. Sodium starch glycolate, croscarmellose sodium, pre-gelatinized starch, PVP, cPVP and Lactose Fast Flo® are excipients that carry trace level of nitrate or nitrite impurities" [79]. Another potential source of trace levels of nitrates and nitrites is processing water, It is worth noting that nitrates do not N-nitrosate directly; rather, reduction to nitrites is first required, and such reduction processes, while known to potentially occur in vivo, do not readily occur upon aging of pharmaceutical DPs [27]; therefore, the focus of a risk assessment should be on nitrites.

Nitrites do not act as direct nitrosating reagents; instead, under acidic conditions (e.g. below pH 4) nitrite converts to HNO_2, and then to N_2O_3, the nitrosating species [80, 81]. Therefore, acidic conditions are favored for the formation of the nitrosating species, while alkaline conditions are favored for reaction with the free base form of the amine to be N-nitrosated.

Tertiary amines are known to nitrosate from reaction with nitrites/nitrous acid with cleavage of one of the alkyl groups (Figure 14.16) [82, 83]. As discussed by Mirvish [84] and documented by Mitch [85], studies indicate that nitrosation rates of tertiary amines are four orders of magnitude slower than for their analogous secondary amines. Interestingly, the stoichiometry for N-nitrosation of tertiary amines was shown by Smith and Loeppky [83] to be 2 : 1, i.e. two molecules of nitrite/nitrous acid per molecule of tertiary amine to yield one molecule of N-nitrosated tertiary amine.

Thus, the risk of N-Nitrosamine formation during storage has become one of a number of risk factors that need to be considered in relation to an N-Nitrosamine risk assessment.

14.12 Conclusions

Understanding the fundamental properties of an API has never been more important than in the context of mutagenic degradants. Although relatively rare, having a degradation pathway that yields a mutagenic degradant can potentially end the development of an otherwise promising candidate drug at an early stage and can be catastrophic if discovered at later stages of development (or in the case of Ranitidine years after launch). It is the authors' views that determination of the fundamental properties of a molecule should be determined as early as possible, to either eliminate a risk or at least identify a risk in such a way that it may be mitigated (e.g. through formulation and/or protective packaging). A combination of predictive modeling and experimental studies, allied to a mutagenicity assessment of degradant structures as described in this chapter affords the best opportunity to do this.

References

1 International Conference on Harmonisation. (2006). Impurities in New Drug Substances. Q3A(R2). Revised Guideline.
2 International Conference on Harmonisation. (2006). Impurities in New Drug Products. Q3B(R2). Revised Guideline.
3 Guideline on the Limits of Genotoxic Impurities (2006). *CPMP/SWP/5199/02, EMEA/CHMP/QWP/251344/2006; Committee for Medicinal Products for Human Use (CHMP).* London: European Medicines Agency (EMEA). 28 June 2006.
4 Committee for Medicinal Products for Human Use Safety Working Party (CHMP) (2008). *Questions and Answers on the CHMP Guideline on the Limits of Genotoxic Impurities.* London: European Medicines Agency (EMEA).
5 International Conference on Harmonisation. (2017). Assessment and Control of DNA reactive (mutagenic) impurities in pharmaceuticals to limit potential carcinogenic risk. M7. Revised guideline (R1).
6 Food and Drug Administration, Center for Drug Evaluation and Research (CDER) (2008). *Genotoxic and carcinogenic impurities in drug substances and drug products: Recommended approaches, Guidance for Industry (draft).* Rockville, MD: U.S. Department of Health and Human Services http://www.fda.gov/downloads/Drugs/GuidanceComplianceRegulatoryInformation/Guidances/ucm079235.pdf.
7 Mueller, L., Mauthe, R.J., Riley, C.M. et al. (2006). A rationale for determining, testing, and controlling specific impurities in pharmaceuticals that possess potential for genotoxicity. *Reg. Toxicol. Pharmacol.* 44: 198–211.
8 Pierson, D.A., Olsen, B.A., Robins, D.K. et al. (2009). Approaches to assessment, testing decisions, and analytical determination of genotoxic impurities in drug substances. *Org. Proc. Res. Dev.* 13 (2): 285–291.
9 Teasdale, A. (ed.) (2010). *Genotoxic Impurities: Strategies for Identification and Control.* Hoboken NJ: Wiley.
10 Szekely, G., Amores de Sousa, M.C., Gil, M. et al. (2015). Genotoxic impurities in pharmaceutical manufacturing: sources, regulation, and mitigation. *Chem. Rev.* 115: 8182–8229.
11 Raillard, S.P. (2012). Control of genotoxic impurities in active pharmaceutical ingredients. *Chim. Oggi/Chem. Today* 30: 1.
12 Callis, C., Bercu, J.P., DeVries, K.M. et al. (2010). Risk assessment of genotoxic impurities in marketed compounds administered over a short-term duration: applications to oncology products and implications for impurity control limits. *Org. Process Res. Dev.* 14: 986–992.
13 McGovern, T. and Jacobson-Kram, D. (2006). Regulation of genotoxic and carcinogenic impurities in drug substances and products. *Trends Anal. Chem.* 25: 790–795.
14 Robinson, D.I. (2010). Control of genotoxic impurities in active pharmaceutical ingredients: a review and perspective. *Org. Process Res. Dev.* 14: 946–959.
15 Snodin, D.J. (2010). Genotoxic impurities: from structural alerts to qualification. *Org. Process Res. Dev.* 14: 960–976.
16 Giordani, A., Kobel, W., and Gally, H.U. (2011). Overall impact of the regulatory requirements for genotoxic impurities. *Eur. J. Pharm. Sci.* 43: 1–15.
17 Raman, N.V.V.S., Prasad, A.V.S.S., and Ratnakar Reddy, K.J. (2011). Strategies for the identification, control and determination of genotoxic impurities in drug substances: a pharmaceutical industry perspective. *Pharm. Biomed. Anal.* 55: 662–667.
18 Dow, L.K., Hansen, M.M., Pack, B.W. et al. (2013). The assessment of impurities for genotoxic potential and subsequent control in drug substance and drug product. *J. Pharm. Sci.* 102: 1404–1418.

19 Dobo, K.L., Greene, N., Cyr, M.O. et al. (2006). The application of structure-based assessment to support safety and chemistry diligence to manage genotoxic impurities in active pharmaceutical ingredients during drug development. *Regul. Toxicol. Pharmacol.* 44: 282–293.

20 Looker, A.R., Ryan, M.P., Neubert-Langille, B.J., and Naji, R. (2010). Risk assessment of potentially genotoxic impurities within the framework of quality by design. *Org. Process Res. Dev.* 14: 1032–1036.

21 McLaughlin, M., Dermenjian, R.K., Jin, Y. et al. (2015). Evaluation and control of mutagenic impurities in a development compound: purge factor estimate vs measured amounts. *Org. Proc. Res. Dev.* 19: 1531–1535.

22 Betori, R.C., Kallemeyn, J.M., and Welch, D.S. (2015). A kinetics-based approach for the assignment of reactivity purge factors. *Org. Proc. Res. Dev.* 19: 1517–1523.

23 Teasdale, A., Fenner, S., Ray, A. et al. (2010). A tool for the semi-quantitative assessment of potentially genotoxic impurity (PGI) carryover into API using physicochemical parameters and process conditions. *Org. Proc. Res. Dev.* 14 (4): 943–945.

24 Kleinman, M.H., Elder, D., Teasdale, A. et al. (2015). Strategies to address mutagenic impurities derived from degradation in drug substances and drug products. *Org. Proc. Res. Dev.* 19 (11): 1447–1457.

25 Raillard, S.P., Bercu, J., Baertschi, S.W., and Riley, C.M. (2010). Prediction of drug degradation pathways leading to structural alerts for potential genotoxic impurities. *Org. Proc. Res. Dev.* 14: 1015–1020.

26 Baertschi, S.W., De Antonis, D., McKeown, A.P. et al. (2011). Stress testing as a predictive tool for the assessment of potential genotoxic degradants. In: *Pharmaceutical Stress Testing: Predicting Drug Degradation*, 2e (eds. S.W. Baertschi, K.M. Alsante and R.A. Reed). London: Informa Healthcare.

27 Baertschi, S.W., Alsante, K.M., and Reed, R.A. (eds.) (2011). *Pharmaceutical Stress Testing: Predicting Drug Degradation*, 2e. London: Informa Healthcare.

28 Alsante, K.M., Ando, A., Brown, R. et al. (2007). The role of degradant profiling in active pharmaceutical ingredients and drug products. *Adv. Drug Deliv. Rev.* 59: 29–37.

29 Blessy, M., Patel, R.D., Prajapati, P.N., and Agrawal, Y.K. (2014). Development of forced degradation and stability-indicating studies of drugs – a review. *J. Pharm. Anal.* 4 (3): 159–165.

30 International Conference on Harmonisation. (2006). Stability Testing of New Drug Substances and Products. Q1A, Sept (1994), followed by a second revision (R2) in Feb (2003).Guideline.

31 Parenty, A.D.C., Button, W.G., and Ott, M.A. (2013). An expert system to predict the forced degradation of organic molecules. *Mol. Pharm.* 10: 2962–2974.

32 Boyd, D.B. and Sharp, T.R. (2011). The power of computational chemistry to leverage stress testing of pharmaceuticals. In: *Pharmaceutical Stress Testing: Predicting Drug Degradation*, 2e (eds. S.W. Baertschi, K.M. Alsante and R.A. Reed). London: Informa Healthcare.

33 Andersson, T., Broo, A., and Evertsson, E. (2014). Prediction of drug candidates' sensitivity toward autoxidation: computational estimation of C–H dissociation energies of carbon-centered radicals. *J. Pharm. Sci.* 103: 949–1955.

34 Lienard, P., Gavartin, J., Boccardi, G., and Meunier, M. (2015). Predicting drug substances autoxidation. *Pharm. Res.* 32: 300–310.

35 Reynolds, D.W., Facchine, K.L., Mullaney, J.F. et al. (2002). Available guidance and best practices for conducting forced degradation studies. *Pharm. Technol.* 26: 48–54.

36 Kleinman, M.H., Baertschi, S.W., Alsante, K.M. et al. (2014). In silico prediction of pharmaceutical degradation pathways: a benchmarking study. *Mol. Pharm.* 11: 4179–4188.

37 Alsante, K.M., Martin, L., and Baertschi, S.W. (2003). A stress testing Benchmarking study. *Pharm. Technol.* 27 (2): 60–72.

38 Baertschi, S.W., Jansen, P.J., and Alsante, K.M. (2011). Stress testing: a predictive tool. In: *Pharmaceutical Stress Testing: Predicting Drug Degradation*, 2e (eds. S.W. Baertschi, K.M. Alsante and R.A. Reed). London: Informa Healthcare.

39 Olsen, B.A., Sreedhara, A., and Baertschi, S.W. (2018). Impurity investigations by phases of drug and product development. *Trends Anal. Chem.* 101: 17–23.

40 Waterman, K.C., Carella, A.J., Gumkowski, M.J. et al. (2007). Improved protocol and data analysis for accelerated shelf-life estimation of solid dosage forms. *Pharm. Res.* 24 (4): 780–790.

41 Waterman, K.C. and Adami, R.C. (2005). Accelerated aging: prediction of chemical stability of pharmaceuticals. *Int. J. Pharm.* 293: 101–125.

42 Colgan, S.T., Timpano, R.J., Roberts, M. et al. (2014). Opportunities for lean stability strategies. *J. Pharm. Innov.* 9: 259–271.

43 <1079> Good storage and distribution practices for drug products.

44 Waterman, K.C. (2010). *Accelerated Stability Assessment Program (ASAP) Using Science to Set Shelf Life*. Philadelphia, PA: IIR Forced Degradation Forum.

45 Kennon, L.J. (1964). Use of models in determining chemical pharmaceutical stability. *J. Pharm. Sci.* 53 (7): 815–818.

46 MacFaul, P.A., Ruston, L., and Wood, J.M. (2011). Activation energies for the decomposition of pharmaceuticals and their application to predicting hydrolytic stability in drug discovery. *Med. Chem. Commun.* 2: 140–142.

47 Davis, J.S. (1991). *Criteria for Accelerated Stability Testing*. Chicago, IL: FDA/ASQC Seminar.

48 European Medicines Agency (EMA). (2010). Committee for Medicinal Products for Human Use (CHMP) Safety Working Party (SWP) Question & Answers on the CHMP Guideline on the Limits of Genotoxic Impurities. EMA/CHMP/SWB/431994/2007, Rev. 3, 23 September 2010.

49 Behnia, K., Cornelius, G., Wang, J. et al. (2011). Estimation of the extent of in vivo formation of a mutagenic aromatic amine from a potential thyromimetic compound: correlation of in vitro and in vivo findings. *Chem. Res. Toxicol.* 24: 905–912.

50 Zhang, S., Coutant, M., O'Connor, D. et al. (2012). Early development GMPs for small-molecule specifications: an industry perspective (Part V). *Pharm. Tech.* 36 (10).

51 Guidance for Industry cGMP for Phase 1 Investigational Drugs. (2008). https://www.fda.gov/downloads/Drugs/GuidanceComplianceRegulatoryInformation/Guidances/UCM070273.pdf.

52 INDs for Phase 2 and Phase 3 Studies Chemistry, Manufacturing, and Controls Information. (2003). https://www.fda.gov/downloads/Drugs/GuidanceComplianceRegulatoryInformation/Guidances/UCM070567.pdf.

53 FDA (1999). INDs for phase II and III studies of drugs, including specified therapeutic biotechnology derived products. *Federal Register (Notices)* 64: 19543–19544.

54 International Conference on Harmonisation. (2006). Quality Risk Management Q9, 2006.

55 Baertschi, S.W., Clapham, D., Foti, C. et al. (2013). Implications of in-use photostability: proposed guidance for photostability testing and labeling to support the administration of photosensitive pharmaceutical products, part 1: drug products administered by injection. *J. Pharm. Sci.* 102 (11): 3888–3899.

56 Baertschi, S.W., Clapham, D., Foti, C. et al. (2015). Implications of in-use photostability: proposed guidance for photostability testing and labelling to support the administration of photosensitive pharmaceutical products. Part 2. Topical drug products. *J. Pharm. Sci.* 102 (11): 3888–3899.

57 Allain, L., Baertschi, S.W., Clapham, D. et al. (2016). Implications of in-use photostability: proposed guidance for photostability testing and labelling to support the administration of photosensitive pharmaceutical products. Part 3. Oral drug products. *J. Pharm. Sci.* 105 (5): 1586–1595.

58 Thatcher, S.R., Mansfield, R.K., Miller, R.B. et al. (2001). Pharmaceutical photostability: a technical and practical interpretation of the ich guideline and its application to pharmaceutical stability: part I. *Pharm. Technol.* 25 (3): 98–110.

59 Clapham, D., Templeton, A.C., Klein, L.J., and Kleinman, M.H. (2011). Practical aspects of conducting photostability stress testing. In: *Pharmaceutical Stress Testing*, 2e (eds. S.W. Baertschi, K.M. Alsante and R.A. Reed), 218–232. New York: Informa Healthcare.

60 Turro, N.J., Ramamurthy, V., and Scaiano, J.C. (2010). *Modern Molecular Photochemistry of Organic Molecules*. Herndon, VA: University Science Books.

61 Templeton, A.C., Xu, H., Placek, J., and Reed, R.A. (2005). Implications of photostability on the manufacturing, packaging, storage, and testing of formulated pharmaceutical products. *Pharm. Technol.* 29: 68–86.

62 Boccardi, G. (1994). Autoxidation of drugs: prediction of degradation impurities from results of reactions with radical chain initiators. *Il Farmaco* 49: 431–435.

63 Elder, D.P., Snodin, D., and Teasdale, A. (2010). Analytical approaches for the detection of epoxides and hydroperoxides in active pharmaceutical ingredients, drug products and herbals. *J. Pharm. Biomed. Anal.* 51: 1051–1023.

64 Nelson, E.D., Thompson, G.M., Yao, Y. et al. (2009). Solvent effects on the AIBN forced degradation of cumene: implications for forced degradation practices. *J. Pharm. Sci.* 98 (3): 959–969.

65 Strege, M.A., Osborne, L.M., Hetrick, E.M. et al. (2015). Assessing the risk of formation of potential genotoxic degradation products in a small-molecule kinase inhibitor drug substance and drug product. *Org. Proc. Res. Dev.* 19 (11): 1458–1464.

66 International Conference on Harmonisation, S2(R1). (2011). Guidance on genotoxicity testing and data interpretation for pharmaceuticals intended for human use.

67 Tweats, D., Trunz, B.B., and Torreele, E. (2012). Genotoxicity profile of fexinidazole—a drug candidate in clinical development for human African trypanomiasis (sleeping sickness). *Mutagenesis* 27 (5): 523–532. https://doi.org/10.1093/mutage/ges015.

68 The Pharmaceutical Drug Degradation Database (Pharma D3) is no longer publicly available.

69 http://toxtree.sourceforge.net/ (accessed 16 Febraury 2020).

70 Baertschi, S.W., Bercu, J., Robert, J., et al. (2018). *Presented by Baertschi SW at Lhasa 2018 Pharmaceutical Industry and Regulators Symposium*, 24–25 May, Brasilia, Brazil (Unpublished data).

71 Galloway, S.M., Vijayaraj Reddy, M., McGettigan, K. et al. (2013). Potentially mutagenic impurities: analysis of structural classes and carcinogenic potencies of chemical intermediates in pharmaceutical syntheses supports alternative methods to the default TTC for calculating safe levels of impurities. *Reg. Toxicol. Pharmacol.* 66 (3): 326–335.

72 Guttenplan, J.B. (1987). N-nitrosamines: bacterial mutagenesis and in vitro metabolism. *Mut. Res.* 186: 81–134.

73 EMA to review ranitidine medicines following detection of NDMA, 13 September 2019 EMA/503622/2019. https://www.ema.europa.eu/en/documents/press-release/ema-review-ranitidine-medicines-following-detection-ndma_en.pdf.

74 https://wayback.archive-it.org/7993/20191212142501/https://www.fda.gov/drugs/drug-safety-and-availability/laboratory-tests-ranitidine, accessed Aug 30, 2021.

75 https://www.gov.uk/government/news/zantac-mhra-drug-alert-issued-as-glaxosmithkline-recalls-all-unexpired-stock, accessed Aug 30, 2021.

76 https://www.fda.gov/news-events/press-announcements/fda-requests-removal-all-ranitidine-products-zantac-market, accessed Aug 31, 2021.

77 Searle, A.D., King, F.J., and Urquhart, M.W. (2020). Ranitidine—Investigations into the Root Cause for the Presence of *N*-Nitroso-*N,N*-dimethylamine in ranitidine hydrochloride drug substances and associated drug products. *Org. Proc. Res. Dev.* 24 (12): 2915–2926.

78 https://www.ipec-europe.org/articles/questionnaire-for-excipient-nitrosamines-risk-evaluation.html, accessed Aug 30, 2021.

79 Wu, Y., Levon, J., Narang, A.S. et al. (2011). Reactive impurities in excipients: profiling, identification and mitigation of drug-excipient incompatibility. *AAPS PharmSciTech* 12 (4): 1248–1263.

80 Rostkowska, K. et al. (1998). Formation and metabolism of N-nitrosamines. *Polish J. Env. Stud.* 7 (6): 321–325.

81 Wainright, T. (1986). The chemistry of nitrosamine formation: relevance to malting and brewing. *J. Inst. Brew.* 92: 49–64.

82 23-June-2020, EMA/526934/2019. Lessons learnt from presence of N-nitrosamine impurities in sartan medicines, accessed Aug 30, 2021.

83 Smith, P.A.S. and Loeppky, R.N. (1967). Nitrosative cleavage of tertiary amines. *J. Am. Chem. Soc.* 89: 1147–1157.

84 Mirvish, S.S. (1975). Formation of N-nitroso compounds: chemistry, kinetics, and in vivo occurrence. *Toxicol. Appl. Pharmacol.* 31: 325–351.

85 Mitch, W. (2019). Literature review of nitrosamine and nitramine formation chemistry relevant to carbon capture. https://ccsnorway.com/wp-content/uploads/sites/6/2019/10/nitrosamineandnitramineformationchemistry_yale-1.pdf.

Index

Mutagenic Impurities: Strategies for Identification and Control, First Edition. Edited by Andrew Teasdale.
© 2022 John Wiley & Sons, Inc. Published 2022 by John Wiley & Sons, Inc.

n

O